Lecture Notes in Computer Science 6701

Commenced Publication in 1973
Founding and Former Series Editors:
Gerhard Goos, Juris Hartmanis, and Jan van Leeuwen

Julia Pahl Torsten Reiners Stefan Voß (Eds.)

Network Optimization

5th International Conference, INOC 2011
Hamburg, Germany, June 13-16, 2011
Proceedings

Springer

Volume Editors

Julia Pahl
Stefan Voß
Torsten Reiners
University of Hamburg
Institute of Information Systems
Von-Melle-Park 5, 20146 Hamburg, Germany
E-mail: {pahl, reiners}@econ.uni-hamburg.de,
stefan.voss@uni-hamburg.de

Torsten Reiners
Curtin University
School of Information Systems
Perth, WA 6845, Australia
E-mail: t.reiners@cbs.curtin.edu.au

ISSN 0302-9743 e-ISSN 1611-3349
ISBN 978-3-642-21526-1 ISBN 978-3-642-21527-8 (eBook)
DOI 10.1007/978-3-642-21527-8
Springer Heidelberg Dordrecht London New York

Library of Congress Control Number: 2011928709

CR Subject Classification (1998): C.2, D.4.4, E.1, F.2.2, G.2.2, H.3.5, H.4.3

LNCS Sublibrary: SL 5 – Computer Communication Networks and Telecommunications

Typesetting: Camera-ready by author, data conversion by Scientific Publishing Services, Chennai, India

Printed on acid-free paper

Springer is part of Springer Science+Business Media (www.springer.com)

Preface

The International Network Optimization Conference (INOC) is the conference of the European Network Optimization Group (ENOG) which aims at regularly bringing together experts from different disciplines especially from operations research, graph theory, queuing theory and stochastic search with the main focus on network optimization. The conference is intended to be a forum for researchers and practitioners working in the area of network optimization. Certainly networks may be seen in the widest possible sense. Telecommunication networks on the physical as well as the logical layer have been the focus of INOC-related research from the very beginning. This relates to network design problems, (multi-commodity) flow problems, location and routing. However, INOC does not only deal with telecommunication networks, but with networks in the widest sense. These could be networks for vehicle routing, electricity provision or maritime shipping, just to mention a few.

One might argue whether globalization and (ferocious) competition are at the very heart of what is needed for mankind. However, one might still follow the idea of an ever increasing need for decision support and solutions of network-related problems using latest information technology. This is especially important regarding the value chain and networks of firms. This may draw special attention to the latest developments in technologies and the challenges that come along regarding theoretical as well as practical implications of network optimization.

This volume of the *Lecture Notes in Computer Science* consists of selected papers presented at the INOC 2011, held at the University of Hamburg, Germany, during June 13–16. The INOC 2011 is the successor of a series of scientific conferences on network optimization. The first was held in 2003 aiming at providing a fruitful environment for discussion and exchange of ideas on current and future challenges regarding information science and communication technologies. Previous conferences were held in Pisa (Italy, 2009), Spa (Belgium, 2007), Lisbon (Portugal, 2005) and Evry (France, 2003). These are well documented as follows:

INOC 2003 W. Ben-Ameur and A. Petrowskin, *Proceedings of the International Network Optimization Conference INOC 2003*, ISSN: 1762-5734, Evry (2003).
Special issue of *Annals of Operations Research*, Vol. 146, Issues 1, pp. 1-202 (2006), editors W. Ben-Ameur, D. Bienstock and I. Saniee

INOC 2005 L. Gouveia and C. Mourao, *Proceedings of the International Network Optimization Conference INOC 2005*, Lisbon (2005), ISSN: 1762-5734.
Special issue "Multicommodity Flows and Network Design" of *Networks*, Vol. 49, Issue 1, pp. 1–133 (2007), guest editors L. Gouveia and S. Voß.

INOC 2007 Contributions are available online at http://www.euro-online.org /enog/inoc2007/INOC%202007%20-%20program.htm.
Special issue "Network Optimization" of *Networks*, Vol. 55, Issue 3, pp. 169–297 (2007), editors B. Fortz and L. Gouveia,
INOC 2009 M.G. Scutellà, CD-Rom with proceedings.
Special Issue of *Networks*, in preparation, guest editors L. Gouveia and M.G. Scutellà.

The INOC 2011 was located in *the* most beautiful city in Germany, Hamburg. Hamburg is part of important logistics and telecommunication networks. Among others, we feature two airports within the city, one of the three biggest container ports in Europe belonging to the largest container ports worldwide. As a metropolitan region, numerous trade activities involving multitudes of companies take place every day providing great challenges for optimization up- and downstream along the value chain. That makes Hamburg the perfect place to inspire new ideas in the field of network optimization as well as information and communication sciences.

The contributions presented at the conference as well as the selected papers in these proceedings highlight recent developments in network optimization. We grouped contributions as follows:

- Network design
- Network flows
- Routing and transportation
- Further optimization problems and applications
  - Energy-oriented network design
  - Telecom applications
  - Location
  - Maritime shipping
  - Telecom
  - Graph theory
  - Miscellaneous

Organizing a conference and publishing the proceedings is of course an endeavor involving many people in numerous activities. We first want to thank all authors and presenters for their contributions. Moreover, we greatly appreciate the valuable help and cooperation from the members of the international Program Committee as well as the referees.

June 2011

Julia Pahl
Torsten Reiners
Stefan Voß

Organization

Organization Chair

Julia Pahl	University of Hamburg, Germany
Torsten Reiners	Curtin University, Australia and University of Hamburg, Germany
Stefan Voß	University of Hamburg, Germany

Organizing Committee

Marco Caserta	IE Business School, Madrid, Spain
Silvia Schwarze	University of Hamburg, Germany

Program Committee and Referees

Edoardo Amaldi, Milan, Italy
Anita Amberg, Darmstadt, Germany
Panagiotis Angeloudis, London, UK
Anant Balakrishnan, Austin, USA
Tolga Bektas, Southampton, UK
Michael G.H. Bell, London, UK
Walid Ben-Ameur, Evry, France
Christian Bierwirth, Halle-Wittenberg, Germany
Branko Bubalo, Berlin, Germany
Kai Brüssau, Hamburg, Germany
Christelle Caillouet, Aachen, Germany
Marco Caserta, Madrid, Spain
Grit Claßen, Aachen, Germany
Angel Corberan, Burjassot, Spain
Amaro de Sousa, Aveio, Portugal
Karl F. Dörner, Linz, Austria
Bernard Fortz, Brussels, Belgium
Mathias Fricke, Darmstadt, Germany
Martin Josef Geiger, Hamburg, Germany
Bernard Gendron, Montreal, Canada
Bruce Golden, College Park, USA
Monica Gentili, Salerno, Italy
Stefan Gollowitzer, Vienna, Austria
Luis Gouveia, Lisbon, Portugal
Peter Greistorfer, Graz, Austria
Richard Hartl, Vienna, Austria
Holger Höller, Hamburg, Germany
Dmitry Ivanov, Hamburg, Germany
Arie M.C.A. de Koster, Aachen, Germany
Manuel Kutschka, Aachen, Germany
Martin Labbe, Brussels, Belgium
Gilbert Laporte, Montreal, Canada
Markus Leitner, Vienna, Austria
Stephan Lemkens, Aachen, Germany
Stefan Lessmann, Hamburg, Germany
Janny M.Y. Leung, Hong Kong, China
Ivana Ljubic, Vienna, Austria
Abilio Lucena, Rio de Janeiro, Brazil
Tom Magnanti, Massachusetts, USA
Ridha Mahjoub, Paris, France
Vittorio Maniezzo, Cesena, Italy
Alexander Martin, Erlangen, Germany
Dirk Mattfeld, Braunschweig, Germany
Belén Melián, La Laguna, Spain
Michel Minoux, Paris, France
Eli Olinick, Dallas, USA
Adam Ourou, Issy-les-Moulineaux, France

Table of Contents

Part I: Theoretical Problems / Uncertainty / Graph Theory / Network Design

Part II: Network Flow

Part III: Routing and Transportation

Part IV: Further Optimization Problems and Applications

On the Design of Optical OFDM-Based Networks

Amal Benhamiche[1], Ridha Mahjoub[2], and Nancy Perrot[3]

[1] Orange Labs R&D; CORE-TPN, 38-40 rue du Général Leclerc,
92794 Issy-Les-Moulineaux Cedex 9, France
amal.benhamiche@orange-ftgroup.com

[2] Université Paris-Dauphine, LAMSADE, CNRS,
Place du Maréchal de Lattre de Tassigny 75775 Paris Cedex 16, France
mahjoub@lamsade.dauphine.fr

[3] Orange Labs R&D; CORE-TPN, 38-40 rue du Général Leclerc,
92794 Issy-Les-Moulineaux Cedex 9, France
nancy.perrot@orange-ftgroup.com

Abstract. In this paper, we are interested in the Optical Multi-Band Network Design. This problem consists, given the physical layer of an optical network and a set of traffic demands, in designing a virtual layer and grooming the traffic demands on virtual links called subbands, then to determine the number of subbands and the wavelength to assign for each subband of the virtual layer. We first propose a node-arcs, and arc-paths integer linear programming formulations for the problem, then we describe the column generation procedure for solving the linear relaxation of the 0-1 arc-paths formulations.

1 Introduction

User demand in traffic is steadily increasing and it is necessary to upgrade the transmission capacity of networks with the emerging high capacitated *WDM* (*Wavelength Division Multiplexing*) systems. The existence of physical phenomena also called transmission impairments [3] that affect the optical fibers, highlights the difficulty of setting up high capacitated wavelengths on long distances. A new technology that may support this evolution is currently under review and would lead to a network architecture that enables the routing of traffic rates up to 100 Gb/s for long distances. This technology is called OFDM (*Orthogonal Frequency Division Multiplexing*) and is based on the division of each wavelength of the network in many *subbands*, this is known as an *Optical Multi-Band OFDM Network*. This property enables to efficiently use the huge capacity of a wavelength and thus to significantly reduce the number of resources required to satisfy the traffic demands of the users.

In this paper, we are interested in the problem of designing an optical WDM layer based on the OFDM technology, with considering the traffic grooming particularity of WDM networks. This is a network design problem which consists in grooming and routing the traffic demands in a multi-band layer and assigning a wavelength for each subband of that layer so that the total cost is minimum. We first give a node-arcs, and an arc-paths integer linear programming formulations for the problem. Then we discuss some columns generation procedure for the arc-paths formulation. The network design problem received a growing interest and many variants of the problem have been

J. Pahl, T. Reiners, and S. Voß (Eds.): INOC 2011, LNCS 6701, pp. 1–6, 2011.

presented and solved by different approaches. Some authors have focused on the multi-layer architecture of these networks [2,5], and proposed many efficient branch-and-cut algorithms, based on polyhedral study [1,4]. In [6], authors have proposed some decomposition methods for a variant of the network design problem that is the Grooming, Routing and Wavelength Assignment problem. Our model is distinguished from the previous ones in that we consider in addition to the traffic routing, the assignment of a physical path to each installed subband, and other specific engineering constraints.

This paper is organized as follows. In Section 2, we present the *Optical Multi-Band Network Design* problem and we give two ILP formulations for the problem. In Section 3, we describe the column generation procedure for solving the relaxation of the 0-1 arcs-paths formulation. We finaly give some concluding remarks in Section 4.

2 The Optical Multi-Band Networks Design Problem

2.1 General Statement

An *optical network* has a *virtual layer* and a *physical layer*. The virtual layer is composed of several ROADMs[1] interconnected by *virtual links*. The physical layer is composed of *transmission nodes* interconnected by *physical links* each of which contains two optical fibers. Each multiplexer in the virtual layer is associated with a transmission node in the physical layer. And every link in the virtual layer consists in one or several OFDM subbands. Each subband of a link in the virtual layer corresponds to a path in the physical layer between the transmission nodes associated with the ends of the subband. The *Optical Multi-band Network Design* problem *(OMBNDP)* consists, given a physical layer of an optical network, where each physical link has a certain capacity, and a set of traffic demands, to design a virtual layer, and to determine for each virtual link, the number of subbands and the wavelength to be assigned to each subband.

In terms of graphs, the problem can be represented as follows. Consider a directed graph $G_1 = (V_1, A_1)$ that represents the physical layer of an optical network, where V_1 is the set of nodes and A_1 is the set of arcs. Each node $v \in V_1$ corresponds to a transmission node. The graph G_1 is such that if there is an arc (i,j) between two nodes i and j of V_1, there is also an arc (j,i) between j and i. Each arc $a \in A_1$ corresponds to an optical fiber, so that the traffic can be routed in both directions between i and j. Each fiber can support at most $|\Omega|$ wavelengths[2]. Every wavelength has a capacity C and can be divided into $|B|$ different subbands. The capacity of a subband can not exceed a certain value denoted by c_{max}.

Let K be a set of traffic demands in G_1. Each demand $k \in K$ has an origin node $o_k \in V_1$, a destination node $d_k \in V_1$ and a traffic amount D_k. The *(OMBND)* problem is to find a directed multigraph $G_2 = (V_2, A_2)$ corresponding to a virtual layer of the OFDM-based network, where V_2 and A_2 are the sets of nodes and arcs, respectively. Each node $v' \in V_2$ in G_2 corresponds to an ROADM and is associated to a node $v \in V_1$ in the graph G_1. Each arc $e \in A_2$ is composed of one or more subbands $b_i \in B$. Even if the traffic demands of K have their origin and destination nodes in the graph G_1, they have to be routed in the graph G_2.

[1] *Reconfigurable Optical Add/Drop Multiplexers.*

[2] In practice, an optical fiber may contain either 8, 32 or 80 wavelengths.

If we suppose that G_2 is complete on the set of nodes V_2, the problem is to remove a subset of nodes and a subset of arcs and find the smallest graph G_2, detemine the number of subbands to install on each virtual arc and the wavelength to be assigned to each subband, so that :

- For each demand k, there exists a path in G_2 between nodes o'_k and d'_k,
- There exists a path in G_1, associated to each subband b_i of the arc $e = (u', v') \in A_2$ between nodes u and v of V_1,
- A unique wavelength is assigned to each subband in a virtual link,
- The total cost of the virtual layer design is minimum, given that each subband b_i is associated a cost denoted by $\gamma(b_i)$.

In what follows we will first introduce some necessary notations, in order to give two integer linear programming formulations.

2.2 Node-Arcs ILP Formulation

Let $y^{eb\omega}$ be a binary variable that takes the value of 1 if a subband $b \in B$ is installed on a virtual link $e \in A_2$ and assigned a wavelength $\omega \in \Omega$ and 0 otherwise. Let $f_a^{eb\omega}$ be a binary variable that takes the value of 1 if the subband $b \in B$ of the virtual link $e \in A_2$ is assigned the wavelength $\omega \in \Omega$, while the associated path in G_1 uses the arc $a \in A_1$, and 0 otherwise. We finaly denote by $x^k_{eb\omega}$, a binary variable that takes the value of 1 if the traffic demand $k \in K$ uses the subband $b \in B$ of the virtual link $e \in A_2$ that is assigned the wavelength $\omega \in \Omega$ for its routing, and 0 otherwise. Let's denote by $m_1 = |A_1|$ and $m_2 = |A_2|$ the number of arcs in G_1 and G_2, respectively. And let $\Gamma(s)^+$, $s \in V_1 \cup V_2$ (resp. $\Gamma(s)^-$) be the set of arc outgoing from the node s (resp. incoming to s). A formulation for (OMBND) is:

$$Min \sum_{e\in A_2}\sum_{\omega\in\Omega}\sum_{b\in B}\gamma(b)y^{eb\omega}$$

$$\sum_{a\in\Gamma^+(s)} f_a^{eb\omega} - \sum_{a\in\Gamma^-(s)} f_a^{eb\omega} = \begin{cases} -y^{eb\omega}, \ if\ s' = u', \\ 0, \ if\ s' \neq u', v', \\ y^{eb\omega}, if\ s' = v', \end{cases} \quad \forall e \in A_2, \forall b \in B, \forall \omega \in \Omega, \forall s \in V_1, \tag{1}$$

$$\sum_{e\in A_2}\sum_{\omega\in\Omega}\sum_{b\in B} f_a^{eb\omega} \leq |\Omega||B|, \quad \forall a \in A_1, \tag{2}$$

$$\sum_{e\in A_2} f_a^{eb\omega} \leq 1, \quad \forall b \in B, \forall \omega \in \Omega, \forall a \in A_1, \tag{3}$$

$$\sum_{\omega\in\Omega}\sum_{b\in B}\sum_{e\in\Gamma^+(s')} x^k_{eb\omega} - \sum_{\omega\in\Omega}\sum_{b\in B}\sum_{e\in\Gamma^-(s')} x^k_{eb\omega} = \begin{cases} -1, \ if\ s' = d'_k, \\ 0, \ if\ s' \neq o'_k, d'_k, \\ 1, \ if\ s' = o'_k, \end{cases} \quad \forall k \in K, \forall s' \in V_2, \tag{4}$$

$$\sum_{k\in K} D^k x^k_{eb\omega} \leq c_{max} y^{eb\omega}, \quad \forall b \in B, \omega \in \Omega, e \in A_2, \tag{5}$$

$$y^{eb\omega} \in \{0,1\}, \quad \forall b \in B, \forall \omega \in \Omega, \forall e \in A_2, \tag{6}$$

$$f_a^{eb\omega} \in \{0,1\}, \quad \forall b \in B, \forall \omega \in \Omega, \forall e \in A_2, \forall a \in A_1, \tag{7}$$

$$x^k_{eb\omega} \in \{0,1\}, \quad \forall b \in B, \forall \omega \in \Omega, \forall e \in A_2, \forall k \in K. \tag{8}$$

In this formulation, there are $|B||\Omega|m_2$ binary design variables, $|B||\Omega|m_1m_2$ binary flow variables for the physical layer and $|B||\Omega||K|m_2$ binary flow variables for the virtual layer. The objective is to minimize the total cost of installing the required number of subbands on the arcs of the virtual layer. The three first constraints are flow constraints for each subband b assigned the wavelength ω. They ensure that a path in G_1 is associated to each subband of G_2. Constraints (2) and (3) ensure that the total number of subbands associated to each arc of G_1 cannot exceed the allowed capacity of that arc. Constraints (4) and (5) are flow constraints for each commodity k. They require that the traffic is not splitted on several paths, and the total flow on each subband cannot exceed the capacity installed on that subband.

2.3 Arcs-Paths ILP Formulation

It is also possible to formulate the problem using path variables as follows. We define, like for the node-arcs formulation the binary variables $y^{eb\omega}$ for the design of the virtual layer. Let's denote by P_e, $e=(u',v')\in A_2$, the set of paths in G_1 between u and v of V_1. We define the binary variables $f^{eb\omega}(p)$ that takes the value 1 if the path $p\in P_e$ is associated to the subband $b\in B$ that is established on $e\in A_2$ and assigned the wavelength $\omega\in\Omega$, and 0 otherwise. We denote by P_k, $k\in K$ the set of candidate paths in G_2 for routing the demand k. Let $x^k(\pi)$ be a binary variable that takes the value 1 if the path $\pi\in P_k$ is selected to route the demand k, and 0 otherwise.

Let's define the coefficients $\tau=(\tau_e^a(p), a\in A_1, e\in A_2, p\in P_e)$ and $\varphi=(\varphi_k^{eb\omega}(\pi), e\in A_2, b\in B, \omega\in\Omega, k\in K, \pi\in P_k)$ as binary parameters that indicate whether the arc a belongs to the path p, and the subband b of the link e that is assigned the wavelength ω that belongs to the path π. The (OMBND) problem is equivalent to the following 0-1 formulation:

$$Min \sum_{e\in A_2}\sum_{\omega\in\Omega}\sum_{b\in B}\gamma(b)y^{eb\omega}$$

$$\sum_{p\in P_e} f^{eb\omega}(p)=y^{eb\omega}, \qquad \forall e\in A_2, \forall b\in B, \forall\omega\in\Omega \tag{9}$$

$$\sum_{e,b,\omega}\sum_{p\in P_e}\tau_e^a(p)f^{eb\omega}(p)\leq|\Omega||B|, \qquad \forall a\in A_1, \tag{10}$$

$$\sum_{e\in A_2}\sum_{p\in P_e}\tau_e^a(p)f^{eb\omega}(p)\leq 1, \qquad \forall b\in B, \forall\omega\in\Omega, \forall a\in A_1 \tag{11}$$

$$\sum_{\pi\in P_k}x^k(\pi)=1, \qquad \forall k\in K \tag{12}$$

$$\sum_{\pi\in P_k}\sum_{k\in K}D^k\varphi_k^{eb\omega}(\pi)x^k(\pi)\leq c_{max}y^{eb\omega}, \qquad \forall e\in A_2, \forall b\in B, \forall\omega\in\Omega, \tag{13}$$

$$y^{eb\omega}, f^{eb\omega}\in\{0,1\}, \qquad \forall e\in A_2, b\in B, \omega\in\Omega, \tag{14}$$

$$x^k\in\{0,1\}, \qquad \forall k\in K. \tag{15}$$

In this formulation, the constraints are similar to the node-arcs formulation and their number is of order $O(V_1^2)$. This arc-paths formulation may contain an exponential number of variables, and the integer column generation procedure is an appropriate method to deal with this kind of formulations. In what follows, we describe the pricing problem associated to the linear relaxation of the formulation.

3 The Column Generation Procedure

In this section, we attempt to describe the column generation procedure that is used to solve the linear relaxation of the formulation (9)-(15). We solve iteratively the problem with a subset of columns. The remaining formulation is called Restricted Master Problem (RMP), and we search the missing columns with negative reduced cost by solving two pricing sub-problems. The (RMP) is given by the linear relaxation of (9)-(15), restricted to a subset of paths $J = J_1 \cup J_2$. J_1 is a subset of paths in the graph G_1 between the pairs of nodes (u,v), u, $v \in V_1$, corresponding to the the end nodes of the arcs $e = (u',v') \in A_2$. J_2 is a subset of paths in the graph G_2 between the pairs of nodes (o'_k, d'_k) corresponding to the source and destination nodes of each traffic demand k.

Let's denote by $\alpha^{eb\omega}$, β_a, $\lambda_a^{b\omega}$, μ^k and $\delta^{eb\omega}$, the dual variables associated to the constraints of the linear relaxation of (RMP). These dual variables allow to express the reduced costs given by (16) and (17)

$$-(\alpha^{eb\omega} + \sum_{a \in A_1} \tau_e^a(p)\beta_a + \sum_{a \in A_1} \tau_e^a(p)\lambda_a^{b\omega}) \tag{16}$$

$$-(\mu^k + \sum_{e \in A_2} \sum_{b \in B} \sum_{\omega \in \Omega} D^k \varphi_k^{eb\omega} \delta^{eb\omega}) \tag{17}$$

associated to the variable x^k. Both of the two associated pricing problems are *shortest path problems*, in the graphs G_1 and G_2, respectively, with specific weigths on the arcs of A_1, and specific weigths on each subband b of the arcs of A_2.

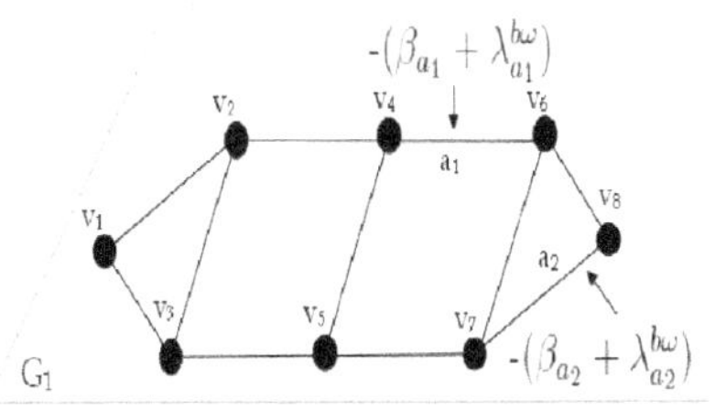

Fig. 1. The reduced cost on the arcs of G_1

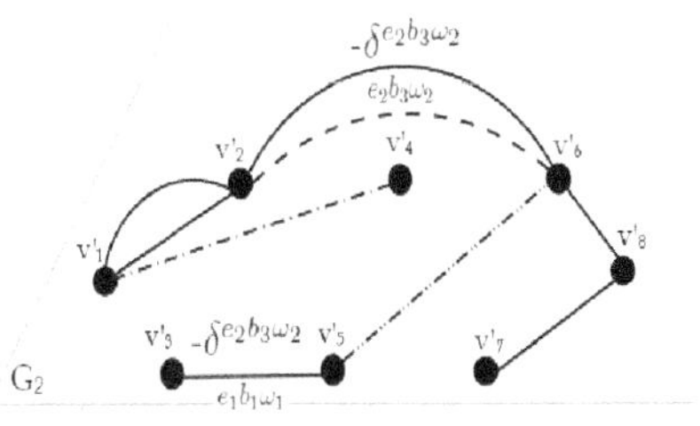

Fig. 2. The reduced cost on the arcs of G_2

The dual variables β_a, $\lambda_a^{b\omega}$ and $\delta^{eb\omega}$ associated to the constraints (10), (11) and (13), respectively are positive. The dual variables $\alpha^{eb\omega}$ and μ^k, associated to the constraints (10) and (12) respectively, are not necessarily positive as they arise from equalities. However, they express a constant cost that does not impact on the shortest path computation (see Fig. 1 and Fig. 2).

4 Conclusion

In this paper, we have considered the Optical Multi-Band Network Design problem. We have given two ILP formulations for the problem and described the column generation procedure for solving its linear relaxation. We have implemented the column generation procedure for the arc-paths formulation and provided some numerical results for both formulations. This work still in a preliminary phase, however some interesting topics would be to performe a polyhedral study of the problem, and to propose efficient algorithms in order to solve the problem for real instances.

References

1. Bienstock, D., Chopra, S., Günlück, O., Tsai, C.Y.: Minimum Cost Capacity Installation for Multicommodity Network Flows. Mathematical Programming 81, 177–199 (1998)
2. Borne, S., Gourdin, E., Liau, B., Mahjoub, A.R.: Design of survivable IP-over-optical networks. Ann. Oper. Res. 146(1), 41–73 (2006)
3. Buchali, F., Dischler, R.: Optical OFDM: A promising High-Speed Optical Transport Technology. Bell Labs Technical Journal 14(1), 125–146 (2009)
4. Dahl, G., Martin, A., Stoer, M.: Routing through Virtual Paths in Layered Telecommunication Networks. Operations Research 47(5), 693–702 (1999)
5. Orlowski, S., Koster, A.M.C.A., Raack, C., Wessäly, R.: Two-layer Network Design by Branch-and-Cut featuring MIP-based Heuristics. ZIB-Report 06-47 (November 2006)
6. Vignac, B., Vanderbeck, F., Jaumard, B.: Reformulation and Decomposition Approaches for Traffic Routing in Optical Networks. INRIA research report 00392256 (August 2009)

An Exact Algorithm for Robust Network Design

Christoph Buchheim[1], Frauke Liers[2], and Laura Sanità[3]

[1] Technische Universität Dortmund, Fakultät für Mathematik,
Vogelpothsweg 87, 44227 Dortmund, Germany
christoph.buchheim@math.tu-dortmund.de

[2] Universität zu Köln, Institut für Informatik, Pohligstraße 1, 50969 Köln, Germany
liers@informatik.uni-koeln.de

[3] Institute of Mathematics, École Polytechnique Fédérale de Lausanne,
1015 Lausanne, Switzerland
laura.sanita@epfl.ch

Abstract. Modern life heavily relies on communication networks that operate efficiently. A crucial issue for the design of communication networks is robustness with respect to traffic fluctuations, since they often lead to congestion and traffic bottlenecks. In this paper, we address an NP-hard single commodity robust network design problem, where the traffic demands change over time. For k different times of the day, we are given for each node the amount of single-commodity flow it wants to send or to receive. The task is to determine the minimum-cost edge capacities such that the flow can be routed integrally through the net at all times. We present an exact branch-and-cut algorithm, based on a decomposition into biconnected network components, a clever primal heuristic for generating feasible solutions from the linear-programming relaxation, and a general cutting-plane separation routine that is based on projection and lifting. By presenting extensive experimental results on realistic instances from the literature, we show that a suitable combination of these algorithmic components can solve most of these instances to optimality. Furthermore, cutting-plane separation considerably improves the algorithmic performance.

1 Introduction

Communication networks play a fundamental role in every-day life. Due to the huge growth of telecommunications services in the last years, the development of efficient methods for an optimal design of such networks is nowadays a crucial research area.

In a standard network design problem, we are given a network represented by a graph with non-negative costs on the edges, and we aim at routing a set D of demands through the network at minimum cost. However, since in practical settings the set of demands is often subject to uncertainty and may vary with time, more accurate models recently have been defined in the literature. In particular, there is a well-studied class of *robust* network design problems, which assumes to have as input a family $\mathcal{D}$ of possible sets of demands to be routed, instead of just one set. The aim is to install minimum cost capacities such that every set of demands $D \in \mathcal{D}$ can be suitably routed. Robust network design problems of this kind have received a lot of attention in the network design community, see e.g. [7,5,14,12,16,26] and the recent survey of Chekuri [10].

J. Pahl, T. Reiners, and S. Voß (Eds.): INOC 2011, LNCS 6701, pp. 7–17, 2011.

In this paper, we focus on a *single commodity* robust network design (RND) problem. As an example, suppose that some clients wish to download some program stored on several servers. For a client, it is not important which server he or she is downloading from, as long as the demand is satisfied. Still, at different times of the day (e.g. morning/afternoon/evening), the demands may change (e.g. different clients show up), and we would like to design a network that is able to route all flow in all different scenarios.

Formally, we are given an undirected graph $G = (V,E)$ with costs $c_{ij} \geq 0$ for every edge $\{i,j\} \in E$, and k sets $\{D_1,\dots,D_k\}$ of demands. A set D_t, also called a traffic matrix, specifies for each node $u \in V$ a value $b_u^t \in \mathbb{Z}$ of flow that the node wants to send ($b_u^t < 0$) or to receive ($b_u^t > 0$); one may also have $b_u^t = 0$. The goal is to install integral min-cost capacities $u \in \mathbb{Z}^E$ such that each traffic matrix D_t can be (non-simultaneously) integrally routed on G without exceeding the capacity.

Note that, if we have only one traffic matrix (i.e. $k = 1$), then the problem is just a min-cost single commodity flow problem, the so-called transshipment problem, and it is therefore easily solvable in polynomial time (see e.g. [11]). In contrast, whenever we take into account more scenarios, the problem becomes NP-hard, already for $k = 3$ [27] (the complexity is open for $k = 2$).

To the best of our knowledge, no exact methods are available in the literature for this problem so far. In this work, we provide a branch-and-cut algorithm, based on the natural flow formulation strengthened by generating local cuts, revisited according to [8]. We test our algorithm on a wide set of realistic instances, and show that in this application local cuts significantly improve the computational time to find an optimal integral solution.

Related Works: Robust network design problems with a family $\mathcal{D}$ of demand sets are widely studied in the literature.

A popular model is the one introduced by [7], where the family $\mathcal{D}$ is described by a polyhedron. In this setting, a well-known polyhedral set of traffic matrices is the *hose model*, defined by [12,16]. In fact, this model is the basis of one of the most important RND problem, namely the Virtual Private Network Design problem [20,19]. For RND problems with a polyhedral set of traffic matrices, many exact algorithms [5,14] as well as approximation algorithms [13,17,21] have been developed.

Whenever the set $\mathcal{D}$ of traffic matrices given in input is a finite list, as in our setting, the problem is a network synthesis problem with non-simultaneous flows. This problem has two main applications [23]: the first one, that we discussed in the previous section, is related to the design of a network with time-varying demands. One reason for considering this model, is that network operators may have historical data on which they can rely to construct an explicit list of traffic matrices to take into account. Here the number k of different sets can be assumed reasonably small, but typically we have multiple sources/destinations. The second application is related to the design of *survivable networks*. Here the number k is large, since it is equal to the number of edges of the graph, but we have a single source and a single destination at the time.

The first application has been studied in more detail in the multi-commodity case (i.e. when each traffic matrix specifies a demand for every pair of nodes). This problem is NP-hard already for $k = 2$ [27]. In this setting, although k can be assumed to be small, a flow formulation would use different flow variables for every pair of nodes and every scenario, which make exact approaches based on flow formulations more difficult to solve in practice when comparing it to our single commodity setting. Some heuristics are given in [23]. Still, the problem can be approximated within a factor of $O(\log |V|)$ using metric embedding techniques [15,27].

In survivable network design, we have a demand $r(i,j)$ for every edge $\{i,j\} \in E$ representing the flow that needs to be re-routed in case the edge $\{i,j\}$ fails, and the problem is to install capacity in order to non-simultaneously route each $r(i,j)$. In this setting, every flow is a single-commodity flow with exactly one source and one destination, but the number k is equal to the number of edges in the graph, therefore in a flow formulation we may again have order of $|V|^2$ different flow variables. The study of this problem was started by [18], who provided combinatorial algorithms for finding an optimal fractional solution in unit-cost metric graphs. Later on, some people studied the polyhedral structure of the problem (see e.g. [25,24]), and exact approaches for special classes of graphs (see e.g. [25,28]).

Interestingly, there is a 2-approximation for the survivable network design problem due to [22] based on an iterative rounding technique. Although our RND problem is also a single-commodity flow problem, we note that our setting has more sources and destinations, which is different from survivable network design. This may make the problem harder to approximate. In fact, the 2-approximation algorithm of [22] does not apply in our case (see [27] for more details), and it is an interesting open question to find a constant factor approximation for the single commodity RND problem.

2 Problem Formulation

We are concerned with the problem of assigning minimum cost edge capacities such that k different flows can be non-simultaneously integrally routed through the network. Clearly, once we compute k flows which realize the demands of the k traffic matrices in input, the capacity which needs to be installed on an edge is just as large as the maximum amount of flow routed along it for all matrices.

For the matrix D_t, let the variable f_{ij}^t model the amount of flow that is routed along edge $\{i,j\}$ in the direction from i to j. Then our optimization problem, which from now on we simply call the RND problem, can be formulated as follows:

$$\begin{aligned} \min \textstyle\sum_{\{i,j\}\in E} c_{ij} \max(f_{ij}^1 + f_{ji}^1, \ldots, f_{ij}^k + f_{ji}^k) & \\ \textstyle\sum_{j:\{j,i\}\in E} f_{ji}^t - \sum_{j:\{i,j\}\in E} f_{ij}^t = b_i^t \quad & \text{for } i \in V,\ t = 1,\ldots,k \\ f_{ij}^t \in \mathbb{Z}_+ \quad & \text{for } \{i,j\} \in E,\ t = 1,\ldots,k \end{aligned} \tag{1}$$

Considering the above set of constraints with linear cost functions, effective algorithms exist for determining optimum flows. However, in this formulation of the RND problem, the cost function is non-linear in the flow variables, which prevents their applicability. Trivially, this non-linear formulation can be linearized by introducing a capacity variable u_{ij} for each edge $\{i,j\} \in E$ that models the maximum amount of flow sent along the edge for all matrices:

$$
\begin{array}{lll}
\min \sum_{\{i,j\}\in E} c_{ij}u_{ij} & & \\
\sum_{j:\{j,i\}\in E} f^t_{ji} - \sum_{j:\{i,j\}\in E} f^t_{ij} = b^t_i & \text{for } i \in V,\ t = 1,\dots,k & \\
\qquad u_{ij} \geq f^t_{ij} + f^t_{ji} & \text{for } \{i,j\} \in E,\ t = 1,\dots,k & (2)\\
\qquad f^t_{ij} \geq 0 & \text{for } \{i,j\} \in E,\ t = 1,\dots,k & \\
\qquad u_{ij} \in \mathbb{Z}_+ & \text{for } \{i,j\} \in E &
\end{array}
$$

Note that in the above formulation we relaxed the integrality constraints for the flow variables. In fact, once an integral feasible capacity vector u is given, one can easily compute integral flows realizing our demands by solving k different flow problems, one for each traffic matrix D_t, with the given capacities u_{ij}. The existence of integral flows is guaranteed since we are dealing with single commodity flows (see e.g. [11]).

The linear relaxation of (2) is the LP at the basis of our branch-and-cut algorithm.

3 Preprocessing

We can preprocess a given RND instance by decomposing the network into biconnecting components. A biconnected component is a maximal connected subgraph such that the removal of any of its nodes does not destroy its connectedness. Any connected graph decomposes into its biconnected components, which are connected to each other by so-called cut vertices.

It is easy to see that the RND instance can be solved for each of the network's biconnected components independently as follows. There exists at least one of them, say $C = (V_C, E_C)$, that contains only one cut vertex $v \in V_C$. All flow into and out of C has to be routed through v. Therefore, we can decompose the RND instance on G into an RND instance on C and an RND instance on a graph G' which is the union of all components different from C. Note that v is included in both C and G'. In the RND instance on C, the demand of the cut vertex is set to $b_v = \sum_{u\in V\setminus V_C} b_u$. In the other instance, we set $b_v = \sum_{u\in V_C} b_u$. The demands for all the other nodes are left unchanged.

Applying the same arguments to G' recursively and appropriately choosing the demands of the cut vertices, the RND problem on G can be reduced to RND problems on its biconnected components, which have a smaller size. The partial optimal solutions for different biconnected components can trivially be combined to an optimum solution for the whole network.

4 Primal Heuristics

Within a branch-and-cut approach, it is important to design so-called primal heuristics that try guessing good feasible solutions in order to derive upper bounds on the optimum value. A standard approach finds feasible solutions by appropriately rounding the optimal solutions of the LP relaxations.

For the RND problem, suppose we have solved an LP relaxation at some node in the branch-and-bound tree, and let (f^*, u^*) be the optimum solution. We compute a feasible solution in three steps. First, we define $\overline{u} \in \mathbb{Z}^E$ to be the vector with entries $\overline{u}_e = \lceil u^*_e \rceil$, for all $e \in E$. Second, we determine k integral flows that satisfy the k different traffic matrices and respect the capacities given by $\overline{u}$. By construction such flows exist, and we compute them by solving a minimum-cost flow problem for each traffic matrix, with a randomly chosen linear objective function. Finally, the k flows are combined to an RND solution by determining for each edge the actual capacity u^{prim} necessary to route the k flows. Note that some entries of the vector u^{prim} could be strictly smaller than the corresponding entries of the vector $\overline{u}$.

In computing the k flows, we use the same cost function for all matrices, as the same edges should be preferred for each of the k flows, in order to keep the values of the capacity variables low. The cost function is chosen randomly in order to have the chance of generating different solutions in each iteration.

In the next lemma, we give an upper bound on the quality of a feasible solution $(f^{\text{prim}}, u^{\text{prim}})$ obtained by this procedure. More specifically, we relate it to the value of an optimum feasible solution contained in the subtree of the corresponding node of the branch-and-bound tree. If the node is not the root, we call such a solution local as it is optimum under the constraints given by the branching decisions.

Lemma 1. *The distance of a (local) optimum RND solution to the feasible solution $(u^{\text{prim}}, f^{\text{prim}})$ generated in the primal heuristic is at most $c^\top(\overline{u} - u^*)$.*

Proof. Let $(f^{\text{loc}}, u^{\text{loc}})$ be a (local) optimum solution that is feasible for RND. Clearly, $c^\top u^{\text{prim}} \leq c^\top \overline{u}$ and $c^\top u^{\text{loc}} \geq c^\top u^*$. This implies $c^\top u^{\text{prim}} - c^\top u^{\text{loc}} \leq c^\top \overline{u} - c^\top u^*$.

5 Separation of Target Cuts

For designing an effective branch-and-cut algorithm, it is essential to separate strong cutting planes so that branching is rarely necessary. In [8], the separation of target cuts was introduced as a variant of the local cuts by Applegate et al. [6]. No predescribed structure is imposed on the generated cutting planes. Furthermore, their separation is a general procedure that can be applied in various contexts. For the separation, the problem is first projected into a low-dimensional space. Let $\overline{P}$ denote the convex hull of all projections of feasible solutions. Let x^* be the point to be separated and $\overline{x}^*$ its projection. The separation problem for x^* and the polytope P in question is solved heuristically by generating a facet separating $\overline{x}^*$ from $\overline{P}$, if it exists. Such a facet can be found by determining an optimal extremal solution of a linear optimization problem whose size is linear in the number of vertices of $\overline{P}$ and the extreme rays. Let $\overline{q}$ belong to the relative interior of $\overline{P}$. The cut-generation LP is of the form

$$\begin{array}{lll} \max & a^\top(\overline{x}^* - \overline{q}) & \\ \text{s.t.} & a^\top(\overline{x}_i - \overline{q}) \leq 1 & \text{for all vertices } \overline{x}_i \text{ of } \overline{P} \\ & a^\top(\overline{x}_i - \overline{q}) \leq 0 & \text{for all extreme rays } \overline{x}_i \text{ of } \overline{P} \\ & a \in \mathbb{R}^r & \end{array} \tag{3}$$

If the optimum value of (3) is larger than 1, the inequality $a^\top(x - \overline{q}) \leq 1$ is violated by $\overline{x}^*$. Furthermore, it is proven in [8] that an optimum solution of (3) defines a facet of $\overline{P}$. In case (3) is unbounded, then $a^\top(x - \overline{q}) = 0$ is a valid equation for $\overline{P}$ and violated by $\overline{x}^*$, where a is an unbounded ray. Finally, the inequality is lifted to become valid (not necessarily facet-defining) for P. In [9], target cuts were successfully used for solving several constrained binary quadratic optimization problems.

5.1 Choice of the Projection

Choosing good projections is crucial for the success of the target-cut separation. In most optimization problems defined on a graph $G = (V,E)$, the polytope $\overline{P}$ is either determined through an orthogonal projection onto some subgraph of G, or through shrinking subsets of nodes or edges in G. The resulting graph is denoted by $\overline{G} = (\overline{V}, \overline{E})$.

For the RND problem, the polytope P in the original variable space is the convex hull of all feasible solutions (f,u) of problem (2). An orthogonal projection onto some subgraph that only contains a subset of nodes is not useful. Indeed, suppose $\overline{G}$ is obtained through an orthogonal projection such that for an edge $\{v_1, v_2\} \in E$ it is $v_1 \in \overline{V}$ but $v_2 \notin \overline{V}$. Then, $\overline{P}$ also contains vectors (f,u) for which f does not need to satisfy the flow-conservation constraints for v_1 because (positive or negative) excess flow at v_1 could be annihilated in G by routing flow along $\{v_1, v_2\}$. Therefore, most of the structure of the RND polytope is lost when using such a projection.

In contrast, we iteratively choose an edge $\{v_1, v_2\} \in E$ randomly and shrink it by identifying the nodes v_1 and v_2. Loops are deleted, multiple edges are replaced by one edge, the demand of the resulting supernode is set to $b_{v_1} + b_{v_2}$. The corresponding entries in the optimum solution x^* of the relaxation are summed up. We shrink until the number of (super-)nodes in the shrunk graph is equal to the value of a fixed parameter c that specifies the size of $\overline{G}$. The latter is also called a *chunk*.

Let $\overline{P}$ be the convex hull of the vertices in the projected space obtained through shrinking edges as outlined above. Clearly, the vectors in $\overline{P}$ being projections of feasible solutions of (2) need to satisfy the flow conservation constraints on $\overline{G}$. It is easy to see that $\overline{P}$ is again the convex hull of feasible solutions of problem (2), but defined on $\overline{G}$.

Finally, a different parameter l specifies the number of traffic matrices that should be taken into account. In case this number is smaller than the original number k of matrices, we randomly choose a subset of them.

The target-cut separation routine now determines facets of $\overline{P}$ that are violated by $\overline{x}^*$, if they exist. These inequalities need to be lifted to become valid for P. We use standard lifting procedures. First, we argue that an inequality valid for a subset of scenarios remains valid if all scenarios are addressed. Indeed, as the capacity variables are not bounded from above in the RND model, their coefficients are necessarily nonpositive in any valid inequality. As the capacity values can only increase when enlarging the set of traffic matrices, an inequality that is valid for a subset remains valid when all matrices are considered. An inequality is then iteratively lifted to an inequality valid for P by simultaneously unshrinking the graph. The shrinking steps are undone one after the other, in reverse order of the shrinking procedure. Suppose some loop was deleted when shrinking an edge $\{v_1, v_2\}$. Furthermore, also suppose some multiple edges were replaced by a single edge e. We obtain the lifted inequality for the graph in which nodes v_1 and v_2 are unshrunk as follows. The coefficient corresponding to the loop edge is set to zero. Let $\overline{a}_e$ be the coefficient of the single edge that represents multiple edges. As the flow along e in $\overline{G}$ can now be split along multiple edges, $\overline{a}_e$ is set as coefficient for each of these. The coefficient of $\{v_1, v_2\}$ is set to zero. All other coefficients remain unchanged. It is easy to see that iteratively applying this lifting and unshrinking procedure yields an inequality valid for P.

5.2 Cut Generation

In our implementation, target cuts are generated by delayed column generation [8]. Starting from a small subset of feasible points in $\overline{P}$, the remaining points of $\overline{P}$ are generated only if necessary. More precisely, a candidate target cut is computed considering the initial set of points, then it is checked whether this inequality is violated by some point in $\overline{P}$ not generated yet. If so, the new point is added and a new candidate cut is computed. To check whether $\overline{P}$ contains a point violating the given cut, we need a so-called oracle that solves the RND problem on the shrunk graph $\overline{G}$. This is done by applying a branch-and-cut algorithm to the MIP model (2).

In our application, the delayed column generation approach is crucial. While for binary problems it might be feasible to completely enumerate all integer points in $\overline{P}$ at once, at least in small dimensions, this is practically not possible in the presence of general integer variables, since the number of these points could become much larger. As described in [8], the delayed column generation procedure can be applied even if the set of initial points is low-dimensional. However, to avoid dealing with numerical problems and to speed up the cut generation process significantly, we always start from a full-dimensional polyhedron $\overline{P}_0$, which is computed as sketched in the following.

First we compute any feasible solution (f,u) for the RND problem on the shrunk graph $\overline{G}$ and set $\overline{P}_0 = \{(f,u)\}$. Such a solution (f,u) can be computed efficiently as a composition of arbitrary feasible solutions for the single matrices, computing appropriate values for the variables u_{ij} in the end. Next, we determine any cycle basis of $\overline{G}$. For each cycle C in the basis and for each of the l matrices considered (recall that we may select a subset of the k matrices), we add the incidence vector of C to the entries of f corresponding to the chosen matrix and adjust the u-entries. The result is a new feasible vector in $\overline{P}$, which we add to $\overline{P}_0$. All vectors added in this way are affinely independent. If $\overline{G}$ has $\bar{n}$ nodes and $\bar{m}$ *directed* edges (counted in both directions (i,j) and (j,i)), the cycle basis contains $\bar{m}-\bar{n}+1$ elements, so the current polytope $\overline{P}_0$ has dimension $l(\bar{m}-\bar{n}+1)$. Additionally, we add an unbounded direction to $\overline{P}_0$ for each variable u_{ij}, since increasing u_{ij} preserves feasibility. Equivalently, in the cut generation LP (3) we may enforce that the coefficient of u_{ij} is non-positive. The dimension of $\overline{P}_0$ increases to $l(\bar{m}-\bar{n}+1)+\frac{1}{2}\bar{m}$. Finally, for each vertex in $\overline{G}$, we have a valid flow conservation constraint (and all but one of these equations are independent). If a is the coefficient vector of any such constraint, we add the unbounded directions a and $-a$ to $\overline{P}_0$. Considering (3), this implies that all generated target cuts will be orthogonal to all flow conservation constraints. The final dimension of $\overline{P}_0$ is $l(\bar{m}-\bar{n}+1)+\frac{1}{2}\bar{m}+l(\bar{n}-1) = (l+\frac{1}{2})\bar{m}$, which means $\overline{P}_0$ is full-dimensional.

If a target cut is found, i.e., if $\bar{x}^* \notin \overline{P}$, we try to compute further target cuts by a reoptimization approach: in (3), we choose the first non-zero coefficient in the last generated cut and fix it to zero. Then we reoptimize (3), using delayed column generation again if necessary, and continue fixing coefficients until no violated target cut is found.

6 Computational Results on Realistic Instances

We implemented an exact branch-and-cut algorithm based on the ideas outlined in the previous sections, using the optimization tool SCIL [4] in combination with ABACUS 3.0 [1], LEDA 6.1 [2], and CPLEX 12.1 [3]. The executable is run on 2.3 GHz machines with a limit of four hours CPU time for each job. Furthermore, as the program is a 32-bit executable, a maximum of 4 GB of memory can be addressed. Parameters controlling the target-cut separation are the chunk size c and the number of traffic matrices l used in the target-cut separation. In particular, for $c = 0$, no separation is performed. We evaluate our method on 1120 realistic network topologies from the literature [5]. For each of these instances, $k = 2,3,4,5$ random traffic scenarios are added. Furthermore, for each network topology and each choice of k, we randomly choose the percentage p of terminals, i.e., nodes with non-zero demand, as $p = 25,50,75,100\%$. For very small instances, we did not use $p = 25\%$. Altogether, these are 1120 different instances. In Table 1, we report the distribution of the instance sizes. As the instances from [5] strongly vary in size, they are grouped with respect to the number of nodes in the network in bins of size 150. Average node and edge numbers of the instances in the respective groups are also given.

Table 1. Distribution of sizes for the realistic instances, grouped in bins by the number of nodes $|V|$ in the network

bin	$\|V\|$avg	$\|E\|$avg	# instances
$0 \le \|V\| \le 149$	34.31	55.48	612
$150 \le \|V\| \le 299$	201.31	383.57	268
$300 \le \|V\| \le 449$	352.00	578.80	240

Within the time and memory limits, 94% of the instances could be solved to optimality, even without separation. Using target cuts separation, we can further increase the number of instances that can be solved to optimality. The fact that almost all of the instance set can be solved shows the effectiveness of our approach. In order to evaluate the computational results in more detail, we first assess the quality of the primal heuristic. For the instances that could be solved to optimality without separation, we report the distance of the optimum solution to the feasible RND solution generated from the first LP relaxation. More specifically, we determine the relative gap g in %, i.e., $g = 100 \cdot \frac{x^{\text{prim}} - x^{\text{opt}}}{x^{\text{opt}}}$. In 24% of the cases, $g < 0.1\%$, in 45% of the cases, $g < 1\%$, and in 55% of the cases, $g > 10\%$. In the worst case, the gap is not larger than 61%.

In Table 2, we report results for solving the instances to optimality. Results are presented separately for each number l of scenarios, grouped with respect to the number of nodes in the network. As the running time usually increases for chunk sizes larger than 4, we restrict ourselves to smaller chunks and therefore use the parameter choices $(c,l) = (0,0),(3,2),(3,3),(4,2),(4,3)$, where the case $(0,0)$ means that no separation takes place. Numbers are only shown for instances in which the number k of scenarios is at least as large as l. For each parameter choice, we report in the first column the

Table 2. Experimental results for the realistic instances from the literature. From top to bottom, the number of scenarios increases from $k = 2$ to $k = 5$. For each parameter choice in the separation, we first report the number of instances solved, followed by the number of instances that were not solved due to the time limit (t) or due to memory constraints (m), respectively. The following columns show the average number of subproblems and the average cpu time in seconds for solving the remaining instances.

$\|V\|$	(0,0)					(3,2)					(3,3)					(4,2)					(4,3)				
	slvd	t	m	subs	CPU	slvd	t	m	subs	CPU	slvd	t	m	subs	CPU	slvd	t	m	subs	CPU	slvd	t	m	subs	CPU
$0 \le \|V\| \le 149$	152	0	1	373.09	1.40	152	0	1	116.78	1.20						153	0	0	62.34	3.46					
$150 \le \|V\| \le 299$	63	1	3	4416.32	112.01	66	0	1	1272.03	133.91						65	0	2	932.94	331.68					
$300 \le \|V\| \le 449$	56	0	4	1093.98	1.64	60	0	0	3301.40	85.96						59	1	0	1370.51	228.32					
$0 \le \|V\| \le 149$	149	0	4	627.54	3.66	150	0	3	323.63	3.28	150	0	3	506.13	12.85	152	0	1	135.61	9.95	151	0	2	146.58	49.46
$150 \le \|V\| \le 299$	59	0	9	7764.75	57.86	61	3	4	237.79	8.42	61	3	4	458.61	67.93	61	3	4	371.62	120.32	59	5	4	409.24	511.14
$300 \le \|V\| \le 449$	57	0	3	8225.11	139.17	60	0	0	3780.33	108.29	58	2	0	3344.52	262.48	59	1	0	1928.53	336.32	56	3	1	938.36	522.55
$0 \le \|V\| \le 149$	146	0	7	844.09	7.53	147	0	6	409.96	7.29	148	0	5	1554.40	50.00	150	0	3	383.42	26.49	150	1	2	421.49	123.11
$150 \le \|V\| \le 299$	61	0	6	4403.41	23.66	62	0	5	1685.76	59.78	62	1	4	1277.21	166.56	60	3	4	1981.53	657.47	58	5	4	426.67	539.23
$300 \le \|V\| \le 449$	57	0	3	1723.42	3.65	58	0	2	579.90	17.66	58	0	2	630.38	34.01	59	1	0	1319.71	220.21	58	2	0	719.55	380.44
$0 \le \|V\| \le 149$	150	0	3	994.91	10.17	150	0	3	441.43	7.39	150	0	3	475.33	15.94	152	0	1	256.52	16.80	150	0	3	250.77	56.18
$150 \le \|V\| \le 299$	57	0	10	8762.09	99.26	61	2	4	4940.90	172.72	57	6	4	4005.89	409.68	58	5	4	2309.07	551.97	55	8	4	584.24	661.29
$300 \le \|V\| \le 449$	50	0	10	7774.72	69.81	55	0	5	6373.80	180.31	55	2	3	6450.38	421.98	54	5	1	3126.33	384.23	51	9	0	1148.20	734.27

number of instances that could be solved to optimality for a specific choice of separation parameters, followed by the number of instances that could not be solved due to time or memory constraints, respectively. Typically, for an instance that could not be solved due to memory constraints, a number of subproblems in the order of 10^5 was generated. The following columns show the average number of subproblems in the branch-and-bound tree and the average cpu time in seconds of the instances that could be solved to optimality. Interestingly, many instances in Table 2 can be solved within a few minutes only. On average, instances with only two scenarios are computationally easier than those with a larger number of scenarios, as can be expected. Furthermore, on average the difficulty often increases with increasing network sizes.

Clearly, target-cut separation considerably improves the performance of the algorithm. Whereas several instances cannot be solved to optimality without separation, the number of unsolved instances is never worse and often better if target-cut separation is used. Usually, when reaching the time limit, the unsolvable instances have a huge number of open subproblems, often in the range of 10^5 or more. Thus, these instances are too difficult for the corresponding algorithmic setting. The fact that the algorithm with target cut separation can solve a larger number of instances to optimality shows the effectiveness of the separation. Furthermore, instances solvable without separation can be solved considerably faster and with a smaller number of subproblems when target cuts are separated.

We conclude that our approach can solve to optimality most of the realistic instances that we have at hand. Whereas instances defined on small networks should probably be solved without separation, in many cases target cut separation leads to faster solution of instances or even makes it possible to solve otherwise unsolvable instances.

Acknowledgement

Financial support from the German Science Foundation (DFG) is acknowledged under contracts Bu 2313/1–2 (Christoph Buchheim) and Li 1675/1–2 (Frauke Liers). Laura Sanità was supported by the Swiss National Science Foundation within the project "Robust Network Design".

References

1. ABACUS – A Branch-And-CUt System, `http://www.informatik.uni-koeln.de/abacus`
2. LEDA – Library of Efficient Data Types and Algorithms, `http://www.algorithmic-solutions.com`
3. ILOG CPLEX 12.1 (2009), `http://www.ilog.com/products/cplex`
4. SCIL 0.9 (2011), `http://www.scil-opt.net`
5. Altin, A., Amaldi, E., Belotti, P., Pinar, M.: Provisioning virtual private networks under traffic uncertainty. Networks 49(1), 100–115 (2007)
6. Applegate, A., Bixby, R., Chvátal, V., Cook, W.: TSP cuts which do not conform to the template paradigm. In: Jünger, M., Naddef, D. (eds.) Computational Combinatorial Optimization. LNCS, vol. 2241, pp. 261–304. Springer, Heidelberg (2001)
7. Ben-Ameur, W., Kerivin, H.: Routing of uncertain demands. Optimization and Engineering 3, 283–313 (2005)

8. Buchheim, C., Liers, F., Oswald, M.: Local cuts revisited. Operations Research Letters 36(4), 430–433 (2008)
9. Buchheim, C., Liers, F., Oswald, M.: Speeding up IP-based algorithms for constrained quadratic 0–1 optimization. Mathematical Programming (Series B) 124(1-2), 513–535 (2010)
10. Chekuri, C.: Routing and network design with robustness to changing or uncertain traffic demands. SIGACT News 38(3), 106–128 (2007)
11. Chvátal, V.: Linear programming. Series of books in the mathematical sciences. W.H. Freeman, New York (1983)
12. Duffield, N., Goyal, P., Greenberg, A., Mishra, P., Ramakrishnan, K., van der Merwe, J.: A flexible model for resource management in virtual private networks. Proceedings of SIGCOMM 29, 95–108 (1999)
13. Eisenbrand, F., Grandoni, F., Oriolo, G., Skutella, M.: New approaches for virtual private network design. SIAM Journal on Computing, 706–721 (2007)
14. Erlebach, T., Rüegg, M.: Optimal bandwidth reservation in hose-model VPNs with multi-path routing. Proceedings of INFOCOM 4, 2275–2282 (2004)
15. Fakcharoenphol, J., Rao, S., Talwar, K.: A tight bound on approximating arbitrary metrics by tree metrics. In: Proceedings of the 35th Annual ACM Symposium on Theory of Computing, pp. 448–455 (2003)
16. Fingerhut, J., Suri, S., Turner, J.: Designing least-cost nonblocking broadband networks. Journal of Algorithms 24(2), 287–309 (1997)
17. Fiorini, S., Oriolo, G., Sanità, L., Theis, D.: The VPN problem with concave costs. SIAM Journal on Discrete Mathematics, 1080–1090 (2010)
18. Gomory, R., Hu, T.: Multi-terminal network flow. SIAM Journal on Applied Mathematics 9, 551–570 (1961)
19. Goyal, N., Olver, N., Shepherd, B.: The VPN conjecture is true. In: Proceedings of STOC, pp. 443–450 (2008)
20. Gupta, A., Kleinberg, J., Rastogi, R., Yener, B.: Provisioning a virtual private network: A network design problem for multicommodity flow. In: Proceedings of STOC, pp. 389–398 (2001)
21. Gupta, A., Kumar, A., Roughgarden, T.: Simpler and better approximation algorithms for network design. In: Proceedings of STOC, pp. 365–372 (2003)
22. Jain, K.: A factor 2 approximation algorithm for the generalized Steiner network problem. In: Proceedings of the 39th Annual Symposium on Foundations of Computer Science (FOCS) (1998)
23. Labbe, M., Séguin, R., Soriano, P., Wynants, C.: Network synthesis with non-simultaneous multicommodity flow requirements: Bounds and heuristics (1999)
24. Magnanti, T., Raghavan, S.: Strong formulations for network design problems with connectivity requirements. Networks 45, 61–79 (2005)
25. Magnanti, T., Wang, Y.: Polyhedral properties of the network restoration problem with the convex hull of a special case. Tech. rep., Operations Research Center, MIT (1997)
26. Minoux, M.: Robust network optimization under polyhedral demand uncertainty is NP-hard Discrete Applied Mathematics, 597–603 (2010)
27. Sanità, L.: Robust Network Design. Ph.D. Thesis, Università Sapienza di Roma (2009)
28. Sridhar, S., Chandrasekaran, R.: Integer solution to synthesis of communication networks. Mathematics of Operations Research 3, 581–585 (1992)

SRG-Disjoint Design with Dedicated and Shared Protection

Bernardetta Addis[1], Giuliana Carello[2], and Federico Malucelli[2]

[1] Università degli studi di Torino, Corso Svizzera 185, Torino
addis@di.unito.it
[2] Politecnico di Milano, via Ponzio 34, Milano
{carello,malucell}@elet.polimi.it

Abstract. We consider a one layer network design problem in the presence of Shared Risk Groups (SRGs). We can define an SRG as a set of links of the logical layer that simultaneously fail in the case of a failure of a lower layer. We propose a mathematical model for the dedicated protection and two alternative models for the shared protection, all based on flow variables. In addition we propose a simple constructive heuristic. We report some computational results that compare the outcome of the models (upper and lower bounds provided by a commercial software) and those of the heuristic.

1 Introduction and Problem Statement

The design of resilient networks is a crucial problem in telecommunications and has obtained much attention by the optimization community. The most traditional setting considers link failures. The link capacity must be allocated at minimum cost so that in case of single or multiple link failures the demands can still be routed. When single failure protection is accounted for, for each demand the capacity on a pair of disjoint paths (primary and backup) from the origin to the destination nodes must be allocated. If the protection is of dedicated type, the capacity must be reserved to the demand on both paths, thus the capacity allocated on each link must consider the sum of flows on the primary and backup paths using that link. However, this kind of protection can lead to an excessive capacity allocation. Indeed if two primary paths are completely disjoint, they can share the capacity allocated on common portions of the backup paths, since the event that both demands need to be rerouted on the backup paths never occurs. This is the so called *shared protection*.

Due to the diffusion of multilayer networks, protection mechanisms, even in the dedicated case, are required to consider more complex features. The links of the logical layer network correspond to complex objects in lower layers. This means that in case of a failure in the lower layers more than one link is affected in the upper one. This complexity is captured by the so called SRGs. We can define a SRG as the set of links of the logical layer, where the design must be decided, that simultaneously fail in the case of a failure of the lower layer. Notice that one link of the logical layer, being made by more entities of the lower layers, may belong to more than one SRGs, as it may fail upon the failing of any of the entities defining it. SRGs may be associated to

J. Pahl, T. Reiners, and S. Voß (Eds.): INOC 2011, LNCS 6701, pp. 18–23, 2011.

single layer networks, as well, as they are related to the physical layout of the network elements. In the presence of SRG, protections (both dedicated or shared) must consider this additional complexity that usually spoils the network structure of the problem as one SRG can include apparently uncorrelated links.

Resilient network design has been widely addressed. In [5] a review on different resilient network design problems is proposed together with path based models: failure scenarios (or states) are taken into account, which, as SRGs, can be used to represent multiple simultaneous failures. The complexity of path-based survivability mechanism is addressed in [6]. Problems arising in the SRG network design context have been considered in several recent works as for example [3] and [4] where some paths and cut problems on colored graphs are analyzed from the computational complexity and approximability point of view. In [7] a network design of resilient networks with SRG is considered. The authors propose a very complex mathematical model that includes many technological aspects. Often SRG related problems arise in two layer networks, as shown in [1] and [2], where SRGs are tackled considering an explicit knowledge of the physical layer affecting them.

In this work we consider a one layer network design problem in the presence of SRGs. We propose one mathematical model for the dedicated protection and two alternative models for the shared protection, all based on flow variables. In addition we propose a simple constructive heuristic. We report some computational results that compare the outcome of models (upper and lower bounds provided by a commercial software) and those of the heuristic.

The network is represented by a directed graph $G = (N, A)$. The set of arcs corresponds to the set of potential links where capacity must be allocated. A set of traffic demands K is given. Each traffic demand is defined by a triple (o_k, t_k, d_k) where o_k and t_k are the origin and the destination of d_k units of flow.

The set $\mathscr{S}$ of SRGs is also given. Each SRG $s \in \mathscr{S}$ is a subset of A. To guarantee resilience to single arc failure, each arc represents a SRG, as well. We assume that the arc capacity is provided by installing transportation channels. Each transportation channel provides a capacity of λ flow units. The cost of installing one transportation channel on arc (i, j) is c_{ij}.

The reliable network design problem consists in finding, for each demand k, two paths from the origin o_k to the destination d_k (unsplit routing) disjoint on the SRGs and allocate the capacity on the arcs at minimum cost. The capacity allocation depends on the type of applied protection. In the case of dedicated protection, the capacity to be allocated on a link (i, j) amounts to the sum of flows (primary or backup) traversing the link. In the case of shared protection, the capacity allocated on link (i, j) is given by the sum of flows of the primary paths traversing the link plus the maximum of the backup paths flows of demands that do not share SRGs on their primary paths.

2 Proposed Approaches: Models and Heuristic

Let us consider the end-to-end dedicated protection case. The problem can be modelled using two sets of binary variables, describing the nominal and backup paths of a demand k, x_{ij}^k and y_{ij}^k, which are equal to one if the nominal path – backup path respectively–

of demand k is routed on arc (i,j). Beside, an integer variable z_{ij} gives the number of channels installed on arc (i,j). The model is the following:

$$\min \sum_{(i,j)\in A} c_{ij}z_{ij} \tag{1}$$

$$\sum_{(j,i)\in A} x^k_{ji} - \sum_{(i,j)\in A} x^k_{ij} = b^k_i \qquad \forall i \in N, k \in K \tag{2}$$

$$\sum_{(j,i)\in A} y^k_{ji} - \sum_{(i,j)\in A} y^k_{ij} = b^k_i \qquad \forall i \in N, k \in K \tag{3}$$

$$\sum_{k\in K} d_k \left(x^k_{ij} + y^k_{ij}\right) \le \lambda z_{ij} \qquad \forall (i,j) \in A \tag{4}$$

$$x^k_{ij} + y^k_{hl} \le 1 \ \forall k \in K, s \in \mathscr{S}, (i,j), (h,l) \in s \tag{5}$$

$$x^k_{ij}, y^k_{ij} \in \{0,1\}, z_{ij} \in \mathbb{Z}_+ \qquad \forall (i,j) \in A, k \in K \tag{6}$$

where $b^k_i = -1$ if $i = o_k$, 1 if $i = t_k$ and 0 otherwise. Objective function (1) guarantees the minimum installation costs. Constraints (2) (and (3)) guarantee that there exists one and only one primary (and backup, respectively) path for each traffic demand. Constraints (5) force the two paths to be SRG disjoint for each demand. Equations (4) are arc dimensioning constraints, while (6) are integrality constraints.

Let us consider now the shared protection case. The problem can be modelled using x^k_{ij} and y^k_{ij} binary variables, and integer variables z_{ij}, with the same meaning of the above model. Besides, a binary variable g^{sk}_{ij} is introduced, which is equal to one if the demand k is re-routed on arc (i,j) in case fault associated to SRG s occurs. Objective function (1) is kept, as well as constraints (2), (3) and (5). The following constraints are added to force the dimensioning of each link:

$$x^k_{hl} + y^k_{ij} - 1 \le g^{sk}_{ij}, \ \forall s \in \mathscr{S}, \forall (h,l) \in s, \forall (i,j) \in A : (i,j) \neq (h,l), (l,h) \tag{7}$$

$$\sum_{k\in K} d_k \left(x^k_{ij} + g^{sk}_{ij}\right) \le \lambda z_{ij}, \qquad \forall (i,j) \in A, \forall s \in \mathscr{S} \tag{8}$$

Constraints (7) force the value of variable g^{sk}_{ij}, while constraints (8) guarantee that the installed capacity on each arc is enough to manage all the traffic demands whose primary path is routed on the arc, plus the amount of re-routed traffic due to the worst fault condition.

An alternative formulation uses a binary variable v_{ks} which is equal to one if demand k is affected by SRG s, and continuous variables r^{ij}_{ks}, which represent the amount of traffic of demand k which must be rerouted on arc (i,j) if SRG s occurs. In such formulation constraints (7) and (8) are replaced by:

$$\Sigma_{(i,j)\in s} x^k_{ij} \le |s| v_{ks} \qquad \forall s \in \mathscr{S}, k \in K \tag{9}$$

$$r^{ij}_{ks} \ge d_k \left(v_{ks} + y^k_{ij} - 1\right) \qquad \forall s \in \mathscr{S}, k \in K, (i,j) \in A \tag{10}$$

$$\Sigma_{k\in K} r^{ij}_{ks} + \Sigma_{k\in K} d_k x^k_{ij} \le \lambda z_{ij} \ \forall (i,j) \in A, s \in \mathscr{S}. \tag{11}$$

Greedy heuristic The problems are solved via a greedy approach integrated in a Multistart. The greedy approach sequentially routes the demands and dimensions the network. Following an ordered set of demands, which can be randomly changed to provide a multistart approach, the algorithm considers one by one the demands and routes them as follows:

- Evaluate incremental costs, i.e. the increase in the capacity installation cost which the demand will cause if routed on the arc: for each arc the capacity installation cost is computed assuming that the considered demand is routed on the arc. The incremental cost are given by the difference between such cost and the current one.
- Find minimum incremental cost pair of primary and backup paths solving suitable ILP models.
- Route demand on graph and dimension link capacity.

3 Computational Results

The models and the proposed heuristic have been tested on a set of a randomly generated instances with 10 nodes. The networks have been generated with GT Internetwork Topology Models code ([8]). The set of demands and the SRGs have been randomly generated. The number of SRGs is four for all the instances. Each instance is named as $\mathrm{i}\_|A| - index\_|K|$, where $|A|$ is the number of arcs, $|K|$ the number of demands, and *index* is the index of the randomly generated topology. For each network we consider the dedicated protection and the shared protection cases.

Models have been solved with CPLEX 11.0.1 with a time limit of 3600 seconds, heuristic has been implemented on C++: both CPLEX and heuristic, with 30 multistart iterations, have been run on a Xeon processor at 2.0GHz with 4Gb RAM.

First we compare the results obtained with CPLEX with those obtained by the heuristic on the dedicated protection instances: results are reported in Table 1. For each instance, four columns are devoted to CPLEX (lower bound LB, upper bound UBC, gap computed as (UBC-LB)/LB in percentage, and computational time) and four are devoted to the heuristic results (UBH, gap with respect to CPLEX UB computed as (UBH-UBC)/UBC, gap with respect to LB computed as (UBH-LB)/LB and computational time). CPLEX manages to solve to optimality 7 instances – those with 5 demands – over 12, in reasonable CPU time (up to about 260 s. in the worst case). The gap is limited for the five instances for which optimality is not proved. The average gap is about 4% and it is about 11% in the worst case. The heuristic algorithm provides gaps with respect to the integer solution of about 14% on the average, which increase up to about 22% in the worst case. Although the computational time increases on the 40 demands instances, however it does never exceed one minute and is limited for the instances with five demands.

The shared protection case is reported in Table 2, where results obtained with the first model (including constraints (7) and (8)) are compared to the ones obtained with the second one (including constraints (9), (10) and (11)). The improvement in the lower bound, with respect to the continuous relaxation, and the gap between upper and lower bound after one hour of computation are compared. Finally we compare upper bound obtained solving the two models with CPLEX with heuristic method results. The last

Table 1. Results on instances with dedicated protection

Instance	CPLEX LB	UB	gap	CPU time	Heuristic UB	gap UB	gap LB	CPU time
i_17_5	7506	7506	0.00%	0.33	7700	2.58%	2.58%	3.74
i_17_40	25350.47	25353	0.01%	263.23	28332	11.75%	11.76%	21.59
i_45-0_5	3348	3348	0.00%	8.19	3693	10.30%	10.30%	5.67
i_45-0_40	9342.85	10265	9.87%	t.l.	12569	22.45%	34.53%	45.16
i_45-1_5	3975	3975	0.00%	2.02	4435	11.57%	11.57%	5.63
i_45-1_40	12815.21	14102	10.04%	t.l.	17223	22.13%	34.39%	45.17
i_45-2_5	4311	4311	0.00%	4.73	4941	14.61%	14.61%	5.62
i_45-2_40	13931.12	15514	11.36%	t.l.	19013	22.55%	36.48%	47.90
i_45-3_5	2939	2939	0.00%	1.35	2940	0.03%	0.03%	5.67
i_45-3_40	9928.96	10765	8.42%	t.l.	12836	19.24%	29.28%	44.91
i_45-4_5	4238	4238	0.00%	1.94	4843	14.28%	14.28%	5.64
i_45-4_40	14474.29	15827	9.35%	t.l.	18966	19.83%	31.03%	44.89
average			4.09%	40.26		14.28%	19.24%	23.47

Table 2. Results on instances with shared protection

instance	CPLEX first model LB impr	gap	CPLEX second model LB impr	gap	gap	Heur. UB gap first model	second model	CPU time
i_17-4_5	82.34%	0.00%	82.58%	0.00%	0.00%	12.50%	12.50%	8.90
i_17-4_40	39.53%	22.42%	39.66%	22.79%	2.44%	7.86%	10.22%	57.26
i_45-0_5	75.53%	67.76%	73.59%	60.69%	7.35%	2.56%	15.56%	25.55
i_45-0_40	17.45%	421.21%	12.72%	169.00%	5.60%	-36.99%	29.32%	295.47
i_45-1_5	67.12%	75.24%	66.30%	73.13%	2.43%	13.81%	18.06%	26.48
i_45-1_40	21.57%	318.82%	17.20%	135.75%	5.29%	-27.51%	35.96%	305.48
i_45-2_5	66.46%	93.08%	65.82%	81.20%	1.87%	4.78%	13.77%	31.47
i_45-2_40	15.29%	269.14%	11.99%	144.03%	3.79%	-20.47%	25.04%	291.01
i_45-3_5	75.54%	57.30%	71.65%	83.10%	13.71%	61.93%	61.21%	25.97
i_45-3_40	20.68%	345.47%	17.63%	171.94%	3.70%	-28.02%	22.44%	285.00
i_45-4_5	63.84%	94.90%	62.73%	94.99%	2.98%	11.16%	14.52%	31.06
i_45-4_40	-	-	14.83%	131.16%		-	36.39%	284.08
average	49.58%	160.49%	44.73%	97.31%		0.14%	24.58%	138.98

column of the table gives the heuristic computational time. Both models manage to prove optimality in only one instance, and the first one cannot find any integer solution for one instance. The number of demands affects significantly the final gap, which are considerable. The continuous relaxation seems quite poor for both models. The comparison between the gap, given by (first model LB-second model LB)/(first model LB), shows that the first model seems to provide slightly better lower bounds. However, such model provides quite poor upper bounds. In fact the heuristic approach can find better upper bounds in 4 instances, while it always finds worse upper bounds that those provided by the second model. The heuristic computational time are reasonable, never rising above 6 minutes.

Beside the 10 node instances, we tested the models and the heuristics on 20 node instances. Preliminary results show that the model can solve 3 out of 20 instances for the dedicated protection case in one hour, and the gap is slightly increased with respect to 10 node instances. The heuristic CPU time increases of about three times on average and it rises up to about 15 minutes in the worst case. We run preliminary tests for the shared protection on the most promising model: none of the instance can be solved to optimality, and the model run out of memory for the more dense instances. The heuristic does not suffer memory problems, but its computational time increases significantly.

Computational experiments show that such models cannot solve the instances of both problems even with a small number of nodes. On the other side, the simple multistart approach, even if providing integer solutions in reasonable time, is affected by sensible gaps. As future developments, both improvements in the formulations and in the heuristics are worth investigating.

References

1. Borne, S., Gourdin, E., Liau, B., Mahjoub, A.R.: Design of Survivable IP-over-Optical Networks. Annals of Operations Research 146, 41–73 (2006)
2. Borne, S., Gourdin, E., Klopfenstein, O., Mahjoub, A.R.: The Multilayer Capacitated Survivable IP Network Design Problem: valid inequalities and Branch-and-Cut. In: Proceeding of International Network Optimization Conference (INOC 2009), Pisa, Italy (April 2009)
3. Coudert, D., Datta, P., Rivano, H., Voge, M.-E.: Minimum color problems and shared risk resource group in multilayer networks, Rapport de Recherche ISRN I3S/RR-2005-37-FR
4. Datta, P., Somani, A.K.: Diverse routing for shared risk resource groups (SRRG) failures in WDM Optical networks. In: BROADNETS 2004 (2004)
5. Piòro, M., Medhi, D.: Routing, flow and capacity design in communication and computer networks. Morgan Kaufman, San Francisco (2004)
6. Orlowski, S., Piòro, M.: On the complexity of column generation in survivable network design with path-based survivability mechanisms, ZIB-Report (2008)
7. Shen, L., Yang, X., Ramamurthy, B.: Shared risk link group (SRLG)-diverse path provisioning under hybrid service level agreements in wavelength-routed optical mesh networks. IEEE/ACM Transactions on networking 13(4), 918–931 (2005)
8. `http://www.cc.gatech.edu/fac/Ellen.Zegura/gt-itm/`

Improved Formulations for the Ring Spur Assignment Problem

Paula Carroll[1], Bernard Fortz[2], Martine Labbé[2], and Seán McGarraghy[1]

[1] Quinn and Smurfit Schools of Business,
University College Dublin, Ireland
[2] Département d'Informatique,
Faculté des Sciences, Université Libre de Bruxelles

Abstract. We present two complete integer programming formulations for the ring spur assignment problem. This problem arises in the design of next generation telecommunications networks. We analyse and compare the formulations in terms of compactness, the resulting LP bound and results from a branch and cut implementation. We present our conclusions with computational results.

1 Introduction

We address a new problem, the *Ring Spur Assignment Problem* (RSAP), introduced in [3,4]. The problem is motivated by a practical situation: a network operator seeks to identify an economical fault tolerant *Next Generation Network* (NGN) topology that can be overlaid on existing physical infrastructure. This problem arose in discussions with an industry partner who wished to identify a survivable backbone topology design in the physical network layer as part of an overall network upgrade plan. The operator wished to achieve this by exploiting existing spare capacity with no (or minimal) further capital investment. At about the same time the Irish Government proposed a project to build a backbone network on the infrastructure of its local agencies aiming to connect EU funded *Metropolitan Area Networks* (MANs) which aim to provide broadband access in all areas of the country. The solution we propose would be suitable for both the industry partner and the government project. In both instances, we seek to mine some value from the existing infrastructure and harness the benefits from emerging technology.

In Section 2, we describe the RSAP in detail, explain how it relates to problems previously addressed in the literature and an initial decomposition approach to solving the problem. In Sections 3 and 4 we describe two new complete integer programming formulations for the RSAP. In Section 5 we compare the formulations. Finally, we present computational results in Section 6 and our conclusions in Section 7.

2 Survivable Network Design Problems

Fibre Optic cable used in NGNs allows speeds in the Tb/s range, higher than traditional copper. However the higher the speed (bandwidth), the greater the

J. Pahl, T. Reiners, and S. Voß (Eds.): INOC 2011, LNCS 6701, pp. 24–36, 2011.

loss if an individual cable or piece of equipment develops a fault. Thus, it has become mandatory for backbone networks to be designed with survivability in mind. Despite the *Dial before you dig* campaigns of many utility companies, it is still a frequent occurrence for cables to be accidently cut as noted in [11]. Survivability issues are reviewed in detail in [10,12].

Synchronous Digital Hierarchy (SDH) is a transmission standard which allows for ease of access to individual channels. SDH, also known as Synchronous Optical NETwork (SONET), provides a fast managed response to failures and so can provide the survivability protection required by modern sparse networks. SDH promotes the use of Self Healing Rings (SHRs) to increase network reliability. Wavelength Division Multiplexing (WDM) is used on Fibre Optic networks to achieve even higher bandwidths. In [16], the authors give an introduction to optical networking issues and indicate that Internet Protocol (IP) over WDM may be the preferred option for NGNs. This protocol can be implemented over the physical SDH layer. The use of WDM introduces additional graph colouring type problems. Wavelength conversion equipment must be installed where traffic is routed across neighbouring rings [1].

The Two Connected Network with Bounded Rings problem (2CNBR) and its variants that arise in the design of SDH/SONET and WDM networks are described in [6,7,8]. This problem concerns designing a minimum cost network where at least two node-disjoint, or alternatively edge-disjoint, paths exist between every pair of nodes. Each edge of the network belongs to at least one cycle whose length (number of edges) is bounded by a given constant. The ring bound is imposed to ensure the quality of the telecommunications signal.

The NP-Hard SDH Ring Assignment Problem (SRAP) is discussed in [9]. They describe the SRAP as a high level design problem that seeks to identify which SHR rings should be built; they choose to minimise network costs by minimising the number of disjoint rings while satisfying customer demand and satisfying a common ring capacity. A special ring, called the federal ring, of the same capacity as the other rings in the network, interconnects the other rings. Another formulation of SRAP as a set partitioning model with additional knapsack constraints is given in [14]. We also mention the Ring Star problem described by [13]. A Ring Star is used to connect terminals to concentrators where not all nodes are required to be 2-connected.

2.1 The Ring Spur Assignment Problem

We now describe the RSAP in detail. Communities of interest, defined in [5] as geographically close nodes that have high traffic demands between them, are identified and their traffic demands are estimated. If such communities can be clustered on node disjoint rings, no wavelength conversion is required, eliminating the cost of wavelength conversion and/or opto-electronic conversion equipment for intra-ring demand; this is an important cost consideration in any network upgrade plan. We call these rings *local rings.*

Local rings can then be connected by a special ring, which we call the *tertiary ring*, often called the federal or backbone ring in the literature. *Tertiary* is a legacy naming convention used by this operator to signify the highest level in the physical infrastructure. The tertiary ring facilitates inter-ring demand; wavelength converters are required where local rings connect to the tertiary ring.

So far the problem described is similar to the SRAP problem; we are identifying rings that can carry the estimated demand. However, as shown in [4], in some real world instances, no SRAP solution is possible. We note that in the SRAP, demand pairs are grouped together so that there is as little inter-ring traffic as possible subject to capacity constraints on rings. The SRAP problem addresses how to design a ring based network by selecting the link capacities to install. In contrast, in the RSAP, we assess an existing network and the problem is to impose a ring topology over existing links at the logical level.

As an alternative, where no SRAP solution exists, we allow locations that have insufficient spare capacity or no possible physical route due to limitations of geography, to be connected to SHRs by spurs off the local rings. Spur nodes must be connected to a local ring by a single edge, i.e., we do not allow a chain of edges to connect spur nodes. We call this problem the Ring Spur Assignment Problem. A solution to the RSAP is a set of disjoint bounded ring stars interconnected by a tertiary ring. In [4] a branch-and-cut algorithm to identify the local ring spur partitions is described. In [3] an integer programming formulation for the RSAP is described. They describe a heuristic procedure that decomposes the RSAP into a subproblem of first finding a minimum cost local ring/spur topology and a second subproblem of finding a minimum cost tertiary ring to interconnect the local ring spur partitions. Computational results on benchmark problems of up to 65 nodes and 108 links are presented. However, the formulation proposed in [3] cannot be used for an exact approach, as the decisions for the local topology and the tertiary ring are explicitly decomposed in two subproblems.

In Sections 3 and 4, we propose two complete IP formulations that can be used to identify the optimal local ring (spur) partitions and tertiary interconnection ring without the need for decomposition. A problem instance is specified by:

- an undirected graph $G = (V, E)$ defined on a set V of nodes, labelled from 1 to n where $n = |V|$, and a set of undirected edges E; the underlying set A of oriented arcs contains, for each edge $\{i, j\} \in E$, two arcs (i, j) and (j, i), one in each direction;
- a non-negative installation cost $c_{ij} \geq 0$ for each ring edge $\{i, j\} \in E$ and the corresponding spur arcs (i, j) and (j, i).

Since we wish to foster high resilience by having locations assigned to rings where possible, we assign a sufficiently high weight, b, to links that are spurs. We use a similar approach to [3] to quantify a penalty weighting value in terms of other network parameters sufficient to ensure the creation of ring solutions if they exist. For simplicity, we set the coefficient of each arc $(i, j) \in A$, to be bc_{ij} in our objective function, i.e., the cost of using a spur edge is the network cost of that edge, c_{ij}, multiplied by the penalty weighting value of b for the network.

3 A Formulation Based on Ring Representatives

We describe the first of our complete formulations for the RSAP. We first concentrate on the local rings and come back to the tertiary ring later. The formulation uses variables to assign nodes and edges to rings, and a third set of variables for spurs. The main problem with such a choice of variables is that it induces a lot of symmetry in the formulation, as rings are interchangeable. One way to break this inherent symmetry is to use the same kind of symmetry breaking that was used e.g. in [2] for graph colouring problems. Each ring is designated by a representative node belonging to it. If we decide that the representative node is the node with smallest index in the ring, the symmetry is completely broken. Note that since each ring is composed of at least 3 nodes, only nodes $k \in K := \{1, \dots, n-2\}$ can be ring representatives.

Let x_{ijk} be a binary variable equal to 1 if and only if edge $\{i, j\}$ appears on ring k, and equal to 0 otherwise; i.e. both i and j are assigned to the same ring k. Also, we implicitly assume that $i < j$ when writing an edge $\{i, j\}$. For each arc $(i, j) \in A$, let y_{ij} be a binary variable equal to 1 if vertex i is assigned to vertex j as a spur; we set $y_{ii} = 1$ for any vertex i that is on a ring. Let z_{ik} be a binary variable equal to 1 if vertex i is assigned to ring k as a ring node.

Let α_{ik} be a binary variable equal to 1 if and only if node i represents ring k on the Tertiary ring, 0 otherwise. Let β_{ij} be a binary variable equal to 1 if and only if edge $\{i, j\}$ appears on the Tertiary ring, and equal to 0 otherwise.

With these sets of variables, the problem can be formulated as follows:

$$(F1) \quad \min \sum_{\{i,j\} \in E} \sum_{k \in K} c_{ij} x_{ijk} + \sum_{(i,j) \in A} bc_{ij} y_{ij} + \sum_{\{i,j\} \in E} c_{ij} \beta_{ij} \tag{1}$$

$$\text{s.t.} \quad \sum_{k=1}^{i} z_{ik} = y_{ii} \qquad \forall\, i \in V \tag{2}$$

$$\sum_{j=1}^{n} y_{ij} = 1 \qquad \forall\, i \in V \tag{3}$$

$$\sum_{j \in \mathrm{adj}(i), j > i} x_{ijk} + \sum_{j \in \mathrm{adj}(i), j < i} x_{jik} = 2 z_{ik} \qquad \forall\, i \in V, k \in K \tag{4}$$

$$\sum_{l \in \mathrm{adj}(i), l < i} x_{lik} + \sum_{l \in \mathrm{adj}(i), l > i, l \neq j} x_{ilk} \geq x_{ijk} \qquad \forall\, \{i, j\} \in E, k \in K \tag{5}$$

$$\sum_{l \in \mathrm{adj}(j), l < j, l \neq i} x_{ljk} + \sum_{l \in \mathrm{adj}(j), l > j} x_{jlk} \geq x_{ijk} \qquad \forall\, \{i, j\} \in E, k \in K \tag{6}$$

$$\sum_{i \in V, i > k} z_{ik} \leq 7 z_{kk} \qquad \forall\, k \in K \tag{7}$$

$$z_{ik} \leq z_{kk} \qquad \forall\, i \in V, k \in K \tag{8}$$

$$\sum_{k \in K} x_{ijk} + y_{ji} \leq y_{ii} \qquad \forall\, i \in V, j \in adj(i) \tag{9}$$

$$\sum_{k \in K} x_{ijk} + y_{ij} + y_{ji} \leq 1 \qquad \forall \{i,j\} \in E \quad (10)$$

$$\sum_{i \in V, i \geq k} \alpha_{ik} \geq z_{kk} \qquad \forall k \in K \quad (11)$$

$$\alpha_{ik} \leq z_{ik} \qquad \forall i \in V, k \in K \quad (12)$$

$$\sum_{j \in V, i<j} \beta_{ij} + \sum_{j \in V, j<i} \beta_{ji} = 2 \sum_{k \in K} \alpha_{ik} \qquad \forall i \in V \quad (13)$$

$$\sum_{l \in \mathrm{adj}(i), l<i} \beta_{il} + \sum_{l \in \mathrm{adj}(i), l>i, l \neq j} \beta_{li} \geq \beta_{ij} \qquad \forall \{i,j\} \in E \quad (14)$$

$$\sum_{l \in \mathrm{adj}(j), l<j, l \neq i} \beta_{lj} + \sum_{l \in \mathrm{adj}(j), l>j} \beta_{jl} \geq \beta_{ij} \qquad \forall \{i,j\} \in E \quad (15)$$

$$\beta_{ij} + y_{ij} + y_{ji} \leq 1 \qquad \forall \{i,j\} \in E \quad (16)$$

$$x_{ijk}, y_{ij}, z_{ik}, \alpha_{ik}, \beta_{ij} \in \{0,1\} \qquad \forall \{i,j\} \in E, k \in K \quad (17)$$

We use constraints (2) to link the y and z variables. Constraints (3) ensure every node is assigned. Constraints (4) say that every node i on ring k has exactly two incident local ring edges on ring k. Constraints (5) and (6) are dis-aggregated edge connectivity constraints and ensure that the head (tail) of a ring edge each have an incident ring edge. This form of connectivity constraints ensure there are at least three edges (three nodes) on each local ring. Rings are restricted to having no more than eight nodes by the Ring Bound Constraints (7), if ring k is active, node k must be active and no more than seven other nodes. Constraints (8) ensure node i is assigned to ring k if and only if node k is assigned to ring k. Constraints (9) say that j is assigned to i as a spur or adjacent to i on a ring if and only if i is a ring node. Constraints (10), which are only needed in the relaxed LP, limit the usage of any edge (or its corresponding arcs) to 1. The next set of constraints define the tertiary ring. Constraints (11) ensure every active local ring k is represented on the tertiary ring by at least one of its nodes. Constraints (12) connect the α to the z variables and ensure that node i only represents ring k if it is active on ring k. Constraints (13) ensure that nodes on the tertiary ring are two connected while constraints (14) and (15) are similar to the local ring edge dis-aggregated connectivity constraints (5) and (6). Again, these connectivity constraints ensure there are at least three edges on the tertiary ring ring. Constraints (16) ensure that no spur edge is used as a tertiary ring edge (this ensures the robustness of the tertiary ring) while constraints (17) are the binary integer constraints.

We note that the formulation thus far would allow subtours on both the local and tertiary rings and that there are an exponential number of subtour elimination constraints (SECs) that would need to be added to eliminate all such subtours. In Section 6 we describe our implementation of a cutting plane algorithm that adds SECs. Where a subtour is detected on ring k we add the following modified versions of the usual SECs.

$$\sum_{\{i,j\}\in E(S)} x_{ijk} \le (|S|-1)z_{kk} \quad \forall\ S \subset V, k \notin S \tag{18}$$

Eq. (18) says that the set of edges of the subset S cannot all be on ring k. Indeed, we can add such a constraint for all $k \le \min(S)$, the minimum index of the subset S. For example, if the edges of the subset $S := \{5, 6, 7\}$ form a local ring, they must be on ring 5 and cannot be on ring 4 or lower. Also note, since constraints (5) and (6) force the number of nodes on an active local ring to be at least three, we consider subtours on $S \subset V$ where $|S| \ge 3$.

We must also add SECs for subtours on the tertiary ring. If a subtour consisting of subsets S_a and S_b is detected, since we do not know which local ring edges will be nonzero in the optimal solution, it is possible that the optimal tertiary ring could consist of all edges of one tertiary ring subtour, either S_a or S_b. However, not all edges of both subtours could be non-zero in the optimal solution. We can use an aggregated version of the usual SECs as follows:

$$\sum_{\{i,j\}\in E(S_a)\cup E(S_b)} \beta_{ij} \le |S_a \cup S_b| - 2 \quad \forall \text{ Tertiary Subtours } S_a \text{ and } S_b \tag{19}$$

The $F1$ model has more decision variables than that described by [3] but we can perform significant pre-processing since not all decision variables can exist. As noted already, the nodes of highest and second highest index in G can never be assigned as ring indices (since a ring must have three nodes). Therefore the maximum value of k is $n-2$. In addition, for each node, its maximum k value is itself. For example, if node 20 is a ring node, $i = 20$ and k for this node is ≤ 20. We can reduce this value further by looking at the indices of the nodes adjacent to i. For example, if i is exactly 2-connected, the maximum k value for x_{ijk} and z_{ik} is $\min(i, \textit{lowest index of adjacent node})$. If node 20 is exactly 2-connected and adjacent to nodes 5 and 25, the maximum ring index for node 20 is 5. So only these decision variables need to be created in $F1$.

We note also that we only need to create ring bound constraints where required. For example, if $k = n-2$, only the highest indexed node, the second highest and k^{th} indexed node in G can be assigned to this ring, giving a ring of size 3, therefore the ring bound constraint is redundant in this instance.

4 An Extended Formulation

Another possible IP formulation for the RSAP is the following: Let x_{ijk} be a binary variable equal to 1 if and only if edge $\{i, j\}$ appears on ring k, and equal to 0 otherwise; i.e. both i and j are assigned to the same ring k where k, the ring designate, is the lowest indexed node on the ring similar to the approach in $F1$ i.e., $k \le i$ and $k \le j$ and $k \in K := \{1, \ldots, n-2\}$. In our extended formulation, we combine the use of the y_{ij} and z_{ik} and variables and use one variable w_{ijk}. For each arc $(i, j) \in A$, let w_{ijk} be a binary variable equal to 1 if vertex i is assigned to vertex j as a spur on ring k; we set $w_{iik} = 1$ for any vertex i that is assigned as a ring node on ring k where $i \ge k$.

We use tertiary ring decision variables similar to those in $F1$. Let α_{ik} be a binary variable equal to 1 if and only if node i represents ring k on the Tertiary ring, 0 otherwise. Let β_{ij} be a binary variable equal to 1 if and only if edge $\{i,j\}$ appears on the Tertiary ring, and equal to 0 otherwise. The second formulation is as follows:

$$(F2) \quad \min \quad \sum_{\{i,j\}\in E}\sum_{k=1}^{n-2} c_{ij}x_{ijk} + \sum_{(i,j)\in A}\sum_{k=1}^{n-2} bc_{ij}w_{ijk} + \sum_{\{i,j\}\in E} c_{ij}\beta_{ij} \tag{20}$$

$$\text{s.t.} \quad \sum_{j=1}^{n}\sum_{k=1}^{j} w_{ijk} = 1 \qquad \forall\, i \in V \tag{21}$$

$$\sum_{j\in \text{adj}(i),j>i} x_{ijk} + \sum_{j\in \text{adj}(i),j<i} x_{jik} = 2w_{iik} \qquad \forall\, i \in V, k \in K \tag{22}$$

$$\sum_{l\in \text{adj}(i),l<i} x_{ilk} + \sum_{l\in \text{adj}(i),l>i,l\neq j} x_{lik} \geq x_{ijk} \qquad \forall\, \{i,j\} \in E, k \in K \tag{23}$$

$$\sum_{l\in \text{adj}(j),l<j,l\neq i} x_{ljk} + \sum_{l\in \text{adj}(j),l>j} x_{jlk} \geq x_{ijk} \qquad \forall\, \{i,j\} \in E, k \in K \tag{24}$$

$$\sum_{i=1,i>k}^{n} w_{iik} \leq 7w_{kkk} \qquad \forall\, k \in K \tag{25}$$

$$w_{iik} \leq w_{kkk} \qquad \forall\, i \in V, k \in K \tag{26}$$

$$\sum_{j\geq k,j\neq i}^{n-2} w_{ijk} \leq w_{kkk} \qquad \forall\, i \in V, k \in K \tag{27}$$

$$x_{ijk} + w_{jik} \leq w_{iik} \qquad \forall\, i \in V, k \in K, j \in adj(i) \tag{28}$$

$$x_{ijk} + w_{ijk} + w_{jik} \leq w_{kkk} \qquad \forall\, \{i,j\} \in E, k \in K \tag{29}$$

$$\sum_{k\in K} (x_{ijk} + w_{ijk} + w_{jik}) \leq 1 \qquad \forall\, \{i,j\} \in E \tag{30}$$

$$\sum_{i\in V,i\geq k} \alpha_{ik} \geq w_{kkk} \qquad \forall\, k \in K \tag{31}$$

$$\alpha_{ik} \leq w_{iik} \qquad \forall\, i \in V, k \in K \tag{32}$$

$$\sum_{j\in V,i<j} \beta_{ij} + \sum_{j\in V,j<i} \beta_{ji} = 2\sum_{k\in K} \alpha_{ik} \qquad \forall\, i \in V \tag{33}$$

$$\sum_{l\in \text{adj}(i),l<i} \beta_{il} + \sum_{l\in \text{adj}(i),l>i,l\neq j} \beta_{li} \geq \beta_{ij} \qquad \forall\, \{i,j\} \in E \tag{34}$$

$$\sum_{l\in \text{adj}(j),l<j,l\neq i} \beta_{lj} + \sum_{l\in \text{adj}(j),l>j} \beta_{jl} \geq \beta_{ij} \qquad \forall\, \{i,j\} \in E \tag{35}$$

$$\beta_{ij} + \sum_{k \in K} w_{ijk} + \sum_{k \in K} w_{jik} \leq 1 \qquad \forall \{i,j\} \in E \qquad (36)$$

$$x_{ijk}, w_{ijk}, w_{jik}, \alpha_{ik}, \beta_{ij} \in \{0,1\} \qquad \forall \{i,j\} \in E, k \in K \qquad (37)$$

Constraints (21) ensure every node $i \in V$ is assigned while constraints (22) ensure that every ring node i on ring k is incident with two edges on the ring. Constraints (23) and (24) are dis-aggregated connectivity constraints for the head and tail of each ring edge, constraints (25) are the ring bound constraints. Constraints (26) only allow node i to be assigned as a ring node on ring k if k is a ring node while the set (27) ensure that node i is only assigned as a spur to node j on ring k if node k is a ring node. Constraints (28) allow node j to be assigned as a spur or adjacent to node i on ring k if i is assigned as a ring node to ring k while constraints (29) and (30) focus on the use of the edge $\{i, j\}$ and allow the edge or its corresponding arcs to be assigned to ring k iff node k is assigned to ring k. Constraints (31) to (36) are the tertiary ring constraints and are similar in style to those of the $F1$ formulation and lastly, (31) are the binary integer constraints.

This model also allows for subtours which can be eliminated on the local rings as follows:

$$\sum_{\{i,j\} \in E(S)} x_{ijk} \leq (|S| - 1) w_{kkk} \quad \forall S \subset V, k \notin S \qquad (38)$$

As before in Eq. (18), $|S| \geq 3$. Subtours on the tertiary ring can be eliminated by adding constraints given by Eq (19) described in the $F1$ formulation.

5 Comparison of the Formulations

Having described the two complete formulations, $F1$ and $F2$, we now wish to determine which is the more promising for implementation.

Let us first consider the size of the formulations (not taking into account subtour elimination constraints). Both formulation have $O(n^3)$ variables: $F1$ and $F2$ share $O(n^3)$ x-variables. In $F1$, there are moreover $O(n^2)$ y and z-variables, while in $F2$ there are $O(n^3)$ w-variables. Asymptotically, $F2$ thus tends towards having twice as many variables as $F1$.

The constraints in both formulations are mostly similar. There is a one-to-one correspondence between all constraints of $F1$ and $F2$, with the exception of (2) that appear only in $F1$, (27) that appear only in $F2$, and each constraint of (9) and (10) in $F1$ is replaced by $|K|$ constraints (28) and (29), plus constraints (30), in $F2$. So there are $O(n^3)$ constraints in both formulations, but the total number of constraints is higher in $F2$.

Computational results for the LP bound achieved by relaxing the integer requirements are described in detail in Sect. 6. They show that there is no empirical difference in the LP bound form these two formulations. The question of this equivalence of the LP relaxations is still open, but it is quite easy to show that $F2$ is at least as strong as $F1$.

Indeed, consider a point (x, w, α, β) satisfying (21)-(36). By defining y and z such that

$$z_{ik} = w_{iik}, \ \forall i \in V, \ k \in K,$$
$$y_{ij} = \sum_{k \in K} w_{ijk}, \ \forall i, j \in V, \ k \in K,$$

by substitution into (21)-(26) and (30)-(36), we immediately obtain that (x, y, z, α, β) satisfies (3)-(8) and (10)-(16). By definition of y and z, (2) is obviously satisfied.

Now summing constraints (28) over all $k \in K$, and substituting the summed w by y, we obtain (9) and we can conclude that (x, y, z, α, β) is valid for the linear relaxation of $F1$.

6 Computational Results

The $F1$ and $F2$ IP formulations were implemented with code written in ANSI C, using Xpress-MP suite 7.0 with Xpress-BCL version 4.2.0 Builder Component library routines and Xpress-Optimizer 20.00.05, and run on a 32 bit Toshiba Satellite Pro with Intel Dual Core Pentium 1.86GHz processors and 1G of RAM under Windows Vista.

The test data used was SNDlib [15], since it provides many real world problem instances with both a network model and a set of demand requirements. In problem instances that have zero pre-installed link capacity, we install one unit of the lowest capacity available at the costs specified to allow us to test our algorithm. Two problems, janos-us-ca and zib54 were integer infeasible for local ring spur partitions, so were omitted from further testing. These two networks have a small number of nodes of very high degree making them unsuitable for the RSAP topology.

Looking at an example of the fractional decision variables from problem giul39 shown in Table 1, we see the same overall result achieved by the two formulations. For example, Node 12 is assigned as a ring node to both rings 1 and 11 by both formulations. Node 30 is partially assigned as a ring node in both formulations. However the sum of the value used by $F2$ to assign node 30 to rings 1, 2 and 6 is the same as the value used by $F1$ to assign node 30 to ring 1. Both formulations assign node 30 equally to the other listed rings.

Both formulations were run without XpressMP heuristics or tightening of the formulation. A summary of the LP bounds is shown in Table 2. This table shows from left to right the problem name and size (in terms of number of nodes, edges), the LP objective function value from $F1$ followed by the LP objective function value from $F2$. As can be seen, there is insufficient difference in the results from the $F1$ and $F2$ LP bounds to perform a signed rank comparison.

Table 1. Fractional Decision Variable examples

$F1$	$F1$ Value	$F2$	$F2$ Value
$y_{12,12}$	1.00	-	-
$z_{12,1}$	0.818182	$w_{12,12,1}$	0.818182
$z_{12,11}$	0.181818	$w_{12,12,11}$	0.181818
$y_{30,30}$	0.849774	-	-
$z_{30,1}$	0.183816	$w_{30,30,1}$	0.085306
$z_{30,13}$	0.031779	$w_{30,30,13}$	0.031779
$z_{30,16}$	0.074714	$w_{30,30,16}$	0.074714
$z_{30,18}$	0.00344	$w_{30,30,18}$	0.00344
-	-	$w_{30,30,2}$	0.045455
$z_{30,20}$	0.018479	$w_{30,30,20}$	0.018479
$z_{30,24}$	0.29953	$w_{30,30,24}$	0.29953
$z_{30,4}$	0.238015	$w_{30,30,4}$	0.238015
-	-	$w_{30,30,6}$	0.053055

Table 2. Formulation LP Bound comparison

Problem	Size n, e	LP Obj $F1$	LP Obj $F2$
dfn-bwin	10, 45	91,499.29	91,499.29
pdh	11, 34	1,176,327.71	1,176,327.71
di-yuan	11, 42	378,142.86	378,142.86
dfn-gwin	11, 47	13,813.33	13,813.33
polska	12, 18	2,989.68	2,989.68
atlanta	15, 22	36,790,750.00	36,790,750.00
new york	16, 49	1,370,720.00	1,370,720.00
ta1	24, 51	8,712,415.62	8,712,415.62
france	25, 45	11,575.00	11,575.00
janos-us	26, 42	13,218.15	13,218.15
norway	27, 51	462,419.27	462,419.27
sun	27,51	462.42	462.42
nobel-eu	28, 41	179,567.32	179,567.32
cost266	37, 57	7,615,390.48	7,615,390.48
giul39	39, 86	735.72	735.72
pioro40	40, 89	6,114.33	6,114.33
germany	50, 88	341,960.77	341,960.77
ta2	65, 108	21,345,067.48	21,345,067.48

The software used, XpressMP, works to a precision of six decimal places but any difference below this precision level is inconsequential in any case. These results suggest that the LP bound provided by $F2$ is not stronger than the $F1$ bound.

We also ran a branch and cut implementation for $F1$ and $F2$ adding SECs as necessary. The relaxed LP of each model was solved, subtours were detected

Table 3. Branch and cut comparison

Problem	Size	F1 IP Obj	F1 SECs	F1 Nodes	F1 time(s)	F2 IP Obj	F2 SECs	F2 Nodes	F2 time(s)
dfn-bwin	10,45	105,810	11	35	0.99	105,810	10	68	1.24
pdh	11,34	1,355,139	16	67	0.70	1,355,139	15	64	0.96
di-yuan	11,42	412,300	0	5	0.21	412,300	0	2	0.28
dfn-gwin	11,47	15,724	3	13	0.42	15,724	4	13	0.53
polska	12,18	3,487	3	19	0.35	3,487	4	15	0.36
atlanta	15,22	55,452,500	14	65	0.90	55,452,500	30	58	1.23
new york	16, 49	1,512,400	185	252	4.82	1,512,400	241	307	8.30
ta1	24,51	11,867,165	2,801	3,941	62.98	11,867,165	3,713	3,951	103.02
france	25,45	16,800	33,534	13,583	447.28	16,800	41,678	12,674	276.48
janos-us	26,42	16,259	1,768	4,626	123.28	16,259	1,617	3,848	106.28
norway	27,51	572,210	791	1,617	68.91	572,210	1,793	2,510	270.76
sun	27,51	572	4,988	5,870	156.55	572	2,093	3,470	297.92
nobel-eu	28,41	-	7,422	12,134	3,600	-	3,356	5,424	3,600
cost266	37, 57	11,432,070	14,167	40,353	1,430.38	11,432,070	4,482	19,360	953.17
giul39	39, 86	990	34,592	14,194	3,600	1,122	11,095	10,944	3,600
pioro40	40, 89	12,283	43,027	47,561	3,600	11,405	14,189	17,856	3,600
germany	50, 88	580,080	46,871	19,296	3,600	582,230	19,984	8,797	3,600
ta2	65, 108	-	8,547	329	3,600	-	3,048	3,126	3,600

using a modified version of the Stoer Wagner Min Cut Algorithm and added to the problems. The Stoer Wagner min cut algorithm is described in [17]. We allowed a maximum run time of one hour and report the best integer solution found by that time for the larger problems. We confirmed for both $F1$ and $F2$ that they produce the desired topology. Results are shown in Table 3. This table shows from left to right the problem name and size (in nodes, edges), the next four columns show the results for $F1$: the IP objective function value, the number of SECs added by the branch and cut algorithm, the number of nodes in the branch and bound tree and the time (seconds). The following four columns show the results for $F2$.

We note that for problem instances solved in less than one hour, $F1$ had a faster run time in most cases. We see for the larger problems that in general for $F1$, more nodes are evaluated within the time limit and produces a better IP bound. Neither formulation found a solution for the largest problem *ta2* within the time limit or for *nobel-eu*.

Figure 1 (left) shows the resulting topology for $F1$ on the 50 node *germany* problem, and (right), the topology from $F2$. The tertiary ring edges are shown with heavy black dashed lines, local rings as coloured lines and spur arcs as dashed red lines. For clarity, we omit any edges of the graph that are not part of the solution topology.

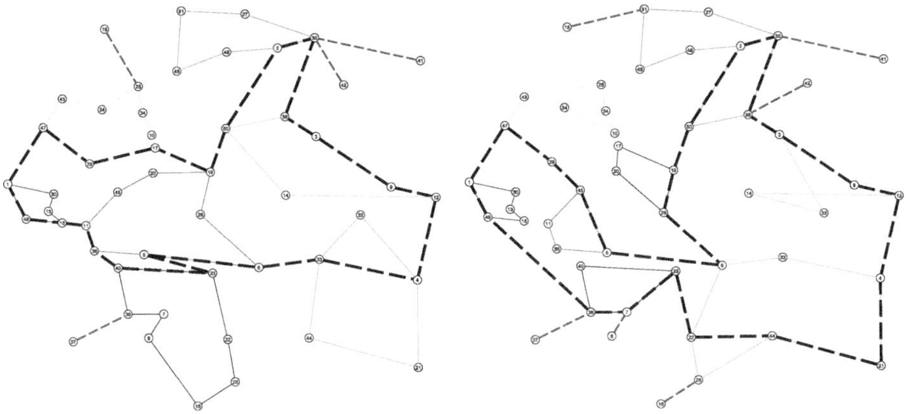

Fig. 1. Left: $F1$ *germany* solution . Right: $F2$ *germany* solution.

7 Conclusion

We have presented two new formulations for the Ring Spur Assignment Problem. The main difference between the two formulations is that decisions for spurs are dis-aggregated by ring in the second formulation, leading to an extended formulation. Numerical experiments tend to show that the two linear relaxations are equivalent, and the compact formulation allows problems to be solved faster.

In our future work, we will try to tighten the linear relaxation by finding new classes of valid inequalities. In particular, it would be interesting to find out if some stronger inequalities can be found for the extended model, as it inherently contains more structure than the compact one.

Acknowledgment

The first author thanks the Quinn and Smurfit Schools of Business at UCD for their support. Bernard Fortz and Martine Labbé were supported by an "Actions de Recherche Concertées" (ARC) projet of the "Communauté française de Belgique" and a FRFC program "Groupe de Programmation Mathémathique".

References

1. Bayvel, P.: Future high-capacity optical telecommunication networks. Philosophical Transactions - A - Mathematical Physical and Engineering Sciences 358(1765), 303 (2000)
2. Campêlo, M., Campos, V.A., Corrêa, R.C.: On the asymmetric representatives formulation for the vertex coloring problem. Discrete Applied Mathematics 156(7), 1097–1111 (2008)

3. Carroll, P., McGarraghy, S.: An algorithm for the ring spur assignment problem. In: Papova, N., O'hEigheartaigh, M. (eds.) Proceedings of the International Eugene Lawler PhD Summer School 2009 held at WIT, Ireland, June 6-10, pp. 180–197. Scientific Computing, WIT (June 2009)
4. Carroll, P., McGarraghy, S.: Investigation of the ring spur assignment problem. In: Bigi, G., Frangioni, A., Scutellà, M. (eds.) Proceedings of the 4th International Network Optimization Conference (INOC 2009), April 26-29, pp. MB1–3. INOC, Pisa (2009)
5. Cosares, S., Deutsch, D.N., Saniee, I., Wasem, O.J.: Sonet toolkit: A decision support system for designing robust and cost-effective fiber-optic networks. Interfaces 25(1), 20–40 (1995)
6. Fortz, B., Labbé, M.: Two-connected networks with rings of bounded cardinality. Computational Optimization and Applications 27(2), 123–148 (2004)
7. Fortz, B., Mahjoub, A.R., McCormick, S.T., Pesneau, P.: Two-edge connected subgraphs with bounded rings: Polyhedral results and Branch-and-Cut. Mathematical Programming 105(1), 85–111 (2006)
8. Fortz, B., Soriano, P., Wynants, C.: A tabu search algorithm for self-healing ring network design. European Journal of Operational Research 151(2), 280–295 (2003)
9. Goldschmidt, O., Laugier, A., Olinick, E.V.: SONET/SDH ring assignment with capacity constraints. Discrete Applied Mathematics 129(1), 99–128 (2003)
10. Grover, W., Doucette, J., Clouqueur, M., Leung, D., Stamatelakis, D.: New options and insights for survivable transport networks. IEEE Communications Magazine 40(1), 34–41 (2002)
11. Grover, W.: Mesh Based Sruvivable Networks, Options and Strategies for Optical, MPLS, Sonet and ATM Networking. Prentice Hall, Englewood Cliffs (2003)
12. Kerivin, H., Mahjoub, A.: Design of survivable networks. Networks 46(1), 1–21 (2005)
13. Labbé, M., Laporte, G., Martin, I., Salazar-Gonzalez, J.: The Ring Star Problem: Polyhedral analysis and exact algorithm. Networks 43(3), 177–189 (2004)
14. Macambira, E., Maculan, N., de Souza, C.: A column generation approach for SONET ring assignment. Networks 47(3), 157–171 (2006)
15. Orlowski, S., Pióro, M., Tomaszewski, A., Wessäly, R.: SNDlib 1.0–Survivable Network Design Library. Networks 55(3), 276–286 (2010), `http://www3.interscience.wiley.com/journal/122653325/abstract`
16. Papadimitriou, G., Obaidat, M., Pomportsis, A.: Advances in Optical Networking. Int. J. Commun. Sys. 15, 101–113 (2001)
17. Stoer, M., Wagner, F.: A simple min-cut algorithm. Journal of the ACM 44(4), 585–591 (1997)

A Chance-Constrained Model and Cutting Planes for Fixed Broadband Wireless Networks

Grit Claßen[1], David Coudert[2], Arie M.C.A. Koster[1], and Napoleão Nepomuceno[3]

[1] Lehrstuhl II für Mathematik, RWTH Aachen University, 52056 Aachen, Germany
{classen,koster}@math2.rwth-aachen.de
[2] Project-team Mascotte, I3S (CNRS/UNS) INRIA Université de Nice Sophia, Sophia Antipolis, France
david.coudert@inria.fr
[3] Institut for Matematik og Datalogi, Syddansk Universitet, Campusvej 55, DK-5230 Odense
napoleao@imada.sdu.dk

Abstract. In this paper, we propose a chance-constrained mathematical program for fixed broadband wireless networks under unreliable channel conditions. The model is reformulated as an integer linear program and valid inequalities are derived for the corresponding polytope. Computational results show that by an exact separation approach the optimality gap is closed by 42 % on average.

1 Introduction

Fixed broadband wireless (FBW) communications is a promising technology for delivering private high-speed data connections by means of microwave radio transmission [2]. Microwave, in the context of this work, refers to terrestrial point-to-point digital radio communications, usually employing highly directional antennas in clear line-of-sight and operating in licensed frequency bands. The rapid and relatively cheap deployment is especially interesting for emerging countries and remote locations as well as for private and isolated networks in urban areas (e.g., connected hospitals, parts of a harbour) where classical copper/fiber lines are too costly [9]. In contrast to wired networks, the capacity of a microwave link is not constant, but depends on the used modulation scheme, which in turn depends on the condition of the channel. Varying channel conditions result in varying link capacities.

In this paper, we extend our earlier study [3] of planning FBW networks under unreliable channel conditions. We restate a chance-constrained optimization model and, for the case where the outage probabilities are independent, an integer linear programming (ILP) formulation (Section 2). We generalize two classes of cutset-based valid inequalities (Section 3) and propose to separate them exactly. Preliminary computational results confirm the importance of these cuts (Section 4).

2 Mathematical Formulation

The minimum cost design of a fixed broadband wireless network can be formulated as follows, cf. [4] for technical details. The network's topology is modeled as a digraph $G = (V, E)$ with V, the set of radio base stations and E, the set of directional

J. Pahl, T. Reiners, and S. Voß (Eds.): INOC 2011, LNCS 6701, pp. 37–42, 2011.

microwave radio links. The traffic requirements are modeled by a set K. For each $k \in K$, s^k denotes the origin, t^k the destination, and $d^k \geq 0$ the expected demand.

For each microwave link $uv \in E$, the capacity is basically determined by the channel bandwidth (e.g., 7 MHz, 28 MHz) and the modulation scheme (e.g., 16-QAM, 128-QAM) used to transmit data. Where exactly one channel bandwidth has to be chosen at design stage, adaptive modulation is performed at runtime, depending on the channel conditions, i.e., if the receiving base station observes a deterioration in signal quality, the modulation scheme is lowered to avoid outage of the link.

Let W_{uv} be the set of bandwidth choices available for arc $uv \in E$. The choice to operate link $uv \in E$ at bandwidth b^w_{uv}, $w \in W_{uv}$, implies a cost c^w_{uv}. The modulation scheme is modeled with a random variable η^w_{uv} with (known) discrete probability, representing the number of bits per symbol of the current modulation scheme. The capacity of a microwave link is basically given by the product of b^w_{uv} and η^w_{uv}.

Given an infeasibility tolerance $\varepsilon > 0$, our aim is to design a minimum cost network such that its capacity is sufficient with a probability of at least $1-\varepsilon$. This joint chance constraint reads

$$\mathscr{P}\left(\sum_{k\in K} d^k f^k_{uv} \leq \sum_{w\in W_{uv}} \eta^w_{uv} b^w_{uv} y^w_{uv} \qquad \forall uv \in E\right) \geq 1-\varepsilon \tag{1}$$

with binary decision variables y^w_{uv} indicating whether bandwidth $w \in W_{uv}$ is chosen for arc $uv \in E$ and flow variables f^k_{uv} denoting the fraction of demand d^k, $k \in K$, routed on arc $uv \in E$.

For independent random variables η^w_{uv}, we can reformulate the left hand side of (1) as the product of probabilities by introducing the following notation: For arc $uv \in E$, let M^w_{uv} be the set of modulations in case of bandwidth choice $w \in W_{uv}$ with, for $m \in M^w_{uv}$, b^{wm}_{uv} the resulting capacity. Given $uv \in E$, $w \in W_{uv}$, and $m \in M^w_{uv}$, let ρ^{wm}_{uv} be the probability that the link is operated at modulation m or higher.

Now, we may assume that each link is operated at a chosen modulation (or higher) as long as the overall probability of the assumptions is at least $1-\varepsilon$. For this, the binary decision variables y obtain a new index m. The minimum cost fixed broadband wireless network design problem then reads:

$$\min \sum_{uv\in E} \sum_{w\in W_{uv}} \sum_{m\in M^w_{uv}} c^w_{uv} y^{wm}_{uv} \tag{2a}$$

$$s.t. \sum_{u\in V: vu\in E} f^k_{vu} - \sum_{u\in V: uv\in E} f^k_{uv} = \begin{cases} 1, & \text{if } v = s^k, \\ -1, & \text{if } v = t^k, \\ 0, & \text{otherwise} \end{cases} \qquad \forall v \in V, k \in K \tag{2b}$$

$$\sum_{w\in W_{uv}} \sum_{m\in M^w_{uv}} y^{wm}_{uv} = 1 \qquad \forall uv \in E \tag{2c}$$

$$\sum_{k\in K} d^k f^k_{uv} \leq \sum_{w\in W_{uv}} \sum_{m\in M^w_{uv}} b^{wm}_{uv} y^{wm}_{uv} \qquad \forall uv \in E \tag{2d}$$

$$\prod_{uv\in E} \Big(\sum_{w\in W_{uv}} \sum_{m\in M^w_{uv}} \rho^{wm}_{uv} y^{wm}_{uv} \Big) \geq 1-\varepsilon \tag{2e}$$

$$f^k_{uv} \in [0,1], y^{wm}_{uv} \in \{0,1\} \tag{2f}$$

Besides the total bandwidth cost function (2a) and the flow conservation constraints (2b), constraints (2c) ensure that exactly one bandwidth-modulation pair is chosen. Constraint (1) is now equivalently modeled in the link capacity constraints (2d) and in the solution confidence constraint (2e). (2d) ensure that all demands on one link can be fulfilled by the chosen bandwidth-modulation pair, whereas (2e) guarantees that the confidence of the solutions is at least $1-\varepsilon$.

Note that we assume explicitly a hypothesis on the modulation scheme in constraints (2d). Obviously, for a given link and bandwidth, the lower the modulation scheme is, the lower the assumed capacity and the higher the probability that the effective capacity supports the routed traffic. In other words, more conservative hypotheses on the modulation schemes lead to more reliable solutions.

Constraint (2e) can be easily linearized: By employing monotonicity of logarithmic functions and because the logarithm of a product equals the sum of the logarithms, (2e) is equivalent to

$$\sum_{uv\in E} \log\left(\sum_{w\in W_{uv}} \sum_{m\in M_{uv}^{w}} \rho_{uv}^{wm} y_{uv}^{wm}\right) \geq \log(1-\varepsilon). \tag{3}$$

By constraints (2c), exactly one of the sum elements within each logarithmic function will be nonzero. Hence, (3) is equivalent to

$$\sum_{uv\in E} \sum_{w\in W_{uv}} \sum_{m\in M_{uv}^{w}} \log(\rho_{uv}^{wm}) y_{uv}^{wm} \geq \log(1-\varepsilon). \tag{4}$$

3 Valid Inequalities

Constraints (2b), (2c), and (2d) define a classical network design problem studied intensively in the literature, see [10] and the references therein. In particular, *cut-based inequalities* have been proven to be effective to enhance the performance of ILP solvers [1]. Let $S \subset V$ be a proper and nonempty subset of the nodes V and $\overline{S} = V \setminus S$ its complement. The set $E(S,\overline{S}) := \{uv \in E : u \in S, v \in \overline{S}\}$, i.e., the set of arcs from S to $\overline{S}$ defines a *cutset*. Similarly, let $K(S,\overline{S}) := \{k \in K : s^k \in S, t^k \in \overline{S}\}$ be the set of demands originating in S and terminating in $\overline{S}$. Finally, let $d(S,\overline{S}) := \sum_{k\in K(S,\overline{S})} d^k$. An appropriate aggregation of constraints (2b), (2d), and nonnegativity of the variables results in the following *base cutset inequality*:

$$\sum_{uv\in E(S,\overline{S})} \sum_{w\in W_{uv}} \sum_{m\in M_{uv}^{w}} b_{uv}^{wm} y_{uv}^{wm} \geq d(S,\overline{S}) \tag{5}$$

Chvátal-Gomory (CG) rounding yields two classes of valid inequalities.

Cutset Inequalities. By dividing (5) by $a \in \{b_{uv}^{wm} : uv \in E(S,\overline{S}), w \in W_{uv}, m \in M_{uv}^{w}\}$ and rounding up both sides, the well-known *cutset inequalities* [10] are obtained:

$$\sum_{uv\in E(S,\overline{S})} \sum_{w\in W_{uv}} \sum_{m\in M_{uv}^{w}} \left\lceil \frac{b_{uv}^{wm}}{a} \right\rceil y_{uv}^{wm} \geq \left\lceil \frac{d(S,\overline{S})}{a} \right\rceil \tag{6}$$

Shifted Cutset Inequalities. Instead of applying CG-rounding directly, we can first shift the coefficients of (5). Given a cutset $E(S,\overline{S})$, let $a_{uv} = \min_{w \in W_{uv}} \min_{m \in M_{uv}^w} b_{uv}^{wm}$ for $uv \in E(S,\overline{S})$. By (2c) and $a(S,\overline{S}) := \sum_{uv \in E(S,\overline{S})} a_{uv}$, (5) can be rewritten as:

$$\sum_{uv \in E(S,\overline{S})} \sum_{w \in W_{uv}} \sum_{m \in M_{uv}^w} (b_{uv}^{wm} - a_{uv}) y_{uv}^{wm} \geq d(S,\overline{S}) - a(S,\overline{S}) \tag{7}$$

Now, let $a' \in \{b_{uv}^{wm} - a_{uv} : uv \in E(S,\overline{S}), w \in W_{uv}, m \in M_{uv}^w\}$. By CG-rounding, we obtain the following *shifted cutset inequalities*:

$$\sum_{uv \in E(S,\overline{S})} \sum_{w \in W_{uv}} \sum_{m \in M_{uv}^w} \left\lceil \frac{b_{uv}^{wm} - a_{uv}}{a'} \right\rceil y_{uv}^{wm} \geq \left\lceil \frac{d(S,\overline{S}) - a(S,\overline{S})}{a'} \right\rceil \tag{8}$$

It can be shown that (6) and (8) define facets of the convex hull of feasible solutions under certain conditions (beyond the scope of this paper).

4 Computational Results

Setting. We have performed preliminary computational experiments on a 5×5 grid network ($|V| = 25$, $|E| = 80$, $|K| = 50$) based on [8]. We consider two bandwidth choices for each link: 7 MHz (28 MHz) with cost 1000 (6000) using the 128-QAM (256-QAM) scheme, with an availability of 99.9 %. In fading conditions, these links will use the 16-QAM (32-QAM) scheme (with 100 % availability).

By assuming the same availability for radio links using the highest modulation scheme and under the hypothesis that the lowest modulation scheme guarantees an availability of 100 % (independent of the bandwidth), we can replace (4) by

$$\sum_{uv \in E} \sum_{w \in W_{uv}} y_{uv}^{w2} \leq \left\lfloor \frac{\log(1-\varepsilon)}{\log(\rho)} \right\rfloor =: N \tag{9}$$

where ρ is the availability probability of the highest modulation scheme. Note that a larger infeasibility tolerance ε implies a larger value N, i.e., the reliability of the solutions decreases. We consider $N = 10$ ($\varepsilon = 0.01$), $20, \ldots, 80$ (no reliability).

All computations are performed with CPLEX 12.2 [6] on a Linux machine with 2.67 GHz Intel Xeon X5650 processor and 12 GB RAM.

Optimality gap closed. In this study, we limit ourselves to a comparison of the optimality gap with/without separation of violated cutset inequalities (6) and/or shifted cutset inequalities (8). To this end, the separation of these inequalities is done exactly by an auxiliary ILP (details omitted, cf. e.g., [5,7]). The cutset inequalities are separated only in the root node of the branch-and-bound tree.

As a reference, we consider the optimality gap, i.e., the difference between LP relaxation and best known solution (computed by CPLEX with a time limit of 12 h, optimal for $N = 10, 60, 70, 80$). Fig. 1 shows the optimality gap closed, i.e., the percental reduction of the optimality gap at the end of the root node. For the results in Fig.1(a), we disabled the internal cuts of CPLEX and separated (i) cutset inequalities (6), (ii) shifted

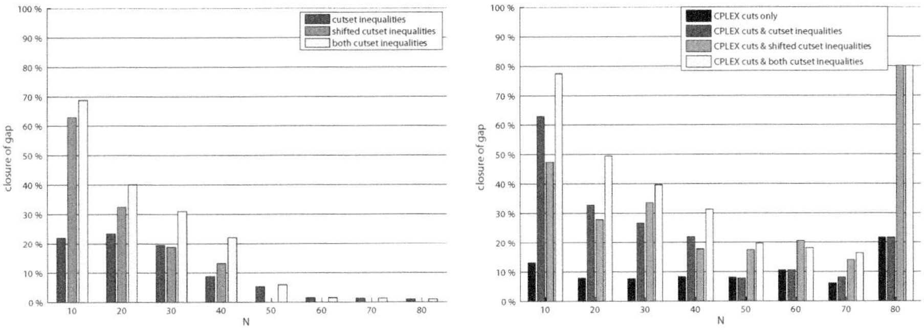

(a) Cutset inequalities only

(b) CPLEX cuts and cutset inequalities

Fig. 1. Optimality gap closed

cutset inequalities (8), and (iii) both. Inequalities (8) close the gap with 16 % on average, whereas inequalities (6) close only 10 %. Obviously, the optimality gap is closed most by the combination of (6) and (8) (up to 69 % for $N = 10$ and 21 % on average). With increasing N, the closure of the optimality gap decreases with hardly any closure from $N = 60$. We conjecture that inequalities (6) and (8) are less likely violated since the constraints (2d) are less restrictive.

In Fig. 1(b), we enabled the internal cuts of CPLEX. The optimality gap closed by internal cuts is only 10 % on average compared to 42 % by the combination of internal cuts and cutset inequalities (6) and (8). Note that also CPLEX can separate cutset inequalities [1]: only for $N = 80$ and (8), some multi-commodity flow (MCF) cuts are found. In case both types are separated, on average 54 violated inequalities are found (17 of type (6) and 37 of type (8)). Again, for increasing N the optimality gap closed decreases, except for $N = 80$ where 80 % of the gap is closed (due to the MCF cuts). For $N = 60$, the optimality gap is closed less by the combination of (6) and (8) than only by the shifted cutset inequalities (8). Such a phenomenon can occur due to varying internal CPLEX cuts.

5 Concluding Remarks

In this paper, we have presented a chance-constrained programming approach for the assignment of bandwidth in reliable fixed broadband wireless networks. We have proposed cutset inequalities and shifted cutset inequalities to enhance the computability of this problem. In our computational studies, we have discussed the optimality gap closed and compared the performance of the different cutset inequalities with and without internal CPLEX cuts. The results show that by the combination of the cutset and the shifted cutset inequalities, the optimality gap is closed by 42 % on average if the internal cuts for CPLEX are enabled.

As future work, we intend to investigate more realistic network topologies, different probability models and the reliability regarding traffic fluctuations.

Acknowledgement

This paper results from a research cooperation which was promoted by the PROCOPE program, a bilateral program funded by the German Academic Exchange Service with funds of the BMBF and by the French Ministry of Foreign Affairs. Additionally, this work has been supported by ANR DIMAGREEN and ECOSCELLS, the PATHFINDER project, Région PACA, and SME 3ROAM as well as the excellence initiative of the German federal and state governments and by the UMIC Research Centre at RWTH Aachen University. The fourth author gratefully acknowledges the financial support by the Villum Kann Rasmussen foundation.

References

1. Achterberg, T., Raack, C.: The MCF-separator: detecting and exploiting multi-commodity flow structures in MIPs. Mathematical Programming Computation 2, 125–165 (2010)
2. Anderson, H.: Fixed Broadband Wireless System Design, 1st edn. John Wiley & Sons, Chichester (2003)
3. Claßen, G., Coudert, D., Koster, A., Nepomuceno, N.: Bandwidth assignment for reliable fixed broadband wireless networks. In: IEEE WoWMoM 2011. IEEE, Lucca (2011)
4. Coudert, D., Nepomuceno, N., Rivano, H.: Power-efficient radio configuration in fixed broadband wireless networks. Computer Communications 33(8), 898–906 (2010)
5. Fischetti, M., Lodi, A., Salvagnin, D.: Just MIP it! Ann. Info. Systems 10, 39–70 (2009)
6. IBM – ILOG: CPLEX Optimization Studio 12.2, `http://www.ilog.com/products/cplex`
7. Koster, A.M.C.A., Kutschka, M., Raack, C.: Cutset inequalities for robust network design. In: Pahl, J., Reiners, T., Voß, S. (eds.) INOC 2011. LNCS, vol. 6701. Springer, Heidelberg (2011)
8. Larsson, T., Yuan, D.: An augmented lagrangian algorithm for large scale multicommodity routing. Computational Optimization and Applications 27(2), 187–215 (2004)
9. Lehpamer, H.: Microwave transmission networks: planning, design, and deployment. McGraw-Hill, New York (2010)
10. Raack, C., Koster, A.M.C.A., Orlowski, S., Wessäly, R.: On cut-based inequalities for capacitated network design polyhedra. Networks 57, 141–156 (2011)

Formulations and Branch-and-Cut Algorithm for the K-rooted Mini-Max Spanning Forest Problem

Alexandre Salles da Cunha[1], Luidi Simonetti[2], and Abilio Lucena[3]

[1] Universidade Federal de Minas Gerais,
Departamento de Ciência da Computação
acunha@dcc.ufmg.br
[2] Universidade Federal Fluminense,
Instituto de Computação
luidi@ic.uff.br
[3] Universidade Federal do Rio de Janeiro,
Programa de Engenharia de Sistemas e Computação
abiliolucena@globo.com

Abstract. In this paper, we discuss two Integer Programming Formulations for the K-rooted Mini-max Spanning Forest Problem. In the first, connectivity is reinforced through Generalized Subtour Breaking inequalities while the second uses Directed cutset constraints. We implement a Branch-and-cut method based on the first formulation that also computes combinatorial lower bounds from the literature and implements a Linear Programming based multi-start heuristic. Our computational results suggest that the Linear Programming lower bounds compare favorably to combinatorial lower bounds. Instances generated as suggested in the literature were solved easily by the algorithms proposed in this study.

1 Introduction

Let $G = (V,E)$ be a connected undirected graph with set of vertices $V = \{1,\ldots,n\}$ and set of edges E ($m = |E|$) with no loops nor multiple edges. A forest is an acyclic subgraph of G that consists of a set of mutually disjoint trees and a spanning forest is a forest where all vertices of V are included in one of its trees. Given that costs $\{c_{ij} \geq 0 : \{i,j\} \in E\}$ are assigned to the edges of E, the cost $w(T_k)$ of a tree $T_k = (V_k,E_k)$ is given by $\sum_{\{i,j\}\in E_k} c_{ij}$. Given a set of roots $\{r_1,\ldots,r_K\} \subset V$, a $K-$rooted forest is a spanning forest of G with K trees $\{T_k = (V_k,E_k) : k = 1,\ldots,K\}$, each one rooted at $r_k \in V_k$. In the K-rooted Mini-max Spanning Forest Problem (K-MMSFP), the objective is to find a K-rooted forest such that the cost of the most expensive of its trees, $w := \max\{w(T_k) : k = 1,\ldots,K\}$, is minimized.

Applications of K-MMSFP (see [8,5] for details) arise when one aims at designing reliable communication networks to serve a set of stations from two or more gateway points. The communication network is a spanning forest, where each station is connected, directly or undirectly, to one of the gateways (roots). The edges chosen in the solution must be such that the costs (which represent a measure of communication failure) of the trees are not only small but balanced as well. For applications of K-MMSFP in other domains, like for example, in supply chain networks, see Huang and Liu [2] and Zhou et al. [10].

J. Pahl, T. Reiners, and S. Voß (Eds.): INOC 2011, LNCS 6701, pp. 43–50, 2011.

K-MMSFP was introduced by Yamada et al. [8], when the problem was proven to be NP-hard when $K \geq 2$. Note that 1-MMSFP is the Minimal Spanning Tree Problem, for which several polynomial time algorithms, Kruskal's [3] for instance, are known. In another contribution, Yamada et al. [9] proposed a Branch-and-bound method where three types of combinatorial lower bounds were evaluated. The algorithm implements a depth-first search and performs branching on edges, given more priority to branch on the least expensive ones.

Recently, Mekking and Volgenant [5] improved on the Branch-and-bound algorithm in [9]. Instead of branching on edges, the new algorithm branches on vertices. The new branching scheme was motivated by the fact that, for solving K-MMSFP, it suffices to have an optimal assignment of vertices to roots. In addition, branching on edges has very little impact on a solution when the graph density is high. Consequently, combinatorial lower bounds tend to grow slowly and the search tree tend to be not balanced in edge based Branch-and-bound algorithms. Because of that, the algorithm in [5], which uses the same three combinatorial bounds mentioned previously, significantly outperforms the method in [9].

Yamada et al. [9] pointed out that a prospective approach to the exact solution of K-MMSFP would be Branch-and-cut methods (see Padberg and Rinaldi [6]), which are based on the polyhedral structure of the problem. In this paper, we investigate one of such methods. We discuss Integer Programming formulations for K-MMSFP and implement a Branch-and-cut method for one of them. Our preliminary computational results suggest that the Branch-and-cut method introduced here compares favorably to the method in [5].

The remaining of the paper is organized as follows. In Section 2, we present two Integer Programming formulations for K-MMSFP. A Branch-and-cut algorithm based on one of them is described in Section 3, together with a Linear Programming based multi-start heuristic. Computational results are discussed in 4, where we also conclude the paper and offer some conclusions.

2 Integer Programming Formulations

In order to present the first Integer Programming formulation for K-MMSFP, consider the following decision variables: (1) $y_i^k \in \{0,1\}$, which assumes value 1 if vertex i belongs to tree T_k (0, otherwise); (2) $x_{ij}^k \in \{0,1\}$ to select edges in each tree (assuming value 1 if edge $\{i,j\}$ belongs to tree T_k and 0, otherwise) and, finally, (3) $w \geq 0$, to denote the cost of the most expensive tree.

In what follows, given sets $M \subseteq E$ and $Q \subseteq V$, define $x^k(M) := \sum_{\{i,j\} \in M} x_{ij}^k$ and $y^k(Q) := \sum_{i \in Q} y_i^k$. Assume that $x = (x^1, \ldots, x^K), y = (y^1, \ldots, y^K)$, $\mathbb{B} = \{0,1\}$ and that $\mathbb{R}$ denotes the set of real numbers. A formulation for K-MMSFP is given by:

$$\min \ \left\{ w : (x,y,w) \in (\mathbb{R}^{Km}, \mathbb{B}^{Kn}, \mathbb{R}) \cap P_U \right\}, \tag{1}$$

where polytope P_U is defined by:

$$w \geq \sum_{\{i,j\} \in E} c_{ij} x_{ij}^k, \ k = 1, \ldots, K, \tag{2a}$$

$$\sum_{k=1}^{K} x^k(E) = n - K, \tag{2b}$$

$$x^k(E(S)) \leq y^k(S \setminus \{j\}), \;\; S \subset V, S \neq \emptyset, j \in S, \;\; k = 1, \ldots, K, \tag{2c}$$

$$\sum_{k=1}^{K} y_i^k = 1, \;\; \forall i \in V \setminus \{r_1, \ldots, r_K\} \tag{2d}$$

$$y_{r_k}^k = 1, \;\; k = 1, \ldots, K, \tag{2e}$$

$$y_i^k \geq 0, \;\; i \in V \setminus \{r_1, \ldots, r_k\}, k = 1, \ldots, K, \tag{2f}$$

$$x_{ij}^k \geq 0, \;\; \{i, j\} \in E, k = 1, \ldots, K. \tag{2g}$$

Constraint (2b) imposes that $n - K$ edges must be selected in any spanning forest of G. Generalized Subtour Breaking constraints (GSEC) (2c) guarantee that the subgraph of G implied by the vertices of V connected to root r_k must be connected and cycle-free. Constraints (2d) guarantee that each vertex of V must be assigned to exactly one tree. Because of (2b)-(2e), we have that $x^k(E) = y^k(V) - 1, \forall k = 1, \ldots, K$. Finally, constraints (2a) assure that w must value at least the cost of the most expensive tree.

One important aspect about polytope P_U is that it may have extreme points with $0 - 1$ entries for y and fractional entries for x variables. To illustrate, in Figure 1(a) - 1(d), we indicate three extreme points of P_U with the same value of w and node-to-root assignments, but with different values for the x variables. In the figures, the two roots are indicated by filled circles and costs are placed close to each edge in Figure 1(a). Figure 1(b) indicates an integral extreme point whereas Figures 1(c) and 1(d) depict two extreme points of P_U with fractional x entries. In all cases, corresponding y entries are binary. All x values are indicated close to the edges in Figures 1(b)–1(d) (fractional x entries are highlighted with dotted lines). Note that, for all points, $w = 25$. In Figure 1(b), the left-most tree is the one that defines the value of w, while for the right-most tree, inequality (2a) is slack. For the other extremes points, inequality (2a) is tight for both values of K. Finally, consider the forest that would be obtained by replacing the edge with cost 12 by the edge with cost 16 in the right-most tree in Figure 1(b). Note that a linear convex combination of these two points (with $\frac{1}{2}'s$ as weights) would lead to $w = 26$ and, therefore, the point indicated in Figure 1(c) is not such convex combination.

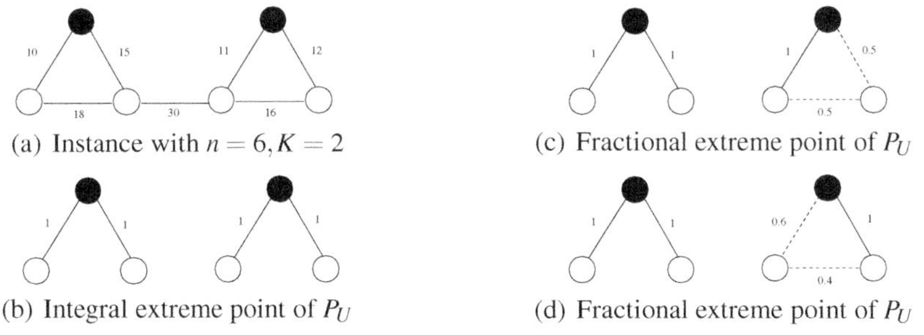

(a) Instance with $n = 6, K = 2$
(b) Integral extreme point of P_U
(c) Fractional extreme point of P_U
(d) Fractional extreme point of P_U

Fig. 1. A 2-MMSFP instance and three extreme points for P_U. Filled circles indicate the two roots.

In the formulation above, though, we do not need to impose x variables to be integer constrained, since whenever an extreme point of P_U having integral y vector is found to be an optimal solution to a Linear Programming (LP) relaxation to K-MMSFP, the corresponding Branch-and-bound node does not require branching. Such observation is obvious whenever x entries are also binary. So let $(\overline{x}, \overline{y}, \overline{w})$ denote an extreme point of P_U that solves the LP relaxation to K-MMSFP in a given node, such that $\overline{x} \notin \mathbb{B}^{Kn}$. Let $\overline{K} \subseteq \{1, \ldots, K\}$ denote the set of indices of trees whose $\overline{x}$ entries are not integer. Since $\overline{y} \in \mathbb{B}^{Kn}$, all we need to do to solve the node is to compute a Minimal Spanning Tree for the vertices assigned to each root $r_k \in \overline{K}$. For $k \in \{1, \ldots, K\} \setminus \overline{K}$, the support graph associated to $(\overline{x}^k, \overline{y}^k)$ already defines a tree. After all optimal trees are available, the most expensive one is retrieved as the solution to that node, which, thus, can be pruned by optimality.

Another way to formulate K-MMSFP is to consider the problem in a directed graph $D = (V, A)$ $(A = \{(i, j) \cup (j, i) : \{i, j\} \in E\})$ obtained from G by duplicating each of its edges $\{i, j\}$ into two arcs $(i, j), (j, i)$ with the same edge cost. For convenience, we also denote the cost of arc $(i, j) \in A$ by c_{ij}. In this case, a solution for K-MMSFP can be sought as a collection of K arborescences, each one directed out of one of the roots $r_1, \ldots, r_K$. Each vertex of V must be spanned by exactly one of such arborescences.

To formulate the problem in D, let us use variables $z_{ij}^k \in \{0, 1\} : (i, j) \in A, k = 1, \ldots, K$, to select the arcs in each arborescence (z_{ij}^k assumes value 1 if arc (i, j) belongs to the arborescence rooted at r_k, 0 otherwise). As before, given $L \subseteq A$, let $z^k(L) := \sum_{(i,j) \in L} z_{ij}^k$. Given any set $W \subset V$, $W \neq \emptyset$, let $\overline{W} = V \setminus W$ and let $(W, \overline{W}) := \{(i, j) \in A : i \in W, j \notin W\}$ denote the arcs in the cut implied by W. The Directed Cutset formulation for K-MMSFP is given by:

$$\min \ \left\{ w : (z, y, w) \in (\mathbb{R}^{2Km}, \mathbb{B}^{Kn}, \mathbb{R}) \cap P_D \right\}, \tag{3}$$

where polytope P_D is implied by (2d)-(2f) and

$$w \geq \sum_{(i,j) \in A} c_{ij} z_{ij}^k, \ \ k = 1, \ldots, K, \tag{4a}$$

$$z^k((V \setminus \{i\}, \{i\})) = y_i^k, \ \ i \in V \setminus \{r_1, \ldots, r_K\}, k = 1, \ldots, K, \tag{4b}$$

$$z^k((W, \overline{W})) \geq y_i^k, \ \ W \subseteq V \setminus \{i\}, r_k \in W, k = 1, \ldots, K. \tag{4c}$$

$$z_{ij}^k \geq 0, \ \ (i, j) \in A, k = 1, \ldots, K. \tag{4d}$$

Directed cutset constraints (4c) guarantee that the solution is connected and cycle free. Constraints (4b) impose that there must be exactly one edge incident to vertex i if it is spanned by the arborescence rooted at r_k. As in the case of P_U, P_D may also have extreme points with binary y entries and fractional z values. For the same reasons, we do not need to impose z variables to be integer constrained.

Following results in Goemans [1] (see also Wolsey and Magnanti [4]), it is not difficult to show that P_U and P_D are polytope-wise equivalent and, thus, provide the same Linear Programming bounds for K-MMSFP. In the following, we describe a Branch-and-cut algorithm based on the undirected formulation. As our computational experiments will show later on, for the instances considered in our test bed, the bounds implied by P_U are stronger than those implied by the three combinatorial lower bounds proposed by Yamada et al. [9] and later improved by Mekking and Volgenant [5].

3 A Branch-and-Cut Algorithm Based on the Undirected Formulation

In this section, we provide the main implementation details on our Branch-and-cut (BC) algorithm for K-MMSFP, based on formulation P_U. The algorithm was implemented with calls to XPRESS MIP solver (release 19.00) callback routines. All pre-processing, heuristics and separation of valid inequalities implemented by XPRESS were turned off. BC implements a *best-first* search policy. Apart from that, default XPRESS settings were used.

BC starts solving the LP relaxation

$$\min\ w:\ (x,y,w) \in \overline{P}_U, \tag{5}$$

where polytope $\overline{P}_U$ is given by (2a)-(2b),(2d)-(2g) and:

$$x_{ij}^k \leq y_i^k,\ k = 1,\ldots,K, \{i,j\} \in E, \tag{6a}$$

$$x_{ij}^k \leq y_j^k,\ k = 1,\ldots,K, \{i,j\} \in E. \tag{6b}$$

Let $(\overline{x},\overline{y},\overline{w}) \in \overline{P}_U$ be the solution to (5) and $\overline{G}_k = (\overline{V}_k,\overline{E}_k) : k = 1,\ldots,K$ be the subgraph of G implied by $(\overline{x}^k,\overline{y}^k)$ ($\overline{V}_k := \{i \in V : \overline{y}_i^k > 0\}$ and $\overline{E}_k := \{\{i,j\} \in E : \overline{x}_{ij}^k > 0\}$). If for all $k = 1,\ldots,K$, the vector $(\overline{x}^k,\overline{y}^k)$ is integer and if there is no GSEC (2c) violated by $(\overline{x}^k,\overline{y}^k)$, $(\overline{x},\overline{y},\overline{w})$ solves (1). Otherwise, we attempt to strengthen $\overline{P}_U$, appending violated GSECs to it.

The exact separation of GSECs can be carried out efficiently, through max-flow (min-cut) computations, in $O(Kn^4)$ time complexity (see [7] and [4] for details). Despite that, in practice, we found advantageous to separate GSECs through the following heuristic. For each value of k, we sort the edges in $\overline{E}_k$ in a non-increasing order of their $\overline{x}_{ij}^k$ values. Then, we find a forest of maximum cardinality of $\overline{G}_k$, using Kruskal's algorithm, giving preference to include edges with higher values of $\overline{x}_{ij}^k$. Each edge included during Kruskal's method merges two sets of vertices into a a new connected component in the forest being built. We check for violated GSECs for such connected components generated after each edge inclusion, until a forest of maximum cardinality has been found. In our implementation, all GSECs violated by at least 10^{-3} are appended into $\overline{P}_U$, and a new LP is formulated and re-optimized. If no GSECs are found to be violated through the separation heuristic outlined above, we branch on y variables.

In our BC implementation, before evaluating the LP relaxation at each node, we first compute the combinatorial lower bounds by Mekking and Volgenant [5]. These bounds, though weaker than the LP relaxation bounds implied by P_U, are very fast to be evaluated. Therefore, we only compute the LP bounds in a node if none of the combinatorial bounds allow to early prune that node. In this sense, BC can be seen as an hybrid algorithm, since the two types of lower bounds are used during the search. In another perspective, BC can be seen as an improvement over the Branch-and-bound algorithm in [5] since it uses LP information to bound, to choose variables on which to branch and the order in which nodes should be investigated.

In order to obtain valid upper bounds for K-MMSFP, we implemented a multi-start Linear Programming heuristic (LPH), that works as follows. Assume that, in a given

Branch-and-cut node, $(\overline{x},\overline{y},\overline{w})$ denotes the solution to the LP relaxation at hands, when no more violated GSECs were found. Due to (2d) and (2e)-(2f), the idea of LPH is to take the values $\{\overline{y}_i^k : k = 1,\ldots,K\}$ as probabilities that vertex i will be connected to each of the roots $r_1,\ldots,r_K$. Accordingly, LPH consists of randomly choosing which root will be assigned to each vertex, according to probabilities given by $\{\overline{y}_i^k : i \in V, k = 1,\ldots,K\}$. Once the node-to-root assignments are chosen, we run Kruskal's algorithm for each set of vertices that are assigned to the same root. Attempting to improve the upper bounds, LPH is called 10 times at the end of each node in the enumeration tree.

4 Preliminary Computational Results and Conclusions

In this section, we discuss results for 2-MMSFP. The only instance available to us from previous studies is instance $K_{20,46}$ (introduced in [8], defined on a planar graph, with $n = 20$ and $m = 46$). Thus, to evaluate BC, additional instances of various types were generated: two dimensional Euclidean and random costs instances, defined on dense and on sparse graphs. For each cost type and graph density, bipartite graphs were considered as well. Euclidean instances considered here have vertices' coordinates randomly chosen in the interval $[0,1000]$. Edge costs correspond to the Euclidean distance between their endpoints rounded down. Random cost instances were generated by assigning randomly chosen integers in the interval $[0,1000]$ to edge costs. For each value of n, instance type and graph density, 3 instances were generated. For each instance, we evaluated BC with the following $n-1$ root choices: $r_1 = 1$ and $r_2 = i, i = 2,\ldots,n$ (see Section 5 in [5] for the reasoning why we have chosen such roots).

All algorithms introduced in this paper were implemented in C and computational experiments were conducted in a Intel XEON processor, running at 2.0Ghz, with 8Gbytes of RAM memory. Computational results reported in Mekking and Volgenant [5] were obtained in a machine with a speed comparable to a Intel Pentium III with 900 Mhz. The algorithms in [5] were coded in Delphi Pascal, usually slower than C. In addition, apart from instance $K_{20,46}$, instances considered here and in [5] are not the same, though were generated by the same guidelines. Thus, a direct comparison between computing times spent by BC and by the Node Based Branch-and-bound algorithm (NBA) in [5] seems hard to be established with accuracy. We therefore only briefly discuss trends in the growth of computing times of both algorithms as n increases.

In Table 1, we compare BC with NBA for instance $K_{20,46}$, for various root choices (indicated in the first column in the table). In the next two columns, we report on the number of nodes required by NBA and BC, followed by the time needed by each algorithm to solve each instance (in seconds). In the subsequent three columns, we present lower bounds for 2-MMSFP: the two strongest lower bounds in [5] (under headings LB1, LB2) and the LP bound implied by P_U, w_{P_U} (we do not report the third lower bound in [5] since for the root node, such bound is always 0). In the last column, we present the value of the optimal objective function, w^*.

As it can be observed, for the root choices considered in the table, w_{P_U} is always stronger than LB1 and LB2. For instances of this size and density, it seems that NBA

Table 1. NBA and BC computational results for instance $K_{20,46}$

	# nodes		CPU time (s)		Root lower bounds			
r_1, r_2	NBA	BC	NBA	BC	LB1	LB2	w_{P_U}	w^*
1, 20	298	31	0.0	0.09	799	631	803.5507	855
2, 19	253	39	0.0	0.07	799	618	801.0098	848
3, 18	359	43	0.0	0.09	799	631	800.7628	848
4, 17	351	39	0.0	0.08	799	631	800.2410	848
5, 16	287	45	0.0	0.09	799	574	799.2847	848
6, 15	269	37	0.0	0.09	799	584	801.0098	848
7, 14	300	41	0.0	0.08	799	660	801.0098	848
8, 13	362	37	0.0	0.08	773	624	797.3352	852
9, 12	319	37	0.0	0.07	807	611	807.7487	848
10, 11	307	29	0.0	0.07	807	635	811.6621	852

outperforms BC. However, $K_{20,46}$ is indeed very easy to solve. In our view, conclusions about the merits of BC should be drawn from its capability of solving larger and harder test instances, presented next.

In Table 2, we present BC results for Euclidean and random costs instances, for various graph densities. In the first two columns, we report instance data (n or $\frac{n}{2}, \frac{n}{2}$ for bipartite graphs and graph densities from 100% to 25%). The next results in the table are the average and the maximum times (in seconds) needed by BC to solve the instances. Similar entries are given on the sequence, for the average and the maximum number of nodes. In the last column, we report the average duality gap at the root node (the relative distance to the optimal value, when the separation heuristic was used to compute the lower bound).

On the average, for Euclidean instances defined on complete graphs with $n = 50, 80$, NBA in [5] needed to evaluate 1.89×10^5 and 4.17×10^6 nodes, respectively, and took, 41.4 and 2313.3 seconds to solve similar sets of instances. BC, on the other hand, took 7.3 and 36.2 seconds, respectively. From $n = 50$ to $n = 80$, computing times increased 5 times for BC and 56 times for NBA. The number of nodes investigated in the search is, on the average, 3 orders of magnitude lower for BC. For bipartite complete instances with 50 vertices, NBA took 19.3 seconds and investigated 9.66×10^4 nodes while BC needed, on the average, to investigate 136 nodes in 2.5 seconds.

To conclude, BC was able to easily solve instances with up to 80 vertices of various different types. It seems that the smaller rate of growth in the computing times for BC is not related only to the computational power available nowadays. This seems to be true since, for the instances considered here, LP bounds dominate combinatorial bounds from the literature and the LP based heuristic performs very well. We plan to implement a Branch-and-cut algorithm based on the directed formulation and to proceed with the polyhedral study started here.

Table 2. BC average computational results for 2-MMSFP Euclidean and random cost instances

Euclidean instances						
		CPU time (s)		# nodes		root
n	den (%)	avg	max	avg	max	gap (%)
50	100	7.3	24.2	207.8	627	2.82
	75	4.3	14.1	190.6	581	2.68
	50	2.7	12.9	193.8	803	2.56
	25	7.0	44.9	643.2	3539	2.38
25,25	100	2.5	16.6	136.6	771	2.08
	75	1.7	5.7	167.4	569	1.60
	50	2.3	19.4	220.0	621	2.19
	25	1.7	14.0	288.6	2051	1.33

Euclidean instances						
		CPU time (s)		# nodes		root
n	den (%)	avg	max	avg	max	gap (%)
80	100	36.2	107.5	319.6	981	1.52
	75	50.5	235.1	542.6	2329	1.93
	50	17.9	123.9	347.p	1881	1.44
	25	27.2	504.2	600.3	6705	1.33
40,40	100	26.1	1065.6	375.8	9117	1.36
	75	12.3	31.6	383.7	731	1.52
	50	16.1	225.1	565.5	6493	2.59
	25	20.3	524.5	824.5	9279	2.97

Random cost instances						
		CPU time (s)		# nodes		root
n	den (%)	avg	max	avg	max	gap (%)
50	100	4.2	46.9	178.9	2279	3.36
	75	2.2	9.5	140.7	373	2.19
	50	1.5	10.2	121.3	593	2.59
	25	1.9	16.5	221.4	1919	2.39
25,25	100	2.7	85.5	310.7	7827	3.38
	75	1.6	8.0	208.4	763	2.84
	50	1.1	4.8	137.8	557	2.05
	25	0.7	3.9	158.2	699	2.41

Random cost instances						
		CPU time (s)		# nodes		root
n	den (%)	avg	max	avg	max	gap (%)
80	100	30.7	533.9	302.9	6217	1.63
	75	18.8	269.2	311.8	4597	1.49
	50	42.6	494.3	876.0	9291	2.06
	25	25.3	345.24	1056.7	9021	2.25
40,40	100	28.9	433.2	598.4	8921	1.88
	75	19.1	218.1	541.1	4267	1.93
	50	14.1	112.1	576.6	2665	1.76
	25	6.7	55.0	421.9	2911	1.82

Acknowledgements

Alexandre Salles da Cunha is partially funded by CNPq grant 302276/2009-2 and FAPEMIG PRONEX APQ-01201-09. Abilio Lucena is partially funded by CNPq grant 310561/2009-4. Luidi Simonetti is partially funded by FAPERJ grant E-26/111.819/2010 and CNPq grant 483.243/2010-8.

References

1. Goemans, M.X.: The Steiner Polytope and Related Polyhedra. Mathematical Programming 63, 157–182 (1994)
2. Huang, B., Liu, N.: Bi-level Programming Approach to Optimizing a Logistic Distribution Network with Balancing Requirements. Transportation Research Record (1894), 188–197 (2004)
3. Kruskal, J.: On the Shortest Spanning Subtree of a Graph and the Traveling Salesman Problem. Proceedings of the American Mathematical Society 7, 48–50 (1956)
4. Magnanti, T.L., Wolsey, L.: Optimal Trees. In: Ball, O., et al. (eds.) Handbooks in OR and MS, vol. 7, pp. 503–615. North-Holland, Amsterdam (1995)
5. Mekking, M., Volgenant, A.: Solving the 2-rooted mini-max spanning forest problem by branch-and-bound. European Journal of Operational Research 203(1), 50–58 (2010)
6. Padberg, M.W., Rinaldi, G.: A Branch-and-Cut algorithm for resolution of large scale of Symmetric Traveling Salesman Problem. SIAM Review 33, 60–100 (1991)
7. Padberg, M.W., Wolsey, L.: Trees and cuts. Annals of Discrete Mathematics 17, 511–517 (1983)
8. Yamada, T., Takahashi, H., Kataoka, S.: A heuristic algorithm for the mini-max spanning forest problem. European Journal of Operational Research 91(3), 565–572 (1996)
9. Yamada, T., Takahashi, H., Kataoka, S.: A branch-and-bound algorithm for the mini-max spanning forest problem. European Journal of Operational Research 101(1), 93–103 (1997)
10. Zhou, G., Min, H., Gen, M.: The balanced allocation of customers to multiple distribution centers in the supply chain network: A genetic algorithm approach. Computers and Industrial Engineering (43), 251–261 (2002)

Negative Cycle Separation in Wireless Network Design

Fabio D'Andreagiovanni[1], Carlo Mannino[2], and Antonio Sassano[3]

[1] Konrad-Zuse-Zentrum für Informationstechnik Berlin (ZIB),
Takustr. 7, D-14195 Berlin, Germany
d.andreagiovanni@zib.de

[2] Department of Computer and System Sciences,
Sapienza Università di Roma, via Ariosto 25, 00185 Roma, Italy
mannino@dis.uniroma1.it

[3] Department of Computer and System Sciences,
Sapienza Università di Roma, via Ariosto 25, 00185 Roma, Italy
sassano@dis.uniroma1.it

Abstract. The Wireless Network Design Problem (WND) consists in choosing values of radio-electrical parameters of transmitters of a wireless network, to maximize network coverage. We present a pure 0-1 Linear Programming formulation for the WND that may contain an exponential number of constraints. Violated inequalities of this formulation are hard to separate both theoretically and in practice. However, a relevant subset of such inequalities can be separated more efficiently in practice and can be used to strengthen classical MILP formulations for the WND. Preliminary computational experience confirms the effectiveness of our new technique both in terms of quality of solutions found and provided bounds.

1 Introduction

Wireless networks have shown a rapid growth over the past two decades and now play a key role in new generation telecommunications networks. Scarce radio resources, such as frequencies, have rapidly became congested and the need for more effective design methods arose. A general planning problem consists in establishing the radio-electrical parameters (e.g., power emission and frequency) of the transmitters of a wireless network so as to maximize the overall network coverage. To present our original contribution, in this paper we focus only on establishing power emissions. This is actually a basic problem in all wireless planning contexts that can be easily extended by introducing additional elements, such as frequencies [4,5].

For our purposes, a wireless network can be described as a set of transmitters B distributing a telecommunication service to a set of receivers T. Each transmitter $b \in B$ emits a radio signal with power $p_b \in [0, P_{\max}]$. The power $p_b(t)$ that receiver t gets from transmitter b is proportional to the emitted power p_b by a factor $a_{tb} \in [0,1]$, i.e. $p_b(t) = a_{tb} \cdot p_b$, commonly called *fading coefficient*. Among the signals received from transmitters in B, receiver t can select a *reference signal* (or *server*), which is the one carrying the service. All the other signals are interfering.

A receiver t is regarded as served by the network, specifically by server $\beta \in B$, if the ratio of the serving power to the sum of the interfering powers (*signal-to-interference*

J. Pahl, T. Reiners, and S. Voß (Eds.): INOC 2011, LNCS 6701, pp. 51–56, 2011.

ratio or *SIR*) is above a threshold δ [11], (*SIR threshold*), whose value depends on the technology and the desired quality of service:

$$\frac{a_{t\beta} \cdot p_{\beta}}{\mu + \sum_{b \in B \setminus \{\beta\}} a_{tb} \cdot p_b} \geq \delta \tag{1}$$

where the system noise $\mu > 0$ is assimilated to an interfering signal with fixed (very low) power emission.

For every $t \in T$, we have one inequality of type (1) for each potential server $\beta \in B$: in particular, we denote by $SIR(t,b)$ the inequality (1) associated with receiver t and server b. Receiver t is served if at least one of these inequalities is satisfied or, equivalently, if the following disjunctive constraint is satisfied:

$$\bigvee_{\beta \in B} \left(a_{t\beta} \cdot p_{\beta} - \delta \cdot \sum_{b \in B \setminus \{\beta\}} a_{tb} \cdot p_b \geq \delta \cdot \mu \right) \tag{2}$$

Each linear inequality of the above disjunction is obtained by simple algebra from the SIR expression (1).

If each receiver $t \in T$ is associated to a value $r_t > 0$ that expresses revenue obtained by serving t, the *Wireless Network Design Problem (WND)* consists in setting the power emission of each transmitter $b \in B$ and the server of each receiver in $t \in T$ with the aim of maximizing the overall revenue of served receivers.

2 A Pure 0-1 Linear Programming Formulation for the WND

The WND is often approached by solving a suitable Mixed-Integer Linear Program (MILP): first, a binary variable x_{tb} is introduced for every $t \in T$, $b \in B$, with $x_{tb} = 1$ if and only if b serves t; then, variables x_{tb} are used to replace each disjunction (2) with a set of $|B|$ linear constraint, that, however, include large positive constants, the notorious *big-M coefficients* [5,7]. The (linear) objective function aims to maximize the overall revenue from coverage, i.e. $\max \sum_{t \in T} \sum_{b \in B} r_t \cdot x_{tb}$ and requires the additional constraints:

$$\sum_{b \in B} x_{tb} \leq 1 \qquad t \in T \tag{3}$$

to ensure that each receiver is associated to at most one server. A vector $x \in \{0,1\}^{T \times B}$ satisfying (3) is a *server assignment*.

The resulting MILP presents severe drawbacks, highlighted in several works, e.g. [5,7,8]. First, the coefficients in the SIR inequalities may vary over a very wide range, with differences up to 10^{12} or even larger. This makes the constraint matrix very ill-conditioned and the solutions returned by solvers are often inaccurate and may contain errors. Also, the presence of big-M terms results in weak bounds thus leading to very large search trees. To tackle these problems a number of different approaches were recently proposed. For a comprehensive introduction to these related works, we refer the reader to [4,5,7].

In this paper, we propose an alternative pure 0-1 Linear Programming formulation for the WND, whose defining inequalities are linear constraints in the assignment variables

x_{tb}. Such inequalities are thus valid for all the formulations that are derived from the previously introduced MILPs and can be included to strengthen them.

Let now $\tilde{x} \in \{0,1\}^{T \times B}$ be a server assignment and let Σ denote the set of all the SIR inequalities $SIR(t,b)$ and the lower and upper bounds constraints $0 \leq p_b \leq P_{\max}$ on power emissions. With $\tilde{x}$ we associate the subsystem $I(\tilde{x})$ of SIR inequalities (1) whose corresponding variables $\tilde{x}_{tb}$ are activated, i.e:

$$I(\tilde{x}) = \{SIR(t,b) \in \Sigma : \tilde{x}_{tb} = 1\}$$

It is easy to check if $I(\tilde{x})$, extended with lower and upper bounds on the variables p_b, is feasible. If this is the case, all of the assigned testpoints can actually be served by the network, and we say that x is a *feasible server assignment.*

At this point, we can restate the WND as the problem of finding a feasible server assignment that maximizes the revenue function. To this aim, a simple characterization of all the feasible server assignments goes as follows. Denote by IS the set of subsystems $I(x)$ such that *x is not feasible*. Then $\tilde{x} \in \{0,1\}^{T \times B}$ is a feasible server assignment if and only if $\tilde{x}$ satisfies the following system of linear inequalities:

$$\sum_{(t,b) \in I} \tilde{x}_{tb} \leq |I| - 1 \qquad \forall\, I \in IS \tag{4}$$

The above system is in general very large and the inequalities must be generated dynamically. Unfortunately, the separation of violated inequalities (4) is hard, both theoretically and in practice [2,10]. Moreover, it may entail some of the numerical difficulties associate with the MILP formulations for the WND. Still, a relevant subset of these inequalities can be separated more effectively, as we describe next.

To this end, we proceed in a similar way to [8]. Namely, we generate a new system Σ' obtained from Σ by substituting each (1) with the system:

$$\frac{a_{t\beta} \cdot p_\beta}{a_{tb} \cdot p_b} \geq \delta \qquad \forall\, b \in B \setminus \{\beta\} \tag{5}$$

where, to simplify the notation, we assume that B also contains the noise μ as a fictitious transmitter with fixed power emission. It is not difficult to see that Σ' is a relaxation of Σ and every infeasible subsystem of Σ' corresponds to an infeasible subsystem of Σ. Basically, this relaxation corresponds to considering a receiver as served if the power emission of its server suffices to contrast each interferer individually and the thermal noise. Or, alternatively, if its *best server* is "stronger" than its *strongest interferer*. In [8] the authors show that, in most cases of practical interest, this is indeed a good approximation of the original SIR constraint.

By assuming $p_b \in [\varepsilon, P_b]$, with $\varepsilon > 0$ very small, and by taking the logarithm[1] of both left and right hand side multiplied by 10, the system Σ' can be rewritten as:

$$q_b - q_\beta \leq w^t_{\beta b} \qquad t \in T, \beta \in B, b \in B \setminus \{\beta\} \tag{6}$$

where $q_b = 10 \log_{10} p_b$ for all $b \in B$ and $w^t_{\beta b} = \lceil 10(\log_{10} a_{t\beta} - \log_{10} a_{tb} - \log_{10} \delta) \rceil$, extended with the lower and upper bounds $10 \log_{10} \varepsilon \leq q_b \leq 10 \log_{10} P_b$, for all $b \in B$.

[1] This corresponds to rewriting all quantities in dB format.

In this way, the system Σ' is transformed into a *system of difference inequalities* (lower and upper bounds can be easily represented in this form as well), where each constraint (6) is associated with a server β and a receiver t and thus with an assignment variable $x_{t\beta}$.

Now, given a generic system of difference constraints Σ^d:

$$(i) \qquad t_v - t_u \leq l_{uv},\ (u,v) \in A \tag{7}$$

where $t \in \mathrm{IR}^A$ and $l \in Z^A$, we can consider the associated weighted directed graph $G = (V,A)$, with weight function l. Then, it is well known that every infeasible subsystem of (7) contains (the constraints corresponding to) the arcs of a negative directed cycle of G [9]. Also, denoting by $x \in \{0,1\}^A$ the incidence vector of (the arcs corresponding to) a feasible subsystem of Σ^d, then x is the set of solutions to:

$$\begin{aligned} (i) \qquad & \Sigma_{uv \in C}\, x_{uv} \leq |C| - 1,\ C \in \mathscr{C}^- \\ & x \in \{0,1\}^A \end{aligned} \tag{8}$$

where $\mathscr{C}^-$ is the set of negative directed cycles of G.

In [6] we develop an exact approach to the separation of violated inequalities (8.i). The resulting algorithm can be used to separate the violated inequalities associated with the system (6) (including upper and lower bounds on the q variables expressed as difference inequalities) which correspond to negative directed cycles in the associated directed graph. One of these cycles C corresponds to a subset of constraints of (6) associated with the pairs $I_C = \{(\beta_1,t_1),\ldots,(\beta_m,t_m)\} \subseteq B \times T$ (plus possibly some lower and upper bound constraints).

One can show that $\beta_1 \neq \beta_2 \neq \ldots \neq \beta_m$ and $t_1 \neq t_2 \neq \ldots \neq t_m$ and the valid constraint:

$$\sum_{(t,b) \in I_C} x_{tb} \leq |I_C| - 1 \tag{9}$$

may be added to the formulation. In our preliminary results, however, we limit to consider cycle inequalities with $|C| = 2$ and separate them by enumeration.

3 Preliminary Computational Results

We test the performance of our new approach to WND on a set of 15 realistic instances, developed with the Technical Strategy & Innovations Unit of British Telecom Italia (BT Italia SpA). All the instances refer to a *Fixed WiMAX Network* [1], deployable in an urban residential area and consider various scenarios with up to $|T| = 529$ receivers, $|B| = 36$ transmitters, $|F| = 3$ frequencies, $|H| = 4$ burst profiles (see Table 1). We remark that the experiments refer to a formulation that extends the basic one considered in Section 1, by including frequency channels and modulation schemes as additional decision variables. Such formulation is denoted by *(BM)* and captures specific features of so-called *Next Generation Networks* like WiMAX [1]. For a detailed description of (BM), we refer the reader to [3].

Table 1. Comparisons between (BM) and (PI) with and without valid inequalities (8)

ID	|T|	|B|	|F|	|H|	(BM)			(S-BM)			(PI)		(S-PI)	
					UB_0	$\|T^*\|$	gap%	UB_0	$\|T^*\|$	gap%	UB_0	$\|T^*\|$	UB_0	$\|T^*\|$
I1	100	12	1	1	98.36	66 (70)	29.43	96.78	66 (70)	27.68	90.77	75	90.23	75
I2	169	12	1	1	165.47	97	59.81	163.15	97	57.39	153.12	101	152.37	101
I3	196	12	1	1	193.61	102 (105)	77.87	192.02	102 (105)	75.11	179.35	108	177.92	108
I4	225	12	1	1	219.76	92	81.13	218.36	92	79.36	202.44	92	201.54	92
I5	289	12	1	1	287.20	76 (77)	195.44	287.20	76 (77)	194.92	274.62	85	274.13	85
I6	361	12	1	1	352.01	126 (132)	154.87	350.43	140	138.76	337.22	156	336.46	156
I7	400	18	1	1	397.21	166	132.01	396.79	166	131.32	386.07	184	384.95	184
I8	400	18	3	4	400.00	356	12.36	400.00	356	12.36	396.53	372	395.80	372
I9	441	18	3	4	441.00	266 (270)	63.33	441.00	266 (270)	63.33	438.28	295	437.52	295
I10	484	27	3	4	484.00	120 (122)	296.72	484.00	120 (122)	296.72	479.10	242	478.68	242
I11	529	27	3	4	529.00	77	587	529.00	77	587	523.15	168	521.76	168
I12	400	36	1	4	398.04	72 (74)	287.30	396.93	77 (78)	264.85	389.61	102	389.14	102
I13	441	36	1	4	433.21	184	131.03	431.42	184	129.77	414.93	194	413.78	194
I14	484	36	1	4	482.78	209	108.31	481.66	209	107.56	472.44	251	471.58	251
I15	529	36	1	4	517.89	98 (105)	226.44	516.14	114	198.57	503.32	232	502.67	232

For each instance, we present preliminary computational results obtained by solving the big-M formulation (BM) and its corresponding Power-Indexed formulation (PI) [5]. We consider (BM) and (PI) formulations with and without the valid inequalities (8) obtained for $|C| = 2$. Formulations strengthened through (8) are distinguished by adding *S*-, i.e. *(S-BM)* and *(S-PI)*.

Experiments are run by imposing a time limit of 1 hour and by using a machine with a 1.80 GHz Intel Core 2 Duo processor and 2 GB of RAM. Table 1 reports the performance of the four considered formulations over the set of WiMAX instances. We solve (BM) and (S-BM) by direct application of IBM ILOG Cplex 11.1 and we report i) the upper bound UB_0 obtained at node 0 of the branch-and-bound tree, ii) the value $|T^*|$ of the best solution found within the time limit and iii) the final integrality gap *gap%*. The presence of two values in some lines of the column $|T^*|$ of (BM) indicates that the coverage plans returned by Cplex contain errors and some receivers are actually not covered. We instead solve (PI) and (S-PI) by the incremental algorithm WPLAN described in [5] and we report i) the upper bound UB_0 obtained at node 0 when considering the basic set of power levels, and ii) the value $|T^*|$ of the best solution found by WPLAN within the time limit.

By adding the new valid inequalities (8) for $|C| = 2$, in most cases stronger bounds are obtained at node 0 and smaller integrality gaps are reached within the time limit. In particular, the benefits are particularly evident in the case of the big-M formulation: in three cases, namely I6, I12, I15, the value of the best solution is increased, even eliminating coverage errors (I6, I15).

Acknowledgement

This work was partially supported by the *German Federal Ministry of Education and Research* (BMBF), project *ROBUKOM*, grant 03MS616E, and by the *Italian Ministry of Education, University and Research* (MIUR), project *Design, optimization and analysis of next-generation converged networks*, grant PRIN 2008LLYXFS.

References

1. Andrews, J.G., Ghosh, A., Muhamed, R.: Fundamentals of WiMAX. Prentice Hall, Upper Saddle River (2007)
2. Amaldi, E., Kann, V.: The complexity and approximability of finding maximum feasible subsystems of linear relations. Theor. Comput. Sci. 147(1-2), 181–210 (1995)
3. D'Andreagiovanni, F., Mannino, C.: An Optimization Model for WiMAX Network Planning. In: Zhang, Y. (ed.) WiMAX Network Planning and Optimization, pp. 369–386. Auerbach Publications, Boca Raton (2009)
4. D'Andreagiovanni, F.: Pure 0-1 Programming Approaches to Wireless Network Design. Ph.D. Thesis, Sapienza Università di Roma, Rome, Italy (2010)
5. D'Andreagiovanni, F., Mannino, C., Sassano, A.: GUB Covers and Power-Indexed Formulations for Wireless Network Design. Technical Report vol. 2(14), Department of Computer and System Sciences, Sapienza Università di Roma, Rome, Italy (2010)
6. D'Andreagiovanni, F., Mannino, C., Sassano, A.: Separating Negative Cycle Inequalities (2010) (in preparation)
7. Kennington, J., Olinick, E., Rajan, D.: Wireless Network Design: Optimization Models and Solution Procedures. Springer, Heidelberg (2010)
8. Mannino, C., Mattia, S., Sassano, A.: Wireless Network Design by Shortest Path. Comp. Opt. and Appl. (2009), doi: 10.1007/s10589-009-9264-3
9. Nehmauser, G., Wolsey, L.: Integer and Combinatorial Optimization. John Wiley & Sons, Hoboken (1988)
10. Pfetsch, M.: The Maximum Feasible Subsystem Problem and Vertex-Facet Incidence of Polyhedra. Ph.D. Thesis, Technische Universität Berlin, Berlin, Germany (2002)
11. Rappaport, T.S.: Wireless Communications: Principles and Practice, 2nd edn. Prentice Hall, Upper Saddle River (2001)

A Node Splitting Technique for Two Level Network Design Problems with Transition Nodes

Stefan Gollowitzer[1], Luís Gouveia[2], and Ivana Ljubić[1]

[1] Department of Statistics and Operations Research, University of Vienna, Austria
{stefan.gollowitzer,ivana.ljubic}@univie.ac.at

[2] Departamento de Estatística e Investigação Operacional - Centro de Investigação Operacional, Faculdade de Ciênçias, Universidade de Lisboa, Portugal
legouveia@fc.ul.pt

Abstract. The Two Level Network Design (TLND) problem arises when local broadband access networks are planned in areas, where no existing infrastructure can be used, i.e., in the so-called *greenfield deployments*. Mixed strategies of Fiber-To-The-Home and Fiber-To-The-Curb, i.e., some customers are served by copper cables, some by fiber optic lines, can be modeled by an extension of the TLND. We are given two types of customers (primary and secondary), an additional set of Steiner nodes and fixed costs for installing either a primary or a secondary technology on each edge. The TLND problem seeks a minimum cost connected subgraph obeying a tree-tree topology, i.e., the primary nodes are connected by a rooted primary tree; the secondary nodes can be connected using both primary and secondary technology. In this paper we study an important extension of TLND in which additional transition costs need to be paid for *intermediate facilities* placed at the transition nodes, i.e., nodes where the change of technology takes place. The introduction of transition node costs leads to a problem with a rich structure permitting us to put in evidence reformulation techniques such as modeling in higher dimensional graphs (which in this case are based on a node splitting technique). We first provide a compact way of modeling intermediate facilities. We then present several generalizations of the facility-based inequalities involving an exponential number of constraints. Finally we show how to model the problem in an extended graph based on node splitting. Our main result states that the connectivity constraints on the splitted graph, projected back into the space of the variables of the original model, provide a new family of inequalities that implies, and even strictly dominates, all previously described cuts. We also provide a polynomial time separation algorithm for the more general cuts by calculating maximum flows *on the splitted graph*. We compare the proposed models both theoretically and computationally.

1 Introduction

The Two Level Network Design (TLND) problem arises in the topological design of hierarchical communication, transportation, and electric power distribution networks. Probably the most important application of TLND is in the context of telecommunication networks, where networks with two cable technologies, fiber optic and copper, are built. In local broadband access networks, if the Fiber-To-The-Home (FTTH) strategy

J. Pahl, T. Reiners, and S. Voß (Eds.): INOC 2011, LNCS 6701, pp. 57–70, 2011.

is used, every customer is provided with a distinct fiber optic connection. A cheaper strategy is Fiber-To-The-Curb (FTTC), where the part of the access network closest to the customer uses copper cables and facilities are installed, to convert optical to electrical signals and vice versa. In greenfield deployments, i.e. where there is no existing infrastructure, a mixed strategy of FTTC and FTTH is often preferable. In such a case, telecommunication companies distinguish between *primary* and *secondary customers*. The switching centers, important infrastructure nodes and small businesses are considered as primary customers (i.e., those to be served by fiber optic connections). Single households are not considered as being consumers of a high potential and hence they only need to be supplied using copper cables. The secondary technology is much cheaper, but the guaranteed quality of the connections and bandwidth is significantly below the quality provided by the primary one.

A large body of work has been done for the TLND and its variations (see below). In this study we incorporate two realistic features that have not yet been considered in previous studies of the TLND. Firstly, none of the previous approaches on TLND considers the cost of establishing *intermediate facilities* at *transition nodes*, i.e., nodes in which the change of technology takes place. Typically, at transition nodes, expensive switching devices need to be installed and the respective costs should not be neglected. Secondly, the previous work on the TLND is based on the assumption that all nodes in the network belong to the customer set. We relax this assumption, allowing the existence of *Steiner* nodes as well. We call the new problem the Two Level Network Design Problem with Intermediate Facilities (TLNDF).

This important problem generalizes problems with tree-star and star-tree topologies, like e.g., connected facility location, hierarchical network design, Steiner trees or uncapacitated facility location. We consider an extended graph, where the installation of facilities is modeled as arcs. We show that connectivity constraints on this splitted graph, projected back into the space of the variables of the original model, provide a new family of inequalities that implies, and even strictly dominates, all previously described constraints. We also provide a polynomial time separation algorithm for the more general inequalities by calculating maximum flows *on the splitted graph*. Finally, our computational study demonstrates the efficiency and practical applicability of the new inequalities.

Problem Definition. We consider the following *generalization* of the two level network design problem:

Definition 1 (TLNDF). *We are given an undirected graph $G = (V, E)$ with a root $r \in V$ and a set of customers $R \subseteq V \setminus \{r\}$. To each edge $e \in E$ we associate two installation costs, $c_e^1 \geq c_e^2 \geq 0$. These correspond to the* primary *and* secondary technology, *respectively. The set of customers, R, is partitioned into the set of* primary *and* secondary customers *P and S, respectively. Our goal is to determine a cost-minimal* subtree *of G satisfying the following properties:*

(P) *each* primary node *in P is connected to the root node by a path that consists of primary edges only,*

(S) *each* secondary node *in S is connected to the root by a path consisting of primary and/or secondary edges,*

(F) *facility opening costs $f_i \geq 0$ are payed for each transition node $i \in V$ and*
(E) *on each edge $e \in E$ at most one of technologies is installed.*

Several observations can be made about the solution space of this problem: i) Since $c_e^1 - c_e^2 \geq 0$, there always exists an optimal solution which is a Steiner tree with a *tree-tree topology*, i.e., it is composed of a rooted subtree of primary edges (*primary subtree*) and a union of subtrees of secondary edges (*secondary subtrees*). Each secondary subtree is rooted in a (transition) node of the primary subtree. ii) If facility opening costs are the same for all facility locations, any leaf of the primary subtree will be a primary node. Otherwise, if facility opening costs are location-dependent, placing facilities at locations of Steiner nodes may provide cheaper solutions, i.e., a leaf of the primary subtree may be any node from $V \setminus \{r\}$. iii) This general definition also covers the case in which potential facility locations are a true subset of V (which can be modeled by setting $f_i := \infty$ for the non-facility locations).

As noted before, the problem discussed here incorporates two new features compared to the original definition in [2], see also [6]. First, we need to consider additional *transition costs* due to the presence of two technologies on the network. The second new feature is that we allow arbitrary subsets of $V \setminus \{r\}$ to be considered as the customer set. This is because in practical applications nodes like street intersections need to be considered, too. Following the spanning tree definition of *multi-level network design* problems in [1], the TLND problem with Steiner nodes can also be seen as a *three-level network design* problem in which the Steiner nodes are assigned to the third group of customers and the installation costs for the third technology are set to zero.

Literature Review. The concept of two level network design problems (more precisely, *two-level spanning trees*) has been developed in the 80's and early 90's. The *hierarchical network design* problem, in which $R = V \setminus \{r\}$ and $|P| = 2$, was the "initial" variant of the TLND introduced by [5]. This problem was later generalized by [6] for $|P| > 2$. [2] have proposed several network flow based models for this latter problem setting and have compared the linear programming bounds of the proposed formulations. In [1], the authors have tested a dual ascent method on the model with the strongest linear programming bound. A more recent approach based on a different formulation is described in [10]. The TLND problem belongs to a class of problems with a tree-tree topology. The reader is referred to [8] where several variants of related problems such as star-tree, tree-star and star-star problems as well as other variants of tree-tree problems are described.

The previous studies on TLND do not incorporate additional constraints. As far as we know, the three exceptions are described in [9,11,7]. In the first one, the authors considered the TLND with *weighted hop constraints* defined as follows: given natural numbers w_1 and w_2, our goal is to construct a two-level minimum spanning tree such that for each node k, the unique path from the root to k contains a weighted number of primary and secondary edges (with weights w_1 and w_2, respectively) which does not exceed H. In [11] the two-level minimum spanning tree problem with *secondary distance constraints* stating that each secondary node must not be too far from the primary network, is considered. In the latter work [7], the authors studied the connected facility location problem (ConFL) wich is a TLNDF variant with a tree-star configuration. In [16] a hop constrained variant of connected facility location has been studied.

For a literature overview on *capacitated network design problems* with two technologies, we refer to a recent work of [4], where a problem has been studied with capacities on edges and with fixed installation and non-linear flow costs. In [13] a *survivable network design problem* with two technologies and facility nodes has been studied.

2 MIP Formulations for the Two Level Network Design Problem

In this section we describe cut based formulations for the TLNDF. We start by giving a formulation of the original TLND problem *without* modeling the facility opening costs.

Directed Graphs. It is well known that for rooted spanning or Steiner tree problems, models with a stronger linear programming bound are obtained by solving the problem on a directed graph (see, e.g., [17]). Thus, we will work on a directed graph $G = (V,A)$ that is obtained from the original undirected graph $G = (V,E)$ as follows: For each edge $e = \{i,j\} \in E$ we include two arcs ij and ji in A with the same cost of the original edge. Since we are modeling an arborescence directed away from the root node, edges $\{r,j\}$ are replaced by a single arc rj only.

To model the TLND problem, we will use the following binary variables:

$$x^1_{ij} = \begin{cases} 1, & \text{if the primary cable technology is installed on arc } ij \\ 0, & \text{otherwise} \end{cases} \qquad \forall ij \in A$$

$$x^2_{ij} = \begin{cases} 1, & \text{if the secondary cable technology is installed on arc } ij \\ 0, & \text{otherwise} \end{cases} \qquad \forall ij \in A, j \notin P$$

Observe that in any feasible solution there will be no secondary arcs entering a primary node (i.e., $x^2_{ij} = 0$, whenever $j \in P$). Therefore, variables corresponding to such arcs will not be considered in our models. However, to simplify the notation, we will allow them in the indexation of the summation terms.

For any $W \subset V$ we denote its complement set by $W^c = V \setminus W$. For any $M,N \subset V$, $M \cap N = \emptyset$, denote the induced cut set of arcs by $(M,N) = \{ij \in A \mid i \in M, j \in N\}$. In particular, let $\delta^-(W) = (W^c,W)$ and $\delta^-(i) = (V \setminus \{i\},\{i\})$. For a set of arcs $\hat{A} \subseteq A$, we will write $x^\ell(\hat{A}) = \sum_{ij \in \hat{A}} x^\ell_{ij}$, for $\ell = 1,2$, and $(x^1 + x^2)(\hat{A}) = \sum_{ij \in \hat{A}} x^1_{ij} + x^2_{ij}$.

The examples described in the next sections use the following symbols: ■ represents the root node, ○ represents a Steiner node. □ represents a primary customer, △ represents a secondary customer. Whenever we solve a problem as the Steiner tree problem, terminals are denoted by ◇.

2.1 Modeling the TLND Problem

The following formulation models the TLND with the set of primary nodes P (that may also be an empty set), and the set of secondary nodes S *without* facility opening costs.

$$(TLND) \quad \min \sum_{ij \in A} (c^1_{ij} x^1_{ij} + c^2_{ij} x^2_{ij})$$

$$x^1(\delta^-(W)) \geq 1 \qquad \forall W \subseteq V \setminus \{r\}, W \cap P \neq \emptyset \qquad \text{(x1)}$$

$$(x^1 + x^2)(\delta^-(W)) \geq 1 \qquad \forall W \subseteq V \setminus \{r\},\ W \cap S \neq \emptyset \tag{x12}$$

$$(x^1 + x^2)(\delta^-(i)) \leq 1 \qquad \forall i \in V \tag{1}$$

$$x^1_{ij}, x^2_{ij} \in \{0,1\} \quad \forall ij \in A \tag{2}$$

The *primary connectivity constraints* (x1) ensure that for every primary node i, there is a path between r and i containing only primary arcs. The *secondary connectivity constraints* (x12) ensure that every secondary node is connected to the root by a path containing primary and/or secondary arcs. The in-degree constraints (1) ensure that the overall solution is a subtree and they are redundant if the edge costs are non-negative.

This gives a valid model for the TLND. In [1,2] a directed MIP formulation based on network flows has been presented. It is easy to show that the set of feasible solutions of the LP-relaxation of the *TLND* model is the projection onto the space of (x^1, x^2) variables of this flow model. This result follows immediately from the max-flow min-cut theorem. Thus, the two models produce the same linear programming bound.

2.2 Modeling Facility Opening Costs

At each node in which a change of technology takes place, expensive facilities (e.g., multiplexors, splitters) need to be installed. In order to model these facility opening costs, we will use variables z_i:

$$z_i = \begin{cases} 1, & \text{if a facility is installed in node } i \\ 0, & \text{otherwise} \end{cases} \qquad \forall i \in V$$

For a set $W \subseteq V$, we will write $z(W) = \sum_{i \in W} z_i$.

Basic Coupling Constraints. To ensure that a facility is open, whenever a change of technology takes place, we request that every secondary arc $jk \in A$ used in a solution is either preceded by another secondary arc entering node j, or there is an open facility at node j. These constraints are an adaptation of degree-inequalities proposed by [14] for the Steiner tree problem. Our problem can then be modeled as follows:

$$(TLNDF) \quad \min \sum_{ij \in A} (c^1_{ij} x^1_{ij} + c^2_{ij} x^2_{ij}) + \sum_{i \in V} z_i f_i$$

$$z_j + \sum_{ij \in A, i \neq k} x^2_{ij} \geq x^2_{jk} \qquad \forall jk \in A, k \notin P \tag{3}$$

$$z_i \in \{0,1\} \quad \forall i \in V \tag{4}$$

$$\text{(x1)}, \text{(x12)}, (1), (2)$$

In this model, the indegree constraints (1) are not redundant even if the arc- and facility costs are non-negative. These constraints namely prevent building of secondary cycles that would satisfy (3) without opening a facility at position j. Together with connectivity constraints (x12), the *basic coupling constraints* (3) guarantee that if a facility is installed at node j, then j is the root of a secondary subtree. This model does not prevent from opening facilities along a secondary path, but this will never be the case in an optimal solution.

Generalized $x^2 - z$ Coupling Constraints. One can generalize the coupling constraints (3) in the following way: Let k be any secondary customer or Steiner node and $W \subseteq V \setminus \{k\}$. Then, the *generalized $x^2 - z$ coupling constraints* can be written as follows:

$$z(W) + x^2(W_k^c, W) \geq x^2(W, \{k\}) \quad \forall k \in V \setminus (P \cup \{r\}), \quad W \subseteq V \setminus \{k\}, W_k = W \cup \{k\}. \tag{5}$$

Note that if $W = \{j\}$, for $j \neq k$, we obtain (3). In [9], the authors consider a similar generalization technique of degree-constraints by [14] in the context of the two-level minimum spanning tree problem with weighted hop constraints. The formulation obtained by replacing constraints (3) by (5) is denoted by $TLNDF^+$.

The convex hull of feasible LP-solutions of $TLNDF^+$ is (for some instances even strictly) contained in the polytope defined by the LP-relaxation of the $TLNDF$ model.

Lemma 1. *Let $\mathscr{P}_{TLNDF^+}$ and $\mathscr{P}_{TLNDF}$ denote the polytopes associated with LP-relaxations of models TLNDF$^+$ and TLNDF, respectively. Then, $\mathscr{P}_{TLNDF^+} \subseteq \mathscr{P}_{TLNDF}$ and there exist instances for which the strict inequality holds.*

Proof. Constraints (3) are contained in the set (5): (3) for $jk \in A, k \notin P$ is derived from (5) for $W = \{j\}$. Figure 1 shows an example where the strict inequality holds: Consider the LP-optimal solution for $TLNDF$ in which $x^1_{r2} = x^1_{24} = 1$ and $x^2_{45} = x^2_{57} = x^2_{46} = x^2_{67} = z_4 = 0.5$. $\upsilon_{LP}(TLNDF) = 3.25$ but constraint (5) is violated for $W = \{4,5,6\}$ and $k = 7$, so $\upsilon_{LP}(TLNDF^+) = 3.5 > \upsilon_{LP}(TLNDF)$. □

Constraints (5) can be rewritten in several equivalent ways which permit an easier comparison with other inequalities. In fact, by adding $x^2(W^c, k)$ to both sides, we can rewrite (5) as follows:

$$z(W) + x^2(\delta^-(W_k)) \geq x^2(\delta^-(k)) \quad \forall k \notin P, \forall W \subseteq V \setminus \{k\}, W_k = W \cup \{k\}. \tag{x2-z}$$

Generalized $x^1 - z$ Coupling Constraints. We have shown how to relate variables z and x_2. We show next how to relate variables z and x_1: For a given $k \in S$ and $W = V \setminus \{k\}$ we can rewrite inequalities (x2-z) as $z(V \setminus \{k\}) \geq x^2(\delta^-(k))$. By using the in-degree constraint (1), we obtain:

$$z(V \setminus \{k\}) + x^1(\delta^-(k)) \geq 1 \quad \forall k \in S$$

The latter constraints can be generalized for subsets $W \cap S \neq \emptyset$ in the following way:

$$z(W^c) + x^1(\delta^-(W)) \geq 1 \quad \forall W \subseteq V \setminus \{r\}, W \cap S \neq \emptyset \tag{x1-z}$$

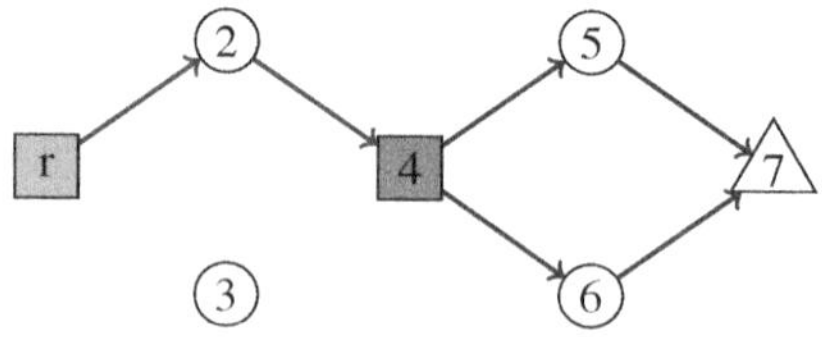

Fig. 1. Example for Lemma 1. All primary arc costs are 1, secondary arc and facility costs are 1/2.

These new inequalities describe the fact that for any subset W containing a secondary node, either there is a primary path between a node from W and r, or there is an open facility in the complementary set W^c.

Observation 1. *The set of inequalities* (x1-z) *cannot replace the coupling constraints* (3) *in the model TLNDF, i.e.* (x1-z) *are not sufficient for modeling the TLND problem with facility nodes. However,* (x1-z) *can be used to strengthen the model TLNDF*$^+$.

We denote the model $TLNDF^+$ extended by (x1-z) as $TLNDF^+_{x1-z}$.

Next, we will show that connectivity constraints (x1), (x12) and both groups of generalized coupling constraints are special cases of a more general group of constraints. These can be derived if we model the problem in a new graph obtained by node-splitting as described below.

3 The Node-Splitting Model

We can model the TLNDF problem as the Steiner arborescence problem in a slightly modified graph $G_{NS} = (V_{NS}, A_{NS})$ with the root r' and the set of terminals R_{NS}, as follows:

$$
\begin{aligned}
V_{NS} &:= V' \cup V'' \cup S \text{ where} & A_{NS} &:= A' \cup A'' \cup A_z \cup A_S \text{ where}\\
V' &:= \{i' \mid i \in V\}, & A' &:= \{i'j' \mid ij \in A\},\\
V'' &:= \{i'' \mid i \in V\}, & A'' &:= \{i''j'' \mid ij \in A\},\\
& S \text{ is the set of secondary nodes;} & A_z &:= \{i'i'' \mid i \in V\},\\
R_{NS} &:= P' \cup S \text{ where} & A_S &:= \{i'i \mid i' \in V', i \in S\}\\
P' &= \{i' \mid i \in P\}; & & \cup \{i''i \mid i'' \in V'', i \in S\}.
\end{aligned}
$$

The graph G_{NS} is composed of several components: i) a subgraph $G' = (V', A')$ which corresponds to the primary network (it contains nodes and arcs that may be included in the primary subtree); ii) a subgraph $G'' = (V'', A'')$ that corresponds to the secondary network (it contains nodes and arcs that may be contained in the secondary subtrees); iii) arcs linking nodes in G' to the corresponding copy in G'' and representing potential facilities and iv) another copy of the secondary nodes with arcs incoming from their representatives in graphs G' and G'' (see Figure 2). Arc costs C_{ij}, $ij \in A_{NS}$ are assigned accordingly to the arcs in G', G''. The arcs linking the two subgraphs are assigned costs $C_{i'i''} := f_i$, for all $i \in V$. To the arcs of the set A_S costs of zero are assigned.

If, for a primary node $i \in P$, its copy $i'' \in V''$ belongs to the optimal solution, there will be no ingoing arcs into i'' (with the only exception of $i'i''$). Therefore, we can reduce the size of G_{NS}, by removing all ingoing arcs of primary nodes in V''. This corresponds to setting $x^2_{ij} := 0$ for all $ij \in A$ such that $j \in P$, as already described in Section 2. Observe that we need a third copy of secondary nodes in G_{NS}, namely the set S, since it is not clear for secondary nodes whether they will be connected within the primary or the secondary subtree.

To provide an ILP model, we assign binary variables X_{ij} to all arcs $ij \in A_{NS}$. Denote by $X(\delta^-(\tilde{W}))$ the sum of X variables that correspond to the arcs in the directed cut

$(\tilde{W}^c, \tilde{W})$ in G_{NS}. Based on the classical cut set model for Steiner trees (cf. [3]) we derive the following ILP formulation:

$$(SA) \quad \min \sum_{ij \in A_{NS}} C_{ij} X_{ij} \tag{6}$$

$$\text{s.t.} \quad X(\delta^-(\tilde{W})) \geq 1 \qquad \forall \tilde{W} \subseteq V_{NS} \setminus \{r'\}, \tilde{W} \cap R_{NS} \neq \emptyset \tag{7}$$

$$\sum_{ij \in A} (X_{i'j'} + X_{i''j''}) \leq 1 \qquad \forall j \in V \tag{8}$$

$$X_{ij} \in \{0,1\} \quad \forall ij \in A_{NS} \tag{9}$$

Constraints (7) are the classical connectivity cuts, inequalities (8) state that of all ingoing edges of both copies of a node in G at most one is allowed to be open.

Lemma 2. *The TLNDF problem can be modeled as the Steiner arborescence problem with additional degree constraints on some node pairs on the graph G_{NS} with the root r' and terminal set R_{NS}.*

Proof. We map each binary solution of formulation *SA* into the variable space of *TLNDF* as follows: $X_{i'j'} \rightarrow x^1_{ij}$, $X_{i''j''} \rightarrow x^2_{ij}$ and $X_{i'i''} \rightarrow z_i$. Let now $\mathbf{X}$ be an LP optimal solution for *SA*. The mapping of $\mathbf{X}$ then satisfies all constraints of *TLNDF*: Connectivity cuts (7) imply (x1) and (x12), together with degree constraints (8) they ensure (1). Finally, constraints (3) are also satisfied since we have:

$$z_j + \sum_{ij \in A, i \neq k} x^2_{ij} = X_{j'j''} + \sum_{i \neq k} X_{i''j''} \geq X_{j''k''} = x^2_{jk}.$$

The last inequality is implied by (7) and (8). □

Let $Proj_{x^1,x^2,z}(\mathscr{P}_{SA})$ denote the projection of the polytope obtained as the convex hull of the LP-solutions of the *SA* formulation into the space of (x^1, x^2, z) variables. In this projection, we set $x^1_{ij} := X_{i'j'}$, $x^2_{ij} := X_{i''j''}$ for all $ij \in A$ and $z_i := X_{i'i''}$ for all $i \in V$.

Theorem 1. *The SA formulation is at least as strong as the previously defined formulation $TLNDF^+_{x1-z}$, i.e., $Proj_{x^1,x^2,z}(\mathscr{P}_{SA}) \subseteq \mathscr{P}_{TLNDF^+_{x1-z}}$.*

To prove this result, we need to analyze the cut set inequalities defined in the *SA* model and their projection onto the original graph G.

Lemma 3. *Cut set inequalities* (7) *such that $\delta^-(\tilde{W}) \cap A_S \neq \emptyset$ are redundant in the model SA.*

Proof. Consider a cut set $\tilde{W} \subseteq V_{NS} \setminus \{r'\}$, $\tilde{W} \cap S \neq \emptyset$, such that $\delta^-(\tilde{W}) \cap A_S \neq \emptyset$. We will show that in that case, $X(\delta^-(\tilde{W})) \geq 1$ is dominated by another cut set inequality $X(\delta^-(\tilde{U})) \geq 1$ where $\tilde{U}$ is defined as stated below. We need to distinguish the following two cases:

i) If for all $i \in S \cap \tilde{W}$, $i'i \in \delta^-(\tilde{W})$ and $i''i \notin \delta^-(\tilde{W})$, a dominating cut is given for $\tilde{U} = \tilde{W} \cup \bigcup_{i \in \tilde{W}} \{i'\}$.

ii) For all other $\tilde{W}$ the dominating cut is obtained by removing nodes $i \in S$ from $\tilde{W}$ if $i'' \in \tilde{W}$ and $i' \notin \tilde{W}$ and adding nodes i' and i'' to $\tilde{W}$ for $i \in S \cap \tilde{W}$ such that $i', i'' \notin \tilde{W}$. □

We will refer to the cut set inequalities such that $\delta^-(\tilde{W}) \cap A_S = \emptyset$ as the *non-dominated cut set inequalities*.

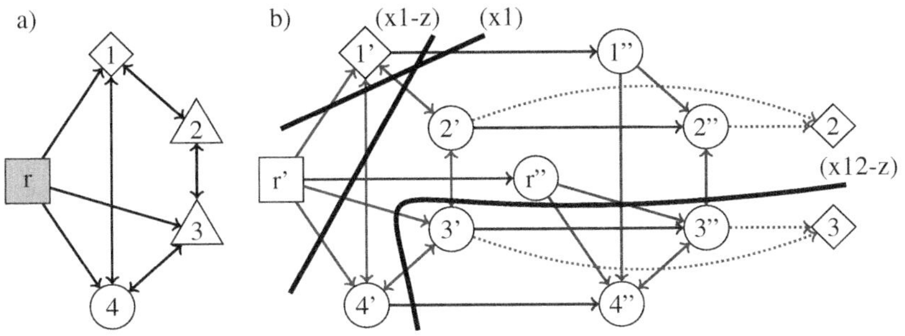

Fig. 2. a) Instance of *TLNDF*; b) Transformed instance and illustration of cuts (x1) for $W = \{1\}$, (x1-z) for $W = \{2,3,4\}$ and cuts (x12-z) for $W' = \{3\}$ and $W'' = \{3,4\}$

Generalized Cut Set Constraints. We will now define the generalized cut set constraints for the TLNDF that are obtained by projecting the non-dominated inequalities among the ones in (7) into the space of (x_1, x_2, z). For an arbitrary cut set $\tilde{W} \subset V_{NS} \setminus \{r'\}$, $\tilde{W} \cap R_{NS} \neq \emptyset$, let us denote the projected subsets of the original graph G as follows:

$$W' = \{i \in V \mid i' \in \tilde{W}\} \text{ and } W'' = \{i \in V \mid i'' \in \tilde{W}\}$$

Then, the projected cut set inequalities (7), that we will refer to as *generalized cut set constraints*, can be written as:

$$x^1(\delta^-(W')) + x^2(\delta^-(W'')) + z(W'' \setminus W') \geq 1 \quad r \notin W', W' \cap W'' \cap S \neq \emptyset \text{ or } W' \cap P \neq \emptyset. \qquad \text{(x12-z)}$$

Observe that all the previously studied inequalities are special cases of this constraint (see Figure 2):

i) If $W'' = \emptyset$, we obtain the primary connectivity constraints (x1).
ii) If $W' = W''$, we obtain the secondary connectivity cuts (x12).
iii) If $W'' = V$, we obtain the generalized coupling constraints (x1-z).
iv) For a given $k \in S$, and a subset $W \subseteq V \setminus \{r,k\}$, the generalized (x2-z) constraint corresponds to (x12-z) for $W' = \{k\}$, $k \in S$, and $W'' = W \cup \{k\}$.

This implicitly proves Theorem 1, i.e., the projection of every feasible LP-solution of the formulation *SA* is also feasible to $TLNDF^+_{x1-z}$.

We conclude this section by noting that even more general classes of inequalities can be obtained by considering non-trivial cases in which $W' \cap W'' \neq \emptyset, W', W''$.

Lemma 4. *The generalized connectivity constraints* (x12-z) *can be separated in polynomial time.*

To separate the constraints (x12-z), one needs to apply the max-flow algorithm on the splitted graph G_{NS} as described in the next section.

4 Computational Study

In this section we report details of the implementation of our Branch-and-Cut algorithm, how we derived the set of benchmark instances and the computational results.

4.1 The Branch-and-Cut Algorithm

To implement our models, we used the Gurobi [12] Branch-and-Cut framework, version 3.0.2. All experiments were performed on a Intel Core2 Quad 2.33 GHz machine with 3.25 GB RAM, where each run was performed on a single processor.

Initialization. As the Gurobi MIP solver requires a compact model for initialization, we used the following Miller-Tucker-Zemlin connectivity constraints (10)-(13) and trivial degree-constraints (14):

$$u_r = 1 \qquad (10)$$

$$|V|(x^1_{ij}+x^2_{ij})+(|V|-2)(x^1_{ji}+x^2_{ji})+u_i-u_j \leq |V|-1 \quad ij \in A, j \notin P \qquad (11)$$

$$|V|(x^1_{ij})+(|V|-2)(x^1_{ji})+u_i-u_j \leq |V|-1 \quad ij \in A, j \in P \qquad (12)$$

$$\sum_{ij \in A: i \neq k} x^1_{ij} \geq x^1_{jk} \quad j \neq r \qquad (13)$$

$$\sum_{ij \in A} x^1_{ij} \geq z_j \quad j \in F \setminus \{r\} \qquad (14)$$

In addition, our model comprises in-degree constraints (1) and coupling constraints (3).

Separation
Separating (x1) **and** (x12) **Cuts:** We separate violated cut set inequalities (x1),(x12) and (x12-z) in every node of the the Branch-and-Bound tree (BnB). To obtain inequalities (x1), we solve a maximum flow problem on the graph $G = (V,A)$. The capacities on each arc are set to the value of the x^1-variable for the respective arc in the current fractional solution. Cut set inequalities (x12) are obtained in a similar fashion. The capacities are equal to the sum of variables x^1 and x^2 for each arc.

Separating (x12-z) **Cuts:** To obtain violated constraints of the largest and strongest group (x12-z), we solve the maximum flow problem on the *splitted graph* G_{NS}. The weights for arcs in A', A'' and A_z are set to the value of the corresponding variable in the current fractional solution. For arcs in A_S, the weight is set to 1, as cuts containing these arcs are dominated by others (cf. Lemma 3).

General Settings: To improve the computational efficiency of our separation, we search for nested and minimum cardinality cuts. To do so, all capacities in the respective graph are increased by some $\varepsilon > 0$. Thus, every detected violated cut contains the least possible number of arcs. The LP is resolved after adding at most 50 violated inequalities of type (x1), (x12) or (x12-z). Finally, we randomly permute the order in which customers are chosen to find violated cuts. To ensure comparability, we fix the seed value for the computations reported in Section 4.3.

Primal Heuristic. We use a primal heuristic (PH) to find incumbent solutions. The PH is entirely carried out on the graph G_{NS}. It consists of the following steps:

1. **Construct primary subtree:** Primary customers are connected to the root node by the arcs in the shortest path to the copy of that customer in V'. For all nodes taken into the primary subtree, ingoing secondary arcs are removed.

2. **Construct secondary subtree:**
   (a) Using zero costs on all arcs in the primary subtree, the shortest paths $P(i')$ and $P(i'')$ from the root to $i' \in V'$ and $i'' \in V''$ are calculated for all $i \in S$. Let $H'(i) = |P(i'') \cap A'|$ and $H''(i) = |P(i'') \cap A''|$.
   (b) Let $Q = S$. For all $i \in Q$ such that $H''(i) = 0$ add $P(i')$ and remove i from Q.
   (c) Sort Q according to (H', H'') in decreasing order and repeat until $Q = \emptyset$: Add $P(i'')$ and remove i from Q.
3. **Pruning of primary subtree:** Superfluous *leaves* are iteratively removed from the primary subgraph: Secondary customers, that are part of the primary and a secondary subtree in which no facility is installed as well as Steiner nodes are removed.
4. **Pruning and repairing of secondary subtree:** Superfluous nodes are removed from the secondary subgraph and infeasible parts of the solution repaired: Steiner node leaves and secondary customer leaves in V'' are iteratively removed, if their respective copy in the primary subtree is used. For each secondary customer with in- and out-going arcs in both A' and A'', we remove the ingoing arcs in A'' and open a facility at this node.

We use the information from the current best LP solution to adjust the weights for calculating the shortest paths. We set the weight w for an arc in G_{NS} to $(1-v)c$ where v is the corresponding variable and c is the initial cost.

4.2 Instances

For our computational study we transform instances of the Steiner tree problem (STP) using the following procedure:

- First, 30% of STP terminals are chosen as primary customers, the remaining 70% are defined as secondary customers. The primary customer with the lowest index is set as root node.
- The Steiner nodes in the STP instance are Steiner nodes in the TLNDF instance.
- As potential facility nodes we chose the root node, primary and secondary customers.
- Primary edge costs equal edge costs of the STP instance. For each secondary edge e, the cost c_e^2 is defined as qc_e^1, where q is uniformly randomly chosen from $[0.25, 0.5]$.
- Facility opening costs are uniform and equal 0.5 times the average primary edge costs.

The parameters for generating instances have been carefully chosen so that trivial solutions (e.g., optimal solutions that do not contain secondary subtrees) are avoided. The sets B, C, D and E of the Steinlib library [15] have been used in our computational study.

4.3 Results

We compared the computational performance of three different settings (two of which using cuts derived from the splitted graph):

Table 1. Computational comparison of three different branch-and-cut settings

	$\lvert V\rvert$	$\lvert E\rvert \leq$	$\lvert T\rvert \leq$	#*OPT* i)	ii)	iii)	t [s] i)	ii)	iii)
c01-10	500	1000	250	**10**	**10**	**10**	**46.52**	69.54	73.48
c11-20	500	12500	250	1	3	**6**	55.84	56.53	**20.86**
d01-10	1000	2000	500	**10**	9	**10**	**180.61**	253.66	271.11
d11-20	1000	25000	500	2	2	**4**	172.38	110.89	**39.19**
e01-10	2500	5000	1250	**6**	5	5	191.71	81.68	**38.60**
e11-20	2500	62500	1250	2	**3**	**3**	178.42	179.29	**80.16**
	$\lvert V\rvert$	$\lvert E\rvert \leq$	$\lvert T\rvert \leq$	avg *gap*[%] i)	ii)	iii)	iii) #*OPT* found in $\leq$ 1h	$\leq$ 2h	$\leq$ 24h
c01-10	500	1000	250	**0.00%**	**0.00%**	**0.00%**	10	10	10
c11-20	500	12500	250	5.93%	3.88%	**1.05%**	7	9	10
d01-10	1000	2000	500	**0.00%**	**0.00%**	**0.00%**	10	10	10
d11-20	1000	25000	500	4.19%	4.17%	**0.75%**	6	6	9
e01-10	2500	5000	1250	**0.11%**	0.15%	0.20%	6	8	10
e11-20	2500	62500	1250	5.78%	**5.34%**	5.35%	3	5	5

i) Model *TLNDF*, in which the basic coupling constraints (3) are inserted at once and the (x1) and (x12) cuts are separated within the branch-and-bound (BnB) tree.

ii) In the second setting, after all violated (x1) cuts have been detected, (x12) are separated. Finally, after no more violated (x1) and (x12) cuts can be found, generalized connectivity constraints (x12-z) are separated.

iii) In the third setup, we refrained from separating inequalities (x12), i.e., after no more violated (x1) cuts can be found, generalized connectivity constraints (x12-z) are separated.

In a preliminary test we tested our three approaches on the instances of set B. The maximum runtime was 5.28, 8.11 and 4.47 seconds respectively. As a consequence we only give detailed results for the larger sets C,D and E.

In Table 1 we show the key figures of our computational study. The first column indicates the group of instances, in columns 2, 3 and 4 we state the (maximum) number of nodes, edges and terminals (i.e. union of primary and secondary customers) of the largest instance of each group, respectively. In the third segment of the upper part we show the number of instances in this group solved to optimality within 1000 seconds of running time. The last segment shows the average running times for the subset of instances that was solved to optimality by all three approaches within 1000 seconds. In segment three of the lower part we state the average gaps of each instance group after 1000 seconds of running time. In segment four we report the number of optimal solutions found by approach iii) within 1h, 2h and 24h, respectively. From the number of instances solved to optimality and the average running times one can see, that for sparse graphs (`.1-10`) the approach based only on connectivity cuts (x1) and (x12) is competitive to the generalized cut set constraints. For denser graphs (`.11-20`) the two new approaches (namely ii) and iii) involving (x12-z) cuts) perform much better:

Table 2. Average running times vs. graph density and vs. number of terminals, respectively. Values are normalized according to the first column in each segment.

	$\|E\|/\|V\|$				$\|T\|$				
	1.25	2	5	25	5	10	$\frac{1}{6}\|V\|$	$\frac{1}{4}\|V\|$	$\frac{1}{2}\|V\|$
c01-20	1.0	2.5	18.3	206.7	1.0	1.0	12.2	23.0	100.4
d01-20	1.0	1.3	6.1	158.8	1.0	1.6	43.3	80.4	143.3
e01-20	1.0	345.7	699.9	1195.6	1.0	2.1	296.7	503.1	256.1

For instances with few arcs, there is little difference in the LP bounds provided by the models with and without constraints (x12-z). Constraints (x1) and (x12) are cheaper to separate, but as the instances grow larger and denser, the advantage of better LP bounds provided by cuts (x12-z) outweighs this.

Table 2 illustrates how the running time performance of the approach iii) depends on the graph density (the second segment) and on the number of terminals (the third segment). Instances C, D, and E have been divided into groups according to their density ($|E|/|V|$) and the number of terminals ($|T|$), respectively. We observe that the average running times increase exponentially with the density and the number of terminals.

5 Conclusions

For the TLNDF we have introduced several new families of valid inequalities combining network design and facility location variables. The so-called generalized cut inequalities (x12-z) are the strongest among those inequalities and can be derived from a cut-set model for Steiner arborescence applied on a splitted graph. We have seen that the separation of (x12-z) cuts is not only computationally tractable, but it also outperforms the standard compact approach of modeling facility nodes. Finally, we have tested our approach on a set of 78 benchmark instances with up to 2500 nodes and 62500 edges. We have been able to solve 60 (66) instances to provable optimality in less than 1h (2h). From the remaining 12 instances 6 were solved optimally after 1 day and for 6 we obtained solutions less than 2% from optimum.

Acknowledgements

The authors would like to thank the referees for their comments and suggestions. Luis Gouveia was funded by Centro de Investigação Operacional [MATH-LVT-Lisboa-152]. Ivana Ljubić was partially supported by the Austrian Science Fund (FWF): [T334].

References

1. Balakrishnan, A., Magnanti, T.L., Mirchandani, P.: A dual-based algorithm for multi-level network design. Management Science 40(5), 567–581 (1994)
2. Balakrishnan, A., Magnanti, T.L., Mirchandani, P.: Modeling and heuristic worst-case performance analysis of the two-level network design problem. Management Science 40(7), 846–867 (1994)

3. Chopra, S., Gorres, E.R., Rao, M.R.: Solving the Steiner Tree Problem on a Graph Using Branch and Cut. INFORMS Journal on Computing 4(3), 320–335 (1992)
4. Costa, A.M., França, P.M., Lyra Filho, C.: Two-level network design with intermediate facilities: An application to electrical distribution systems. Omega 39(1), 3–13 (2011)
5. Current, J.R., ReVelle, C.S., Cohon, J.L.: The hierarchical network design problem. European Journal of Operational Research 27(1), 57–66 (1986)
6. Duin, C., Volgenant, T.: The multi-weighted Steiner tree problem. Annals of Operations Research 33(6), 451–469 (1991)
7. Gollowitzer, S., Ljubić, I.: MIP models for connected facility location: A theoretical and computational study. Computers & Operations Research 38(2), 435–449 (2011)
8. Gourdin, E., Labbé, M., Yaman, H.: Telecommunication and location - a survey. In: Drezner, Z., Hamacher, H. (eds.) Facility Location: Applications and Theory. Springer, Heidelberg (2001)
9. Gouveia, L., Janssen, E.: Designing reliable tree networks with two cable technologies. European Journal of Operational Research 105(3), 552–568 (1998)
10. Gouveia, L., Telhada, J.: The multi-weighted Steiner tree problem: A reformulation by intersection. Computers & Operations Research 35(11), 3599–3611 (2008)
11. Gouveia, L., Telhada, J.: Distance-constrained hierarchical networks. In: Proceedings of International Network Optimization Conference, INOC 2005, pp. B2.416–B2.421 (2005)
12. Gurobi: Gurobi Optimization, `http://www.gurobi.com/`
13. Jongh, A.D., Gendreau, M., Labbé, M.: Finding disjoint routes in telecommunications networks with two technologies. Operations Research 47(1), 81–92 (1999)
14. Khoury, B.N., Pardalos, P.M., Du, D.Z.: A test problem generator for the Steiner problem in graphs. ACM Trans. Math. Softw. 19(4), 509–522 (1993)
15. Koch, T., Martin, A., Voß, S.: SteinLib: An updated library on steiner tree problems in graphs. Tech. rep., Konrad-Zuse-Zentrum für Informationstechnik Berlin (2000), `http://elib.zib.de/steinlib`
16. Ljubić, I., Gollowitzer, S.: Modeling the hop constrained connected facility location problem on layered graphs. In: Proceedings of the International Symposium on Combinatorial Optimization (ISCO). Electronic Notes in Discrete Mathematics, vol. 36, pp. 207–214 (2010)
17. Magnanti, T., Wolsey, L.: Optimal trees. In: Handbook in Operations Research and Management Science, pp. 503–615 (1995)

The Two Level Network Design Problem with Secondary Hop Constraints

Stefan Gollowitzer[1], Luís Gouveia[2], and Ivana Ljubić[3]

[1] Department of Statistics and Operations Research, University of Vienna, Austria
stefan.gollowitzer@univie.ac.at

[2] Departamento de Estatística e Investigação Operacional - Centro de Investigação Operacional, Faculdade de Ciênçias, Universidade de Lisboa, Portugal
legouveia@fc.ul.pt

[3] Institute of Computer Graphics and Algorithms, Vienna University of Technology, Austria
ivana@ads.tuwien.ac.at

Abstract. The Two Level Network Design problem asks for a cost-minimal Steiner subtree of a given graph $G = (V,E)$ that connects all *primary customers* using a *primary technology* only, and all *secondary customers* using either the primary or the *secondary technology*. Thereby, the secondary technology is cheaper but less reliable and hence, hop constraints on the length of each secondary path are imposed. In addition, in some applications facility opening costs need to be paid for *transition nodes*, i.e., for nodes where the change of technology takes place. We consider various MIP models for this new problem and derive a new class of strong inequalities that we call *generalized cut-jump constraints*. We also show that these inequalities can be obtained by projecting the cut-set formulation obtained on a graph in which we split the potential facility locations and introduce layers for installing the secondary technology.

1 Introduction

Two Level Network Design (TLND) problems with a tree structure arise when local broadband access networks are planned in areas, where no existing infrastructure can be used, i.e., in so-called greenfield deployments (see, e.g., [1]). Topological network design for mixed strategies of Fiber-To-The-Home and Fiber-To-The-Curb, i.e., some customers are served by copper cables, some by fiber optic lines, can be modeled by an extension of the TLND. Since the secondary technology (e.g., copper) is usually much cheaper but also less reliable than the primary technology (e.g. optical fiber), hop-constraints on the length of secondary paths need to be introduced.

Definition 1 (TLNDSH). *We are given an undirected graph $G = (V,E)$ with a root r and a set of customers $R \subseteq V \setminus \{r\}$. To each edge $e \in E$, two installation costs $c_e^1 \geq c_e^2 \geq 0$ are associated. These correspond to the* primary *and* secondary technology, *respectively. The primary edges are more reliable, and hence, more expensive. The set of customers, R, is partitioned into two subsets P and S. We are also given facility opening costs $f_i \geq 0 \forall i \in V$, and a hop limit H. Our goal is to determine a cost-minimal* subtree *of G satisfying the following properties:*

J. Pahl, T. Reiners, and S. Voß (Eds.): INOC 2011, LNCS 6701, pp. 71–76, 2011.

(Psub) *The subgraph made of primary edges is a tree rooted at r interconnecting the primary nodes in P.*

(S) *Each* secondary node *in S is connected to the root by a path consisting of primary and/or secondary edges.*

(F) *Facility opening costs are payed for* transition nodes $i \in V$, *i.e., nodes where change of technology takes place,*

(H) *for each secondary node v, the number of secondary edges on the path between v and r may not exceed H, and*

(E) *on each edge $e \in E$ at most one of technologies is installed.*

We observe that, since the edge costs are non-negative, there always exists an optimal solution which is a Steiner tree interconnecting the nodes from $R \cup \{r\}$. Furthermore, since $c_e^1 - c_e^2 \geq 0$, this Steiner tree is composed of a subtree of primary edges (*primary subtree*) and a union of subtrees of secondary edges (*secondary subtrees*). Each secondary subtree is rooted in a node of the primary subtree. Due to facility opening costs and hop-constraints, every node from V may be a leaf of the primary subtree.

The TLND with secondary hop constraints has not yet been studied in the literature. Minimum spanning trees with two technologies and *secondary distance constraints* have been introduced in [6]. In [4] the TLND is considered with weighted hop constraints defined as follows: the goal is to construct a two level minimum spanning tree such that for each node k, the r-k path contains a weighted number of primary and secondary edges (with weights w_1 and w_2, respectively) which does not exceed H (for given $w_1, w_2, H \in \mathbb{N}$).

2 MIP Models

There are two challenges in modelling the TLNDSH problem: a) facility nodes and b) hop constraints. MIP formulations focused on modelling facility nodes are studied in in [3]. In this short abstract, we mainly concentrate on modeling the hop constraints on secondary trees. The models will be derived on graph $G = (V,A)$ with the set of directed arcs $A = \{ij \mid \{i,j\} \in E,\ j \neq r\}$ and with $c_{ij}^k = c_{ji}^k = c_e^k$ for $k = 1,2$ and $e = \{i,j\}$. We use sets $I_H := \{1,2,\ldots,H\}$ and $I_H^0 := I_H \cup \{0\}$. For any $W \subset V$ we denote its complement set by $W^c = V \setminus W$. For any $M,N \subset V$, $M \cap N = \emptyset$, we denote the induced cut set of arcs by $(M,N) = \{ij \in A \mid i \in M, j \in N\}$. In particular, let $\delta^-(W) = (W^c,W)$, $\delta^+(W) = (W,W^c)$ and $\delta^-(i) = (V \setminus \{i\},\{i\})$. The following binary variables are used in our models: Variables x_{ij}^1 (x_{ij}^2) will indicate if the primary (secondary) technology is installed on $ij \in A$, while z_i will be one if a facility is installed on $i \in V$. For a set of arcs $\hat{A} \subseteq A$, we write $x^\ell(\hat{A}) = \sum_{ij \in \hat{A}} x_{ij}^\ell$, for $\ell = 1,2$, and $(x^1+x^2)(\hat{A}) = \sum_{ij \in \hat{A}} x_{ij}^1 + x_{ij}^2$. For $W \subseteq V$ we define $z(W) = \sum_{i \in W} z_i$.

A Cut-Jump Model. The first formulation uses a generalization of the well known *jump constraints* that are crucial for modelling problems with network design constraints [2]. Let $[S_0,S_1,\ldots,S_{H+1}]$ be a partition of V, such that the root node $r \in S_0$ and $S_{H+1} \cap S \neq \emptyset$ (observe that some of sets S_i, for $i \in \{1,\ldots,H\}$ may also be empty). We call $J_H = J(S_0,S_1,\ldots,S_{H+1}) = \bigcup_{(i,j):i<j-1} [S_i,S_j]$ where $[S_i,S_j] = \{uv \in A : u \in S_i, v \in S_j\}$ a H-jump. In fact, without loss of generality, we can consider only $S_{H+1} = \{i\}$ for $i \in S$.

Letting J_H denote the set of all possible H-jumps, we derive the following formulation for the TLNDSH:

$$(JUMP) \quad \min \sum_{ij \in A} (c^1_{ij} x^1_{ij} + c^2_{ij} x^2_{ij}) + \sum_{i \in V} f_i z_i$$

$$x^2(J_H) + x^1(\delta^+(S_0)) \geq 1 \qquad \forall J_H : \{r\} \cup P \subseteq S_0,\ S_{H+1} = \{i\},\ i \in S \qquad (1)$$

$$z_j + \sum_{ij \in \delta^-(j), i \neq k} x^2_{ij} \geq x^2_{jk} \qquad \forall jk \in A : k \notin P \qquad (2)$$

$$x^1(\delta^-(W)) \geq 1 \qquad \forall W \subseteq V \setminus \{r\},\ W \cap P \neq \emptyset \qquad (3)$$

$$x^1(\delta^-(W)) \geq z_j \qquad \forall W \subseteq V \setminus \{r\} : j \in W \qquad (4)$$

$$(x^1 + x^2)(\delta^-(i)) \leq 1 \qquad \forall i \in V \qquad (5)$$

$$x^1_{ij}, x^2_{ij}, z_i \in \{0,1\} \quad \forall ij \in A, \forall i \in V \qquad (6)$$

The new *cut-jump* constraints (1) state that each secondary customer is connected to the root by primary or secondary edges, whereas the length of secondary edges along the path does not exceed H.

Lemma 1. *Formulation JUMP is valid for the TLNDSH.*

Proof. To show that the property (S) is satisfied, assume that there exists a secondary customer i that is not connected to r. Let C be the set of nodes belonging to the same connected component as i. We now set $S_{H+1} = \{i\}$, $S_H = C$, $S_0 = V \setminus (C \cup \{i\})$, and $S_h = \emptyset$, for $h \in \{1, \dots, H-1\}$. Then constraint (1) is violated, which is a contradiction. Constraints (5) ensure that each node has an in-degree of at most 1. This guarantees that no two opposing arcs are installed at the same time, and that a primary and a secondary arc cannot enter the same node. Inequalities (2) are an adaptation of degree-inequalities proposed by [7] for the Steiner tree problem. They force opening a facility in each node whose secondary out-degree is positive, and there are no ingoing secondary arcs. Hence, together with (5) and (1), (2) ensure that the property (F) is fulfilled.

We now show property (Psub), i.e., that the primary subnetwork is a subtree rooted in r. Due to the cut set (3), primary customers are connected to the root using primary arcs only. Assume a solution contains an infeasible path. The case of two or more subpaths of each technology is impossible due to connectivity cuts (4). So the path consists of two primary subpaths connected by a secondary subpath: But then inequality (1) is violated for the following case: Nodes in the primary subpath containing the root node are in S_0; the nodes of the secondary subpath are in set S_1; all but the last node in the remaining primary subpath are in S_2 and the last node is in S_{H+1}.

Finally, assume that there exists a binary solution (x^1, x^2, z) not satisfying property (H). That is, there is a customer $i \in S$ which is connected to r by a path in which more than H secondary arcs are used. Without loss of generality we can assume that there are exactly $H+1$ arcs in this path. Denote the nodes on the secondary path as $v_1, \dots, v_{H+2}$. Let S_0 denote the set of nodes on that path that are adjacent to primary arcs only. We now consider the following jump-set: $J_H = \{S_0, \{v_1\}, \dots, \{v_{H+1}\}, \{v_{H+2}\}\}$. Obviously, for J_H defined in this way, constraints (1) are violated, which is a contradiction. □

3 Layered Graph Models for TLNDSH

In this section we show how to model the TLNDSH as a Steiner Arborescence problem with additional degree constraints in an adequate layered graph. Similar ideas have been proposed for the hop-constrained spanning tree problem [5] and the hop-constrained connected facility location problem [8]. We first consider a layered graph model without node splitting and then a layered graph model with node splitting.

3.1 Solving the TLNDSH as a Steiner Arborescence Problem with Facility and Node-Degree Constraints

A first layered graph $G^L = (V^L, A^L)$ with the set of terminals denoted by R^L is constructed as follows:

$$
\begin{aligned}
V^L &:= \bigcup_{h=0}^{H} V^h \text{ where} & A^L &:= \bigcup_{h=0}^{H} A^h \text{ where}\\
V^0 &:= \{i^0 \mid i \in V\}, & A^0 &:= \{i^0 j^0 \mid ij \in A\},\\
V^h &:= \{i^h \mid i \in V \setminus P\} \ \forall h \in I_{H-1}, & A^h &:= \{i^{h-1} j^h \mid ij \in A, j \notin P\} \ \forall h \in I_{H-1},\\
V^H &:= \{i^H \mid i \in S\}, & A^H &:= \{i^h i^H \mid i \in S, h \in I^0_{H-1}\}\\
R^L &:= \{i^0 \in V_0 \mid i \in P\} \cup V^H \text{ and} & &\cup \{i^{H-1} j^H \mid j^H \in V^H, ij \in A\}.
\end{aligned}
$$

The costs of the arcs are set as follows: i) arcs in A^0 are assigned the primary edge costs; ii) arcs in A^h for $h \geq 1$ are assigned secondary edge costs, with only exception of the arcs of type (i^h, i^H), $h \in I^0_{H-1}$, that are assigned a cost of zero. Figure 1a) illustrates a small segment taken from a layered graph obtained this way.

We associate binary variables to the arcs in A^L as follows: variable x^1_{ij} to arc $ij \in A^0$, variable z_i to node $i \in V^0$, variable x^{2h}_{ij} to arc $i^{h-1} j^h \in A^h \forall h \in I_H$, and variable x^{2h}_{ii} to arc $i^h i^H \in A^H$. Let $\delta^-(W) = (V^L \setminus W; W)$ denote a directed cut in G^L and $X(\delta^-(W))$ the sum of all x-variables corresponding to arcs in that cut. Then, the model reads:

$$
\begin{aligned}
(SA_z) \quad \min \sum_{ij \in A} c^1_{ij} x^1_{ij} &+ \sum_{ij \in A} \sum_{h=1}^{H} c^2_{ij} x^{2h}_{ij} + \sum_{i \in V} f_i z_i\\
\text{s.t.} \quad X(\delta^-(W)) &\geq 1 & \forall W \subseteq V^L \setminus \{r\}, W \cap R^L \neq \emptyset \quad (7)\\
z_i &\geq x^{21}_{ij} & \forall i^0 j^1 \in A^1 \quad (8)\\
\sum_{ij \in \delta^-(i)} (x^1_{ij} + \sum_{h=1}^{H} x^{2h}_{ij}) &\leq 1 & \forall j \in V \quad (9)\\
x^1_{ij}, x^{2h}_{ij}, z_i &\in \{0,1\} & \forall ij \in A, h \in I^0_H, \forall i \in V \quad (10)
\end{aligned}
$$

Consider an optimal binary solution of the SA_z formulation. Removing the arcs in A^H and ignoring the indices h of the remaining arcs, one obtains a two-level Steiner tree with secondary depth less or equal than H in the original graph and with the cost obtained as a sum of facility opening, primary and secondary edge costs. In other words, variables x^{2h} can be projected into the original variable space as follows: $x^2_{ij} = \sum_{h=1}^{H} x^{2h}_{ij} \forall ij \in A$, while the variables x^{2h}_{ii} are ignored.

Lemma 2.

$$\upsilon_{LP}(SA_z) \geq \upsilon_{LP}(JUMP).$$

Proof. We prove the claimed relation by showing that an LP-optimal solution $(\mathbf{x}, \mathbf{z})$ of SA_z can be projected onto an LP-feasible solution $(\bar{\mathbf{x}}, \bar{\mathbf{z}})$ of formulation *JUMP*: For an LP-optimal solution $z_i = \max_j x^{21}_{ij}$ holds $\forall i \in V^0$. Together with inequalities (7) this implies

$$X(\delta^-(i^0)) \geq \begin{cases} x^1_{ij} \\ x^{21}_{ij} \end{cases} \qquad \forall i^0 \in V^0 \text{ and} \tag{11}$$

$$X(\delta^-(i^h)) \geq x^{2,h+1}_{ij} \qquad \forall i^h \in V^h, h \in I_{H-1}. \tag{12}$$

Then inequalities (7) imply constraints (4) and (3). By summing up (12) and (8) we derive (2). Inequalities (9) imply (5). (8) together with (12) guarantee (4). Finally each constraint in (1) is the projection of constraint (7) where

$$W = \{i^0 \mid i \in S_0^c\} \cup \{i^k \mid i \in \bigcup_{k=h}^{H} S_k, k \in I_H\} \subset V^L \setminus \{r\}. \qquad \square$$

3.2 Generalized Cut-Jump Inequalities

We describe how to eliminate facility constraints (8) from the Steiner arborescence formulation in order to obtain even a stronger formulation. We do so by splitting the nodes in V^0 into one primary and one facility copy and obtain a new layered graph $\bar{G}^L = (\bar{V}^L, \bar{A}^L)$. The set of nodes in $\bar{G}^L$ is then $\bar{V}^L = V^L \cup V^z$ where $V^z = \{i^z \mid i \in V\}$. We adapt the sets of arcs in G^L by replacing A^1 by $\bar{A}^1 := \{i^z j^1 \mid ij \in A\}$ and adding $A^z := \{i^0 i^z \mid i \in V\}$. The costs of arcs in A^h, for $h \in I_H \setminus \{1\}$ are set as above. Arcs of A^z are split facility nodes and therefore $c_{i^0 i^z} := f_i \forall i \in V$. Figure 1 illustrates this transformation.

Associating binary variables z_i to arc $i^0 i^z \in A^z$ and using the variables introduced for formulation SA_z we can model the TLNDSH as Steiner Arborescence problem with node degree constraints on $\bar{G}^L$. We denote it by SA.

In the proof of Lemma 2 we have seen that the cut-jump inequalities are a special case of Steiner cuts in the layered graph G^L. We now show how a more general family of stronger inequalities can be derived from Steiner cuts in the split layered graph $\bar{G}^L$.

Looking at the valid Steiner cuts in $\bar{G}^L$ we can derive two different types of *generalized cut-jump constraints* (GCJ) as follows:

a) Let S_0' be a superset of S_0. Then, the following cuts are valid:

$$x^1(\delta^+(S_0')) + x^2(J) + z(S_0' \setminus S_0) \geq 1 \quad \forall \{r\} \cup P \subseteq S_0 \subset S_0', S_{H+1} = \{i\}, i \in S,\ i \notin S_0' \tag{13}$$

b) Let S_0' be a subset of S_0. Then the following cuts are valid for TLNDSH:

$$x^1(\delta^+(S_0')) + x^2(J) \geq 1 \quad \forall \{r\} \cup P \subseteq S_0, S_{H+1} = \{i\}, i \in S, S_0' \subset S_0 \tag{14}$$

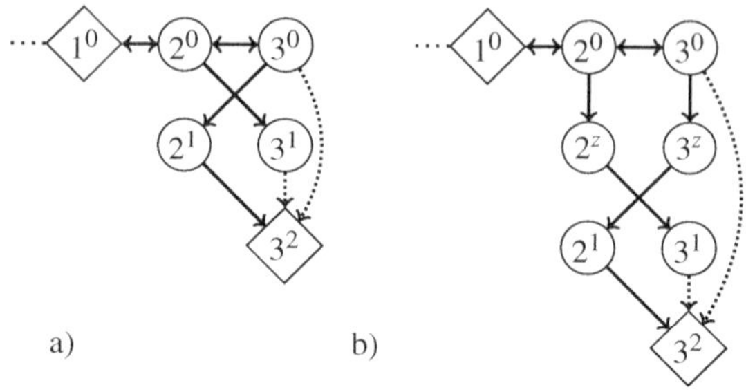

Fig. 1. Illustration of the transformation from layered graph G^L to $\bar{G}^L$ for $H = 2$. Node 1 is a primary customer, thus the copy indexed by 0 is a terminal. Node 2 is a Steiner node and node 3 is a secondary customer, thus the copy with index 2 is a terminal as well. Dotted arcs are the ones in A^H with a cost of 0.

In our future work we intend to examine the (generalized) cut-jump inequalities more completely, in terms of studying their practical usefulness as well as in finding a suitable and intuitive interpretation for them. We also plan to implement a branch-and-cut algorithm involving the (generalized) cut-jump inequalities and to investigate the practical importance of the layered graph models when compared to other, weaker formulations.

Acknowledgement

The authors like to thank the referees for their useful comments and suggestions. Luis Gouveia was funded by Centro de Investigação Operacional [MATH-LVT-Lisboa-152].

References

1. Balakrishnan, A., Magnanti, T.L., Mirchandani, P.: Modeling and heuristic worst-case performance analysis of the two-level network design problem. Management Science 40(7), 846–867 (1994)
2. Dahl, G.: Notes on polyhedra associated with hop-constrained paths. Operations Research Letters 25(2), 97–100 (1999)
3. Gollowitzer, S., Gouveia, L., Ljubić, I.: A node splitting technique for two level network design problems with transition nodes. Tech. rep. (2011)
4. Gouveia, L., Janssen, E.: Designing reliable tree networks with two cable technologies. European Journal of Operational Research 105(3), 552–568 (1998)
5. Gouveia, L., Simonetti, L., Uchoa, E.: Modeling hop-constrained and diameter-constrained minimum spanning tree problems as Steiner tree problems over layered graphs. Mathematical Programming, 1–26
6. Gouveia, L., Telhada, J.: Distance-constrained hierarchical networks. In: Proceedings of International Network Optimization Conference, INOC 2005, pp. B2.416–B2.421 (2005)
7. Khoury, B.N., Pardalos, P.M., Du, D.Z.: A test problem generator for the Steiner problem in graphs. ACM Trans. Math. Softw. 19(4), 509–522 (1993)
8. Ljubić, I., Gollowitzer, S.: Modeling the hop constrained connected facility location problem on layered graphs. In: Proceedings of the International Symposium on Combinatorial Optimization (ISCO). Electronic Notes in Discrete Mathematics, vol. 36, pp. 207–214 (2010)

Spanning Trees with Generalized Degree Constraints Arising in the Design of Wireless Networks

Luís Gouveia[1], Pedro Moura[1], and Amaro de Sousa[2]

[1] CIO-DEIO, Bloco C6-Piso 4, Faculdade de Ciências da Universidade de Lisboa, Cidade Universitária, Campo Grande, 1749-016 Lisboa, Portugal
{legouveia,pmmoura}@fc.ul.pt

[2] Instituto de Telecomunicações, Universidade de Aveiro, 3810-193 Aveiro, Portugal
asou@ua.pt

Abstract. In this paper we describe a minimum spanning tree problem with generalized degree constraints which arises in the design of wireless networks. The signal strength on the receiver side of a wireless link decreases with the distance between transmitter and receiver. In order to work properly, the interference on the receiving part of the link must be under a given threshold. In order to guarantee this constraint, for each node we impose a degree constraint that depends on the "length" of the links adjacent to the corresponding node, more precisely, nodes adjacent to long links must have a smaller degree and vice-versa. The problem is complicated by considering different signal strengths for each link. Increasing the strength in a link increases the cost of the link. However, it also reduces the maximum allowed degree on its end nodes. We create two models using adequate sets of variables, one may be considered an extended version of the other, and relate, from a theoretical perspective, the corresponding linear programming relaxations.

1 Introduction

In this paper we consider a wireless network design problem that generalizes a problem previously defined and studied in [4] (see also [2,3]). These problems also generalize the well-known degree constrained spanning tree problem (see [1] and the references inside) in the sense that they consider node degree dependent costs and more complicated degree constraints (the constraint on the degree of a node depends on the edges adjacent to it in the solution). Section 2 describes and motivates the new problem. Section 3 describes several models for the problem.

2 Description and Motivation of the Problem

In point-to-point wireless networks, each network connection is implemented through a point-to-point wireless transmission system (wireless link, for short) composed by a pair of transmitter/receiver antennas and signal processing units (one at each side of the connection) working on a frequency channel, chosen from a possible set of channels. Thus, consider an undirected graph $G = (V, E)$ where $V = \{1, \ldots, n\}$ is the set of

J. Pahl, T. Reiners, and S. Voß (Eds.): INOC 2011, LNCS 6701, pp. 77–82, 2011.

network nodes and $E \subseteq V^2$ is the set of edges $\{i, j\}$, representing each network connection. A network node with wireless links for different neighbour nodes must use different frequency channels. In most wireless technologies, due to the scarcity of the spectrum, there is a limited set of available frequency channels and many of them are partially overlapped between each other. Therefore, in a node using partially overlapped channels to different neighbour nodes, part of the transmitted signal on one channel is added as interference to the received signal on the other channel. Note that the signal strength on the receiver side of a wireless link decreases with the distance between transmitter and receiver antennas due to attenuation and other propagation effects. In order to work properly, the received signal must be such that the signal-to-interference-and-noise ratio (SINR) on the receiver is above a required threshold. Therefore, the coverage of a wireless link, i.e., the maximum distance between antennas that make the link work properly, depends on the amount of interference introduced by the other frequency channels on its end nodes. When a given wireless link cannot meet the required SINR threshold, we can consider three possible cases.

Case 1. *Several costs / A single maximum degree parameter*: In this case, we assume that pairs of nodes with higher distance have more expensive wireless links, with a higher power transmission, in order to raise the SINR over the required threshold. A parameter D is set as the maximum degree for each node (based on the available frequency channels) and, for each pair of nodes i and j with a distance equal to d_{ij}, a cost value $c_{\{i,j\}}$, which depends on d_{ij}, is defined as the least cost wireless link that can still provide the required SINR whatever the degree of its end nodes is. This is the case adopted in [4] where 3 types of wireless links were considered, each one with a different coverage and cost.

Case 2. *A single cost / Several maximum degree parameters*: In this case, we assume that there is only one type of wireless link with an associated cost value c and such that it is not used when the required SINR threshold is not met. For each pair of nodes i and j with a distance equal to d_{ij}, a degree parameter $D_{\{i,j\}}$, which depends on d_{ij}, is defined as the maximum degree of both i and j such that interference does not jeopardize the required SINR threshold for the wireless link to work properly.

Case 3. *Several costs / Several maximum degree parameters*. In this case, we assume that there are T types of wireless links with associated (increasing) costs $f_t, 1 \leq t \leq T$, and the degree of its endnodes depends on the type of link installed. Consider a pair of nodes i and j with a distance equal to d_{ij}. We define $D^t_{\{i,j\}}$, which depends on d_{ij}, as the maximum degree on both nodes i and j, if we install a wireless link of type t between these two nodes. Then, we can install a higher cost wireless link, allowing both nodes to have higher degrees, or install a lower cost wireless link constraining the degrees of nodes i and j to be lower. That is, we have a cost model for a wireless link to be installed between two nodes, i and j, which not only depends on the distance between those two nodes, but also depends on the degree that nodes i and j will have in the solution of the problem. Then, for each pair of nodes i and j we define a cost $c^m_{\{i,j\}}$, which gives the cost of the cheapest cost wireless link that can be used, assuming that m is the maximum of the degrees of nodes i and j.

Note that cases 1 and 2 are particular cases of case 3. In fact, if we consider just one type of wireless link we obtain case 2. Also, case 1 is a particular case of case 3 when we assume that all types of wireless links allow the degree of its end nodes to be the maximum degree D *i.e.*, the degree parameters $D^t_{\{i,j\}}$ are equal to D for all pairs of nodes i and j and all types t of wireless links. In the next section we describe several models for this more general case 3.

3 Formulations

In this section we describe two integer linear formulations for the problem. Consider binary variables $x_{\{i,j\}}$ indicating whether edge $\{i,j\} \in E$ is selected, as well as binary variables y_i^d indicating whether node $i \in V$ has degree equal to $d \in \{1,\dots,D\}$ in the solution. These variables were used in the models introduced in the works [2,3,4] where problems with non linear costs associated to the node degrees were studied. The two models studied in this paper use the previous two sets of variables. They differ, however, on the set of variables that characterize the type of links to be installed.

3.1 Model (P_1)

Besides the two sets of variables x and y, model (P_1) also uses binary variables $v^m_{\{i,j\}}$ indicating whether the edge $\{i,j\} \in E$ is selected and the maximum degree between nodes i and j is m (with $m = 2,\dots,D$). Clearly, these variables are not defined for $m = 1$, since we cannot have an edge where the degree of both endpoints is equal to 1. The problem can then be formulated as (P_1) (we denote by $E(i) \subseteq E$ the set of edges incident on node i). The objective cost function is straightforward.

$$(P_1)\ min \quad \sum_{\{i,j\}\in E}\sum_{m=2}^{D} c^m_{\{i,j\}} \cdot v^m_{\{i,j\}} \tag{1}$$

$$s.to: \quad \{\,\{i,j\} \in E : x_{\{i,j\}} = 1\} \text{ is a SpTree} \tag{2}$$

$$\sum_{d=1}^{D} d \cdot y_i^d = \sum_{\{i,j\}\in E(i)} x_{\{i,j\}} \qquad i \in V \tag{3}$$

$$\sum_{d=1}^{D} y_i^d = 1 \qquad i \in V \tag{4}$$

$$x_{\{i,j\}} = \sum_{m=2}^{D} v^m_{\{i,j\}} \qquad \{i,j\} \in E \tag{5}$$

$$v^m_{\{i,j\}} \le y_i^m + y_j^m \qquad \{i,j\} \in E; m = 2,\dots,D \tag{6}$$

$$2 \cdot v^m_{\{i,j\}} \le \sum_{d=1}^{m} (y_i^d + y_j^d) \qquad \{i,j\} \in E; m = 2,\dots,D-1 \tag{7}$$

$$x_{\{i,j\}} \in \{0,1\} \qquad \{i,j\} \in E \tag{8}$$

$$v^m_{\{i,j\}} \in \{0,1\} \qquad \{i,j\} \in E; m = 2,\dots,D \tag{9}$$

$$y_i^d \in \{0,1\} \qquad i \in V; d = 1,\dots,D \tag{10}$$

Constraints (2), stating that the solution is a Spanning Tree, are given in a generic form and can be written in several ways (see [5]). Constraints (3) and (4) define the degree variables y_i^d and guarantee that $y_i^d = 1$ *iff* the number of edges adjacent to node i is equal to d. Constraints (5) link the two sets of edge variables, $x_{\{i,j\}}$ and $v_{\{i,j\}}^m$, stating that, if edge $\{i,j\}$ is selected, then the maximum between the degrees of its endnodes must be a value in $\{2,\ldots,D\}$. Constraints (6) and (7) link the node variables y_i^d with the edge variables $v_{\{i,j\}}^m$: for a given edge $\{i,j\}$, constraints (6) guarantee that if $v_{\{i,j\}}^m = 1$ then one of the nodes i or j must have a degree equal to m, and constraint (7) guarantees that neither one of these nodes has a degree greater than m. Constraints (8)-(10) define the domain of the variables.

The variables $v_{\{i,j\}}^m$ are sufficient to describe the objective function of the problem since the extra index indicates the maximum degree of the endpoints associated to each edge. In the next subsection we create a model with edge variables having two extra indexes, associated to the degrees of each endpoint. We will show that these extra variables, although leading to a model with more variables, permit us to write a model with fewer constraints since it is easier (we need fewer constraints) to relate the new variables with the degree variables y_i^d. Furthermore, with the new set of variables we can derive, hopefully strong, valid inequalities.

3.2 Model (P_2)

Besides the $x_{\{i,j\}}$ and y_i^d variables, model (P_2) also uses binary variables $z_{\{i,j\}}^{pq}$, indicating whether the edge $\{i,j\} \in E$ is selected and degree(i) = p and degree(j) = q. Again, these variables are not defined for $(p,q) = (1,1)$. Before describing the new model, we note that the two sets of edge variables, $v_{\{i,j\}}^m$ and $z_{\{i,j\}}^{pq}$, can be related as follows

$$v_{\{i,j\}}^m = \sum_{q=1}^{m} z_{\{i,j\}}^{mq} + \sum_{p=1}^{m-1} z_{\{i,j\}}^{pm} \qquad \{i,j\} \in E; m = 2,\ldots,D \tag{11}$$

$$(P_2)\ min \quad \sum_{\{i,j\}\in E} \sum_{m=2}^{D} c_{\{i,j\}}^m \cdot \left(\sum_{q=1}^{m} z_{\{i,j\}}^{mq} + \sum_{p=1}^{m-1} z_{\{i,j\}}^{pm} \right) \tag{12}$$

$$s.to: \quad \{\ \{i,j\} \in E : x_{\{i,j\}} = 1\} \text{ is a SpTree} \tag{2}$$

$$\sum_{d=1}^{D} d \cdot y_i^d = \sum_{\{i,j\}\in E(i)} x_{\{i,j\}} \qquad i \in V \tag{3}$$

$$\sum_{d=1}^{D} y_i^d = 1 \qquad i \in V \tag{4}$$

$$x_{\{i,j\}} = \sum_{p=1}^{D} \sum_{q=1}^{D} z_{\{i,j\}}^{pq} \qquad \{i,j\} \in E \tag{13}$$

$$p \cdot y_i^p = \sum_{\{i,j\}\in E(i)} \sum_{q=1}^{D} z_{\{i,j\}}^{pq} \qquad i \in V; p = 1,\ldots,D \tag{14}$$

$$x_{\{i,j\}} \in \{0,1\} \qquad \{i,j\} \in E \tag{8}$$

$$y_i^d \in \{0,1\} \qquad i \in V; d = 1,\ldots,D \qquad (10)$$
$$z_{\{i,j\}}^{pq} \in \{0,1\} \qquad \{i,j\} \in E; p,q = 1,\ldots,D \qquad (15)$$

The objective function follows straightforwardly from the objective function of the previous model and the linking constraints (11). Note the constraints (14) linking the degree variables with the new link variables, which are much easier to write in this model. These constraints state that, if the degree of node i is p then, in the solution, exactly p edges are incident in that node, whatever the degree of node j is (for $p = 1$, the summation on q starts at 2). Note that under (13), constraints (3) for a given node i, can be obtained by adding constraints (14) for all $p = 1,\ldots,D$ and for the same i. Thus constraints (14) are a disaggregation of (3) and the latter can be omitted from the integer model. However, we will come back again to the weaker constraints (3) since we will see later that we can obtain a different valid model for the problem where we can use the weaker but more compact set (3) rather than the stronger but less compact set (14). For the moment, we also point out that there is no dominance relationship between the linear programming relaxations of the two models (P_1) and (P_2).

As we have stated before, another advantage of using the new variables is that we can write, hopefully strong, valid inequalities such as,

$$\sum_{q=1}^{D} z_{\{i,j\}}^{pq} \leq y_i^p \qquad i \in V, \{i,j\} \in E(i); p = 1,\ldots,D \qquad (16)$$

The valid inequalities (16) state that if edge $\{i,j\}$ is in the solution and node i has degree equal to p, whatever the degree on node j is, then the corresponding y variable associated to node i and degree p must be equal to 1. We do not need to consider inequalities (16) for $p = 1$ because these are implied by constraints (14) for node i and $p = 1$. Denoting by (P_2^*) the model obtained by adding the inequalities (16) to model (P_2) as well as the definitional constraints (11) (these do not improve the linear programming bound) we can prove the following result.

Proposition 1. *The projection of the set of feasible solutions of the linear programming relaxation of (P_2^*) on the subspace defined by the variables x, y and v is contained in the set of feasible solutions of the linear programming relaxation of the model (P_1).*

In the proof of Proposition 1 (not presented here), we did not make use of constraints (14) of model (P_2^*). In fact, it is not difficult to see that, in the presence of the new constraints (16), we still obtain a valid model by using only the weaker constraints (3) instead of both (3) and (14) constraints. We denote by (P_2^{**}) this model.

Preliminary results show that in almost any case, model (P_2) performs better than model (P_1). By including the valid inequalities (16), model (P_2^*) is able to reduce the gaps of model (P_2) in most of the cases. The results of model (P_2^{**}) show that the presence of constraints (14) in the model is of significance since the lower bounds obtained with this model are considerably larger than the ones obtained with the model (P_2^*).

Acknowledgement

Sponsored by FCT, research grants POCTI-ISFL-1-152 and PTDC/EIA/64772/2006.

References

1. Cunha, A., Lucena, A.: Lower and upper bounds for the Degree-constrained Minimum Spanning Tree Problem. Networks 50(1), 55–66 (2007)
2. Duhamel, C., Gouveia, L., Moura, P., Souza, M.: Models and Heuristics for the k-degree constrained minimum spanning tree problem with node-degree costs. To appear in Networks (2011)
3. Gouveia, L., Moura, P.: Spanning Trees with Node Degree Dependent Costs and Knapsack Reformulations. Electronic Notes in Discrete Mathematics 36, 985–992 (2010)
4. Gouveia, L., Moura, P., Sousa, A.: Prize collecting Steiner trees with node degree dependent costs. Computers & Operations Research 38(1), 234–245 (2010)
5. Magnanti, T., Wolsey, L.: Optimal Trees. In: Ball, M.O., Magnanti, T.L., Monma, C.L., Nemhauser, G.L. (eds.) Handbooks in Operational Research and Management Science (1995)

Reformulation by Intersection Method on the MST Problem with Lower Bound on the Number of Leaves

Luís Gouveia and João Telhada

Centro de Investigação Operacional, Universidade de Lisboa
{legouveia,joao.telhada}@fc.ul.pt

Abstract. We consider a variant of the the minimum spanning tree with a constraint imposing a minimum number of leaves. This paper is motivated by the computational results taken from a small set of instances with an enhanced directed model where it is shown that the corresponding linear programming bound values strongly depend on the choice of the root node. Thus, we present a new formulation for this problem that is based on "intersecting" all the rooted tree models at the same time. We will show that the linear programming bound of the new model is, in general, substantially better than the linear programming bound obtained by the best directed model. The computational results indicate that the model is too large to solve efficiently medium sized instances. In order to overcome this disadvantage, we present an iterative procedure that starts with a single rooted model and sequentially adds other rooted models. The idea of this method is to obtain an intermediate intersection model (that is, a model where only some of the rooted models are considered in the intersection) and such that the corresponding linear programming bound will be close to the bound obtained by the model which results from intersecting all the rooted models. The computational results show that the iterative procedure is worth using and should be further investigated when using the reformulation by intersection for other problems.

1 Introduction

Consider a graph $G = (V,E)$ with costs c_{ij} associated to each edge. In this paper we consider a variant of the the minimum spanning tree which includes a constraint imposing a minimum number of leaves. Let L represent the lower bound imposed on the number of leaves. This problem was first studied by Luís Gouveia and Lucinda Fernandes [1], who have also presented some models for the related maximum leaf spanning tree problem (see [2,3,6] for more work on the problem). The problem studied earlier was defined as having a root node and the leaf constraint was defined only for the remaining nodes. The problem studied here is "unrooted" and all the nodes are considered for the leaf constraint.

It is well known that, for tree design problems, models with a better linear programming (LP) relaxation are obtained by selecting a root node and modelling the solution as a directed tree (see, e.g., [7]). This paper is motivated by the computational results taken from a small set of instances with an enhanced directed model. The results show that the corresponding LP bound values strongly depend on the choice of the root node. Thus, following [4], we present a new formulation for this problem that is based on

J. Pahl, T. Reiners, and S. Voß (Eds.): INOC 2011, LNCS 6701, pp. 83–91, 2011.

the technique of reformulation by intersection. More precisely, we describe a model that considers all the rooted tree models at the same time linked by adequate coupling constraints (the "intersection" constraints). We will show that the LP bound of the new model is, in general, better than the LP bound obtained by the best directed model.

Using previous results known from Kipp Martin [5] (see also [4]), the intersection model, although relatively huge, does not need to include the exponentially sized set of subtour elimination constraints (SEC) that are required in each single rooted model, since those constraints are redundant (this is one of the important outcomes of the intersection technique). However, the computational results still indicate that the model is too large to solve medium sized instances (e.g., more than 60 nodes).

In order to overcome the large number of constraints and variables ($o(n^3)$) still remaining in the model, we also present an iterative procedure that starts with a single rooted model and sequentially adds other rooted models. The idea of this method is to obtain an intermediate intersection model (that is, a model where only some of the rooted models are considered in the intersection) and such that the corresponding LP bound will be close to the bound obtained by the model which results from intersecting all the rooted models. The computational results show that the iterative procedure is worth using and should be further investigated when using the reformulation by intersection for other problems.

2 The Directed Model and the Reformulation by Intersection

We use, as a starting point, a directed formulation that is similar to the one proposed by Luís Gouveia and Lucinda Fernandes [1]. We denote the root node by r, and consider the binary variables $x(i,j)$ which indicate whether arc (i,j) is in the solution and binary variables $y(i)$ which indicate whether node i is a leaf (degree 1 in the solution).

We use $x(S,T)$ to denote the summation of variables associated with arcs whose start node is in S and whose end node is in T. That is,

$$x(S,T) = \sum_{(i,j)\in A: i\in S, j\in T} x(i,j)$$

For the particular case when S and T are the same, we use an abbreviated notation $x(S)$. Similarly for the summation of leaf variables associated with nodes in a certain set S,

$$y(S) = \sum_{i\in S} y(i)$$

Let $d(i)$ denote the degree of node i and $\delta(i)$ represent the set of nodes adjacent to node i. We denote by (**Sub$^{\mathbf{r}}$**) the directed model presented below:

$$x(V-\{i\},\{i\}) = 1, \forall i \in V-\{r\} \tag{1}$$

$$x(U) \leq |U|-1, \forall U \subseteq V-\{r\}, |U| \geq 2 \tag{2}$$

$$x(\{i\}, V-\{i\}) \leq (d(i)-1)(1-y(i)), \forall i \in V-\{r\} \tag{3}$$

$$x(\{r\}, V-\{r\}) \leq 1+(d(r)-1)(1-y(r)) \tag{4}$$

$$y(V) \geq L \tag{5}$$

$$y(i) \in \{0,1\}, \forall i \in V \tag{6}$$

$$x(i,j) \in \{0,1\}, \forall (i,j) \in A \tag{7}$$

Constraints (1), (2) and (7) ensure that vector x defines a directed tree with root node r. The cardinality constraint (5) gives the lower bound on the number of leaves and constraints (3) and (4) link arc and leaf variables. Observe that constraints for the root node, (4), need to be different from the ones considered for other nodes, (3), since in the directed model the leaf variables are defined in terms of the outgoing arcs on each node. Note that constraints (3) and (4) only guarantee that the set of nodes with leaf variables equal to one is a subset of the leaf nodes in any feasible solution. To guarantee that leaf variables exactly define the set of leaves in a spanning tree, other constraints such as a set of constraints given in [Luís Gouveia and Lucinda Fernandes] [1] need to be included in the model. However, they are not necessary for modelling this problem since the cardinality lower bound (5) guarantees that at least L such variables will be equal to one.

Since this model is based on a procedure of choosing a node to be the root, we can write different directed models by selecting different root nodes. The following result, not proven here, states that the LP bound provided by the directed model is independent of choice on the root node.

Result 1: $v(\mathbf{Sub}^{r_1}{}_L) = v(\mathbf{Sub}^{r_2}{}_L), \forall r_1, r_2 \in V$

Some computational results also show that the LP bound of this formulation is, in general, still far from the optimal value. In order to tighten the LP relaxation of the model, we add a set of constraints proposed by [Lucena et al.] [6] for the problem of determining spanning trees with a maximum number of leaves.

$$x(i,j) + y(i) \leq 1, \forall (i,j) \in A : i \neq r \tag{8}$$

For a given arc (i,j), these constraints state that either the arc is in the solution or node i is a leaf, but not both. As before, we need to consider a slightly different set of constraints for the root node with a similar interpretation.

$$x(r,j) + y(r) + y(j) \leq 2, \forall j \in \delta(r) \tag{9}$$

We denote by ($\mathbf{Sub+^r}$) the formulation obtained by adding constraints (8) and (9) to model ($\mathbf{Sub^r}$). Results taken from a few instances indicate that the LP gaps substantially decrease by using this additional set of linking constraints. More interesting for the development of the paper is that the LP bounds are no longer independent of the choice of the root node. This suggests the use of the so-called reformulation by intersection (see [4]) leading to an enlarged model which considers the models for all possible roots combined together by suitable coupling constraints.

Consider the undirected variables $u(i,j)$ indicating whether edge $\{i,j\}$ is in the solution, directed variables $z(i,j,r)$ indicating whether arc (i,j) is in the directed tree rooted in node r, as well as the following constraints linking the directed variables with the undirected variables.

$$u(i,j) = z(i,j,r) + z(j,i,r), \forall \{i,j\} \in E(V - \{r\}), r \in V \tag{10}$$

$$u(r,j) = z(r,j,r), \forall j \in \delta(r), r \in V \tag{11}$$

Constraints (10) indicate that an edge $\{i,j\}$ is in the solution if and only if for each root r, the edge is used in any of the two possible directions, (i,j) or (j,i). Constraints (11) have a similar interpretation but use the fact that for any root r, arcs with the root as end node are not allowed.

Set notation, as introduced before for variables x and y, are also considered for the new variables. Furthermore, we use $z^r(S,T)$ to represent the summation of the new rooted arc variables associated with root r and with arcs whose start node is in S and whose end node is in T.

The intersection model includes constraints from all directed models, as well as constraints (10) and (11) which guarantee that all directed trees have common edges.

$$z^r(V-\{i\},\{i\}) = 1, \forall i \in V-\{r\}, r \in V \tag{12}$$
$$z^r(\{i\},V-\{i\}) \leq (d(i)-1)(1-y(i)), \forall i \in V-\{r\}, r \in V \tag{13}$$
$$z^r(\{r\},V-\{r\}) \leq 1+(d(r)-1)(1-y(r)), \forall r \in V \tag{14}$$
$$z(i,j,r)+y(i) \leq 1, \forall (i,j) \in A : i \neq r, r \in V \tag{15}$$
$$z(r,j,r)+y(r)+y(j) \leq 2, \forall j \in \delta(r), r \in V \tag{16}$$
$$y(V) \geq L \tag{5}$$
$$u(i,j) = z(i,j,r)+z(j,i,r), \forall \{i,j\} \in E(V-\{r\}), r \in V \tag{10}$$
$$u(r,j) = z(r,j,r), \forall j \in \delta(r), r \in V \tag{11}$$
$$y(i) \in \{0,1\}, \forall i \in V \tag{6}$$
$$z(i,j,r) \in \{0,1\}, \forall (i,j) \in A, r \in V \tag{17}$$
$$u(i,j) \in \{0,1\}, \forall \{i,j\} \in E \tag{18}$$

Let this model be denoted by **Int**(V), where V indicates that all possible rooted models are being considered. Note that this model does not include the subtour elimination constraints from each directed model, since they are redundant in the model. This fact was first shown by Kipp Martin [5] where he shows that the projection of the LP relaxation of the system defined by (12), (10), (11), (17) and (18) into the space defined by the undirected variables $u(i,j)$ gives a complete description of the spanning tree polytope defined by:

$$u(E) = |V|-1 \tag{19}$$
$$u(S) \leq |S|-1, \forall S \subset V : |S| > 2 \tag{20}$$
$$u(i,j) \in \{0,1\}, \forall \{i,j\} \in E \tag{18}$$

Constraints (20) are usually designated by subtour elimination constraints (SECs).

We obtained a model that is compact and that can be used directly by any MIP package. The results given in Section 4 show that the LP bounds of the new model are significantly better than the ones obtained by using the single rooted models and in several cases, the bound given by the intersected model is better than the bound obtained by the best single rooted model.

Still, the model remains huge in the number of variables and constraints ($O(n^3)$), and is difficult to use for solving medium sized instances (e.g., more than 60 nodes).

As mentioned in the introduction, this suggests the use of an iterative procedure where directed models associated with different root nodes are added in an iterative fashion.

3 The Iterative Procedure

We start by examining the intermediate model in the iterative procedure, more precisely the model that is obtained by intersecting some, but not all, of the directed models. For simplicity, we denote by R the current root node set. Clearly, the intermediate model may not even be valid for the corresponding integer problem since it lacks all the constraints associated to the variables $z(i,j,k)$ with k not in R.

The main idea is to know how to augment an intermediate model by selecting a new root and adding the corresponding constraints. This augmentation process will iteratively improve the LP bound given by the resulting model and is divided into two phases: i) adding implicitly violated SECs and ii) adding implicitly violated leaf linking constraints (15).

3.1 Phase 1: Adding Violated SECs

In a given intermediate model, some of the SECs may be not redundant and their inclusion may be needed to make the model valid. In the next result, we consider the intersected model restricted to a subset of root nodes and show which SECs are redundant in the model. The proof is similar to the proof given in [4] for the multi-weighted Steiner tree case.

Result 2: Let (u, z^r, y) be a feasible solution for the model $\mathbf{Int}(R)$. Then, subtour elimination constraints, for subsets that intersect R, are redundant.

Proof: Consider a root node $r \in R$, which also belongs to subset U. Then,

$$u(U) = z^r(U) = \sum_{i \in U - \{r\}} z^r(U - \{i\}, \{i\}) \leq \sum_{i \in U - \{r\}} z^r(V - \{i\}, \{i\}) = |U| - 1$$

This proves the redundancy when root node belongs to U. On the other hand, if $r \in R \cap U^C$, by intersection constraints,

$$x^r(U) = x^{r'}(U),$$

with $r' \in R \cap U$, and this proves the result. □

The previous result means that subtour elimination constraints on the undirected variables for subsets U included in $V \backslash R$ may not be redundant in the model. Thus, the augmenting step is based on: i) finding SECs such that U is included in $V \backslash R$ and that are violated by the current solution and, consequently, ii) choose an adequate root k in order to guarantee that those constraints are no longer violated. The previous result shows that by selecting a root k (and adding the corresponding equations (12), (10), (11) and (17) to the model) included in $V \backslash R$ will guarantee that SECs for sets of node including k will then become satisfied.

To find such a violated constraint and a corresponding node k, we solve a max flow problem between a source node r in R (it is easy to show that finding a violated SEC does not depend on the choice of a source node in R) and a node k not in R. If the value of the maximum flow is less than 1, then a cut $[V \backslash S, S]$ with r in $V \backslash S$ and k in S is violated. Using the indegree constraint (12) for the nodes in S and the equality constraints (10) and (11) for the same k, we obtain a violated SEC for the set S. By the previous result, if we add node k to the root set R (meaning that we add all constraints involving variables $z(i,j,k)$ for the chosen node k), that constraint, as well as all SECs including node k, will be satisfied by the new model. Note that some of these constraints, the ones with set of nodes intersecting R were already satisfied. However, the new SECs associated to sets $U \subseteq V \backslash R$, with $k \in R$, will now become satisfied.

We repeat this procedure for all nodes k not in R and the process is repeated for the new LP solution. The process stops when violated SECs are no longer found by the max flow routine, or when all roots have been included in R.

3.2 Phase 2: Adding Violated Leaf Linking Constraints

Let R^* be the set of roots obtained when the process previously described stops. We know that all SECs are implicitly satisfied. However, it may happen (confirmed by our computational experiments) that the current LP bound is not equal to the LP bound obtained by the model with all roots included, $\mathbf{Int}(V)$, and, in some cases, the bounds are still significantly different. This is explained by the fact that model $\mathbf{Int}(V)$ also includes, for all possible roots, the constraints linking the two sets of variables, (15) and (16). The iterative process previously described only seeks for violated SECs and will produce a bound at least equal to the optimal minimum spanning tree value but not necessarily a bound equal to the LP bound produced by the intersection model with all roots for the problem with the additional constraints. Thus, some of the models with roots not in R^* may still help to increase the lower bound.

To choose a root k to include in the intersection model in order to improve the LP bound, we use the following heuristic approach. Given the optimal solution (u^*, z^*, y^*) for an intermediate relaxed intersection model, and for each k not in R, we evaluate feasible upper and lower bound values, $\text{LB}(i,j)$ and $\text{UB}(i,j)$, for the new arc variables $z(i,j,k)$, in the following way:

1. for any j and k we set $z(k,j,k) := u^*(k,j)$ (in this case, $\text{UB}(k,j) = \text{LB}(k,j) = u^*(k,j)$);
2. for (i,j) such that $i, j \neq k$, we define upper bounds for each variable $z(i,j,k)$ given by $\text{UB}(i,j) := \min\{u^*(i,j); 1\text{-}y^*(i)\}$ (this follows from the equalities (10) and the inequalities (15));
3. for (i,j) such that $i, j \neq k$, we define lower bounds for each variable $z(i,j,k)$ given by $\text{LB}(i,j) := u^*(i,j)\text{-}\text{UB}(j,i)$ (this also follows from equalities (10));
4. for (i,j) such that $i, j \neq k$, we may improve the lower bounds for each variable $z(i,j,k)$ by setting $\text{LB}(i,j) := \min\{\text{LB}(i,j), 1 - \sum_{l \neq i} \text{LB}(l,j)\}$ (this follows from the indegree constraints (12)).

Then, we search for feasibility by checking if lower and upper bounds on the variables are consistent with the indegree constraints (12). If that consistency does not hold, the assignment of new $z(i,j,k)$ values is not possible to achieve with the incumbent solution (u,x,y). Due to the fact that all SECs are satisfied (and thus, it would be possible to find $z(i,j,k)$ values to satisfy equalities (12), (10) and (11)) we conclude that some of the linking constraints (15) and (16) for that given root node k must be violated and we include k in the set of roots.

4 Computational Results

We performed some computational experiments to evaluate the proposed approach. We generated random instances with 30, 40, 50 and 60 nodes, and with densities of 15% and 50%. For each pair number of nodes / density, 5 instances were generated and the reported results are averages based on those 5 instances. All results for linear relaxations

Table 1. Results obtained for the rooted models

	L=20						L=25					
	Sub+^{r}			**Sub^r**		Opt	**Sub+^{r}**			**Sub^r**		Opt
Instance	r	LP	Gap	LP	Gap		r	LP	Gap	LP	Gap	
1	1	3337,5	4,31%	3062	12,21%	3488	1	4752,049	7,31%	3062	40,28%	5127
	2	3339,678	4,25%	3062	12,21%	3488	2	4781,595	6,74%	3062	40,28%	5127
	3	3312	5,05%	3062	12,21%	3488	3	4744,692	7,46%	3062	40,28%	5127
	4	3334,321	4,41%	3062	12,21%	3488	4	4754,212	7,27%	3062	40,28%	5127
	5	3339,588	4,25%	3062	12,21%	3488	5	4813,2	6,12%	3062	40,28%	5127
	21	**3348,455**	**4,00%**	3062	12,21%	3488	**12**	**4844,909**	**5,50%**	3062	40,28%	5127
2	**1**	**2042,5**	**2,18%**	1939	7,14%	2088	1	2331,207	3,79%	1939	19,98%	2423
	2	2027,363	2,90%	1939	7,14%	2088	2	2308,625	4,72%	1939	19,98%	2423
	3	2030,675	2,75%	1939	7,14%	2088	3	2311,879	4,59%	1939	19,98%	2423
	4	2040,667	2,27%	1939	7,14%	2088	4	2332,469	3,74%	1939	19,98%	2423
	5	2039,112	2,34%	1939	7,14%	2088	5	2330,909	3,80%	1939	19,98%	2423
	6	2026,441	2,95%	1939	7,14%	2088	**11**	**2335,457**	**3,61%**	1939	19,98%	2423
3	1	18447,2	0,53%	18175	2,00%	18546	1	20607,88	1,99%	18175	13,56%	21027
	2	18473,5	0,39%	18175	2,00%	18546	2	20722,4	1,45%	18175	13,56%	21027
	3	18467	0,43%	18175	2,00%	18546	3	20679,67	1,65%	18175	13,56%	21027
	4	18473,5	0,39%	18175	2,00%	18546	4	20779,63	1,18%	18175	13,56%	21027
	5	18424,5	0,66%	18175	2,00%	18546	5	20636,33	1,86%	18175	13,56%	21027
	8	**18484**	**0,33%**	18175	2,00%	18546	**15**	**20866**	**0,77%**	18175	13,56%	21027
4	1	7342,375	6,81%	6335	19,60%	7879	**1**	**12304,07**	**6,73%**	6335	51,98%	13192
	2	7320,613	7,09%	6335	19,60%	7879	2	12278,12	6,93%	6335	51,98%	13192
	3	7425,25	5,76%	6335	19,60%	7879	3	12042,1	8,72%	6335	51,98%	13192
	4	7441,364	5,55%	6335	19,60%	7879	4	11980,23	9,19%	6335	51,98%	13192
	5	7207,753	8,52%	6335	19,60%	7879	5	11946,08	9,44%	6335	51,98%	13192
	10	**7493,031**	**4,90%**	6335	19,60%	7879	6	11921,72	9,63%	6335	51,98%	13192
5	1	5730,115	2,90%	5252	11,00%	5901	1	7866,412	7,88%	5252	38,49%	8539
	2	5786,25	1,94%	5252	11,00%	5901	2	7881,959	7,69%	5252	38,49%	8539
	3	5789,25	1,89%	5252	11,00%	5901	3	7851,442	8,05%	5252	38,49%	8539
	4	5803,833	1,65%	5252	11,00%	5901	4	7890,556	7,59%	5252	38,49%	8539
	5	5821,833	1,34%	5252	11,00%	5901	5	7791,143	8,76%	5252	38,49%	8539
	25	**5858,25**	**0,72%**	5252	11,00%	5901	**25**	**7931,452**	**7,11%**	5252	38,49%	8539

Table 2. Comparison between single rooted models and the intersection model

	L=20						L=25					
Instance	r	Sub+	Gap	Int	Gap	Opt	r	Sub+	Gap	Int	Gap	Opt
1	21	3348,455	4,00%	3481,5	0,19%	3488	12	4844,909	5,50%	4978	2,91%	5127
2	1	2042,5	2,18%	2080	0,38%	2088	11	2335,457	3,61%	2378,039	1,86%	2423
3	8	18484	0,33%	18494,5	0,28%	18546	15	20866	0,77%	21027	0,00%	21027
4	10	7493,031	4,90%	7852,5	0,34%	7879	1	12304,07	6,73%	12857,27	2,54%	13192
5	25	5858,25	0,72%	5901	0,00%	5901	25	7931,452	7,11%	8176,108	4,25%	8539

were performed with CPLEX 11.2 and the iterative process was implemented with the BCL language in a machine with an Intel T9400@2.53GHz, with 4Gb of RAM.

Table 1 presents results for (**Sub**$^{\mathbf{r}}$) and (**Sub+**$^{\mathbf{r}}$) in 5 instances with 30 nodes, a density of 50%, and considering L to be equal to 20 and 25. The first column indicates to which instance the figures correspond. For each instance, and for each value of L considered, results are presented for 6 possible roots. Among those 6 roots is the one that produced the best result for **Sub+**$^{\mathbf{r}}{}_{L}$, which is highlighted in bold text.

One can check the improvement in the LP bound which results from including the additional linking constraints, (15) and (16), which lead to model (**Sub+**$^{\mathbf{r}}$). In the case of $L = 25$, that improvement is very significant. A second observation is on the value of the LP bound provided by model (**Sub+**$^{\mathbf{r}}$) which strongly depends on the choice of the root (see, e.g., instance 4 with $L = 20$).

Table 2 shows results for the same instances reported in table 1. As before, the first column indicates the instance and, in the next columns, two sets of results are presented, for each of the cases $L = 20$ and $L = 25$. The figures illustrate the improvement

Table 3. Results obtained for the intersection model and the iterative process

			$\mathbf{Int}(V)_L$		Iterative Process					
$\|V\|$	D	L	LP Value	CPU (in sec.)	♯ subtours	♯ leaves	Value	CPU (in sec.)	Gap	♯ opt.
30	0,15	20	9811,80	0,992	1,6	20,6	9811,80	1,804	0,00%	4
30	0,5	20	7561,90	3,342	3,4	12,4	7561,90	4,027	0,24%	1
30	0,5	25	9883,28	4,116	12,6	17,4	9883,28	14,103	2,31%	1
40	0,15	25	12795,60	8,302	1,6	6,4	12795,60	0,571	0,00%	5
40	0,15	30	15119,58	10,248	1,6	25	15119,58	7,524	0,08%	3
40	0,5	25	11912,90	31,358	3,6	10,6	11912,90	4,325	0,00%	4
40	0,5	30	12448,68	41,978	4,2	24	12448,68	22,225	0,02%	3
40	0,5	35	14470,28	42,856	21,8	18,2	14470,28	106,200	2,28%	0
50	0,15	35	13709,13	46,780	1,4	41,2	13709,13	14,656	0,17%	2
50	0,15	40	16286,66	74,998	10,4	39,6	16286,66	47,188	2,78%	0
50	0,5	35	16394,45	85,384	9,6	35,2	16394,45	37,525	0,05%	3
50	0,5	40	18188,58	109,544	33	17	18188,58	206,134	1,38%	0
50	0,5	45	24035,60	122,764	36,2	13,8	24035,60	1517,208	7,37%	0
60	0,15	40	21875,27	80,572	2	40,4	21875,27	15,550	0,04%	4
60	0,15	45	23792,56	142,726	10,8	41,2	23792,56	57,054	0,65%	0
60	0,15	50	27383,87	101,648	25,6	34,4	27383,87	206,640	4,56%	0
60	0,5	40	11249,77	321,006	10,2	41,4	11249,77	336,243	0,06%	2
60	0,5	45	11610,97	357,530	12,4	47,6	11610,97	896,300	0,16%	1

obtained by the intersection model compared with the best rooted model, in terms of the corresponding LP relaxations. In fact, even for two instances it was possible to reach optimality with the intersection model.

The results obtained by the intersection model and the iterative process are presented in table 3. The first three columns indicate the characteristics of the networks and the bound considered for the required number of leaves. The fourth and fifth columns indicate the average results of the LP value and respective CPU time given by the intersection model. The next four columns show the results given by the iterative process, including the average number of roots considered in the first phase, the average number of roots considered in the second phase, the lower bound obtained at the end of the procedure, as well as the corresponding CPU time. To better evaluate the relative quality of these two lower bounding procedures, in the last two columns we report the average gap and the number of integer optimal solutions achieved within each set of 5 instances. Note that, since the lower bound given by the two methods was equal for all cases, we include only one column for each of these two values.

Observe that, in many cases, the iterative process is able to achieve the same lower bound with much less computational effort. For example, this happens for instances with 40 nodes and a density of 50%, in the case of L being 25 or 30. At present, we are running the proposed approach for larger instances (e.g., 80 nodes). The results obtained by the iterative procedure suggest that it may be worth using this type of approach for other related network optimization problems.

Acknowledgments

We would like to thank the important suggestions made by the anonymous referee, which improved in a considerable way this work. The authors also wish to acknowledge the financial support of Centro de Investigação Operacional, through project MATH-LVT-Lisboa-152.

References

1. Fernandes, L.M., Gouveia, L.: Minimal spanning trees with a constraint on the number of leaves. European Journal of Operational Research 104, 250–261 (1998)
2. Fujie, T.: An Exact Algorithm for the Maximum Leaf Spanning Tree Problem. Computers & Operations Research 30(13), 1931–1944 (2003)
3. Fujie, T.: The Maximum-Leaf Spanning Tree Problem: Formulations and Facets. Networks 43(4), 212–223 (2004)
4. Gouveia, L., Telhada, J.: The Multi Weighted Steiner Tree Problem: A Reformulation by Intersection. Computers & Operations Research 35(11), 3599–3611 (2008)
5. Kipp Martin, R.: Using separation algorithms to generate mixed integer model reformulations. Operations Research Letters 10(3), 119–128 (1991)
6. Lucena, A., Maculan, N., Simonetti, L.: Reformulations and Solution Algorithms for the Maximum Leaf Spanning Tree Problem. Computational Management Science 7(3), 289–311 (2010)
7. Magnanti, T.L., Wolsey, L.: Optimal Trees. In: Ball, M.O., Magnanti, T.L., Monma, C.L., Nemhauser, G.L. (eds.) Network Models, Handbooks in Operations Research and Management Science, vol. 7, pp. 503–615. North-Holland, Amsterdam (1995)

A Polyhedral Approach for Solving Two Facility Network Design Problem

Faiz Hamid and Yogesh K. Agarwal

Indian Institute of Management Lucknow, Prabandh Nagar,
Off-Sitapur Road, Lucknow 226013, India
{faiz,yka}@iiml.ac.in

Abstract. The paper studies the problem of designing telecommunication networks using transmission facilities of two different capacities. The point-to-point communication demands are met by installing a mix of facilities of both capacities on the edges to minimize total cost. We consider 3-partitions of the original graph which results in smaller 3-node subproblems. The extreme points of this subproblem polyhedron are enumerated using a set of proposed theorems. We introduce a new approach for computing the facets of the 3-node problem based on polarity theory after obtaining the extreme points. The facets of the subproblem are then translated back to those of the original problem using an extended version of a previously known theorem. We have tested our approach on several randomly generated and real life networks. The computational results show that 3-partition facets reduce the integrality gap by approximately 30-50% compared to that provided by 2-partition facets. Also there is a substantial reduction in the size of the branch-and-bound tree if these facets are used.

1 Introduction and Problem Formulation

We consider the *network design problem* (NDP) or *network loading problem* (NLP) which involves determining the mix of facilities of two capacities (*high* and *low*) on the edges of a given graph in order to satisfy the point-to-point demands at minimum cost. Applications of this problem and its variants arise frequently in the telecommunications industry for both service providers and their customers. The tariffs of these facilities are complex and offer strong economies of scale. The problem of designing a network becomes "hard" to solve when more than one type of facility is involved due to complexity of the cost structure.

In [9], the authors consider a single facility NDP, develop facets and completely characterize the convex hulls of the feasible solutions for two subproblems. The polyhedral properties of a single facility NDP are also studied by [1]. [10] study polyhedral properties of the two-facility undirected NDP and introduce three basic classes of valid inequalities: cut inequalities, 3-partition inequalities and arc residual capacity inequalities. [2] discusses the polyhedral properties of the flow formulation for a special NLP over a "cutset". [6] compare cutting plane algorithms based on the flow and capacity formulation. A new family of mixed-integer rounding inequalities is proposed in [7]. [4] presents a cutting plane algorithm based on cut inequalities.

J. Pahl, T. Reiners, and S. Voß (Eds.): INOC 2011, LNCS 6701, pp. 92–97, 2011.

[5] study polyhedra based on directed demand, flow costs, and existing capacities. In [3], a new class of tight metric inequalities is introduced, that completely characterize the convex hull of the integer feasible solutions of the NDP. A detailed discussion on polyhedral terminology can be found in [11].

The multicommodity capacitated NDP we consider is defined on an undirected network $G = (V, E)$, where V is the set of nodes and E the set of edges. The communication demands between origin-destination pairs are represented by the set of commodities K. Each commodity $k \in K$ has demand d_k that must flow between the origin $O(k)$ and the destination $D(k)$. We assume the larger modularity size to be an integral multiple of the smaller one (a realistic assumption). By rescaling demands, the smaller modularity size can be made equal to 1. Installing one LC facility on edge (i,j) provides one unit capacity at a cost a_{ij}. Whereas, installing a single HC facility on edge (i,j) provides C units of capacity at a cost b_{ij}. The two-facility network design problem (TFNDP) is formulated mathematically as the following mixed-integer programming model:

$$\text{Minimize} \sum_{(i,j)\in E} (a_{ij}x_{ij} + b_{ij}y_{ij}) \tag{1}$$

subject to

$$\sum_{j\in V} f_{ji}^k - \sum_{j\in V} f_{ij}^k = \begin{cases} -d_k \text{ if } i = O(k) \\ d_k \text{ if } i = D(k) \\ 0 \text{ otherwise} \end{cases} \quad \forall i \in V, \forall k \in K \tag{2}$$

$$\sum_{k\in K} (f_{ij}^k + f_{ji}^k) \le x_{ij} + Cy_{ij} \quad \forall (i,j) \in E \tag{3}$$

$$x_{ij}, y_{ij} \ge 0 \text{ and integer } \forall (i,j) \in E$$

$$f_{ij}^k, f_{ji}^k \ge 0 \; \forall (i,j) \in E, \forall k \in K$$

In the above formulation there are two kinds of variables: integral *capacity* variables x_{ij} and y_{ij} that define the number of LC and HC facilities loaded on the edge (i,j), and continuous *flow* variables f_{ij}^k that model the flow of commodity k on edge (i,j) in the direction i to j. Constraints (2) correspond to the flow conservation constraints for each commodity at each node. Capacity constraints (3) model the requirement that the total flow (in both directions) on an edge cannot exceed the capacity installed on that edge.

2 Solution Approach and Strategy

We consider the projection of the original problem polyhedron from the plane of integer and continuous to pure integer variables, eliminating the continuous variables corresponding to commodity flows. By virtue of Japanese Theorem [8], any feasible solution of this projection polyhedron is guaranteed to permit a feasible multicommodity flow for the original problem. The facial structure of the projection polyhedron is studied and facets are derived. According to a theorem proposed in [1], a facet inequality of the k-node subproblem resulting from a k-partition translates into a facet of the original problem for single facility NDP. We extend this theorem for the TFNDP and use it to translate the facets of 3-node TFNDP to those of the original TFNDP. The strategy that we adopt to generate facets of the TFNDP is discussed in detail in this section.

2.1 Shrinking the Original NDP Graph

First we shrink the original NDP graph by considering the partition of the set of nodes into three subsets $\{V_1, V_2, V_3\}$. Nodes present within the same subset of the partition merge to a single node. The demand pairs from V_i to V_j gets aggregated.

Each partition gives a different 3-node subproblem of the same original NDP. Therefore, the number of 3-node subproblems increases exponentially with the increase in number of nodes. For smaller size networks we can enumerate all the partitions (exact separation) and check all of them in each iteration if they are violated or not. However, for larger problems we use a separation heuristic based on neighborhood search, discussed below, to generate only violated 3-partitions.

Separation Heuristic - We start with a randomly generated 3-partition and evaluate all its neighbors obtained by shifting and exchange of one node from each subset. The neighbor with the maximum (demand-capacity) violation is selected as the next solution. The process is repeated until there is no further increase in the violation. The search can be made more intense by increasing the number of random restarts.

2.2 Enumeration of the Extreme Points of the Subproblem Polyhedron

After obtaining a 3-partition we systematically enumerate all the extreme points of this 3-node problem polyhedron. Using a set of theorems we can identify whether a given solution is an extreme point or not. Some of these theorems are presented below. It is found that there are maximum of 126 extreme points of this polyhedron depending on the demands between the three nodes.

Theorem 1. *A non minimal solution cannot be an extreme point.*

Theorem 2. *Simultaneous diversion of traffic from two or more edges will not produce an extreme point.*

Theorem 3. *Let $\delta_{ij} = \max(s_{ij}, x_{ij})$ where s_{ij} is the amount of spare capacity available on edge (i, j) of the 3-node problem and x_{ij} is the number of LC facilities installed on the edge. A solution will not be an extreme point solution if $\delta_{ij} > 0 \;\forall\; (i, j)$.*

2.3 Computation of Facets of the Subproblem

We use polarity theory [11] to find only the most violated facet of a 3-node problem polyhedron, P, avoiding generation of all the facets. To the best of our knowledge this approach has not so far been used to generate facets. The extreme points of the polar polyhedron Π give the facets of polyhedron P and vice-versa. This was done by solving the following subproblem:

$$
\begin{aligned}
z_p = \ & \text{Minimize } \alpha\pi \\
& \text{subject to } \pi x^k \geq 1, \ k \in K \\
& \qquad\qquad\quad \pi_i \geq 0, \ i = 1, \ldots, 6
\end{aligned}
$$

where, α is the aggregated capacity across the subsets of a given 3-node problem, π the vector of the coefficients of the violated facet being generated and x^k the k-th extreme point of the polyhedron P. The constraints correspond to the extreme points of the polyhedron P obtained in previous step. This is a simple linear programming problem with a maximum of 126 constraints over six variables. If $z_p < 1$, a violated facet is found for the given partition with respect to the given capacity aggregation.

We now present the algorithm to solve the problem.

Step 1. Solve LP relaxation.
Step 2. Generate 3-node subproblem.
Step 3. List extreme points.
Step 4. Compute violated facet.
Step 5. Translate the facets to that of original NDP.
Step 6. Add it to the LP problem and re-solve.
Step 7. Go back to Step 2 to generate a new 3-node subproblem. If all the partitions have been scanned for generating violated facets, stop.

3 Computational Study

We have tested our approach on several randomly generated networks and also on problem instances available at *www.sndlib.zib.de*. The routing cost and pre-installed capacity, if present, in the SND.LIB problem instances were ignored for our purpose. The smaller size networks (8 to 10 nodes) we considered had fully connected topology and were solved to optimality using exact separation. However, for larger size networks the separation was done using the heuristic mentioned in Section 2.1. The algorithm was implemented in Visual C++ 2008 and callable library of CPLEX 12.1 was used for optimization. The CPLEX generated cuts were suppressed while running branch-and-bound (B&B). All computations were carried out on a Pentium 4 processor with 3.0 GHz clock speed.

The value of C, ratio of HC and LC facility cost (HC/LC), and the demand between node-pairs were treated as input parameters for the experiments. Our objective of this study was to observe the effectiveness of the 3-partition facets in getting tighter lower bound compared to that provided by 2-partition facets as described in [10]. The performance measures used were (1) bound value for LP relaxation with 2-partition facets (Z_{LP2F}), (2) bound value for LP relaxation with 3-partition facets (Z_{LP3F}), (3) percentage gap reduction ($= (Z_{LP3F} - Z_{LP2F})/(Z_{IP} - Z_{LP2F}) \times 100$), (4) percentage node count reduction of the B&B tree of IP with 2-partition facets (IP2F) due to introduction of 3-partition facets and (5) optimality gap at the termination of the B&B.

From the computational results (Tables 1, 2 and 3) it is evident that the 3-partition facets reduce the integrality gap by approximately 30-50% compared to that provided by 2-partition facets. Also there is a substantial reduction in the size of the branch-and-bound tree if these facets are used.

Table 1. Average performance measures on smaller networks — fully connected topology, solved to optimality using exact separation

S. No.	Nodes	Edges	C	HC/ LC	Demand Min	Demand Max	Z_{LP} %	Z_{LP2F} %	Z_{LP3F} %	Gap Redc.%	B&B Node Count IP2F	B&B Node Count IP3F	B&B Node Count Redc.%
1	8	28	10	3.5	0.5	2.0	64.1	81.8	91.2	51.7	669	218	67.4
2	9	36	10	3.5	0.5	1.5	59.2	78.9	87.3	39.8	3451	1207	65.0
3	9	36	10	3.5	0.5	2.5	66.5	84.3	92.0	49.0	9905	2476	75.0
4	9	36	10	3.5	0.5	3.5	73.1	85.7	91.9	43.4	19690	6512	66.9
5	10	45	20	8.0	0.5	1.5	43.8	70.7	84.4	46.8	35560	9323	73.8
6	10	45	20	11.0	0.5	1.5	51.7	78.4	90.8	57.6	406581	133607	67.1
7	10	45	20	14.0	0.5	1.5	65.8	88.8	98.5	86.4	338385	21998	93.5

Table 2. Average performance measures on larger networks — Nodes = 25, Edges = 50, C = 10, Min Demand = 0.05, Max CPU Time = 1800 sec, solved using heuristic separation

S. No.	HC/ LC	Dem Max	Z_{LP} %	Z_{LP2F} %	Z_{LP3F} %	Gap$^{\natural}$ Redc.%	B&B Node Count IP2F	B&B Node Count IP3F	CPU Time(sec) 2F	CPU Time(sec) 3F	Opt. Gap % 2F	Opt. Gap % 3F
1	3.5	1.0	87.4	95.2	96.7	31.3	72195	46826	1800	1694	1.5	0
2	3.5	3.0	96.2	98.8	99.2	33.3	180983	44505	1800	499	0.4	0
3	3.5	5.0	97.9	99.4	99.6	33.3	133146	57055	921	484	0	0
4	5.0	1.0	88.9	94.1	95.7	27.1	123199	62342	1800	1800	2.0	1.1
5	5.0	3.0	96.7	98.6	98.9	21.4	200448	152813	1800	1800	0.3	0.2
6	5.0	5.0	98.0	99.3	99.4	15.0	197992	181155	1800	1800	0.2	0.1

$^{\natural}$ Gap Redc. % is computed w.r.t. best known integer solution.

Table 3. Performance measures on SND.LIB problem instances — Max CPU Time = 3600 sec

Prob Name	Nodes	Edges	Z_{LP} %	$Z^{\sharp}_{LP2F}$ %	$Z^{\S}_{LP3F}$ %	Gap$^{\natural}$ Redc%	B&B Node Count IP2F$^{\sharp}$	B&B Node Count IP3F§	CPU Time(sec) 2F$^{\sharp}$	CPU Time(sec) 3F§	Opt. Gap % 2F$^{\sharp}$	Opt. Gap % 3F§
Atlanta	15	22	93.0	97.9	99.2	61.9	1368	85	3	2	0.0	0.0
NewYork†	16	49	7.0	63.8	79.0	42.0	588680	472523	3600	3600	16.0	3.2
Norway	27	51	36.0	73.6	87.8	53.8	82618	13602	3600	1028	9.3	0.0
Cost266‡	37	57	67.5	85.4	89.8	30.1	48219	17614	3600	3600	5.9	5.5

$^{\natural}$ Gap Redc. % is computed due to polar 3-partition facets over 2-partition facets w.r.t. best known integer solution.
† With 3-partition facets, the problem could be solved to optimality in 1.2 hrs. However, the problem with 2-partition facets had an optimality gap of 12.7% even after 3 hrs.
‡ After 3 hrs, the optimality gap for problems with 2- and 3-partition facets were 4.2% and 2.5%, respectively.

4 Conclusion

In this paper we addressed the problem of network design with facilities of two different capacities. We fully characterized and enumerated all the extreme points of the 3-node problem polyhedron. A new approach for computing facets is introduced and a new family of facets is identified. The 3-partition based facets strengthen the linear programming formulation to a great extent. Computational results show that these facets significantly reduce the integrality gap and also the size of the branch-and-bound tree. Thus our approach provides both a very good lower bound and a starting point for branch-and-bound. The results assure that our approach can be an effective tool for solving real life problems. The future work is to develop viable heuristic based on this approach to solve larger real life problems. We are also trying to explore whether this approach can be extended to 4-partition based facets.

References

1. Agarwal, Y.K.: k-Partition-based facets of the network design problem. Networks 47(3), 123–139 (2006)
2. Atamtürk, A.: On capacitated network design cut-set polyhedra. Mathematical Programming B 92(3), 425–437 (2002)
3. Avella, P., Mattiab, S., Sassanob, A.: Metric inequalities and the network loading problem. Discrete Optimization 4(1), 103–114 (2007)
4. Barahona, F.: Network design using cut inequalities. SIAM Journal on Optimization 6(3), 823–837 (1996)
5. Bienstock, D., Günlük, O.: Capacitated network design - polyhedral structure and computation. INFORMS Journal on Computing 8(3), 243–259 (1996)
6. Bienstock, D., Chopra, S., Günlük, O., Tsai, C.: Minimum cost capacity installation for multicommodity network flows. Mathematical Programming 81(2), 177–199 (1998)
7. Günlük, O.: A branch-and-cut algorithm for capacitated network design problems. Mathematical Programming A 86(1), 17–39 (1999)
8. Lomonosov, M.V.: Combinatorial approaches to multiflow problems. Discrete Applied Mathematics 11(1), 1–93 (1985)
9. Magnanti, T.L., Mirchandani, P.: Shortest paths, single origin-destination network design and associated polyhedra. Networks 23(2), 103–121 (1993)
10. Magnanti, T.L., Mirchandani, P., Vachani, R.: Modeling and solving the two-facility capacitated network loading problem. Operations Research 43(1), 142–157 (1995)
11. Nemhauser, G.L., Wolsey, L.A.: Integer and combinatorial optimization. Wiley, New York (1988)

FTTH Network Design under OA&M Constraints

Matthieu Chardy and Cedric Hervet

Orange Labs, 38-40 rue du Général Leclerc 92125 Issy-les-Moulineaux, France
{matthieu.chardy,cedric.hervet}@orange-ftgroup.com

Abstract. Due to the emergence of bandwidth-requiring services, telecommunication operators are being compelled to renew their fixed access network, most of them favouring the Fiber To The Home (FTTH) technology. For long, network design strategies have been driven by mere deployment CAPital EXpenditures (CAPEX). Today however, the feedback and the experience gathered from the management of former networks strongly push for the consideration of other sources of cost for the design of networks. This paper focuses on the optimization of FTTH networks deployment under Operations, Administration and Maintenance (OA&M) considerations. Mixed integer formulations are first argued for the modelling of these decision problems. Then, numerical tests performed on real-life data prove the efficiency of branch and bound approaches for such models. Assessment of the economic impact of OA&M considerations is also made.

1 Introduction

For the past few decades, the business model of telecommunication operators has been based on the design of innovative value-added services that require high level of bandwidth. The consequent need for bandwidth upgrade in networks has been handled in different manners according to the type of network (core versus access networks, mobile versus fix networks). Like most telecommunication operators concerned with fixed access networks, Orange has adopted the FTTH technology with a specific point to multipoint architecture: the Passive Optical Networks (PON). Beyond technical challenges, tremendous amounts of money are at stake, and that makes its deployment a major issue for the coming years [7]. Literature related to optical access network design is quite abundant [1,3,4,5,6]. However, to our knowledge, all the models tackle the decision from a mere CAPEX costs perspective (costs of the equipments, costs of the deployment intervention resources, etc.) whereas major sources of costs are not direct one-shot CAPEX costs but indirect recurrent costs linked to OA&M [2]. This includes, but is not limited to Information System costs, monitoring and supervision costs, preventive and curative maintenance costs. This paper focuses on the decision problem of FTTH PON network optimization with the aim of taking into account several engineering rules motivated by OA&M considerations. The remaining of this paper is organised as follows. Section 2 is dedicated to the mathematical modelling of PON network design problems. Numerical results assessing branch and bound solving approaches and the impact of OA&M on CAPEX costs are presented in Section 3, before concluding in Section 4.

J. Pahl, T. Reiners, and S. Voß (Eds.): INOC 2011, LNCS 6701, pp. 98–104, 2011.

2 Mathematical Modelling

2.1 K-Level PON Network Design

PON are specific multi-level point to multipoint architectures dedicated to optical fibers. Precisely, their basic principle is the following: each optical fiber originating from the core[1] network (the optical entry point of the core network being called a NRO) will go through a sequence of passive network elements called optical splitters before reaching multiple subscriber households in the end. When going through a splitter, an incoming fiber is "divided/multiplied" into several outgoing fibers, this multiplicative factor being a positive integer (called capacity of the splitter). We stress the fact that all fibers coming out from a splitter are not necessarily used. Finally, in the following, the number of levels of a PON architecture will refer to the size of the sequence of splitters. For the sake of clarity, an illustration is given in Figure 1.

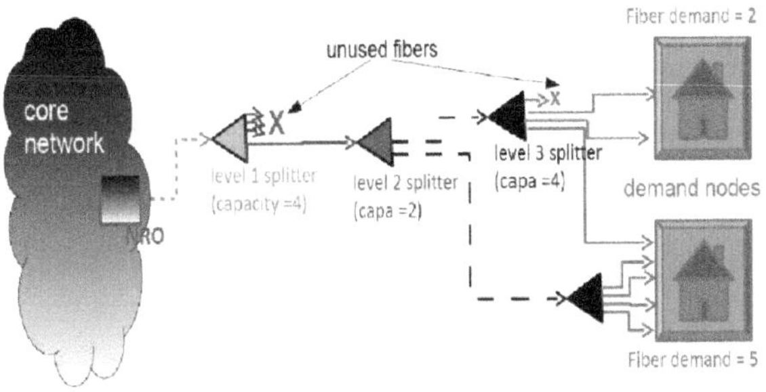

Fig. 1. Example of a PON architecture with three levels of splitters with respective capacities 4, 2 and 4

In such context, the PON design problem consists of delivering fibers to a set of located households and can be seen as a joint problem of spitters (of each level) location and fibers (of each level) routing. We take as basis the model introduced by [5,6] for a two-level architecture and, due to space limitations, refer to these articles for detailed aspects. We first propose to extend this formulation to a K-level architecture. Let $G = (V,E)$ be an undirected graph representing an existing infrastructure (e.g. the legacy copper network) and b_e, $e \in E$ be the remaining capacity (in number of fibers). Concerning the demand for fibers, let $\mathscr{D} \subset V$ denote the set of demand nodes with an associated demand a_i, $\forall i \in \mathscr{D}$ (we assume that individual household demands for one fiber have already been aggregated by building or neighborhood). As decision variables, we denote by $\mathbf{z}_i^k$ the number of splitters of level $k = 1..K$ installed at site $i \in V$, knowing that only the set of sites $\mathscr{S}^k \subset V$ is eligible to store level-k splitters, of capacity m^k and unitary cost C^k. Likewise, we denote by $\mathbf{f}_{ij}^k$ the number of fibers of level $k = 1..K+1$ going from site i to site j ($\mathbf{u}_i^k$ being the number of unused fibers), of unitary routing

[1] A core network is the central part of an end to end network, generally interconnecting large cities.

cost d_{ij}^k. The K-level PON design problem can be formulated as follows (denoted as $\mathscr{PON}^K$):

$$
\begin{aligned}
\min_{\mathbf{f},\mathbf{z}} \quad & \sum_{k=1..K} \sum_{i\in\mathscr{S}^k} C^k \mathbf{z}_i^k + \sum_{[i,j]\in E} \sum_{k=1..K+1} d_{ij}^k (\mathbf{f}_{ij}^k + \mathbf{f}_{ji}^k) \\
\text{such that :} \quad & \\
& \sum_{j|[i,j]\in E} \mathbf{f}_{ji}^1 = \mathbf{z}_i^1 + \sum_{j|[i,j]\in E} \mathbf{f}_{ij}^1 && \forall i \in \mathscr{S}^1 && (1)\\
& \sum_{j|[i,j]\in E} \mathbf{f}_{ji}^k + m^{k-1}\mathbf{z}_i^{k-1} = \mathbf{z}_i^k + \sum_{j|[i,j]\in E} \mathbf{f}_{ij}^k + \mathbf{u}_i^k && \forall k=2..K,\ \forall i \in \mathscr{S}^k && (2)\\
& \sum_{j|[i,j]\in E} \mathbf{f}_{ji}^{K+1} + m^K \mathbf{z}_i^K = a_i + \sum_{j|[i,j]\in E} \mathbf{f}_{ij}^{K+1} + \mathbf{u}_i^{K+1} && \forall i \in \mathscr{D} && (3)\\
& \sum_{k=1..K+1} (\mathbf{f}_{ij}^k + \mathbf{f}_{ji}^k) \le b_{ij} && \forall [i,j] \in E && (4)\\
& \mathbf{z}_i^k,\ \mathbf{u}_i^{k+1} = 0 && \forall k=1..K,\ \forall i \notin \mathscr{S}^k && (5)\\
& \mathbf{z}_i^k,\ \mathbf{u}_i^{k+1}, \mathbf{f}_{ij}^k \in \mathbb{N} && \forall k=1..K,\ \forall i \in \mathscr{S}^k,\ \forall [i,j] \in E &&
\end{aligned}
$$

2.2 PON Network Design under OA&M Constraints

We propose models for PON design problems taking OA&M concerns into account. As we think that OA&M costs are not comparable with deployment CAPEX costs, we favor an integration of such considerations through additional constraints to $\mathscr{PON}^K$.

OA&M and the "Splitter Delocation" Rule: Considering both network administration (ease of fault detection) and customer relationship (equity in user experience), network managers appreciate that subscribers of the same area (building or neighborhood) have the same "connection". This is classically achieved by mono-routing strategies in multiflow-based network design problems. When dealing with PON architectures, such a strategy would imply that both optical routes and location of splitters be the same and would be detrimental in terms of costs. Therefore, they push for the following compromise: when a demand at a given demand node exceeds a certain value (called delocation threshold) then all clients of this node must be "served" by fibers of the last[2] levels initiated by splitters located at the demand node. Let $\{e_0 = +\infty, e_1, ..., e_K\}$ be the set of delocation thresholds (positive integers sorted in decreasing order). For any demand node $i \in \mathscr{D}$, let k_i^{max} denote the maximum index k such that the threshold e_k exceeds the demand value a_i ($k_i^{max} = \max_{k=1,...,K}(a_i < e_k)$), then the "splitter delocation" rule can be formulated as follows:

$$
\begin{cases}
\mathbf{f}_{[i,i]}^{k+1} = \mathbf{z}_i^{k+1} & \forall k = (k_i^{max}+1),...,K, \text{ with the convention } \mathbf{z}_i^{K+1} = a_i \\
\mathbf{z}_i^k = \left\lceil \frac{a_i}{\Pi_{j=k}^K m^j} \right\rceil & \forall k = (k_i^{max}+1),...,K
\end{cases}
\qquad (6)
$$

OA&M and the "Household Grouping" Rule: PON deployment strategies driven by CAPEX costs can lead to an important scattering of the splitters among the eligible

[2] "last" is to be defined demand node by demand node, according to the demand value.

storing sites. Be they CAPEX costs optimum, those deployment schemes should be avoided regarding the future maintenance costs (technicians rounds). Therefore, considering PON deployments, network managers push for the following requirement: a site can be chosen for storing splitters of a given level only if these installed splitters "deliver" fibers for a minimum number of households (called household grouping threshold for this level). Let us introduce the household grouping thresholds for level $k = 1..K$, denoted by HG^k and let $\mathbf{v}_i^k$ be a binary variable equal to 1 if the site $i \in \mathscr{S}^k$ is opened to splitters of level k, 0 otherwise. The OA&M household grouping constraints can thus be formulated by means of the following logical constraints:

$$\left(\mathbf{v}_i^k = 1\right) \Rightarrow \left(HG^k \leq \prod_{j=k}^{K} m^j \mathbf{z}_i^k\right) \forall k = 1,...,K, \forall i \in \mathscr{S}^k \tag{7}$$

Proposition 1. *Constraints (7) can be linearized as follows:*

$$\begin{cases} \mathbf{z}_i^k \leq M_i^k \mathbf{v}_i^k & \forall k = 1..K,\ i \in \mathscr{S}^k \text{ , with } M_i^k = \begin{cases} \sum\limits_{j \neq i} b_{ij},\ k = 1 \\ m^{k-1} M_i^{k-1},\ \forall k \geq 2 \end{cases} \\ \left\lceil \dfrac{HG^k}{\prod_{j=k}^{K} m^j} \right\rceil \mathbf{v}_i^k \leq \mathbf{z}_i^k\ \ \forall k = 1..K,\ i \in \mathscr{S}^k \end{cases} \tag{8}$$

Furthermore, such constraints can lead to the installation of "artificial" splitters, these splitters being unused but installed simply in order to satisfy the previous constraints. We thus have to introduce the following constraints:

$$\mathbf{u}_i^k \leq m^k - 1 \forall k = 1,...,K, \forall i \in \mathscr{S}^k \tag{9}$$

In the following, $\mathscr{PON}_{OAM}^K$ refers to the $\mathscr{PON}^K$ model with (6), (8) and (9).

3 Numerical Tests

Tests are performed on 10 real-life instances. These are selected so as to be representative of two types of areas where FTTH PON deployment is impending: first local areas of very high density of population (Net_1-Net_5) and second, local areas of moderate density of population (Net_6-Net_{10}). Features of these instances are synthesized in columns "instance" of Table 1. Objectives of these tests are first to assess the efficiency of branch and bound approaches for $\mathscr{PON}_{OAM}^K$. Then, we give an insight into the CAPEX overcosts implied by OA&M. Note that CPLEX is used for MIP solving.

3.1 Numerical Results

In this section, results are presented for the whole set of instances, on the basis of a 2-level architecture with $m^1 = m^2 = 8$. The computation time limit is set to 1 hour. Furthermore, we instantiate OA&M rules as follows: first, the "splitter delocation" rule with the thresholds set $\{e_0 = e_1 = +\infty,\ e_2 = 1\}$, which means that all the level-2 splitters are forced to be located at the demand node they serve, and that the location of

Table 1. Computational results

Instance				PON^K problem			PON^K_{OAM} problem			Overcost (%)
Name	$\|V\|$	$\|E\|$	Demand	UB^K	LB^K	gap^K	UB^K_{OAM}	LB^K_{OAM}	gap^K_{OAM}	$(\frac{UB^K_{OAM}-LB^K}{UB^K_{OAM}})$
Net_1	52	65	1828	**32933**	32933	0.00	**36294**	36294	0.00	9.26
Net_2	235	313	12604	243182	242711	0.19	256723	256414	0.12	5.46
Net_3	322	421	16068	315282	314503	0.25	333294	332362	0.28	5.64
Net_4	392	537	22326	440180	438797	0.31	461903	460952	0.21	5.00
Net_5	229	297	11946	230444	230010	0.19	243974	243667	0.13	5.72
Net_6	602	730	8778	185222	183494	0.93	208813	204615	2.01	12.13
Net_7	234	297	2345	48593	48199	0.81	56011	55481	0.95	13.95
Net_8	955	1173	14000	296608	293222	1.14	334011	328842	1.55	12.21
Net_9	449	562	5652	115188	114030	1.01	130866	129241	1.24	12.87
Net_10	117	160	1147	22948	22731	0.94	**25029**	25029	0.00	9.18

the level-1 splitters remains unconstrained (whatever the demand value); second, the "household grouping" rule with the set $\{HG^1 = 200, HG^2 = 1\}$. As for results, three indicators are given in each column "PON^K_{OAM} problem" and "PON^K problem" (given as a reference): "UB" refers to the best solution found and "gap" (in %) to the relative difference between the best upper and lower bound (noted "LB").

Our first observation is that branch and bound approaches prove efficient for solving $\mathscr{PON}^K_{OAM}$ (mean final gap of 0.6%), despite hardly ever proving the optimality of the best solution found (only on the smallest instance of each category). Second, results suggest that the introduction of OA&M constraints has little impact on practical complexity (the two problems having the same theoretical one). Concerning overcosts, we observe a mean "Max overcost" of 9.1% for this OA&M setting (note that this noticeably varies with the type of area: mean ratio of 0.5). This suggests that OA&M considerations do not induce too overcostly deployments. This is analysed next.

3.2 Impact of OA&M Constraints

We base this sensitivity analysis on the OA&M setup previously defined. Computation time is limited to 10 minutes. We focus on instances Net_4 and Net_9 (1 of each category), for which we observe the evolution of the overcost with respect to (i) the architecture for the "splitter delocation" (SD) rule, (ii) the grouping threshold for the "household grouping" (HG) rule (see. respectively Figures 2 and 3). These results first enable us to highlight the partial contribution of each rule and the fact that, within the previous setting, the SD rule is far more detrimental (in terms of CAPEX overcosts) than the HG rule. Concerning the SD rule, two main observations are drawn: first, the induced overcosts are almost linear with respect to the level-2 splitters capacity; second, a drastic difference exists between the two (type of) instances, suggesting that the SD rule has a reasonable impact on overcosts for areas of high density ($< 10\%$ even when $m^2 = 32$) whereas a potentially highly detrimental one on those of lower density. This leads us to recommend further analysis before imposing it for operational deployments on areas of moderate population density. Concerning the HG rule, we first highlight the moderate we first highlight the moderate overcosts induced, whatever the type of instance ($< 10\%$

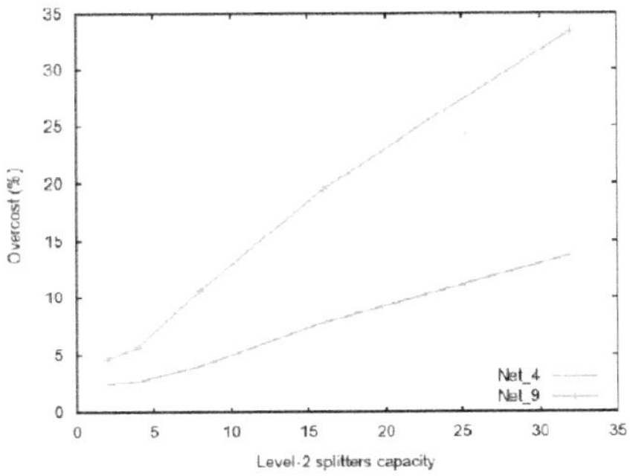

Fig. 2. Impact of the SD rule according to PON architectures (the cumulative splitting ratio is 1:64 and the HG rule is relaxed)

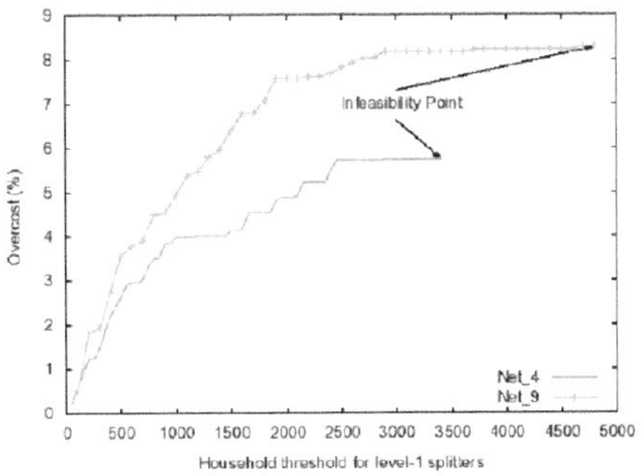

Fig. 3. Impact of the HG rule according to the household threshold of level-1 splitters (the SD rule is relaxed)

even for thresholds > 3500). Consequently, and due to the strictly concave curves, we recommend operational units not to hesitate considering quite high thresholds (between 1500 and 2000), thinking of maintenance round costs savings due to the decrease in the number of splitters storing sites. Second we stress the fact that very high thresholds lead to problem infeasibility, which was expected regarding the capacity constraints.

4 Conclusions

This paper focuses on FTTH PON design problems, which are of major importance for telecommunication operators. Mixed integer formulations have been proposed, with the aim of introducing specific constraints derived from OA&M considerations. Numerical tests performed on real-life data prove the efficiency of branch and bound approaches and a short insight into the overcosts induced by OA&M enables us to draw initial operational recommendations, whose confirmation and extension appear a natural and necessary prospect for this work.

References

1. Sherali, H.D., Lee, Y., Park, T.: New modelling approaches for the design of local access transport area networks. European Journal of Operational Research 127, 94–108 (2000)
2. Boutaba, R., Xiao, J.: Network Management: State of the Art. In: Proceedings of the World Computer Congress, pp. 127–146 (2002)
3. Lee, Y., Kim, Y., Han, J.: FTTH-PON Splitter Location-Allocation Problem. In: Proceedings of the Eighth INFORMS Telecommunications Conference (2006)
4. Li, J., Shen, G.: Cost Minimization Planning for Greenfield Passive Optical Networks. Journal of Optical Communications and Networking 1(1), 17–29 (2009)
5. Trampont, M.: Modélisation et optimisation du déploiement des réseaux de télécommunications: application aux réseaux d'accès cuivre et optique (in French). PhD Thesis, CEDRIC (2009)

6. Chardy, M., Costa, M.-C., Faye, A., Trampont, M.: Optimizing the deployment of a multilevel optical FTTH network. Submitted in EJOR (2011)
7. Chicheportiche, O.: Fibre optique : Orange et le gouvernement réaffirment leurs ambitions (in French) (2011), available via ZDNET.fr.
`http://www.zdnet.fr/actualites/fibre-optique-orange -et-le-gouvernement-reaffirment-leurs-ambitions-39758023.htm`
(cited February 4, 2011)

Introducing the Virtual Network Mapping Problem with Delay, Routing and Location Constraints

Johannes Inführ and Günther R. Raidl

Institute of Computer Graphics and Algorithms, Vienna University of Technology, Favoritenstraße 9-11, 1040 Vienna, Austria
{infuehr,raidl}@ads.tuwien.ac.at

Abstract. Network virtualization is a main paradigm of Future Internet research. It allows for automatic creation of virtual networks with application specific resource management, routing, topology and naming. Since those virtual networks need to be implemented by means of the underlying physical network, the Virtual Network Mapping Problem (VNMP) arises. In this work, we introduce the Virtual Network Mapping Problem with Delay, Routing and Location Constraints (VNMP-DRL), a variant of the VNMP including some practically relevant aspects of Virtual Network Mapping that have not been considered before. We describe the creation of a benchmark set for the VNMP-DRL. The main goal was to include VNMP-DRL instances which are as realistic as possible, a goal we met by using parts of real network topologies to model the physical networks and by using different classes of virtual networks to model possible use-cases, instead of relying on random graphs. As a first approach, we solve the VNMP-DRL benchmark set by means of a multicommodity flow integer linear program.

1 Introduction

Network virtualization has been identified as a main paradigm of Future Internet research [3,4] because it helps to overcome the ossification of the internet [15]. In this context, ossification means that it is very hard or even impossible to replace or fundamentally change a widespread technology, such as internet protocols. With the help of virtualization, such changes can be implemented in an incremental and non-disruptive manner. Another viewpoint is that virtualization is not only useful to switch technologies, but allows the coexistence of different technologies with different tradeoffs, each targeting a different user group. Network virtualization techniques have already been successfully used in scientific network testbeds such as GENI [2], PlanetLab [8] or G-Lab [17]. Its area of application is the splitting of a shared underlying network infrastructure (substrate) into virtual networks (slices), which are under the full control of different research groups for their experiments. The properties of the slices can be controlled by the experimenters and are not simply the result of the used substrate network. It is even possible to implement custom resource management, routing and naming on a per slice basis. One main idea of Future Internet research is that these mechanisms can be directly transferred to the internet, to be able to create application specific virtual networks, specifically tailored to each application. Network virtualization also enables

J. Pahl, T. Reiners, and S. Voß (Eds.): INOC 2011, LNCS 6701, pp. 105–117, 2011.

application specific choice of internet service provider, depending on the network characteristics of those providers and the application's requirements.

In such a scenario, the question naturally arises of how to map a set of virtual networks, each with its specific performance requirements, onto the existing network, which is the core of the Virtual Network Mapping Problem (VNMP). As is often the case, the problems are hidden in the details. For the VNMP, there is no standard set of requirements of virtual networks or properties of the substrate network, or even a clearly specified aim. For our work, we use the common properties of bandwidth (supplied by links of the substrate network and required by links of virtual networks) and CPU power (supplied by the substrate nodes and required by the virtual network nodes to implement custom protocols). In addition, we use communication delay on arcs (transporting data across a link in the substrate network incurs a delay and each virtual link has a specified maximum allowed value for such delay) and routing capacity on nodes (in most cases, routers can not route the full bandwidth with which they are connected). One additional class of constraints we consider is the possible placement of virtual nodes. In practice, users of a virtual network are located at specific positions in the substrate network and cannot be relocated to positions where it would be more suitable. Our goal will be to use the cheapest subset of substrate resources to satisfy all virtual network demand. We call this problem the Virtual Network Mapping Problem with Delay, Routing and Location constraints (VNMP-DRL).

Formally, the VNMP-DRL is defined as follows: We are given a directed multigraph $G = (V,A)$ with node set V and arc set A representing the substrate network. Additionally given is the available CPU power of a substrate node $c_i \in \mathbb{N}^+$, $\forall i \in V$, the amount of bandwidth units that can be routed by a substrate node $r_i \in \mathbb{N}^+$, $\forall i \in V$, the cost of using a substrate node as host for virtual nodes $p_i^V \in \mathbb{N}^+$, $\forall i \in V$, the delay of a substrate arc $d_e \in \mathbb{N}^+$, $\forall e \in A$, the available bandwidth of a substrate arc $b_e \in \mathbb{N}^+$, $\forall e \in A$ and the cost of using a substrate arc to implement virtual connections $p_e^A \in \mathbb{N}^+$, $\forall e \in A$. The slices are given by (the components of) the directed graph $G' = (V',A')$ (the virtual network graph), with node set V' and arc set A'. Associated with the slices is the required CPU power by a virtual node $c_k \in \mathbb{N}^+$, $\forall k \in V'$, the allowed delay on the path implementing a virtual connection $d_f \in \mathbb{N}^+$, $\forall f \in A'$ and the required bandwidth on the path implementing a virtual connection $b_f \in \mathbb{N}^+$, $\forall f \in A'$. The set $M \subseteq V' \times V$ defines the allowed mappings between virtual and substrate nodes. The functions $s : A \cup A' \to V \cup V'$ and $t : A \cup A' \to V \cup V'$ associate each arc of G and G' with their source and target nodes respectively. The objective is to find an assignment of the virtual nodes to substrate nodes (subject to the allowed mappings and CPU constraints) and for each virtual arc a path in the substrate network from the location of the virtual source node to the location of the virtual target node in the substrate network (subject to routing, bandwidth and delay constraints), so that the total cost of used substrate nodes and arcs is minimized.

2 Related Work

On the topic of network virtualization, its application and available technologies, see [6] for a survey. The VNMP has been solved in diverse variants [7,10,11,14,16,19,20,22] and under various names (Virtual Network Mapping, Virtual Network Assignment, Network Testbed Mapping) in the literature. The solution methods that have been applied to VNMP variants include (quadratic) mixed integer programming [7,10,14], approximation algorithms [10], simulated annealing [16], distributed algorithms [11], multicommodity flow algorithms [19,20] or algorithms especially tailored to the considered problem variant [22].

One type of virtual network demand considered by nearly all works is the required bandwidth, but how it is taken into account varies. One method is to use traffic bounds to describe a whole range of bandwidth requirements that all have to be feasibly routed (e.g. [10,14]), another is to specify the node-to-node communication demand in the form of a traffic matrix (e.g. [19]). If another requirement is taken into account, it is usually the required CPU processing power of each virtual node (e.g. [20]).

In most cases, the aim of optimization is a tradeoff between the cost of the mapping and load balancing on the substrate nodes ([7,11]) but also other aspects are taken into consideration, such as reliability requirements [20], configuration costs [14] or node and link stress (the amount of virtual nodes or links mapped to a single substrate node or link) [22]. The considered substrate sizes vary between 20 [14] and 100 [22] nodes and are either real topologies or generated by tools such as GT-ITM [21]. The requested virtual networks are mostly random graphs and consist of about ten nodes. All the cited works use undirected or directed graphs to model the substrate and virtual networks, to the best of our knowledge this is the first work to consider substrate multigraphs. We chose multigraphs because there may be multiple connections between nodes of the substrate networks with different characteristics, and one has to be able to represent those in a natural manner.

3 The Model

In this section we present a multicommodity flow based mixed integer programming formulation for the VNMP-DRL. It utilizes the decision variables $x_{ki} \in \{0,1\}$, $\forall k \in V'$, $\forall i \in V$ to indicate where the virtual nodes are located in the substrate graph and $y_e^f \in \{0,1\}$, $\forall f \in A'$, $\forall e \in A$ to indicate if a virtual connection is implemented by using a substrate connection. Further auxiliary decision variables are $z_i^f \in \{0,1\}$, $\forall f \in A'$, $\forall i \in V$ to indicate that a substrate node is used to route a virtual connection, $u_i^V \in \{0,1\}$, $\forall i \in V$ to indicate that a substrate node hosts at least one virtual node and $u_e^A \in \{0,1\}$, $\forall e \in A$ to indicate that a substrate arc is used for at least one virtual connection.

Now follows the multicommoditiy flow (MCF) based integer linear programming model we use to solve the VNMP-DRL.

$$\text{(FLOW)} \quad \min \quad \sum_{i \in V} p_i^V u_i^V + \sum_{e \in A} p_e^A u_e^A \tag{1}$$

$$\sum_{(k,i) \in M} x_{ki} = 1 \qquad \forall k \in V' \tag{2}$$

$$\sum_{e \in A | t(e) = i} y_e^f + x_{s(f)i} - \sum_{e \in A | s(e) = i} y_e^f - x_{t(f)i} = 0 \qquad \forall i \in V,\ \forall f \in A' \tag{3}$$

$$\sum_{e \in A | t(e) = i} y_e^f + x_{s(f)i} \leq z_i^f \qquad \forall i \in V,\ \forall f \in A' \tag{4}$$

Table 1. Summary of the used variables, constants and functions of the MCF formulation of the VNMP-DRL ($i \in V$, $e \in A$, $k \in V'$, $f \in A'$, $l \in A \cup A'$)

Symbol	Meaning	Symbol	Meaning	Symbol	Meaning
$G(V,A)$	Substrate graph	$G'(V',A')$	Virtual graph	x_{ki}	Map node k to i
c_i	Av. CPU	c_k	Req. CPU	y_e^f	Use arc e for f
d_e	Delay	d_f	Max. allowed delay	z_i^f	Use node i for f
b_e	Av. bandwidth	b_f	Req. bandwidth	u_i^V	Use node i
r_i	Av. routing capacity	M	Set of allowed mappings	u_e^A	Use arc e
p_i^V	Node price	$s(l)$	Source node of arc l		
p_e^A	Arc price	$t(l)$	Target node of arc l		

$$\sum_{(k,i)\in M} c_k x_{ki} \le c_i \qquad \forall i \in V \tag{5}$$

$$\sum_{f\in A'} b_f z_i^f \le r_i \qquad \forall i \in V \tag{6}$$

$$\sum_{f\in A'} b_f y_e^f \le b_e \qquad \forall e \in A \tag{7}$$

$$\sum_{e\in A} d_e y_e^f \le d_f \qquad \forall f \in A' \tag{8}$$

$$x_{ki} \le u_i^V \qquad \forall i \in V,\ \forall k \in V' \tag{9}$$

$$y_e^f \le u_e^A \qquad \forall e \in A,\ \forall f \in A' \tag{10}$$

$$x_{ki} = 0 \qquad \forall (k,i) \in (V' \times V)\setminus M \tag{11}$$

$$x_{ki} \in \{0,1\} \qquad \forall (k,i) \in M \tag{12}$$

$$y_e^f \in \{0,1\} \qquad \forall e \in A,\ \forall f \in A' \tag{13}$$

$$z_i^f \in \{0,1\} \qquad \forall i \in V,\ \forall f \in A' \tag{14}$$

Equalities (2) ensure that each virtual node is mapped to exactly one substrate node, subject to the mapping constraints. The flow conservation constraints (3) make sure that for each virtual connection there is a connected path in the substrate network. Linking constraints (4) make certain that variables z_i^f are equal to one when the corresponding node is used to route the traffic of a particular virtual connection. Inequalities (5)–(8) ensure that the solutions are valid with regard to CPU, routing capacity, bandwidth and delay constraints. Linking constraints (9) and (10) force variables u_i^V and u_e^A to be (at least) one when the corresponding substrate node or arc is used by any virtual node or arc. Constraints (11) exclude any forbidden mappings from the solution. Note that while the model only includes integrality constraints for x_{ki}, y_e^f and z_i^f (12)–(14), constraints (9) and (10) together with the objective function (1) also cause variables u_i^V and u_e^A to be integral (and binary). Table 1 gives a short reference of the used variables, constants and functions.

4 Generating the Benchmark-Instances

This section describes how the VNMP-DRL benchmark set was created, by first illustrating how the components of a benchmark instance were created and then how they were combined to form a complete instance. The main goal was to create hard VNMP-DRL instances which are as realistic as possible, to serve as a common basis for the comparison of different VNMP-DRL solution approaches. The benchmark set can be obtained at [13].

4.1 Substrate Network

The substrate networks were based on real internet topology maps. The base data came from the Rocketfuel project [18] and the scan-lucent map (the union of the topologies measured by the SCAN [9] and Lucent [5] internet mapping projects). To create networks of the required size, nem-0.9.6 was used to extract subgraphs which retain the main characteristics of the source graph. In the cases where that was not possible (when the required network size exceeded 30 percent of the source network size), nodes with in- and out-degree of at most one were randomly deleted until the target size was reached. In absence of such nodes, random nodes were deleted. No measures to ensure connectivity were taken because the source networks were not always connected themselves.

Since the Rocketfuel graphs (rf) have assigned latencies to their arcs, those were used as delay values d_e. For the graphs generated from the scan-lucent (sl) map delay values were chosen uniformly at random between 1 and 10. Arcs which were the only connection of a node were assigned a bandwidth b_e of 25. The other arcs were assigned a bandwidth value of 25 times the minimum of the in-degree of their source node and the out-degree of their target node, but at least 25. The routing capacity r_i of a node was calculated as the minimum of the sum of the incoming and outgoing bandwidth. The available CPU capacity of a node c_i was set to be the same as the routing capacity. The costs of nodes and edges are chosen uniformly at random between 1 and 20.

Figure 1 shows a generated substrate graph based upon the rf1221 topology of size 20.

4.2 Slices

In this work, we used four different slice-types to represent possible use-cases in the virtual network setting with different sets of requirements regarding needed bandwidth, needed processing power per node and maximum delay. Those types were web slices to represent general http-traffic, stream slices to represent video streaming, P2P slices to represent P2P networks and VoIP slices to represent voice chat.

Web Slice. The general characteristics of web slices were chosen to be low bandwidth requirements, short delays (for fast responses) and no special CPU requirements. Web slices were modeled by a star graph, where the central node represented a web server and the leafs were the users. All edges were assigned a bandwidth requirement of 1 and a maximum allowed delay of 25. All nodes were assigned a CPU requirement of 1, except

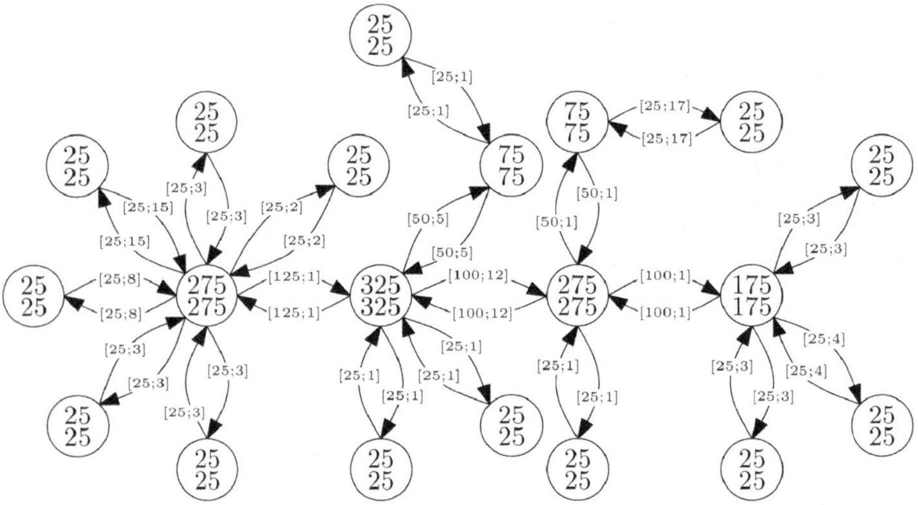

Fig. 1. Example substrate graph of size 20, generated from the rf1221 topology map. The node labels are the routing capacity r_i and the available CPU c_i, the arc labels are the bandwidth b_e and the delay values d_e.

the root node which was assigned the sum of the outgoing bandwidth as CPU requirement. The allowed mapping calculated for web slices placed the leaf nodes of the star at a random location at the edge of the substrate network (which was defined as the set of nodes with minimal degree), while the central node was placed at a random location at the core of the network (which was defined as not being the edge). Figure 2a shows a generated web slice of size 5.

Stream Slice. The general characteristics of stream slices were chosen to be medium to high bandwidth requirements, no relevant delay bounds and 3 units of CPU processing power per routed bandwidth. The idea of the stream slices was to use a random tree graph, where the root node is the source of the video stream, the leafs are the customers receiving specific channels of the video stream and the intermediate nodes split the video stream and forward only the channels which are watched by the customers. The stream splitting is the reason for the high CPU requirements in relation to bandwidth units. This is an example of a customized routing protocol which delivers more features than currently possible. The delay bound of the arcs of the stream slices was set to 1000, which effectively means that they are not relevant. The number of channels in the video stream was chosen uniformly at random between 10 and 20, while the total bandwidth requirement of the stream was chosen from a discrete $N(5,1)$ distribution, but at least 3 and at most 7. Each child node in the stream network only received a random fraction of the channels the parent received (between 0.3 and 1), but it was made sure that all channels that a node received were forwarded. The bandwidth requirements of each arc were set to $\lceil(\text{bandwidthPerChannel} * \text{forwardedChannels}\rceil$ and the CPU requirement of each node to three times the received bandwidth. The calculated allowed mapping for stream slices placed the root node at a random location at the core of the substrate network and the leaf nodes at a random location at the edge of the substrate network.

The leaf nodes were placed in a way so that the distance between siblings was less or equal to four hops in the substrate network. The placement of the intermediate node was not constrained, i.e. they were allowed to be mapped anywhere. Figure 2b shows a generated stream slice of size 5.

P2P Slice. The general characteristics of P2P slices were chosen to be medium bandwidth requirements, no relevant delay bounds and medium CPU requirements. The network structure of a P2P slice was generated by the small_world_iterator of the boost graph library [1], after which every arc was duplicated and reversed. The bandwidth requirement of each arc was chosen uniformly at random between 1 and 3, the delay bound was set to 1000 and the CPU requirement of the nodes was chosen uniformly at random between 1 and 5. The CPU requirement was chosen independent of the routed bandwidth to model that some traffic may be encrypted or compressed and therefore has a higher computational demand. All nodes of the P2P slice were only allowed to be mapped to one random node at the edge of the substrate network. Figure 2c shows a generated P2P slice of size 5.

VoIP Slice. The general characteristics of VoIP slices were chosen to be medium bandwidth requirements, medium delays and high CPU requirements. The networks structure of VoIP slices was generated in the same way as P2P slices. The bandwidth requirement of each arc was chosen uniformly at random between 1 and 3 and the delay bound was set to 50. The CPU requirement was set to the minimum of incoming and outgoing bandwidth, so that "super-nodes" (which route a lot of VoIP traffic) have high CPU requirements. Figure 2d shows a generated VoIP slice of size 5.

4.3 VNMP-DRL Instance

To generate a complete VNMP-DRL instance, first a specific topology map is used to create a substrate graph (in conjunction with its associated costs) of the required size. Then the virtual graph is build by adding slices (of random type and size between 10 and 20 percent of the substrate graph size) to the virtual graph, until the problem becomes "too hard" (see next paragraph). If that happens, the last added slice is removed and a new one is generated and added. If this process fails to find an addable slice 40 times in a row, the generation process is finished. From this generated instance, five variants are constructed by only using 50 to 90 percent of the added slices (in ten percent increments, each variant includes all the slices used by a smaller variant), so the complete generation process creates six problem instances of incremental difficulty.

We tried different criteria to define "too hard", for instance only solving the LP-relaxation of FLOW and rejecting virtual graphs which cause the relaxation to be unsolvable or only calculating the root node of the branch-and-bound tree and rejecting virtual graphs which are by then proven to be unsolvable. However, preliminary runs showed that the instances that were generated in this way were either extremely easy to solve to optimality (without branching) or provably unsolvable. Especially for larger instance sizes the fraction of generated unsolvable instances became unmanageable. The hardness definition we used in the end was that if CPLEX 12.2 [12] is not able to find an integer solution after 300 seconds once in five tries, the virtual graph is rejected as

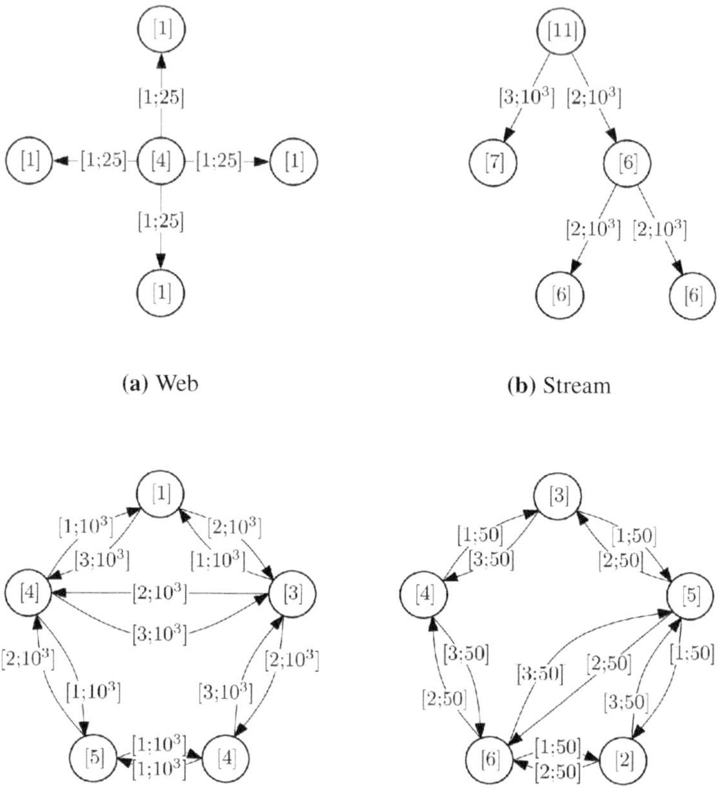

(a) Web **(b)** Stream

(c) P2P **(d)** VoIP

Fig. 2. Examples of generated slices of size 5

"too hard". In each of the five trials, the sequence in which the slices are added to the virtual graph is permuted, because preliminary runs showed that in some cases a bad sequence can double the time needed to solve the LP relaxation and that time is missed afterwards when trying to find an integer solution.

The used substrate sizes for instance creation were 20, 30, 40, 50, 70 and 100. For each of the seven topology maps (six rf and one sl) the problem generation procedure described previously was used ten times, so in total 420 instances per size were created, with the exception of substrate size 100, for which only 300 instances were created because two of the rocketfuel topology maps contain less than 100 nodes.

5 Computational Results

All results presented in this section have been achieved by using CPLEX 12.2 to solve the (FLOW) problem. Each computational experiment has been performed on one core of an Intel Xeon E5540 multi-core system with 2.53 GHz and 3 GB RAM per core. A time limit of 10000 seconds was used. The reported gaps are the optimality gaps calculated by CPLEX.

5.1 Created Benchmark-Set

The basic properties of the generated benchmark instances are shown in Table 2. It can be seen that while the substrate grows, the number of slices contained in the virtual network graph does not grow and for the sizes 70 and 100 is even less than the number of slices of instances of size 20. The total size of the virtual network however does not shrink. From this we can conclude that because larger instances include larger slices (keep in mind that slices are created with a size between 10 and 20% of the substrate size), fewer slices can be mapped onto the substrate network, even though it is bigger and should be able to carry more slices. Also noteworthy is the slice composition development. From the description of the instance generation procedure, one would expect an equal number of slice types in every virtual network graph. It seems to be the case that it is harder to find P2P and VoIP slices that can be mapped, than it is for the other slice types. For VoIP slices, this effect gets worse as the substrate size grows. Also finding mappable Web slices gets harder with increasing substrate size. The reason for this behaviour could be, that Stream slices are not delay bound, while Web and VoIP slices are, so when the substrate grows, so does the average delay between two points at the edge of the substrate network and so paths within the delay bounds become more scarce.

The influence of the choice of topology source for the substrate network on the generated benchmark instances can be seen in Table 3. Note that the average substrate size for the rf1775 and rf3967 instances is smaller than those of the other instances because those topologies were too small to create instances of size 100. It can be seen that the sl instances are the source of the sparsest substrate graphs in the benchmark set. This leads to the highest number of slices in the virtual network graph, but web and stream slices

Table 2. Overview of the properties of the created benchmark instances: Average size of substrate graph $G = (V, A)$, average size of virtual network graph $G' = (V', A')$, average number of slices contained in G' (#S) and average fraction of slice types contained in the virtual network graph (Web, Stream, P2P, VoIP)

Size	$\overline{\lvert V \rvert}$	$\overline{\lvert A \rvert}$	$\overline{\lvert V' \rvert}$	$\overline{\lvert A' \rvert}$	$\overline{\#S}$	$\overline{\text{Web}}$	$\overline{\text{Stream}}$	$\overline{\text{P2P}}$	$\overline{\text{VoIP}}$
20	20	53.3	72.4	81.9	14.5	0.36	0.36	0.13	0.15
30	30	87.3	101.1	117.9	20.2	0.31	0.37	0.19	0.14
40	40	126.9	104.8	145.7	18.7	0.28	0.42	0.18	0.12
50	50	172.9	111.8	174.2	16.7	0.27	0.44	0.19	0.11
70	70	253.1	100.8	140.9	10.5	0.25	0.54	0.14	0.07
100	100	386.4	119.0	156.9	8.7	0.19	0.59	0.16	0.06

Table 3. Influence of the chosen topology source on the created benchmark instances: Average size of substrate graph $G = (V,A)$, average size of virtual network graph $G' = (V',A')$, average number of slices contained in G' (#S) and average fraction of slice types contained in the virtual network graph (Web, Stream, P2P, VoIP)

Top. source	$\overline{\lvert V \rvert}$	$\overline{\lvert A \rvert}$	$\overline{\lvert V' \rvert}$	$\overline{\lvert A' \rvert}$	$\overline{\#S}$	$\overline{\text{Web}}$	$\overline{\text{Stream}}$	$\overline{\text{P2P}}$	$\overline{\text{VoIP}}$
rf1221	51.7	164.7	120.5	163.3	17.4	0.34	0.43	0.12	0.11
rf1239	51.7	202.0	75.8	95.3	11.1	0.35	0.44	0.10	0.12
rf1755	42.0	141.2	108.6	147.0	18.4	0.31	0.42	0.14	0.13
rf3257	51.7	197.4	76.5	112.9	11.3	0.29	0.38	0.19	0.14
rf3967	42.0	140.4	93.7	141.6	16.1	0.07	0.56	0.26	0.12
rf6461	51.7	207.3	73.3	113.1	11.0	0.10	0.54	0.29	0.07
sl	51.7	125.0	157.2	176.4	21.7	0.47	0.37	0.09	0.07

are highly overrepresented. Also note that it seems to be very hard to find mappable web slices when using rf3967 and rf6461 as substrate topology source, even though it is not problematic for all other topology sources.

5.2 Solving the VNMP-DRL

The results of solving the benchmark instances with the FLOW formulation are summarized in Table 4. It can be seen that in general, the generated instances are not very hard to solve, with at least 74.3% instances solved to optimality per substrate size. Interestingly, the hardest instances turned out to be those of size 50 (or 70 when regarding the average gap). In total, the results show that up to substrate sizes of 100 nodes it is possible to solve the VNMP-DRL to proven optimality within one hour on average.

Table 5 shows the influence of the chosen source topology on the solution characteristics of the instances. Unsurprisingly, instances based on the sl topology map are the easiest to solve, as the substrate graphs are very sparse, which reduces the routing and mapping possibilities. But also the rf instances vary a lot, which indicates that the hardness of the VNMP-DRL is very sensitive to the choice of the substrate topology.

Table 4. Results of solving the VNMP-DRL with the FLOW formulation: Number of instances (# Inst.), fraction of instances solved to optimality (# Opt [%]), average gap, average number of branch-and-bound nodes and average required CPU time

Size	# Inst.	# Opt [%]	$\overline{\text{gap}}$ [%]	$\overline{\text{BB-Nodes}}$	$\overline{\text{t}}$ [s]
20	420	100.0	0.00	53.5	8
30	420	93.8	0.12	1352.3	758
40	420	85.7	0.36	1955.3	1842
50	420	74.3	0.52	1176.8	3320
70	420	82.9	0.60	420.0	2465
100	300	90.7	0.20	218.1	1859

Table 5. Influence of the substrate topology source: Number of instances (# Inst.), fraction of instances solved to optimality (# Opt [%]), average gap, average number of branch-and-bound nodes and average required CPU time

Top. source	# Inst.	#Opt [%]	gap [%]	BB-Nodes	t [s]
rf1221	360	87.8	0.19	961.2	1781
rf1239	360	94.2	0.14	193.0	902
rf1755	300	80.7	0.70	2336.7	2736
rf3257	360	85.3	0.29	400.9	2087
rf3967	300	73.7	0.79	2153.1	3145
rf6461	360	89.2	0.17	644.8	1650
sl	360	100.0	0.00	24.7	20

Table 6. Influence of pS: Number of instances (# Inst.), fraction of instances solved to optimality (# Opt [%]), average gap, average number of branch-and-bound nodes and average required CPU time

pS	# Inst.	#Opt [%]	gap [%]	BB-Nodes	t [s]
0.5	400	97.2	0.05	547.1	555
0.6	400	95.5	0.07	766.8	862
0.7	400	91.2	0.17	912.7	1449
0.8	400	85.8	0.33	941.3	1935
0.9	400	81.8	0.44	1026.1	2397
1	400	75.0	0.77	1175.4	3010

The influence of the fraction of the set of slices (denoted as pS), created during instance creation, on the hardness of the created instances is presented in Table 6. It can be seen that the complexity of the instances can be selected very well by an appropriate choice of pS. From another point of view this means that the more slices are added, the harder the instance gets. This seems like an obvious statement, but there were some instances in the benchmark set for which the pS=0.8 or pS=0.9 variants were harder to solve than the pS=1 variant, probably because the pS=1 variants were so densely packed with slices that the number of routing possibilities was greatly reduced.

6 Conclusion and Future Work

In this work, we introduced the Virtual Network Mapping Problem with Delay, Routing and Location constraints (VNMP-DRL). We presented a method for creating realistic benchmark instances for this problem and solved those instances by means of a multi-commodity flow based integer linear program. Even this simple approach was able to solve more than 74% of problem instances to proven optimality in less than one hour on average. The biggest influence on instance hardness was shown to be the topology map used to create the instance. Larger problem instances were not harder to solve than smaller instances on average, instances of size 100 were about as easily solved as instances of size 40. We were able to show that the fraction of slices used of the complete set of slices found during instance creation can be used to control instance hardness.

Future work will include improvements of the instance creation method, so that larger instances are actually harder than smaller instances and scaling the substrate network size up to 1000 nodes. Another venue is the application of more advanced ILP solution methods like branch-and-price to improve solution time and quality and eventually of heuristic methods to solve large problem instances. One simplifying assumption of VNMP-DRL was that the set of slices is known in advance, so future research could target the online aspect of this problem. Another interesting research direction could be analyzing the effect of rising delays with rising link utilization, which is currently not modelled.

Acknowledgement

We would like to thank Kurt Tutschku and David Stezenbach for their invaluable support and helpful discussions. This research was funded by the WWTF, project number ICT10-027.

References

1. Boost C++ Libraries, `http://www.boost.org`
2. GENI.net Global Environment for Network Innovations, `http://www.geni.net`
3. Anderson, T., Peterson, L., Shenker, S., Turner, J.: Overcoming the Internet impasse through virtualization. Computer 38(4), 34–41 (2005)
4. Berl, A., Fischer, A., de Meer, H.: Virtualisierung im future internet. Informatik-Spektrum 33, 186–194 (2010)
5. Burch, H., Cheswick, B.: Mapping the internet. Computer 32(4), 97–98, 102 (1999)
6. Chowdhury, N.M.K., Boutaba, R.: A survey of network virtualization. Computer Networks 54(5), 862–876 (2010)
7. Chowdhury, N., Rahman, M., Boutaba, R.: Virtual network embedding with coordinated node and link mapping. In: INFOCOM 2009, pp. 783–791. IEEE, Los Alamitos (April 2009)
8. Chun, B., Culler, D., Roscoe, T., Bavier, A., Peterson, L., Wawrzoniak, M., Bowman, M.: Planetlab: an overlay testbed for broad-coverage services. SIGCOMM Comput. Commun. Rev. 33, 3–12 (2003)
9. Govindan, R., Tangmunarunkit, H.: Heuristics for internet map discovery. In: Proceedings of Nineteenth Annual Joint Conference on the IEEE Computer and Communications Societies, INFOCOM 2000, vol. 3, pp. 1371–1380. IEEE, Los Alamitos (March 2000)
10. Gupta, A., Kleinberg, J., Kumar, A., Rastogi, R., Yener, B.: Provisioning a virtual private network: a network design problem for multicommodity flow. In: Proceedings of the Thirty-Third Annual ACM Symposium on Theory of Computing, STOC 2001, pp. 389–398. ACM, New York (2001)
11. Houidi, I., Louati, W., Zeghlache, D.: A distributed virtual network mapping algorithm. In: IEEE International Conference on Communications, ICC 2008, pp. 5634–5640 (May 2008)
12. IBM ILOG: CPLEX 12.2, `http://www-01.ibm.com/software/integration/optimization/cplex-optimizer`
13. Inführ, J., Raidl, G.R.: The VNMP-DRL benchmark set, `http://www.ads.tuwien.ac.at/projects/optFI/`
14. Lu, J., Turner, J.: Efficient mapping of virtual networks onto a shared substrate. Tech. rep., Washington University in St. Louis (2006)

15. National Research Council: Looking Over the Fence at Networks. National Academy Press (2001)
16. Ricci, R., Alfeld, C., Lepreau, J.: A solver for the network testbed mapping problem. SIGCOMM Comput. Commun. Rev. 33(2), 65–81 (2003)
17. Schwerdel, D., Günther, D., Henjes, R., Reuther, B., Müller, P.: German-lab experimental facility. In: Berre, A., Gémez-Pérez, A., Tutschku, K., Fensel, D. (eds.) FIS 2010. LNCS, vol. 6369, pp. 1–10. Springer, Heidelberg (2010)
18. Spring, N., Mahajan, R., Wetherall, D.: Measuring ISP topologies with rocketfuel. In: Proceedings of the 2002 Conference on Applications, Technologies, Architectures, and Protocols for Computer Communications, SIGCOMM 2002, pp. 133–145. ACM, New York (2002)
19. Szeto, W., Iraqi, Y., Boutaba, R.: A multi-commodity flow based approach to virtual network resource allocation. In: Global Telecommunications Conference, GLOBECOM 2003, vol. 6, pp. 3004–3008. IEEE, Los Alamitos (2003)
20. Yeow, W.L., Westphal, C., Kozat, U.: Designing and embedding reliable virtual infrastructures. In: Proceedings of the Second ACM SIGCOMM Workshop on Virtualized Infrastructure Systems and Architectures, VISA 2010, pp. 33–40. ACM, New York (2010)
21. Zegura, E., Calvert, K., Bhattacharjee, S.: How to model an internetwork. In: Proceedings IEEE of Fifteenth Annual Joint Conference of the IEEE Computer Societies, Networking the Next Generation, INFOCOM 1996, vol. 2, pp. 594–602 (March 1996)
22. Zhu, Y., Ammar, M.: Algorithms for assigning substrate network resources to virtual network components. In: Proceedings of 25th IEEE International Conference on Computer Communications, INFCOM 2006, pp. 1–12 (2006)

Cutset Inequalities for Robust Network Design

Arie M.C.A. Koster[1], Manuel Kutschka[1], and Christian Raack[2]

[1] RWTH Aachen University, Lehrstuhl II für Mathematik, Wüllnerstr. 5b, 52062 Aachen, Germany
{koster,kutschka}@math2.rwth-aachen.de
[2] Zuse Institute Berlin (ZIB), Takustr. 7, 14195 Berlin, Germany
raack@zib.de

Abstract. Robust optimization is an emerging approach in integer programming taking data uncertainty into account. Recently, the adjustable Γ-robustness approach of Bertsimas and Sim [3, 4] has been applied to network design problems. In this paper, we consider this so-called Γ-robust network design problem. We investigate its polyhedral structure and present robust versions of the cutset inequalities as well as computational results on their effectiveness.

1 Introduction

The classical network design problem aims at finding a cost-minimal network configuration, i. e., integer link capacities and a routing hosting all traffic demands. Given one traffic matrix only, this problem has been studied extensively in the literature, see [14] and the references therein. Being already a complex task, it does not take traffic fluctuations in real-life networks into account. Instead of accepting expensive and resource-inefficient network designs due to (highly) overestimated traffic values, we consider in this paper the Γ-robust network design problem (Γ*-RNDP*) which has been applied successfully, e. g., by [10, 11] following [4] and [2]. In this approach, two values are given for every demand: a (nominal) default value and a peak value. Further, at most Γ demands are assumed to be at their peaks simultaneously in realistic traffic scenarios.

This paper complements [11] introducing the Γ-robust model and [10] analyzing the design w. r. t. real-life traffic measurements. Here, we focus on the effectiveness of the robust version of the well-known cutset inequalities.

In Section 2, we describe the Γ-RNDP and investigate the Robust-Γ Cutset Polyhedron in Section 3. Next in Section 4, we evaluate the effectiveness of separating valid inequalities for this polyhedron, including an exact separation for the natural extension. We close with concluding remarks in Section 5.

2 Γ-Robust Network Design

The classical network design problem is described on an undirected graph $G = (V,E)$ representing the network topology. Capacity can be installed in batches of $C > 0$ units on each link $e \in E$ with costs κ_e per batch. For every commodity k in a set K of point-to-point demands, a (bifurcated) routing has to be defined from source $s^k \in V$ to target

J. Pahl, T. Reiners, and S. Voß (Eds.): INOC 2011, LNCS 6701, pp. 118–123, 2011.

$t^k \in V$ such that its traffic volume d^k can be carried. This problem can be formulated as integer linear program:

$$\min \sum_{e \in E} \kappa_e x_e \tag{1a}$$

$$s.t. \sum_{j \in V: ij \in E} (f_{ij}^k - f_{ji}^k) = \begin{cases} 1 & i = s^k \\ -1 & i = t^k \,, \\ 0 & \text{else} \end{cases} \quad \forall i \in V, k \in K \tag{1b}$$

$$\sum\nolimits_{k \in K} d^k (f_{ij}^k + f_{ji}^k) \leq C x_e, \quad \forall e \in E \tag{1c}$$

$$f, x \geq 0, \; x \in \mathbb{Z}^{|E|} \tag{1d}$$

where x_e denotes the number of batches installed and f_{ij}^k the flow percentage.

In practice, actual traffic demands d^k fluctuate and are not known a-priori. Robust network design addresses this issue by taking data uncertainty into account. Following [3, 4], (1c) is reformulated by assuming the existence of a default value $\bar{d}^k > 0$ and a deviation $\hat{d}^k > 0$ for every $k \in K$ such that the actual demand $d^k \in [0, \bar{d}^k + \hat{d}^k]$. Note that these intervals do not have an impact compared to the original proposed symmetric intervals $[\bar{d}^k - \hat{d}^k, \bar{d}^k + \hat{d}^k]$. Given a parameter $\Gamma \in \{0, 1, \ldots, |K|\}$, the Γ-robust network design problem is to find a minimum-cost installation of capacities such that a routing exists not exceeding the link capacities if at most Γ commodities are at their peaks simultaneously. By varying Γ, we may adjust the robustness. Based on [3, 4], a *compact* linear formulation of the Γ-RNDP can be obtained by LP duality (introducing new dual variables p_e^k and π_e, cf. [11]), replacing (1c) by

$$\Gamma \pi_e + \sum\nolimits_{k \in K} \bar{d}^k (f_{ij}^k + f_{ji}^k) + \sum\nolimits_{k \in K} p_e^k \leq C x_e, \quad \forall ij \in E \tag{2a}$$

$$-\pi_e + \hat{d}^k (f_{ij}^k + f_{ji}^k) - p_e^k \leq 0, \quad \forall ij \in E, k \in K \tag{2b}$$

$$p, \pi \geq 0 \tag{2c}$$

3 Valid Inequalities

In deterministic network design, cutset inequalities are of particular importance, see [14] and the references therein. In this section, we generalize the cutset inequality to robust network design and provide a complete description in a particular case.

The Robust-Γ Cutset Polyhedron. We consider a proper and nonempty subset S of the nodes V and the corresponding cutset $\delta(S) := \{ij \in E : i \in S, j \notin S\}$ and denote by $Q_S \subseteq K$ the subset of commodities with source s^k and target t^k in different shores of the cut. W. l. o. g., we assume $Q_S \neq \emptyset$, $s^k \in S$ for all $k \in Q_S$, and denote by $\bar{d}_S := \sum_{k \in Q_S} \bar{d}^k > 0$ the aggregated *default cut-demand* with respect to S.

Let $\rho : Q_S \mapsto \{1, \ldots, |Q_S|\}$ be a permutation of the commodities $k \in Q_S$ such that $\hat{d}^{\rho^{-1}(1)} \geq \hat{d}^{\rho^{-1}(2)} \geq \ldots \geq \hat{d}^{\rho^{-1}(|Q_S|)}$. Let $J = \{-\Gamma, \ldots, |Q_S| - \Gamma\}$. We define $Q_i := \{k \in Q_S : \rho(k) \leq i + \Gamma\}$ for $i \in J$ as the commodities corresponding to the $i + \Gamma$ largest $\hat{d}^k$, and $d_i := \bar{d}_S + \hat{d}(Q_i)$. Aggregating (2a), adding (2b) for $e \in \delta(S)$ and $k \in Q_i \subseteq Q_S$, $Q \neq \emptyset$, and relaxing the backward flow variables results in

$$Cx(\delta(S)) + i\pi(\delta(S)) \geq d_i. \tag{3}$$

Next, we consider $X_\Gamma(S) = \{(x,\pi) \in \mathbb{Z}_+^{|\delta(S)|} \times \mathbb{R}_+^{|\delta(S)|} : (x,\pi) \text{ satisfies (3) } \forall i \in J\}$ and the polyhedron defined by $\text{conv}(X_\Gamma(S))$. Every valid inequality for $\text{conv}(X_\Gamma(S))$ is also valid for the Γ-robust formulation (2).

We define $r(d,c) := d - c(\lceil \frac{d}{c} \rceil - 1)$ as the remainder of the division of d by c with $r(d,c) = c$ in case c divides d. Setting $r_i := r(d_i, C)$ and applying Mixed Integer Rounding to (3) results in

$$r_i x(\delta(S)) + \max(0,i)\pi(\delta(S)) \geq r_i \left\lceil \frac{d_i}{C} \right\rceil. \tag{4}$$

In particular, for $i = 0$ this inequality reduces to

$$x(\delta(S)) \geq \left\lceil \frac{d_0}{C} \right\rceil \tag{5}$$

which defines a facet of $\text{conv}(X_\Gamma(S))$ if and only if $r_0 < C$. This generalizes the classical cutset inequality [12], stating that the capacity on the cut should be at least the default cut demand plus the Γ largest deviations among Q_S.

Introducing index sets $J_- = \{-\Gamma, \ldots, -1\}$ and $J_+ := \{1, \ldots, |Q_S| - \Gamma\}$, we consider two arbitrary base constraints corresponding to $i, j \in J_-$ (or $i, j \in J_+$) with $i < j$ in the following. The valid inequalities defined below cut off fractional points (x,π) of the LP relaxation with $x \in [\lfloor b_{i,j} \rfloor, \lceil b_{i,j} \rceil]$ where $b_{i,j} := (jd_i - id_j)/((j-i)C)$ denotes the x-value at the intersection of the two base inequalities. We define $r_{i,j} := r(jd_i - id_j, (j-i)C)$.

Lemma 1. *For $i, j \in J_-$ with $i < j$, and for $k, \ell \in J_+$ with $k < \ell$, the following inequalities are valid for* $\text{conv}(X_\Gamma(S))$*:*

$$(-Cj + r_{i,j})x(\delta(S)) - ij\pi(\delta(S)) \geq r_{i,j} \lceil b_{i,j} \rceil - jd_i \tag{6}$$

$$(Ck + r_{k,\ell})x(\delta(S)) + k\ell\pi(\delta(S)) \geq r_{k,\ell} \lceil b_{k,\ell} \rceil + kd_\ell \tag{7}$$

Only linearly many of the inequalities (6) and (7) are non-redundant. To identify these, we define the function $\pi(k,x) := (d_k - Cx)/k$ for all $k \in J_- \cup J_+$ and $x \in \mathbb{R}_+$.

Lemma 2. *Let $a \in \mathbb{Z}_+$, $i = \arg\min_{\alpha \in J_-} \pi(\alpha, a+1)$, $j = \arg\min_{\alpha \in J_-} \pi(\alpha, a)$, $k = \arg\max_{\alpha \in J_+} \pi(\alpha, a)$, and $\ell = \arg\max_{\alpha \in J_+} \pi(\alpha, a+1)$. Then the following holds*

1. *If $i \neq j$, inequality* (6) *for i, j dominates all other inequalities* (6) *on* $[a, a+1]$.
2. *If $i = j$, base inequality* (3) *dominates all inequalities* (6) *on* $[a, a+1]$.
3. *If $k \neq \ell$, inequality* (7) *for k, ℓ dominates all other inequalities* (7) *on* $[a, a+1]$.
4. *If $k = \ell$, base inequality* (3) *dominates all inequalities* (7) *on* $[a, a+1]$.

For the deterministic model, ineq. (5) completely describes the cutset polyedron.

Theorem 1. *If $|\delta(S)| = 1$, then inequalities* (3)–(7) *completely describe* $\text{conv}(X_\Gamma(S))$.

Analog to the deterministic case, it can be shown that these inequalities are facet-defining for the general robust network design polytope under mild conditions [14].

Separation. To separate inequalities (4)–(7) we consider the following heuristic dating back to [5, 9] and used by [14] for the deterministic model (1): by contracting edges with large slacks in (2a) and (2b), we shrink the network until 5 nodes or only edges with positive slacks are left. Then, we enumerate up to $|V|^2$ cuts in the resulting graph and separate violated inequalities. Further, violated inequalities are separated for all cuts $\delta(S)$ with $|S| = 1$.

In addition, we implemented an ILP based exact algorithm to separate inequalities (5). We define binary variables β_i $(i \in V)$ with $\beta_i = 1$ iff $i \in S$, α^k with $\alpha^k = 1$ iff $k \in Q_S$, γ^k with $\gamma^k = 1$ iff commodity $k \in Q_S$ deviates from its default, and z_{ij} $(ij \in E)$ with $z_{ij} = 1$ iff $ij \in \delta(S)$. In addition, let d determine the worst-case total demand value crossing the cut, and let R be the right-hand side value of the corresponding cutset inequality (5). We minimize the feasibility of (5) such that a negative objective value yields a violated cut. Then, the separation problem reads

$$
\begin{aligned}
\min\ & \textstyle\sum_{ij\in E} z_{ij} - R && && (8a)\\
s.t.\ & \max\{\beta_{s^k} - \beta_{t^k}, \beta_{t^k} - \beta_{s^k}\} & \le \alpha^k & \le \min\{\beta_{s^k} + \beta_{t^k}, 2 - \beta_{s^k} - \beta_{t^k}\} \quad \forall k \in Q && (8b)\\
& \max\{\beta_i - \beta_j, \beta_j - \beta_i\} & \le z_{ij} & \le \min\{\beta_i + \beta_j, 2 - \beta_i - \beta_j\} \quad \forall ij \in E && (8c)\\
& d/C & \le R & \le (d + C - 1)/C && (8d)\\
& \textstyle\sum_{k\in Q} \gamma^k & \le \Gamma & && (8e)\\
& & d & = \textstyle\sum_{k\in Q} (\bar{d}^k \alpha^k + \hat{d}^k \gamma^k) && (8f)\\
& & \gamma^k & \le \alpha^k \quad \forall k \in Q && (8g)\\
& \alpha^k, \beta_i, \gamma^k, z_{ij} \in \{0,1\}, rhs \in \mathbb{Z}, d & \ge 0 & \ \forall k \in Q, \forall ij \in E, \forall i \in V && (8h)
\end{aligned}
$$

with (8b), (8c), (8g) defining the dependencies between α^k, β_i, γ^k, and z_{ij}, and (8d) guaranteeing the round-up of R. The total demand d is calculated by (8f), the number of deviating commodities limited by (8e).

4 Computations

In this section, we summarize the results of our computational study on the impact of separating violated inequalities (5)–(7) on the integrality gap closed.

Test instances. We consider live traffic data of the U. S. Internet2 Network (ABILENE) [1] ($|V| = 12$, $|E| = 15$, $|K| = 66$), the European research backbone GÉANT [8] ($|V| = 22$, $|E| = 36$, $|K| = 461$), the German research backbone [7] mapped on the networks GERMANY17 ($|V| = 17$, $|E| = 26$, $|K| = 136$), and GERMANY50 [13] ($|V| = 50$, $|E| = 89$, $|K| = 1044$). Traffic measurements with a granularity of 5 minutes (ABILENE, GERMANY17, GERMANY50) or 15 minutes (GÉANT) are given. We consider one (two) time period(s) of 1 day (week) each for GERMANY17 and GERMANY50 (ABILENE and GÉANT). All demands are scaled such that the maximum total demand of each network equals 1 Tbps. For each network and each point-to-point demand the

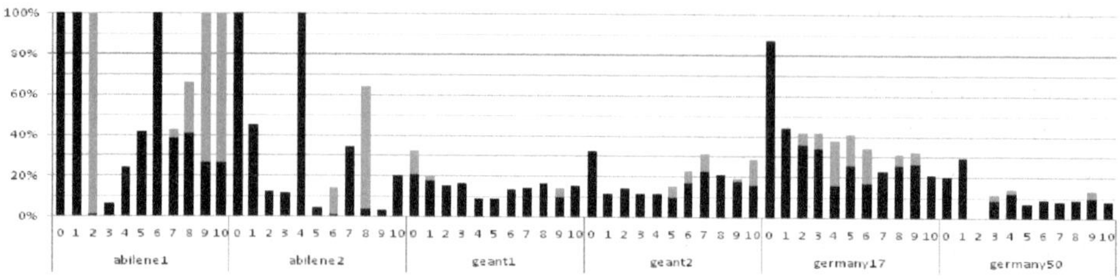

Fig. 1. Additional gap closed by separating (5)–(7) with setting (ii) [black] and setting (iii) [grey]

arithmetic mean (95%-percentile) of the corresponding traffic matrices is taken as default (peak) value. We set $C = 40$ Gbps.

Setting. We implemented model (2) in C++ using IBM ILOG CPLEX 12.1 [6]. We added a separator as described in Section 3. It is called at the root node of the branch-and-cut tree. All computations were done on a Linux machine with 2.93 GHz Intel Xeon W3540 CPU, 12 GB RAM, and a time limit of 12 h per instance.

We compare three settings: solving the Γ-RNDP with (i) CPLEX alone, (ii) additional heuristical separation of (5)–(7) in a branch-and-cut, and (iii) an additional exact separation if none has been found in (ii). In all settings, only the root node is solved and the achieved bounds on the optimal value are evaluated.

Results. Figure 1 shows the additional optimality gap closed (i. e., the percentual decrease in the gap between best known/optimal primal bound and dual bound by separating (5)–(7)) for each network, $\Gamma = 0, 1, \ldots, 10$, and settings (ii) (black bars) and (iii) (black+grey bars). The additional gap closed seems to be more related with the network (topology) than with Γ, e. g., by trend separating (5)–(7) is more effective for ABILENE or GERMANY17 than for GÉANT or GERMANY50. Still, for most networks and values Γ the gap can be closed by at least 10%. The added value of exact separation is unpredictable. For ABILENE1 the gap can be closed completely in most cases.

The additional computational effort in (ii) and (iii) results in increased average root node solving times of 2.8 resp. 3.1 times the one in (i). However, in a full cut-and-branch approach without node limit, the separation yields a significant speed-up in the geometric mean (e. g., 42,6% for GEANT1, or 64,8% for GEANT2).

Among all separated cuts, 99.9% are of type (5) and 0.1% are of type (4). No violated cuts of types (6) and (7) have been found by the heuristic.

In summary, the separation of violated robust cutset inequalities is beneficial, speeding-up the solving process by closing the optimality gap in the root node by 10% or more already.

5 Concluding Remarks

In this paper, we have considered the Γ-robust network design problem (Γ-RNDP) presented by [11]. We have shown a generalization of the well-known class of cutset inequalities and given a complete non-redundant description in a particular case. Further, we have presented their effectiveness in a computational study with realistic real-life traffic data of several research backbone networks using both an heuristical and an exact ILP-based separation approach.

In the future, the polyhedral structure of the Γ-RNDP has to be studied further to identify more classes of valid (and facet-defining) inequalities.

Acknowledgement

This work was supported by the Federal Ministry of Education and Research (BMBF grant 03MS616A, project *ROBUKOM - Robust Communication Networks*, www.robukom.de

References

1. Abilene Internet2 Network, www.internet2.edu/network
2. Altin, A., Amaldi, E., Belotti, P., Pinar, M.Ç.: Provisioning virtual private networks under traffic uncertainty. Networks 49(1), 100–155 (2007)
3. Bertsimas, D., Sim, M.: Robust discrete optimization and network flows. Math. Prog. Ser. B 98, 49–71 (2003)
4. Bertsimas, D., Sim, M.: The Price of Robustness. Operations Research 52(1), 35–53 (2004)
5. Bienstock, D., Chopra, S., Günlük, O., Tsai, C.Y.: Minimum cost capacity installation for multicommodity network flows. Math. Prog. 81, 177–199 (1998)
6. IBM CPLEX version 12.1 (2009), http://www.ibm.com
7. Deutsche Forschungsnetz (DFN), http://www.dfn.de
8. GÉANT – the pan-european research and education network (2000–2005), http://www.geant.net
9. Günlük, O.: A branch and cut algorithm for capacitated network design problems. Math. Prog. 86, 17–39 (1999)
10. Koster, A.M.C.A., Kutschka, M., Raack, C.: On the robustness of optimal network designs. In: Proceedings IEEE International Conference on Communications, ICC 2011 (to appear 2011)
11. Koster, A.M.C.A., Kutschka, M., Raack, C.: Towards Robust Network Design using Integer Linear Programming Techniques. In: Proceedings Next Generation Internet, NGI 2010. IEEE Xplore (2010), http://dx.doi.org/10.1109/NGI.2010.5534462
12. Magnanti, T.L., Mirchandani, P.: Shortest paths, single origin-destination network design and associated polyhedra. Networks 33, 103–121 (1993)
13. Orlowski, S., Wessäly, R., Pióro, M., Tomaszewski, A.: SNDlib 1.0–Survivable Network Design Library. Networks 55, h276–h286 (2009)
14. Raack, C., Koster, A.M.C.A., Orlowski, S., Wessaely, R.: On cut-based inequalities for capacitated network design polyhedra. Networks 57, 141–156 (2010)

Stabilized Branch-and-Price for the Rooted Delay-Constrained Steiner Tree Problem

Markus Leitner, Mario Ruthmair, and Günther R. Raidl

Institute of Computer Graphics and Algorithms, Vienna University of Technology,
Favoritenstraße 9-11, 1040 Vienna, Austria
{leitner,ruthmair,raidl}@ads.tuwien.ac.at

Abstract. We consider the rooted delay-constrained Steiner tree problem which arises, e.g., in the design of centralized multicasting networks where quality of service constraints are of concern. We present a mixed integer linear programming formulation based on the concept of feasible paths which has already been considered in the literature for the spanning tree variant. Solving its linear relaxation by column generation has, however, been regarded as computationally not competitive. In this work, we study various possibilities to speed-up the solution of our model by stabilization techniques and embed the column generation procedure in a branch-and-price approach in order to compute proven optimal solutions. Computational results show that the best among the resulting stabilized branch-and-price variants outperforms so-far proposed methods.

1 Introduction

When designing a communication network with a central server broadcasting or multicasting information to all or some of the participants of the network, some applications such as video conferences require a limitation of the maximal delay from the server to each client. Beside this delay-constraint minimizing the cost of establishing the network is in most cases an important design criterion. As another example, consider a package shipping organization with a central depot guaranteeing its customers a delivery within a specified time horizon. Naturally the organization aims at minimizing the transportation costs but at the same time has to hold its promise of being in time. Such network design problems can be modeled using an NP-hard combinatorial optimization problem called *rooted delay-constrained Steiner tree problem* (RDCSTP) [14]. The objective is to find a minimum cost Steiner tree on a given graph with the additional constraint that the total delay along each path from a specified root node to any other required node must not exceed a given delay bound.

More formally, we are given an undirected graph $G = (V,E)$ with node set V, edge set E, a fixed root node $s \in V$, a set $T \subseteq V \setminus \{s\}$ of terminal or required nodes, a set $S = V \setminus (T \cup \{s\})$ of optional Steiner nodes, a cost function $c : E \rightarrow \mathbb{Z}^+$, a delay function $d : E \rightarrow \mathbb{Z}^+$, and a delay bound $B \in \mathbb{Z}^+$. A feasible solution to the RDCSTP is a Steiner tree $G_S = (V_S, E_S)$, $s \in V_S$, $T \subset V_S \subseteq V$, $E_S \subseteq E$, satisfying the constraints $\sum_{e \in p_S(t)} d_e \leq B$, $\forall t \in T$, where $p_S(t) \subseteq E$ denotes the edge set of the unique path from root s to terminal t. An optimal solution $G_S^* = (V_S^*, E_S^*)$ is a feasible solution with minimum costs $c(G_S^*) = \sum_{e \in E_S^*} c_e$.

J. Pahl, T. Reiners, and S. Voß (Eds.): INOC 2011, LNCS 6701, pp. 124–138, 2011.

After discussing existing related work in Section 2 we describe a mixed integer linear programming (MIP) formulation involving exponentially many path variables as well as its solving by branch-and-price in Section 3. Section 4 details two different column generation stabilization techniques. Computational results in Section 5 show that the best among the resulting stabilized branch-and-price variants outperforms so-far proposed methods. We conclude in Section 6 and sketch potential future work.

2 Previous and Related Work

Kompella et al. [14] introduced the RDCSTP, proved its NP-hardness and presented a construction heuristic based on the algorithm by Kou et al. [15] for the Steiner tree problem (STP) on graphs. Manyem et al. [21] showed that the problem is not in APX. There are many recent publications dedicated to this problem and its more special variants. Several metaheuristics have been applied to the RDCSTP, such as tabu-search [29], GRASP [30,33], path-relinking [11], variable neighborhood descent (VND) [24], and variable neighborhood search (VNS) [33]. A hybrid algorithm in [34] combines scatter search with tabu-search, VND, and path-relinking. More heuristic approaches can be found for the spanning tree variant with $T = V \setminus \{s\}$, e.g. a GRASP and a VND in [26] and an ant colony optimization and a VNS in [27]. Furthermore, preprocessing methods are presented in [27] to reduce the size of the input graph significantly in order to speed up the solving process.

Exact methods based on integer linear programming (ILP) have been explored by Leggieri et al. [16] who describe a compact extended node-based formulation using lifted Miller-Tucker-Zemlin inequalities. Since the used Big-M inequalities usually yield rather weak linear programming (LP) relaxation bounds this formulation is improved by directed connection cuts. Several ILP approaches for the spanning tree variant have been examined by Gouveia et al. in [12] based on a path formulation solved by three different methods. Standard column generation turns out to be computationally inefficient while a Lagrangian relaxation approach together with a fast primal heuristic exhibits better performance. The third approach reformulates the constrained shortest path problem for each node on a layered graph and solves it using a multi commodity flow (MCF) formulation. Since the size of the layered graph and therefore the efficiency of the according model heavily depends on the number of achievable discrete delay values, this approach can in practice only be used for instances in which this number is quite restricted. Additionally an MCF model usually suffers in practice from the huge amount of flow variables used, altogether leading to a slow and memory-intensive solving process. Nevertheless solving these layered graph models turned out to be highly effective on certain classes of instances. In [28] not just the constrained shortest path problem but the whole RDCSTP is modeled on a layered graph which reduces to solving the classical STP on this graph. The acyclicity of the layered graph allows to eliminate sub-tours using a compact formulation for the STP without additional variables. However, the well-known directed cut formulation on this graph with an exponential number of constraints yields a tighter or at least equal LP bound than all other known formulations for the RDCSTP. This result was shown by Gouveia et al. [13] for the *hop-constrained minimum spanning tree problem* where $d_e = 1$, $\forall e \in E$, and can be generalized to the RDCSTP in a natural way. To overcome the issue of an excessive number of layers in case of a huge set of achievable delay values, a strategy based on iteratively solving smaller layered graphs is presented in [28] obtaining lower and upper bounds to the optimal costs. By successively extending these smaller graphs appropriately, the bounds are tightened to finally converge to an optimal solution. In practice, this approach usually yields very small gaps even on instances where the directed cut formulation on the layered graph is not able to derive an optimal LP value.

Recently, we [20] proposed stabilized column generation approaches for the RDCSTP. The current article significantly extends this work by embedding column generation in a branch-and-bound approach to compute proven optimal solutions, describing an additional pricing strategy and many other aspects in more detail. We also present more extensive results including a comparison to the above mentioned layered graph approaches.

3 Branch-and-Price

In this section we present the details of a branch-and-price approach for solving the RDCSTP, which is based on a MIP formulation utilizing variables corresponding to feasible paths for each terminal. This model is a straightforward modification of the one discussed by Gouveia et al. [12] for the spanning tree variant of the RDCSTP, i.e. for the case of $T = V \setminus \{s\}$. Our directed formulation uses an arc set A containing an arc (s, j) for each edge $\{s, j\} \in E$ incident to the root node and two oppositely directed arcs (i, j), (j, i) for all other edges $\{i, j\} \in E$, $i, j \neq s$. Note that we assume the edge cost and delay functions to be correspondingly defined on the set of arcs, too, i.e. $c_{ij} = c_e$ and $d_{ij} = d_e$, $\forall (i, j) \in A$, $e = \{i, j\} \in E$.

The *integer master problem* (IMP) defined by (1)–(6) is based on variables $x_{ij} \in \{0,1\}$, $\forall (i, j) \in A$, indicating which arcs are included in the directed solution. Each such directed solution must form an outgoing arborescence rooted at node s. We further use path variables $\lambda_p \in \{0,1\}$, $\forall p \in P = \bigcup_{t \in T} P_t$, where $P_t \subseteq 2^A$ is the set of feasible paths from the root node s to terminal t. A path $p \in P_t$ to terminal $t \in T$, which is represented by its arc set, is feasible if and only if it satisfies the delay bound, i.e. $\sum_{(i,j) \in p} d_{ij} \leq B$. Variable λ_p is set to one if path $p \in P$ is realized. Dual variables are given in parenthesis in model (1)–(6).

$$\text{(IMP)} \quad \min \sum_{(i,j) \in A} c_{ij} x_{ij} \tag{1}$$

$$\text{s.t.} \quad \sum_{p \in P_t} \lambda_p \geq 1 \qquad (\mu_t) \quad \forall t \in T \tag{2}$$

$$x_{ij} - \sum_{p \in P_t | (i,j) \in p} \lambda_p \geq 0 \qquad (\pi^t_{ij}) \quad \forall t \in T,\ \forall (i, j) \in A \tag{3}$$

$$\sum_{(i,j) \in A} x_{ij} \leq 1 \qquad (\gamma_j) \quad \forall j \in V \tag{4}$$

$$x_{ij} \in \{0,1\} \qquad \forall (i, j) \in A \tag{5}$$

$$\lambda_p \geq 0 \qquad \forall p \in P \tag{6}$$

The convexity constraints (2) ensure that at least one path is realized for each terminal, while the coupling constraints (3) link paths to the corresponding arc variables. Inequalities (4) restrict the in-degree of each node and thus together with inequalities (2) and (3) ensure that the directed solution is an arborescence with root s. Given strictly positive edge costs, removing inequalities (4) would also yield a valid model. We did nevertheless include them to stay consistent with the model by Gouveia et al. [12]. Further note that only lower bounds are given for variables λ_p, $\forall p \in P$, in inequalities (6). These variables will become automatically integral due to the remaining inequalities.

Since the number of feasible paths for each terminal $t \in T$ and thus the total number of variables in the model is in general exponentially large, we cannot solve the IMP directly. Hence we embed delayed column generation – see e.g. [5,7] – in a branch-and-bound procedure to solve the IMP, i.e. we apply branch-and-price. Branching is performed on arc variables x_{ij}, $\forall (i,j) \in A$. The *restricted master problem* (RMP) which then needs to be solved in each node of the branch-and-bound tree is defined by considering only a subset $\tilde{P}_t \subseteq P_t$, $\tilde{P}_t \neq \emptyset$, $\forall t \in T$, of path variables and by replacing the integrality conditions on arcs (5) by $x_{ij} \geq 0$, $\forall (i,j) \in A$. Further variables are added on demand according to the solution of the pricing subproblem which will be discussed in the following.

3.1 The Pricing Subproblem

Let $\tilde{P}$ denote the set of paths for which corresponding variables have already been included in the RMP. We further denote by $\mu_t \geq 0$, $\forall t \in T$, the dual variables associated to the convexity constraints (2) and by $\pi^t_{ij} \geq 0$, $\forall t \in T$, $\forall (i,j) \in A$, the dual variables associated to the coupling constraints (3). In the pricing subproblem, we need to identify at least one path variable λ_p, $p \in P \setminus \tilde{P}$, yielding negative reduced costs $\bar{c}_p = -\mu_t + \sum_{(i,j)\in p} \pi^t_{ij}$ or prove that no such variable exists.

Thus, the pricing subproblem is formally defined as

$$(t^*, p^*) = \text{argmin}_{t\in T, p\in P_t} - \mu_t + \sum_{(i,j)\in p} \pi^t_{ij}. \tag{7}$$

It can be solved by computing a resource constrained shortest path on a graph (V,A) with non-negative arc costs π^t_{ij}, $\forall (i,j) \in A$, for each terminal $t \in T$. Computing a minimum cost resource constrained shortest path between two nodes is NP-hard in the weak sense [10] and can thus be solved in pseudo-polynomial time, see [9] for a survey. We use the dynamic programming based algorithm from [12] in our implementation which has computational complexity $O(B \cdot |A|)$.

For solving the RMP of the currently considered branch-and-bound tree node, we need to add path variables and resolve the RMP as long as at least one path variable $p \in P$ with negative reduced costs $\bar{c}_p$ exists. We compare two pricing strategies, which both require a single run of the dynamic program from [12] for each terminal $t \in T$. In the first approach we add the variable corresponding to the cheapest feasible path for each terminal in case it has negative reduced costs. Thus at most $|T|$ path variables are added in each pricing iteration. The second approach follows [12] and potentially adds multiple path variables for a single terminal in each iteration: We consider all nodes $v \in V$ adjacent to terminal t and all delay values $b = 1, \ldots, B - d_{vt}$ for which a path from s to v in conjunction with arc (v,t) is a feasible path to t. In case a shortest path p to v of total delay b, $0 < b \leq B - d_{vt}$, exists and $p' = p \cup \{(v,t)\}$ yields negative reduced costs, the corresponding variable is added to the RMP.

4 Column Generation Stabilization

It is well known that basic column generation based approaches typically suffer from computational instabilities often leading to long running times. Vanderbeck [31] describes five major causes of these instabilities including primal degeneracy, the tailing-off, and the heading-in effect. In order to improve the efficiency of such methods, several so-called column generation stabilization techniques aiming to reduce the effects of these problems have been proposed. These can be classified into problem specific techniques, such as the usage of dual-optimal inequalities [3,6] and problem independent approaches, see e.g. [2,25,32,23]. The latter are often based on the concept of stability centers which are current estimates of good dual variable values. The boxstep method [22] restricts each dual variable value to a small trust region around its current stability center. Other methods penalize deviations from the current stability center, e.g. by using piecewise linear penalty functions [3,8]. Except for the weighted Dantzig-Wolfe decomposition approach, these so far proposed stabilization methods, however, usually need to add additional constraints and variables to the RMP. Recently, we proposed a stabilization technique [18,17,19] which does not modify the RMP, but is based on using alternative dual-optimal solutions within the pricing subproblem. This method turned out to significantly accelerate the column generation process for a survivable network design problem by reducing the necessary number of iterations as well as the total number of included variables.

In the following we will show how this technique can be applied to the RDCSTP before we discuss two alternative stabilization approaches based on piecewise linear penalty functions. The latter two will then be used to compare our method.

4.1 Alternative Dual-Optimal Solutions

In order to describe our stabilization approach, we briefly discuss the dual of the RMP (8)–(13); primal variables are given in parenthesis.

$$\max \sum_{t \in T} \mu_t + \sum_{j \in V} \gamma_j \tag{8}$$

$$\text{s.t.} \quad \sum_{t \in T} \pi_{ij}^t + \gamma_j \leq c_{ij} \qquad (x_{ij}) \quad \forall (i,j) \in A \tag{9}$$

$$\mu_t - \sum_{(i,j) \in p} \pi_{ij}^t \leq 0 \qquad (\lambda_p) \quad \forall t \in T,\ \forall p \in \tilde{P}_t \tag{10}$$

$$\mu_t \geq 0 \qquad \forall t \in T \tag{11}$$

$$\pi_{ij}^t \geq 0 \qquad \forall t \in T,\ \forall (i,j) \in A \tag{12}$$

$$\gamma_j \leq 0 \qquad \forall j \in V \tag{13}$$

Inequalities (9) are capacity constraints imposing upper bounds on the sum of dual values $\sum_{t\in T}\pi^t_{ij}$ for each arc $(i,j)\in A$, while inequalities (10) ensure that the sum of dual arc costs π^t_{ij} along each included path is at least μ_t.

Let (μ^*,π^*,γ^*) denote the current dual solution computed by an LP solver for the RMP and $A'=\{(i,j)\in A \mid \nexists p\in\tilde{P}:(i,j)\in p\}$ be the set of arcs which are not used in any of the so far included path variables. Since only the capacity constraints (9) are relevant for these arcs, any dual variable values $\pi^{t\,*}_{ij}\geq 0$ are optimal as long as $\sum_{t\in T}\pi^{t\,*}_{ij}\leq c_{ij}-\gamma_j^*, \forall(i,j)\in A'$, holds. In case the capacity constraints are not binding, it is further possible to increase dual variable values $\pi^{t\,*}_{ij}$, $t\in T$, for arcs $(i,j)\in A\setminus A'$ while maintaining dual optimality.

Let $\delta_{ij}=c_{ij}-\gamma_j-\sum_{t\in T}\pi^{t\,*}_{ij}, \forall(i,j)\in A$, denote the slack of each capacity constraint (9). Then, obviously any values $\pi^t_{ij}\geq\pi^{t\,*}_{ij}, \forall(i,j)\in A, \forall t\in T$, are dual-optimal as long as $\sum_{t\in T}\pi^t_{ij}\leq\sum_{t\in T}\pi^{t\,*}_{ij}+\delta_{ij}$, $\forall(i,j)\in A$, holds. Note that state-of-the-art LP solvers usually yield minimal optimal dual variable values, i.e. $\pi^{t\,*}_{ij}=0$, $\forall t\in T$, $\forall(i,j)\in A'$. Based on these observations our stabilization approach aims to choose alternative dual-optimal solutions by distributing the slack δ_{ij} to the relevant dual variables $\pi^{t\,*}_{ij}$, $\forall t\in T$. We expect that increasing the dual variable values resulting in higher arc costs in the pricing subproblem facilitates the generation of meaningful path variables. One main advantage of choosing such alternative dual-optimal solutions for solving the pricing subproblem is that on the contrary to most other stabilization approaches we do not modify the RMP or increase its size by adding further variables or constraints.

Our first strategy is based on simply distributing the potential increase δ_{ij} equally among all relevant dual variables, i.e. we use alternative dual variables $\bar{\pi}^t_{ij}=\pi^{t\,*}_{ij}+\frac{\delta_{ij}}{|T|}$, $\forall t\in T$, $\forall(i,j)\in A$. In our previous work for a survivable network design problem [18,17,19], however, it turned out to be beneficial to initially use different dual-optimal solutions, one for each terminal t, and let them finally converge towards $\bar{\pi}^t_{ij}$, $\forall t\in T$, $\forall(i,j)\in A$. Given an exogenous parameter $Q\geq 2$, denoting a total number of major iterations, the approach is iterated with parameter $q=1,\dots,Q$, indicating the current major iteration. Thus, parameter q is initially set to one (first major iteration) and gradually incremented by one in case no negative reduced cost path has been found. Let $t'\in T$ be the terminal currently considered in the pricing subproblem. Then the resulting dual variable values, which are denoted by $\hat{\pi}^t_{ij}$, $\forall t\in T$, $\forall(i,j)\in A$, are defined as follows:

$$\hat{\pi}^t_{ij}=\begin{cases}\pi^{t\,*}_{ij}+\frac{\delta_{ij}}{|T|}+\frac{Q-q}{Q-1}\left(\delta_{ij}-\frac{\delta_{ij}}{|T|}\right) & \text{if } t=t'\\ \pi^{t\,*}_{ij} & \text{otherwise}\end{cases}, \quad \forall t\in T,\ \forall(i,j)\in A \qquad (14)$$

This approach divides the interval $\left[\frac{\delta_{ij}}{|T|},\delta_{ij}\right]$ into $Q-1$ equally sized sub-intervals defining the dual variable values used for each value of q, $1\leq q\leq Q$. Note that for each terminal $t\in T$ the resulting vector $\hat{\pi}$ is a dual-optimal solution. After Q major iterations, i.e. if $q=Q$, $\hat{\pi}^t_{ij}=\bar{\pi}^t_{ij}$ holds for each terminal t, i.e. we essentially use the same dual solution for all terminals. Thus, we can terminate the column generation process of the current node if $q=Q$ and no path variables have been added. Since most path variables are usually already generated in the root node of the branch-and-bound tree, we do not

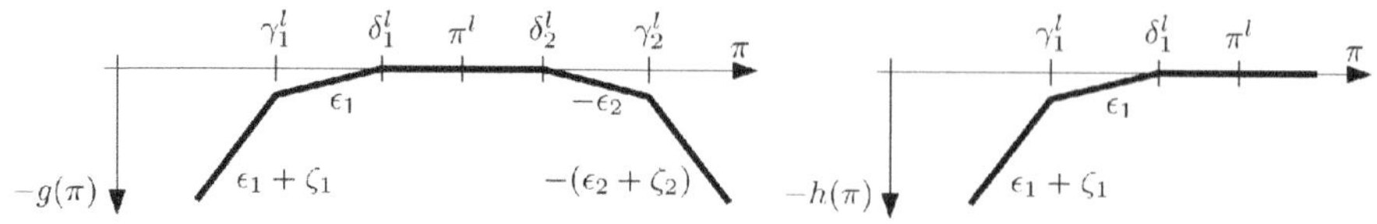

Fig. 1. 5-Piecewise and 3-Piecewise Linear Dual Penalty Functions $g(\pi)$ and $h(\pi)$

reinitialize parameter q. Hence, dual variable values $\bar{\pi}_{ij}^t$, $\forall t \in T$, $\forall (i,j) \in A$, are used in all further nodes of the branch-and-bound tree.

4.2 Piecewise Linear Stabilization

As mentioned before, successful stabilization techniques are often based on penalizing deviations from a current stability center by adding a stabilization term to the primal problem, i.e. a penalty function to the dual problem. Amor et al. [4] compared the performance of a Bundle-type approach and k-piecewise linear penalty functions using three and five pieces, respectively. They concluded that using five-piecewise linear penalty functions as originally proposed in [2] yields good results if all parameters are chosen carefully. We also adopted this approach in order to compare its performance to the previously described stabilization technique based on alternative dual-optimal solutions. Given the current stability center $\pi^l \in \mathbb{R}_+^{|T|\cdot|A|}$ of major iteration $l \in \mathbb{N}$, $l \geq 1$, and a correspondingly defined penalty function $g(\pi)$ – see Figure 1 – dual variable values outside the trust region $[\delta_1^l, \delta_2^l]$ are penalized according to ζ_1, ε_1, ε_2, and ζ_2, respectively.

Let $\pi_{ij}^{t\,*}$, $\forall t \in T$, $\forall (i,j) \in A$, denote the dual variable values when the column generation approach on the penalized model at major iteration l terminates. If there exists at least one dual variable value in the penalized region, we need to update the stability center according to the current dual solution, i.e. $\pi^{l+1} = \pi^*$, correspondingly set γ_1^{l+1}, δ_1^{l+1}, δ_2^{l+1}, and γ_2^{l+1}, and continue the column generation process. As has been shown previously [2] this process, which needs to be repeated until each dual variable value lies within an unpenalized region, terminates after finitely many steps yielding the LP relaxation of the current branch-and-bound node.

In our case, however, preliminary tests with various settings showed that due to a typically large number of relatively time consuming updates of the stability center this concept does not seem to pay off. Since the analysis in Section 4.1 shows that high dual variable values facilitate the generation of reasonable path variables, we further apply a second variant of this concept where only dual variable values smaller than δ_1^l are penalized using the penalty function $h(\pi)$ shown in Figure 1.

5 Computational Results

All described variants of the branch-and-price approach – denoted by BP throughout all tables – have been implemented in C++ using ZIB SCIP 2.0.1 [1] with IBM CPLEX 12.2 as embedded LP solver. We further decided to additionally test corresponding

Table 1. Median CPU-Times in Seconds for CG with Different Pricing Strategies and Stabilization Techniques based on Alternative Dual-Optimal solutions

				OPT					MPT					CG_G	Lag_G
dual solution				π^*	$\bar{\pi}$	$\hat{\pi}$			π^*	$\bar{\pi}$	$\hat{\pi}$			-	-
Set	B	$\|V\|$	$\|E\|$	-	-	Q=2	Q=5	Q=10	-	-	Q=2	Q=5	Q=10	-	-
C2	3	41	279	1	1	1	2	2	**0**	**0**	1	1	1	2	4
	5	41	321	7	4	5	5	6	3	**2**	**2**	3	4	173	46
	7	41	321	25	7	8	6	8	11	**3**	4	4	5	3658	76
	9	41	321	62	10	10	8	9	33	7	6	**5**	6	8367	64
E2	3	41	597	5	5	5	7	8	**2**	**2**	**2**	4	5	13	12
	5	41	680	229	72	93	47	54	166	55	43	**22**	31	10045	208
	7	41	680	10000	983	989	246	243	10000	1600	871	113	**102**	10149	205
	9	41	680	10000	1326	1434	229	131	10000	3119	2142	657	**110**	10162	243
C100	100	41	561	173	36	35	31	31	41	**9**	10	**9**	12	10026	809
	150	41	572	808	61	71	48	46	118	24	**14**	16	16	10034	544
	200	41	572	3245	105	97	62	60	567	32	30	22	**21**	10061	711
	250	41	572	8742	103	113	64	63	3137	40	24	**17**	21	10076	1066
E100	100	41	651	520	82	93	56	54	201	62	26	**17**	19	1033	976
	150	41	672	3814	286	278	170	131	2911	376	227	126	**67**	10106	1817
	200	41	672	10000	1869	1501	325	192	10000	4098	1626	238	**158**	10096	2972
	250	41	672	10000	1589	1851	439	201	10000	10000	3453	734	**159**	10104	4008
C1000	1000	41	572	138	47	38	37	31	17	**7**	9	12	15	8186	668
	1500	41	589	648	74	63	64	60	115	**22**	29	**22**	28	10024	942
	2000	41	589	1730	136	144	131	90	599	80	47	38	**36**	10037	2389
	2500	41	589	6952	141	145	96	91	1336	56	47	**45**	54	10037	1256
E1000	1000	41	632	387	82	74	66	58	183	58	41	28	**27**	10065	2846
	1500	41	668	2830	268	343	268	148	2413	348	154	99	**69**	10031	3041
	2000	41	668	10000	671	1038	265	**177**	10000	1516	765	232	180	10083	5882
	2500	41	668	10000	863	1035	420	316	10000	4121	1365	315	**278**	10070	5726
T10	16	96	469	1	1	1	1	1	**0**	**0**	**0**	**0**	**0**	-	-
	30	100	932	23	6	6	6	6	3	**1**	**1**	**1**	2	-	-
	50	100	1269	188	17	18	18	16	15	**1**	**1**	2	2	-	-
	100	100	1695	2173	38	40	37	31	125	**2**	**2**	3	4	-	-
T30	16	98	482	3	3	3	3	4	**1**	**1**	**1**	2	2	-	-
	30	100	932	68	26	25	24	24	17	**4**	**4**	6	8	-	-
	50	100	1269	663	97	92	88	77	118	13	**12**	14	15	-	-
	100	100	1695	9973	345	370	311	269	1906	32	38	**27**	28	-	-
T50	16	99	486	7	6	5	6	7	3	**2**	**2**	4	5	-	-
	30	100	932	114	45	43	36	40	38	**11**	**11**	13	18	-	-
	50	100	1269	1128	159	167	135	140	249	30	**28**	31	38	-	-
	100	100	1695	10000	693	766	725	511	7739	192	106	112	**92**	-	-
T70	16	99	487	11	8	8	9	12	4	**3**	4	6	8	-	-
	30	100	932	177	60	62	59	56	51	**19**	20	24	28	-	-
	50	100	1269	1706	247	250	193	171	438	**67**	68	70	**67**	-	-
	100	100	1695	10000	1179	1036	1041	822	9017	402	436	**240**	249	-	-
T99	16	100	490	17	11	13	15	19	7	**4**	7	10	12	-	-
	30	100	932	320	108	98	83	93	147	49	**42**	45	50	-	-
	50	100	1269	2952	409	343	292	263	858	159	134	**111**	124	-	-
	100	100	1695	10000	1584	1759	1574	1275	10000	1166	1120	835	**629**	-	-

pure column generation – denoted by CG in the following – implemented by solely using CPLEX in order to analyze an eventually existing overhead of SCIP compared to CPLEX. Each computational experiment has been performed on a single core of an Intel Xeon E5540 processor with 2.53 GHz and 3 GB RAM per core. An absolute time limit of 10000 CPU-seconds has been applied to all experiments.

First, we tested our approaches on instances originally proposed by [12] for the spanning tree variant of the RDCSTP, i.e. $T = V \setminus \{s\}$. The three main instance sets R, C and E each have different graph structures defined by their edge cost functions: R has random edge costs, C and E both have Euclidean costs fixing the source s near the center and near the border, respectively. Each main instance set consists of different subsets of five complete input graphs with 41 nodes varying in the number of possible discrete edge delay values; e.g. C100 denotes the set of instances with 100 different integer delay values: $d_e \in \{1,\ldots,100\}, \forall e \in E$. Additionally we ran tests on instance sets Tα consisting of 30 randomly generated complete graphs with $|V| = 100$ where α denotes the number of terminal nodes $|T|$. Here all delays and costs are uniformly distributed in $\{1,\ldots,99\}$. For each instance set we tested our approaches on different delay bounds B. Since we consider these larger randomly generated instances Tα and since all instances with random costs from [12] could be solved relatively fast, we do not report here our detailed results for the latter. All preprocessing methods described in [27] are used to reduce the input graphs prior to solving. To build an initial set of paths a simple construction heuristic is applied on Steiner tree instances: the delay constrained shortest paths from the root to all terminal nodes are iteratively added to the tree dissolving possible cycles. On instances where $T = V \setminus \{s\}$ we apply the Kruskal-based heuristic followed by VND as introduced in [26].

Table 1 details median CPU-times in seconds for determining the LP relaxation of the IMP by unstabilized column generation, denoted by π^*, and when using stabilization based on alternative dual-optimal solutions $\bar{\pi}$ and $\hat{\pi}$, respectively. Here, OPT denotes the first described pricing strategy where at most one path per terminal is added in each iteration and MPT the one potentially adding multiple paths for a single terminal, compare Section 3.1. We further report average numbers of nodes $\overline{|V|}$ and edges $\overline{|E|}$ for each instance set after preprocessing. Finally, average CPU-times in seconds for the conceptually identical column generation approach by [12] – denoted as $\mathrm{CG_G}$ – as well as the Lagrangian approach $\mathrm{Lag_G}$ from the same authors are given in Table 1. The results of the latter two have, however, been computed on a different hardware and are thus not directly comparable. We observe that MPT outperforms OPT in almost all cases. Hence, we do not consider OPT in all further experiments. We further conclude that all stabilization strategies based on alternative dual-optimal solutions significantly outperform standard column generation. Note that already our unstabilized column generation variant needs significantly less iterations than the conceptually identical one discussed by [12]. We believe that next to different CPLEX versions these differences mainly come from choosing a better set of initial path variables, more sophisticated graph preprocessing, and the fact that we use the dual simplex algorithm which turned out to perform better than the primal one in our case.

Table 2 shows more detailed results for the variants of column generation using alternative dual-optimal solutions. Next to median CPU times in seconds (t_{total}), numbers of

Table 2. Median CPU-times, Median Times for Reaching the LP Value, Average Pricing Iterations, and Included Path variables for CG and Stabilization based on Alternative Dual-Optimal Solutions

		t_{total} [s]			t_{best} [s]			Iterations			Variables		
dual solution		$\bar{\pi}$	$\hat{\pi}$		$\bar{\pi}$	$\hat{\pi}$		$\bar{\pi}$	$\hat{\pi}$		$\bar{\pi}$	$\hat{\pi}$	
Set	B	-	Q=5	Q=10	-	Q=5	Q=10	-	Q=5	Q=10	-	Q=5	Q=10
C2	3	**0**	1	1	**0**	**0**	**0**	**10**	32	48	697	648	**641**
	5	**2**	3	4	**1**	**1**	2	**29**	63	89	2645	2024	**1901**
	7	**3**	4	5	**2**	3	**2**	**58**	82	122	4812	3312	**3014**
	9	7	**5**	6	2	**1**	**1**	**104**	106	130	7252	4700	**3808**
E2	3	**2**	4	5	**1**	2	2	**18**	46	74	1814	1639	**1596**
	5	55	**22**	31	54	**12**	16	**122**	154	203	11413	7740	**7186**
	7	1600	113	**102**	1597	100	**86**	383	**342**	365	36396	17908	**13459**
	9	3119	657	**110**	2746	45	**43**	570	565	**495**	59749	33377	**18123**
C100	100	**9**	**9**	12	5	**2**	**2**	**33**	68	101	16788	12828	**10783**
	150	24	**16**	**16**	16	**3**	**3**	**48**	79	113	33938	28465	**23247**
	200	32	22	**21**	15	**2**	**2**	**65**	104	131	52273	39777	**33140**
	250	40	**17**	21	17	**2**	**2**	**63**	89	132	56258	43697	**40100**
E100	100	62	**17**	19	34	**8**	9	**43**	90	127	27737	15702	**12177**
	150	376	126	**67**	184	25	**9**	**59**	114	147	75041	40916	**28441**
	200	4098	238	**158**	4069	**71**	82	**89**	149	194	142512	58547	**45980**
	250	10000	734	**159**	7804	76	**50**	**98**	194	201	209531	99451	**61488**
C1000	1000	**7**	12	15	6	**2**	3	**24**	60	97	30898	26540	**23374**
	1500	**22**	**22**	28	11	**4**	6	**33**	72	110	69064	59589	**53193**
	2000	80	38	**36**	52	10	**9**	**44**	81	118	103734	86713	**75027**
	2500	56	**45**	54	45	**6**	**6**	**42**	89	124	127440	112045	**100876**
E1000	1000	58	28	**27**	50	**6**	**6**	**31**	83	112	43492	29744	**24004**
	1500	348	99	**69**	303	22	**19**	**54**	119	151	116867	78501	**59886**
	2000	1516	232	**180**	740	**13**	14	**83**	141	162	239176	148387	**113407**
	2500	4121	315	**278**	2128	**18**	19	**92**	136	190	334410	187341	**156620**
T10	16	**0**	**0**	**0**	**0**	**0**	**0**	**11**	28	43	348	306	**290**
	30	**1**	**1**	2	**0**	**0**	**0**	**19**	37	59	1400	1221	**1083**
	50	**1**	2	2	**0**	**0**	**0**	**20**	42	63	2427	2153	**1894**
	100	**2**	3	4	**0**	1	1	**20**	42	66	3456	3249	**2779**
T30	16	**1**	2	2	**0**	**0**	1	**14**	39	61	982	852	**787**
	30	**4**	6	8	2	**1**	**1**	**29**	61	88	4946	4300	**3876**
	50	**13**	14	15	5	**2**	3	**43**	73	108	11489	10370	**9303**
	100	32	**27**	28	13	**4**	5	**44**	81	120	23889	23430	**20989**
T50	16	**2**	4	5	**1**	2	2	**15**	44	69	1458	1239	**1144**
	30	**11**	13	18	6	**5**	**5**	**32**	67	101	7271	6404	**5566**
	50	**30**	31	38	15	**5**	6	**50**	86	122	18202	16611	**15131**
	100	192	112	**92**	87	**11**	13	**63**	107	148	46145	43165	**40350**
T70	16	**3**	6	8	**2**	3	3	**15**	50	76	1864	1611	**1492**
	30	**19**	24	28	14	**7**	10	**30**	74	109	9269	7999	**7012**
	50	**67**	70	**67**	32	**11**	12	**50**	97	140	23640	22124	**19302**
	100	402	**240**	249	263	34	**33**	**74**	127	173	65179	63348	**57621**
T99	16	**4**	10	12	**3**	4	**3**	**16**	52	76	2397	2049	**1887**
	30	49	**45**	50	33	**15**	19	**33**	82	115	12063	9991	**8717**
	50	159	**111**	124	91	**20**	28	**54**	110	153	33106	27500	**24996**
	100	1166	835	**629**	750	73	**67**	**88**	160	202	99801	88283	**78012**

Table 3. Median Total CPU-Times and Updates of the Stability Center for CG and Piecewise Linear Stabilization Techniques

		t_{total} [s]	3PL								5PL							
			t_{total} [s]				$\overline{\text{Updates}}$				$\overline{t_{\text{total}}}$ [s]				$\overline{\text{Updates}}$			
		π^*	sml		lrg		sml		lrg		sml		lrg		sml		lrg	
Set	B	-	S_1	S_2	S_1	S_2	S_1	S_2	S_1	S_2	S_1	S_2	S_1	S_2	S_1	S_2	S_1	S_2
C100	100	56	204	255	138	173	18	23	9	11	10000	10000	10000	10000	51	78	48	47
	150	133	674	493	575	470	15	20	8	11	10000	10000	10000	10000	39	62	22	33
	200	1056	2482	2554	3008	2728	18	20	10	11	10000	10000	10000	10000	24	40	14	22
	250	3485	10000	9474	10000	7350	16	20	10	12	10000	10000	10000	10000	15	24	9	14
E100	100	463	692	627	522	639	19	24	10	12	10000	10000	10000	10000	35	62	18	29
	150	4018	10000	10000	10000	10000	18	19	9	10	10000	10000	10000	10000	21	35	10	17
	200	9137	10000	10000	10000	10000	7	7	4	4	10000	10000	10000	10000	14	25	6	11
	250	9861	10000	10000	10000	10000	5	6	3	3	10000	10000	10000	10000	9	15	4	7
T10	30	4	6	6	5	5	3	3	2	2	52	79	21	34	26	50	11	20
	50	17	24	28	21	22	3	3	2	2	242	322	102	137	29	56	12	22
T30	30	20	43	51	39	40	12	15	7	8	2281	2797	1250	1525	105	181	50	85
	50	137	353	355	301	325	12	17	6	9	10000	10000	8409	9961	89	113	64	87
T50	30	49	132	135	98	117	22	29	11	16	10000	10000	6586	7294	87	126	84	136
	50	364	986	957	695	856	24	33	12	17	10000	10000	10000	10000	28	47	18	29

pricing iterations (Iterations) and total number of included path variables (Variables) we also report median CPU times for finding the correct LP value (t_{best}), i.e. the remaining time is needed for proving this value. We do not report on our experiments for Q=2 in Table 2 since we already observed from Table 1 that Q=2 is not competitive compared to Q=5 and Q=10, respectively. From these results we conclude that using $\hat{\pi}$ clearly outperforms $\bar{\pi}$ regarding the total CPU-time as well as the time for finding the correct LP value. Especially for harder instances – i.e. those requiring more time – Q=10 performs significantly better than the other configurations. Using $\bar{\pi}$ usually leads to a smaller number of total pricing iterations than all variants of $\hat{\pi}$ while the latter reduce the total number of included variables. Hence, we observe that $\hat{\pi}$ allows for finding more meaningful path variables already in the beginning of the column generation process. Since the best performing variants – i.e. using $\hat{\pi}$ and $Q \in \{5, 10\}$ – exhibit a quite significant tailing-off effect there is potential for further possible improvement e.g. by computing additional (Lagrangian) bounds or performing early branching in branch-and-price.

Next, we analyze and compare the performance of the two column generation variants using piecewise linear stabilization terms. Table 3 reports median CPU-times in seconds and performed updates of the stability centers for different instance sets from both types. 3PL denotes the approach penalizing only small values and 5PL the full approach using a 5-piecewise linear penalty function. For both variants the initial stability center is chosen according to the first strategy for using alternative dual-optimal solutions, i.e. $\pi^1 = \bar{\pi}$, and we tested various settings of the inner trust region radius $\pi_{ij}^{t\ l} - \delta_{ij_1}^{t\ l} = \delta_{ij_2}^{t\ l} - \pi_{ij}^{t\ l} = r_{\text{i}} \frac{c_{ij}}{|T|}$ and the outer trust region radius $\pi_{ij}^{t\ l} - \gamma_{ij_1}^{t\ l} = \gamma_{ij_2}^{t\ l} - \pi_{ij}^{t\ l} = r_{\text{o}} \frac{c_{ij}}{|T|}$, $\forall t \in T$, $\forall (i,j) \in A$, respectively. For 5PL we used symmetric penalty functions, i.e. $\varepsilon = \varepsilon_1 = \varepsilon_2$ and $\zeta = \zeta_1 = \zeta_2$, and identical penalty slopes for all dimensions. Among the various tested configurations we report here experiments with smaller (sml) and larger (lrg) trust region radii $r_{\text{i}} = 1$, $r_{\text{o}} = 3$ and $r_{\text{i}} = 2$, $r_{\text{o}} = 6$, respectively. Furthermore, we tested penalty slopes S_1 where $\varepsilon = 0.3$ and $\zeta = 1$, and S_2 where $\varepsilon = 0.5$ and $\zeta = 1.5$.

Table 4. Number of Instances Solved to Proven Optimality, Average Optimality Gap, Root Node Gap, Number of Branching Nodes, Median Total CPU-Time, and Median Time for Solving the Root Node for BP Compared to Layered Graph Approaches SL and DL

		Σ Opt			$\overline{\text{gap}}$ [%]			$\overline{t_{\text{total}}}$ [s]			$\overline{\text{gap}_{\text{root}}}$ [%]	$\overline{\text{Nodes}}$	$\overline{t_{\text{root}}}$ [s]	
Set	B	SL	DL	BP	SL	DL	BP	SL	DL	BP	BP	BP	BP	CG
C2	3	**5**	**5**	**5**	**0.0**	**0.0**	**0.0**	**0**	**0**	1	0.0	1.0	1	1
	5	**5**	**5**	**5**	**0.0**	**0.0**	**0.0**	**0**	3	5	0.7	5.0	3	4
	7	**5**	**5**	**5**	**0.0**	**0.0**	**0.0**	**1**	5	9	1.0	13.0	5	5
	9	**5**	**5**	**5**	**0.0**	**0.0**	**0.0**	**3**	12	8	0.6	5.0	5	6
E2	3	**5**	**5**	**5**	**0.0**	**0.0**	**0.0**	**0**	1	4	0.0	1.0	4	5
	5	**5**	**5**	**5**	**0.0**	**0.0**	**0.0**	**3**	31	34	2.5	5.2	25	31
	7	**5**	**5**	**5**	**0.0**	**0.0**	**0.0**	**18**	522	743	2.3	14.2	122	102
	9	**5**	**5**	4	**0.0**	**0.0**	0.1	**48**	893	914	2.3	7.6	149	110
C100	100	**5**	**5**	**5**	**0.0**	**0.0**	**0.0**	130	33	**9**	0.1	4.2	9	12
	150	**5**	**5**	**5**	**0.0**	**0.0**	**0.0**	1383	208	**18**	0.0	1.0	18	16
	200	3	**5**	**5**	0.9	**0.0**	**0.0**	7552	1219	**21**	0.1	3.8	21	21
	250	1	**5**	**5**	10.7	**0.0**	**0.0**	10000	1415	**35**	0.1	2.2	20	21
E100	100	**5**	**5**	**5**	**0.0**	**0.0**	**0.0**	1212	766	**22**	0.9	2.4	20	19
	150	0	3	**5**	6.8	0.3	**0.0**	10000	8299	**122**	0.3	4.6	70	67
	200	0	0	**4**	13.1	2.6	**0.1**	10000	10000	**453**	0.8	9.0	361	158
	250	0	0	**4**	12.2	3.0	**0.1**	10000	10000	**993**	0.9	8.0	303	159
C1000	1000	4	**5**	**5**	1.8	**0.0**	**0.0**	2509	20	**13**	0.0	1.0	13	15
	1500	0	**5**	**5**	12.2	**0.0**	**0.0**	10000	175	**40**	0.2	2.6	33	28
	2000	0	4	**5**	100.0	**0.0**	**0.0**	10000	5203	**75**	0.6	19.8	61	36
	2500	0	**5**	**5**	100.0	**0.0**	**0.0**	10000	2163	**53**	0.0	3.0	53	54
E1000	1000	1	**5**	**5**	11.3	**0.0**	**0.0**	10000	1296	**24**	0.1	1.4	24	27
	1500	0	3	**5**	34.8	0.7	**0.0**	10000	3127	**63**	0.3	4.2	63	69
	2000	0	1	**5**	100.0	1.3	**0.0**	10000	10000	**223**	0.5	2.2	223	180
	2500	0	1	**4**	100.0	2.2	**0.1**	10000	10000	**1588**	0.6	9.4	642	278
T10	16	**30**	**30**	**30**	**0.0**	**0.0**	**0.0**	**0**	**0**	0	0.1	1.3	0	0
	30	**30**	**30**	**30**	**0.0**	**0.0**	**0.0**	3	4	**1**	0.1	1.5	1	2
	50	**30**	**30**	**30**	**0.0**	**0.0**	**0.0**	16	16	**2**	0.2	1.2	2	2
	100	**30**	**30**	**30**	**0.0**	**0.0**	**0.0**	106	44	**3**	0.0	1.1	3	4
T30	16	**30**	**30**	**30**	**0.0**	**0.0**	**0.0**	**0**	1	2	0.2	1.5	2	2
	30	**30**	**30**	**30**	**0.0**	**0.0**	**0.0**	**4**	8	8	0.0	1.5	8	8
	50	**30**	**30**	**30**	**0.0**	**0.0**	**0.0**	25	52	**21**	0.5	2.7	17	15
	100	**30**	29	**30**	**0.0**	0.2	**0.0**	186	103	**43**	0.9	4.2	39	28
T50	16	**30**	**30**	**30**	**0.0**	**0.0**	**0.0**	**0**	1	4	0.1	1.6	5	5
	30	**30**	**30**	**30**	**0.0**	**0.0**	**0.0**	**5**	15	17	0.5	6.3	15	18
	50	**30**	**30**	**30**	**0.0**	**0.0**	**0.0**	**34**	67	52	0.8	6.1	39	38
	100	**29**	27	28	0.2	0.2	**0.1**	319	505	**217**	0.9	6.6	151	92
T70	16	**30**	**30**	**30**	**0.0**	**0.0**	**0.0**	**0**	2	8	0.1	1.8	8	8
	30	**30**	**30**	**30**	**0.0**	**0.0**	**0.0**	**5**	24	33	0.7	8.1	27	28
	50	**30**	**30**	**30**	**0.0**	**0.0**	**0.0**	**34**	84	76	0.5	3.5	66	67
	100	**28**	26	27	**0.1**	0.3	0.6	**388**	519	426	1.1	6.9	315	249
T99	16	**30**	**30**	**30**	**0.0**	**0.0**	**0.0**	**1**	2	11	0.5	4.0	11	12
	30	**30**	**30**	**30**	**0.0**	**0.0**	**0.0**	**7**	26	54	0.9	14.6	42	50
	50	**30**	**30**	29	**0.0**	**0.0**	0.1	**36**	76	167	0.5	7.4	117	124
	100	**28**	**28**	26	0.3	**0.2**	0.3	**298**	675	867	0.8	4.5	516	629

In Table 3 we observe that 3PL and 5PL are clearly not competitive to all approaches based on alternative dual-optimal solutions. Although the number of updates of the stability center was not too high, the additional overhead due to these updates and the additional variables and constraints in the model lead to long running times which are usually even higher than those of unstabilized column generation. As we could not identify better parameter values in further tests we conclude that using piecewise linear stabilization does not seem to be promising for the RDCSTP.

Table 4 compares number of instances solved to proven optimality, average remaining gaps, and median CPU-times in seconds for the full branch-and-price approach using alternative dual-optimal solutions $\hat{\pi}$ and ten major iterations (Q=10) to the theoretically stronger static (SL) and dynamic (DL) layered graph approach from [28] (rerun with CPLEX 12.2). SL starts with the same primal solution as BP whereas DL does not use any initial heuristics here. Directed connection cuts are separated in SL and DL only for the instance sets from [12]; Tα instances are solved faster when omitting them. We additionally report the average number of considered branch-and-bound nodes, the average integrality gap at the root node and the median time needed for solving the root node of the branch-and-bound tree for BP as well as the corresponding CPU-time of the column generation approach.

We conclude that due to the stabilization based on alternative dual-optimal solutions the proposed branch-and-price approach outperforms both layered graph approaches in many cases and performs particularly good when the delay bound is not too strict as well as when the relative number of terminal nodes is not too high. Thus we consider it a good complement to the layered graph approaches from [28].

6 Conclusions and Future Work

In this paper we presented a branch-and-price approach for the RDCSTP. Column generation stabilization methods based on alternative dual-optimal solutions and piecewise linear penalty functions have been applied to accelerate the approach. We further compared the performance of two different pricing strategies. We conclude that when using stabilization based on alternative dual-optimal solutions our method outperforms so-far proposed exact methods for the RDCSTP in many cases and allows for computing proven optimal solutions solutions to medium sized instances within reasonable time. In future, we want to compare our approach to further stabilization techniques such as e.g. interior point stabilization [25] or weighted Dantzig-Wolfe decomposition [32,23] as well as combine promising aspects of different stabilization techniques potentially yielding an additional speed-up. We further want to study the impact of different possibilities for choosing an initial set of columns as well as aim at reducing the tailing-off effect.

References

1. Achterberg, T.: SCIP: Solving constraint integer programs. Mathematical Programming Computation 1(1), 1–41 (2009)
2. Amor, H.B., Desrosiers, J.: A proximal trust-region algorithm for column generation stabilization. Computers & Operations Research 33, 910–927 (2006)

3. Amor, H.B., Desrosiers, J., Carvalho, J.M.V.: Dual-optimal inequalities for stabilized column generation. Operations Research 54(3), 454–463 (2006)
4. Amor, H.B., Desrosiers, J., Frangioni, A.: On the choice of explicit stabilizing terms in column generation. Discrete Applied Mathematics 157, 1167–1184 (2009)
5. Barnhart, C., Johnson, E.L., Nemhauser, G.L., Savelsbergh, M.W.P., Vance, P.H.: Branch-and-price: Column generation for solving huge integer programs. Operations Research 46, 316–329 (1998)
6. de Carvalho, J.M.V.: Using extra dual cuts to accelerate convergence in column generation. INFORMS Journal on Computing 17(2), 175–182 (2005)
7. Desaulniers, G., Desrosiers, J., Solomon, M.M. (eds.): Column Generation. Springer, Heidelberg (2005)
8. du Merle, O., Villeneuve, D., Desrosiers, J., Hansen, P.: Stabilized column generation. Discrete Mathematics 194(1-3), 229–237 (1999)
9. Dumitrescu, I., Boland, N.: Improved preprocessing, labeling and scaling algorithms for the weight-constrained shortest path problem. Networks 42(3), 135–153 (2003)
10. Garey, M.R., Johnson, D.S.: Computers and Intractability: A Guide to the Theory of NP-Completeness. W. H. Freeman, New York (1979)
11. Ghaboosi, N., Haghighat, A.T.: A path relinking approach for delay-constrained least-cost multicast routing problem. In: 19th IEEE International Conference on Tools with Artificial Intelligence, pp. 383–390 (2007)
12. Gouveia, L., Paias, A., Sharma, D.: Modeling and solving the rooted distance-constrained minimum spanning tree problem. Computers & Operations Research 35(2), 600–613 (2008)
13. Gouveia, L., Simonetti, L., Uchoa, E.: Modeling hop-constrained and diameter-constrained minimum spanning tree problems as Steiner tree problems over layered graphs. Mathematical Programming, 1–26 (2010)
14. Kompella, V.P., Pasquale, J.C., Polyzos, G.C.: Multicasting for multimedia applications. In: Eleventh Annual Joint Conference of the IEEE Computer and Communications Societies, INFOCOM 1992, pp. 2078–2085. IEEE, Los Alamitos (1992)
15. Kou, L., Markowsky, G., Berman, L.: A fast algorithm for Steiner trees. Acta Informatica 15(2), 141–145 (1981)
16. Leggieri, V., Haouari, M., Triki, C.: An exact algorithm for the Steiner tree problem with delays. Electronic Notes in Discrete Mathematics 36, 223–230 (2010)
17. Leitner, M., Raidl, G.R.: Strong lower bounds for a survivable network design problem. In: Haouari, M., Mahjoub, A.R. (eds.) ISCO 2010. Electronic Notes in Discrete Mathematics, vol. 36, pp. 295–302. Elsevier, Amsterdam (2010)
18. Leitner, M., Raidl, G.R., Pferschy, U.: Accelerating column generation for a survivable network design problem. In: Scutellà, M.G., et al. (eds.) Proceedings of the International Network Optimization Conference 2009, Pisa, Italy (2009)
19. Leitner, M., Raidl, G.R., Pferschy, U.: Branch-and-price for a survivable network design problem. Tech. Rep. TR 186–1–10–02, Vienna University of Technology, Vienna, Austria (2010), submitted to Networks
20. Leitner, M., Ruthmair, M., Raidl, G.R.: Stabilized column generation for the rooted delay-constrained Steiner tree problem. In: VII ALIO/EURO – Workshop on Applied Combinatorial Optimization, Porto, Portugal (May 2011)
21. Manyem, P., Stallmann, M.: Some approximation results in multicasting. Tech. Rep. TR-96-03, North Carolina State University (1996)
22. Marsten, R.E., Hogan, W.W., Blankenship, J.W.: The BOXSTEP method for large-scale optimization. Operations Research 23(3), 389–405 (1975)
23. Pessoa, A., Uchoa, E., de Aragão, M., Rodrigues, R.: Exact algorithm over an arc-time-indexed formulation for parallel machine scheduling problems. Mathematical Programming Computation 2(3), 259–290 (2010)

24. Qu, R., Xu, Y., Kendall, G.: A variable neighborhood descent search algorithm for delay-constrained least-cost multicast routing. In: Stützle, T. (ed.) LION 3. LNCS, vol. 5851, pp. 15–29. Springer, Heidelberg (2009)
25. Rousseau, L.M., Gendreau, M., Feillet, D.: Interior point stabilization for column generation. Operations Research Letters 35(5), 660–668 (2007)
26. Ruthmair, M., Raidl, G.R.: A kruskal-based heuristic for the rooted delay-constrained minimum spanning tree problem. In: Moreno-Díaz, R., Pichler, F., Quesada-Arencibia, A. (eds.) EUROCAST 2009. LNCS, vol. 5717, pp. 713–720. Springer, Heidelberg (2009)
27. Ruthmair, M., Raidl, G.R.: Variable neighborhood search and ant colony optimization for the rooted delay-constrained minimum spanning tree problem. In: Schaefer, R., Cotta, C., Kołodziej, J., Rudolph, G. (eds.) PPSN XI. LNCS, vol. 6239, pp. 391–400. Springer, Heidelberg (2010)
28. Ruthmair, M., Raidl, G.R.: A layered graph model and an adaptive layers framework to solve delay-constrained minimum tree problems. In: Fifteenth Conference on Integer Programming and Combinatorial Optimization (IPCO XV) (2011) (to appear)
29. Skorin-Kapov, N., Kos, M.: The application of Steiner trees to delay constrained multicast routing: a tabu search approach. In: Proceedings of the 7th International Conference on Telecommunications, vol. 2, pp. 443–448 (2003)
30. Skorin-Kapov, N., Kos, M.: A GRASP heuristic for the delay-constrained multicast routing problem. Telecommunication Systems 32(1), 55–69 (2006)
31. Vanderbeck, F.: Implementing mixed integer column generation. In: Desaulniers et al [7], pp. 331–358
32. Wentges, P.: Weighted Dantzig-Wolfe decomposition for linear mixed-integer programming. International Transactions in Operational Research 4(2), 151–162 (1997)
33. Xu, Y., Qu, R.: A GRASP approach for the delay-constrained multicast routing problem. In: Proceedings of the 4th Multidisplinary International Scheduling Conference (MISTA4), pp. 93–104 (2009)
34. Xu, Y., Qu, R.: A hybrid scatter search meta-heuristic for delay-constrained multicast routing problems. Applied Intelligence, 1–13 (2010)

A Heuristic Algorithm for a Prize-Collecting Local Access Network Design Problem

Ivana Ljubić[1], Peter Putz[2], and Juan-José Salazar-González[3]

[1] Institute of Computer Graphics and Algorithms,
Vienna University of Technology, Austria
ivana@ads.tuwien.ac.at
[2] Department of Statistics and Operations Research, University of Vienna, Austria
p.putz@univie.ac.at
[3] DEIOC, Universidad de La Laguna, Tenerife, Spain
jjsalaza@ull.es

Abstract. This paper presents the main findings when approaching an optimization problem proposed to us by a telecommunication company in Austria. It concerns deploying a broadband telecommunications system that lays optical fiber cable from a *central office* to a number of *end-customers*, i.e., *fiber to the home* technology. End-customers represent buildings with apartments and/or offices. It is a capacitated network design problem that requires an installation of optical fiber cables with sufficient capacity to carry the traffic from the central office to the end-customers at minimum cost. This type of problem arises in the design of a Local Access Network (LAN) and in the literature is also named Single-Source Network Loading. In the situation motivating our work the network does not necessarily need to connect all customers (or at least not with the best available technology). Instead, some nodes are *potential* customers. The aim is to select the customers to be connected to the central server and to choose the cable capacities to establish these connections. The telecom company takes the strategic decision of fixing a percentage of customers that should be served, and aims for minimizing the total cost of the network proving this minimum service. For that reason the underlying problem is called the *Prize-Collecting* LAN problem (PC-LAN). We propose a sophisticated heuristic to solve real-world instances with up to 86 000 nodes and around 1 500 potential customers.

1 Introduction

This paper models the fiber-to-the-building/fiber-to-the-home strategy applied by telecommunication companies when designing the new generation of local access networks. *Customer nodes* are associated to physical locations representing buildings, business locations or single households. There are three features associated to each potential customer node. One is called *prize* and it represents the number of subscribers (e.g., apartments and/or offices) in the building. Another is called *demand* and gives the number of optical fibers required by the potential customer. The third feature is called *cost* and represents the setup cost

J. Pahl, T. Reiners, and S. Voß (Eds.): INOC 2011, LNCS 6701, pp. 139–144, 2011.

of installing a device if the customer is connected and receives its service. The available hardware (i.e., splitter devices) determines the demand and the cost of each customer node. Usually, several splitter devices with various splitting ratios (e.g., 1:4, 1:16, 1:32) are available. Their costs obey economies of scales. For example, to connect 16 subscribers, a device must be installed that costs €2000 and 1 optical fiber should come in that building. To connect a building with 17 subscribers, a device that costs €3000 and 2 fibers are needed and this larger device is sufficient to support up to 32 subscribers. We are not allowed to connect only a fraction of subscribers at the customer node. Instead, decisions have to be made whether all subscribers or none of them are going to be served. This allows for preprocessing of input data and exact calculation of customers' demands and the corresponding set-up costs.

The customers selected for being served must be connected by cables to the server (the center). The company provides different types of cables. Each type of cable is characterized by two features. One is its capacity and represents the number of optical fibers. The other is its cost. Clearly, for connecting two sites one may need several cables. Each combination of cables leads to a *module* with a given *capacity* and *cost*. The capacity of a module is simply the sum of the fibers included in the cables. The cost of a module is the sum of the cable costs plus the installation on the roads taking into account the length. The goal is to decide the modules to be installed so that the whole demand can be routed through the network at minimum cost.

The flow between the center and a customer is allowed to split apart, thus we are speaking of a *bifurcated* formulation. An optimal solution of the problem is not necessarily a tree in the graph. Obviously, if there is only one module per edge providing sufficient capacity to route the total flow through it, then the optimal solution will be a tree.

There are works in the literature focusing on other aspects of fiber-to-the-home (see e.g. [5] where more than one server in two layers are considered). The literature also contains several approaches for solving problems arising in LAN design. See e.g. recent works in [3,4] or [2] for a more general multiple-source multiple-sink setting. However, the previously published articles deal with problems where all customers must be served. In this work we develop a sophisticated heuristic for the Prize-Collecting variant of the problem.

2 Problem Definition

We are given an undirected and connected graph $G = (V, E)$ with a node $r \in V$ representing the center (central server, access to the backbone network,...) and a set of potential customers $D \subseteq V \setminus \{r\}$. To each potential customer $k \in D$, a positive demand d_k, a positive prize p_k and a positive setup cost c_k are assigned. We denote by p_0 the minimum customer prizes the company wants to collect.

It is possible to install combinations of different cable types with positive costs and capacities on every edge. Each combination is called a *module*. We assume that modules $N_e = \{n_1, n_2, \dots, n_{|N_e|}\}$ are given for each edge, with capacities

$u_{e,n} \in \mathbb{R}_{>0}$ and costs $c_{e,n} \in \mathbb{R}_{>0}$ for each $1 \leq n \leq |N_e|$. Module indices are sorted such that $u_{e,n} < u_{e,n+1}$. The optimization problem of our interest is the selection of the customers to be served, the single-source multiple-sink routing, and the edge capacity design. To this end we allow an installation of *at most one module* on every edge.

We now show a Single Commodity Flow (SCF) formulation of the PC-LAN problem. This and other models have been derived from the ones given in [3] for the problem where all customers must be in the network. We make use of a binary variable y_k to model whether a potential customer k must be served or not. The SCF formulation uses a bidirected problem representation and models the flow on every arc as the total amount of flow routed from the center towards the customers. The arc set is denoted by A. To model a non-decreasing step cost function on every edge, binary variables need to be used. Binary variables $x_{a,n}$ decide whether the module n shall be installed on the arc a, whereas flow variables $f_a \geq 0$ describe the amount of flow on arc $a \in A$. Then the model is:

$$\min \sum_{a \in A} \sum_{n \in N_a} c_{a,n} x_{a,n} + \sum_{k \in D} c_k y_k$$

subject to

$$\sum_{a \in \delta^+(i)} f_a - \sum_{a \in \delta^-(i)} f_a = \begin{cases} -d_i y_i, & i \in D \\ \sum_{k \in D} d_k y_k, & i = r \\ 0, & \text{otherwise} \end{cases} \qquad \forall i \in V$$

$$\sum_{k \in D} p_k y_k \geq p_0$$

$$\sum_{n \in N_a} x_{a,n} \leq 1 \qquad \forall a \in A$$

$$0 \leq f_a \leq \sum_{n \in N_a} u_{a,n} x_{a,n} \qquad \forall a \in A$$

$$x_{a,n} \in \{0,1\} \qquad \forall a \in A,\ \forall n \in N_a$$

$$y_k \in \{0,1\} \qquad \forall k \in D.$$

3 Heuristic

We now describe the main elements of the heuristic approach that we are developing to solve large-sized instances of the PC-LAN problem. In a first phase we select a set of customers. In a second phase we construct the network iteratively by using shortest path calculations on the graph with adapting edge weights. In a third phase, we locally improve the solution. The three phases are repeated within an reactive search framework.

Selection of Customers: Given pre-installed capacities z_e on the edges, we sort the potential customers according to the ratio between the prize of the customer and the setup cost plus the cost of the shortest path from the center to the customer site. The calculation of the shortest paths is described below. Customers are then chosen so that they satisfy the coverage constraint $\sum_{k \in D} p_k y_k \geq p_0$.

Network Construction: The input to the network construction heuristic is the undirected graph G, a vector $b \in \mathbb{R}_{\geq 0}^{|V|}$ with $b_k = d_k y_k$ for a potential customer $k \in D$ and $b_i = 0$ otherwise. The heuristic subsequently modifies b and creates a flow vector $f \in \mathbb{R}_{\geq 0}^{|E|}$ and a design vector $z \in \mathbb{N}^{|E|}$, where z_e denotes the module to be installed on the edge e and 0 if no module is installed. We randomly pick a node v with positive $b_v > 0$ and define the edge weights for the shortest path problem as $w_e = c_{e,n_e(f_e+b_v)} - c_{e,z_e}$. Here n_e represents a function that gives the *most appropriate module* for some required capacity, i.e. the cheapest module with sufficient capacity, or simply the largest module if there is no module with sufficient capacity. More formally, we have $n_e(U) = \arg\min_{\{n \in N_e | u_{e,n} \geq U\}} c_{e,n}$ if $\exists n \in N_e | u_{e,n} \geq U$, and $n_e(U) = |N_e|$ otherwise. Hence w_e is the cost for expanding the installation on e from the currently selected module z_e, to the module $n_e(f_e + b_v)$. An edge e is *saturated* if we already use the largest module $z_e = |N_e|$ and all available capacity is being used $u_{e,z_e} = f_e$. Once an edge is saturated it is not considered in subsequent computations. Now that we are given a node v and edge weights w we can compute the shortest path $P = \langle (r, i_1), (i_1, i_2), \ldots, (i_m, v) \rangle$ from r to v in G. We transport the demand along this path $f'_e = f_e + b_v$ for all $e \in P$, and make the necessary installation $z'_e = n_e(f'_e)$. We mark this customer as done $b'_v = 0$ and start the next iteration.

If the edge $e = (i, j)$ on P is about to be saturated in the current iteration, we transport the maximum possible amount and leave the appropriate demands on each ends of the arc. More formally, if $u_{e,n_e(f_e+b_v)} < f_e + b_v$ then we take the maximum $u = u_{e,n_e(f_e+b_v)} - f_e$, transport it along the edge $f'_e = f_e + u$ and change the demand vector to $b'_i = b_i + u$ and $b'_j = b_j + b_v - u$. Now the edge is saturated and we continue with the next node.

After the construction has produced a feasible solution we take z and y to define input parameters for a minimum cost flow problem. To this end, for each edge e with $z_e > 0$ the capacity is set to u_{e,z_e} and the cost is set to $c_{e,z_e}/u_{e,z_e}$. The minimum cost flow problem is solved and produces a new flow vector f'. Using f' we define a new design z' with $z'_e \leq z_e$ for all e.

Local Improvement: Given a feasible solution associated to the vector (z, y, f), we attempt the following local improvement. Pick the edge e maximizing c_{e,z_e} or $c_{e,z_e}/f_e$, alternating between these two criteria. The flow f is split into a set of paths. Take all those paths that use the edge e and remove all the flow along these paths from f to produce a smaller flow f'. The values of f' define pre-installed modules z_e on the edges. The customers whose flows have been just canceled out are removed. Now we re-enter the selection phase and choose the remaining set of customers to ensure the coverage constraint. This is done repeatedly to eventually get a better solution. The procedure stops after 10 edges have been considered without improving the objective value.

Reactive Search (RS): To enable a clustering of demands we do not install the modules along the whole path right to the center but instead stop at some earlier node j. Two criteria are used to select j: (1) j is the first node with a positive demand $b_j > 0$ encountered along the path, or (2) j is at most m edges away from v. Observe that the demands are clustered if criteria (1) is applied and m is set to a small number. The idea of this clustering is to merge customers that are close to each other with respect to the stepwise edge cost function.

To decide whether we activate or not criteria (1) and the best value m for criteria (2) we employ a learning adaptation mechanism known as *Reactive Search Optimization* [1]. The Network Construction procedure is applied repeatedly. Initially five prespecified settings are used. Then we go from a diversification of the settings towards an intensification, i.e. from randomly perturbed settings towards settings that have produced the best objective values so far.

Primal Heuristic: In order to use the previous framework within a branch-and-cut, for a given fractional solution $(\bar{x}, \bar{y})$ we: a) sort the customers according to the $\bar{y}_k$ values and b) pre-install capacities z_e according to the $\bar{x}_{a,n}$ values before entering the network construction phase.

4 Results

We have implemented the heuristic on a computer Intel Xeon 2.6 Ghz with 3 Gb RAM. In addition we extended the branch-and-cut framework from [3] using CPLEX 12.1. We initiated our evaluation on the small instances in [4]. Most of these instances can be solved to optimality which allows to measure the quality of the heuristic approach. To produce instances for the PC-LAN problem from the 60 instances used in [4] we have defined $c_k = d_k/2$, $p_k = d_k$, $p_0 = 0.7 \sum_{k \in D} p_k$.

Table 1 displays results of our experiments. Each line shows the average values over 5 instances. The instances are grouped according to the number $|V|$ of nodes (i.e., 20, 30 or 40), the position of the center (central or random) and the demand level (high or low). $|N|$ denotes the average number of modules available over all edges. The columns UB0 refer to the initial heuristic approach where the RS is executed until 50 consecutive iterations without improvement. The columns UB1 refers to the primal heuristic applying the RS on the fractional solutions generated while solving the root node. The RS is repeated while there is an improvement in the solution. The columns BEST refer to the best known solution obtained by running the branch-and-cut framework with our primal heuristic within 1000 seconds. Over the 60 instances this best known was optimal in 42 cases. The gap in UB0 and UB1 is computed over the solution obtained in BEST, while the gap in BEST is between the best upper and lower bounds. Based on this experiment the construction heuristic UB0 already produces good results on these small instances. To conclude, Table 2 shows the performance of the approach on the three real world instances. The branch-and-cut framework did not solve the root node within two hours. It remains an open question whether the presented gaps are due to the lower bounds. Still the feasible solutions are of reasonable quality for practical purpose.

Table 1. Results for instances from [4]

$\lvert V\rvert$	$\lvert E\rvert$	$\lvert D\rvert$	$\lvert N\rvert$			UB0		UB1		BEST	
						gap0	time0	gap1	time1	gap	time
20.0	40.2	9.2	27.6	c	h	1.2	0.1	0.8	4.3	0.0	18.5
20.0	40.2	9.0	12.6	c	l	0.0	0.0	0.0	3.9	0.0	17.7
20.0	39.8	9.2	23.2	r	h	0.5	0.0	0.1	4.3	0.0	64.3
20.0	39.8	10.0	13.0	r	l	3.5	0.0	0.1	3.9	0.0	22.0
30.0	58.4	15.8	47.4	c	h	0.3	0.2	0.3	10.6	0.2	489.8
30.0	58.4	16.0	24.2	c	l	0.2	0.1	0.2	8.9	0.0	467.8
30.0	59.2	12.2	33.4	r	h	0.1	0.1	0.1	6.1	0.0	182.5
30.0	59.2	14.4	22.2	r	l	0.1	0.1	0.1	7.3	0.0	46.2
40.0	80.0	19.2	49.4	c	h	0.6	0.2	0.4	17.7	0.4	968.8
40.0	80.0	19.2	27.4	c	l	1.1	0.1	1.0	13.1	0.5	665.5
40.0	80.6	18.2	46.2	r	h	1.1	0.2	1.0	13.1	0.7	905.8
40.0	80.6	20.6	31.4	r	l	0.9	0.1	0.8	14.9	0.4	876.3

Table 2. Results for real-world instances

$\lvert V\rvert$	$\lvert E\rvert$	$\lvert D\rvert$	$\lvert N\rvert$	UB0	time0	UB1	gap1
86 745	116 750	1 157	9.0	2 881 000	1 517	2 814 139	50
48 247	65 304	720	9.0	1 499 750	290	1 499 750	26
77 329	107 696	1 498	9.9	3 280 373	2 361	3 280 373	54

Acknowledgement

This work was partially supported by the bilateral program "Acciones Integradas" between Austria (E15/2010) and Spain (AT2009-0021).

References

1. Battiti, R., Brunato, M., Mascia, F.: Reactive Search and Intelligent Optimization. Operations Research/Computer Science Interfaces Series, vol. 45. Springer, Heidelberg (November 2008)
2. Frangioni, A., Gendron, B.: 0-1 reformulations of the multicommodity capacitated network design problem. Discrete Applied Mathematics 157(6), 1229–1241 (2009)
3. Ljubić, I., Putz, P., Salazar, J.J.: Exact approaches to the single-source network loading problem. Networks (2011) (to appear)
4. Salman, F.S., Ravi, R., Hooker, J.N.: Solving the Capacitated Local Access Network Design Problem. INFORMS Journal on Computing 20(2), 243–254 (2008)
5. Gualandi, S., Malucelli, F., Sozzi, D.: On the design of the next generation access networks. In: Lodi, A., Milano, M., Toth, P. (eds.) CPAIOR 2010. LNCS, vol. 6140, pp. 162–175. Springer, Heidelberg (2010)

The Two Layer Network Design Problem

Sara Mattia

Technische Universität Dortmund, Fakultät für Mathematik, Vogelpothsweg 87, 44227 Dortmund, Germany
sara.mattia@math.tu-dortmund.de

Abstract. A two layer network design problem arising from the design of an optical network is considered. Two versions of the optical network design problem are studied and polyhedral results for the capacity formulation of the problems are presented.

1 Introduction and Motivation

An optical network is a two layer network where the lower layer is called *physical* or *optical* while the upper layer (service layer) is called *logical*. The nodes of the logical layer are a subset of the nodes of the physical layer (usually the assumption that the two sets coincide is made). The edges of the physical network corresponds to fiber cables, the edges of the logical network, or *lightpaths*, are logical connections corresponding to physical paths. Given the physical and the logical network, both with capacity installation costs for the edges, and given a set of point-to-point traffic demands (or *commodities*) in the logical network, the two layer network design problem consists of choosing minimum costs integer capacities for the edges of both layers so that all the demands can be routed simultaneously on the network.

Several problems related to the design of an optical network have been studied in the literature. See for example [10,4,5,6,22,15,12,11,19] and references therein. In [10] the authors present polyhedral results for a two layer problem where only the dimensioning of the logical layer is required, while the physical capacities are given. In [9] a problem with survivability requirements is considered. In [5] a branch-and-cut-and-price approach is proposed. In [22] heuristics to be used within a branch-and-cut algorithm are discussed. In [15] the authors use single layer cuts to solve a survivable two layer network design problem. In [11,19] computational results for the capacity formulation of the two layer network design problem with and without survivability requirements are presented, but no polyhedral results are given. For a survey on technical details of optical networks and on optimization problems arising from telecommunications networks see [23,24].

While the polyhedron of single layer network design problem has received a lot of attention in the literature (see for example [17,18,8,13,7,2,1,3] and references therein), the two layer network design problem has received only a limited attention and, to our knowledge, no polyhedral result is known for the capacity formulation of the problem addressed in this paper. Here two versions of the problem are considered. The two models depend on the definition of the lightpaths and have already been considered in the

J. Pahl, T. Reiners, and S. Voß (Eds.): INOC 2011, LNCS 6701, pp. 145–149, 2011.

literature, see for example [22,15,11,19] and references therein, but no polyhedral contribution has been given for them so far. In [22,15,19] and references therein several variants of the two layer network design problem using the first model are addressed, but the facet defining status of the inequalities used in the computational experiments is not proved. In [11,19] the authors present computational experiments based on a Benders decomposition approach for the two layer network design problem with and without survivability requirements using the second model, but no polyhedral contribution is given. In this paper the polyhedral properties of the capacity formulation of both problems are studied. The paper is organized as follows. In Section 2 we present the mathematical formulation of the considered problems, in Sections 3 polyhedral results are given.

2 Formulation of the Problem

Depending on the definition of the lightpaths, two different models can be used to formulate the problem. If the physical path corresponding to the lightpaths is known in advance, then we have an *explicit* lightpath model. If the paths corresponding to the lightpaths are not known in advance and they have to be computed during the optimization process, then we have an *implicit* lightpath model. In the second case the assumption that the cost of a lightpath does not depend on its routing is made. For applications of the explicit model see for example [22,15,19] and references therein. For applications of the implicit model see for example [11,19] and references therein.

Let $G(V,E)$ be an undirected graph without loops and parallel edges corresponding to the physical network and let $H(V,L)$ be an undirected graph corresponding to the logical network. Capacity on logical and physical edges can only be installed in batches, let $B \geq 1$ be the capacity batch for physical edges while the batch for the logical edges is set to one. Let c^E_e and c^L_ℓ be the cost of installing a capacity batch on edge $e \in E$ and $\ell \in L$, respectively. Let K be the set of commodities, where each commodity k is a triple (s^k, t^k, d^k), i.e. s^k is the source node, t^k is the destination node and d^k is the demand. For the explicit model, let L_e be the set of lightpaths using physical edge $e \in E$. Let x_e be an integer variable representing the number of capacity batches installed on physical edge $e \in E$, and let y_ℓ be an integer variable representing the number of capacity batches installed on logical edge $\ell \in L$.

If we use the explicit model, the capacity formulation of the problem is the following (see [19]).

$$(ECF) \quad \min \sum_{e \in E} c^E_e x_e + \sum_{\ell \in L} c^L_\ell y_\ell$$

$$\sum_{\ell=(i,j) \in L} \mu_{ij} y_\ell \geq \sum_{k \in K} \pi^\mu_k d^k \qquad \mu \geq 0 \tag{1}$$

$$\sum_{\ell \in L_e} y_\ell \leq B x_e \qquad e \in E \tag{2}$$

$$x \in \mathbb{Z}^{|E|}_+, y_\ell \in \mathbb{Z}^{|L|}_+$$

Constraints (1) are metric inequalities [21,14], constraints (2) are capacity constraints for the physical layer. For the implicit model, the capacity formulation is the following (see [11] and references therein).

$$(ICF)\ \min \sum_{e\in E} c_e^E x_e + \sum_{\ell\in L} c_\ell^L y_\ell$$

$$\sum_{\ell=(i,j)\in L} \mu_{ij} y_\ell \geq \sum_{k\in K} \pi_k^\mu d^k \qquad \mu \geq 0 \tag{3}$$

$$\sum_{e\in E} \eta_e B x_e \geq \sum_{\ell\in L} \pi_\ell^\eta y_\ell \qquad \eta \geq 0 \tag{4}$$

$$x \in \mathbb{Z}_+^{|E|}, y_\ell \in \mathbb{Z}_+^{|L|}$$

Constraints (3) and (4) are metric inequalities for the logical and the physical layer.

Given a graph R, let us denote by Met_R the set generated by all non-zero metrics on R. As pointed out in [16], μ vectors corresponding to metrics are enough to guarantee feasibility, therefore we can restrict to $\mu \in Met_H$ for constraints (1) and (3), and to $\eta \in Met_G$ for constraints (4).

3 The Polyhedron

Let *ELPol* be the convex-hull of integer feasible solutions of (ECF) and let *ILPol* be the convex-hull of integer feasible solutions of (ICF).

Theorem 1. *ELPol and ILPol are full-dimensional.*

A edge $e \in E$ is a physical bridge if its removal physically disconnects at least one origin-destination pair in the physical network. A edge $\ell \in L$ is a logical bridge if its removal disconnects at least one origin destination pair in the logical network.

Theorem 2. *Inequalities $x_e \geq 0$ and $y_\ell \geq 0$ are facet-defining both for ELPol and for ILPol if and only if e is not a physical bridge and ℓ is not a logical bridge.*

Theorem 3. *Let $a^T x + g^T y \geq b$ be a valid inequality for ELPol or ILPol, then $a \geq 0$.*

Theorem 4. *Let $a^T x + g^T y \geq b$ be a facet-defining inequality for ILPol, then $a \in Met_G$.*

Theorem 5. *Let $a^T x + g^T y \geq b$ be a valid inequality for ELPol or ILPol having $a = 0$, then $g \geq 0$.*

Let $SL(H,K)$ be the convex-hull of integer feasible solutions of the single layer problem corresponding to the logical layer. Any inequality that is valid for $SL(H,K)$ is also valid for *ELPol* and *ILPol*, moreover the following result holds.

Theorem 6. *All facet-defining inequalities of $SL(H,K)$ are facet-defining for ELPol and ILPol.*

Let $\mu \in Met_H$, we define R_μ as $R_\mu = \min\{\mu^T y : y \in SL(H,K)\}$. Inequalities $\mu^T y \geq R_\mu$ are called *tight metric inequalities* [3].

Theorem 7. *All facet-defining inequalities $a^T x + g^T y \geq b$ for ELPol or ILPol having $a = 0$ are tight metric inequalities.*

Let $P = \{P^1, \ldots, P^s\}$ be an s-partition of the nodes, i.e. $P^i \subseteq V$ for all $i \in \{1, \ldots, s\}$, $\cup_{i=1}^s P^i = V$, $P^i \cap P^j = \emptyset$ for all $i, j \in \{1, \ldots, s\}$, $i \neq j$. Let $G(P, E_P)$ and $H(P, L_P)$ be the physical and logical graphs obtained shrinking all the nodes in a subset into a single node. The parallel edges are replaced by a single edge for the implicit model, they are not replaced by a single edge to avoid losing the correspondence between logical and physical edges for the explicit one. Any inequality that is valid for the s-node problem is valid for the original problem (in the implicit case each edge must be replaced by the sum of the parallel edges that were shrunk). The following theorem holds.

Theorem 8. *Let $a^T x + g^T y \geq b$ a facet-defining inequality for the s-node problem, then it is facet-defining for ELPol and ILPol.*

Theorem 9. *Inequality (2) $\sum_{\ell \in L_e} y_\ell \leq B x_e$ is facet-defining for ELPol.*

Let $\{S, V - S\}$ be a 2-partition of the nodes and let $\Delta(S) = \left\lceil \sum_{k \in \delta^K(S)} d^k \right\rceil$ where $\delta^K(S)$ is the set of the demands having s^k and t^k is different subsets of the partition.

Theorem 10. *The inequality:*

$$\sum_{e \in \delta^E(S)} x_e \geq \lceil \Delta(S)/B \rceil \tag{5}$$

is facet-defining for ILPol if and only if: (i) $\Delta(S) > 0$, (ii) the subsets are connected both in the physical and in the logical network, (iii) $\Delta(S)/B$ is not integer.

Theorem 11. *Let $S \subseteq V$. If S and $V - S$ are connected then inequality $\sum_{e \in \delta^E(S)} B x_e \geq \sum_{\ell \in \delta^L(S)} y_\ell$ is facet-defining for ILPol.*

It is easy to see that the physical connection of the subsets is also a necessary condition for $B \sum_{e \in \delta^E(S)} x_e \geq \sum_{\ell \in \delta^L(S)} y_\ell$ to be facet-defining. The proofs of the above results can be found in [20].

References

1. Agarwal, Y.: K-partition-based facets of the network design problem. Networks 47(3), 123–139 (2006)
2. Atamtürk, A.: On capacitated network design cut-set polyhedra. Mathematical Programming 92, 425–437 (2002)
3. Avella, P., Mattia, S., Sassano, A.: Metric inequalities and the network loading problem. Disc. Opt. 4, 103–114 (2007)
4. Banerjee, D., Mukherjee, B.: Wavelength-routed optical networks: Linear formulatation, resource budgeting tradeoffs, and a reconfiguration study. IEEE Transactions on Networking 8(5), 598–607 (2000)
5. Belotti, P., Koster, A.M.C.A., Orlowski, S.: A cut&branch&price approach to two-layer network design. In: The 8th INFORMS Telecommunications Conference, Dallas, TX (2006)

6. Belotti, P., Malucelli, F.: Multilayer network design: a row-column generation algorithm. In: Proceedings of the 2nd International Network Optimization Conference (INOC 2005), Lisbon, Portugal, vol. 3, pp. 422–427 (March 2005)
7. Bienstock, D., Chopra, S., Günlük, O., Tsai, C.Y.: Mininum cost capacity installation for multicommodity flows. Mathematical Programming 81, 177–199 (1998)
8. Bienstock, D., Günlük, O.: Capacitated network design - polyhedral structure and computation. INFORMS Journal on Computing 8, 243–259 (1996)
9. Borne, S., Gourdin, E., Liau, B., Mahjoub, A.: Design of survivable IP-over-optical networks. Ann. Oper. Res. 146, 41–73 (2006)
10. Dahl, G., Martin, A., Stoer, M.: Routing through virtual paths in layered telecommunication networks. Operations Research 47(5), 693–702 (1999)
11. Fortz, B., Poss, M.: An improved benders decomposition applied to a multi-layer network design problem. Oper. Res. Lett. 37(5), 777–795 (2009)
12. Gouveia, L., Patricio, P., de Sousa, A.: Hop-constrained node survivable network design: An application to MPLS over WDM. Networks and Spatial Economics 8(1), 3–21 (2008), `http://ideas.repec.org/a/kap/netspa/v8y2008i1p3-21.html`
13. Günlük, O.: A branch-and-cut algorithm for capacitated network design problems. Mathematical Programming 86, 17–39 (1999)
14. Iri, M.: On an extension of the max-flow min-cut theorem to multicommodity flows. Journal of the Operations Research Society of Japan 13, 129–135 (1971)
15. Koster, A., Orlowski, S., Raack, C., Baier, G., Engel, T.: Single-layer Cuts for Multi-Layer Network Design Problems, ch. 1, pp. 1–23. Springer, Heidelberg (2008), selected proceedings 9th INFORMS Telecommunications Conference
16. Lomonosov, M.: Combinatorial approaches to multiflow problems. Disc. Appl. Math. 11, 1–93 (1985)
17. Magnanti, T.L., Mirchandani, P., Vachani, R.: The convex hull of two core capacitated network design problems. Mathematical Programming 60, 233–250 (1993)
18. Magnanti, T.L., Mirchandani, P., Vachani, R.: Modeling and solving the two-facility capacitated network loading problem. Operations Research 43, 142–157 (1995)
19. Mattia, S.: Solving survivable two-layer network design problems by metric inequalities. Computational Optimization and Applications (2010), doi: 10.1007/s10589-010-9364-0
20. Mattia, S.: A polyhedral study of the capacity formulation of the multilayer network design problem. Optimization Online 2011-03-2947 (2011)
21. Onaga, K., Kakusho, O.: On feasibility conditions of multicommodity flows in network. IEEE Trans. Circuit Theory 18(4), 425–429 (1971)
22. Orlowski, S., Koster, A., Raack, C., Wessäly, R.: Two-layer network design by branch-and-cut featuring MIP-based heuristics. In: Proceedings of the INOC 2007, Spa, Belgium (2007), `http://www.poms.ucl.ac.be/inoc2007/Papers/author.89/paper/paper.89.pdf`
23. Orlowski, S., Wessäly, R.: An integer programming model for multi-layer network design. ZIB Preprint ZR-04-49, Konrad-Zuse-Zentrum für Informationstechnik Berlin (December 2004), `http://www.zib.de/PaperWeb/abstracts/ZR-04-49`
24. Yuan, D.: An annotated bibliography in communication network design and routing. Ph.D. thesis, Institute of Technology, Linköpings Universitet (2001)

Affine Recourse for the Robust Network Design Problem: Between Static and Dynamic Routing

Michael Poss[1] and Christian Raack[2]

[1] Department of Computer Science, Faculté des Sciences,
Université Libre de Bruxelles, Brussels, Belgium
mposs@ulb.ac.be

[2] Zuse Institute Berlin (ZIB), Takustr. 7, D-14195 Berlin, Germany
raack@zib.de

Abstract. Affinely-Adjustable Robust Counterparts are used to provide tractable alternatives to (two-stage) robust programs with arbitrary recourse. We apply them to robust network design with polyhedral demand uncertainty, introducing the affine routing principle. We compare the affine routing to the well-studied static and dynamic routing schemes for robust network design. It is shown that affine routing can be seen as a generalization of the widely used static routing still being tractable and providing cheaper solutions. We investigate properties on the demand polytope under which affine routings reduce to static routings and also develop conditions on the uncertainty set leading to dynamic routings being affine. We show however that affine routings suffer from the drawback that (even strongly) dominated demand vectors are not necessarily supported by affine solutions.

1 Introduction

In the classical deterministic network design problem, a set of point-to point commodities with known demand values is given, and capacities have to be installed on the network links at minimum cost such that the resulting capacitated network is able to accommodate all demands simultaneously by a multi-commodity flow. In practice however, exact demand values are usually not known at the time the design decisions must be made. Robust optimization overcomes this problem by explicitly taking into account the uncertainty of the data introducing so-called uncertainty sets. A solution is said to be feasible if it is feasible for all realizations of the data in a predetermined uncertainty set $\mathscr{D}$. Introducing even more flexibility, two-stage robust optimization allows to adjust a subset of the problem variables only after observing the actual realization of the data. [3]. In fact, it is natural to apply this two-stage approach to network design since very often first stage capacity design decisions are made in the long term while the actual routing is adjusted based on observed user demands. This second stage adjusting procedure is called *recourse* which in the context of network design relates to what is known as traffic engineering. Unrestricted second stage recourse in robust network design is called *dynamic routing*, see [5]. Given a fixed design, the commodity routing can be changed arbitrarily for every realization of the demands. In [5] it is shown that

J. Pahl, T. Reiners, and S. Voß (Eds.): INOC 2011, LNCS 6701, pp. 150–155, 2011.

allowing for dynamic routing makes robust network design intractable. Already deciding whether or not a fixed capacity design allows for a dynamic routing of demands in a given polytope is $\mathcal{NP}$-complete (on directed graphs).

This paper is motivated by the scarcity of works using affine routing. Following [3], we introduce affine routing as a generalization of static routing allowing for more routing flexibility but still yielding polynomially solvable robust counterparts, (in opposition to the $\mathcal{NP}$-hard scheme from [9], among others). In this context affine routing provides a tractable alternative in between static and dynamic routing. Affine routing has been used implicitly already in [8] for a robust network design problem with a particular uncertainty set. The contributions of this paper consist of a theoretical and empirical study of network design under the affine routing principle for general polyhedral demand uncertainty sets $\mathcal{D}$. Section 2 introduces the mathematical models and defines formally static, affine and dynamic routings. In Section 3 we present our main results. Proofs are omitted due to space restrictions. We also conducted numerical comparisons of static, affine and dynamic routings, which are not presented due to space restrictions.

2 Robust Network Design with Recourse

We are given a directed graph $G = (V,A)$ and a set of commodities K. A commodity $k \in K$ has source $s(k) \in V$, destination $t(k) \in V$, and demand value $d^k \geq 0$. A *flow* for k is a vector $f^k \in \mathbb{R}_+^A$ satisfying:

$$\sum_{a\in\delta^+(v)} f_a^k - \sum_{a\in\delta^-(v)} f_a^k = d^k \psi_{vk} \quad \text{for all } v \in V, \tag{1}$$

where $\delta^+(v)$ and $\delta^-(v)$ denote the set of outgoing arcs and incoming arcs at node v, respectively. For node $v \in V$ and commodity $k \in K$ we set $\psi_{vk} := 1$ if $v = s(k)$, $\psi_{vk} := -1$ if $v = t(k)$, and $\psi_{vk} := 0$ else. Flows are non-negative. A *multi-commodity flow* is a collection of flows, one for each commodity in K. A *circulation* (or cycle-flow) is a vector $g \in \mathbb{R}^A$ satisfying

$$\sum_{a\in\delta^+(v)} g_a - \sum_{a\in\delta^-(v)} g_a = 0 \quad \text{for all } v \in V. \tag{2}$$

A circulation is not necessarily non-negative. A value $g_a < 0$ can be seen as a flow from the head of arc a to its tail. We call a circulation g *non-negative* if $g \geq 0$ and *positive* if additionally $g \neq 0$. Notice that any two flows $\hat{f}^k, f^k$ for k only differ by a circulation, that is, there always exists a circulation g such that $\hat{f}^k = f^k + g$.

In the sequel we assume that $d \in \mathcal{D} \subset \mathbb{R}^K$ with $\mathcal{D}$ being a polytope. A capacity allocation $x \in \mathbb{R}_+^A$ is said to *support* the set $\mathcal{D}$ if there exists a routing f serving $\mathcal{D}$ such that for every $d \in \mathcal{D}$ the corresponding multi-commodity flow $f(d)$ is not exceeding the arc-capacities given by x. The problem can be written as the following (semi-infinite) linear program:

$$(RND) \quad \min \sum_{a\in A} \kappa_a x_a$$

$$\sum_{a\in\delta^+(v)} f_a^k(d) - \sum_{a\in\delta^-(v)} f_a^k(d) = d^k \psi_{vk}, \quad v \in V, k \in K, d \in \mathscr{D} \tag{3}$$

$$\sum_{k\in K} f_a^k(d) \leq x_a, \quad a \in A, d \in \mathscr{D} \tag{4}$$

$$f_a^k(d) \geq 0, \quad a \in A, k \in K, d \in \mathscr{D} \tag{5}$$

$$x_a \geq 0, \quad a \in A,$$

where $\kappa_a \in \mathbb{R}_+$ is the cost for installing one unit of capacity on arc $a \in A$.

Most authors ([2,1], among others) use a simpler version of (RND) introducing a restriction on the second stage recourse known as static routing (also called oblivious routing). Each component $f^k : \mathscr{D} \to \mathbb{R}_+^A$ is forced to be a linear function of d^k:

$$f_a^k(d) := y_a^k d^k \qquad a \in A, k \in K, d \in \mathscr{D}. \tag{6}$$

By combining (6) and (3) it follows that the multipliers $y \in \mathbb{R}_+^{A\times K}$ define a multi-commodity (percentage) flow. For every $k \in K$, the vector $y^k \in \mathbb{R}_+^A$ satisfies (1) setting $d^k = 1$. The flow y is called a *routing template* since it decides, for every commodity, which paths are used to route the demand and what is the percental splitting among these paths.

Ben-Tal *et al.* [3] introduce Affine Adjustable Robust Counterparts restricting the recourse to be an affine function of the uncertainties. Applying this framework to (RND) means restricting f^k to be an affine function of *all* components of d giving

$$f_a^k(d) := f_a^{0k} + \sum_{h\in K} y_a^{kh} d^h \geq 0, \qquad a \in A, k \in K, d \in \mathscr{D}, \tag{7}$$

where $f_a^{0k}, y_a^{kh} \in \mathbb{R}$ for all $a \in A, k, h \in K$, also see [8]. In what follows, a routing f serving $\mathscr{D}$ and satisfying (7) for some vectors f^0 and y is called affine. We see that affine routing generalizes static routing allowing for more flexibility in reacting to demand fluctuations, but it is not as flexible as dynamic routing.

The difficulty of model (RND) is that it contains an infinite number of inequalities. However, it is easy to see that (RND) can be discretized by restricting the model to the extreme demand scenarios that correspond to vertices of $\mathscr{D}$ (for all three routing schemes).

3 Affine Routings

In this section, we study properties and consequences of the affine routing principle. First we show that affine routing has a nice interpretation as paths and cycles:

Lemma 1. *Let $\mathscr{D}$ be a demand polytope and let $(f^0, y) \in \mathbb{R}^{A\times K} \times \mathbb{R}^{A\times K\times K}$ be an affine routing serving $\mathscr{D}$. If $\mathscr{D}$ is full-dimensional, then $y^{kk} \in \mathbb{R}^A$ is a routing template for $k \in K$ and $f^{0k} \in \mathbb{R}^A, y^{kh} \in \mathbb{R}^A$ are circulations for every $k, h \in K$ with $k \neq h$.*

We illustrate in Example 1 that affine routing can be as good as dynamic routing in terms of the cost for capacity allocation and that f^0 and y^{kh} may not describe circulations when $\mathscr{D}$ is not full-dimensional.

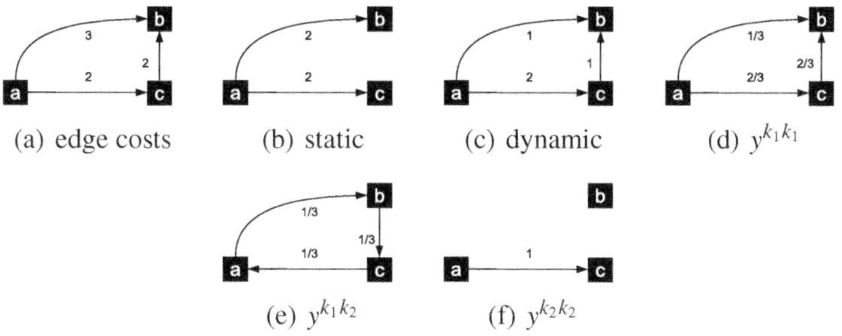

Fig. 1. Static, dynamic, and affine recourse

Example 1. Consider the network design problem for the graph depicted in Figure 1(a) with two commodities $k_1 : a \to b$ and $k_2 : a \to c$. The uncertainty set $\mathscr{D}$ is defined by the extreme points $d_1 = (2,1), d_2 = (1,2)$ and $d_3 = (1,1)$, and the capacity unitary costs are the edge labels of Figure 1(a). Edge labels from Figure 1(b) and 1(c) represent optimal capacity allocations with static and dynamic routing, respectively. They have costs of 10 and 9, respectively. Then, Figure 1(d)-1(f) describes coefficients y^{kh} for an affine routing feasible for the capacity allocation 1(c). If we remove $d_3 = (1,1)$ from the set of extreme points, the dimension of the uncertainty set reduces to 1. The affine routing prescribed by $y_{ac}^{k_2k_2} = 1$, $f_{ab}^{0k_1} = 3$ and $y_{ab}^{k_1k_2} = -1$ serves all demands in the convex hull of $d_1 = (2,1)$ and $d_2 = (1,2)$ but f^{0k_1} and $y^{k_1k_2}$ do not describe a circulation.

Compact reformulations. Reformulating by dualizing constraints is a standard technique in robust optimization that results in so-called *robust counterparts*. Applying this technique to (RND) with affine routing yields the following result.

Proposition 1. *If $\mathscr{D}$ is a full-dimensional polytope and either the number of its vertices or the number of its facets is polynomial in $(|A|,|V|,|K|)$, then (RND) with the affine recourse* (7) *can be solved in polynomial time in $(|A|,|V|,|K|)$.*

Proposition 1 implies that given a capacity allocation x, the existence of an affine routing can be answered in polynomial time as long as $\mathscr{D}$ can be described in a compact way, which is also true in the static case but is in contrast to the $\mathscr{NP}$-complete results for dynamic routing [5].

Domination of demands. For static and dynamic routings, not all extreme points of $\mathscr{D}$ have to be considered in a discretization of $\mathscr{D}$. Given two demands vectors d_1 and d_2, Oriolo [6] says that d_1 *dominates* d_2 if any $x \in \mathbb{R}_+^A$ supporting d_1 also supports d_2. Hence, removing dominated (extreme) points from $\mathscr{D}$ is not changing the problem in the dynamic case. Oriolo defines similarly the concept ot *total domination* for static routing. For general affine routings, however, there is no notion of domination of demands:

Proposition 2. *Let $d_1, d_2 \in \mathbb{R}_+^K$, $d_1 \neq d_2$. There exists (x, f^0, y) that satisfies* (3)–(5),(7) *for d_1 but not for d_2.*

Relation to static routing. Notice that if a flow f^k for k contains a positive circulation, that is, there exists a positive circulation g such that $f^k - g$ is a flow for k then f^k can be reduced to $f^k - g$ without changing the flow balance at $s(k)$ and $t(k)$. Moreover, the percental splitting among the used paths is unchanged. In this spirit we call any routing f *cycle-free* if for all $d \in \mathscr{D}$ and all commodities $k \in K$ the commodity flows do not contain positive circulations. Of course every optimal capacity allocation has a *cycle-free* (static, affine, or dynamic) routing.

Let e^k be the k-th unit vector in $\mathbb{R}^K_+$ and $\mathscr{D}^k_0$ be the set obtained from $\mathscr{D}$ by removing $d \in \mathscr{D}$ with $d^k > 0$, that is, $\mathscr{D}^k_0 := \{d \in \mathscr{D} : d^k = 0\}$. We can prove the following:

Proposition 3. *Let $\mathscr{D}$ be a demand polytope. If $0 \in \mathscr{D}$ and for each $k \in K$ there is $\varepsilon_k > 0$ such that $\varepsilon_k e^k \in \mathscr{D}$, then all cycle-free affine routings serving $\mathscr{D}$ are static.*

Proposition 4. *Let $\mathscr{D}$ be a demand polytope and let G be acyclic. If $\dim(\mathscr{D}^k_0) = |K| - 1$ for all $k \in K$, then all cycle-free affine routings serving $\mathscr{D}$ are static.*

Theorem 1. *Let $\mathscr{D}$ be a demand polytope. If all cycle-free affine routings serving $\mathscr{D}$ are static then $\dim(\mathscr{D}^k_0) = |K| - 1$ for all $k \in K$.*

Combining Proposition 4 with Theorem 1, we have complectly described polytopes for which cycle-free affine routings and static routings are equivalent, assuming that G is acyclic. However, we can show that Proposition 4 is wrong for general graphs.

Relation to dynamic routing. Theorem 3 identifies demand polytopes for which affine routing is no better than static routing. However, we saw in Example 1 that affine routing may also perform as well as dynamic routing does, yielding strictly cheaper capacity allocations. For general robust optimization problems, [4] show that affine recourse is equivalent to dynamic recourse when $\mathscr{D}$ is a simplex. Here we show that in the context of robust network design this condition is also necessary.

Theorem 2. *Given a demand polytope $\mathscr{D}$, all dynamic routings serving $\mathscr{D}$ are affine routings if and only if $\mathscr{D}$ is a simplex.*

An insight into the numerical computations. We compared numerically static, affine, and dynamic routings on networks *giul39*, *janos-us*, and *sun* from SNDlib [7]. These experiments showed that the static-dynamic gap is small (usually below 10%) and that solutions based on affine routings tend to be as cheap as those based on dynamic routings. In addition, affine routing enables us to solve problems with more commodities and more complex uncertainty polytopes than using dynamic routing.

References

1. Altin, A., Amaldi, E., Belotti, P., Pinar, Ç.: Provisioning virtual private networks under traffic uncertainty. Networks 49(1), 100–115 (2007)
2. Ben-Ameur, W., Kerivin, H.: New economical virtual private networks. Communications of the ACM 46(6), 69–73 (2003)
3. Ben-Tal, A., Goryashko, A., Guslitzer, E., Nemirovski, A.: Adjustable robust solutions of uncertain linear programs. Math. Prog. 99(2), 351–376 (2004)

4. Bertsimas, D., Goyal, V.: On the power and limitations of affine policies in two-stage adaptive optimization. Math. Prog., 1–41 (2011) (in press)
5. Chekuri, C., Shepherd, F.B., Oriolo, G., Scutellà, M.G.: Hardness of robust network design. Networks 50(1), 50–54 (2007)
6. Oriolo, G.: Domination between traffic matrices. Mathematics of Operations Research 33(1), 91–96 (2008)
7. Orlowski, S., Pióro, M., Tomaszewski, A., Wessäly, R.: SNDlib 1.0–Survivable Network Design Library. Networks 55(3), 276–286 (2010)
8. Ouorou, A., Vial, J.P.: A model for robust capacity planning for telecommunications networks under demand uncertainty. In: 6th Intl. Workshop on Design and Reliable Communication Networks, pp. 1–4 (2007)
9. Scutellà, M.G.: On improving optimal oblivious routing. Operations Research Letters 37(3), 197–200 (2009)

On the Weight-Constrained Minimum Spanning Tree Problem

Agostinho Agra[1], Adelaide Cerveira[2], Cristina Requejo[1], and Eulália Santos[3]

[1] CIDMA and Department of Mathematics, University of Aveiro, 3810-193 Aveiro, Portugal
{aagra,crequejo}@ua.pt

[2] CIO and Department of Mathematics, University of Trás-os-Montes and Alto Douro, 5001-801 Vila Real, Portugal
cerveira@utad.pt

[3] CIDMA and School of Technology and Management, Polytechnic Institute of Leiria, 2411-901 Leiria, Portugal
eulalia.santos@ipleiria.pt

Abstract. We consider the weight-constrained minimum spanning tree problem which has important applications in telecommunication networks design. We discuss and compare several formulations. In order to strengthen these formulations, new classes of valid inequalities are introduced. They adapt the well-known cover, extended cover and lifted cover inequalities. They incorporate information from the two subsets: the set of spanning trees and the knapsack set. We report computational experiments where the best performance of a standard optimization package was obtained when using a formulation based on the well-known Miller-Tucker-Zemlin variables combined with separation of cut-set inequalities.

1 Introduction

Consider an undirected complete graph $G=(V,E)$, with node set $V=\{0,1,\ldots,n-1\}$ and edge set $E=\{\{i,j\},\ i,j\in V, i\neq j\}$. Associated with each edge $e=\{i,j\}\in E$ consider nonnegative integer costs c_e and nonnegative integer weights w_e. The Weight-constrained Minimum Spanning Tree problem (WMST) is to find a spanning tree $T=(V,E_T)$, $E_T\subseteq E$, in G of minimum cost $C(T)=\sum_{e\in E_T} c_e$ and with total weight $W(T)=\sum_{e\in E_T} w_e$ not exceeding a given limit H. This combinatorial optimization problem is weakly NP-hard [1].

The WMST is known under several different names. It was first mentioned in [1] as the *MST problem subject to a side constraint*. In this paper the authors propose an exact algorithm that uses a Lagrangian relaxation combined with a branch and bound strategy. A similar approach can also be found in [10]. Approximation algorithms were developed in [9,4,3]. In [3] the results in [9] are improved. A branch-and-bound algorithm for the weight-constrained maximum spanning tree problem was developed in [11].

The WMST appears in several real applications and the weight restrictions are mainly concerned with a limited budget on installation/upgrading costs. A classical application arises in the areas of communication networks and network design, in which information is broadcast over a minimum spanning tree. The upgrade and/or the design of the physical system is usually restricted to a pre-established budget.

J. Pahl, T. Reiners, and S. Voß (Eds.): INOC 2011, LNCS 6701, pp. 156–161, 2011.

In this paper we intend to fill a gap, i.e., the lack of research on formulations and valid inequalities for the WMST problem. Firstly we discuss extended formulations that are adapted from formulations for the Minimum Spanning Tree problem: the multicommodity flow formulation and formulations based on the well-known Miller-Tucker-Zemlin (MTZ) inequalities. These formulations are compared, from the computational point of view, with the classical cut-set formulation for the MST. Computational experiments show that interesting results can be obtained when a MTZ based reformulation [2] is combined with separation over the cut-set inequalities. Secondly, we discuss valid inequalities for the set of feasible solutions that take into account properties from the two subsets, the knapsack set and the set of spanning trees, simultaneously. These inequalities adapt for the WMST problem the well-known cover, extended cover and lifted cover inequalities.

In Section 2 we discuss formulations for the WMST problem while in Section 3 we discuss valid inequalities. In Section 4 we report some computational experiments.

2 Formulations

A natural way to formulate the WMST problem is to use a formulation for the Minimum Spanning Tree (MST) problem [6] and add the weight constraint $\sum_{(i,j)\in A} w_{ij}x_{ij} \leq H$.

It is well-known (see [6]) that oriented formulations (based on the underlying directed graph) for the MST lead, in general, to tighter formulations (formulations whose lower bounds provided by the linear relaxations are closer to the optimum values). Thus, in this section we consider the corresponding directed graph, with root node 0, where each edge $e = \{0, j\} \in E$ is replaced with arc $(0, j)$ and each edge $e = \{i, j\} \in E, i \neq 0$, is replaced with two arcs, (i, j) and (j, i), yielding the arc set $A = \{(i, j),\ i \in V \setminus \{0\}, j \in V, i \neq j\}$. These arcs inherit the cost and weight of the ancestor edge.

The two classical formulations on the space of the original variables (the binary variables x_{ij}, for all $(i, j) \in A$, indicating whether arc (i, j) is chosen or not) for the WMST, one using the circuit elimination inequalities and the other the cut-set inequalities to ensure connectivity/prevent circuits can be considered. The linear relaxation of both models provide the same bound [6]. We use the formulation with the cut-set inequalities, Cut-Set formulation, denoted by CS. As the number of cut-set inequalities increases exponentially with the size of the model, these inequalities are introduced in the model as cuts using separation. However, it is well-known that in order to ensure connectivity/prevent circuits, instead of using one of those families of inequalities with an exponential number of inequalities, one can use compact extended formulations. That is the case of the well-known Multicommodity Flow (MF) formulation where connectivity of the solution is ensured through the flow conservation constraints together with the connecting constraints [6] and the case of the well-known MTZ formulation where connectivity of the solution is ensured through the node position variables [2].

As stated above, all these four formulations can be used directly to formulate the WMST by adding the weight constraint. Next we propose one more extended formulation, based on the MTZ variables, requiring some additional elaboration. In addition to the binary variables x_{ij} defining the topology of the solution, we consider variables p_i, $i \in V$, which specify the weighted-position of node i in the tree, i.e. the sum of the weights of the arcs in the path between the root node and node i. The Weighted Miller-Tucker-Zemlin (WMTZ) formulation is as follows:

$$\min \sum_{(i,j)\in A} c_{ij}x_{ij}$$

$$\text{s.t.} \sum_{i\in V} x_{ij} = 1 \qquad j \in V \setminus \{0\} \tag{1}$$

$$\sum_{(i,j)\in A} w_{ij}x_{ij} \leq H \tag{2}$$

$$w_{ij}x_{ij} + p_i \leq p_j + H(1 - x_{ij}) \qquad (i,j) \in A \tag{3}$$

$$0 \leq p_i \leq H \qquad i \in V \tag{4}$$

$$x_{ij} \in \{0,1\} \qquad (i,j) \in A. \tag{5}$$

Constraints (1) ensure that there is one arc incident to each node, with the exception of the root node. Constraint (2) is the weight constraint. Constraints (4) impose bounds on variables p_i. Constraints (3) prevent circuits and act as the well-known subtour elimination constraints given in [7] for the Traveling Salesman Problem: adding (3) for a circuit $\mathscr{C}$ ($x_{ij} = 1, (i,j) \in \mathscr{C}$) one obtains $\sum_{(i,j)\in\mathscr{C}} w_{ij} \leq 0$. On the other hand for any feasible weighted tree one can always find values for $p_j, \forall j \in N$, such that (3) and (4) are satisfied. Setting p_j to the weight of the path from the root node to any node j, ($p_j = p_i + w_{ij}$ for all (i,j) such that $x_{ij} = 1$ and $p_0 = 0$), then constraints (4) and (3), for all (i,j) such that $x_{ij} = 0$, are implied by the knapsack constraint (2).

Following Gouveia [2], constraints (3) can be lifted into several sets of inequalities. Computational results indicate that among all the lifted inequalities better computational results are obtained when the following inequalities

$$(H - w_{ji})x_{ji} + w_{ij}x_{ij} + p_i \leq p_j + H(1 - x_{ij}) \qquad (i,j) \in A \tag{6}$$

are incorporated in the formulation. Thus, henceforward we consider the WMTZ formulation with inequalities (3) replaced by (6).

3 Valid Inequalities

In order to strengthen the formulations presented in the previous section we discuss classes of valid inequalities.

We denote by X the set of feasible solutions to WMST. Set X can be regarded as the intersection of two well-known sets: $X = X_T \cap X_K$, where X_T is the set of spanning trees and X_K is the binary knapsack set defined by (2) and (5). Valid inequalities for X_T and valid inequalities for X_K are valid for X. While the polyhedral description of the convex hull of X_T, $P_T = conv(X_T)$ is well-known, see [6], for the polyhedral characterization of the convex hull of knapsack sets, $P_K = conv(X_K)$, only partial descriptions are known. This polyhedron is probably one of the most combinatorial optimization polyhedra studied. Kaparis and Letchford [5] present a very complete study on separation of valid inequalities for P_K. As in general $P = conv(X)$ is strictly included in $P_T \cap P_K$, there are fractional solutions that cannot be cut off by valid inequalities derived for P_T

or P_K. Hence, here we focus on valid inequalities derived for P that take into account properties from the two sets, simultaneously.

We call a set $C \subset E$ a Tree-Completion (TC) cover if for every spanning tree $T = (V, E_T)$ such that $C \subset E_T$, $W(T) = \sum_{e \in E_T} w_e > H$.

Proposition 1. *Given any TC cover C, the tree completion cover inequality (TCCI) $\sum_{e \in C} x_e \leq |C| - 1$ is valid for X.*

It can be checked that every cover inequality is a Subtour breaking Constraint (SC), or a TCCI or it is dominated by a SC or a TCCI. As for cover inequalities, TCCI are in general weak. One possible approach to strengthen these inequalities is by lifting. Given a TCCI, $\sum_{e \in C} x_e \leq |C| - 1$, a valid inequality $\sum_{e \in C} x_e + \sum_{e \in E \setminus C} \beta_e x_e \leq |C| - 1$, with $\beta_e \geq 0, e \in E \setminus C$, is called a lifted TCCI (LTCCI). One first approach to lift a TCCI is to adapt the well-known extended cover inequalities.

Proposition 2. *Let $C \subset E$ be a TC cover and let $C' = \{e \in E | w_e \geq \max\{w_f : f \in C\}$ and $C \cup \{e\}$ forms a cycle$\}$. The extended TCCI (ETCCI), $\sum_{e \in C} x_e + \sum_{e \in C'} x_e \leq |C| - 1$, is valid for X.*

LTCCIs can also be obtained via sequential lifting where the coefficients β_e are computed one at a time. Given a TCCI, $\sum_{e \in C} x_e \leq | C | -1$, and a LTCCI $\sum_{e \in C} x_e + \sum_{e \in R} \beta_e x_e \leq |C| - 1$, one can lift x_f, with $f \in E \setminus (C \cup R)$, by computing β_f, such that: $0 \leq \beta_f \leq f(C, R, \beta) = \min\{| C | -1 - \sum_{e \in C} x_e - \sum_{e \in R} \beta_e x_e : x \in X, x_f = 1\}$.

In order to derive LTCCIs no variables are fixed a priori (no restrictions are considered). However, a new class of lifted inequalities can be derived by the "usual" lifting procedure where the lifting is done by fixing the value of a set of variables, deriving a valid inequality for the restricted set that results from the variable fixing, and then sequentially lift each variable whose value has been fixed. We call the inequalities obtained by this procedure the Generalized Lifted TCCI Inequalities (GLI). To derive a GLI one fix a set E_0 of variables to zero, a set E_1 of variables to one, then generate a LTCCI for the restricted set ($X^R = X \cap \{x : x_e = 0, e \in E_0, x_e = 1, e \in E_1\}$) and lift sequentially all the variables with null coefficient in the LTCCI.

4 Computational Experiments

To compare the proposed formulations and test the valid inequalities, a test set of instances was generated. The costs and weights of a first test set with up to 100 nodes, were generated based on Pisinger's [8] spanner instances. In a second set of instances, with nodes from 150 to 300, the costs were based on Euclidean distances while weights were randomly generated in $[1, 100]$. In all instances H was fixed to the average between the minimum and the maximum weight spanning tree. All the tests were run on a Intel(R) Core(TM)2Duo CPU 2.00GHz with 1.99Gb of RAM and using the optimization software Xpress 7.1. To solve large size instances we focused on the comparison of the two hybrid procedures that result from the MTZ and WMTZ formulations

Table 1. Average gaps and average running times with formulations MTZ, WMTZ, MF and CS and the two hybrid procedures MTZ+C and WMTZ+C

$\|V\|$	LPgap MTZ	time MTZ	LPgap WMTZ	time WMTZ	LPgap MF	time MF	LPgap CS	time CS	LPgap MTZ+C	time MTZ+C	LPgap WMTZ+	time WMTZ+C
10	10.3	0.08	10.3	0.09	1.2	0.02	1.2	0.01	1.2	0.08	1.2	0.01
20	7.4	0.84	7.4	0.99	1.0	0.72	1.0	0.12	1.0	0.05	1.0	0.05
40	3.7	2236	3.7	2276	0.5	12.3	0.5	71.2	0.5	0.27	0.5	0.12
60	2.2	4335	2.2	4329	0.5	144.2	0.5	6269	0.5	0.6	0.5	0.36
80	1.7	4655	1.7	2868	1.0	306	1.0	*	1.0	1.46	1.0	1.25
100	1.3	4333	1.3	4326	0.3	3517	0.3	*	0.3	4329	0.3	2173
150									0.03	4.5	0.03	3.6
200									0.04	13.0	0.04	11.6
250									0.04	37.2	0.04	25.0
300									0.05	44.2	0.05	41.4

strengthened at the root node of the Branch and Bound with the cut-set inequalities, MTZ+C and WMTZ+C. Using the two hybrid procedures we solved all the generated instances up to 300 nodes. Table 1 gives the average integrality gap and the average running times (in seconds) for a set of 5 instances for each number of vertices. An asterisk means that some of the 5 instances were not solved within a time limit of one day.

For valid inequalities we tested separation heuristics for TCCIs, ETCCIs, LTCCIs and GLIs. Since the tested separation heuristics for GLIs provided no improvement when compared to the LTCCIs case, we do not discuss here the separation of GLIs. For separation of TCCIs we: (i) sort the edges accordingly to a given order; (ii) following that order, include one edge at a time into set C and check whether C defines a TC cover, and, if so, checks if the TCCI cuts off the fractional solution; (iii) if no cuts were found, and there are different orders to be tested, then return to (i) and use the next order to find cuts, otherwise STOP. For Step (i) we tested four (non-increasing) orderings of the edges based on the values: x_e^* (fractional solution); w_e; $(1-x_e^*)/w_e$ and $w_e x_e^*$. Although not reported here, tests using only one ordering showed that the best bounds were obtained using $w_e x_e^*$. Separation of ETCCIs was done similarly. When a cut from a TCCI is found we lift the TCCI into an ETCCI. For LTCCIs, when a cut from a TCCI is found we lift the variables accordingly to the order used to find the TCCI. The lifting coefficient of x_e is given by $\beta_e = \max\{0, \lceil \underline{f}(C,R,\beta) \rceil\}$ where $\underline{f}(C,R,\beta)$ denotes the value of the linear relaxation of $f(C,R,\beta)$.

For a selected set of 21 instances from 10 to 80 nodes we obtained an average integrality gap of 2.1%, 0.94%, 0.9% and 0.75% with the linear relaxation (LP), LP with TCCI cuts, LP with ETCCI cuts and LP with LTCCI cuts, respectively. To obtain the linear relaxation we used the MF formulation. For the linear relaxation of $f(C,R,\beta)$ we used the WMTZ formulation.

Acknowledgement

This work was partially supported by the Portuguese Foundation for Science and Technology, co-financed by the European Community Fund FEDER/POCI 2010.

References

1. Aggarwal, V., Aneja, Y.P., Nair, K.P.K.: Minimal spanning tree subject to a side constraint. Computers and Operations Research 9, 287–296 (1982)
2. Gouveia, L.: Using the Miller-Tucker-Zemlin constraints to formulate a minimal spanning tree problem with hop constraints. Computers and Operations Research 22(9), 959–970 (1995)
3. Hassin, R., Levin, A.: An efficient polynomial time approximation scheme for the constrained minimum spanning tree problem using matroid intersection. SIAM Journal on Computing 33(2), 261–268 (2004)
4. Hong, S.P., Chung, S.J., Park, B.H.: A fully polynomial bicriteria approximation scheme for the constrained spanning tree problem. Operations Research Letters 32, 233–239 (2004)
5. Kaparis, K., Letchford, A.: Separation algorithms for 0-1 knapsack polytopes. Mathematical Programming 124(1–2), 69–91 (2010)
6. Magnanti, T., Wolsey, L.: Optimal trees. In: Ball, M., Magnanti, T., Monma, C., Nemhauser, G. (eds.) Network Models. Handbooks in Operations Research and Management Science, vol. 7, pp. 503–615. Elsevier Science Publishers, North-Holland, Amsterdam (1995)
7. Miller, C., Tucker, A., Zemlin, R.: Integer programming formulations and travelling salesman problems. Journal of the Association for Computing Machinery 7, 326–329 (1960)
8. Pisinger, D.: Where are the hard knapsack problems? Technical Report 2003/08, DIKU, University of Copenhagen, Denmark (2003)
9. Ravi, R., Goemans, M.: The constrained minimum spanning tree problem. In: Karlsson, R., Lingas, A. (eds.) SWAT 1996. LNCS, vol. 1097, pp. 66–75. Springer, Heidelberg (1996)
10. Shogan, A.: Constructing a minimal-cost spanning tree subject to resource constraints and flow requirements. Networks 13, 169–190 (1983)
11. Yamada, T., Watanabe, K., Kataoka, S.: Algorithms to solve the knapsack constrained maximum spanning tree problem. International Journal of Computer Mathematics 82, 23–34 (2005)

The Minimum Connected Dominating Set Problem: Formulation, Valid Inequalities and a Branch-and-Cut Algorithm

Luidi Simonetti[1], Alexandre Salles da Cunha[2], and Abilio Lucena[3]

[1] Universidade Federal Fluminense, Instituto de Computação
luidi@ic.uff.br
[2] Universidade Federal de Minas Gerais, Departamento de Ciência da Computação
acunha@dcc.ufmg.br
[3] Universidade Federal do Rio de Janeiro, Programa de Engenharia de Sistemas e Computação
abiliolucena@globo.com

Abstract. We consider the minimum connected dominating set problem. We present an integer programming formulation and new valid inequalities. A branch-and-cut algorithm based on the reinforced formulation is also provided. Computational results indicate that the reinforced lower bounds are always stronger than the bounds implied by the formulation from which resulted one of the best known exact algorithms for the problem. In some cases, the reinforced lower bounds are stronger than those implied by the strongest known formulation to date. For dense graphs, our algorithm provides the best results in the literature. For sparse instances, known to be harder, our method is outperformed by another one. We discuss reasons for that and how to improve our current computational results. One possible way to achieve such goals is to devise specific separation algorithms for some classes of valid inequalities introduced here.

1 Introduction

Let $G = (V, E)$ be a connected undirected graph with a set of vertices $V = \{1, \ldots, n\}$ and edges E ($m = |E|$). Given $i \in V$, assume that $\Gamma_i \subseteq V$ denotes the union of $\{i\}$ with the set of vertices of V that share an edge with i. A set $W \subseteq V$ is a dominating set of G if, for every $i \in V$, there is $k \in \Gamma_i \cap W$. Given any set W, let $E(W) = \{\{i, j\} \in E : i, j \in W\}$ be the subset of edges of E with both endpoints in W. A dominating set W is connected if the subgraph $G_W = (W, E(W))$ of G is connected. In the Minimum Connected Dominating Set Problem (MCDSP), one wishes to find a connected dominating set of G with minimum cardinality. MCDSP was shown to be NP-hard in [6].

MCDSP is closely related to the Maximum Leaf Spanning Tree Problem [7] (MLSTP), which consists in finding a spanning tree of G with as many leaves as possible. It should be clear that, given a dominating set W, a spanning tree $T(W) = (W, E_T)$ of G_W could be found efficiently. Such a tree could be enlarged into a spanning tree T of G, where all vertices in $V \setminus W$ are leaves. Thus, for every connected dominating set W of G, a spanning tree of G with at least $n - |W|$ leaves could be efficiently found. In particular, if W is a minimum connected dominating set, a spanning tree of G with the maximum possible number of leaves results from the procedure outlined above.

J. Pahl, T. Reiners, and S. Voß (Eds.): INOC 2011, LNCS 6701, pp. 162–169, 2011.

Applications for MCDSP (and MLSTP, too) arise, e.g., in the design of ad-hoc wireless sensor networks, where network topologies may change dynamically [1] as well as the design of fiber optics networks where regenerators of information may be required at some vertices [3]. Polyhedral investigations for MLSTP were conducted in [4] and exact algorithms have been proposed in [5,9]. Two *integer programming* (IP) formulations were proposed for MLSTP in [9]. MCDSP has been tackled by approximation algorithms in [7,11]. Although not explicitly stated, the underlying problem associated with the Regenerator Location Problem (RLP) in [3] is a MCDSP.

Here we present an IP formulation, valid inequalities and a *branch-and-cut* (BC) algorithm [12] for MCDSP. Exploring specific properties of MCDSP, new valid inequalities are proposed. Lower bounds obtained by the reinforced formulation significantly improves on the original bounds. For some instances, such bounds are stronger than the best known bounds reported in [9]. Preliminary results obtained with the BC algorithm implemented here suggest that the formulation might be promising. Though, our current BC implementation is outperformed by the overall best exact algorithm for MLSTP in [9], for sparse instances.

The paper is organized as follows. In Section 2, we present the IP formulation for MCDS. We discuss a BC algorithm in Section 3. Computational results are reported in Section 4. We conclude the paper in Section 5.

2 Integer Programming Formulation

In order to present an IP formulation for MCDSP, let us use the following decision variables: $y_i \in \{0,1\}$, $i \in V$: to select which vertices are to be included ($y_i = 1$) or not ($y_i = 0$) in the dominating set; $x_e \in \{0,1\}$, $e \in E$: to choose edges that guarantee that the dominating set is indeed connected. In what follows, assume that $\mathbb{B} = \{0,1\}$ and that $\mathbb{R}$ denotes the set of real numbers. An IP formulation for MCDSP is:

$$\min \left\{ \sum_{i \in V} y_i : (x,y) \in \mathscr{R}_0 \cap (\mathbb{R}^m_+, \mathbb{B}^n) \right\}, \tag{1}$$

where polyhedral region $\mathscr{R}_0$ is implied by:

$$\sum_{e \in E} x_e = \sum_{i \in V} y_i - 1, \tag{2a}$$

$$\sum_{e \in E(S)} x_e \leq \sum_{i \in S \setminus \{j\}} y_i, \quad S \subset V, j \in S \tag{2b}$$

$$\sum_{j \in \Gamma_i} y_j \geq 1, \quad i \in V \tag{2c}$$

$$x_e \geq 0, \quad e \in E \tag{2d}$$

$$0 \leq y_i \leq 1, \quad i \in V. \tag{2e}$$

The idea behind the formulation above is to use variables x to select edges that guarantee that a spanning tree must be found in the subgraph $G_W = (W, E(W))$, implied by a dominating set W. More precisely, constraint (2a) guarantees that the number of selected edges is exactly one unity less than the number of vertices in a connected dominating set. Generalized Subtour Breaking Constraints (GSEC) (2b) guarantee that the selected edges imply a tree. Constraints (2c), on the other hand, make sure that the set of vertices selected in a solution is dominating.

Formulation $\mathscr{R}_0$ embeds two basic structures. On the one hand, constraints (2a), (2b), (2d) – (2e) fully characterize the tree polytope of G [10]. The other structure, implied by constraints (2c) and (2e), is that of the Covering Problem. Facet defining inequalities for the covering polytope are widely recognized as difficult to separate [2]. In the sequence, we show how formulation $\mathscr{R}_0$ can be strengthened by other means, using problem specific arguments.

Firstly, we claim that (2c) can be lifted to

$$\sum_{j\in\Gamma_i} y_j - \sum_{e\in E(\Gamma_i)} x_e \geq 1, \forall i \in V. \tag{3}$$

To show that (3) is valid for MCDSP, consider a connected dominating set W. Since $|W \cap \Gamma_i| \geq 1$ and since the edges in $E(W)$ selected to span the set must imply a tree, we have that the number of selected edges in $E(\Gamma_i)$ must be at most $|\Gamma_i| - 1$. Otherwise, at least one cycle must be included in the solution.

Constraints (3) can also be viewed as a strengthened version of the Generalized Subtour Breaking constraints (2b). To verify that, let $S = \Gamma_i$, for which the corresponding GSEC reads as: $\sum_{e\in E(\Gamma_i)} x_e \leq \sum_{j\in\Gamma_i\setminus\{k\}} y_j, k \in \Gamma_i$. Since at least one vertex in Γ_i must be chosen, the latter can be replaced by the stronger form $\sum_{e\in E(\Gamma_i)} x_e \leq \sum_{j\in\Gamma_i} y_j - 1$, which is precisely (3). The previous observation leads to a lifting of (2b) as follows. Assume that, for any given way, we can certify that, out of those vertices in a particular given set $C \subset V$, at least one vertex must be included in a connected dominating set. Clearly, for sets satisfying such property, GSECs (2b) can be replaced by the stronger version:

$$\sum_{e\in E(C)} x_e \leq \sum_{j\in C} y_j - 1. \tag{4}$$

To present another valid inequality for MCDSP, assume that given $S \subset V$, $S \neq \emptyset$, $\Gamma(S) := \bigcup_{i\in S} \Gamma_i$ and that $(S, V\setminus S) := \{\{i,j\} \in E : i \in S, j \notin S\}$ denotes the edges in the cut implied by S. Whenever $\Gamma(S) \neq V$ and $\Gamma(V\setminus S) \neq V$, at least one edge in $(\Gamma(S), V \setminus \Gamma(S))$ must be chosen. This is true since the vertices in a connected dominating set cannot be exclusively confined to S or to $V \setminus S$. Mathematically, we have that:

$$\sum_{e\in(S,V\setminus S)} x_e \geq 1, \forall S \subset V : \Gamma(S) \neq V, \Gamma(V\setminus S) \neq V. \tag{5}$$

Therefore, a strengthened formulation $\mathscr{R}_1$ for MCDSP can be obtained by constraints (2a)-(2b),(2d),(2e), (3),(4) and (5).

3 Branch-and-Cut Algorithm

In this section, we provide the main implementation details on a preliminary BC algorithm for MCDSP, based on formulation $\mathscr{R}_1$. The algorithm was implemented with calls to `XPRESS` MIP solver (release 19.00) callback routines. The algorithm implements a *best-first* search strategy. All other default `XPRESS` settings were used.

The algorithm starts solving the LP relaxation

$$\min \sum_{i \in V} y_i : \ (x,y) \in \mathscr{P}, \tag{6}$$

where polytope $\mathscr{P}$ is given by (2a),(2d),(2e), (3) and $x_{ij} \leq y_i, \{i,j\} \in E$, as well as $x_{ij} \leq y_j, \{i,j\} \in E$.

Let $(\bar{x},\bar{y}) \in \mathscr{P}$ be the solution to (6) and $\overline{G} = (\overline{V},\overline{E})$ be the subgraph of G implied by $(\bar{x},\bar{y})$ ($\overline{V} := \{i \in V : \bar{y}_i > 0\}$ and $\overline{E} := \{\{i,j\} \in E : \bar{x}_{ij} > 0\}$). If $(\bar{x},\bar{y})$ is integer and if there is no GSEC (2b) violated by $(\bar{x},\bar{y})$, $(\bar{x},\bar{y})$ solves (1). Otherwise, we attempt to reinforce $\mathscr{P}$, appending violated valid inequalities to it.

The exact separation of GSECs can be carried out efficiently, through max-flow (min-cut) computations, in $O(n^4)$ [13]. Despite that, in our current implementation, we do not use any exact separation algorithm. Here, the algorithm of [13] is only used to present a tighter approximation of the bounds implied by $\mathscr{R}_1$. In practice, in our BC method, we found advantageous to separate GSECs only through the following heuristic. We sort the edges in $\overline{E}$ in non-increasing order of their $\bar{x}_{ij}$ values. Then, we find a forest of maximum cardinality of $\overline{G}$, using Kruskal's algorithm [8], giving preference to include edges with higher values of $\bar{x}_{ij}$. Each edge included during Kruskal's method merges two sets of vertices into a new connected component in the forest being built. We check for violated GSECs for the connected components generated after each edge inclusion, until a forest of maximum cardinality has been found. If no violated GSECs are found with the separation heuristic, we branch on y variables.

Whenever (with the GSEC separation heuristic outlined above) we find a set C whose corresponding GSEC is violated, we attempt to lift the inequality to (4), by checking whether at least one vertex in C must be in a connected dominating set for G. One simple test allowing to conclude so is the following. Whenever $\Gamma(V \setminus C) \neq V$, (4) holds for C. Apart from that, specific separation algorithms for inequalities (4) are not yet implemented as well as separation algorithms for inequalities (5).

Initial valid upper bounds for MCDSP are computed with the dynamic greedy heuristic introduced in [9]. It is oriented towards generating spanning trees of G with as many leaves as possible or, equivalently, to enforce that these trees contain as few inner vertices as possible. The heuristic works with two sets: $\mathscr{D}$, to represent vertices in a connected dominating set and $\mathscr{L}$, to represent those vertices which have at least one neighbor in $\mathscr{D}$. The procedure is initialized by $\mathscr{D} = \{v\}$ and $\mathscr{L} = \Gamma_v \setminus \{v\}$ for any $v \in V$. The basic operation is to try to push vertices from $\mathscr{L}$ into $\mathscr{D}$, until a connected dominating set is found. Assuming that i is moved from $\mathscr{L}$ to $\mathscr{D}$, in the next iteration we have: $\mathscr{D} \leftarrow \mathscr{D} \cup \{i\}$ and $\mathscr{L} \leftarrow \mathscr{L} \setminus \{i\} \cup (\Gamma_i \setminus \mathscr{D})$. Preference is given to include in $\mathscr{D}$ vertices with as many neighbors as possible, not already included in $\mathscr{L}$. The heuristic stops when $V = \mathscr{L} \cup \mathscr{D}$, when $\mathscr{D}$ represents a connected dominating set and $\mathscr{L}$ denotes the set of leaf implying vertices in the tree.

4 Preliminary Computational Results

The algorithms introduced here were implemented in C and all computational testings were conducted in a Intel XEON machine running at 2Ghz, with 8 Gbytes of RAM

Table 1. MCDSP lower bounds and Branch-and-cut results

		Lower Bounds				OPT	Branch-and-cut			$t_{DGR}(s)$
n	density (%)	$\mathscr{R}_0$	$\mathscr{R}_1$	DGR	STR		$t(s)$	BLB	BUB	
30	10	8.60	14.40	14.12	14.34	15	0.01	15	15	0.01
	20	5.09	6.18	5.68	6.49	7	0.02	7	7	0.10
	30	2.92	3.60	3.05	3.50	4	0.05	4	4	0.03
	50	1.95	2.38	1.86	2.11	3	0.04	3	3	0.08
	70	1.36	1.83	1.30	2.00	2	0.02	2	2	0.01
50	5	15.55	31.00	31.00	31.00	31	0.02	31	31	0.01
	10	9.17	10.68	10.37	11.15	12	0.42	12	12	0.36
	20	4.76	5.24	4.88	5.52	7	0.66	7	7	1.32
	30	3.28	3.69	3.26	3.93	5	0.25	5	5	1.21
	50	1.98	2.44	1.82	2.20	3	0.25	3	3	0.51
	70	1.45	1.84	1.31	2.00	2	0.29	2	2	0.04
70	5	17.10	26.31	25.29	26.44	27	1.42	27	27	0.26
	10	9.82	11.23	10.90	11.40	13	34.29	13	13	4.73
	20	4.92	5.37	5.12	5.63	7	2.16	7	7	16.30
	30	3.27	3.62	3.20	3.86	5	1.00	5	5	2.90
	50	2.05	2.44	1.95	2.05	3	0.70	3	3	1.33
	70	1.43	1.91	1.35	2.00	2	0.79	2	2	1.92
100	5	18.00	21.63	20.79	22.04	24	342.25	24	24	12.50
	10	10.05	10.98	10.62	11.07	13	32.11	13	13	9.36
	20	5.24	5.52	5.15	5.62	8	174.93	8	8	86.16
	30	3.37	3.74	3.33	-	6	193.65	6	6	258.15
	50	2.10	2.51	1.97	-	4	35.41	4	4	132.55
	70	1.45	1.94	1.36	2.05	3	12.03	3	3	154.10
120	5	19.12	22.74	22.48	22.87	35	-	24.34	26	2.65
	10	9.79	10.66	10.33	10.87	13	-	12.79	15	65.49
	20	5.14	5.35	5.07	-	8	610.89	8	8	393.47
	30	3.40	3.76	3.31	-	6	475.54	6	6	653.70
	50	1.99	2.49	1.37	2.15	4	168.55	4	4	815.64
	70	1.44	1.92	-	-	3	31.67	3	3	356.31
150	5	19.60	21.72	21.35	21.94	26	-	23.74	27	2954.00
	10	10.27	10.69	10.56	10.84	14	-	12.47	15	3247.89
	20	5.05	5.37	4.95	-	9	-	6.97	9	-
	30	3.42	3.81	3.33	-	6	1954.00	6	6	2317.35
	50	1.98	2.47	1.90	-	4	481.61	4	4	2756.36
	70	1.44	1.99	1.37	-	3	43.75	3	3	1828.86
200	5	20.35	22.52	22.17	22.69	-	-	23.78	29	-
	10	10.16	10.53	10.39	-	-	-	11.9	16	-
	20	4.95	5.26	4.87	-	-	-	6.41	9	-
	30	3.35	3.77	3.23	-	-	-	4.56	7	-
	50	2.01	2.53	1.93	-	4	2249.43	4	4	20155.00
	70	1.44	2.00	1.37	2.03	3	271.91	3	3	8154.13

memory. In order to evaluate BC and the quality of the lower bounds of the proposed reinforced model, we considered those MCDSP instances introduced in [9], for which $n \in \{30, 50, 70, 100, 120, 150, 200\}$ and graph densities range from 5% to 70%. For each value of n and graph density, BC was allowed to run for at most 3600 seconds, after which, the problem is left unsolved.

Detailed computational results are reported in Table 1. In the first two columns of the table, we report n and the graph density. In the subsequent four columns, we present lower bounds for MCDSP: the lower bound implied by $\mathscr{R}_0$, an approximation of the lower bound implied by $\mathscr{R}_1$, followed by the lower bounds implied by the Directed Graph Reformulation (DGR) and by the Steiner Arborescence Reformulation (STR) in [9]. In the next column, we report the optimal objective function value (OPT) for the instance, whenever available (an entry "-" in that column indicates that the corresponding optimal value is yet unknown). In the next three columns, we present results for our BC method: $t(s)$, the time (in seconds) it takes to solve the instance and the best lower (BLB) and upper (BUB) bound found after the search. Whenever BC does not solve the instance within the imposed time limit, an entry "-" is given in that column. In the last column of the table, we report $t_{DGR}(s)$, the time (in seconds) needed by the BC algorithm in [9] (based on DGR) to solve the instance. These times were obtained in a machine whose speed is directly comparable to the one used in our testings. We do not compare our method with the BC algorithm in [3], since instances in that study were not available to us.

As can be seen from Table 1, optimizing over $\mathscr{R}_1$ allows significantly stronger lower bounds, when compared to the bounds implied by $\mathscr{R}_0$. We should recall that the bounds reported in column $\mathscr{R}_1$ are not the true bounds provided by that formulation, since inequalities (4) and (5) are not separated yet (only GSECs were separated exactly, for the sake of approximating the bounds given by $\mathscr{R}_1$). In many cases, these approximations are stronger than the best lower bounds in [9], given by STR. Note also that the new reinforced lower bounds are always at least as strong as the bounds implied by DGR, the formulation from which resulted the best exact algorithm for MSLTP in [9].

For small instances, with up to 70 vertices, the BC algorithm implemented here is always faster than the BC algorithm based on DGR. Similar conclusions can be drawn for larger dense instances. However, despite the potential benefits of optimizing over $\mathscr{R}_1$, for sparse larger instances, the BC procedure implemented here is outperformed by the one based on DGR. For example, note that when $n = 120$ and graph density is 10% or less, the algorithm implemented here failed on solving the problem within the imposed time limit, whereas the algorithm in [9] solved such instances quite easily, in less than 66 seconds. One possible explanation is the following. Cuts based on GSECs are usually denser than directed cutset counterparts. As a consequence, solving the LP relaxations involved in the formulation discussed here is likely to be more time consuming than solving DGR relaxations.

5 Conclusions

Our computational results indicate that the (approximated) lower bounds implied by the reinforced formulation are stronger than the bounds given by the formulation from which resulted one of the best exact solution approaches in MCDSP literature. Our method is at least as fast as the Branch-and-cut algorithm in [9] for dense graphs. Despite the potential benefits of optimizing over a better approximation of the convex hull of feasible solutions, our Branch-and-cut method is outperformed by the best solution algorithm in [9] for sparse problems.

We believe that our computational results could be improved significantly, if specific separation algorithms for the new valid inequalities were implemented. In addition, the valid inequalities suggested here for an undirected formulation for MCDSP have direct counterparts for MCDSP formulated in a directed graph. Such inequalities should be easier to separate over a directed structure. Consequently, a Branch-and-cut algorithm based on a directed version of the formulation provided here could also benefit from the stronger lower bounds introduced in this study. Benefits of such directed Branch-and-cut implementation could arise also from other two different sources. Firstly, directed cutsets are typically sparser than GSECs. Consequently, the time needed to solve the Linear Programming relaxations may decrease. Secondly, directed cutsets can be separated in $O(n^3)$ time, while GSECs take $O(n^4)$, in the worst case.

Acknowledgements

Alexandre Salles da Cunha is partially funded by CNPq grant 302276/2009-2 and FAPEMIG PRONEX APQ-01201-09. Abilio Lucena is partially funded by CNPq grant 310561/2009-4. Luidi Simonetti is partially funded by FAPERJ grant E-26/111.819/2010 and CNPq grant 483.243/2010-8.

References

1. Balasundaram, B., Butenko, S.: Graph domination, coloring and cliques in telecommunications. In: Handbook of Optimization in Telecommunications, pp. 865–890. Springer, Heidelberg (2006)
2. Borndörfer, R.: Aspects of Set Packing, Partitioning and Covering. Ph.D. thesis, Konrad-Zuse-Zentrum fur Informationstechnik, Berlin (1998)
3. Chen, S., Ljubić, I., Raghavan, S.: The regenerator location problem. Networks 55(3), 205–220 (2010)
4. Fujie, T.: The Maximum-leaf Spanning Tree Problem: formulations and facets. Networks 43(4), 212–223 (2004)
5. Fujie, T.: An exact algorithm for the Maximum-leaf Spanning Tree Problem. European Journal of Operational Research 104, 250–261 (2003)
6. Garey, M.R., Johnson, D.S.: Computers and Intractability: A Guide to the Theory of NP-Completeness (1979)
7. Guha, S., Khuller, S.: Approximation algorithms for connected dominating sets. Algorithmica 20(4), 374–387 (1998)
8. Kruskal, J.: On the Shortest Spanning Subtree of a Graph and the Traveling Salesman Problem. Proceedings of the American Mathematical Society 7, 48–50 (1956)
9. Lucena, A., Maculan, N., Simonetti, L.: Reformulation and solution algorithms for the maximum leaf spanning tree problem. Computational Management Science 7, 289–311 (2010)

10. Magnanti, T.L., Wolsey, L.: Optimal Trees. In: Ball, O., et al. (eds.) Handbooks in OR and MS, vol. 7, pp. 503–615. North-Holland, Amsterdam (1995)
11. Marathe, M.V., Breu, H., Hunt III, H.B., Ravi, S.S., Rosenkrantz, D.J.: Simple heuristics for unit disc graphs. Networks 25, 59–68 (1995)
12. Padberg, M.W., Rinaldi, G.: A Branch-and-Cut algorithm for resolution of large scale of Symmetric Traveling Salesman Problem. SIAM Review 33, 60–100 (1991)
13. Padberg, M.W., Wolsey, L.: Trees and cuts. Annals of Discrete Mathematics 17, 511–517 (1983)

Multilayer Survivable Optical Network Design

Sylvie Borne[1], Virginie Gabrel[2], Ridha Mahjoub[2], and Raouia Taktak[2]

[1] Institut Galilée, Avenue J.B. Clément 93430 Villetaneuse, France
sylvie.borne@lipn.univ-paris13.fr

[2] LAMSADE, Université Paris Dauphine, Place du Maréchal de Lattre de Tassigny 75775 Paris Cedex 16, France
{gabrel,mahjoub,taktak}@lamsade.dauphine.fr

Abstract. With the explosive growth of traffic data, telecommunication networks have evolved toward a model of high-speed IP routers interconnected by intelligent optical core networks. This IP-over-optical architecture is particularly considered as an important opportunity for telecommunication carriers who want to vary services and add more multimedia applications.

In our work, we are interested in the problem of survivability in multilayer IP-over-optical networks. Given a set of traffic demands for which we know a survivable logical routing in the IP layer, our purpose is to determine the corresponding survivable topology in the optical layer. We show that the problem is NP-hard even for one demand. We formulate the problem in terms of $0-1$ linear program based on path variables. We discuss the pricing problem and prove that it reduces to a shortest path problem. Using this, we propose a Branch-and-Price algorithm. Some preliminary computational results are also discussed.

1 Introduction

Telecommunication networks have witnessed within the past years an explosive growth of traffic data. This rapid evolution has induced a need to a new promising architecture that enable an efficient management of huge amount of data. Telecommunication networks have hence evolved toward a multilayer architecture consisting of high-speed routers interconnected by intelligent optical core networks. The IP-over-WDM networks are composed of a virtual (IP/MPLS) layer over a physical (WDM) layer. Multilayer Network Design problems has recently interested many researchers [3,5]. Moreover, survivability of this networks has become unavoidable in order to ensure a continuously routing of data in case of failures [2].

The problem that we are studying belongs actually to the multilayer survivability context. Consider an IP-over-WDM network consisted of an IP/MPLS layer over a WDM layer. The logical layer is composed of IP routers which are interconnected by virtual links and the optical layer consists of a number of Optical Cross Connects OXC interconnected by physical links. To each IP router corresponds an OXC. Consider also a set of demands and for each demand two node-disjoint paths routing it in the virtual layer. Finally, for each physical link we associate a cost corresponding to its cost of installation. The Multilayer Survivable Optical Network Design (MSOND) problem is to find, for each demand, two elementary node-disjoint physical paths routing it in

J. Pahl, T. Reiners, and S. Voß (Eds.): INOC 2011, LNCS 6701, pp. 170–175, 2011.

the optical layer going in order through the OXCs corresponding to the routers in the logical paths and such that the total cost of installation is minimum. Apart from the importance of MSOND in the telecommunication context, our problem is very interesting and raised from challenging classical problems. In fact, for a single demand (Single Commodity MSOND or SC-MSOND), the problem can be seen as a Steiner cycle that should visits specific nodes with some precedence constraints between them. This is in a close relationship with classical problems such as the shortest path with specified nodes [4], the Steiner cycle [6] and the travelling salesman problem with precedence constraints [1].

The paper is organized as follows. In the following section we give some definitions and notations that are necessary for the sequel. In Section 3, we prove that MSOND is NP-hard even for a single commodity. Section 4 will be devoted to present the path formulation, discuss the corresponding pricing problem and give some preliminary results. We conclude in Section 5 by some future works and perspectives.

2 Notations

We associate to the logical layer an undirected graph $G_1 = (V_1, E_1)$ where nodes correspond to routers and edges to possible links between these routers. We associate to the optical layer an undirected graph $G_2 = (V_2, E_2)$ where nodes correspond to the OXCs and edges to the physical links between these OXC. To every router $v_i \in V_1$ we associate an OXC $w_i \in V_2$. We assume that between nodes of G_1 there exist traffic demands. Let us denote by K the set of these demands. Denote by (O_k, D_k) the pair of routers origin-destination for $k \in K$ and by O'_k and D'_k the corresponding OXCs in the optical layer. Let $L^1_k = (v^{1,1}_k, ..., v^{1,j}_k, ..., v^{1,l_{1,k}}_k)$ and $L^2_k = (v^{2,1}_k, ..., v^{2,j}_k, ..., v^{2,l_{2,k}}_k)$ be the two paths routing demand $k \in K$ in G_1. These paths pass through terminal routers $v^{i,j}_k$ for which are associated terminal optical end-nodes OXCs $w^{i,j}_k (k \in K, i \in \{1,2\}, j = 1, ..., l_{1,k} + l_{2,k})$. Denote by T_k the set of these *terminals*. The other nodes in V_2 are called *Steiner nodes* for the demand $k \in K$ and are denoted $S_k = V_2 \backslash T_k$. Denote by $\mathscr{T}_k = \{T^q_k, q = 1, ..., n_k, n_k = l_{1,k} + l_{2,k} - 2, k \in K\}$ the set of sections between the different pairs of terminals OXC. A graph $G^{q,k}$ is the induced graph obtained from G_2 by deleting all terminals T_k of the demand k but extremities of section q or T^q_k. Graph G_2 is assumed to be complete with infinite capacities on the edges. Let $c(e) > 0$ be the cost of an edge $e \in E_2$.

3 Complexity

In this section, we study the complexity of the problem MSOND. We are in particular interested in the problem SC-MSOND (case of $|K| = 1$). The SC-MSOND can be defined as follows:

Input: an undirected graph $G' = (V', E')$, a cost $w'_e \geq 0$ associated to each $e' \in E'$ and $T' = (v_1, ..., v_l)$ terminals.

Output: An elementary cycle going in order through the terminals T' such that the total cost is minimum.

The corresponding decision problem is to find if there exists an elementary cycle going in order through the terminals T' such that the total cost is at most equal to a positive integer U'. Recall that T' constitutes the terminals corresponding actually to the source, the destination and intermediary nodes used in the two paths between the source and the destination. Since the two given paths are vertex disjoint, we have always $|T'| \geq 3$.

Theorem 1. *SC-MSOND Problem is NP-hard.*

Proof. We prove that the decision problem associated to SC-MSOND is NP-hard by proposing a polynomial reduction from the decision problem associated to Weighted Min-Sum Vertex Disjoint Paths WMSVDP proved to be NP-hard in [7,8]. This problem can be defined as follows:

Input: an undirected graph $G = (V,E)$, a cost $w_e \geq 0$ associated to each $e \in E$ and $T = \{(s_i,t_i) \in V, i = 1,...,k\}$ pairs of origin-destination, we assume that k is fixed and greater than or equal to 3.

Output: Does it exist k vertex disjoint paths $P_1,...,P_k$, P_i is a path from s_i to t_i, $i = 1,...,k$ such that the total cost is at most equal to a positive integer U.

Consider an instance (G,W,T) of the WMSVDP. We construct from (G,W,T) an instance (G',W',T') of MSOND as follows. We add to a copy of the graph G, k vertices $u_1,...,u_k$ and $2k$ edges $\{t_i,u_i\},\{u_i,s_{i+1}\}, i = 1,...,k$ $(s_1 = s_{k+1})$. Denote E_u the added edges. Let $w'_e = w_e$ if $e \in E$ and 0 otherwise (see Figure 1). Finally we set $T' = (s_1,t_1,u_1,s_2,...,s_j,t_j,u_j...,s_k,t_k,u_k,s_1)$ the terminals.

In the following we show that there exist k vertex disjoint paths between the pairs of T in G such that the total cost is at most equal to U if and only if there exists in G' an elementary cycle going in order through the terminals T' such that the total cost is at most equal to U.

Consider first a solution of WMSVDP in G with a total cost $C \leq U$. The solution consists of k vertex disjoint paths between the pairs $(s_i,t_i), i = 1,...,k$. These paths plus the set of edges E_u constitute by construction an elementary cycle in G' going in order through the terminals of T'. And since, the weights of all edges in E_u is equal to 0, the cost of the cycle is equal to C which is at most equal to U. Consider now an elementary cycle in G' going in order through the terminals T' with a total cost $C' \leq U$. Consider the sections between the terminals $(s_i,t_i), i = 1,...,k$. Since the cycle is elementary, these sections are

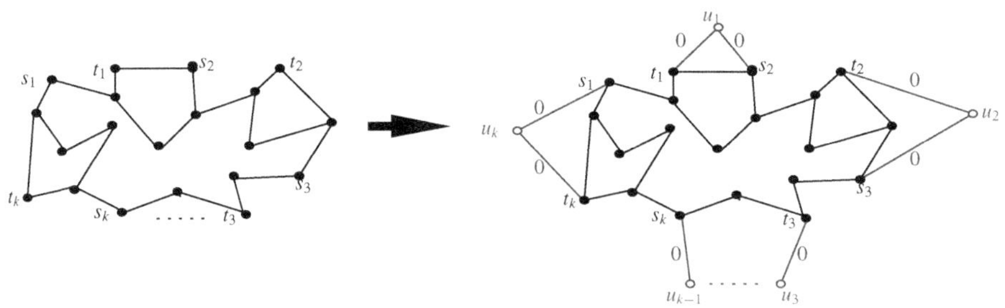

Fig. 1. WMSVDP reduction to SC-MSOND

vertex disjoint. Moreover, as the weights of all edges in E_u are 0, the total weight of the sub-paths between $(s_i,t_i), i=1,...,k$ is exactly equal to C' which is at most equal to U.

Corollary 1. *Since SC-MSOND is a particular case of MSOND, MSOND is NP-hard.*

4 Path Formulation

We denote by P_k^q the set of paths routing the section q of demand k calculated in the reduced graph $G^{q,k}$ previously defined. We associate for each path $p \in P_k^q$ a binary variable $x_p^{q,k}$ which takes 1 if $p \in P_k^q$ is selected to rout section q of demand k and 0 otherwise. Let $y_e = 1$ if the edge $e \in E_2$ is installed and 0 if not. We define coefficients $a = (a_p^{q,k}, k \in K, q \in T_k, p \in P_k^q)$ and $b = (b_p^{q,k}, k \in K, q \in T_k, p \in P_k^q)$ as follows. $a_p^{q,k}(w)$ characterize the degree of a vertex w in a path p routing section q of demand k: it is equal to 1 if w is one of the extremities of section q, 2 if w belongs to p and 0 otherwise. $b_p^{q,k}(e)$ designs the belonging of an edge e to the path p routing section q of demand k: it is equal to 1 if e belongs p and 0 otherwise. The MSOND problem is equivalent to the following $0-1$ linear program.

$$\min \sum_{e \in E_2} c(e) y_e$$

$$\sum_{p \in P_k^q} x_p^{q,k} = 1 \qquad \forall k \in K, \forall q \in \mathscr{T}_k \tag{1}$$

$$\sum_{q \in \mathscr{T}_k} \sum_{p \in P_k^q} a_p^{q,k}(w) x_p^{q,k} \leq 2 \qquad \forall w \in V_2, \forall k \in K \tag{2}$$

$$\sum_{p \in P_k} b_p^{q,k}(e) x_p^{q,k} \leq y_e \qquad \forall e \in E_2, \forall k \in K, \forall q \in \mathscr{T}_k \tag{3}$$

$$0 \leq x_p^{q,k} \leq 1 \qquad \forall k \in K, \forall q \in \mathscr{T}_k, \forall p \in P_k^q \tag{4}$$

$$x_e^{q,k} \in \{0,1\} \qquad \forall k \in K, \forall q \in \mathscr{T}_k, \forall p \in P_k^q \tag{5}$$

$$0 \leq y_e \leq 1 \qquad \forall e \in E_2 \tag{6}$$

$$y_e \in \{0,1\} \qquad \forall e \subset E_2 \tag{7}$$

Constraints (1) ensure routing of the demands through terminals with respect to the order constraints since paths are calculated in reduced graphs. Constraints (2) ensure the elementarity and disjunction of the two paths. Constraints (3) force routing variables to be equal to 0 if design variables are equal to 0 as well. Finally, constraints (4), (6) and (5), (7) represent, respectively, the trivial and integrity constraints.

4.1 Pricing Problem

Let us denote by $\pi^{q,k}$, λ_w^k and $\beta_e^{q,k}$ the dual variables associated respectively with constraints (1), (2) and (3), with respect to primal variable $x_p^{q,k}$. The reduced cost of the variable $x_p^{q,k}$ is given by $R_p^{q,k} = -(\pi^{q,k} + \sum_{w \in V_2} \lambda_w^k a_p^{q,k}(w) + \sum_{e \in E_2} b_p^{q,k}(e) \beta_e^{q,k})$. Here, the pricing problem is to find, for each section q of a demand k, a path of P_k^q such as

$R_p^{q,k} = min_{p\prime \in P_k^q} R_{p\prime}^{q,k}$ and $R_p^{q,k} < 0$. This can be seen as a shortest path problem in a reduced graph $G^{k,q}$ with weights λ_w^k on vertices and $\beta_e^{q,k}$ on edges. λ_w^k can be after split and hence weights are then only on edges. As dual variables λ_w^k and $\beta_e^{q,k}$ are negative, edge weights are non negative and the shortest path pricing problem can be solved in polynomial time.

4.2 Preliminary Results

We compute two relaxations of the previous program. The first relaxing both integer constraints and the second relaxing only x variables integrality. Results are reported in Table 1. The columns represent the numbers respectively of nodes in G_2, nodes in G_1, demands, generated paths for the first and second relaxations and finally the gaps of these relaxations comparing to the optimal value obtained by a Branch-and-Cut algorithm based on a cut formulation of the problem. The results show that branching only on y variables is interesting for small instances but is inefficient for larger ones. In addition, both relaxations are weak with a mean gap of near to 25% and have to be strengthened mainly by identifying and adding new valid inequalities.

Table 1. Preliminary computational results

V_1	V_2	K	paths1	paths2	gap1	gap2
6	4	4	55	65	21.26	0.00
7	4	3	99	202	21.26	0.00
7	4	5	147	227	21.26	0.00
8	5	4	340	1533	21.44	0.00
8	5	6	214	908	27.16	0.00
10	7	8	2260	18947	12.57	0.00
12	10	8	11163	348697	24.71	0.00
12	10	12	2619	195999	22.63	100.00

V_1	V_2	K	paths1	paths2	gap1	gap2
13	8	12	2474	228436	27.29	100.00
14	12	18	3620	2071	20.16	17.89
16	13	16	4890	4324	25.68	24.88
16	13	18	8071	5179	26.17	24.73
17	15	18	10403	7773	24.72	23.81
17	15	20	11474	6377	30.68	29.61
18	15	25	18478	8044	25.08	23.03
20	17	25	23168	14506	25.43	25.04

5 Conclusion

In this paper, we study the problem of Multilayer Survivable Optical Networks Design. We prove that this problem is NP-hard and we propose a path-based formulation to it. We discuss the corresponding pricing problem and give some preliminary computational results for two relaxations of the formulation. Current experimentations concern the test of different branching rules to achieve the Branch-and-Price algorithm. These results will be shown later.

References

1. Balas, E., Fischetti, M., Pulleyblank, W.R.: The precedence-constrained asymmetric traveling salesman polytope. Mathematical Programming 68(1), 241–265 (1995)
2. Borne, S., Gourdin, E., Liau, B., Mahjoub, A.R.: Design of survivable IP-over-optical networks. Annals of Operations Research (146), 41–73 (2006)

3. Dahl, G., Martin, A., Stoer, M.: Routing through virtual paths in layered telecommunication networks. Operations Research 47(5), 693–702 (1999)
4. Laporte, G., Mercure, H., Nobert, Y.: Optimal tour planning with specified nodes. RAIRO, rech. Oplle/ Opns. Res. 18, 203–210 (1984)
5. Orlowski, S., Raack, C., Koster, A.M.C.A., Baier, G., Engel, T., Belotti, P.: Branch-and-Cut Techniques for Solving Realistic Two-Layer Network Design Problems. In: Graphs and Algorithms in Communication Networks, pp. 95–118. Springer, Heidelberg (2010)
6. Salazar-González, J.J.: The steiner cycle polytope. European Journal of Operational Research 147, 671–679 (2003)
7. Eilam-Tzoreff, T.: The disjoint shortest paths problem. Discrete Applied Mathematics 85, 113–138 (1998)
8. Zhang, P., Zhao, W.: On the complexity and Approximation of the Min-Sum and Min-Max Disjoint Paths Problems. In: Combinatorics, Algorithms, Probabilistic and Experimental Methodologies (Book Chapter), pp. 70–81 (2007)

Hop-Level Flow Formulation for the Hop Constrained Survivable Network Design Problem

Ridha Mahjoub[1], Luidi Simonetti[2], and Eduardo Uchoa[2]

[1] Université Paris-Dauphine, Place du Maréchal De Lattre de Tassigny, 75775, Paris Cedex 16, France
mahjoub@lamsade.dauphine.fr
[2] Universidade Federal Fluminense, Rua Passo da Pátria, 156 CEP 24210-240, Niterói, Brazil
luidi@ic.uff.br, uchoa@producao.uff.br

Abstract. The HSNDP consists in finding a minimum cost subgraph containing K edge-disjoint paths with length at most H joining each pair of vertices in a given demand set. The only formulation found in the literature that is valid for any K and any H is based on multi-commodity flows over suitable layered graphs (Hop-MCF) and has typical integrality gaps in the range of 5% to 25%. We propose a new formulation called Hop-Level-MCF (in this short paper only for the rooted demands case), having about H times more variables and constraints than Hop-MCF, but being significantly stronger. Typical gaps for rooted instances are between 0% and 6%. Some instances from the literature are solved for the first time.

1 Introduction

Let $G = (V, E)$ be an undirected graph with n vertices, numbered from 0 to $n-1$, and m edges with non-negative costs c_e, $e \in E$; $D \subseteq V \times V$ be a set of demands; and $K \geq 1$ and $H \geq 2$ be natural numbers. The Hop-constrained Survivable Network Design Problem (HSNDP) consists in finding a subgraph of G with minimum cost containing, for each demand $d = (u, v) \in D$, K edge-disjoint (u, v)-paths with at most H edges. If all demands have a common vertex, w.l.o.g. the vertex 0, we say that the demands are *rooted*, otherwise they are *unrooted*. When $|D| = 1$, the HSNDP is polynomial for $H \leq 3$ and NP-hard for $H \geq 4$ (see [1]). When the cardinality of D is not constrained, the problem is NP-hard even if D is rooted, $K = 1$ and $H = 2$ (see [3]). In this short paper we only consider the case of rooted demands.

2 Hop Multi-Commodity Flow Formulation (Hop-MCF)

An extended formulation was recently proposed for the general HSNDP [2]. As D here is assumed to be rooted, a demand $(0, d) \in D$ will be identified by its destination vertex d. Let $V' = V - \{0\}$ and $E' = E \setminus \delta(0)$, where $\delta(i)$ represents the set of edges adjacent to a vertex i. For each demand $d \in D$, define the hop layered directed graph $G_H^d = (V_H^d, A_H^d)$, where $V_H^d = \{(0,0)\} \cup \{(i,h) : i \in V'; 1 \leq h \leq H-1\} \cup \{(d,H)\}$. Assuming that G is a complete graph, $A_H^d =$

J. Pahl, T. Reiners, and S. Voß (Eds.): INOC 2011, LNCS 6701, pp. 176–181, 2011.

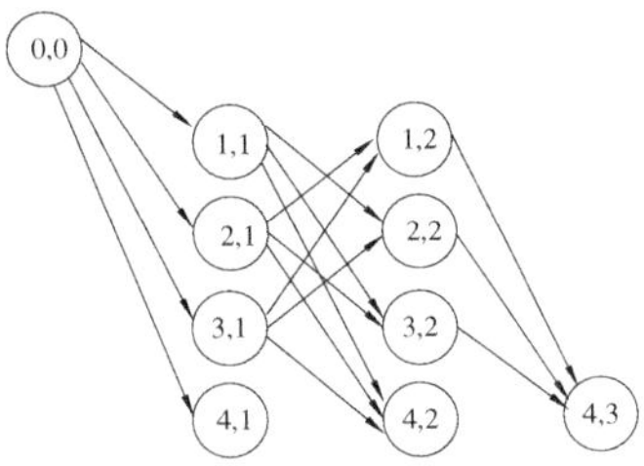

Fig. 1. Example of auxiliary graph G_H^d: G complete, $n = 5$, $d = 4$, and $H = 3$

$$\{[0,j,1] = [(0,0),(j,1)] : j \in V'\}$$
$$\cup \{[i,j,h] = [(i,h-1),(j,h)] : i,j \in V' - \{d\}, i \neq j; 2 \leq h \leq H-1\}$$
$$\cup \{[i,d,h] = [(i,h-1),(d,h)] : i \in V' - \{d\}; 2 \leq h \leq H\}.$$

Each arc in A_H^d is identified by a triple $[i,j,h]$, giving its origin, destination and hop. When G is not complete, if $(i,j) \notin E$, arcs of form $[i,j,h]$ and $[j,i,h]$ are omitted from A_H^d. Figure 1 depicts an example of such auxiliary network. For each $d \in D$, and for each arc $[i,j,h]$ in A_H^d, define binary flow variables f_{ij}^{dh}. For each edge (i,j) in E, define design binary variables x_{ij}. Let $\delta^-(i,h,d)$ and $\delta^+(i,h,d)$ denote, respectively, the set of arcs in A_H^d entering and leaving vertex (i,h). The Hop-MCF formulation follows:

$$\min \sum_{(i,j) \in E} c_{ij} x_{ij} \tag{1}$$

$s.t.$

$$\sum_{a \in \delta^-(i,h,d)} f_a - \sum_{a \in \delta^+(i,h,d)} f_a = 0 \quad d \in D; (i,h) \in V_H^d, i \notin \{0,d\} \tag{2}$$

$$\sum_{h=1}^{H} \sum_{a \in \delta^-(d,h,d)} f_a = K \quad d \in D \tag{3}$$

$$f_{0j}^{d1} \leq x_{0j} \quad d \in D; (0,j) \in \delta(0) \tag{4}$$

$$\sum_{h=2}^{H-1} (f_{ji}^{dh} + f_{ij}^{dh}) \leq x_{ij} \quad d \in D; (i,j) \in E' \setminus \delta(d) \tag{5}$$

$$\sum_{h=2}^{H} f_{jd}^{dh} \leq x_{jd} \quad d \in D; (j,d) \in \delta(d) \setminus \delta(0) \tag{6}$$

3 Hop-Level Multi-Commodity Flow Formulation (HL-MCF)

It is well-known that directed formulations of network design problems, when available, are much stronger that their undirected counterparts. The relative weakness of known HSNDP formulations, including Hop-MCF, is related to the impossibility of directing the solutions, since both orientations of an edge can be used in the paths for different demands. The proposed formulation tries to remedy this difficulty by introducing the concept of *solution level*. Given a solution T, we can partition V into $L+2$ levels, according to their distances to 0 in T. In the rooted case, L is set as equal to H, in the unrooted case (not presented here) L is usually greater than H. Level 0 only contains vertex 0; level l, $1 \leq l \leq L$, contains vertices with distance l; and level $L+1$ contains the vertices that are not connected to 0 in T. Besides variables x, HL-MCF also has:

- Binary variables w_i^l, $i \in V', 1 \leq l \leq L+1$, indicating that vertex i is in level l; constant w_0^0 is defined as 1.
- Binary variables $y_{ij}^{l_1 l_2}$ indicating that edge (i,j) belongs to T, i is in level l_1 and j in level l_2. For each $(0,j) \in \delta(0)$ there is a single variable y_{0j}^{01}. Each $e=(i,j) \in E'$ is associated with a set of $3(L-1)$ variables $\{y_{ij}^{ll} : 1 \leq l \leq L-1\} \cup \{y_{ij}^{l(l+1)}, y_{ji}^{l(l+1)} : 1 \leq l \leq L-1\}$.
- Binary flow variables $g_{ij}^{dhl_1l_2}$ associated to $|D|$ auxiliary hop-level networks.

The x and (w,y) variables are linked by the following constraints:

$$\sum_{l=1}^{L+1} w_i^l = 1 \quad i \in V' \tag{7}$$

$$w_j^1 = y_{0j}^{01} = x_{0j} \quad (0,j) \in \delta(0) \tag{8}$$

$$\sum_{l=1}^{L-1} y_{ij}^{ll} + \sum_{l=1}^{L-1} (y_{ij}^{l(l+1)} + y_{ji}^{l(l+1)}) = x_{ij} \quad (i,j) \in E' \tag{9}$$

$$\begin{aligned} y_{ij}^{11} + y_{ij}^{12} &\leq w_i^1 \\ y_{ij}^{11} + y_{ji}^{12} &\leq w_j^1 \end{aligned} \quad (i,j) \in E' \tag{10}$$

$$\begin{aligned} y_{ij}^{ll} + y_{ij}^{l(l+1)} + y_{ji}^{(l-1)l} &\leq w_i^l \\ y_{ij}^{ll} + y_{ji}^{l(l+1)} + y_{ij}^{(l-1)l} &\leq w_j^l \end{aligned} \quad (i,j) \in E'; l = 2,\ldots,L-1 \tag{11}$$

$$\begin{aligned} y_{ji}^{(L-1)L} &\leq w_i^L \\ y_{ij}^{(L-1)L} &\leq w_j^L \end{aligned} \quad (i,j) \in E' \tag{12}$$

$$w_i^l \leq \sum_{(j,i) \in \delta(i), j \neq 0} y_{ji}^{(l-1)l} \quad i \in V'; l = 2,\ldots,L \tag{13}$$

It can be checked that for any fixed binary solution x there is a single (w,y) solution that satisfies (7–13), that solution is binary and every vertex is assigned to a single level. However, a fractional x usually forces a (w,y) solution that splits vertices and edges into different levels. In order to profit from that splitting, for each $d \in D$, we define hop-level directed graphs $G_{HL}^d = (V_{HL}^d, A_{HL}^d)$, where $V_{HL}^d = \{(0,0,0)\} \cup \{(i,h,l) : i \in V'; 1 \leq h \leq H-1; h \leq l \leq L\} \cup \{(d,H,l) : 1 \leq l \leq L\}$, and $A_{HL}^d =$

$$\begin{aligned} &\{[0,j,1,0,1] = [(0,0,0),(j,1,1)] : j \in V'\} \\ \cup\ &\{[i,j,h+1,l,l'] = [(i,h,l),(j,h+1,l')] : i,j \in V', i \neq d; 1 \leq h \leq H-2; \\ &\quad 1 \leq l \leq h; max(l-1,1) \leq l' \leq l+1\} \\ \cup\ &\{[i,d,H,l,l'] = [(i,H-1,l),(d,H,l')] : i \in V', i \neq d; \\ &\quad 1 \leq l \leq L-1; max(l-1,1) \leq l' \leq l+1\} \end{aligned}$$

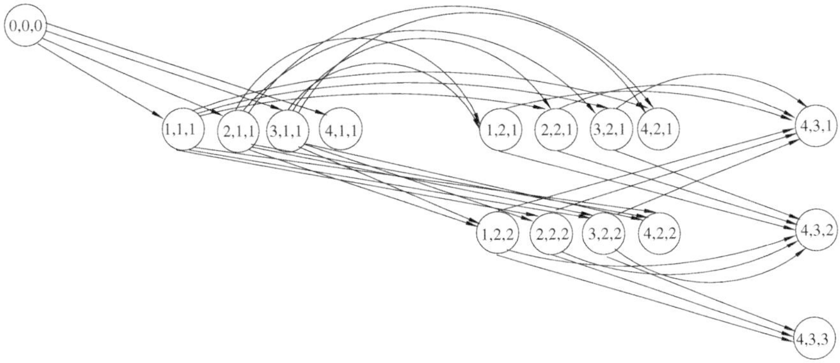

Fig. 2. Example of auxiliary graph G^d_{HL}: G complete, $n = 5$, $d = 4$, and $H = L = 3$

Again, if G is not complete, the arcs corresponding to missing edges are removed. Each arc in A^d_{HL} is identified by a tuple $[i, j, h, l_1, l_2]$, giving its origin, destination, hop, origin level and destination level. For each such arc, we define a binary flow variable $g_{ij}^{dhl_1l_2}$. Let $\delta^-(i,h,l,d)$ and $\delta^+(i,h,l,d)$ denote, respectively, the set of arcs in A^d_H entering and leaving vertex (i,h,l). The new formulation HL-MCF is (1), subject to (7)–(13) and to the following constraints:

$$\sum_{a\in\delta^-(i,h,l,d)} g_a - \sum_{a\in\delta^+(i,h,l,d)} g_a = 0 \quad d \in D; (i,h,l) \in V^d_{HL}, i \notin \{0,d\} \tag{14}$$

$$\sum_{h=1}^{H} \sum_{a\in\delta^-(d,h,l,d)} g_a = K.w^l_d \quad d \in D;\ 1 \le l \le L \tag{15}$$

$$g_{0j}^{d101} \le y_{0j}^{01} \quad d \in D; (0,j) \in \delta(0) \tag{16}$$

$$\sum_{h=l+1}^{H-1} (g_{ji}^{dhll} + g_{ij}^{dhll}) \le y_{ij}^{ll} \quad d \in D; (i,j) \in E' \setminus \delta(d); 1 \le l \le L-2 \tag{17}$$

$$\sum_{h=l+2}^{H-1} g_{ji}^{dh(l+1)l} + \sum_{h=l+1}^{H-1} g_{ij}^{dhl(l+1)} \le y_{ij}^{l(l+1)} \quad d \in D; (i,j) \in E' \setminus \delta(d); 1 \le l \le L-2 \tag{18}$$

$$\sum_{h=l+1}^{H} g_{jd}^{dhll} \le y_{jd}^{ll} \quad d \in D; (j,d) \in \delta(d) \setminus \delta(0); 1 \le l \le L-2 \tag{19}$$

$$\sum_{h=l+1}^{H} g_{jd}^{dhl(l+1)} \le y_{jd}^{l(l+1)} \quad d \in D; (j,d) \in \delta(d) \setminus \delta(0); 1 \le l \le L-2 \tag{20}$$

The HL-MCF formulation has $O(|D|.H.L.m)$ variables and $O(|D|.H.L.n)$ constraints, an increase by a factor of $L = H$ (in both dimensions) with respect to Hop-MCF. It can be proved that HL-MCF is at least as strong as Hop-MCF in terms of bounds provided by their linear relaxations.

4 Computational Experiments

The experiments were performed with CPLEX 12.1 MIP solver over a single core of an Intel i5 2.27GHz CPU. The first tests were on the rooted instances used in [2], complete graphs with 21 vertices associated to random points in a square, euclidean distances. The root vertex is in the center on instances TC-5 and TC-10, and on a corner on instances TE-5 and TE-10. The numbers 5 and 10 refer to the number of demands. We also added instances TC-20 and TE-20 with 20 demands. The average gaps of formulations Hop-MCF and HL-MCF are listed in Table 1.

- Formulation HL-MCF is very strong when $K = 1$ or $H = 2$, all problems are solved to optimality with almost no branching. The rooted case with $K = 1$ is equivalent to the hop-constrained Steiner tree problem, for which very strong formulations are already known [3]. But for $H = 2$ and $K > 1$, HL-MCF is much stronger than any other known formulation.
- When $K = 2$ and $H = 3$, HL-MCF is much stronger than Hop-MCF, the decreased gaps more than compensate for having to solve larger LPs. For example, instance TE-20 can be solved to optimality in 162 seconds using HL-MCF, but can not be solved in 1 hour with Hop-MCF.
- When $K = 2$ and $H = 4$ or when $K = 3$ and $H = 3$, HL-MCF is significantly stronger than Hop-MCF. However, the decreased gaps and smaller enumeration trees are roughly compensated by the burden of the larger LPs; the overall results are comparable.
- In the remaining three cases, HL-MCF is only slightly stronger than Hop-MCF, which performs much better on solving those instances to optimality.

Table 1. Average percentage gaps on instances TC-5, TC-10, TC-20, TE-5, TE-10, TE-20

	K	$H = 2$	$H = 3$	$H = 4$	$H = 5$
Hop	1	14.99	23.91	25.82	26.94
HL		0.00	0.00	0.83	1.93
Hop	2	12.92	13.13	13.05	9.60
HL		0.40	2.83	5.62	5.08
Hop	3	7.38	7.64	7.00	6.01
HL		0.08	3.27	5.20	5.27

The remaining experiments were performed on rooted instances from [4], defined over complete graphs with up to 40 vertices, for cases $K = 2$ and $H = 2,3$. Table 2 compares the performance of CPLEX MIP solver over formulations Hop-MCF and HL-MCF, in terms of duality gap, number of nodes in the branch tree and total time to solve the instance. The root gaps of the branch-and-cut in [4] for some instances are also presented.

Table 2. Results on the rooted instances from Huygens et al.[4] ($K = 2$)

		$H = 2$							$H = 3$						
n	$\|D\|$	HL-MCF			Hop-MCF			[4]	HL-MCF			Hop-MCF			[4]
		Gap	Nds	T(s)	Gap	Nds	T(s)	Gap	Gap	Nds	T(s)	Gap	Nds	T(s)	Gap
20	5	0.00	1	0.12	6.30	2	0.03	0.9	3.23	6	2.75	8.15	12	0.66	5.2
	10	0.00	1	0.12	12.73	22	0.44	6.8	4.09	16	10.6	12.52	181	26.0	6.8
	15	0.00	1	0.13	12.90	55	0.40	9.2	3.82	141	76.6	16.33	6788	1635	-
30	8	0.22	2	1.05	9.28	5	0.99	3.3	5.28	14	25.2	10.75	53	8.45	5.3
	15	0.00	1	0.37	14.27	36	0.44	7.4	4.23	86	217	18.36	7026	2110	-
	22	0.08	1	0.99	18.31	685	8.29	-	3.37	1608	6754	22.01	-	>24h	-
40	10	0.00	1	0.97	12.64	22	0.46	7.4	2.24	5	100	8.84	117	43.7	7.4
	20	0.00	1	1.09	14.96	263	5.47	-	6.23	6819	21866	18.06	-	>24h	-
	30	0.00	1	0.51	16.60	8456	198	-	4.11	23014	83150	20.98	-	>24h	-
Avg.		0.03	1	0.59	13.11	1061	23.8	5.8	4.07	3523	12466	15.11	-	-	6.18

References

1. Bley, A., Neto, J.: Approximability of 3- and 4-hop bounded disjoint paths problems. In: Eisenbrand, F., Shepherd, F.B. (eds.) IPCO 2010. LNCS, vol. 6080, pp. 205–218. Springer, Heidelberg (2010)
2. Botton, Q., Fortz, B., Gouveia, L., Poss, M.: Benders decomposition for the hop-constrained network design problem, Optimization on-line, report ULB (2010)
3. Gouveia, L., Simonetti, L., Uchoa, E.: Modeling hop-constrained and diameter-constrained minimum spanning tree problems as Steiner tree problems over layered graphs. Mathematical Programming, Online first (2009)
4. Huygens, D., Labbé, M., Mahjoub, A.R., Pesneau, P.: Two-edge connected hop-constrained network design problem: valid inequalities and branch-and-cut. Networks 49, 116–133 (2007)

Maximum Delay Computation under Traffic Matrix Uncertainty and Its Application to Interdomain Path Selection

Isabel Amigo[1], Sandrine Vaton[3], Thierry Chonavel[3], and Federico Larroca[2]

[1] Telecom Bretagne, France, and Facultad de Ingeniería Eléctrica, Universidad de la República, Uruguay
maria.amigo@telecom-bretagne.eu
[2] Facultad de Ingeniería Eléctrica, Universidad de la República, Uruguay
[3] Telecom Bretagne, France

Abstract. One of the most important problems when deploying interdomain path selection with quality of service requirements is being able to rely the computations on metrics that hold for a long period of time. Our proposal for achieving such assurance is to compute bounds on the metrics, taking into account the uncertainty on the traffic demands. In particular, we will explore the computation of the maximum end-to-end delay of traversing a domain considering that the traffic is unknown but bounded. Since this provides a robust quality of service value for traversing the Autonomous System (AS), without revealing confidential information, we claim that the bound can be safely conceived as a metric to be announced by each AS to the entities performing the path selection, in the process of interdomain path selection. We show how the maximum delay value is obtained for an interdomain bandwidth demand and we propose an exact method for solving the optimization problem. Simulations with real data are also presented.

1 Introduction

New Internet market proposals are emerging, mainly due to the new technological offers and the positioning towards them of all the involved actors [23]. Value-added services with real time requirements, such as videoconferencing and telepresence, are showing up as the new stars of the Internet for Service Providers (SPs) and Customers. The Network Providers interests are moving towards obtaining new revenues and business opportunities out of these new services, that rely completely on their deployed network infrastructure. Network Providers must be able to assure Quality of Service (QoS), so as to fulfil the Customers expectations and to be able to trade among SPs, for instance by means of Service Level Agreements (SLAs), which means, in turn, that SPs can offer better services to their Customers.

This arising scenario becomes more complex when the services provided traverse several domains, or Autonomous Systems (ASs), in its way from the SP

J. Pahl, T. Reiners, and S. Voß (Eds.): INOC 2011, LNCS 6701, pp. 182–195, 2011.

to the Customer. In this case, QoS must be provided all throughout the path, involving different ASs, which raises several technical, economical and political issues. Concerning the technical aspect, achieving scalability, preserving confidentiality and providing interoperability is of paramount importance in any technical solution [25].

In the framework of an alliance of ASs, carriers work together in order to achieve a common interest. In this scenario QoS values related to each domain are exchanged, and Traffic Engineering decisions are taken according to them. Different mechanisms have been proposed for the selection and establishment of interdomain QoS constrained tunnels, that mainly rely on RSVP-TE [4] and the PCE architecture [1] (e.g. [22,6,21]). These mechanisms are based on metrics announced by each AS but they do not specify how to compute such metrics. The complexity resides mainly in the fact that the announced metrics have to hold for some period of time, ideally as long as the service is provided. Hence, ASs must be able to provide QoS values that are guaranteed to hold for a certain period of time.

We shall put our focus on point-to-point services with QoS requirements. In this case the service may be abstracted to a QoS guaranteed tunnel (for instance an MPLS tunnel [10]). The path traversed by the tunnel must fulfil the QoS parameters required by the service.

In particular, our attention will be focused on those services for which available bandwidth and end-to-end delay are critical parameters. The end-to-end delay is compounded of the sum of the delays introduced by each transit AS and the terminal ones, from source to destination. As illustrated in Fig. 1, where we show a situation with two terminal ASs and one transit AS, the delay in each of the ASs depends on the traffic already present in the AS (t_* flows in Fig. 1), the topology, the routing configuration, and the traffic coming from the new tunnel (flow u in Fig. 1).

But, why do not simply advertise an instantaneous value of the metric? The main fact that makes an instantaneous value of the metric a non appropriate one is the existence of uncertainty, which implies that the value can change in

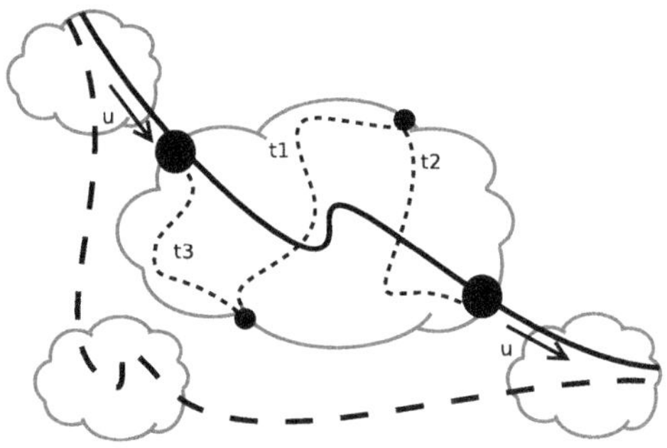

Fig. 1. Scenario

the immediate future. We could, however, follow a dynamic approach, in which network state is continuously monitored and metric value is updated. These reactive approaches make it possible to tightly follow the variations of the traffic but they require a monitoring infrastructure to be present and some sophisticated algorithms to process the measurement data. Moreover, reactive approaches are able to detect variations in the traffic demand such as abrupt changes but they are not able to forecast them [8]. On the contrary, proactive mechanisms provide pessimistic values of QoS metrics but they are able to provide metrics values which are likely to hold for a given period of time since in that case uncertainty is taken proactively into account. In this work we will use the robust approach, in which a bound for the metric is provided.

In this context, uncertainty can be classified into two types: network state uncertainty and traffic uncertainty. Uncertainty in network state refers to the situation where the topology changes or is partially known. This may be due to information arriving out of date or not synchronized to the entity performing the computation, or simply to link failures. In the literature some approaches have been proposed for performing QoS routing under this kind of uncertainty [19,12,18]. However, in the present paper we will assume that the topology does not change, and considering this uncertainty is left for future work.

On the other hand, we will consider uncertainty in the traffic. This refers to the fact that the flows traversing the domain are not perfectly known. This can be due to the fact that changes occur rather frequently. The reason of these changes may be several, for instance, external routing modification, the presence of unexpected events such as network equipment failures outside the domain, large-volume network attacks or flash crowd occurrences [24].

In summary, we shall focus on the computation of a bound for the end-to-end delay of traversing an AS, from a given Origin to a Destination node, as a function of the AS parameters we mentioned before: the routing configuration, the traffic demands and the traffic injected through the new tunnel. We will consider the situation where traffic variation is the principal cause of delay variation and we will assume that the topology and the routing configuration are fixed. However, we will consider that traffic is non-static, and that it is contained in a so-called uncertainty set [5]. The question of how to choose this set is discussed later in the paper.

2 Problem Statement

In this section we formally present the problem of finding the maximum end-to-end delay experienced by a bounded amount of traffic traversing an AS through a particular path. As mentioned before, we will consider that traffic varies within an uncertainty set. First, let us introduce the notations that are going to be used throughout the paper and state some assumptions.

2.1 Assumptions and Notations

The network is compounded of n nodes and of a set L of links, $L = \{l_1 \dots l_{|L|}\}$, where the notation $|\cdot|$ refers to the cardinality of the set. Traffic demands will be represented by the so-called traffic matrix $TM = \{tm_{i,j}\}$, where $tm_{i,j}$ is the amount of traffic from node i to node j. We shall use as well the term Origin Destination (OD) flows to refer to them. We reorder every traffic demand and rewrite the OD flows ($tm_{i,j}$) in vector form as t, $t = \{t_k\}$, $k = 1 \dots n(n-1)$. The amount of traffic coming from the interdomain injected into the new tunnel will be u.

The link load Y is a vector containing in the i-th entry the load on link i without considering u. With these definitions we can see that $Y = R.t$ where R, a $|L| \times m$ matrix ($m = n(n-1)$), is the routing matrix, which means that $\{R_{i,j}\} = 1$ if flow j traverses link i, and 0 otherwise.

The flow that carries u will traverse the AS from an origin to a destination node following a certain path. We will call this path P. We will equally refer to the set of links that belong to that path as P, in this case P is a subset of L.

The mean link delay is approximated by the M/M/1 model, that is to say $D_l = \frac{K}{c_l - y_l}$, where c_l is the capacity of the link l and K the mean packet size. We then approximate the mean delay of a path by the sum of the delays of the links it traverses:

$$Delay_P = \sum_{l \in P} \frac{K}{c_l - y_l}. \tag{1}$$

The propagation delay may be ignored in our formulation since it does not change with the load and may be added as a constant later on. Moreover, the M/M/1 model is used for illustrative purposes only. In fact, any convex function may be used instead. See [17] on how to obtain a good approximation of the delay function based on measurements. We will as well ignore the constant K in the following formulations, for the sake of notations simplification.

2.2 Modelling Traffic Uncertainty

As mentioned above, we will not make any assumptions on the traffic matrix except that it always belongs to a certain uncertainty set. In particular we will follow the approach presented in [5] and define the uncertainty set as a polytope formed by the result of the intersection of several half-spaces. Consequently, all constraints can be written as $A \times t \leq b$, where A is a certain matrix which can be defined after different models, and b is a given bound. We now present four examples of polytope definition.

The Hose Model. This particular case of the general polytope definition was presented in [9] in the context of VPN services specification. It establishes that the input and output total traffic on each node is bounded. That is to say: $\sum_i tm_{i,j} \leq b_j^+$ and $\sum_i tm_{j,i} \leq b_j^- \ \forall \ i \in N$, $j \in \{N \setminus i\}$, where b_j^-, b_j^+ are given bounds on the total ingress and egress traffic and N is the set of network nodes.

Links Capacity Model. This model results of the application of bounds on the total traffic traversing the different links of the network, $y_i \leq b_i$. These constraints can also be written as $R^h \times t \leq b$, where $b = \{b_i\}$ are historical maximums taken for instance form measurements, and R^h is the routing matrix at the moment when the measurements were taken. This approach is used for example in [13] where a polyhedral definition of the traffic matrix is preferred to its estimation because of non stationarity artifacts and estimation errors.

Known Statistical Values. If mean, variance and covariance values of link loads are known, we can compute the variance ellipsoid as $\{w = \varrho + \alpha \mid \alpha^T \Omega \alpha \leq 1\}$ where ϱ is the expected value of the link loads, and Ω its covariance matrix. Therefore, the variables w describe an ellipsoid. Several half-planes tangent to the ellipsoid can be defined in order to obtain linear constraints. Figure 2 illustrates this example. The polytope can then be written as $A \times R \times t \leq b$, were R is the routing matrix and A and b define the polytope in which the ellipsoid is inscribed.

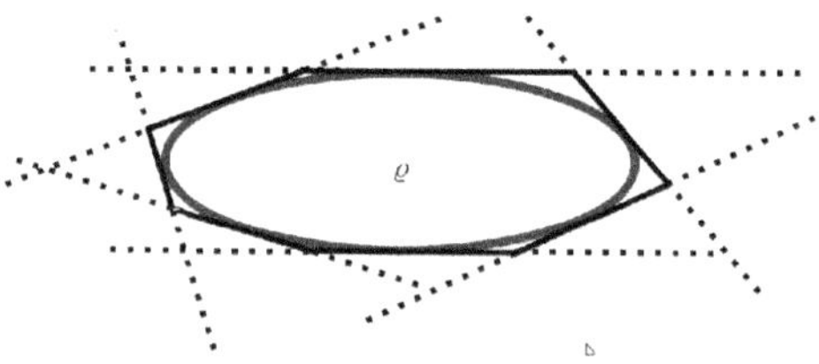

Fig. 2. Example for defining a polytope after known statistical values

Prediction Based Model. This model consists of defining bounds on the value of traffic demands which are based on traffic prediction. The prediction of future demands is based on past observations. For example artificial intelligence methods such as neural networks or time series analysis can be used in order to forecast the future values of the traffic demand; see for example [11] for prediction based on a seasonal ARIMA model.

2.3 Mathematical Formulation

The problem consists on, given a path, computing the maximum end-to-end delay of that path, allowing the traffic matrix t to vary within a polytope. That is to say that we will work with a maximization problem with linear constraints. Let us introduce the m-dimensional column vector w_l, $l \in P$, as $w_l = \{w_{l,i}\} = R_{l,i}/c_l$.

The optimization problem is described by Problem 1, where A and b define the polytope.

Problem 1

$$\begin{aligned} \max_{t} \quad & \sum_{l \in P} 1/c_l \frac{1}{1 - w_l^T t - u/c_l} \\ \text{s.t.} \quad & A \times t - b \leq 0. \end{aligned}$$

Please note that if some additional linear constraints must be taken into account they can be integrated in the definition of the polytope $A \times t \leq b$. Example of such constraints can be $w_l^T t + u/c_l < 1$, for $l \in P$, which simply states that there should be enough link capacity in order to accommodate all the traffic, including the new tunnel.

The objective function in the maximization problem defined by Problem 1 is not a concave function, consequently, the problem is not a convex one. On the contrary, the problem is the maximization of a convex function over a polytope. This is a very difficult problem, all the more so since the objective function is not strictly convex.

Intuitively we can see that the function is not strictly convex due to the difference between the number of links and the number of OD flows. Indeed, while the number of links grows linearly with the number of nodes in the network, the number of OD flows squares with the number of nodes in the network. This means that for different values of the vector t the objective function of Problem 1 can have the same value, while its gradient remains always non-negative.

More formally, we state the following proposition.

Proposition 1. *The function $f(t)$, objective function of Problem 1, is a convex function over the set $S = \{t \in \mathbb{R}^m | A \times t \leq b\}$, but not a strictly convex one.*

Proof. We explore if the following inequality holds [20]

$$f(t_1) \geq f(t_2) + \nabla f(t_2)^T (t_1 - t_2), \; t_1, t_2 \in S. \tag{2}$$

Applying the definition of f to Eq. (2) we obtain the following inequality for $t_1, t_2 \in S$:

$$\sum_{l \in P} \frac{1/c_l}{1 - w_l^T t_1 - u/c_l} \geq \sum_{l \in P} \frac{1/c_l}{1 - w_l^T t_2 - u/c_l} + \sum_{l \in P} \frac{1/c_l \times w_l^T (t_1 - t_2)}{(1 - w_l^T t_2 - u/c_l)^2}. \tag{3}$$

Let us now define $g_l(t)$, an auxiliary function in order to simplify the notations, as

$$g_l(t) = 1 - w_l^T t - u/c_l, \; t \in S. \tag{4}$$

Substituting the latter definition in Eq. (3) and performing some regular math operations we obtain the following inequality

$$\sum_{l \in P} \frac{(g_l(t_2) - g_l(t_1))^2}{g_l(t_1) g_l(t_2)^2} \geq 0, \; t_1, t_2 \in S. \tag{5}$$

Each term on Inequality (5) is either zero or greater than zero for all $t_1, t_2 \in S$. Therefore, the function f is convex over S. It remains to show if the function is strictly convex or not. Which is equivalent to showing if there exist t_1 and t_2 $\in S$ such that $< w_l, t_2 - t_1 >$ is equal to zero for all $l \in P$, that is to say, having all vectors w_l, $l \in P$ orthogonal to the vector $(t_2 - t_1)$, or not. Since the vectors w_l do not form a basis of $\mathbb{R}^m$ it is possible to find t_1 and $t_2 \in S$ such that their difference is orthogonal to all vectors w_l, $l \in P$. □

Proposition 1 showed that f is a convex function, but not a strictly convex one. However, in the following section we reformulate the problem and show a way to find its solution.

3 Finding the Solution

We now state the problem in a different way which will allow us to find its solution. We aim at formulating the problem in such a way that the objective function is strictly convex and the dimension of the problem is reduced. For doing so we shall decompose the vector t over a particular basis of $\mathbb{R}^m$.

The procedure consists in decomposing the vector t over the vectors w_l, $l \in P$, and their orthogonal complement. We define the matrix W_1 as an m by $|P|$ matrix, whose columns are the vectors w_l, with $l \in P$, and W_2, an m by $m - |P|$ matrix such that it verifies

$$W_1^T \times W_2 = 0. \tag{6}$$

Provided that the columns of W_1 are linearly independent, it can be proven that the columns of the matrix W defined after W_1 and W_2 as

$$W = [W_1 W_2] = [w_1, \ldots, w_l, \ldots, w_{|P|}, \ldots w_m] \tag{7}$$

represent a basis of $\mathbb{R}^m$.

We shall decompose the vector t over the defined basis using the auxiliary variables $x \in \mathbb{R}^{|P|}$ and $h \in \mathbb{R}^{m-|P|}$ as

$$t = W_1 x + W_2 h. \tag{8}$$

By multiplying both sides of Eq. (8) by w_l^T, and using Eq. (6) we obtain

$$w_l^T t = w_l^T W_1 x = v_l^T x, \tag{9}$$

where we have set $v_l^T = w_l^T W_1$, for all $l \in P$. Note that both v_l and x are column vectors of dimension $|P|$.

Equation (9) will directly lead us to rewriting the objective function of Problem 1 as a function of x. We shall now redefine the polytope by writing it in the basis W which leads to defining a new matrix denoted D and computed as $A \times W$. The polytope over the new basis can be compactly written as $D[x^T \ h^T]^T \leq b$.

All in all, Problem 1 can be rewritten in the form of Problem 2. Please note that the objective function depends only on the variable x.

Problem 2

$$\begin{aligned} \max_x \quad & \sum_{l \in P} 1/c_l \frac{1}{1 - v_l^T x - u/c_l} \\ \text{s.t.} \quad & D \begin{pmatrix} x \\ h \end{pmatrix} \le b \end{aligned}$$

Let us call the objective function of Problem 2 as $J(x)$ and the new polytope as V (i.e. $V = \left\{[x^T \ h^T]^T \in \mathbb{R}^m : D[x^T \ h^T]^T \le b\right\}$). Let us as well define the polytope V_x as

$$V_x = \left\{x \in \mathbb{R}^{|P|} \mid \exists \ h \in \mathbb{R}^{m-|P|} : D[x^T \ h^T]^T \le b\right\}. \tag{10}$$

Let $\mathcal{W}_1 = span\{w_1 \ldots w_{|P|}\}$, where span refers to the set of all linear combinations of vectors $w_1 \ldots w_{|P|}$. Clearly V_x is the projection of V onto $\mathcal{W}_1$.

Since V is a convex polytope by definition, it is easy to check that V_x is also a convex polytope. More precisely, V_x is the convex hull of the projection of the extreme points of V onto $\mathcal{W}_1$ [7].

Then, since $J(x)$ does not depend on h, Problem 2 can be represented in the space $\mathcal{W}_1$ as follows:

Problem 3

$$\begin{aligned} \max_x \quad & J(x) \\ \text{s.t.} \quad & x \in V_x. \end{aligned}$$

The following statement summarizes our development of the problem.

Proposition 2. *The optimization problem defined by Problem 1 is equivalent to the one defined by Problem 3.*

We now show that $J(x)$ is a strictly convex function over V_x, which will in turn allow us to prove that the solution of Problem 3 is attained at an extreme point of the polytope V_x.

Proposition 3. *The function $J(x)$, objective function of Problem 2, is a strictly convex function over the set V_x defined as in* (10).

Proof. We define $\lambda_l(x)$ as

$$\lambda_l(x) = (1 - v_l^T x - u/c_l)^{-2}, \ \forall l \in P \tag{11}$$

and the matrix Λ as

$$\Lambda(x) = diag(\lambda_1, \ldots, \lambda_{|P|}). \tag{12}$$

For all $x \in V_x$ and $l \in \{1 \ldots |P|\}$, $\lambda_l(x) > 0$. Thus, $\Lambda(x)$ is a positive-definite matrix[1].

[1] A $n \times n$ real symmetric matrix M is positive-definite if $z^T M z > 0$ for all non-zero vectors z, $z \in \mathbb{R}^n$.

In addition, we can check that $[v_1 \dots v_{|P|}] = W_1^T W_1$ is also a positive-definite matrix. Thus, the Hessian of $J(x)$, which is

$$\nabla^2 J(x) = (W_1^T W_1)\Lambda(x)(W_1^T W_1) \tag{13}$$

is as well a positive-definite matrix. □

We are now able to show that the solution to Problem 3 is attained at an extreme point of V_x.

Theorem 1. *The solution of Problem 3 is attained at an extreme point of the polytope V_x, defined by the set* (10).

Proof. We prove by contradiction that the maximum of $J(x)$ over V_x must be reached at an extreme point of V_x. Since, by Proposition 3, J is a strictly convex function, inequality (14) holds [20].

$$J(\Phi) > J(\theta) + \nabla J(\theta)^T(\Phi - \theta),\ \forall\ \theta, \Phi\ \in V_x. \tag{14}$$

Now, let $\bar{\theta} \in V_x$ be an optimal point of Problem 3. Therefore, $\bar{\theta}$ is a strict maximum, since J is strictly convex, and, for all $\Phi \in V_x \setminus \{\bar{\theta}\}$, we must have:

$$J(\Phi) - J(\bar{\theta}) < 0. \tag{15}$$

Together with inequality (14), we get

$$\nabla J(\bar{\theta})^T(\Phi - \bar{\theta}) < 0,\ \forall \Phi \in V_x \setminus \{\bar{\theta}\}. \tag{16}$$

By contradiction we suppose that $\bar{\theta}$ is not an extreme point of V_x. Then there exists $\mu \in \mathbb{R}^{|P|}$ such that $||\mu|| > 0$ and $\bar{\theta} + \mu$, $\bar{\theta} - \mu \in V_x$. By letting $\Phi = \bar{\theta} - \mu$ and $\Phi = \bar{\theta} - \mu$ at a time, we would get:

$$\nabla J(\bar{\theta})^T \mu < 0 \text{ and } -\nabla J(\bar{\theta})^T \mu < 0, \tag{17}$$

which is not possible. □

Problem 3 allows us to work with a strictly convex function, and to reduce the dimension of the feasible region, in some cases, considerably. According to Preposition 2 along with Theorem 1, finding the extreme points of the polytope V_x renders the solution of Problem 1. Therefore, we need to be able to perform the projection of a polytope, and afterwards enumerate its extreme points. Methods for doing so are available (see for instance [14]), although these can be computationally expensive tasks. In the following section we explore this solution by performing simulations in real topologies.

4 Simulations

To evaluate the proposed method we present some simulation studies. The simulations are carried out using two different research networks. Namely, the Abilene network, whose topology, historical traffic demands and routing matrix are available from [26], and the GÉANT network [3]. All results are computed on a regular computer (Intel Pentium Dual-core 1.86GHz, 2GB of RAM). For computing the polytope projection and enumerating its extreme points we use the MPT library [2] and the ET library [16], distributed along with the former.

4.1 The Abilene Network

The Abilene network consists of 30 internal links and 12 routers, all exchanging traffic among them. Figure 3 shows a traffic trace of Abilene's network. In this example we can see how the traffic matrix is prone to sudden traffic variations. Figure 3(a) shows the traffic for some OD flows corresponding to 2016 consecutive measurements, while Fig. 3(b) shows the link load.

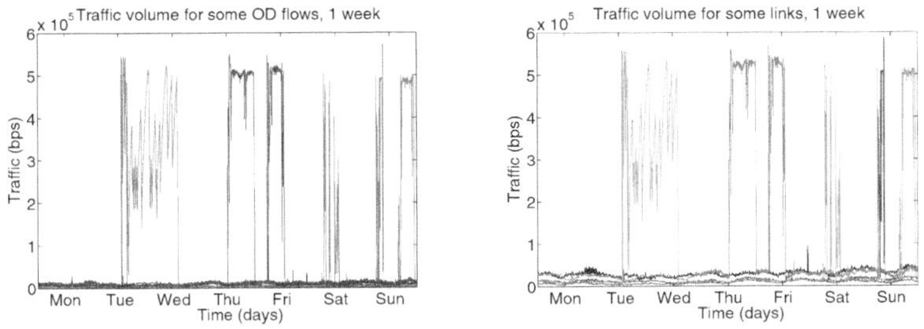

(a) Traffic volume per OD flow (b) Link load

Fig. 3. Example of traffic variation in the Abilene network, one week of traffic

For illustrative purposes we compute results for three different types of services. Namely, a VoIP service with 1 Mbps of bandwidth, a broadcast quality HDTV service with 19.4 Mbps and a VPN service with a demand of 270 Mbps. We compute the maximum delay suffered by a flow traversing the AS through a particular path and carrying each one of these services at a time. The path is chosen arbitrarily, from one origin to one destination node. Please note that this choice and its impact on the delay are out of the scope of the present paper.

In the first place, we define the polytope using the Links Load model. That is to say, the polytope is defined by imposing bounds on each link load, which are based on the maximum values obtained historically.

The values obtained for the defined path and the three services are shown in Fig.4(a) (dotted line) along with the current delay value. The current delay value corresponds to a value obtained instantaneously. For this particular case the maximum delay value is approximately 3 times more than the current one which illustrates the weakness of the current value as a metric on which rely. We will come back to this kind of comparisons later on this section.

In the second simulation, we define the polytope based on the Known Statistical Values model, introduced in Sect. 2.2. We compute the variance ellipsoid using a historical traffic trace (the same trace used for the first simulation) and we approximate the ellipsoid by a polytope, by intersecting several half spaces tangent to it. The maximum delay of traversing the AS is computed for the same path used in the previous simulation.

The results are shown on Fig. 4(a) (dashed line) for a flow traversing the same path as in the previous simulation and carrying the three defined services, one

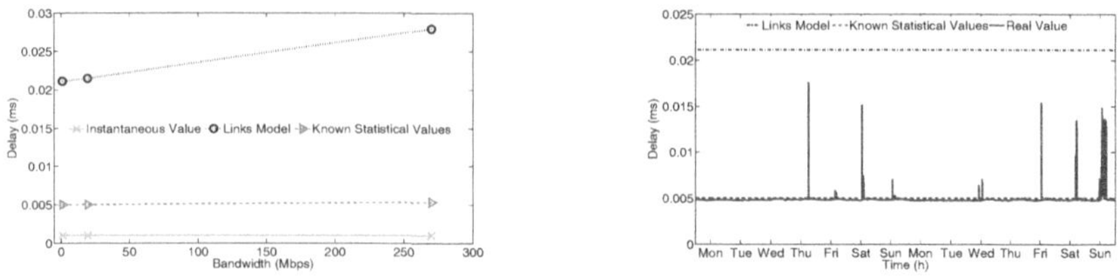

(a) Instantaneous delay value and Maximum delay value for two different polytopes and three bandwidth demands

(b) Maximum value for two different polytopes and real values during two weeks for one bandwidth demand

Fig. 4. Simulations in the Abilene network

at a time. We can see that in this case the bound obtained is smaller than the one obtained in the first place and closer to the instantaneous value.

We now compare the two bounds with the real delay suffered by the path during the two weeks after the computation of the polytopes, in all the cases assuming an interdomain bandwidth demand of 1 Mbps. The results are shown in Fig. 4(b) which illustrates the behaviour of the bounds with respect to the real values. We can see that there is a trade-off between assuring a delay value for most of the time, by using a big polytope, or having a tighter bound most of the time, but having delays that outstrip the bound. Nevertheless, the polytope could be reduced in a safe way if we had additional information, for example by using as well the hose model which imposes bounds to the traffic coming from other clients, which may be limited by a contract and traffic shaping.

The time consumed to perform the computations varied between 48 minutes and 36 hours, which for a moderately sized network is rather high. In fact, even if in several topologies we were able to find the exact solution through these means, it is still an open question whether there exists an algorithm for enumerating all extreme points of a polytope of an arbitrary dimension in running polynomial time [15]. We will, on the next subsection, empirically explore the time consumed by the method in a larger network.

4.2 The GÉANT Network

In order to test the proposed solution on a larger topology, we use the GÉANT network. This network is compounded of 23 nodes and 74 links. Thus, we can define up to 506 independent OD flows. As we have already mentioned the computation complexity of the proposed solution is likely to grow with the dimension of the network (i.e. the number of links in the path and the number of OD flows in the network). The simulations with this network aid as to assessing the performance of the method when the number of OD flows grows. We perform the simulations considering several subsets of OD flows, containing each of them 170, 200, 230 and 260 OD flows. The polytope is defined using the Links Load model and historical data.

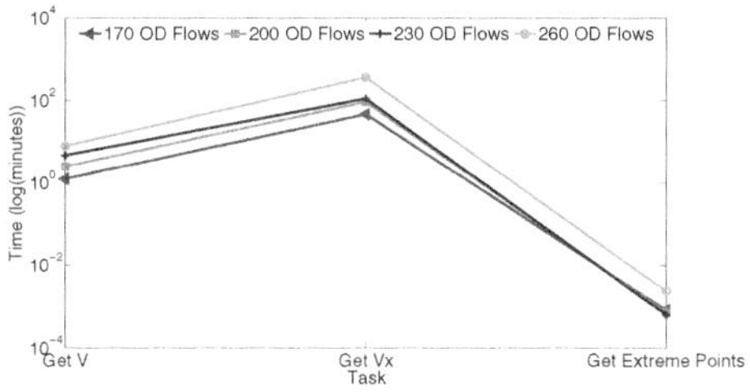

Fig. 5. Computation time as a function of the number of OD flows considered on the GÉANT Network

Figure 5 shows the time consumed by each phase of the procedure, that is to say obtaining the polytope in the new basis, projecting the polytope and finding its extreme points. We can see that in all the cases, when we increase the number of OD flows considered, the task that consumes most of the time is the projection of the polytope.

The procedure has shown rather high computational times, though it was still feasible in all the tests. It is because of this that we think of this method as of great aid when developing approximated, but less time consuming, methods, since it provides the ground truth, thus a validation tool for such methods.

5 Conclusion and Future Work

In this work we have addressed the problem of the existence of uncertainties on the traffic demands in the context of interdomain QoS provisioning. The uncertainty was modeled as a polytope and different examples for building it were mentioned. We have focused our attention on the computation of a robust value of the end-to-end delay of traversing an AS under traffic uncertainty, which means obtaining a value that does not change when traffic demands do so, assuming the demands remain inside the uncertainty set. This bound was conceived as a metric to be used in the interdomain path selection process, since it provides a value that the AS can guarantee for a certain period of time, while it can be advertised without reveling confidential information. The problem was mathematically formulated and an exact solution, based on the projection of the polytope onto a subspace of smaller dimension, was proposed. Simulations with real data were performed and shown.

The theoretical study suggested that the computational times could be rather high, simulations with large network confirmed this. In order to find a remedy to this situation, we are currently studying alternative solutions based on heuristics and numerical approximation methods. The exact method proposed in this paper will be extremely useful as a tool of validation of the approximated solutions. In addition, as future work, we shall address the case of having uncertainty on the AS topology in addition to traffic uncertainty. For instance, taking into account the case of link or node failures, and being able to provide even in those cases a tight end-to-end delay bound.

The Spatially Equitable Multicommodity Capacitated Network Flow Problem

Paolo Dell'Olmo and Antonino Sgalambro

Dipartimento di Scienze Statistiche, Sapienza Università di Roma,
P.le Aldo Moro 5, 00185, Roma
{paolo.dellolmo,antonino.sgalambro}@uniroma1.it

Abstract. In this work we propose a novel approach addressing the need for a spatially equitable distribution of the flows when routing multiple commodities on a capacitated network. In our model, the spatial distribution of the flows is considered by partitioning the area in which the network is embedded by means of a grid of uniform size cells and then computing the impact of the network flows on each cell by a weighted linear combination of the flows interesting each cell. A spatially equitable distribution of the flows is therefore obtained when all the multicommodity demands are satisfied in such a way to minimize the maximum impact registered on the cells of the grid. We refer to this problem as the spatially equitable multicommodity capacitated network flow problem and propose a minimax linear programming formulation. The need to find a proper trade-off between the total routing cost and the spatial equity is treated as well by considering both the objective functions and computing pareto-optimal solutions for the bicriteria optimization problem. Computational results obtained on a real traffic network are presented and discussed in the paper.

1 Introduction

In this paper we consider the problem of distributing network flows in a spatially equitable way while satisfying given multicommodity demands on a capacitated network. This topic was previously treated in the literature concerning hazardous material road trasportation by introducing the concept of spatially dissimilar paths [2] in order to reduce the exposure of the population to the risk associated to accidents. In their approach this issue is addressed by computing a set of k-shortest routes between an origin-destination pair and then seeking for the dissimilar paths by applying a p-dispersion algorithm. The dissimilarity measure in this approach is based on the length of the arcs shared by each couple of paths. In [5] two main contributions are introduced in this field. The first is the use of the *Buffer Zones* of a path, defined as the area obtained by moving a circle along a path whose center is the vehicle while the radius is proportional to the impact area due to possible accidents. Buffer Zones are therefore considered in order to take into account the exposure during a shipment for the population when computing path dissimilarities in the p-dispersion method. The second contribution consists in selecting the routes among a set of pareto-optimal paths according to a multicriteria

J. Pahl, T. Reiners, and S. Voß (Eds.): INOC 2011, LNCS 6701, pp. 196–209, 2011.

shortest path problem considering concurrently the risk and the length of the paths. The Buffer Zones method addresses one of the main drawbacks of the approach previously presented in [2], in which routes may be spatially very close to one another. Nevertheless the effectiveness of the latter approach in obtaining an equitable distribution of the flows from the spatial point of view suffers some limits depending on the topology of the network: in those cases in which the spatial distribution of the arcs on the network area in not uniform, that is in particular the case for urban areas, the overlapping between the Buffer Zones can interest several times the same areas, even if the dimension of the overlapping can by tiny with respect to the total length of the paths. This turns out in providing low similarity measures between the paths, but at the same time it can produce an overload on those shared portions of the networks. In [4] a different approach is considered: in addition to the main objective of selecting a set of paths of minimum total risk, a link-based risk equity is considered as well by imposing a threshold for the maximum risk sustained by the population living in the proximity of each populated link of the network. A path-based mixed integer programming formulation is provided in that paper, together with two heuristic methods and a Lagrangian relaxation providing effective lower bounds on the optimal solution value. In this paper we generalize the scope of the spatially equitable distribution of the flows: the suitability of the spatial dissimilarity of the flows can interest several different aspects in the real applications of network flow models, and is not always limited to the population living at the boundaries of the network roads as assumed in the previous contributions on this topic. In the specific field of transportation networks, e.g. the spatial distribution of the flows influences the concentration of the polluting emissions, the exposure to acoustic noises, as well as the risk associated to accidents occurring when transporting materials that are airborne when released. We propose a novel approach based on an extension of the topology considered in the network optimization problem: the area in which the network is embedded is partitioned by a grid of cells of uniform dimension, and the impact of the flows on each cell is considered to be proportional to the length of the portion of the arcs that interest the cell itself. In this way, it becomes possible to consider an explicit measure for the *cell load* when routing the flows in the context of the multicommodity capacitated network flow problem, and a spatially equitable distribution of the flows can be obtained by minimizing the maximum cell load while satisfying the multicommodity demands and respecting the capacity constraints on the network links. This approach generalizes the contributions in the literature on the link-based balanced network flow problem (see e.g. [1,8]). The remainder of the paper is organized as follows. In Section 2 the spatially equitable multicommodity network flow problem is introduced and formulated as a minimax linear programming problem. In Section 3 the computational results obtained by testing the proposed optimization problem on a real urban transportation network are presented and discussed. In Section 4 we extend the model to the bi-objective case considering concurrently the minimization of the maximum cell load and the total routing costs. Finally, we discuss further lines of research for the proposed optimization problem, aiming at the real suitability of the model for practical purposes.

2 The Spatially Equitable Multicommodity Network Flow Problem

Recall the definition of the capacitated multicommodity network flow problem: we are given a directed graph $G = (V,A)$ with $|V| = n$ and $|A| = m$, with arc capacities $u_{ij} > 0$, $\forall (i,j) \in A$, arc costs $c_{ij} \geq 0$, $\forall (i,j) \in A$, and a set D of commodities, each associated with a certain amount of demand d_i, a source node $s_i \in V$, and a destination node $t_i \in V \setminus \{s_i\}$, $\forall i \in D$. The goal is to minimize the total routing cost $TC = \sum_{k \in D} \sum_{(i,j) \in A} c_{ij} x_{ij}^k$ while satisfying all the demands without violating arc capacity constraints. The capacitated multicommodity network flow problem can be therefore formulated as follows:

$$\min \quad TC = \sum_{k \in D} \sum_{(i,j) \in A} c_{ij} x_{ij}^k \tag{1}$$

$$\text{s.t.} \quad \sum_{j \in FS(i)} x_{ij}^k - \sum_{j \in BS(i)} x_{ji}^k = \begin{cases} d^k, & i = s^k, \forall k \in D \\ 0, & \forall i \in V_{\setminus \{s_k, t_k\}}, \forall k \in D \\ -d^k, & i = t^k, \forall k \in D \end{cases} \tag{2}$$

$$\sum_{k \in D} x_{ij}^k \leq u_{ij} \qquad \forall (i,j) \in A \tag{3}$$

$$x_{ij}^k \geq 0 \qquad \forall (i,j) \in A, \forall k \in D$$

where the minimization of the total routing costs is subject to the satisfaction of the demand and the conservation of the flows (2) and to the constraints on the arc capacities (3). In order to introduce the spatially equitable multicommodity network flow problem, we first consider a set Z of cells of uniform size and dimension overlapped to the underlying network, as depicted in Figure 1. Then, a set of weights $\{w_{ij}^z\}$ can be defined as follows:

$$w_{ij}^z = c_{ij} \cdot l_{ij}^z \qquad \forall z \in Z, \forall (i,j) \in A \tag{4}$$

where l_{ij}^z, defined in $[0,1]$, expresses the fraction of the arc (i,j) that is embedded in the cell z. With this notation in mind, the spatial distribution of the flows in our approach is considered by introducing a linear measure of the impact f_z that the routed flows produce on a certain cell $z \in Z$, namely the *cell load*, as follows:

$$f_z = \sum_{k \in D} \sum_{(i,j) \in A} w_{ij}^z \cdot x_{ij}^k \qquad \forall z \in Z \tag{5}$$

The objective of a spatial equity in the distribution of the multicommodity network flows can be now obtained if we consider the following minimax linear model in which the objective function λ to be minimized is defined as the maximum among the cell loads f_z, $\forall z \in Z$, registered over all the cells, while satisfying all the demands on the capacitated network. We refer to this problem as the *spatially equitable multicommodity capacitated network flow problem* that can be formulated as follows:

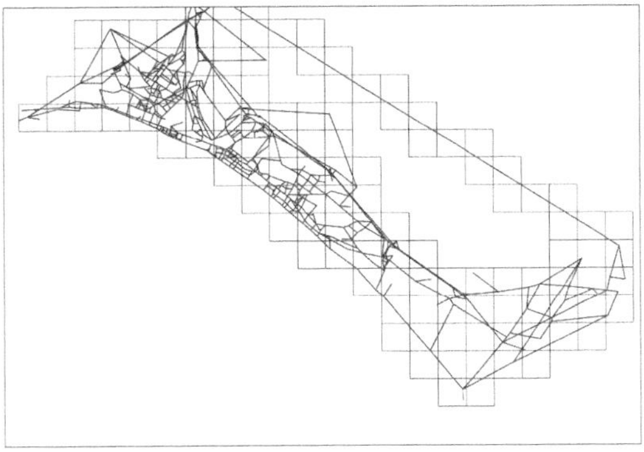

Fig. 1. A grid of uniform cells overlapped to the underlying road transportation network of the city of Salerno, Italy

$$\min \quad \lambda \tag{6}$$

$$\text{s.t.} \sum_{j \in FS(i)} x_{ij}^k - \sum_{j \in BS(i)} x_{ji}^k = \begin{cases} d^k, & i = s^k, \forall k \in D \\ 0, & \forall i \in V_{\setminus \{s_k, t_k\}}, \forall k \in D \\ -d^k, & i = t^k, \forall k \in D \end{cases} \tag{7}$$

$$\sum_{k \in D} x_{ij}^k \leq u_{ij} \qquad \forall (i,j) \in A \tag{8}$$

$$\sum_{k \in D} \sum_{(i,j) \in A} w_{ij}^z x_{ij}^k = f_z \qquad \forall z \in Z \tag{9}$$

$$f_z \leq \lambda \qquad \forall z \in Z \tag{10}$$

$$x_{ij}^k \geq 0 \qquad \forall (i,j) \in A, \forall k \in D \tag{11}$$

$$f_z \geq 0 \qquad \forall z \in Z \tag{12}$$

$$\lambda \geq 0$$

where constraints (9) assign the cell load of each cell z to the variable f_z, while constraints (10) define the maximum cell load $\lambda = max_{z \in Z} f_z$ to be minimized as an objective function.

3 Computational Results on a Real Traffic Network

A set of computational tests was performed in order to verify the effectiveness of the proposed optimization problem with respect to the purposes, considering as a basic instance the real traffic network of the city of Salerno, Italy. The network presents a set of 556 nodes and 1186 arcs, and a set of 100 commodities was generated at pseudo-random with fixed demand equal to 10. The real lengths of the traffic roads were used for the set of arc costs $\{c_{ij}\}$ while the arc capacities $\{u_{ij}\}$ were fixed equal to 100 and 50 in two separate sets of experiments. The considered traffic network is depicted in Figure 1 and was processed by using the open-source Geographical Information System

Table 1. Sketch of the computation of the l_{ij}^z coefficient for some arcs of the network

Arc ID $(i,j) \in A$	Origin node $i \in V$	Destination node $j \in V$	Arc Length (m)	Intersection Zone $(z \in Z)$	Intersection Length (m)	Coefficient l_{ij}^z
1	1	2	223.89	4866	25.76	0.11504
1	1	2	223.89	4867	1.50	0.00671
1	1	2	223.89	4996	86.08	0.38448
1	1	2	223.89	4995	84.58	0.37777
1	1	2	223.89	4997	25.97	0.11600
2	3	4	69.86	4235	26.06	0.37310
2	3	4	69.86	4363	43.79	0.62691
3	5	6	141.65	4368	30.52	0.21545
3	5	6	141.65	4496	52.91	0.37355
3	5	6	141.65	4497	53.28	0.37617
3	5	6	141.65	4625	4.93	0.03484
4	5	7	125.75	4368	70.16	0.55791
4	5	7	125.75	4369	55.59	0.44210
...	...	...	...	...	...	...

(GIS) software Quantum GIS (www.qgis.org). Given a grid of cells Z, the set of coefficients $\{l_{ij}^z\}$ must be preliminarily computed to assess the portion of each arc $(i,j) \in A$ that is embedded in each zone $z \in Z$ in order to compute the set of weights $\{w_{ij}^z\}$, as reported in (4). In the implementation of the model, this task was performed by using the *fTools* plugins in the Quantum GIS environment. More in detail, two overlapped layers are considered in the GIS project: a polygon vector layer containing the cells of the grid and a linear vector layer containing the arcs of the network. The intersection process between the latter and the former layers gives rise to a new linear vector layer (intersection layer) in which each arc of the original network is repeated several times, one for each cell of the grid that contains a portion of the arc itself. It is also possible to compute the length of each portion of arc that belongs to the new intersection layer by exporting the geometrical features of the intersection layer within the underlying database containing the attribute table. Finally, each coefficient l_{ij}^z can be obtained by considering the ratio between the length of the portion of arc $(i,j) \in A$ associated to the zone $z \in Z$ in the intersection layer and the length of the associated arc (i,j) in the linear vector layer containing the original network. A sketch of the results of this geometrical analysis within the GIS environment is presented in Table 1. Since in this computational test we are assuming that the arc costs c_{ij} are equal to the arc lengths, the arc intersection lengths directly coincide with the weights w_{ij}^z. Nevertheless this is not always valid, since arc costs could differ in general from the length of the arcs, depending on the purposes and the field of application of the model.

In Table 2 the computational results are presented at varying the size of the cell side, ranging from 100 to 1000 meters. In this case, cell grids covering the whole network area were filtered keeping only those cells containing at least a portion of road, while the empty cells were disregarded in the model. All the numerical results are expressed by comparing the results provided by the minimization of the maximum cell load λ with those obtained by minimizing the total routing cost function TC. In Table 2 and

Table 2. Comparison of the computational results on different sizes of the cell side for arc capacities equal to 100

Cell side (m)	$\|Z\|$	MCLD (%)	ACLI (%)	RT
100	1284	47.34	2.87	16.51
120	1006	57.85	5.34	10.83
140	799	55.92	16.56	3.38
160	682	58.57	21.54	2.72
180	560	58.88	21.56	2.73
200	486	52.59	23.02	2.28
300	265	53.28	24.18	2.20
400	172	55.84	31.26	1.79
500	122	42.61	20.89	2.04
600	94	43.68	24.76	1.76
700	73	41.89	22.30	1.88
800	59	42.53	26.62	1.60
900	51	41.42	32.45	1.28
1000	45	36.37	21.72	1.67

Table 3 the numerical results are presented as follows. For each cells size, the number of cells containing at least one portion of arcs, namely $|Z|$, is considered. The value *MCLD* (Maximum Cell Load Decrease) expresses the reduction of the maximum cell load that is obtained with the spatially equitable model with respect to the minimum routing cost model, and is computed as follows:

$$MCLD = \frac{\lambda^{TC} - \lambda^*}{\lambda^{TC}}$$

being λ^* the value of the objective function for the minimax model and λ^{TC} the maximum cell load f_z, $z \in Z$, obtained by minimizing the total routing cost function TC. The value *ACLI* (Average Load Cell Increase) expresses the increase of the mean cell load arising from the equitable distribution of the flows in comparison with the minimum routing cost model, and can be expressed as follows:

$$ACLI = \frac{\bar{f_z}^* - \bar{f_z}^{TC}}{\bar{f_z}^{TC}}$$

where $\bar{f_z}^* = \frac{\sum_{z \in Z} f_z}{\|Z\|}$ is the average value of the cell loads obtained with the minimax model and $\bar{f_z}^{TC} = \frac{\sum_{z \in Z} f_z}{\|Z\|}$ is the average value of the cell loads arising from the minimization of the total routing cost function. Finally, the ratio $RT = \frac{MCLD}{ACLI}$ is computed and presented in the last column. In Table 2 the results obtained by fixing all the arc capacities equal to 100 are presented, while Table 3 refers to the case with arc capacities equal to 50.

All the instances were solved at the optimum within a CPU time ranging from some seconds to a few minutes by the IBM ILOG CPLEX 12 software running on a 2.4 GHz intel core i5 with 4Gb DDR Ram. The running time increases with the number

Table 3. Comparison of the computational results on different sizes of the cell side for arc capacities equal to 50

Cell side (m)	$\|Z\|$	MCLD (%)	ACLI (%)	RT
100	1284	69.74	10.23	6.82
120	1006	68.99	10.51	6.56
140	799	60.93	18.33	3.32
160	682	67.92	19.83	3.42
180	560	56.62	17.67	3.20
200	486	61.97	22.29	2.78
300	265	58.77	23.59	2.49
400	172	55.84	31.26	1.79
500	122	37.44	23.67	1.58
600	94	41.10	24.59	1.67
700	73	47.92	18.39	2.60
800	59	38.38	26.25	1.46
900	51	34.13	20.12	1.70
1000	45	35.48	18.20	1.95

of cells associated with each instance. By analyzing the numerical results, it turns out how the increase in the number of cells gives rise to better results, providing a higher relative decrease of the maximum cell load and higher values for the ratio *RT* between *MCLD* and *ACLI* as well. The comparison between the results in Table 2 and Table 3 show the effects of reducing arc capacities from 100 to 50: some of the *MCLD* values improve significantly, but the concurrent higher relative increase in the *ACLI* values influences the *RT* ratios, that are lower for the case with arc capacities equal to 50. The results of the application of our model to the considered traffic network can be graphically observed as well in Figure 2 and Figure 3, in which the intensity of the grayscale represents the cell load value for the cases of the minimum total routing cost variant and the spatially equitable variant of the considered multicommodity capacitated network flow problem.

3.1 Evaluating the Spatial Concentration of the Flows

A further set of experiments was performed in order to verify the outcomes of the proposed optimization problem on the spatial concentration of the flows at varying the size and the number of the cells in the grid. The considered measure for the spatial concentration of the flows is defined as the ratio between the cell load f_z interesting a given cell $z \in Z$ and the area of the cell z. The experimental testbed was built in such a way to guarantee a fair comparison in terms of the total area covered by the grid of cells, independently on the size of the cells, therefore a different approach was adopted with respect to the results presented so far in this section. The whole rectangular area containing the traffic network was partitioned in a grid of $128 \cdot 128$ rectangular cells of uniform size, referred to as the *basic grid*, and composed by an overall number of cells equal to 16384 as presented in Figure 4. The area of each cell in the basic grid is equal to 4895 square meters, while the total area of the grid covers 80199680 square meters.

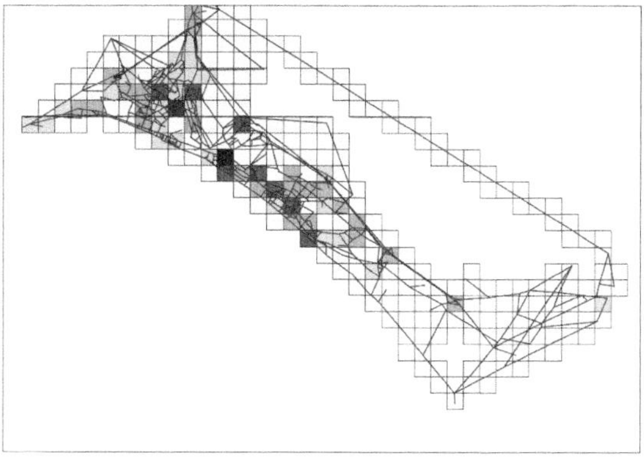

Fig. 2. Graphical representation expressing the cell load values f_z in grayscale when minimizing the total routing costs TC for the grid with 300 meters side cells on the traffic network of Salerno

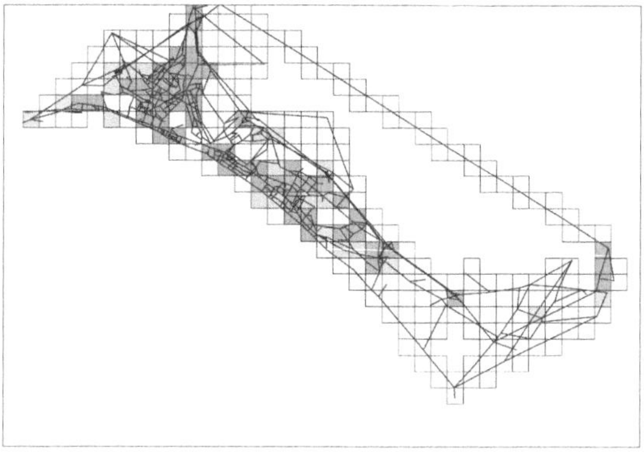

Fig. 3. Graphical representation expressing the cell load values f_z in grayscale when minimizing the maximum cell load $\lambda = \max_{z \in Z} f_z$ for the grid with 300 meters side cells on the traffic network of Salerno

We will refer to those cells as the *basic cells* that will be used as reference area units in the spatial concentration test. In order to perform a fair comparison of the effects of varying the size of the grid cells on the spatial concentration of the flows, we consider different grids composed by a set of *macrocells*, defined as a uniform partition of the basic cells. All the macrocells are therefore composed by a regular number of basic cells and have the same shape. The set of macrocells in a grid is indicated with Z' and indexed by z'. Each macrocell grid covers the same overall area with respect to the basic grid. In particular, in our test we consider two macrocell grids, presented in Figure 5 and in Figure 6. The macrocell grid in Figure 5 (Instance 2) is composed by 4096 macrocells, where each macrocell contains 4 basic cells. The macrocell grid in Figure 6 (Instance 3) is composed by 1024 macrocells, and each macrocell contains 16 basic cells. An additional instance (Instance 4) was considered (see Figure 7) presenting one single

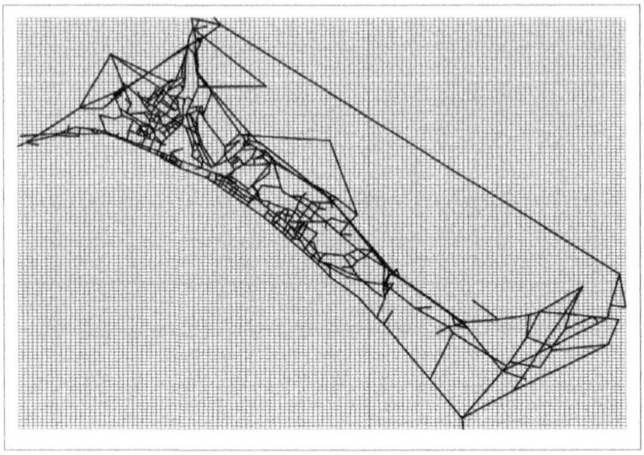

Fig. 4. Instance 1: basic grid composed by 16384 cells overlapped to the traffic network of Salerno

macrocell, containing all the 16384 basic cells. The purpose of the test consists in the minimization of the maximum macrocell load on the considered instances, followed by a comparison among the measures of the spatial concentration on the basic cells for the different cases. The spatially equitable multicommodity capacitated network flow problem is applied on the described cases as follows: the whole set of basic cells is considered for all the four instances as the set Z of the cells in the model (6)-(12), hence the variables f_z still represent the load on each basic cell. The minimization of the maximum macrocell load for each instance is ensured by modifying constraints (10) in such a way that for each macrocell $z' \in Z'$ the sum of the loads f_z for all the basic cells z belonging to the macrocell z' must be less than or equal to λ:

$$\sum_{z \in z'} f_z \leq \lambda \qquad \forall z' \in Z' \tag{13}$$

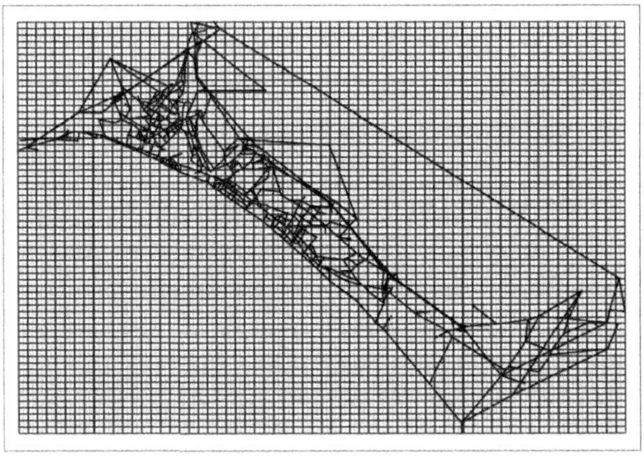

Fig. 5. Instance 2: grid composed by 4096 macrocells each one containing 4 basic cells overlapped to the traffic network of Salerno

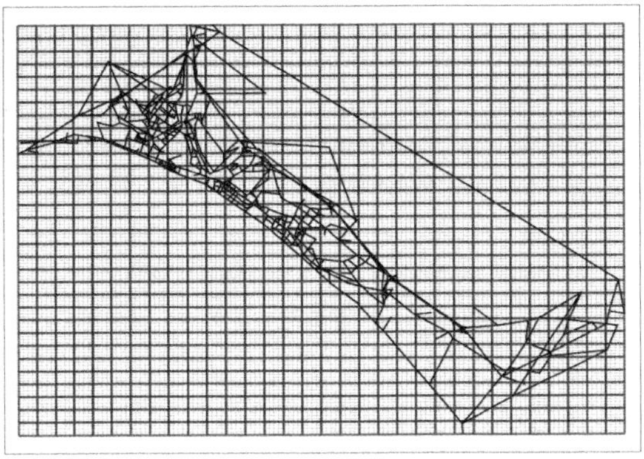

Fig. 6. Instance 3: grid composed by 1024 macrocells each one containing 16 basic cells overlapped to the traffic network of Salerno

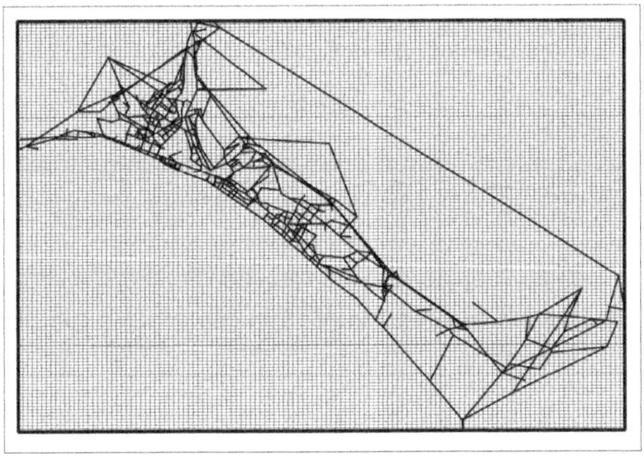

Fig. 7. Instance 4: grid composed by a single macrocell containing all the 16384 basic cells overlapped to the traffic network of Salerno

Thus, an index $SC_{max} = \frac{\max_{z \in Z} f_z}{basic\ cell\ area}$ can be introduced to compare the maximum spatial concentration of the flows on the basic cells arising from the application of the model on different macrocell grids. Similarly, an index $SC_{mean} = \frac{\sum_{z \in Z} f_z}{total\ area}$ can be defined to compare the average spatial concentration of the basic cell loads. In particular, the application of the spatially equitable network flow model to the Instance 4 (Figure 7) corresponds to the minimization of the total routing cost, being the macrocell equal to the whole area embedding the network. The results of the spatial concentration test on the considered instances are presented in Table 4 and Table 5 for the cases with arc capacities equal to 100 and 50 respectively.

The analysis of the spatial concentration indexes suggests very interesting results concerning the effectiveness and suitability of the proposed spatially equitable network flow optimization problem, confirming at the same time the empirical evidence provided by the previous tests. The average spatial concentration increases are very low for

Table 4. Comparison of the spatial concentration indexes for different macrocell grids with arc capacities equal to 100

Instance number	$\|Z'\|$	SC_{mean}	SC_{max}
1	16384	0.02306	2.823
2	4096	0.02827	2.855
3	1024	0.02758	4.939
4	1	0.02282	4.993

Table 5. Comparison of the spatial concentration indexes for different macrocell grids with arc capacities equal to 50

Instance number	$\|Z'\|$	SC_{mean}	SC_{max}
1	16384	0.02508	2.823
2	4096	0.02711	4.300
3	1024	0.02857	5.732
4	1	0.02368	10.325

both the arc capacity settings. The maximum spatial concentration regularly improves with the density of the grid, with a 43% decrease between Instance 4 and Instance 1 for high arc capacities and a 73% decrease between Instance 4 and Instance 1 for low arc capacities. The latter observation suggests in particular the suitability of the model for congested urban areas with reduced arc capacities.

4 A Bi-objective Approach Balancing Spatial Equity and Total Routing Costs

As described in the previous section, depending on the structure of the network and the distribution of the demands, the spatial equitable distribution of the flows yields in general an increase in the total routing costs. Depending on the purposes and the application context of the model, the increase in the total routing cost can be controlled by considering concurrently both the objective functions, namely the minimization of the maximum cell load and the total routing cost. In this case the network optimization problem falls in the class of multi-objective multicommodity network flow problems (see e.g. [6] for a review on this topic), and the proper trade-off between the two criteria can be individuated by studying the pareto efficient solutions through a convex combination of the objective functions at varying the weights, according to the following formulation:

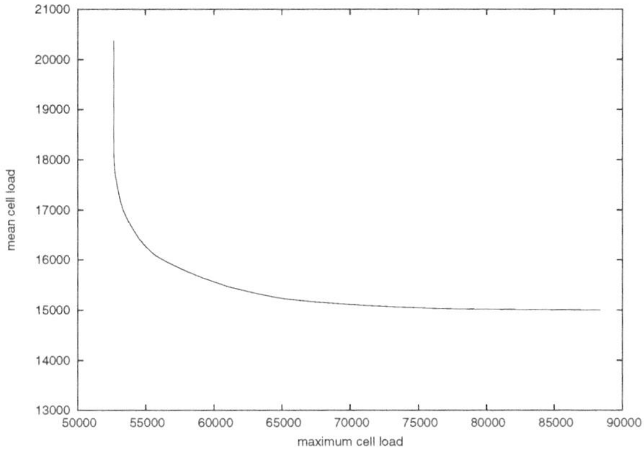

Fig. 8. Numerical results for the mean cell load and the maximum cell load at varying γ_1 and γ_2 coefficients for the traffic network of Salerno with a 500 meters cell side grid

$$\min \quad \gamma_1 \lambda + \gamma_2 \sum_{k \in D} \sum_{(i,j) \in A} c_{ij} x_{ij}^k \tag{14}$$

$$\text{s.t.} \sum_{j \in FS(i)} x_{ij}^k - \sum_{j \in BS(i)} x_{ji}^k = \begin{cases} d^k, & i = s^k, \forall k \in D \\ 0, & \forall i \in V_{\setminus \{s_k, t_k\}}, \forall k \in D \\ -d^k, & i = t^k, \forall k \in D \end{cases} \tag{15}$$

$$\sum_{k \in D} x_{ij}^k \leq u_{ij} \qquad \forall (i,j) \in A$$

$$\sum_{k \in D} \sum_{(i,j) \in A} w_{ij}^z x_{ij}^k = f_z \qquad \forall z \in Z \tag{16}$$

$$f_z \leq \lambda \qquad \forall z \in Z \tag{17}$$

$$x_{ij}^k \geq 0 \qquad \forall (i,j) \in A, \forall k \in D \tag{18}$$

$$f_z \geq 0 \qquad \forall z \in Z \tag{19}$$

$$\lambda \geq 0 \tag{20}$$

In Figure 8 the values of the mean cell load and the maximum cell load are depicted for each optimal solution at varying the convex combination of the γ_1 and γ_2 coefficients for the traffic network of Salerno with a 500 meters cell side grid. The full range of the coefficients was considered in the study, setting $\gamma_2 = 1 - \gamma_1$ and varying γ_1 in $[0, 1]$.

5 Conclusions and Further Research

In this paper we proposed a minimax approach in the field of multicommodity capacitated network flow problems with the aim to provide a spatially equitable distribution of the flows with respect to the area in which the network is embedded. The model is based on the introduction of a grid of cells overlapped to the network area and on the definition of a weighted linear combination of the flows interesting each cell as a measure of the impact of the routed flows on the area underlying the cell itself. A spatially equitable distribution of the flows is obtained by imposing the minimization of the maximum load registered among the whole set of grid cells. The contribution of our approach is twofold: first, it addresses the need for a spatial equitable distribution of the flows avoiding the excessive spatial concentration of the negative outcomes associated to the routed flows, independently on the spatial position of the arcs and the nodes of the network, differently from the previous approaches proposed in the literature in this field that are based on path dissimilarity measures and link load minimization. Second, the arising mathematical programming formulation falls in the class of the linear programming problems, avoiding the need for heuristic approaches in the path selection process that characterizes the previous path-based models based on the discrete p-dispersion optimization problem. The preliminary computational results obtained on a real urban traffic network confirm the suitability and the efficiency of the proposed approach, showing that it is possible to spread the flows in a spatially equitable way with relatively low increases in the value of the total routing cost function. The size of the grid cells influences both the relative value of the maximum cell load decrease and the ratio between the latter and the relative increase of the total routing cost function: better results correspond to dense grids with small size cells. The computational results are also influenced by the capacities of the arcs: the effectiveness of the model in reducing the spatial concentration of the flows is higher for the cases with low arc capacities, suggesting the suitability of the optimization problem for highly congested urban areas. The right choice of a proper trade-off between the spatial equity in the flow distribution and the overall routing cost can be achieved by considering a bi-objective model and selecting a solution among the set of the pareto-optimal solutions obtained by varying the coefficients in the convex combination of the two considered objective functions. The computational results were obtained for all the instances in very low computational times. Still, the size of the cells influences the computational effort required for solving the model at the optimum, being the number of variables and constraints of the LP formulation related to the number of cells considered in each instance. Therefore, in the follow-up of the research on this topic, it can be worth to design efficient algorithms and approximation schemes to solve the spatially equitable multicommodity network flow problem on very large scale instances, that can be useful when applying the model on big urban areas. Further developments of this work from the modeling point of view can be considered assuming that the effects of the flows routed within each cell $z \in Z$ interest also to a certain extent the population belonging to the cells in the neighborhood of the cell z itself. This issue can be easily considered and implemented by adding a weighted sum of the adjacent cell loads on the left hand side of each constraint of the set (10). Possible extensions of the practical suitability of the optimization problem proposed in this paper will be enabled by considering multiple decision maker optimization models as well, in particular in the field of bilevel network design and toll setting problems (see e.g. the contributions in [7,3]).

Acknowledgement

Partial funding for this research has been provided by the projects: PRIN-2007J49483 from the Italian Ministry of University and Research, Sapienza Università di Roma Ateneo Federato Research Fellowship 2009.

References

1. Ahuja, R.K.: The balanced linear programming problem. European Journal of Operational Research 101(1), 29–38 (1997)
2. Akgun, V., Erkut, E., Batta, R.: On finding dissimilar paths. European Journal of Operational Research 121(2), 232–246 (2000)
3. Bianco, L., Caramia, M., Giordani, S.: A bilevel flow model for hazmat transportation network design. Transportation Research Part C: Emerging Technologies 17(2), 175–196 (2009)
4. Carotenuto, P., Giordani, S., Ricciardelli, S.: Finding minimum and equitable risk routes for hazmat shipments. Computers & Operations Research 34(5), 1304–1327 (2007)
5. Dell'Olmo, P., Gentili, M., Scozzari, A.: On finding dissimilar pareto-optimal paths. European Journal of Operational Research 162(1), 70–82 (2005)
6. Hamacher, H.W., Pedersen, C.R., Ruzika, S.: Multiple objective minimum cost flow problems: A review. European Journal of Operational Research 176(3), 1404–1422 (2007)
7. Marcotte, P., Mercier, A., Savard, G., Verter, V.: Toll policies for mitigating hazardous materials transport risk. Transportation Science 43(2), 228–243 (2009)
8. Scutellà, M.G.: A strongly polynomial algorithm for the uniform balanced network flow problem. Discrete Applied Mathematics 81(1-3), 123–131 (1998)

Approximating Minimum Cut with Bounded Size

Giulia Galbiati

Dipartimento di Informatica e Sistemistica, Università degli Studi di Pavia, 27100 Pavia, Italy
giulia.galbiati@unipv.it

Abstract. We present the Minimum Cut with Bounded Size problem and two efficient algorithms for its solution. In this problem we want to partition the n vertices of a edge-weighted graph into two sets S and T, with S including a given source s, T a given sink t, and with $|S|$ bounded by a given threshold B, so as to minimize the weight $\delta(S)$ of the edges crossing the cut (S,T). If B is equal to $n-1$ the problem is well-known to be solvable in polynomial time, but for general B it becomes NP-hard. The first algorithm is randomized and, for each $\varepsilon > 0$, it returns, with high probability, a solution S having a weight within ratio $(1 + \frac{\varepsilon B}{\log n})$ of the optimum. The second algorithm is a deterministic bi-criteria algorithm which can return a solution violating the cardinality constraint within a specified ratio; precisely, for each $0 < \gamma < 1$, it returns a set S having either (1) a weight within ratio $\frac{1}{1-\gamma}$ of the optimum or (2) optimum weight but cardinality $|S| \leq \frac{B}{\gamma}$, and hence it violates the constraint by a factor at most $\frac{1}{\gamma}$.

1 Introduction

Graph cuts are very well-studied objects in computer science, modelling a large variety of basic problems in algorithm design, in combinatorial optimization and in many application areas like communications and electrical networks. Recently unbalanced graph cuts have received some attention [4,5,7]. Combinatorial optimization problems with cardinality constraints have also recently received attention [2,9].

Here we address the following problem: given an undirected graph $G = (V,E)$, with vertex set V of cardinality n, and edge set E where each edge (i,j) has a non-negative weight w_{ij}, and given an integer B, $0 < B < n$, an identified source $s \in V$ and sink $t \in V$, find a cut $(S, V \setminus S)$ separating s from t, with $s \in S$, such that the cardinality of S is not greater than B, and minimizing the weight $\delta(S)$ of the edges crossing the cut, i.e., having one endvertex in S and the other in $V \setminus S$. We call this problem Minimum Cut with Bounded Size (MINCUTBS for short). When B is equal to $n-1$ it is the well known $s-t$ Minimum Cut problem, which is solvable in polynomial time [6], otherwise the problem becomes NP-hard, as explained below.

The MINCUTBS problem, besides the classical applications, has interesting applications in the control of disasters, epidemic outbreaks, etc.. In these contexts node s represents, for instance, the infected node where the disaster arises, and the goal is to minimize the cost of treating the disaster, subject to the request of preserving uninfected both node t and all other nodes except at most B ones. Hence new applications areas include control of disasters in telecommunication networks, in social or biological networks, as well as in the web graph.

J. Pahl, T. Reiners, and S. Voß (Eds.): INOC 2011, LNCS 6701, pp. 210–215, 2011.

In this paper we present two polynomial-time algorithms for solving MINCUTBS. The first algorithm is randomized and, for each $\varepsilon > 0$, it returns, with high probability, a solution S having a weight within ratio $(1+\frac{\varepsilon B}{\log n})$ of the weight of an optimum solution S^*. This algorithm is derived from the algorithm in [5] which solves a problem related to MINCUTBS, that we call Minimum Cut with Exact Size (MINCUTES for short). In MINCUTES one wants to find a cut of minimum weight, where the cardinality of one shore of the cut is required to be exactly equal to a given integer k. In Section 2 we show how this algorithm can be extended to solve our problem and how the approximation results obtained for MINCUTES can also be fully extended to MINCUTBS.

The second algorithm is a bi-criteria algorithm that, for each $0 < \gamma < 1$, returns a set S, with $s \in S, t \notin S$, having either (1) $\delta(S) \leq \frac{1}{1-\gamma}\delta(S^*)$ and $|S| \leq B$ or (2) $\delta(S) \leq \delta(S^*)$ and $|S| \leq \frac{B}{\gamma}$, where S^* denotes an optimum solution for MINCUTBS. Therefore this algorithm can return a solution violating the cardinality constraint within a factor at most $\frac{1}{\gamma}$. This algorithm, described in Section 3, has been inspired by the one in [4] devoted to the solution of a companion problem, that we call Minimum Size with Bounded Cut (MINSIZEBC for short). The latter problem looks for a cut $(S, V \setminus S)$ with $s \in S$, separating s from t and minimizing the cardinality of S among those having the weight $\delta(S)$ of the edges crossing the cut not greater than B. This problem and MINCUTBS obviously have the same formulation in recognition form. Hence, as MINSIZEBC is NP-hard [4], the same is true for MINCUTBS. We show that, quite surprisingly, the fast parametric algorithm in [8], used in [4] for solving MINSIZEBC, can also be used, on the same parameterized network but in a different way, to solve also MINCUTBS. In the same section we highlight the differences and the similarities between our algorithm for solving MINCUTBS and the one in [4] for solving MINSIZEBC.

We conclude this section by saying that throughout the paper we often refer to a cut $(S, V \setminus S)$ simply as S, we denote by $\delta(S)$ the weight of the edges crossing S and by $w(S)$ the sum of the weights of the edges having both endvertices in S. We call a cut separating s from t an $s-t$ cut. Extensive testing of these algorithms is planned.

2 The Randomized Algorithm

We start from the randomized algorithm presented in [5]. Here the authors solve, with high probability, the problem of finding, in an undirected graph having non negative edge weights, a cut S with $|S| = B$ and $\delta(S) \leq (1+\frac{\varepsilon B}{\log n})\delta(S')$, where S' denotes an optimum cut among those having one of the two shores of the cut of cardinality equal to a given integer B, and ε is any fixed positive constant. We show here that their algorithm can be modified to find, with high probability, an $s-t$ cut having the source side S of cardinality at most B and having weight $\delta(S) \leq (1+\frac{\varepsilon B}{\log n})\delta(S^*)$. The algorithm in [5] works in iterations, where each iteration consists of a "contraction stage" followed by a "combining stage". In our modified version, which is described below, we maintain unaltered the "contraction stage" but we modify instead the "combining stage" where, instead of cuts S with $|S| = B$, we consider $s-t$ cuts S with $s \in S$ and $|S| \leq B$. For a full comprehension of what follows the reader is invited to consult [5]. A brief explanation is given below.

Algorithm 1

- initialize n clusters, each containing a distinct vertex of G;
- while G still contains edges do { /* begin contraction stage */
  - . choose at random one edge of G;
  - . contract in a single cluster the two clusters that contain the endvertices of the edge;
  - . remove from G (updating G) self-loops, i.e., edges having both endvertices in the new cluster but retain parallel edges, i.e., edges having the endvertices in different clusters;
    /* the clusters now contain vertices of G having no edges adjacent to any two of them */
  - . let $C = C_1, ..., C_L$ be the current set of clusters, with $s \in C_1$;
    /* end contraction stage - begin combining stage*/
  - . find, if it exists, a subset $Q \subseteq C$ such that, if we let $\widetilde{Q} = \cup_{C_i \in Q} C_i$ then:
    i) $|\widetilde{Q}| \leq B$, $s \in \widetilde{Q}, t \notin \widetilde{Q}$;
    ii) $\sum_{C_i \in Q} \delta(C_i)$ is minimal;
    and compute the weigth of the $s-t$ cut $\widetilde{Q}$; /* end combining stage */

  }
- return the $s-t$ cut $\widetilde{Q}$ of minimum weight among those found in the combining stages.

In the contraction stage an edge of G is chosen at random, it is "contracted", and G is "updated". Precisely, the endpoints of the chosen edge are merged in a newly created vertex, and loops and parallel edges that may result from the contraction are treated as described in the algorithm, where the name cluster (that corresponds to a newly created vertex) is used as a synonym of independent set, i.e., set of vertices having no edge adjacent to any two of them. In the combining stage instead we find, in the graph G of the current iteration, a set of clusters which include s but not t, have at most B vertices, and minimize the sum of the weights of the edges crossing each cluster.

The implementation of this algorithm and the analysis of its performance can be done as in [5], yielding the conclusion expressed in Theorem 1. Two crucial observations allow to reach such conclusion. The first is about the implementation of the combining stage and the running time of the algorithm. The combining stage can be implemented using a dynamic programming table T as in [5], but here each table entry $T(i,k)$, $i = 1,...,L$, $1 \leq k \leq B$ has a different meaning. Precisely, if the clusters are labelled by $1,2,...,L$ in an arbitrary way, with cluster C_1 containing s, then $T(i,k)$ describes the minimal $\sum_{C_j \in Q} \delta(C_j)$, over all subsets Q of the clusters $C_1,...,C_i$ for which it happens that $s \in \widetilde{Q}$, $t \notin \widetilde{Q}$ and $|\widetilde{Q}| \leq B$, with $\widetilde{Q} = \cup_{C_j \in Q} C_j$. It is straightforward to see that each table entry in row i can be computed from the entries in row $i-1$ and the preceding entries in the same row, and that the entries in the first row are easy to initialize. The size of the table is $LB = O(n^2)$ and hence $T(L,B)$, which is the desired output of the combining stage, can be computed in polynomial time. The second observation is about the success probability. It can be shown that Lemma 5 in [5] becomes here the following lemma.

Lemma 1. *For every (not necessarily fixed) $\rho > 0$, Algorithm 1 outputs an $s-t$ cut S having $s \in S$, $|S| \leq B$, and $\delta(S) \leq (1+\rho B)\delta(S^*)$, with probability at least $e^{-2/\rho}$.*

The proof of the correctness of this lemma follows closely the one in [5]; the important thing that must be modified is the cut initially fixed for the analysis, that must be an $s-t$ cut S having $s \in S$, $|S| \leq B$, and optimum weight $\delta(S^*)$.

If now we take $\rho = \frac{\varepsilon}{\log n}$, for any $\varepsilon > 0$, the probability of success of the algorithm becomes at least $n^{-2/\varepsilon}$; hence if we repeat the algorithm $n^{2/\varepsilon}$ times, taking from all the repetitions the cut of minimum weight, the next result follows.

Theorem 1. *For every fixed $\varepsilon > 0$, there is a polynomial-time randomized algorithm that finds, with high probability, a solution S to* MINCUTBS *having a weight $\delta(S)$ within ratio $(1+\frac{\varepsilon B}{\log n})$ from the optimum weight $\delta(S^*)$.*

Notice that if $B = O(\log n)$ this algorithm is a PTAS, but in this case we could solve the problem exactly with simple enumeration. Hence this algorithm becomes interesting when $B = \Omega(\log n)$, where the approximation ratio becomes $O(\frac{B}{\log n})$.

3 The Bi-criteria Algorithm

Now we start the presentation of a deterministic, bi-criteria algorithm for MINCUTBS. This algorithm returns a solution S that either approximates the weight of an optimum solution or violates, within a specified performance guarantee, the constraint on the cardinality bound but has optimum weight. Theorem 2 gives the performance guarantees of this algorithm, which has been inspired by and is based on the one in [4] for solving MINSIZEBC. The results in Theorem 2 show that the algorithm is a $(\frac{1}{1-\gamma}, \frac{1}{\gamma})$ bi-criteria approximation algorithm, if we define an (α,β) bi-criteria approximation algorithm for MINCUTBS to be an algorithm returning a solution S having $\delta(S) \leq \alpha\delta(S^*)$ and $|S| \leq \beta B$. We highlight here differences and similarities between our algorithm and the algorithm in [4] for solving MINSIZEBC. The differences lie in the performance guarantees, since the latter algorithm is a $(\frac{1}{\gamma}, \frac{1}{1-\gamma})$ bi-criteria approximation algorithm, if we define an (α,β) bi-criteria approximation algorithm for MINSIZEBC to be an algorithm returning a solution S having $\delta(S) \leq \alpha B$ and $|S| \leq \beta|\bar{S}|$, where $\bar{S}$ denotes an optimum solution for MINSIZEBC. The similarities are due to the fact that, quite surprisingly, the methodology used in [4] for finding a cut having size and weight within performance guarantees of the corresponding values in an optimum solution $\bar{S}$ for MINSIZEBC, can be modified to find a cut having size and weight within performance guarantees of the corresponding values in an optimum solution S^* for MINCUTBS. Before proceeding with the description of the algorithm we notice that bi-criteria algorithms, even if variously defined, have already appeared in the literature, dealing with problems whose solutions need to be evaluated with respect to two cost criteria [1,3].

Let $0 < \gamma < 1$. The algorithm essentially uses the algorithm in [8] applied to graph G^λ, which is obtained from G by adding, for each vertex $v \in V - \{t\}$, a new edge from

v to t of weight λ (if an edge to t already exists it increases its weight by λ). Obviously the weight of an $s-t$ cut S of G^λ is $\delta(S)+\lambda|S|$ and it is not difficult to see that, as λ increases, the source side of a minimum $s-t$ cut of G^λ decreases in cardinality, until it contains the single vertex s. To reach this conclusion it is enough to add the two inequalities $\delta(S)+\lambda|S| \leq \delta(S')+\lambda|S'|$ and $\delta(S')+\lambda'|S'| \leq \delta(S)+\lambda'|S|$, where S (resp. S') denotes a minimum $s-t$ cut of G^λ (resp. $G^{\lambda'}$), with $\lambda < \lambda'$. The algorithm of [8] has very interesting properties. It starts with the parameter $\lambda = 0$ and in a single run is able to output minimum $s-t$ cuts of G^λ, for all values of $\lambda \geq 0$; the source sides of all these cuts form a finite nested family of vertex sets $S_0 \supset S_1 \supset \ldots \supset S_k$, with S_0 a minimum $s-t$ cut of G and $S_k = \{s\}$; moreover for any sequence of successive sets S_j, S_{j+1}, $j = 0,\ldots,k-1$, there is a value of λ, say λ_j^*, for which both S_j and S_{j+1} are minimum $s-t$ cuts of $G^{\lambda_j^*}$. An important observation concerning the weights of the edges crossing these successive cuts is that, if $i < j$, it can be shown that $\delta(S_i) < \delta(S_j)$.

We now describe our bi-criteria algorithm; of course the computation at step 2 is done with the algorithm in [8].

Algorithm 2

1. form graph G^λ;
2. compute the nested family $S_0 \supset S_1 \supset \ldots \supset S_k$ of the source sides of optimum minimum $s-t$ cuts of G $^\lambda$, for all values of $\lambda \geq 0$;
3. if $|S_0| \leq B$ then return S_0; /* an optimum solution has been found*/
4. find j such that $|S_j| > B \geq |S_{j+1}|$;
5. if $|S_j| \leq \frac{B}{\gamma}$ then return S_j else return S_{j+1}.

The following theorem gives the performance of this algorithm.

Theorem 2. *For each* $0 < \gamma < 1$, *Algorithm 2 outputs an* $s-t$ *cut* S *of* G, *with* $s \in S$, *such that either (i)* $\delta(S) \leq \frac{1}{1-\gamma}\delta(S^*)$ *and* $|S| \leq B$ *or (ii)* $\delta(S) \leq \delta(S^*)$ *and* $|S| \leq \frac{B}{\gamma}$.

Proof. (sketched) If the algorithm returns at Step 3, of course the optimum solution satisfies both (i) and (ii). Otherwise let j be the integer computed at Step 4. By definition $\delta(S^*) \leq \delta(S_{j+1})$, and since S_0 is a minimum $s-t$ cut of G obviously $\delta(S_0) \leq \delta(S^*)$. Hence from the inequality $\delta(S_0) \leq \delta(S^*) \leq \delta(S_{j+1})$ we derive that there exists an integer $i \leq j$ such that $\delta(S_i) \leq \delta(S^*) \leq \delta(S_{i+1})$. It cannot be that $i < j$; otherwise from the inequalities $|S^*| \leq B < |S_j|$ and $\delta(S^*) \leq \delta(S_{i+1}) \leq \delta(S_j)$ it would follow that S^* would have been a better cut than S_j for all values of λ, and the algorithm would have generated S^*, not S_j (and $S^* \neq S_j$ since $S^* \leq B < |S_j|$). Hence $i = j$ and $\delta(S_j) \leq \delta(S^*) \leq \delta(S_{j+1})$. Now, at Step 5 of the algorithm, if $|S_j| \leq \frac{B}{\gamma}$ then the set S_j returned by the algorithm satisfies (ii). It remains to be shown that when $|S_j| > \frac{B}{\gamma}$ then $\delta(S_{j+1}) \leq \frac{1}{1-\gamma}\delta(S^*)$, so that the set S_{j+1} returned by the algorithm satisfies (i) (it does satisfy $|S_{j+1}| \leq B$, from Step 4). This part of the proof is omitted due to space constraints, but can be done along the lines of that of Theorem 2 in [4].

References

1. Angel, E., Bampis, E., Gourvès, L.: Approximation Algorithms for the Bi-criteria Weighted MAX-CUT Problem. Discrete Applied Mathematics 154, 1685–1692 (2006)
2. Bruglieri, M., Horst, M.E., Hamacher, W., Maffioli, F.: An annotated bibliography of combinatorial optimization problems with fixed cardinality constraints. Discrete Applied Mathematics 154(9), 1344–1357 (2006)
3. Dutot, P.-F., Eyraud, L., Mounié, G., Trystram, D.: Bi-criteria Algorithm for Scheduling Jobs on Cluster Platforms. In: ACM Symposium on Parallel Algorithms and Architectures, pp. 125–132 (2004)
4. Hayrapetyan, A., Kempe, D., Pál, M., Svitkina, Z.: Unbalanced Graph Cuts. In: Brodal, G.S., Leonardi, S. (eds.) ESA 2005. LNCS, vol. 3669, pp. 191–202. Springer, Heidelberg (2005)
5. Feige, U., Krauthgamer, R., Nissim, K.: On cutting a few vertices from a graph. Discrete Applied Mathematics 127, 643–649 (2003)
6. Ford, L., Fulkerson, D.: Maximal Flow Through a Network. Can. J. Math. 8, 399–404 (1956)
7. Galbiati, G., Maffioli, F.: Approximation algorithms for maximum cut with limited unbalance. Theoretical Computer Science 385, 78–87 (2007)
8. Gallo, G., Grigoriadis, M.D., Tarjan, R.E.: A fast parametric Maximum Flow Algorithm And Applications. SIAM J. Comput. 18, 30–55 (1989)
9. Svitkina, Z., Fleischer, L.: Submodular Approximation: Sampling-Based Algorithms and Lower Bounds. In: FOCS 2008 (2008)

Lexicographical Minimization of Routing Hops in Telecommunication Networks

Luís Gouveia[1], Pedro Patrício[2], and Amaro de Sousa[3]

[1] CIO and DEIO, Faculdade de Ciências da Universidade de Lisboa, Bloco C6, Piso 4, 1749-016 Lisboa, Portugal
legouveia@fc.ul.pt

[2] CIO and Departamento de Matemática, Universidade da Beira Interior, Rua Marquês d'Ávila e Bolama, 6201-001 Covilhã, Portugal
pedrofp@ubi.pt

[3] Instituto de Telecomunicações, Universidade de Aveiro, Campus Universitário de Santiago, 3810-193 Aveiro, Portugal
asou@ua.pt

Abstract. Consider a telecommunications network with given link capacities and a set of commodities with known demands that must be routed over the network. We aim to determine a single routing path for each commodity such that the whole set of paths does not violate the link capacities and the number of routing hops is minimized in a lexicographical sense, i.e., minimizing the number of paths with the worst number of hops; then, among all such solutions, minimizing the number of paths with the second worst number of hops; and so on. We present two approaches for solving this problem. The first approach is iterative where a sequence of hop constrained problems is solved and the optimal solution value of each iteration defines a new constraint added to the problems of subsequent iterations. The second approach is based on defining a single integer programming model for the whole problem. In this approach, we consider appropriate cost parameters associated with the number of hops of each routing path such that the cost of a path with h hops is higher than the cost of all paths with $h-1$ hops. In both cases, we propose multi-commodity flow and hop-indexed models and compare them both in terms of linear programming relaxation values and in terms of efficiency.

1 Introduction

In a telecommunications network, the minimization of the number of hops in the paths supporting traffic has a positive impact on the performance of the network due to two main reasons: delay and robustness. Concerning delay, a data packet traveling through a set of links from its origin node to its destination node suffers a total delay given by the queuing and transmission delays on intermediate nodes and by the propagation delays on links. Usually, the first two factors are dominant and, in some cases, it is even possible to guarantee a bound on the maximum packet delay which is proportional to the number of routing nodes [15]. Therefore, it is desirable to route demand commodities on the smallest possible number of links. Concerning robustness, when a network element fails, all commodity demands whose routing paths use the failed element are affected

J. Pahl, T. Reiners, and S. Voß (Eds.): INOC 2011, LNCS 6701, pp. 216–229, 2011.

and, therefore, the probability of a commodity being affected by a failure is lower if its routing path has fewer hops.

The problem addressed in this paper arises in the context of a traffic engineering task over pre-dimensioned networks. We consider the networks which are solutions of a network design problem studied in [5], where the aim was to find minimum cost networks guaranteeing the existence of D hop-constrained node disjoint paths for every commodity k in a set of commodities K, accommodating a given estimated demand matrix T. The hop-constrained paths guarantee a worst case delay performance for the network design solution. The D node disjoint paths, together with the demand reserved on each path, guarantee a desired degree of survivability. In this paper we consider the traffic engineering problem for the $D = 1$ case only and the case with $D > 1$ will be addressed as subsequent work.

The constraint on the maximum number of H hops for each path in the network design problem means that some paths may have fewer hops. Moreover, since the estimated demand matrix T that was used in the design problem may be different from the one that should be accommodated by the network when it is put in operation, it becomes interesting to consider the problem of routing a new demand matrix R while optimizing the number of hops of all routing paths.

In [6,7], two hop related objective functions are considered: the minimization of the average number of hops and the minimization of the worst case number of hops. The first objective gives the best overall average performance but it may be unfair for some of the commodities: it leads to solutions where many commodities are supported by routing paths with a number of hops close to the minimum but some commodities may be supported by routing paths having a large number of hops. One way of optimizing the fairness criterion is to consider the minimization of the worst case number of hops (the second objective studied in [6,7]). Nevertheless, this objective does not account for the minimization of the number of hops of the routing paths that can have fewer hops than its worst value. Here, we generalize the "worst case" objective function by minimizing the number of routing hops in a lexicographical sense, i.e., minimizing the number of paths with the worst number of hops; then, among all such solutions, minimizing the number of paths with the second worst number of hops, and so on. As will be seen in the computational results, optimal solutions of this objective function exhibit a very small penalty on the optimal average number of routing hops, which means that we can improve the routing fairness without jeopardizing the overall average performance.

Consider the vector of routing hops $(h_k : k \in K)$ where h_k is the number of hops of the routing path of commodity $k \in K$. Let $[h_k]_{k \in K}$ be the vector obtained by rearranging the previous vector in a non-increasing order. Given two vectors $[a_k]_{k \in K}$ and $[b_k]_{k \in K}$, the first vector is said to be lexicographically smaller than the latter if either $a_1 < b_1$ or there exists an index $l \in \{1, \dots, |K| - 1\}$ such that $a_i = b_i$ for all $i \leq l$ and $a_{l+1} < b_{l+1}$. Our problem can now be stated as follows. Given an undirected graph $G = (V, E)$, representing a network where V and E denote, respectively, the sets of nodes and edges, with installed edge capacities $b_e, e \in E$, a set of commodities K where each commodity $k \in K$ has demand r_k, origin node s_k and destination node t_k, the traffic engineering problem addressed in this work consists of routing the demand matrix R through paths that lexicographically minimize $[h_k]_{k \in K}$. Figure 1 illustrates two feasible solutions considering

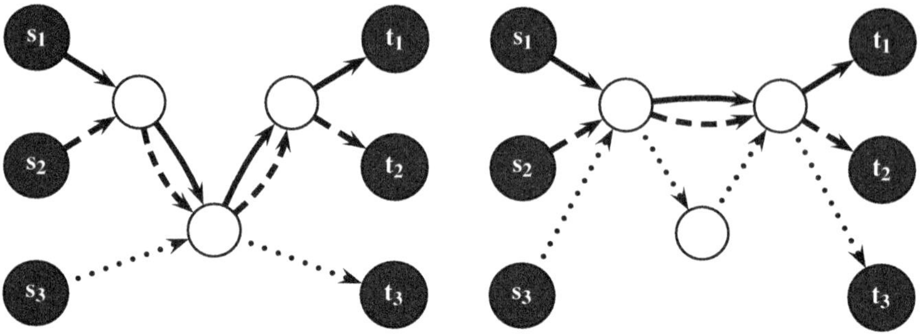

Fig. 1. Example of Two Feasible Solutions

an instance with $|K| = 3$ and $H = 4$, where thick, dashed and dotted lines represent the paths of each commodity. Clearly, the solution represented on the right is the best one, in lexicographic terms, since it has a vector $[a_k]_{k\in K} = \begin{bmatrix} 4 & 3 & 3 \end{bmatrix}$ while the solution represented on the left has a vector $[b_k]_{k\in K} = \begin{bmatrix} 4 & 4 & 2 \end{bmatrix}$. Note that in regard to both the average and maximum number of hops minimization criteria considered in [6,7], these two solutions have the same objective function value.

There are very few references dealing with delay optimization and the reader is referred to [9,13,18] for heuristic techniques dealing with delay related traffic engineering methods. The concept of lexicographical minimization is similar to that of max-min fairness (MMF) previously applied to routing, load balancing and resource allocation on telecommunication networks [11,12,17] (see [16] for more general issues on MMF) but, as far as we are aware, has never been applied to the minimization of the number of routing hops.

We present two approaches for solving this problem. The first approach is iterative where a sequence of hop constrained problems is solved and the optimal solution value of each iteration defines a new constraint that is added on the problems of the subsequent iterations. The second approach is based on defining a single integer programming model to the whole problem. In this second approach, we consider appropriate cost parameters associated to the number of hops of each routing path in such a way that the cost of a path with h hops is higher than the cost of all paths with $h-1$ hops.

This paper is organized as follows. In Section 2, we present the iterative approach based on solving sequentially an auxiliary hop constrained problem. We present two different models for the auxiliary problem and compare their linear programming (LP) relaxations. In Section 3, we describe the single model approach to the problem. In Section 4, we present and discuss the computational results. Finally, Section 5 presents the main conclusions and future work.

2 The Iterative Approach

The reasoning behind this approach is straightforward. Given the lexicographic minimization objective described in the previous section, we start by minimizing the number of commodities with H hops. Letting F^h represent the number of commodities with h hops, we then solve a second problem minimizing the number of commodities with

$H-1$ hops with the constraint that the commodities with H hops do not exceed F^H. The iterative procedure continues until all commodities have been accounted for, that is, $\sum_{h=1,\ldots,H} F^h = |K|$.

Let us denote the auxiliary problem to be solved in each step of the iterative approach by Restricted-Hop(H^*). This problem consists of minimizing the number of commodities with H^* hops (with $H^* < H$) subject to the traffic engineering problem constraints stated in the previous section, together with the extra constraints that guarantee that the number of commodities with $H^*+1,\ldots,$ and H hops is, respectively, not greater than the values $F^{H^*+1},\ldots,$ and F^H . We note that hop-constrained network design problems have been studied in the past in many settings (see, for instance, the surveys in [3] and [10], and several other papers such as [1,5,8,14]). However, this work appears to be the first time where such problems have been studied with this kind of objective. Letting $P(H^*)$ denote a valid model for this auxiliary problem (models for this problem will be described in the following subsections), the iterative approach can be specified as follows.

Initialization:
$F^h = 0, h = 1,\ldots,H$
$H^* = H$
Iteration:
Do while $\sum_{h=1,\ldots,H} F^h < |K|$
Find the optimal solution of $P(H^*)$
$F^{H^*} =$ number of commodities with H^* hops
$H^* = H^* - 1$
End do

Following the model sequence of previous works on network design with hop constraints, we present next two valid models for the Restricted-Hop(H^*) problem. The first one is a straightforward multi-commodity flow model with a cardinality constraint for the hop limit and constraints linking the flow variables with an auxiliary set of variables. The second one uses hop-indexed variables and is based on a better polyhedral representation of the underlying hop-constrained shortest path problem. As we shall show next, the relevant advantage for problems with the proposed type of objective functions is that these hop-indexed variables let us model the objective functions in a straightforward way permitting us to get rid of the auxiliary variables and to obtain models with a stronger LP relaxation.

2.1 The $MCF(H^*)$ Model

Consider binary variables: $y_{ij}^k = 1$ if edge $\{i,j\}$, traversed from i to j, is in the path supporting commodity k, and 0 otherwise ($\forall\{i,j\} \in E; k \in K$), and $u^{hk} = 1$ if commodity k is supported by a path with h hops, and 0 otherwise ($\forall h = 1,\ldots,H; k \in K$).

We denote by $MCF(H^*)$ the following model:

$$Min \sum_{k\in K} u^{H^*k}$$

subject to:

$$\sum_{j\in V} y_{ij}^k - \sum_{j\in V} y_{ji}^k = \begin{cases} 1, \; i = s_k \\ 0, \; i \neq s_k, t_k \\ -1, \; i = t_k \end{cases}, k \in K \tag{1}$$

$$\sum_{(i,j):\{i,j\}\in E} y_{ij}^k \leq H, k \in K \tag{2}$$

$$\sum_{(i,j):\{i,j\}\in E} y_{ij}^k = \sum_{h=1,\ldots,H} hu^{hk}, k \in K \tag{3}$$

$$\sum_{h=1,\ldots,H} u^{hk} = 1, k \in K \tag{4}$$

$$\sum_{k\in K} u^{hk} \leq F^h, h = H^* + 1, \ldots, H \tag{5}$$

$$\sum_{k\in K} r_k \left(y_{ij}^k + y_{ji}^k \right) \leq b_e, e = \{i,j\} \in E \tag{6}$$

$$y_{ij}^k, y_{ji}^k \in \{0,1\}, \{i,j\} \in E; k \in K \tag{7}$$

$$u^{hk} \in \{0,1\}, h = 1, \ldots, H; k \in K \tag{8}$$

For a given $H^* = 1, \ldots, H$, the objective function that we wish to minimize expresses the number of commodities with H^* hops. Constraints (1) are the usual flow conservation constraints, they guarantee a path for each commodity, starting on s_k and ending on t_k, and constraints (2) guarantee that the path supporting each commodity has at most H hops. The linking constraints (3) define the new variables u^{hk} and together with the cardinality constraints (4) state that each commodity has either $1, 2, \ldots,$ or H hops. Constraints (5) guarantee that the number of commodities with more than H^* hops, that is, $H^* + 1, \ldots,$ and H hops (with $H^* < H$) is not greater than $F^{H^*+1}, \ldots,$ and F^H, respectively. Note that, according to the iterative approach described before, when finding the optimal solution of $MCF(H^*)$, the values of $F^{H^*+1}, \ldots,$ and F^H are determined in previous iterations. Finally, constraints (6) guarantee that the total demand that traverses each edge does not exceed the capacity installed in that edge. Constraints (7) and (8) represent the variables domain. Note that in the iterative procedure we may write $F^{H^*} = \sum_{k\in K} u^{H^*k}$.

2.2 The $HOP(H^*)$ Model

We now model the Restricted-Hop(H^*) problem using a different set of (flow) variables, the so-called hop-indexed variables which also indicate the position of an arc in the path: $z_{ij}^{hk} = 1$ if edge $\{i,j\}$ is traversed from i to j in position h in the path supporting commodity k, and 0 otherwise ($\forall \{i,j\} \in E \cup \{t_k, t_k\}; h = 1, \ldots, H; k \in K$).

Note the existence of "loop" variables $z_{t_k t_k}^{hk}$, with $h = 2, \ldots, H$, which are necessary to model paths that have less than H hops (the reader is referred to [4,6,7] for an explanation of these variables). Let $HOP(H^*)$ denote the following model:

$$Min \sum_{k\in K} \sum_{j\in V\setminus\{t_k\}} z_{jt_k}^{H^*k}$$

subject to:

$$\sum_{j \in V} z_{s_k j}^{1k} = 1, k \in K$$
$$\sum_{j \in V} z_{ij}^{h+1,k} - \sum_{j \in V} z_{ji}^{hk} = 0, i \neq s_k; h = 1, \ldots, H-1; k \in K \tag{9}$$
$$\sum_{j \in V} z_{jt_k}^{Hk} = 1, k \in K$$
$$\sum_{k \in K} \sum_{j \in V \setminus \{t_k\}} z_{jt_k}^{hk} \leq F^h, h = H^* + 1, \ldots, H \tag{10}$$
$$\sum_{k \in K} r_k \left(\sum_{h=1,\ldots,H} z_{ij}^{hk} + \sum_{h=1,\ldots,H} z_{ji}^{hk} \right) \leq b_e, e = \{i,j\} \in E \tag{11}$$
$$z_{ij}^{hk}, z_{ji}^{hk} \in \{0,1\}, \{i,j\} \in E; k \in K; h = 1, \ldots, H \tag{12}$$
$$z_{t_k t_k}^{hk} \in \{0,1\}, k \in K; h = 2, \ldots, H \tag{13}$$

For a given $H^* = 1, \ldots, H$, the additional (hop count) information contained in the new variables permits us to minimize the number of paths with H^* hops by simply minimizing the number of edges $\{j, t_k\}$ that are traversed from j to t_k, with $j \neq t_k$, in position H^*. Constraints (9) guarantee the existence of a path from s_k to t_k with at most H hops for every commodity $k \in K$ (note that these constraints contain the loop variables $z_{t_k t_k}^{hk}$ mentioned above). Similarly to constraints (5) of the $MCF(H^*)$ model, constraints (10) state that the number of commodities with $H^* + 1, \ldots,$ and H hops (with $H^* < H$) is not greater than $F^{H^*+1}, \ldots,$ and F^H, respectively (note, however, that as stated for the objective function, the variables $z_{jt_k}^{hk}$ are sufficient for expressing these constraints). As before, note that these values are given by the optimal solutions of the $HOP(h)$ models, with $h = H^* + 1, \ldots, H$. Constraints (11) are capacity constraints and (12) and (13) represent the variables domain. Similarly to the observation made at the end of the previous subsection, note that we may now write $F^{H^*} = \sum_{k \in K} \sum_{j \in V \setminus \{t_k\}} z_{jt_k}^{H^* k}$ in the iterative procedure.

2.3 Comparing the LP Relaxations of $MCF(H^*)$ and $HOP(H^*)$

We will show next that the LP relaxation bound of the $HOP(H^*)$ model is at least as good as the LP relaxation bound of the $MCF(H^*)$ model. For this proof, we introduce the following intermediate hop-indexed model, denoted by $HOPi(H^*)$, that uses the variables u^{hk} needed for the first model but unnecessary for the model with hop-indexed variables:

$$Min \sum_{k \in K} u^{H^* k}$$

subject to:

(9)

$$\sum_{(i,j):\{i,j\}\in E}\ \sum_{h=1,\ldots,H} z_{ij}^{hk} = \sum_{h=1,\ldots,H} hu^{hk}, k \in K \tag{14}$$

$$\sum_{h=1,\ldots,H} u^{hk} = 1, k \in K \tag{4}$$

(5), (11)-(13), (8)

The first result follows from the well known fact (see, *e.g.*, [4]) that the projection in the space of the y_{ij}^k variables of the polyhedron defined by (9), $z_{ij}^{hk}, z_{ji}^{hk} \geq 0$ $(\{i,j\} \in E; k \in K; h = 1,\ldots,H)$, $z_{t_k t_k}^{hk} \geq 0$ $(k \in K; h = 2,\ldots,H)$ and $\sum_{h=1,\ldots,H} z_{ij}^{hk} = y_{ij}^k$ is contained in the polyhedron defined by (1), (2) and $y_{ij}^k, y_{ji}^k \geq 0$ $(\{i,j\} \in E; k \in K)$. In the following, let P_L represent the LP relaxation of model P and $v(P_L)$ its optimal value.

Result 2.1: $v(HOPi(H^*)_L) \geq v(MCF(H^*)_L)$.
In contrast to results obtained from pure network design problems with hop constraints (the objective function involves the minimization of edge costs), for all the instances tested in this work, we have obtained equality between the two LP bounds. These results are not surprising due to the different objective functions and most of all to the fact that the dimensioned networks already allow feasible paths satisfying the hop constraints. Thus, it is quite natural to expect that for many optimal solutions obtained by the model $MCF(H^*)_L$ in our experiments, constraints (2) are non binding. What can be gained by using the hop-indexed variables model is explained in the second result.

To prove this result, we start by adding the constraints

$$\sum_{j\in V\setminus\{t_k\}} z_{jt_k}^{hk} = u^{hk}, h = 1,\ldots,H; k \in K \tag{16}$$

to the intermediate model. Clearly we will obtain a model whose LP relaxation is not worse. We will show that the model with these extra constraints is precisely the model $HOP(H^*)$. Constraints (16) are equivalent to

$$(h-1)\sum_{j\in V\setminus\{t_k\}} z_{jt_k}^{hk} + \sum_{j\in V\setminus\{t_k\}} z_{jt_k}^{hk} = hu^{hk}, h = 1,\ldots,H; k \in K \tag{17}$$

By using the flow conservation constraints (9) for the same j and h, we may rewrite the first expression on the left hand side and then by adding for all h these modified constraints we obtain (14). Thus these constraints are redundant in this augmented model. Finally, by using (16), the u^{hk} variables can be eliminated and we obtain the model $HOP(H^*)$. We have just proven that

Result 2.2: $v(HOP(H^*)_L) \geq v(HOPi(H^*)_L)$.
Computational results given in Section 4 will show that this dominance is strict for several instances. As a corollary we get

Result 2.3: $v(HOP(H^*)_L) \geq v(MCF(H^*)_L)$.

3 The Single Model Approach

In order to model the problem with a single model, let A^h represent the cost of a commodity with h hops such that $A^h > |K|A^{h-1}, h = 2, \ldots, H$. In this way, we ensure that the cost of a path with h hops is higher than the sum of the costs of all paths with $h-1$ hops. As a consequence, we guarantee that for any two vectors $[a_k]_{k\in K}$ and $[b_k]_{k\in K}$ such that either $a_1 < b_1$ or there exists an index $l \in \{1, \ldots, |K|-1\}$ such that $a_i = b_i$ for all $i \leq l$ and $a_{l+1} < b_{l+1}$, $[a_k]_{k\in K}$ is selected as the optimal solution (that is, the lexicographically minimum vector). In Figure 1, letting $A^1 = 1$, and, thus, $A^2 = 4$, $A^3 = 13$ and $A^4 = 40$, the cost of the left solution is equal to 84 whereas the cost of right solution is equal to 66.

Using the same variables defined in the previous section, we present two valid models for our problem, the MCF and HOP models, that are similar to the models discussed in the previous section. For the objective function of these models, the coefficients of the u^{hk} and $z^{hk}_{jt_k}$ variables use the proposed cost structure. Due to the similarity of the models with the ones presented in the previous subsection, we simply present the models and the main result comparing the corresponding LP relaxations.

The MCF Model

$$Min \sum_{k\in K} \sum_{h=1,\ldots,H} A^h u^{hk}$$

subject to:

(1)-(4), (6)-(8)

The HOP Model

$$Min \sum_{k\in K} \sum_{j\in V\setminus\{t_k\}} \sum_{h=1,\ldots,H} A^h z^{hk}_{jt_k}$$

subject to:

(9), (11)-(13)

Using the same reasoning as in Subsection 2.3, we can state the following result

Result 3.1: $v(HOP_L) \geq v(MCF_L)$.
Computational results reported in the next section show that this dominance is strict for several instances.

4 Computational Results

In our computational tests, we have considered the network topologies obtained by solving the network design problem described in [5] applied to the NSFNet and EON backbone optical networks. These instances were determined with $H = 4$, $D = 1$ and different traffic demand matrices T, randomly generated using uniform distributions in the intervals $[0,0.4]$, $[0,0.6]$, $[0,0.8]$, and $[0,1.0]$, where the values are normalized with respect to the capacity of each network link. For the NSFNet network we have considered $|K| = 28$ commodities and, for the EON network, $|K| = 45$ commodities. The models of both approaches presented in this paper were solved through the branch-and-cut algorithm of the CPLEX 10.2 software package. The results were obtained on an Intel Core 2 Duo at 2.0 GHz with 640 MB of RAM.

Similarly to the tests reported in [6,7], here we also consider two main scenarios for the new demands r_k. The first considers that the actual demands are smaller than the initial estimates. In this case, the demands r_k range from 90% to 97.5% of the original demands in 2.5% steps, that is, $r_k = (1-\delta)t_k$, with $\delta = 0.025, 0.05, 0.075, 0.1$. The second scenario assumes that the total demand of the initial estimate is roughly correct but the individual values are different from the initial ones. In this set, each demand r_k is randomly generated with a uniform distribution between $(1-\varepsilon)t_k$ and $(1+\varepsilon)t_k$, with $\varepsilon = 0, 0.05, 0.1, 0.15, 0.2$ (ε represents the maximum relative error of the initial estimated demands t_k). Note that in the first scenario, all problems are feasible but in the second, some instances might be infeasible for $\varepsilon > 0$ (we only report the results of the feasible cases). Table 1 contains the results of the NSFNet network and Table 2 (for demand matrices T in $[0,0.4]$ and $[0,0.6]$) and Table 3 (for demand matrices T in $[0,0.8]$ and $[0,1.0]$) contain the results of the EON network.

These tables show the results of the iterative approach (columns $MCF(H^*)$ and $HOP(H^*)$) and the single model approach (columns MCF and HOP). In the iterative approach, all instances required three iterations, that is, since $H = 4$, both models were solved with $H^* = 4, 3, 2$ (the number of commodities with one hop is automatically determined). Therefore, for each instance and each model, we present three values (in percentage) which are the corresponding three LP relaxation gaps. Below these percentage values, we present the total CPU time, in seconds, taken in solving the LP relaxation of the $MCF(H^*)$ and $HOP(H^*)$ models (given by the sum of the CPU times taken by each iteration) and, separately, the total CPU time taken in finding the integer optimal solution of the same models. Similarly, in the single model approach, for each instance and each model, we present the corresponding LP relaxation gap as well as the CPU time taken in finding the LP relaxation and integer optimal solutions. Concerning the A^h costs, we have considered $A^1 = 1$, $A^2 = 29$, $A^3 = 813$ and $A^4 = 22765$ for the NSFNet network (with $|K| = 28$ and $H = 4$) and $A^1 = 1$, $A^2 = 46$, $A^3 = 2071$ and $A^4 = 93196$ for the EON network (with $|K| = 45$ and $H = 4$).

In all tables we have an extra column C showing the increase (in percentage) on the average number of hops of our solutions over the minimum average number of hops (computed with the model presented in [6,7] which dealt specifically with the average number of hops minimization).

The analysis of the results obtained for the NSFNet and EON networks allows to conclude that, in LP relaxation terms, the dominance of the $HOP(H^*)$ and HOP models with respect to the $MCF(H^*)$ and MCF models, respectively (Results 2.3 and 3.1), is strict for several instances. This dominance is identified by the gap values highlighted in bold in Tables 1, 2 and 3. Regarding the iterative approach results, observe that dominance can be quite significant and also that the corresponding gap reaches 0% for some instances.

Table 1. Computational Results for the NSFNet Network

T		$MCF(H^*)$	$HOP(H^*)$	MCF	HOP	C
[0,0.4]	$\delta = 0.1$	0%; 0%; 0%	0%; 0%; 0%	0%	0%	*0%*
		0.07; 0.23	0.04; 0.1	0.02; 0.04	0.01; 0.02	
	$\delta = 0.075$	0%; 0%; 0%	0%; 0%; 0%	0%	0%	*0%*
		0.07; 0.19	0.05; 0.12	0.01; 0.06	0.02; 0.03	
	$\delta = 0.05$	0%; 0%; 0%	0%; 0%; 0%	0%	0%	*0%*
		0.09; 0.25	0.05; 0.15	0.01; 0.04	0.01; 0.03	
	$\delta = 0.025$	0%; 16.7%; 9.1%	0%; 16.7%; **0%**	15.1%	15.1%	*0%*
		0.06; 0.32	0.03; 0.11	0.01; 0.2	0.01; 0.02	
	$\varepsilon = 0$	0%; 16.7%; 9.1%	0%; 16.7%; **0%**	15.1%	15.1%	*0%*
		0.06; 0.29	0.05; 0.1	0.01; 0.08	0.02; 0.03	
	$\varepsilon = 0.05$	100%; 25%; 15.4%	100%; **0%**; **7.7%**	83.2%	83.2%	*0%*
		0.06; 0.52	0.06; 0.2	0.02; 0.46	0.01; 0.13	
[0,0.6]	$\delta = 0.1$	0%; 7.2%; 2.4%	0%; **6.1%**; **0%**	6.2%	**5.4%**	*0%*
		0.07; 0.2	0.04; 0.06	0.01; 0.03	0.01; 0.02	
	$\delta = 0.075$	0%; 5.7%; 2%	0%; **5.1%**; **0%**	4.9%	**4.5%**	*0%*
		0.06; 0.15	0.03; 0.05	0.01; 0.03	0.01; 0.02	
	$\delta = 0.05$	0%; 20.2%; 9.9%	0%; **20.1%**; **0%**	18.1%	18.1%	*0%*
		0.04; 0.17	0.04; 0.06	0.01; 0.06	0.01; 0.02	
	$\delta = 0.025$	0%; 19.1%; 8.7%	0%; 19.1%; **0%**	17.0%	17.0%	*0%*
		0.05; 0.18	0.03; 0.06	0.01; 0.06	0.01; 0.02	
	$\varepsilon = 0$	0%; 38.5%; 29.2%	0%; 38.5%; **0%**	35.4%	35.4%	*1.9%*
		0.07; 0.16	0.03; 0.06	0.02; 0.09	0.01; 0.03	
[0,0.8]	$\delta = 0.1$	0%; 0%; 15.1%	0%; 0%; 15.1%	1.2%	1.2%	*2%*
		0.05; 0.21	0.03; 0.14	0.01; 0.05	0.01; 0.03	
	$\delta = 0.075$	0%; 0%; 14.6%	0%; 0%; 14.6%	1.2%	1.2%	*2%*
		0.06; 0.27	0.06; 0.07	0.01; 0.04	0.01; 0.04	
	$\delta = 0.05$	0%; 28.6%; 28.4%	0%; 28.6%; **10%**	26.6%	26.6%	*4%*
		0.06; 0.32	0.04; 0.13	0.01; 0.17	0.02; 0.09	
	$\delta = 0.025$	0%; 37.5%; 41.8%	0%; 37.5%; **11.1%**	35.0%	35.0%	*3.9%*
		0.06; 0.9	0.05; 0.14	0.01; 0.94	0.01; 0.09	
	$\varepsilon = 0$	100%; 25%; 19.8%	100%; **0%**; **15.3%**	83.3%	83.3%	*0%*
		0.07; 0.44	0.07; 0.22	0.03; 1.01	0.01; 0.26	
[0,1.0]	$\delta = 0.1$	0%; 4.1%; 2%	0%; **2%**; **1.3%**	3.6%	**1.9%**	*0%*
		0.06; 0.29	0.04; 0.06	0.01; 0.03	0.01; 0.03	
	$\delta = 0.075$	0%; 2.6%; 1.3%	0%; **1.3%**; **0.8%**	2.4%	**1.2%**	*0%*
		0.07; 0.17	0.04; 0.05	0.01; 0.03	0.01; 0.02	
	$\delta = 0.05$	0%; 1.3%; 0.7%	0%; **0.7%**; **0.4%**	1.2%	**0.6%**	*0%*
		0.07; 0.13	0.04; 0.04	0.01; 0.04	0.01; 0.02	
	$\delta = 0.025$	100%; 33.4%; 30.8%	100%; **22.5%**; **23.1%**	81.3%	81.3%	*0%*
		0.07; 0.32	0.05; 0.14	0.01; 0.12	0.01; 0.03	
	$\varepsilon = 0$	100%; 32%; 30.2%	100%; **21.9%**; **23.1%**	81.1%	**80.8%**	*0%*
		0.07; 0.33	0.06; 0.11	0.02; 0.1	0.01; 0.04	

Table 2. Computational Results for the EON Network with T in $[0, 0.4]$ and $[0, 0.6]$

T		$MCF(H^*)$	$HOP(H^*)$	MCF	HOP	C
$[0, 0.4]$	$\delta = 0.1$	0%; 0%; 4% 0.16; 4.64	0%; 0%; 4% 0.1; 0.19	0.4% 0.05; 0.13	0.4% 0.02; 0.05	*0%*
	$\delta = 0.075$	0%; 16.7%; 11.7% 0.16; 0.78	0%; **15.3%**; **8%** 0.11; 0.24	15.5% 0.05; 0.15	**14.3%** 0.02; 0.05	*0%*
	$\delta = 0.05$	0%; 16.7%; 11.3% 0.14; 4	0%; **14%**; **8%** 0.1; 1.21	15.4% 0.03; 0.12	**13.1%** 0.02; 0.05	*0%*
	$\delta = 0.025$	0%; 28.6%; 11.3% 0.15; 8.9	0%; **25.2%**; **4.3%** 0.11; 0.96	26.1% 0.03; 0.67	**23.1%** 0.03; 0.1	*0%*
	$\varepsilon = 0$	100%; 50%; 14.2% 0.17; 10.43	100%; **30.6%**; **4.3%** 0.13; 1.07	89.2% 0.03; 3.11	**88.6%** 0.02; 1.3	*0%*
	$\varepsilon = 0.05$	100%; 50%; 14.5% 0.17; 13.59	100%; **31.2%**; **4.3%** 0.17; 2.61	89.2% 0.06; 8.03	**88.7%** 0.06; 1.01	*0%*
	$\varepsilon = 0.10$	0%; 0%; 4% 0.16; 0.57	0%; 0%; 4% 0.11; 0.25	0.4% 0.04; 0.15	0.4% 0.05; 0.1	*0%*
	$\varepsilon = 0.15$	0%; 16.7%; 11.5% 0.18; 3.4	0%; **14.5%**; **8%** 0.11; 0.28	15.5% 0.05; 0.19	**13.6%** 0.04; 0.05	*0%*
	$\varepsilon = 0.20$	0%; 16.7%; 14.8% 0.16; 2.85	0%; **15.5%**; **11.5%** 0.1; 1.01	15.8% 0.05; 0.24	**14.7%** 0.02; 0.06	*1.2%*
$[0, 0.6]$	$\delta = 0.1$	0%; 22.3%; 10.8% 0.2; 7.76	0%; **19.9%**; **9.1%** 0.12; 1.81	20.4% 0.06; 0.21	**18.3%** 0.02; 0.15	*1.3%*
	$\delta = 0.075$	0%; 21.1%; 10% 0.18; 1.33	0%; **17.1%**; **9.1%** 0.14; 0.49	19.3% 0.05; 0.31	**15.8%** 0.03; 0.13	*1.3%*
	$\delta = 0.05$	0%; 19.6%; 9.3% 0.18; 3.7	0%; **14.5%**; **9.1%** 0.12; 1.56	17.8% 0.06; 0.21	**13.4%** 0.02; 0.08	*1.3%*
	$\delta = 0.025$	0%; 17.3%; 12.2% 0.21; 4.04	0%; **11.9%**; 12.2% 0.12; 2.05	15.7% 0.05; 0.36	**11.3%** 0.02; 0.09	*1.2%*
	$\varepsilon = 0$	0%; 14.9%; 14.5% 0.21; 5.01	0%; **9.5%**; **14.4%** 0.12; 0.41	13.8% 0.06; 0.21	**9.3%** 0.03; 0.1	*0%*
	$\varepsilon = 0.05$	0%; 25.3%; 19.5% 0.19; 5.8	0%; **20.3%**; **17.4%** 0.13; 1.29	23.4% 0.06; 0.93	**19.3%** 0.02; 0.26	*1.2%*
	$\varepsilon = 0.10$	0%; 21.7%; 7.2% 0.17; 1.31	0%; 21.7%; **4.8%** 0.12; 0.78	19.7% 0.08; 0.15	19.7% 0.02; 0.08	*0%*
	$\varepsilon = 0.15$	0%; 19.1%; 13.2% 0.2; 1.53	0%; **15.2%**; **13%** 0.13; 1.34	17.4% 0.07; 0.38	**14.2%** 0.03; 0.13	*2.5%*
	$\varepsilon = 0.20$	0%; 56.6%; 53.9% 0.18; 37.07	0%; **55.4%**; **17.6%** 0.16; 2.1	53.9% 0.06; 6.48	**52.9%** 0.02; 1.67	*2.3%*

Table 3. Computational Results for the EON Network with T in $[0, 0.8]$ and $[0, 1.0]$

T		$MCF(H^*)$	$HOP(H^*)$	MCF	HOP	C
$[0, 0.8]$	$\delta = 0.1$	0%; 36.2%; 9.4%	0%; 36.2%; **0%**	32.9%	32.9%	*0%*
		0.25; 1.93	0.16; 1.79	0.07; 0.73	0.05; 0.78	
	$\delta = 0.075$	0%; 35.6%; 8.3%	0%; **35%**; **0%**	32.2%	**31.5%**	*0%*
		0.22; 2.73	0.15; 0.45	0.08; 0.35	0.06; 0.24	
	$\delta = 0.05$	0%; 35.1%; 12%	0%; **32%**; **5.3%**	31.8%	**28.9%**	*0%*
		0.25; 8.43	0.14; 0.52	0.07; 0.3	0.06; 0.15	
	$\delta = 0.025$	0%; 34.3%; 11%	0%; **29%**; **5.3%**	30.8%	**26.1%**	*0%*
		0.25; 2.8	0.15; 1.58	0.07; 0.29	0.03; 0.09	
	$\varepsilon = 0$	0%; 31.5%; 18.6%	0%; **26.1%**; **14.3%**	28.5%	**23.8%**	*0%*
		0.26; 22.17	0.12; 2.04	0.08; 0.48	0.03; 0.3	
	$\varepsilon = 0.05$	0%; 32%; 9.9%	0%; **26.6%**; **5.3%**	28.6%	**23.8%**	*0%*
		0.22; 20	0.16; 2.15	0.06; 0.63	0.05; 0.09	
	$\varepsilon = 0.10$	0%; 35.2%; 9.3%	0%; 35.2%; **0%**	32.1%	32.1%	*0%*
		0.21; 1.75	0.15; 0.34	0.06; 0.68	0.03; 0.1	
	$\varepsilon = 0.15$	0%; 36.1%; 9.4%	0%; 36.1%; **0%**	32.8%	32.8%	*0%*
		0.2; 0.83	0.12; 0.44	0.06; 0.26	0.03; 0.1	
	$\varepsilon = 0.20$	0%; 31.5%; 15.2%	0%; **29.4%**; **10%**	28.5%	**26.6%**	*0%*
		0.21; 8.25	0.12; 0.56	0.06; 0.39	0.02; 0.16	
$[0, 1.0]$	$\delta = 0.1$	0%; 25%; 11.5%	0%; 25%; **5.6%**	23.2%	23.2%	*0%*
		0.2; 6.64	0.12; 2.41	0.08; 0.59	0.03; 0.15	
	$\delta = 0.075$	0%; 25%; 15%	0%; **24.3%**; **10.5%**	23.2%	**22.6%**	*0%*
		0.27; 1.62	0.13; 2.67	0.08; 0.49	0.02; 0.15	
	$\delta = 0.05$	0%; 40%; 26.3%	0%; **37.7%**; **11.8%**	37.4%	**35.4%**	*0%*
		0.22; 12.5	0.14; 2.67	0.08; 0.71	0.03; 0.62	
	$\delta = 0.025$	0%; 48.8%; 36.3%	0%; **46.7%**; **7.1%**	46.1%	**44.1%**	*0%*
		0.26; 15.75	0.19; 2.97	0.11; 6.35	0.03; 3.85	
	$\varepsilon = 0$	0%; 51.4%; 47.3%	0%; **49.5%**; **14.3%**	48.8%	**47.1%**	*0%*
		0.28; 18.79	0.2; 3.25	0.07; 5.54	0.05; 2.44	
	$\varepsilon = 0.05$	0%; 51.6%; 47%	0%; **49.8%**; **14.3%**	48.9%	**47.3%**	*0%*
		0.33; 10.75	0.16; 1.87	0.06; 1.97	0.05; 1.48	
	$\varepsilon = 0.10$	0%; 14.3%; 11.1%	0%; 14.3%; **10%**	13.3%	13.3%	*0%*
		0.25; 22.08	0.17; 0.49	0.06; 0.38	0.05; 0.26	
	$\varepsilon = 0.15$	0%; 33.2%; 15.8%	0%; **30.8%**; **5.9%**	30.8%	**28.5%**	*0%*
		0.27; 7.05	0.19; 0.54	0.08; 2.56	0.06; 1.25	

Table 4. Average CPU Time (in Seconds) Taken in Finding the Integer Optimal Solution

	$MCF(H^*)$	$HOP(H^*)$	MCF	HOP
NSFNet	0.29	0.1	0.18	0.05
EON	8.02	1.33	1.28	0.51

As for the CPU times, even though the order of magnitude of CPU times of the LP relaxations is rather small, these times are smaller for the $HOP(H^*)$ and HOP models. On the other hand, considering the CPU times taken in finding the integer optimal solutions, the $HOP(H^*)$ and the HOP models are markedly faster. In order to highlight these results, Table 4 shows the CPU time taken in finding the optimal solutions averaged over all NSFNet instances and all EON instances. These results show the superior efficiency of HOP based models over MCF based models and, also, that the single model approach is preferable over the iterative approach.

To conclude this section, we now compare the average number of hops of the paths determined in this work to the optimal values given in column C of all tables. For all instances tested, the average number of hops increased at most 4.0% and for most of the instances the increase was 0%. These results show that the optimal solutions for the lexicographical minimization of routing hops exhibit a very small penalty on the optimal average number of routing hops, which means that we can improve the routing fairness without jeopardizing the overall average performance.

5 Conclusions and Future Work

In this paper, we have addressed the problem of determining a single routing path for each commodity over a telecommunications network aiming to minimize the number of routing hops in a lexicographical sense. We have presented two approaches: an iterative approach and a single model approach. In both cases, we have exploited a multi-commodity flow model and a hop-indexed model. We have shown that the LP bound of the hop-indexed model is at least as good as the multi-commodity model. The computational results showed that this dominance is strict for several instances and, as a consequence, the approaches based on the hop-indexed model were more efficient. The computational results also showed that the single model approach has significantly shorter running times.

As mentioned previously, this work has only addressed the $D = 1$ case (which also explains why all instances are easy to solve) but we intend to address the $D = 2, 3, 4$ cases in the near future. We emphasize, however, that the case with $D = 1$ is of interest alone since the objective function under study is, as far as we know, new in the context of these problems, and the $D = 1$ case is suitable for understanding the behavior of hop-indexed models in the context of more "standard" models as the *MCF* model. Lastly, since the LP gaps are often quite large, we are investigating valid cut inequalities in order to obtain better formulations.

References

1. Ben-Ameur, W., Kerivin, H.: Routing of Uncertain Traffic Demands. Optim. Eng. 6, 283–313 (2005)
2. Botton, Q., Fortz. B., Gouveia, L., Poss, M.: Benders decomposition for the hop constrained survivable network design problem. CIO working paper (March 2010)
3. Dahl, G., Gouveia, L., Requejo, C.: On Formulations and Methods for the Hop-Constrained Minimum Spanning Tree Problem. In: Resende, M., Pardalos, P. (eds.) Handbook of Optimization in Telecommunications, pp. 493–515. Springer, Heidelberg (2006)
4. Gouveia, L.: Using Variable Redefinition for Computing Lower Bounds for Minimum Spanning and Steiner Trees with Hop Constraints. INFORMS J. Comput. 10, 180–188 (1998)
5. Gouveia, L., Patrício, P.F., de Sousa, A.F.: Hop-Constrained Node Survivable Network Design: An Application to MPLS over WDM. Netw. Spat. Econ. 8(1), 3–21 (2008)
6. Gouveia, L., Patrício, P.F., de Sousa, A.F.: Optimal Survivable Routing with a Small Number of Hops. In: Raghavan, S., Golden, G., Wasil, E. (eds.) Telecommunications Modeling, Policy, and Technology. Operations Research/Computer Science Interfaces Book Series, vol. 44, pp. 253–274. Springer, Heidelberg (2008)
7. Gouveia, L., Patrício, P.F., de Sousa, A.F.: Models for Optimal Survivable Routing with a Minimum Number of Hops: Comparing Disaggregated with Aggregated Models. Int. Trans. Oper. Res. (2011), doi: 10.1111/j.1475-3995.2010.00766.x
8. Huygens, D., Labb, M., Mahjoub, A.R., Pesneau, P.: The two-edge connected hop-constrained network design problem: Valid inequalities and branch-and-cut. Netw 49(1), 116–133 (2007)
9. Juttner, A., Szviatovski, B., Mecs, I., Rajko, Z.: Lagrange relaxation based method for the QoS routing problem. In: Infocom Proc., vol. 2, pp. 859–868 (2001)
10. Kerivin, H., Mahjoub, A.R.: Design of Survivable Networks: A Survey. Netw. 46(1), 1–21 (2005)
11. Nace, D., Pióro, M.: Max-Min Fairness And Its Applications to Routing and Load-Balancing in Communication Networks: A Tutorial. IEEE Surv. and Tutor. 10(4), 5–17 (2008)
12. Ogryczak, W., Milewski, M., Wierzbicki, A.: Fair and Efficient Bandwidth Allocation with the Reference Point Methodology. In: INOC Proc. (2007)
13. Orda, A.: Routing with end to end QoS guarantees in broadband networks. IEEE/ACM Trans. Netw. 7(3), 365–374 (1999)
14. Orlowski, S., Wessaely, R.: The Effect of Hop Limits on Optimal Cost in Survivable Network Design. In: Raghavan, S., Anandalingam, G. (eds.) Telecommunications Network Planning: Innovations in Pricing, Network Design and Management. Operations Research/Computer Science Interfaces Book Series, vol. 33, pp. 151–166. Springer, Heidelberg (2006)
15. Parekh, A., Gallager, R.: A Generalized Processor Sharing Approach to Flow Control in Integrated Services Networks: The Multiple Node Case. IEEE/ACM Trans. Netw. 2(2), 137–150 (1994)
16. Radunovic, B., Le Boudec, J.-Y.: A Unified Framework for Max-min and Min-max Fairness with Applications. IEEE/ACM Trans. Netw. 15(5), 1073–1083 (2007)
17. Santos, D., de Sousa, A., Alvelos, F., Pióro, M.: Link Load Balancing Optimization of Telecommunication Networks: a Column Generation based Heuristic Approach. In: Int. Telecommun. In: Netw. Strategy Plan. Symp. Proc., pp. 1–6 (2010)
18. Yuan, X.: Heuristic Algorithms for multiconstrained quality-of-service routing. IEEE/ACM Trans. Netw. 10(2), 244–256 (2002)

A Method for Obtaining the Maximum (δ, η)-Balanced Flow in a Network

Wataru Kishimoto

Graduate School of Advanced Integration Science Chiba University, Chiba-shi, Chiba 263-8522, Japan
wkishi@faculty.chiba-u.jp

Abstract. In a directed flow network we assign capacities on vertices as well as on edges. We consider a (δ, η)-balanced flow problem of single commodity case. A (δ, η)-balanced flow is defined as a flow such that the flow value at each edge is not more than $\delta \cdot f$ and the flow value at each vertex is not more than $\eta \cdot f$, where f is the total amount of the flow. Based on (δ, η)-balanced flow, the (δ, η)-capacity is defined for a mixed cut in a network. A mixed cut in a network is a set of edges and vertices removal of those separates the network. Then the max-flow min-cut theorem for this (δ, η)-balanced flow is proved for the single commodity case in a directed network. The theorem for (δ, η)-balanced flow is not easily to be proved by only applying the max-flow min-cut theorem of ordinary flows. Then we show a method for evaluating the maximum (δ, η)-balanced flow. The algorithm gives the maximum value of (δ, η)-balanced flow between s and t in N with at most $|V| \cdot |E|$ evaluations of maximum flow in a network, where V is the vertex set of N and E is the edge set of N, respectively. Each evaluations of the maximum flow is performed in N with altered capacities on edges and on vertices. We can apply all the results to undirected networks.

1 Introduction

A flow in a network is a non-negative combination of flows along paths. Depending on the paths the flow goes along, an edge failure may cause almost entire loss of the flow. This situation is not preferable for some applications. This motivated the definition of a multi-route flow [16,17,18]. Later, a δ-reliable flow is defined as a generalization of the multi-route flow [19]. A δ-reliable flow is a flow for which an edge failure causes a loss of at most δ of the entire flow. We say it is a δ-condition. Node failures of a flow are considered similarly to edge failures. If we care about only node failures, it is easy to solve the problem as a δ-reliable flow. Also if we require the amount of loss a node failure causes is at most the same value of an edge failure, it is solved easily. In addition to the δ-condition, we admit a node failure to the loss of at most η of the entire flow, where $0 < \delta < \eta \leq 1$. We call it an η-condition. We say that such a flow satisfying both of δ-condition and η-condition is a (δ, η)-balanced flow, or a (δ, η)-reliable flow.

Since $\delta < \eta$, it is not a straight forward extension of a δ-reliable flow, which we can solve by using the same method. We construct a flow with value f satisfying δ-condition, which means at most δf of the flow goes through each edge. Then, we happen to decrease the entire flow to $f' < f$ for satisfying η-condition. Now, for satisfying

J. Pahl, T. Reiners, and S. Voß (Eds.): INOC 2011, LNCS 6701, pp. 230–242, 2011.

δ-condition the flow value at each edge is at most $\delta f' < \delta f$. Therefore, the current flow with valuer f' may not satisfy δ-condition. The flow is decreased again for satisfying δ-condition and then it may not satisfy η-condition, and so on. We have to consider both of δ-condition and η-condition simultaneously for constructing a (δ,η)-balanced flow. We prove a max-flow min-cut theorem for a (δ,η)-balanced flow. The theorem for (δ,η)-balanced flow is not easily to be proved by only applying the max-flow min-cut theorem for ordinary flow. Then we give a method for evaluating a maximum (δ,η)-balanced flow between a pair of vertices s and t in a network. The algorithm gives the maximum value of (δ,η)-balanced flow from s to t in N with at most $|V|\cdot|E|$ evaluations of maximum flow in a network, where V is a vertex set of N and E is an edge set of N, respectively. Each evaluation of the maximum flow is done in N with altered capacities of edges and vertices. We can apply all the results in this paper to undirected networks.

The problem to construct a balanced flow corresponds to, for example, the design of a reliable data communication channel for reliable data flow in a network. Usually, a network is designed to satisfy some specified performance objectives under normal operating conditions, without explicit consideration of network survivability. The network performance in the event of a failure is difficult to predict. Setting the objectives of survivability performance will enable us to ensure that, under given failure scenarios, network performance will not degrade below predetermined levels [21,22]. The design of such a reliable communication channel corresponds to a balanced flow problem in a network. For reliability of network flows we can refer to many researches e.g. [3,4,5,12,20].

Single commodity multiroute flows are studied by [1,16,18]. Multi-commodity multiroute flow problem is considered in [7]. There is a study of multiroute flow problem based on different cuts [8]. Other multiroute flow problems are considered in [6,9,10,15]. [7,8,10] gives good reference of multiroute flow problem.

A multiroute flow corresponds to a generalized fan [11] in graph theory. On the basis of [11], m-route flow is proved to be a δ-reliable flow where $\delta = 1/m$ [17,18]. A δ-reliable flow is an extension of a multiroute flow [19] and is the basis of a (δ,η)-balanced flow we define in this paper.

This paper is organized as follows. In Section 2 the terms used in this paper are defined. In Section 3, the definitions of the (δ,η)-balanced flow and the (δ,η)-capacity of a cut are given. Then the max-flow min-cut theorem for the (δ,η)-balanced flow is covered in Section 4. Section 5 shows the main algorithm for obtaining the maximum (δ,η)-balanced flow between a pair of vertices.

2 Preliminaries

2.1 Flow Network

Let $G = (V,E)$ be a graph with the vertex set V and the edge set E. Then $N = (G,c)$ is a network, where we say c is a capacity function such that $c(x)$ is non-negative real value and indicates the capacity of edge x or that of vertex x in G (i.e. $x \in V \cup E$). We set $c_i(x) = 0$ if $x \notin V \cup E$. If G is directed, N is directed, otherwise N is undirected. Unless stated otherwise, networks we consider are directed.

Definition 1. *Let $N_a = (G_a, c_a)$ and $N_b = (G_b, c_b)$ be two networks. If G_b is a subgraph of G_a and $c_b(x) \le c_a(x)$ for each edge x and for each vertex x of G_b, we say N_b is a subnetwork of N_a.*

Definition 2. *Let $N_1 = (G_1, c_1), N_2 = (G_2, c_2), \cdots, N_n = (G_n, c_n)$ be networks. Let V_i and E_i be the vertex set and the edge set of G_i; let V and E be the vertex set and the edge set of G. The sum of networks $N_1, N_2, \cdots, N_n$ is $N = (G, c)$ such that*

$$\begin{aligned} V &= V_1 \cup V_2 \cup \cdots \cup V_n, \\ E &= E_1 \cup E_2 \cup \cdots \cup E_n, \\ c(x) &= c_1(x) + c_2(x) + \cdots + c_n(x) \quad \textit{for all } x \in E \cup V. \end{aligned}$$

Definition 3. *A network is λ uniform or uniform with value λ if each edge and each vertex has capacity λ.*

Let V_1 be a nonempty subset of vertex-set V of network N, and V_2 be the complement of V_1. In a directed network N, a cut (V_1, V_2) of N is the set of edges, each of which is directed away from a vertex of V_1 and directed toward a vertex of V_2. The set of edges directed away from v is called the out-incidence cut of v, and the set of edges directed into v is the in-incidence cut of v. Then, $I_o(v)$ stands for the out-incidence cut of v and $I_i(v)$ the in-incidence cut of v.

We denote by $\bar{U}$ the complement of a subset U of V. When $s \in V_1$ and $t \in \bar{V_1}$, we say $(V_1, \bar{V_1})$ is a *s*-*t* cut. The sum of edge capacities of an edge set S is called the capacity of the edge set, and is indicated by $c(S)$. The capacity of a cut $(V_1, \bar{V_1})$ is simply represented by $c(V_1, \bar{V_1})$ instead of $c((V_1, \bar{V_1}))$.

The capacity of a vertex also restricts the amount of flows passing through the vertex. Clearly, no vertex, even s nor t can convey a flow with the value greater than $c(I_i(v))$ and $c(I_o(v))$. If the capacity of each vertex v denoted as $c(v)$ satisfies

$$c(v) \ge \max\{c(I_i(v)), c(I_o(v))\},$$

we say c is affluent at v. Moreover, a network $N = (G, c(*))$ is smooth if c is affluent at each vertex in G. In this paper, if c is affluent at v we set

$$c(v) = \max\{c(I_i(v)), c(I_o(v))\}. \tag{1}$$

If N is smooth, we set equation (1) holds at each vertex of N. In a smooth network we can define the value of an ordinary flow between any pair of vertices without care of capacities of vertices in the network.

Definition 4. *In a directed network $N = (G, c(*))$, let V_1 and V_2 be two nonempty proper subsets of the vertex set V of G such that $V_1 \cap V_2 = \emptyset$. The mixed cut is the set consisting of edges each of which is directed away from a vertex in V_1 toward a vertex in V_2, and vertices in $V - V_1 - V_2$. The symbol (V_1, V_2) is used for this mixed cut. The capacity $c(V_1, V_2)$ is defined as*

$$c(V_1, V_2) = \sum_{e \in E(V_1, V_2)} c(e) + \sum_{v \in V(V_1, V_2)} c(v),$$

where $E(V_1, V_2)$ is the set of edges in (V_1, V_2), and $V(V_1, V_2)$ is the set of vertices in (V_1, V_2).

When $s \in V_1$ and $t \in V_2$, we say (V_1,V_2) is an *s-t* mixed cut. If $V_1 \cup V_2 = V$, i.e. $V_2 = \bar{V_1}$, the mixed cut (V_1,V_2) coincides with the ordinary cut $(V_1,\bar{V_1})$.

2.2 Definition of Flow

A flow from a vertex s to a vertex t is usually defined as a function f from the edge set of a network to nonnegative real numbers[2,14]; however, the following definitions are used here. Henceforth, N is assumed to be a directed network.

Definition 5. *A path-flow from a vertex s to a vertex t with the value λ is a uniform network $P = (G_p, c_p)$ of value λ such that G_p is a directed path from s to t. We also say that P is an s-t path-flow.*

Definition 6. *A flow from a vertex s to a vertex t is defined as a network $F = (G_f, c_f)$ which is the sum of s-t path-flows $P_1, P_2, \cdots, P_n$. The value of F is defined as the sum of the values of these path-flows. We also say that F is an s-t flow.*

Note that a path flow and a flow are both smooth networks. Let a flow F be a subnetwork of N; then F is a flow in N. There is no essential difference between the definition 6 and the ordinary one using a flow function f. Let $F = (G,c)$ be a flow from s to t. For each vertex v of the vertices except s and t in F, we have that $c(v) = c(I_i(v)) = c(I_o(v))$. Moreover, we have that $c(s) = c(I_o(v))$, and $c(t) = c(I_i(v))$. In the figure of a flow we omit the value at each vertex.

From the max-flow min-cut theorem, the maximum flow value from s to t is equal to the minimum capacity of *s-t* cuts in N [2,14].

2.3 Cut-Flow

Definition 7. *[18] Let $(V_1,\bar{V_1})$ be a cut of a network N. A cut-flow of $(V_1,\bar{V_1})$ is a subnetwork $F_c = (G_f, c_f)$ of N such that each edge of G_f is in $(V_1,\bar{V_1})$. The value of F_c is defined by $c_f(V_1,\bar{V_1})$.*

Since $F_c = (G_f, c_f)$ is a subnetwork of N, $c_f(e) \leq c(e)$ for each edge e in G_f. The maximum value of cut-flow of $(V_1,\bar{V_1})$ is equal to $c(V_1,\bar{V_1})$. A cut-flow of $(V_1,\bar{V_1})$ can be considered to be a flow between a pair of vertex sets, instead of between a single pair of vertices.

Definition 8. *[18] Let (V_1,V_2) be a mixed cut of a network N. A cut-flow of (V_1,V_2) in N is a subnetwork $F_c = (G_c, c_c)$ of N such that each edge of G_c is an edge in (V_1,V_2) and each vertex of G_c is a vertex in (V_1,V_2). The value of F_c is defined by $c_c(V_1,V_2)$.*

It can be easily seen that the maximum value of a cut-flow of a mixed cut (V_1,V_2) is equal to $c(V_1,V_2)$.

3 (δ,η)-Balanced FLOW

3.1 (δ,η)-Balanced Flow

From now on δ and η always satisfy that $0 < \delta \leq \eta \leq 1$.

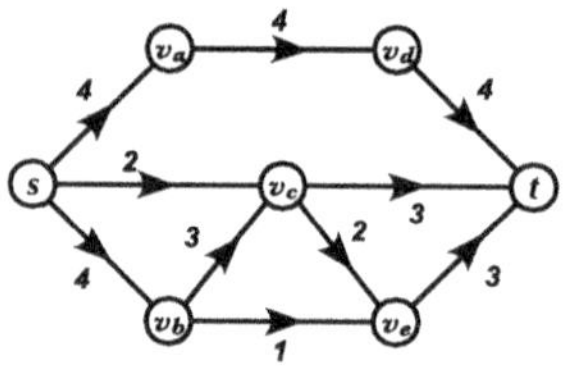

Fig. 1. $(0.4, 0.5)$-balanced flow D_0

Definition 9. *A (δ, η)-balanced flow is a flow $F = (G_F, c_F(*))$ from a vertex s to a vertex t, such that*

$$c_F(e) \leq \delta f,$$

for each edge e in F, and

$$c_F(v) \leq \eta f,$$

for each vertex v in F except s and t where f is the value of F,

A single edge-failure causes, at most, a fraction δ of a (δ, η)-balanced flow to fail, while a single-vertex failure does at most a fraction η of the flow to fail. Therefore, at least $(1-\delta)\cdot$ (value of the flow) survives an edge failure, and at least $(1-\eta)\cdot$ (value of the flow) survives a vertex failure. Even when n edges and m vertices fail, at most $(n\delta + m\eta)\cdot$ (value of a (δ, η)-balanced flow) fails. Thus (δ, η)-balanced flow has a specified reliability against edge and vertex failures.

Example: A smooth network D_0 of figure 1 is a flow with the value 10 from s to t. As denoted before, the capacity of each vertex v satisfies equation (1). The capacity of each edge in D_0 is not greater than 4 ($= 0.4 \times 10$), as well as each capacity of the vertex except s and t in D_0 is not greater than 5 ($= 0.5 \times 10$). Therefore, D_0 is a $(0.4, 0.5)$-balanced flow.

3.2 (δ, η)-Capacity of a Mixed Cut

Definition 10. *A (δ, η)-balanced cut-flow of a mixed cut (V_1, V_2) in N is a cut-flow $F_c = (G_c, c_c(*))$ of (V_1, V_2) in N such that*

$$c_c(e) \leq \delta f_c,$$
$$c_c(v) \leq \eta f_c$$

for each edge e and each vertex v of G_c, where f_c is the value of F_c.

Definition 11. *The (δ, η)-capacity of a mixed cut is the maximum value of (δ, η)-balanced cut-flow of the mixed cut.*

Since the (δ,η)-capacity is the value of a cut-flow, then

$$c_{(\delta,\eta)}(V_1,V_2) \le c(V_1,V_2)$$

where $c_{(\delta,\eta)}(V_1,V_2)$ stands for the (δ,η)-capacity of (V_1,V_2).

Let F be a (δ,η)-balanced flow from s to t in N with the value f. Suppose that a subnetwork F_1 consists of the edges and the vertices of an s-t mixed cut (V_1,V_2) in F and that its capacities of edges and vertices are the same values as in F. Then F_1 is a cut-flow of (V_1,V_2) with the value f_1 ($\ge f$). Since each edge capacity is less than δf ($\le \delta f_1$) and each vertex capacity is less than ηf ($\le \eta f_1$), F_1 is a (δ,η)-balanced cut-flow of (V_1,V_2). Consequently, the value of a (δ,η)-balanced flow from s to t is not greater than the (δ,η)-capacity of (V_1,V_2).

3.3 Evaluation of (δ,η)-Capacity of a Mixed Cut

Let the edges of a mixed cut (V_1,V_2) in $N = (G, c)$ be $e_1, e_2, \cdots, e_p$ and, the vertices of (V_1,V_2) be $v_1, v_2, \cdots, v_q$, such that

$$c(V_1,V_2) = \sum_{i=1}^{p} c(e_i) + \sum_{j=1}^{q} c(v_j); \tag{2}$$

$$c(e_i) \ge c(e_{i+1}), \quad \text{for } 1 \le i \le p-1; \tag{3}$$

$$c(v_j) \ge c(v_{j+1}), \quad \text{for } 1 \le j \le q-1. \tag{4}$$

Let $F = (G', c')$ be a cut-flow of a mixed cut (V_1,V_2) in $N = (G, c)$. Then, F is a (δ,η)-balanced cut-flow of (V_1,V_2) if and only if

$$c'(e_i) \le \min\{c(e_i), \delta c'(V_1,V_2)\}, \quad \text{for } i = 1,2,\cdots,p, \tag{5}$$

$$c'(v_j) \le \min\{c(v_j), \eta c'(V_1,V_2)\}, \quad \text{for } j = 1,2,\cdots,q. \tag{6}$$

We can easily show the following lemmas.

Lemma 1. *A necessary and sufficient condition for a value θ to be the value of a (δ,η)-balanced cut-flow of (V_1,V_2) is*

$$\theta \le \sum_{k=1}^{p} \min\{c(e_k), \delta\theta\} + \sum_{\ell=1}^{q} \min\{c(v_\ell), \eta\theta\}. \tag{7}$$

That is, the value of a (δ,η)-balanced cut-flow θ must satisfy inequality (7). Conversely, if θ satisfies inequality (7), there exists a (δ,η)-balanced cut-flow F with this value θ.

Let θ_{max} is the maximum value of θ that satisfies inequality (7). Since $c_{(\delta,\eta)}(V_1,V_2)$, the (δ,η)-capacity of (V_1,V_2), is the value of a (δ,η)-balanced cut-flow, $c_{(\delta,\eta)}(V_1,V_2)$ satisfies inequality (7). We have

$$\theta_{max} \ge c_{(\delta,\eta)}(V_1,V_2),$$

Since from lemma 1 there exists a (δ,η)-balanced cut-flow with the value θ_{max} of (V_1,V_2) in N,

$$\theta_{max} \le c_{(\delta,\eta)}(V_1,V_2).$$

Consequently, θ_{max} is equal to $c_{(\delta,\eta)}(V_1,V_2)$.

Lemma 2. *Let θ be a value such that $\theta \le \theta_{max}$. Then θ satisfies inequality (7).*

Inequalities (3) and (5) indicate that there is some index i such that the capacity of the cut-flow edges subsequent to i are bounded by $c(e_k)$ and the capacity of ones equal to and preceding i are bounded by $\delta\theta$. Similarly, inequalities (4) and (6) imply that there is some index j such that the capacity of the cut-flow vertices subsequent to j are bounded by $\eta\theta$ and the capacity of ones equal to and preceding j are bounded by $c(v_k)$.

Intuitively thinking, the maximum θ can be obtained by making inequality (7) an equality; this, in turn, causes inequalities (5) and (6) to be equalities. Then, let $\mathscr{I}$ be the set of all pairs of (i,j) such that $i\delta + j\eta < 1$, $i \in \{0,\ldots,p\}$, and $j \in \{0,\ldots,q\}$, and solve the following equation to obtain a θ value for each pair $(i,j) \in \mathscr{I}$

$$\theta = (i\delta + j\eta)\theta + \sum_{k=i+1}^{p} c(e_k) + \sum_{\ell=j+1}^{q} c(v_\ell).$$

Note that if $(i,j) \in \mathscr{I}$ then $\theta \ge 0$. If the obtained θ satisfies

$$c(e_i) \ge \delta\theta \ge c(e_{i+1}), \quad \text{and}$$
$$c(v_j) \ge \eta\theta \ge c(v_{j+1}),$$

then θ also satisfies the equality of inequality (7) and is expected to be the maximum.

On the basis of the results and intuitive discussion above, we describe a method to obtain the (δ,η)-capacity of a cut with $1/\delta + 1/\eta$ (at most $p+q$) evaluations of θ.

Algorithm #1: Evaluation of the (δ,η)-capacity of a mixed cut.
In the following, the sequence $\theta_{i,j}$, for $(i,j) \in \mathscr{I}$, is the θ sequence of (V_1,V_2) defined as follows:

$$\begin{aligned}\theta_{i,j} &= \psi(i,j)\left\{\sum_{k=i+1}^{p} c(e_k) + \sum_{k=j+1}^{q} c(v_k)\right\} \\ &= \psi(i,j)\left\{c(V_1,V_2) - \sum_{k=1}^{i} c(e_k) - \sum_{k=1}^{j} c(v_k)\right\},\end{aligned} \tag{8}$$

where $\psi(i,j) = \frac{1}{1-i\delta-j\eta}$.

1. Set $\beta = 0$.
2. Iterate the next procedure until it stops.
   Procedure A:
   (a) Set α to be the minimum i ($(i,\beta) \in \mathscr{I}$) such that

$$c(e_{i+1}) \le \delta\psi(i,\beta)\left\{\sum_{k=i+1}^{p} c(e_k) + \sum_{\ell=\beta+1}^{q} c(v_\ell)\right\} = \delta\theta_{i,\beta}. \tag{9}$$

(b) If $c(v_{\beta+1}) > \eta\theta_{\alpha,\beta}$, then set β to be the minimum j ($(\alpha, j) \in \mathscr{I}$) such that

$$c(v_{j+1}) \le \eta\psi(\alpha, j)\left\{\sum_{k=\alpha+1}^{p} c(e_k) + \sum_{k=j+1}^{q} c(v_k)\right\} = \eta\theta_{\alpha,j}. \quad (10)$$

(c) If $c(e_{\alpha+1}) \le \delta\theta_{\alpha,\beta}$, then terminate the procedure. Otherwise, go back to (a). (end of procedure)

3. The (δ,η)-index (τ,ρ) is defined as that $\tau = \alpha$, $\rho = \beta$. Then $\theta_{\tau,\rho} = c_{(\delta,\eta)}(V_1,V_2)$ $= \theta_{max}$, the (δ,η)-capacity of (V_1,V_2).

(end of algorithm)

Note that the indices α and β never decrease in the algorithm #1. Therefore, $\theta_{i,j}$ is evaluated $1/\delta$ times in step 2(a) of this algorithm, as well as, $\theta_{i,j}$ is evaluated $1/\eta$ times in step 2(b). This algorithm terminates after at most $1/\delta + 1/\eta$ iterations of step 2.

Example: In a network N of figure 2, let $V_1 = \{s, v_a, v_b\}$, $V_2 = \{v_d, v_e, t\}$. Then $(V_1,V_2) = \{e_1, e_2, v_c\}$. By using the above algorithm, we can evaluate $c_{(0.4,0.5)}(V_1,V_2)$ as follows.

1. Set $\beta = 0$.
2. (a) Since $\theta_{0,0} = 16$, we have $6 \le 0.4 \cdot 16$, that is, $c(e_1) \le \delta\theta_{0,\beta}$. We set $\alpha = 0$.
   (b) Since $c(v_1) = 9$, we have $9 > 0.5 \cdot 16$, that is, $c(v_1) > \eta\theta_{0,0}$. Then $c(v_2) = 0$ and $\theta_{0,1} = 14$. Therefore $0 \le 0.5 \cdot 14$, that is, $c(e_2) \le \delta\theta_{0,\beta}$. We set $\beta = 1$ and iterate the procedure A again.
   (c) Since $c(e_1) = 6$, we have $6 > 0.4 \cdot 14$, that is, $c(e_1) > \delta\theta_{0,1}$.
   (a) Since $\theta_{1,1} = 10$ and $c(e_2) = 1$, we have $1 \le 0.4 \cdot 10$, that is, $c(e_2) \le \delta\theta_{1,1}$. We set $\alpha = 1$.
   (b) Since $c(v_2) = 0$, we have $c(v_2) \le \eta\theta_{1,1}$, that is, $0 \le 0.5 \cdot 10$.
3. The (δ,η)-index (τ,ρ) is defined as that $\tau = 1$, $\rho = 1$. Then $c_{(0.4,0.5)}(V_1,V_2) = \theta_{1,1}$ $= 10$.

Therefore, we obtain a value of 10 as the $(0.4,0.5)$-capacity of (V_1,V_2) with $\tau = 1$ and $\rho = 1$. Since D_0 of figure 1 is a $(0.4,0.5)$-balanced flow with value 10 in N, D_0 is a maximum $(0.4,0.5)$-balanced flow from s to t in N.

Theorem 1 shows that algorithm #1 is correct. We can conclude that the maximum θ satisfies the equality of inequality (7). The maximum θ is also obtained from evaluations of all $\theta_{i,j}$, which needs at most pq times of evaluations of $\theta_{i,j}$.

For the proof of theorem 1 we show some lemmas which are useful for the following sections.

Lemma 3. *1. For any β ($0 \le \beta < 1/\eta$) there always exists a (i,β) ($\in \mathscr{I}$) satisfying inequality (9).*

2. For any α ($0 \le \alpha < 1/\delta$) there always exists a (α,j) ($\in \mathscr{I}$) satisfying inequality (10).

3. There always exists a (α,β) ($\in \mathscr{I}$) satisfying both of inequalities (9) and (10).

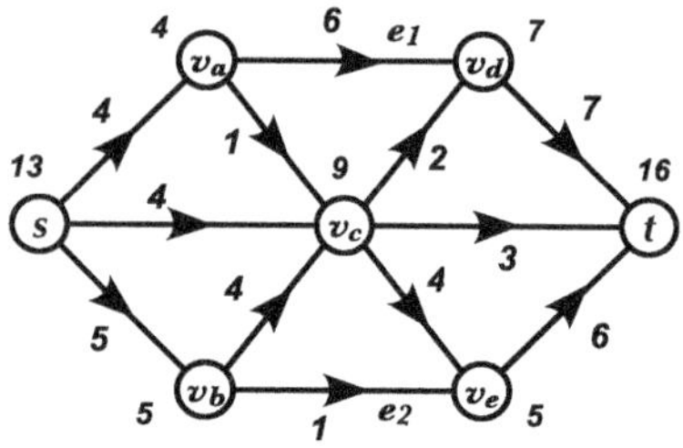

Fig. 2. Network N

Lemma 4. *For $(i,j) \in \mathscr{I}$,*
1. if $c(e_{i+1}) > \delta\theta_{i,j}$

$$c(e_{k+1}) > \delta\theta_{k,j} \quad \text{for } 1 \le k \le i-1,$$

2. if $c(v_{j+1}) > \eta\theta_{i,j}$,

$$c(v_{\ell+1}) > \eta\theta_{i,\ell} \quad \text{for } 1 \le \ell \le j-1.$$

From lemma 4, at least α or β increases at each iteration of step 2 in algorithm #1 until the algorithm stops. Furthermore, from lemma 3, (α,β) is always in finite set $\mathscr{I}$. Algorithm #1 finds (δ,η)-reliable capacity, and terminates in a finite number of iterations. The (δ,η)-index (τ,ρ) is always in $\mathscr{I}$.

Lemma 5. *For any mixed-cut with p edges and q vertices, $\theta_{\tau,\rho} = 0$, if and only if, $(p,q) \in \mathscr{I}$.*

Lemma 6. *Let (τ,ρ) be (δ,η)-index of a mixed cut (V_1,V_2). Then,*

$$c(e_i) > \delta\theta_{\tau,\rho}, \quad \text{for } 1 \le i \le \tau;$$
$$c(v_j) > \eta\theta_{\tau,\rho}, \quad \text{for } 1 \le j \le \rho.$$

From the definitions of τ and ρ, we also have the following inequalities:

$$c(e_i) \le \delta\theta_{\tau,\rho}, \quad \text{for } \tau+1 \le i \le p; \tag{11}$$
$$c(v_j) \le \eta\theta_{\tau,\rho}, \quad \text{for } \rho+1 \le j \le q. \tag{12}$$

Theorem 1. *The output $\theta_{\tau,\rho}$ of algorithm #1 is the (δ,η)-capacity of (V_1,V_2).*

4 Max-Flow Min-Cut Theorem of (δ,η)-Balanced Flow

In this section we prove the max-flow min-cut theorem for (δ,η)-balanced flow.

Lemma 7. *Let (τ,ρ) be (δ,η)-index of a mixed cut (V_1,V_2). For any (i,j) $(\in \mathscr{I})$,*

1. *If* $\tau \le i$ *and* $\rho \le j$,

$$\theta_{i,j} \ge \theta_{\tau,\rho},$$

2. *otherwise,*

$$\theta_{i,j} > \theta_{\tau,\rho}.$$

Using the lemmas proved so far, and the max-flow min-cut theorem for ordinary flow, we can prove the max-flow min-cut theorem of (δ,η)-balanced flow.

Theorem 2. *There exists a* (δ,η)*-balanced flow* $D = (G_D, c_D)$ *with the value* μ *from* s *to* t *in a network* N *if and only if the* (δ,η)*-capacity of any* s-t *mixed cut in* N *is not less than* μ.

5 How to Obtain the Maximum (δ,η)-Balanced Flow

We develop a method for evaluating the maximum value of (δ,η)-balanced flow between a specified pair of vertices. The method consists of calculations of the maximum value of ordinary flows between the pair of vertices.

We recall the set $\mathscr{I}$ of all pairs of (i,j) such that $i\delta + j\eta < 1$, $i \ge 0$, and $j \ge 0$. For $\mathscr{I}$, let $\mathscr{J} = \{\delta i + j\eta : (i,j) \in \mathscr{I}\}$. We denote $\mathscr{J}$ as follows.

$$\mathscr{J} = \{\omega_0(=0), \omega_1, \ldots, \omega_{r-1}\},$$

such that $\omega_{k-1} < \omega_k$ where $r = |\mathscr{J}|$. We define a function $ord_{\delta,\eta}\colon \mathscr{I} \to \{0,1,\ldots,r-1\}$ and $\Psi(k)$ as follows.

$$ord_{\delta,\eta}(i,j) = k \longleftrightarrow \delta i + \eta j = \omega_k, \qquad \Psi(k) = \frac{1}{1-\omega_k}.$$

Note that

$$\Psi(ord_{\delta,\eta}(i,j)) = \frac{1}{1 - i\delta - j\eta} = \psi(i,j).$$

There may be exist (i_1,j_1) and (i_2,j_2) $((i_1,j_1) \ne (i_2,j_2))$ such that $ord_{\delta,\eta}(i_1,j_1) = ord_{\delta,\eta}(i_2,j_2)$. For $k = 0, 1, \ldots, r-1$ we also define $(\alpha(k),\beta(k))$ as an arbitrary $(i,j) \in \mathscr{I}$ such that $ord_{\delta,\eta}(i,j) = k$. Note that

$$\alpha(k)\delta + \beta(k)\eta = \omega_k, \qquad \psi(\alpha(k),\beta(k)) = \Psi(k).$$

For any $k \ge 1$, and a given network $N = (G,c)$, we denote $N_k = (G,c_k)$ as the network in which capacity function c_k is defined for each edge e and for each vertex v in G except s and t as follows.

$$\begin{aligned} c_k(e) &= \min\{c(e), \delta\xi_k\} && \text{for each edge } e \text{ in } G \\ c_k(v) &= \min\{c(v), \eta\xi_k\} && \text{for each vertex } v \text{ in } G \text{ except } s \text{ and } t, \end{aligned}$$

where ξ_k is defined as equation (13) in the following algorithm #2.

Algorithm #2: Evaluation of the maximum value of (δ,η)-balanced flow.
Let s and t be the distinct vertices in a given network N. We will obtain the maximum value of (δ,η)-balanced flow between s and t.

1. Let $N_0 = N$. Evaluate the maximum value μ_0 of the ordinary flows from s to t in N. Set $\xi_1 = \mu_0$, and $k = 1$.
2. Repeat the following procedure B until the procedure stops.
   Procedure B: Calculate the maximum value of the ordinary flows from s to t in N_k. Then set μ_k be the value.

   (a) If $\mu_k = \xi_k$, output ξ_k as the maximum value of (δ, η)-balanced flows between s and t and stop. (end of algorithm #2)
   (b) Otherwise, set

$$\xi_{k+1} = \frac{\mu_k - \omega_k \xi_k}{1 - \omega_k} = \Psi(k)(\mu_k - \omega_k \xi_k)$$
$$= \psi(\alpha(k), \beta(k)) \cdot [\mu_k - \{\alpha(k)\delta + \beta(k)\eta\}\xi_k], \quad (13)$$

   and $k = k + 1$. (end of procedure)

(end of algorithm)

Theorem 3. *Algorithm #2 gives the maximum value of (δ, η)-balanced flow between s and t in N with at most $|V| \cdot |E|$ evaluations of maximum ordinary flow in a network, where V is the vertex set of N and E is the edge set of N, respectively. Each evaluation of the maximum flow is applied in N with altered capacities of edges and vertices.*

6 Example

By using algorithm #2 we evaluate the maximum $(0.4, 0.5)$-balanced flow from s to t in N of figure 2.

1. Let $N_0 = N$. The maximum flow from s to t in N_0 is evaluated as 13. mixed-cut $\left(V_1^{(0)}, V_2^{(0)}\right)$ is a minimum s-t mixed-cut in N_0, where $V_1^{(0)} = \{s\}$ and $V_2^{(0)} = \{v_a, v_b, v_c, v_d, v_e, t\}$. Set $\xi_1 = \mu_0 = 13$, and $k = 1$. Moreover, we have

$$\mathscr{I} = \{(0,0), (1,0), (2,0), (0,1), (1,1),\}, \quad \mathscr{J} = \{0, 0.4, 0.5, 0.8, 0.9\}.$$

2. Procedure B:

   First iteration. Since $\xi_1 = 13$, N_1 is obtained as figure 3. The maximum flow from s to t in N_1 is evaluated as 11.5, and $\mu_1 = 11.5$. mixed-cut $\left(V_1^{(1)}, V_2^{(1)}\right)$ is a minimum s-t mixed-cut in N_1, where $V_1^{(1)} = \{s, v_b\}$ and $V_2^{(1)} = \{v_a, v_d, v_e, t\}$. We have $\mu_1 \neq \xi_1$. Since $\omega_1 = 0.4$, we set

$$\xi_2 = \frac{\mu_1 - \omega_1 \xi_1}{1 - \omega_1} = \frac{11.5 - 0.4 \times 13}{1 - 0.4} = \frac{11.5 - 5.2}{0.6} = 10.5,$$

   and $k = 2$.

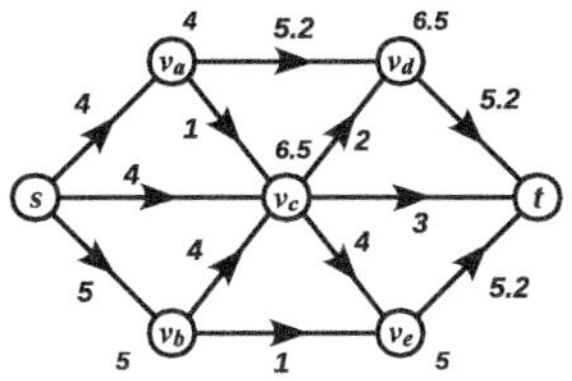

Fig. 3. Network N_1

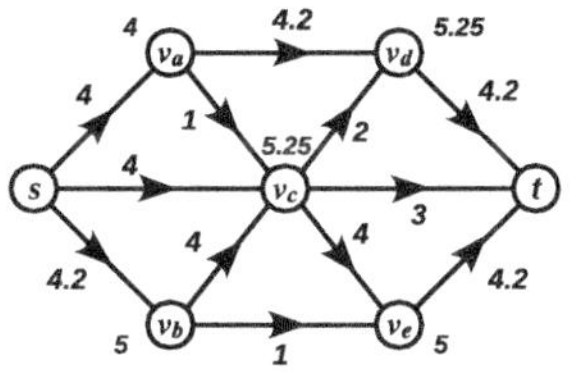

Fig. 4. Network N_2

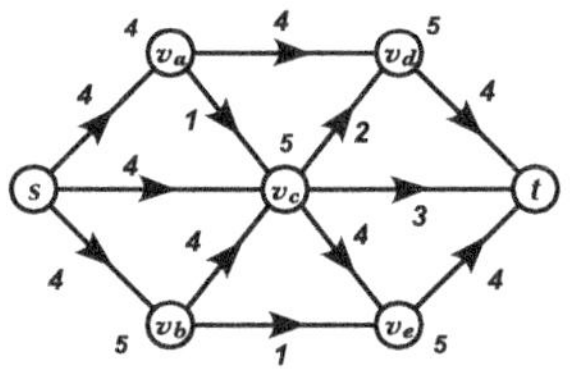

Fig. 5. Network N_3

Second iteration
Since $\xi_2 = 10.5$, N_2 is obtained as figure 4. The maximum flow from s to t in N_2 is evaluated as 10.25, and $\mu_2 = 10.25$. mixed-cut $\left(V_1^{(2)}, V_2^{(2)}\right)$ is a minimum s-t mixed-cut in N_2, where $V_1^{(2)} = \{s, v_b\}$ and $V_2^{(2)} = \{v_a, v_d, v_e, t\}$. We have $\mu_2 \neq \xi_2$. Since $\omega_2 = 0.5$, we set

$$\xi_3 = \frac{\mu_2 - \omega_2\xi_2}{1-\omega_2} = \frac{10.25 - 0.5 \times 10.5}{1-0.5} = \frac{10.25 - 5.25}{0.5} = 10,$$

and $k = 3$.

Third iteration. Since $\xi_3 = 10$, N_3 is obtained as figure 5. The maximum flow from s to t in N_3 is evaluated 10, and $\mu_3 = 10$. mixed-cut $\left(V_1^{(3)}, V_2^{(3)}\right)$ is a minimum s-t mixed-cut in N_3, where $V_1^{(3)} = \{s, v_b\}$ and $V_2^{(3)} = \{v_a, v_d, v_e, t\}$. Since $\mu_3 = \xi_3$, we have $\mu_3 = \xi_3 = 10$ as the maximum value of $(0.4, 0.5)$-balanced flow between s and t.

References

1. Aggarwal, C.C., Orlin, J.B.: On multiroute maximum flows in networks. Networks 39, 43–52 (2002)
2. Ahuja, R.K., Magnanti, T.L., Orlin, J.B.: Network Flows: Theory, Algorithms, and Applications. Prentice Hall, Englewood Cliffs (1993)
3. Alexopoulos, C., Fishman, G.S.: Characterizing Stochastic Flow Networks Using the Monte Carlo Method. Networks 21, 775–798 (1991)
4. Alexopoulos, C., Fishman, G.S.: Sensitivity Analysis in Stochastic Flow Networks Using the Monte Carlo Method. Networks 23, 605–621 (1993)
5. Alexopoulos, C.: A Note on State-Space Decomposition Methods for Analyzing Stochastic Flow Networks. IEEE Trans. Reliability 44, 354–357 (1995)
6. Bagchi, A., Chaudhary, A., Kolman, P., Sgall, J.: A simple combinatorial proof for the duality of multiroute flows and cuts, Technical Report 2004-662, Charles University, Prague (2004)
7. Bruhn, H., Černý, J., Hall, A., Kolman, P., Sgall, J.: Single Source Multiroute Flows and Cuts on Uniform Capacity Networks. Theory of Computing 4(1), 1–20 (2008), `http://www.theoryofcomputing.org/articles/main/v004/a001/`
8. Chekuri, C., Khanna, S.: Algorithms for 2-Route Cut Problems. In: Aceto, L., Damgård, I., Goldberg, L.A., Halldórsson, M.M., Ingólfsdóttir, A., Walukiewicz, I. (eds.) ICALP 2008, Part I. LNCS, vol. 5125, pp. 472–484. Springer, Heidelberg (2008)
9. Du, D.: Multiroute Flow Problem, Ph.D. thesis, The University of Texas at Dallas (2003)
10. Du, D., Chandrasekaran, R.: The multiroute maximum flow problem revisited. Networks 47, 81–92 (2006)
11. Egawa, Y., Kaneko, A., Matsumoto, M.: A mixed version of Menger's theorem. Combinatorica 11, 71–74 (1991)
12. Evans, J.R.: Maximum Flow in Probabilistic Graphs – The Discrete Case. Networks 6, 161–183 (1976)
13. Ford, L.R., Fulkerson, D.R.: Maximum flow through a network. Canad. J. Math. 8, 399–404 (1956)
14. Ford Jr., L.R., Fulkerson, D.R.: Flows in Networks. Princeton University Press, Princeton (1962)
15. Kar, K., Kodialam, M., Lakshma, T.V.: Routing restorable bandwidth guaranteed connections using maximum 2-route flows. IEEE/ACM Transactions on Networking 11, 772–781 (2003)
16. Kishimoto, W., Takeuchi, M., Kishi, G.: Two-Route Flows in an Undirected Flow Network. IEICE Trans. J75–A, 1699–1717 (1992) (in Japanese)
17. Kishimoto, W., Takeuchi, M.: On M-route flows in a network. In: Proceedings of Singapore ICCS/ISITA, vol. 3, pp. 1386–1390 (1992)
18. Kishimoto, W., Takeuchi, M.: On m-Route Flows in a Network. IEICE Trans. J76–A, 1185–1200 (1993) (in japanese)
19. Kishimoto, W.: Reliable Flow with Failures in a Network. IEEE Trans. Reliability 46, 308–315 (1997)
20. Kulkarni, V.G., Adlakha, V.G.: Maximum Flow in Planar Networks with Exponentially Distributed Arc Capacities. Communications in Statistics-Stochastic Models 1, 263–289 (1985)
21. Wu, T.H.: Fiber Network Service Survivability. Artech House, Boston (1992)
22. Zolfaghari, A., Kaudel, F.J.: Framework for Network Survivability Performance. IEEE J. Select. Areas in Commun. 12, 46–51 (1994)

Quickest Cluster Flow Problems on Tree Networks

Kathrin Leiner and Stefan Ruzika

University of Kaiserslautern, Germany
{leiner,ruzika}@mathematik.uni-kl.de

Abstract. In this publication we examine a dynamic network flow problem, called the *quickest cluster flow problem*. This problem is motivated by evacuation planning and yields improved lower bounds on evacuation times. Our approach models people moving in groups rather than individually which can be observed even in the situation of an emergency. To the best of our knowledge, this fact has received little attention in dynamic network flow literature. Interrelations of this new model to existing network flow models like multicommodity flows are pointed out. The quickest cluster flow problem is proven to be NP-hard even on tree networks. We restrict the cluster sizes to pairwise divisible values and obtain an exact greedy-based algorithm.

1 Introduction

Dynamic network flow models can be used for macroscopic modeling of evacuation scenarios. In particular *quickest flow problems*, which can be interpreted as asking for the minimal time horizon needed to evacuate a specified number of people, are very useful in security research. The resulting time computed with a quickest flow algorithm is a lower bound on the time needed to evacuate this number of people in reality, provided the network models the area appropriately [5].

[7] stated that people arriving in groups, e.g. families or cliques of friends, do not split up even in case of an emergency. This gives reason to the approach of improving the lower bounds provided by dynamic network flow algorithms. To this end we model different sizes of (unsplittable) flow units, so-called *clusters*, where the size of the cluster resembles the size of the respective group. This model is called the *cluster flow model* and has been introduced by [4].

Our research in this area is motivated by the evacuation of the area around a soccer stadium located in a densely populated neighborhood. In case of emergency it is not sufficient to only evacuate the stadium itself. In order to avoid blocking of the streets around the stadium, evacuees have to depart to areas of safety at sufficient distance from the stadium. We do not assign specific targets to the evacuees since we model the fact that each person leaves the endangered area as fast as possible in an arbitrary direction.

In an evacuation scenario, people tend to avoid taking detours and choose the shortest path leading to their destination, even if this path is crowded and takes a longer time to traverse [6]. This means there will essentially be a single path to each destination used in an evacuation progress. This observation motivates the examination of *quickest cluster flow problems* on tree networks. More precisely, we consider directed out-trees with a single source (the root node representing the stadium) and multiple sinks (the leaves of the tree representing the areas of safety).

J. Pahl, T. Reiners, and S. Voß (Eds.): INOC 2011, LNCS 6701, pp. 243–255, 2011.

In [4] a polynomial time 2-approximation is derived for the quickest cluster flow problem on general networks with one cluster size and single flow units (clusters of size 1). This approach can easily be generalized to a D-approximation for D different sizes of cluster flow units.

This paper is organized as follows: In Section 2 we state some basic definitions for (dynamic) network flows as well as the notation that will be used in the remainder of the paper. In Section 3 we introduce cluster flow problems, in particular quickest cluster flow problems. We prove NP-hardness of the quickest cluster flow problem even on tree networks. Afterwards we relate the cluster flow problem to other network flow problems. In Section 4 we state an exact pseudopolynomial greedy-based algorithm for the special case of the quickest cluster flow problem on trees with pairwise divisible cluster sizes. This algorithm exploits the tree structure of the network, especially the fact, that the path from the source node to any sink node is unique.

2 Preliminaries and Notation

A *(discrete-time) dynamic network* $N = (G = (V,A),\tau,u)$ is a directed graph G consisting of a node set V with $|V| = n$ and an arc set A with $|A| = m$ as well as a capacity function $u : A \to \mathbb{N}$ and a travel time function $\tau : A \to \mathbb{N}$. The travel time $\tau(a)$ of an arc $a = (v,w) \in A$ is the number of time steps that flow units need to traverse arc a, i.e., a unit of flow starting at node v at time θ will arrive at node w at time $\theta + \tau(a)$. One specific node $s \in V$ is called the *source node* of N. In all problems examined in this paper, there will be a single source node. This node will distribute flow into the network. Analogously, a set of nodes $\mathcal{T} \subset V$ with $|\mathcal{T}| \geq 1$ is called the set of *sink nodes* of N. Only nodes $t \in \mathcal{T}$ may receive flow from the network without sending it onward. For simplicity we assume that the arc set A does not contain any outgoing arcs from sink nodes. The nodes $v \in V \setminus \{\mathcal{T} \cup \{s\}\}$ are called *transshipment nodes*.

In the following we denote by $\delta^+(v)$ the set of all arcs emanating from node $v \in V$. Analogously, $\delta^-(v)$ denotes the set of all arcs entering it. Additionally, for notational ease we define $f(a,\theta) := 0$ for negative values $\theta \in \mathbb{Z}$.

A *(discrete-time) dynamic network flow problem* asks for a *feasible dynamic flow* with certain properties. A dynamic flow $f : A \times \{0,\dots,T\} \to \mathbb{N}$ with *time horizon* $T \in \mathbb{N}$ is given by *flow values* $f(a,\theta)$ that represent the number of flow units travelling from node v to w on arc $a = (v,w)$ starting at time $\theta \in \{0,\dots,T\}$. The dynamic flow f is called *feasible*, if it fulfills the capacity constraints

$$f(a,\theta) \leq u(a) \tag{1a}$$

for each arc $a \in A$ and $\theta \in \{0,\dots,T\}$, the flow balance constraints for transshipment nodes

$$\sum_{a\in\delta^+(v)} f(a,\theta) - \sum_{a\in\delta^-(v)} f(a,\theta-\tau(a)) = 0 \tag{1b}$$

for all $v \in V \setminus \{\mathcal{T} \cup \{s\}\}$ and all time steps $\theta \in \{0,\dots,T\}$, and the flow balance constraints for source and sink nodes

$$\sum_{\theta=0}^{T}\left(\sum_{a\in\delta^+(v)} f(a,\theta) - \sum_{a\in\delta^-(v)} f(a,\theta-\tau(a))\right) = b(v) \tag{1c}$$

where $b(v) \in \mathbb{Z}$ is the *demand* or *supply* of any node $v \in \{\mathcal{T} \cup \{s\}\}$. The source node $s \in V$ has a positive supply value $b(s) > 0$. Analogously, for any $t \in \mathcal{T}$ it holds that $b(t) < 0$, i.e., these nodes request flow units from the network.

In the *quickest flow problem (QFP)*, we are given a discrete-time dynamic network with a source node $s \in V$ and a single sink node $t \in V$ as well as a supply value $b \in \mathbb{N}$ and ask for the minimal time horizon T^* needed to send b units of flow from s to t and the corresponding flow values at each time step $\theta \in \{0,\ldots,T^*\}$. The quickest flow problem can be formulated as follows:

$$T^* = \min \quad T$$

$$\text{s.t.} \quad f(a,\theta) \leq u(a) \quad \forall a \in A, \theta \in \{0,\ldots,T\} \tag{2a}$$

$$\sum_{a\in\delta^+(v)} f(a,\theta) - \sum_{a\in\delta^-(v)} f(a,\theta-\tau(a)) = 0 \quad \forall v \in V \setminus \{\mathcal{T} \cup \{s\}\}, \quad \theta \in \{0,\ldots,T\} \tag{2b}$$

$$\sum_{\theta=0}^{T}\left(\sum_{a\in\delta^+(v)} f(a,\theta) - \sum_{a\in\delta^-(v)} f(a,\theta-\tau(a))\right) = \begin{cases} b & \text{if } v = s \\ -b & \text{if } v = t \end{cases} \tag{2c}$$

$$f(a,\theta) \in \mathbb{N} \quad \forall a \in A, \theta \in \{0,\ldots,T\} \tag{2d}$$

[2] showed that this problem can be solved in strongly polynomial time.

In Section 3 we point out the relationship of cluster flow problems to other network flow problems. Therefore we need to define cluster flow problems also for *static* (non-dynamic) networks. A static network is defined as a network N with cost function $k : A \to \mathbb{Z}$ instead of a travel time function and capacity function $u : A \to \mathbb{N}$ as in the dynamic case. Analogously to dynamic flows, a *feasible static flow* $f : A \to \mathbb{N}$ has to fulfill the (static) capacity constraints

$$f(a) \leq u(a) \tag{3a}$$

for all $a \in A$ and the (static) flow balance constraints

$$\sum_{a\in\delta^+(v)} f(a) - \sum_{a\in\delta^-(v)} f(a) = b(v) \tag{3b}$$

for all $v \in V$. Based on the feasibility constraints (3a) and (3b), the *(static) minimum cost flow problem* is to find a feasible static flow f with minimum cost

$$\sum_{a\in A} k(a)f(a). \tag{3c}$$

We examine network flow problems on special networks called *directed out-tree networks* (or shorter: *tree networks*). A tree network is a network with a single source $s \in V$ and multiple sinks $\mathcal{T} = \{t_1,\ldots,t_{|\mathcal{T}|}\} \subset V$. In this case the underlying graph of the network is a *directed out-tree* rooted at s, i.e., a cycle-free, connected graph with root node s and leaves $t_i \in \mathcal{T}$, where each arc $a = (v,w) \in A$ is oriented such that any s-t_i path for $t_i \in \mathcal{T}$ is directed (cf. [1]).

3 Quickest Cluster Flow Problems

Cluster flow problems are dynamic network flow problems which aim at routing not only flow units of size 1 but of possibly many other sizes that do not split up on their way from source to a sink. In this context each unsplittable unit is called a *cluster*. In an evacuation scenario, a cluster may represent a group of people, for instance a family or a clique of friends that does not split up even in an emergency situation [7]. We focus on dynamic cluster flow problems with single-source and multiple sinks because of their relevance in the context of evacuation problems.

A *(dynamic) cluster flow* consists of flow values $f_d(a,\theta)$ for each cluster size $d \in \mathcal{D} \subset \mathbb{N}$, each arc $a \in A$, and time step $\theta \in \{0,\dots,T\}$. In analogy to general dynamic flows, a cluster flow is called *feasible*, if it fulfills the capacity constraints

$$\sum_{d \in \mathcal{D}} d f_d(a,\theta) \le u(a) \tag{4a}$$

for all $a \in A$ and $\theta \in \{0,\dots,T\}$ as well as the flow balance constraints

$$\sum_{a \in \delta^+(v)} f_d(a,\theta) - \sum_{a \in \delta^-(v)} f_d(a,\theta-\tau(a)) = 0 \tag{4b}$$

for all $v \in V \setminus \{\mathcal{T} \cup \{s\}\}$, $\theta \in \{0,\dots,T\}$, and $d \in \mathcal{D}$ and

$$\sum_{\theta=0}^{T} \left(\sum_{a \in \delta^+(s)} f_d(a,\theta) - \sum_{a \in \delta^-(s)} f_d(a,\theta-\tau(a)) \right) = b_d, \tag{4c}$$

$$\sum_{t \in \mathcal{T}} \sum_{\theta=0}^{T} \left(\sum_{a \in \delta^+(t)} f_d(a,\theta) - \sum_{a \in \delta^-(t)} f_d(a,\theta-\tau(a)) \right) = -b_d \tag{4d}$$

for all $d \in \mathcal{D}$, where b_d is the number of clusters of size d that shall be routed, $s \in V$ is the source node and $\mathcal{T} \subseteq V$ is the set of sink nodes. Constraint (4d) implies that it is not specified, to which one of the sources $t_i \in \mathcal{T}$ a cluster is sent. Constraint (4a) models the fact that the different clusters have to share the capacity on each arc, where a cluster of size d needs d times as much capacity as a single flow unit.

A *(single-source) quickest cluster flow instance* consists of a directed network $N = (G = (V,A),\tau,u)$ with one source $s \in V$ and several sinks $\mathcal{T} = \{t_1,\dots,t_{|\mathcal{T}|}\}$, a set of cluster sizes $\mathcal{D} = \{d_1,\dots,d_{|\mathcal{D}|}\}$, and demand values b_d for all $d \in \mathcal{D}$. Thereby each value b_d gives the number of clusters of size d that should be sent. The *quickest cluster flow problem* asks for a feasible dynamic cluster flow with minimal time horizon T^*. A feasible cluster flow sends each of the b_d clusters of size $d \in \mathcal{D}$ on a single path from the source s to a sink $t_i \in \mathcal{T}$. Recall that it is not specified to which sink node $t_i \in \mathcal{T}$ a cluster is routed. This problem can be formulated as follows:

$$
\begin{aligned}
T^* = \quad & \min \quad T \\
\text{s.t.} \quad & \sum_{d \in \mathcal{D}} d f_d(a,\theta) \le u(a) \quad \forall a \in A, \theta \in \{0,\dots,T\} && \text{(5a)} \\
& \sum_{a \in \delta^+(v)} f_d(a,\theta) - \sum_{a \in \delta^-(v)} f_d(a,\theta-\tau(a)) = 0 \quad \forall v \in V \setminus \{\mathcal{T} \cup \{s\}\}, && \text{(5b)} \\
& \qquad \theta \in \{0,\dots,T\}, d \in \mathcal{D} \\
& \sum_{\theta=0}^{T} \left(\sum_{a \in \delta^+(s)} f_d(a,\theta) - \sum_{a \in \delta^-(s)} f_d(a,\theta-\tau(a)) \right) = b_d \quad \forall d \in \mathcal{D} && \text{(5c)} \\
& \sum_{t \in \mathcal{T}} \sum_{\theta=0}^{T} \left(\sum_{a \in \delta^+(t)} f_d(a,\theta) - \sum_{a \in \delta^-(t)} f_d(a,\theta-\tau(a)) \right) = -b_d \quad \forall d \in \mathcal{D} && \text{(5d)} \\
& f_d(a,\theta) \in \mathbb{N} \quad \forall a \in A,\ \theta \in \{0,\dots,T\},\ d \in \mathcal{D} && \text{(5e)}
\end{aligned}
$$

Notice that this is not a *linear* integer program as the parameter T to be minimized in the objective function appears in the summation limits of constraints (5c) and (5d) and determines the number of constraints of type (5a), (5b), and (5e).

Next, we establish a close relationship between cluster flow problems and multicommodity flow problems. In the *minimum cost multicommodity flow problem* we are given a static network $N = (G = (V,A),k,u)$ and a set of source-sink pairs (*commodities*) $C = \{(s_i,t_i) | i \in \{1,\dots,p\}\}$, for some $p \in \mathbb{N}$, with $s_i, t_i \in V$, each with a specified value b_i of flow units (all of the same size 1) that have to be sent from s_i to t_i. The total amount of flow of all commodities on each arc must not exceed the capacity value on this arc. The minimum cost multicommodity flow problem asks for a multicommodity flow with minimal cost. Let $f_i(a)$ denote the flow value of commodity (s_i,t_i) on arc $a \in A$. Then, the minimum cost integral multicommodity flow problem can be formulated as follows:

$$
\begin{aligned}
\min \quad & \sum_{a \in A} k(a) \sum_{(s_i,t_i) \in C} f_i(a) \\
\text{s.t.} \quad & \sum_{(s_i,t_i) \in C} f_i(a) \le u(a) \quad \forall a \in A && \text{(6a)} \\
& \sum_{a \in \delta^+(v)} f_i(a) - \sum_{a \in \delta^-(v)} f_i(a) = \begin{cases} b_i & \text{if } v = s_i \\ -b_i & \text{if } v = t_i \\ 0 & \text{else} \end{cases} \quad \forall v \in V,\ (s_i,t_i) \in C && \text{(6b)} \\
& f_i(a) \in \mathbb{N} \quad \forall a \in A,\ (s_i,t_i) \in C && \text{(6c)}
\end{aligned}
$$

To relate cluster flows and multicommodity flows, we need to define *static minimum cost cluster flow problems.* A static cluster flow consists of $|\mathcal{D}|$ functions $f_d : A \to \mathbb{N}$ giving the number of clusters of size $d \in \mathcal{D}$ on each arc $a \in A$. The cluster flow is said to be *feasible*, if it fulfills the capacity constraints

$$
\sum_{d \in \mathcal{D}} d f_d(a) \le u(a) \qquad \text{(7a)}
$$

for all $a \in A$ and the flow balance constraints

$$\sum_{a \in \delta^+(v)} f_d(a) - \sum_{a \in \delta^-(v)} f_d(a) = b_d(v) \tag{7b}$$

for all $v \in V$ and $d \in \mathcal{D}$, where $b_d(v) \in \mathbb{Z}$ gives the number of clusters of size $d \in \mathcal{D}$ that the node $v \in V$ receives from or distributes into the network. Note that in this case, it is exactly specified to which sink the clusters have to be routed. By introducing a supersink (cf. [5]), however, we can still model the case with unspecified sink.

Then the minimum cost cluster flow problem can be stated as follows: Given a static network $N = (G = (V, A), k, u)$, find a feasible static cluster flow satisfying the demands at minimal cost

$$\sum_{a \in A} \sum_{d \in \mathcal{D}} k(a)\, d f_d(a). \tag{7c}$$

Note that this problem is already mentioned in [1] as a generalization of multicommodity flow problems. The following proposition substantiates this:

Proposition 1. *The integral single-source minimum cost multicommodity flow problem is a relaxation of the minimum cost cluster flow problem.*

Proof. Consider the static minimum cost cluster flow problem. As we have seen above this problem can be formulated as

$$\min \quad \sum_{a \in A} \sum_{d \in \mathcal{D}} k(a)\, d f_d(a) \tag{8a}$$

$$\text{s.t.} \quad \sum_{a \in \delta^+(v)} f_d(a) - \sum_{a \in \delta^-(v)} f_d(a) = b_d(v) \quad \forall v \in V \tag{8b}$$

$$\sum_{d \in \mathcal{D}} d f_d(a) \le u(a) \quad \forall a \in A \tag{8c}$$

$$f_d(a) \ge 0, \quad \forall a \in A,\ d \in \mathcal{D} \tag{8d}$$

$$f_d \in \mathbb{N} \quad \forall a \in A,\ d \in \mathcal{D}, \tag{8e}$$

where $b_d(v) \in \mathbb{Z}$ is the demand or supply of each node $v \in V$ and each cluster size $d \in \mathcal{D}$.

Set $h_d(a) := d f_d(a)$ for each $a \in A$ and $d \in \mathcal{D}$. Additionally, set $b'_d(v) := d b_d(v)$ for all $v \in V$ and $d \in \mathcal{D}$. Since $f_d(a)$ is the number of clusters of size d on arc $a \in A$, the new variables $h_d(a)$ can be interpreted as the number of single flow units bound in clusters of size d on arc a. This way, constraints (8b) alter to

$$\sum_{a \in \delta^+(v)} \frac{h_d(a)}{d} - \sum_{a \in \delta^-(v)} \frac{h_d(a)}{d} = \frac{b'_d(v)}{d} \tag{8b'}$$

for all $v \in V$ and $d \in \mathcal{D}$. Multiplying each of these constraints by d yields the following equivalent formulation:

$$\min \quad \sum_{a\in A}\sum_{d\in\mathcal{D}} k(a)\, h_d(a) \tag{9a}$$

$$\text{s.t.} \quad \sum_{a\in\delta^+(v)} h_d(a) - \sum_{a\in\delta^-(v)} h_d(a) = b'_d(v) \quad \forall v \in V \tag{9b}$$

$$\sum_{d\in\mathcal{D}} h_d(a) \le u(a) \quad \forall a \in A \tag{9c}$$

$$h_d(a) \ge 0, \quad \forall a \in A,\ d \in \mathcal{D} \tag{9d}$$

$$h_d(a) \in d\mathbb{N} \quad \forall a \in A,\ d \in \mathcal{D} \tag{9e}$$

Relaxing constraints (9e) to

$$h_d(a) \in \mathbb{N} \tag{9e'}$$

for all $a \in A$ and $d \in \mathcal{D}$ leads to a minimum cost integral multicommodity flow formulation (cf. formulation (6)). □

This combinatorial relaxation can be interpreted as dropping the cluster affiliations of the flow units.

Corollary 1. *The integral single-source quickest multicommodity flow problem is a relaxation of the quickest cluster flow problem.*

Proof. A quickest cluster flow problem, in analogy to the general quickest flow problem, can be solved as a minimum cost cluster flow problem in a static network, precisely the time-expanded network with turnstile-costs (cf. [5]). Then the result follows directly from Theorem 1. □

Remark 1. In a single-sink network, the relaxation of a minimum cost cluster flow problem described in the proof of Proposition 1 corresponds to a classical minimum cost flow problem. Because of Corollary 1 this implies that this relaxation of a single-sink quickest cluster flow problem is equivalent to a quickest flow problem.

Theorem 1. *The decision version of the quickest cluster flow problem is NP-complete even on tree networks.*

Proof. Reduction from *bin packing*, which is an NP-complete problem [3].

Bin Packing: **Given:** Set C of L items, sizes $s(c) \in \mathbb{N}$ for all $c \in C$, bin size $B \in \mathbb{N}$, integral number $Q \in \mathbb{N}$.

Question: Is there a partition $C_1 \cup C_2 \cup \cdots \cup C_Q$ of the set C such that $\sum_{c\in C_i} s(c) \le B$ for all $i \in \{1,\dots,Q\}$?

We construct a quickest cluster flow instance in a tree network from a bin packing instance: Graph G has node set $V = \{s, t_1, t_2, \dots, t_Q\}$, where s is the source node, t_i are sink nodes for all $i \in \{1,\dots,Q\}$, and arc set $A = \{a = (s,t) | t \in \{t_1,\dots,t_Q\}\}$ with $u(a) = B$ and $\tau(a) = 1$ for all $a \in A$. Let $\mathcal{D} = \{d_1,\dots,d_{|\mathcal{D}|}\}$ be the set of different item sizes of the set C. For $i \in \{1,\dots,|\mathcal{D}|\}$ let $b_i = |\{c \in C | s(c) = d_i\}|$.

Claim. There is a feasible quickest cluster flow solution with time horizon $T \leq 1$ if and only if the bin packing instance is a YES-instance.

Proof of the Claim. Assume there is a quickest cluster flow solution in the instance constructed from a bin packing instance as described above. Let $C_i := \{c \in C|$ the QCFP-solution routes c to sink $t_i\} \forall i \in \{1,\ldots,Q\}$. Since all clusters are routed exactly once in the cluster flow solution and $T \leq 1$, this is a partition of the set C into Q subsets. Since the capacity $u(a)$ is equal to B and the cluster sizes correspond to the item sizes, it holds that

$$\sum_{c \in C_i} s(c) \leq u(a) = B \tag{10}$$

for all $i \in \{1,\ldots,Q\}$. Because of the partition $C = \{C_1,\ldots,C_Q\}$ we can conclude that the bin packing instance is a YES-instance.

On the other hand, let the partition $C = \{C_1,\ldots,C_Q\}$ with the property

$$\sum_{c \in C_i} s(c) \leq B \tag{11}$$

for all $i \in \{1,\ldots,Q\}$ be a valid partition for the bin packing instance. For $i \in \{1,\ldots,Q\}$ send a cluster of size $s(c)$ to sink t_i for any $c \in C_i$. Since $B = u(a)$ for all $a \in A$ this is a feasible solution for the cluster flow instance. Since we have Q subsets and also Q sinks and the travel time on each arc $a \in A$ is 1, the time horizon T of the constructed solution is 1. □

4 Quickest Cluster Problems with Divisible Cluster Sizes

Since for the general quickest cluster flow problem neither an exact combinatorial algorithm nor an integer programming formulation has been developed so far, we examine the special case of pairwise divisible cluster sizes.

A set $\mathcal{D}$ of different cluster sizes is said to be *pairwise divisible* if and only if it holds that

$$d_i | d_{i+1} \tag{12}$$

for all $i \in \{1,\ldots,|\mathcal{D}|-1\}$, i.e., each cluster size is a divisor of any larger size occurring in the set of clusters.

We show that for the quickest cluster flow problem on tree networks with pairwise divisible cluster sizes a greedy-approach works, i.e., sending the largest clusters as fast as possible, then the second largest with respect to the remaining capacities and so on.

This can be realized by adding a super-sink and then using a quickest flow algorithm for instances with time-dependent capacities, as it is stated in [8], for any $d \in \mathcal{D}$ after scaling the capacities appropriately in each step. This algorithm has a worst-case complexity of $O(nmb_{max}T^2)$, where T is an upper bound on the minimal time horizon and b_{max} is the maximum number of clusters of any size. It has to be executed $\log d_{max}$ times in order to achieve an optimal quickest cluster flow on a tree network with pairwise divisible cluster sizes, resulting in an overall complexity of $O(\log(d_{max})nmb_{max}T^2)$.

In the following we derive a faster algorithm for quickest cluster flow problems on tree networks.

4.1 Assumption

We assume that $b_d > 0$ for all $d \in \mathcal{D}$, i.e., for every cluster size in the list there is actually a cluster to be sent. If this is not the case for any $d \in \mathcal{D}$, the cluster size d can be removed from the set $\mathcal{D}$.

Since in a tree network, every path P between two nodes $v, w \in V$ is unique, the travel time between these nodes is constant. We denote this travel time value, i.e., the sum of the transit times of all arcs contained in the path by $\tau(v,w)$ or $\tau(P)$.

The algorithm starts with a 0-flow and iteratively increments cluster flow of the largest remaining cluster on the currently quickest s-t_i-path P_i. To achieve this, the sinks are ordered with respect to their earliest arrival time τ_i^d currently available for a cluster of size d. In order to obtain these earliest arrival times, the residual capacities u_i^θ of any s-t_i-path P_i starting at time step $\theta \in \mathbb{N}$ are stored in a table. This table is expanded dynamically whenever a new layer is needed. This may occur only once in each iteration. Thus, the size of the table is bounded by $|\mathcal{T}|$ and the number of iterations. If the quickest s-t_i-path P_i is found, the number of clusters to be sent on this path is set to the minimum of the number of remaining clusters of the current size and the number of clusters this path can carry. After increasing flow on some path the residual capacities of the path itself as well as the capacities of all other affected paths are updated. Since in a tree network every s-v-path is unique for $v \in V$, only paths with the same starting time as the one with modified flow are affected. Algorithm 1 formalizes this description.

In the algorithm, setting $f_d(t,\theta)$ to some value $x \in \mathbb{N}$ is understood as follows: If the path from s to t consists of arcs $a_1, a_2, \ldots, a_l \in A$, the flow values on these arcs are set to x in the following way:

$$f_d\left(a_i, \theta + \sum_{j=1}^{i-1} \tau(a_j)\right) = x \tag{13}$$

for all $i \in \{1, \ldots, l\}$.

Theorem 2. *The output of Algorithm 1 is an optimal solution for the quickest cluster flow problem for any input instance on a tree network with pairwise divisible cluster sizes.*

Proof. Let f with time horizon T^* be the dynamic cluster flow output by Algorithm 1. Assume there is another feasible dynamic cluster flow g with time horizon $T < T^*$.

Let $\hat{C}$ be the overall set of clusters that are routed, i.e., $\hat{C}$ consists of b_d clusters of size d for any $d \in \mathcal{D}$. For $c \in \hat{C}$ let $d(c)$ denote the cluster size of c.

Let $C = \{c_1, \ldots, c_l\}$ be the set of clusters that are routed differently by the dynamic cluster flows f and g. Thereby we distinguish only different cluster sizes, not specific clusters of the same size.

Each cluster that is shipped by the flow f such that it arrives at its sink later than time T is contained in the set C, since the time horizon of the flow g is T.

Let (P_i^f, θ_i^f) be the path and the corresponding starting time such that a cluster $c_i \in C$ is routed on P_i^f at time θ_i^f in the cluster flow f. Analogously, let (P_i^g, θ_i^g) denote the path and starting time of cluster c_i in g. Note that by definition of C it holds that $(P_i^f, \theta_i^f) \neq (P_i^g, \theta_i^g)$ for all $c_i \in C$.

Input: Dynamic tree network N, cluster sizes $\mathcal{D} = \{d_1,\dots,d_K\}$, supplies $\{b_{d_1},\dots,b_{d_K}\}$
Output: $f_d(t_i,\theta)$ for $d \in \mathcal{D}$, $t_i \in \mathcal{T}$, $\theta \in \mathbb{N}$ giving a quickest cluster flow for the input instance, optimal time horizon T^*

$\mathcal{D}_{org} := \mathcal{D}$;
$f := 0$;
$T^* = 0$;
$u_i^\theta := \min_{a=(v,w)\in P_i} \{u_a\}$ for $\theta = 0,1,2,\dots; i = 1,\dots,|\mathcal{T}|$;
while $\mathcal{D} \neq \emptyset$ **do**
  $d_{max} := \max\{d \in \mathcal{D}\}$;
  while $b_{d_{max}} > 0$ **do**
    foreach $t_j \in \mathcal{T}$ **do**
      $\theta_j^{d_{max}} := \min_\theta \left\{u_j^\theta \geq d_{max}\right\}$;
      $\tau_j^{d_{max}} := \theta_j^{d_{max}} + \tau(s,t_j)$;
    end
    $\tau_i := \min\left\{\tau_j^{d_{max}} \mid t_j \in \mathcal{T}\right\}$;
    $f_{d_{max}}(t_i,\theta_i^{d_{max}}) := \min\left\{b_{d_{max}}, \left\lfloor \frac{u_i^{\theta_i^{d_{max}}}}{d_{max}} \right\rfloor\right\}$;
    if $T^* < \tau_i$ **then** $T^* := \tau_i$;
    $b_{d_{max}} := b_{d_{max}} - f_{d_{max}}\left(t_i,\theta_i^{d_{max}}\right)$;
    $u_i^{\theta_i^{d_{max}}} := u_i^{\theta_i^{d_{max}}} - d_{max} f_{d_{max}}\left(t_i,\theta_i^{d_{max}}\right)$;
    foreach $t_j \in \mathcal{T} \setminus \{t_i\}$ *with* $P_i \cap P_j \neq \emptyset$ **do**
      $u_j^{\theta_i^{d_{max}}} := \min\left\{u_j^{\theta_i^{d_{max}}}, \min_{a=(v,w)\in P_i\cap P_j}\left\{u(a) - \sum_{d\in\mathcal{D}_{org}} d f_d\left(a, \tau(s,v)+\theta_i^{d_{max}}\right)\right\}\right\}$;
    end
    if $b_{d_{max}} = 0$ **then** $\mathcal{D} := \mathcal{D} \setminus d_{max}$;
  end
end

Algorithm 1. Algorithm for quickest cluster flow problems on trees with pairwise divisible cluster sizes

Following this notation it holds that

$$\max_{c_i\in C}\left\{\theta_i^g + \tau(P_i^g)\right\} \leq T < \max_{c_i\in C}\left\{\theta_i^f + \tau(P_i^f)\right\} = T^*. \tag{14}$$

Consider a rearrangement of the elements of the set C such that $\theta_1^f + \tau(P_1^f) > T$ for $c_1 \in C$. Then $(P_1^f,\theta_1^f) \neq (P_1^g,\theta_1^g)$ and $\theta_1^g + \tau(P_1^g) \leq T$. Since $\theta_1^g + \tau(P_1^g) < \theta_1^f + \tau(P_1^f)$ and due to the fact that the algorithm did not choose the path (P_1^g,θ_1^g) (or an alternative path with equal or lower arrival time) for the cluster c_1, there is another cluster $c_2 \in C$ with $d(c_2) > d(c_1)$ routed on (P_2^f,θ_2^f) in the flow f that blocks some arc necessary to route c_1 on (P_1^g,θ_1^g). There must be at least one such cluster c_2 with $d(c_2) > d(c_1)$, since if $d(c_2) < d(c_1)$, the cluster c_2 would have been routed after routing c_1 by the algorithm, so c_1 would have been assigned this path in f, too. If $d(c_2) = d(c_1)$, however, the cluster c_2 would not be contained in the set C of clusters routed differently in f and g.

It follows that in the flow g, the cluster c_2 is routed on a path (P_2^g, θ_2^g) with $\theta_2^g + \tau(P_2^g) \leq T$. Since the algorithm has chosen (P_2^f, θ_2^f) instead of (P_2^g, θ_2^g), the path (P_2^f, θ_2^f) must have been one of the paths with minimal arrival time available for a cluster of size $d(c_2)$ in that iteration with only clusters of smaller or equal size remaining to be routed.

By the same argumentation as above it follows that if $\theta_2^g + \tau(P_2^g) < \theta_2^f + \tau(P_2^f)$ the capacity of the path (P_2^g, θ_2^g) was occupied by (at least) one cluster c_3 with $d(c_3) > d(c_2)$. If, however, $\theta_2^g + \tau(P_2^g) > \theta_2^f + \tau(P_2^f)$, the capacity of (P_2^g, θ_2^g) can be occupied in f by a cluster c_3 larger than c_2 or by a set of clusters $\{c_{3,1}, \ldots, c_{3,J}\}$ with $d(c_{3,j}) < d(c_2)$ for $j \in \{1, \ldots, J\}$ but $\sum_{j=1}^{J} d(c_{3,j}) \geq d(c_2)$. For each of these clusters $c_{3,j}$ it holds that $d(c_{3,j}) \geq d(c_1)$, since otherwise c_1 would have been routed on this path by the algorithm. Choose any $c_3 \in \{c_{3,1}, \ldots, c_{3,J}\}$.

Carrying on this argumentation, we obtain a sequence $(c_i)_i$ of clusters contained in C with cluster sizes $d(c_j) \geq d(c_1)$ for all $c_j \in (c_i)_i$. This sequence can be chosen such that each $c_j \in C$ is contained at most $p \in \mathbb{N}$ times if $d(c_j) = pd(c_1)$. Since the number of elements in C is finite it follows that there is a final element c_L with size $d(c_L)$ in the sequence $(c_i)_i$. Then no cluster c_{L+1} can be found in the set C that blocks the capacity on the path (P_L^g, θ_L^g) in the flow f. Since $d(c_L) \geq d(c_1)$ and $\theta_L^g + \tau(P_L^g) < \theta_1^f + \tau(P_1^f)$, the path (P_L^g, θ_L^g) would have been chosen to route the cluster c_1 instead of (P_1^f, θ_1^f) in the algorithm. Since this argumentation holds for any $c_1 \in C$ with $\theta_1^f + \tau(P_1^f) > T^*$ there cannot be a feasible dynamic cluster flow with a time horizon smaller than T^*. □

We estimate the worst-case complexity of Algorithm 1: The outer `while`-loop must be executed $|\mathcal{D}|$ times. This number is polynomial in the input size, as the cluster sizes are pairwise divisible; hence, the cardinality of the set of cluster sizes $\mathcal{D}$ can be at most $\log(d_{max})$. The inner `while`-loop is called at most $b_{max} := \max\{b_d | d \in \mathcal{D}\}$ times in each iteration. In this loop we have to examine all sinks (two times), where in the second loop we calculate a minimum over all arcs $a \in A$ and sum over $\log(d_{max})$ values. Hence this results in a worst-case complexity of $O(\log(d_{max}) b_{max} |\mathcal{T}| m) \in O(\log(d_{max}) b_{max} nm)$, which makes it a pseudopolynomial algorithm.

4.2 Examples

In the following we see two examples for which Algorithm 1 will not output an optimal solution. In such cases as non-divisible cluster sizes or different graph topologies, the algorithm can still be used as a heuristic for the quickest cluster flow problem.

Figure 1 shows a tree network, in which Algorithm 1 does not result in a quickest cluster flow for not pairwise divisible cluster sizes $\mathcal{D} = \{d_2 = 3, d_1 = 2\}$ and demands $b_2 = 2, b_3 = 2$. Algorithm 1 might route the two clusters of size 3 to t_1, because this is the fastest path to a sink, resulting in a temporary time horizon of 2. Then no time horizon smaller than 3 can be achieved by routing the clusters of size 2 as fast as possible with respect to the remaining capacities. A smaller time horizon, however, could be achieved by routing one the clusters of size 3 to t_2 instead. Then both clusters of size 2 could be sent to t_1, resulting in an overall time horizon of 2.

In Figure 2 a non-tree network is depicted, where Algorithm 1 does not result in a quickest cluster flow even for pairwise divisible cluster sizes $\mathcal{D} = \{d_2 = 2, d_1 = 1\}$,

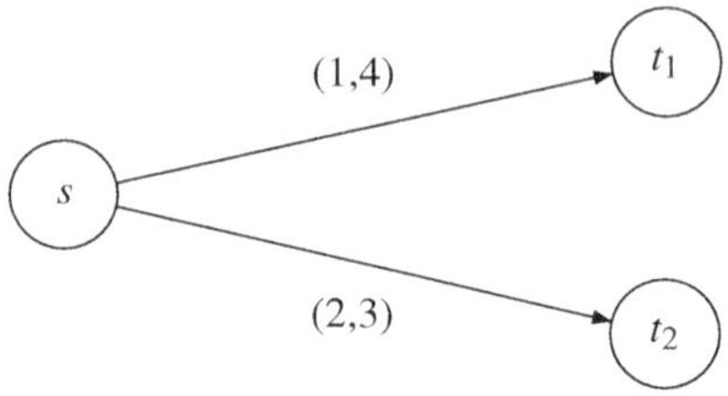

Fig. 1. Example for cluster flows with non-unit clusters. The arc-labels represent transit times and capacities $(\tau(a), u(a))$.

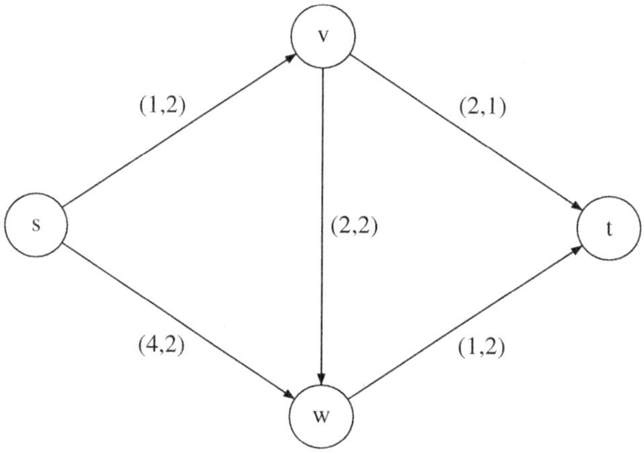

Fig. 2. Example for cluster flows on non-tree networks. The arc-labels represent transit times and capacities $(\tau(a), u(a))$.

$b_2 = 2$, $b_1 = 5$. The algorithm given above will send the larger clusters on the path s-v-w-t, resulting in a temporary time horizon of 5. Then no way of sending the clusters of size 1 yields a time horizon smaller than 7. A better solution consists of sending the clusters of size 2 on the path s-w-t. Then the temporary time horizon is 6 and hence larger than in the previous solution, but the smaller clusters can also be sent within the time horizon of 6, so the overall time horizon is smaller.

5 Conclusions

Macroscopic models based on dynamic network flows are frequently applied in transportation and evacuation planning [5]. The corresponding mathematical programming problems can often be efficiently solved by dedicated algorithms making use of the combinatorial structure. The solutions obtained provide provable system optimal characteristics, e.g., lower bounds on transportation or evacuation times. Increasing the degree of detail covered by the mathematical model allows for more precise statements.

However, there is typically a trade-off between ease of solution and complexity of the model. In this article, we elaborate on the so-called quickest cluster flow problem [4]. In this problem, different kinds of flow commodities which use different amounts of resources available in the network, are taken into account. In evacuation modeling, this may represent different sizes of groups of evacuees which tend to stay together during the egress movement [7]. We relate the quickest cluster flow problem to other existing network flow models and prove NP-hardness. We propose an exact algorithm for some special case on tree networks, prove its correctness, and analyze its running time.

Currently we are working on approximation approaches for different classes of quickest cluster flow problems, for example problems with non-divisible cluster sizes on trees as well as divisible cluster sizes on more general networks. Additionally, we are investigating the complexity status of the problem mentioned in this paper which is still unsolved up to now.

Acknowledgement

This paper is supported in part by the Federal Ministry for Education and Research (Bundesministerium für Bildung und Forschung, BMBF), Project REPKA, under FKZ 13N9961 (TU Kaiserslautern).

References

1. Ahuja, R., Magnanti, T., Orlin, J.: Network Flows. Prentice-Hall, Englewood Cliffs (1993)
2. Burkard, R., Dlaska, K., Klinz, B.: The Quickest Flow Problem. Math. Methods Oper. Res. 37(1), 31–58 (1993)
3. Garey, M., Johnson, D.: Computers and Intractability: A Guide to the Theory of NP-Completeness. Freeman, New York (1979)
4. Hamacher, H., Leiner, K., Ruzika, S.: Quickest Cluster Flow Problems. In: Proceedings of PED 2010 Conference (to appear, 2011)
5. Hamacher, H., Tjandra, S.: Mathematical Modelling of Evacuation Problems–A State of the Art. In: Pedestrian and Evacuation Dynamics, pp. 227–266. Springer, Heidelberg (2002)
6. Helbing, D., Farkas, I., Molnar, P., Vicsek, T.: Simulation of Pedestrian Crowds in Normal and Evacuation Situations. In: Pedestrian and Evacuation Dynamics, pp. 21–58. Springer, Heidelberg (2002)
7. Mawson, A.: Understanding Mass Panic and Other Collective Responses to Threat and Disaster. Psychiatry: Interpersonal and Biological Processes 68(2), 95–113 (2005)
8. Tjandra, S.: Dynamic Network Optimization with Application to the Evacuation Problem. Ph.D. thesis, TU Kaiserslautern (2003)

Strong Duality for the Maximum Borel Flow Problem

Ronald Koch and Ebrahim Nasrabadi

Technische Universität Berlin, Institut für Mathematik, Straße des 17. Juni 136, 10623 Berlin, Germany
{koch,nasrabadi}@math.tu-berlin.de

Abstract. Research on flows over time has been conducted mainly in two separate and independent approaches, namely *discrete* and *continuous* models, depending on whether a discrete or continuous representation of time is used. Recently, Borel flows have been introduced to build a bridge between these two models. In this paper, we consider the maximum Borel flow problem formulated in a network where capacities on arcs are given as Borel measures and storage might be allowed at the nodes of the network. This problem is formulated as a linear program in a space of measures. We define a dual problem and prove a strong duality result. We show that strong duality is closely related to a MaxFlow-MinCut Theorem.

1 Introduction

Network flows over time (also called *dynamic network flows* in the literature) form a fascinating area of study. In contrast to classical *static* flows, transit times are introduced on the arcs to describe how long it takes to traverse an arc. This would imply that flows on arcs are not constant but may change over time. Ford and Fulkerson [5,6] study the *maximum flow over time* problem and show that this problem can be solved efficiently by one minimum cost flow computation on the given network. In the model studied by Ford and Fulkerson [5,6], time is represented in discrete time steps and arc capacities are constant over time. In contrast to this, Anderson, Nash, and Philpott [1] study the maximum flow over time problem in a network with time-varying transit and storage capacities for the case where time is modeled as a continuum. They establish a MaxFlow-MinCut Theorem for the case that transit times are zero and the transit capacities are bounded measurable. This result was later extended to arbitrary transit times by Philpott [8].

Since the seminal research of Ford and Fulkerson in the 1950s, many authors have extensively studied flows over time from different viewpoints, but in two separate ways with respect to time-modeling, leading to discrete and continues models. Fleischer and Tardos [4] point out a close correspondence between these two models. Recently, Koch, Nasrabadi, and Skutella [7] introduced the notion of Borel flows to unify discrete and continuous flows over time into a single model. They establish a MaxFlow-MinCut Theorem for the maximum Borel flow problem.

Developing a duality theory for various classes of optimization problems has received a great deal of attention because of its importance in designing solution algorithms. In the classical static network flows, it is known that the dual problem for the

J. Pahl, T. Reiners, and S. Voß (Eds.): INOC 2011, LNCS 6701, pp. 256–261, 2011.

maximum flow problem corresponds to the cuts and strong duality is equivalent to the MaxFlow-MinCut Theorem. Anderson and Philpott [2] explore this relationship for the maximum flow over time in the continues model for the special case where transit time are zero and the arc and node capacities are piecewise analytic. The aim of this paper is to establish a strong duality result for the maximum Borel flow problem and examine its relationship to the MaxFlow-MinCut Theorem. Due to page limitations, the proofs of lemmas and further details will be presented in the full version of the paper.

2 The Maximum Borel Flow Problem

We consider a directed graph $G = (V, E)$ with *node set* V and *arc set* E. Let $s \in V$ be a *source* and $t \in V$ be a *sink* in G. Each arc $e \in E$ has an associated *transit time* $\tau_e \in \mathbb{R}$ specifying the required amount of time for traveling from the tail to the head of e. More precisely, if flow leaves node v at time θ along an arc $e = (v,w)$, it arrives at w at time $\theta + \tau_e$. Consider the real line $\mathbb{R}$ as the time domain. A *Borel flow* x is defined by a family of Borel measures $x_e : \mathscr{B} \longrightarrow \mathbb{R}_+, e \in E$, where $\mathscr{B}$ is the *Borel σ-algebra on* $\mathbb{R}$. We refer readers to [7] for the motivation of Borel flows. A member $B \in \mathscr{B}$ is called a *Borel set* or *measurable set*. The value $x_e(B)$ gives the amount of flow entering arc e over the Borel set B.

The *maximum Borel flow problem* is to find a Borel flow that maximizes the total flow value subject to arc and node capacity constraints. This problem can be formulated as the following infinite dimensional linear program:

$$\begin{aligned}
\max \quad & \sum_{e\in\delta^+(s)} |x_e| - \sum_{e\in\delta^-(s)} |x_e| \\
\text{s.t.} \quad & \sum_{e\in\delta^+(v)} x_e - \sum_{e\in\delta^-(v)} (x_e - \tau_e) + y_v = 0 \qquad \forall v \in V \setminus \{s,t\}\,, \\
& 0 \le x_e \le u_e \qquad \forall e \in E\,, \\
& 0 \le Y_v \le U_v \qquad \forall v \in V \setminus \{s,t\}\,.
\end{aligned} \tag{MBFP}$$

Here, the value $|x_e|$ denotes the total amount of flow entering arc e, i.e., $|x_e| := x_e(\mathbb{R})$. Hence, the objective function equals the net outflow of s. Furthermore, the arc capacities are given by Borel measures u_e, $e \in E$, and the node capacities are given by right-continuous functions U_v, $v \in V \setminus \{s,t\}$, of bounded variation. We assume that $|u_e| < \infty$ for each arc $e \in E$. The value $u_e(B)$ is an upper bound on the amount of flow that is able to enter arc e over the Borel set B and the value $U_v(\theta)$ is an upper bound on the amount of flow that can be stored at node v at time θ. We assume that there is no initial storage at any node and flow must not remain at any node except s and t. Hence, we must have $U_v(-\infty) := \lim_{\theta \to -\infty} U_v(\theta) = 0 = \lim_{\theta \to -\infty} Y(\theta) =: U_v(\infty)$ for each $v \in V \setminus \{s,t\}$. We notice that $x_e - \tau_e$ is a *shifted measure* defined by $(x_e - \tau_e)(B) = x_e(B - \tau_e)$ for all $B \in \mathscr{B}$, where $B - \tau_e := \{\theta - \tau_e \mid \theta \in B\}$. For each $v \in V \setminus \{s,t\}$, Y_v is a right-continuous function of bounded variation and y_v is a *signed* Borel measure derived from the formula $y_v\big((-\infty, \theta]\big) = Y_v(\theta)$. The value $Y_v(\theta)$ represents the amount of flow stored at node v at the point in time $\theta \in \mathbb{R}$ and the value $y_v(B)$ gives the overall change in storage at v over the Borel set B.

In order to state a MaxFlow-MinCut Theorem, we require a definition of s-t-cuts. For a right-continuous function $M : \mathbb{R} \to \mathbb{R}_{\geq 0}$ of bounded variation, we shall use $M^{\succ 0}$ to denote the set of all points $\theta \in \mathbb{R}$ where M or its left limit is positive at θ, that is, $M^{\succ 0} := \{\theta \in \mathbb{R} \mid M(\theta-) > 0 \text{ or } M(\theta) > 0\}$. Here, $M(\theta-)$ denotes the limit of M at θ from the left, i.e., $M(\theta-) := \lim_{\vartheta \nearrow \theta} M(\vartheta)$. We now define an *s-t Borel cut* $S := (S_v)_{v \in V}$ by measurable sets S_v, one for each $v \in V$ so that $S_s = \mathbb{R}$, $S_t = \emptyset$ and for every node $v \in V \setminus \{s,t\}$ the set $\Gamma_v := S_v \cap U_v^{\succ 0}$ is a countable union of pairwise disjoint intervals.

Let $S = (S_v)_{v \in V}$ be an s-t Borel cut and consider a node v. By definition, Γ_v is expressible as $\bigcup_{i \in J_v} I_{v,i}$, where J_v is a countable set of indices and $I_{v,i}$, $i \in J_v$, are pairwise disjoint intervals. Each interval $I_{v,i}$ is supposed to be inclusion-wise maximal, i.e., there is no interval $I \subseteq \Gamma_v$ strictly containing $I_{v,i}$. Let $\alpha_{v,i}$ and $\beta_{v,i}$ be the left and right boundary of the interval $I_{v,i}$, respectively. An interval $I_{v,i}$ can be of the form $(\alpha_{v,i}, \beta_{v,i})$, $[\alpha_{v,i}, \beta_{v,i})$, $(\alpha_{v,i}, \beta_{v,i}]$, or $[\alpha_{v,i}, \beta_{v,i}]$. Therefore we partition the set J_v of indices into four subsets. Let J_v^1 (J_v^2, J_v^3, and J_v^4) be the set of indices i for which $I_{v,i}$ is open (left-closed & right-open, right-closed & left-open, and closed, respectively). With these constructions, the *capacity* $\text{cap}(S)$ of S is defined by

$$\text{cap}(S) := \sum_{e=(v,w) \in E} u_e\big(S_v \cap (S_w - \tau_e)^c\big) + \sum_{v \in V \setminus \{s,t\}} \left(\sum_{i \in J_v^1 \cup J_v^2} U_v(\beta_{v,i}-) + \sum_{i \in J_v^3 \cup J_v^4} U_v(\beta_{v,i}) \right) . \tag{1}$$

We set the capacity $\text{cap}(S)$ to ∞ if any infinite sum does not converge. We refer to an s-t Borel cut whose capacity is minimum among all s-t Borel cuts as a *minimum Borel cut*. The following theorem is due to [7].

Theorem 1. *The MaxFlow-MinCut Theorem holds for Borel flows, i.e., there exists an s-t-flow over time x and an s-t-cut over time S for which $val(x) = cap(S)$.*

3 Dual Formulation and Strong Duality

In the context of static network flows, the MaxFlow-MinCut Theorem is equivalent to strong duality. Here we wish to establish a similar result for Borel flows. To do this, we need a dual problem for (MBFP). In order to state a dual formulation, we require the concept of a function of *σ-bounded variation*. Let f be a real-valued function on $\mathbb{R}$. The *total variation* of f within a bounded interval $[a,b]$ is defined by

$$V(f;[a,b]) := \sup\left\{ \sum_{i=1}^{n} \big(f(a_i) - f(a_{i-1})\big) \mid \{a_1, \ldots, a_n\} \text{ is a partition of } [a,b] \right\} .$$

The function f is called of *bounded variation* on $[a,b]$ if $V(f,[a,b]) < \infty$. The function f is said to be of *bounded variation* on $\mathbb{R}$ if there exists a constant $K < \infty$ such that $V(M;[a,b]) < K$ for any (bounded) interval $[a,b] \subset \mathbb{R}$.

The function f is said to be of *σ-bounded variation* if it can be decomposed into a countable sum of functions of bounded variation on $\mathbb{R}$. Similarly, f is said to be *σ-monotonic increasing* if it can be decomposed into a countable sum of monotonic increasing and bounded functions on $\mathbb{R}$. These two definitions can be regarded as the extension of finite measures to σ-finite measures and because of that we have used the symbol σ.

It is a well-known result that a function is of bounded variation if and only if it is the difference between two monotonic increasing and bounded functions (see, e.g., Chapter 6 in [3]). This implies that a function is of σ-bounded variation if and only if it is the difference between two σ-monotonic increasing functions. Let $f : \mathbb{R} \to \mathbb{R}$ be a function of σ-bounded variation with $f(\infty) = 0$ where $f(\infty) := \lim_{\theta\to\infty} f(\theta)$. Notice that this limit exists since f is of σ-bounded variation. Then there exist functions $f^{(+)}$ and $f^{(-)}$ (subsequently referred to as the *Jordan decomposition* of f) that are σ-monotonic increasing on f with $f^{(+)}(\infty) = f^{(-)}(\infty) = 0$ and $f(\theta) = f^{(+)}(\theta) - f^{(-)}(\theta)$ for $\theta \in \mathbb{R}$. The functions $f^{(+)}$ and $f^{(-)}$ are called the positive and negative parts of f, respectively.

We now consider a dual problem for (MBFP) as follows:

$$\begin{aligned} \min \quad & \sum_{e\in E} \int_{\mathbb{R}} \rho_e \, du_e + \sum_{v\in V\setminus\{s,t\}} \int_{\mathbb{R}} U_v \, d\pi_v^{(-)} \\ \text{s.t.} \quad & \rho_e(\theta) - \lambda_v(\theta) + \lambda_w(\theta+\tau_e) \geq 0 \quad \forall e=(v,w)\in E,\ \theta\in\mathbb{R}, \\ & \rho_e(\theta) \geq 0 \quad \forall e\in E,\ \theta\in\mathbb{R}, \\ & \lambda_s(\theta) = 1 \quad \forall \theta\in\mathbb{R}, \\ & \lambda_t(\theta) = 0 \quad \forall \theta\in\mathbb{R}, \\ & \pi_v := \lambda_v|_{U_v^{\succ 0}} \text{ of } \sigma\text{-BV on } \mathbb{R} \quad \forall v\in V\setminus\{s,t\}, \end{aligned} \tag{MBFP*}$$

where $\rho_e, e\in E$ and $\lambda_v, v\in V\setminus\{s,t\}$ are measurable functions. We should mention that this problem generalizes the dual problem studied by Anderson and Philpott [2] for the continuous-time maximum flow problem. Note that $\pi_v^{(-)}$ denotes the negative part of π_v and $\pi_v := \lambda_v|_{U_v^{\succ 0}}$ is given for each $v\in V\setminus\{s,t\}$ as follows:

$$\pi_v(\theta) := \begin{cases} \lambda_v(\theta) & \text{if } \theta\in U_v^{\succ 0}, \\ 0 & \text{otherwise}. \end{cases}$$

It follows from this definition that $\pi_v(\infty) = 0$ since $Y_v(\infty) = 0$ due to our assumption. Furthermore, the dual variable $\rho_e, e\in E$ can be eliminated from (MBFP*). In fact, if we know optimal values for the dual variables λ_v, $v\in V$, we can compute the optimal values for $\rho_e, e\in E$ by

$$\rho_e(\theta) = \max\{0, \lambda_v(\theta) - \lambda_w(\theta+\tau_e)\} \qquad \forall e=(v,w)\in E,\ \theta\in\mathbb{R}.$$

The integrals in the objective function of (MBFP*) should be explained. For each $e\in E$ the first integral involves the function ρ_e as the integrand and the measure u_e as the integrator and is understood in the sense of Lebesgue-Stieltjes. Since ρ_e is supposed to be measurable and μ_e is a Borel measure, the first integral is well defined and exists. The second integral for each $v\in V\setminus\{s,t\}$ involves two functions U_v and $\pi_v^{(-)}$ (U_v as the integrand and $\pi_v^{(-)}$ as the integrator) and is regarded as a generalization of the Riemann-Stieltjes integral. More precisely, if $\pi_v^{(-)} = \sum_{i\in\mathbb{N}} \pi_{v,i}^{(-)}$ where $\pi_{v,i}^{(-)}, i\in\mathbb{N}$ are monotonic increasing and bounded, then

$$\int_{\mathbb{R}} U_v \, d\pi_v^{(-)} := \sum_{i \in \mathbb{N}} \int_{\mathbb{R}} U_v \, d\pi_{v,i}^{(-)} .$$

Each left integral of the above equation is treated as the Riemann-Stieltjes integral as developed in [3, Chapter 7]. However, although U_v is supposed to be of bounded variation and right-continuous and $\pi_{v,i}^{(-)}$ is monotonic increasing and bounded, the Riemann-Stieltjes integral $\int_{\mathbb{R}} U_v \, d\pi_{v,i}^{(-)}$ need not exist for some $i \in \mathbb{N}$ as U_v and $\pi_{v,i}^{(-)}$ may have common discontinuous from the left or from the right at some points. In such a case, the integral does not exist (see [3, Theorem 7.29]) and we replace U_v by another function, say $\bar{U}_v$, defined by

$$\bar{U}_v(\theta) := \begin{cases} U_v(\theta-) & \text{if } U_v \text{ is discontinuous at } \theta \text{ and } \pi_{v,i} \text{ is right-continuous at } \theta \, , \\ U_v(\theta) & \text{otherwise} \, . \end{cases}$$

Note that the function $\bar{U}_v$ differs from U_v only at those points θ for which U_v and $\pi_{v,i}^{(-)}$ share a common discontinuity from the left or from the right at θ. Then the integral $\int_{\mathbb{R}} U_v \, d\pi_{v,i}^{(-)}$ is defined by

$$\int_{\mathbb{R}} U_v \, d\pi_{v,i}^{(-)} := \int_{\mathbb{R}} \bar{U}_v \, d\pi_{v,i}^{(-)}$$

Note that the left integral of the above equation is guaranteed to exist (see [3, Theorem 7.29]). In what follows, each integral with a measure as the integrator is regarded in the sense of Lebesgue-Stieltjes and each one with a function as the integrator is regarded in the sense of Riemann-Stieltjes as defined above.

The first result that we would like to have between (MBFP) and (MBFP*) is weak duality. To state this and subsequent results, we introduce some notation. For a given optimization problem (OP), we use the notation $V[\text{OP}]$ to denote its optimal value and use the notation $V[\text{OP}, x]$ to denote the objective function value for a given feasible solution x.

Lemma 1. *Suppose that x is feasible for* (MBFP) *and λ is feasible for* (MBFP*). *Then* $V[(\text{MBFP}), x] \leq V[(\text{MBFP}^*), \lambda]$.

A stronger result than weak duality is to prove the existence of a feasible solution x for (MBFP) and a feasible solution λ for (MBFP*) in which $V[(\text{MBFP}), x] = V[(\text{MBFP}^*), \lambda]$. The arc capacities $u_e, e \in E$ are finite and this guarantees the existence of an optimal solution x, say, for (MBFP). Moreover, by Theorem 1, there exists an s-t Borel cut S for which $\text{val}(x) = \text{cap}(S)$. It thus enough to show that S corresponds to a feasible solution λ with $\text{cap}(S) = V[(\text{MBFP}^*), \lambda]$.

Lemma 2. *Given an s-t Borel cut $S = (S_v)_{v \in V}$, let $\lambda_v : \mathbb{R} \to \mathbb{R}$ be the indicator function of S_v and $\rho_e : \mathbb{R} \to \mathbb{R}$ be the indicator function of $S_v \cap (S_w - \tau_e)^c$ for each arc $e = (v, w) \in E$. Then $\lambda := (\lambda_v)_{v \in V}$, $\rho := (\rho_e)_{e \in E}$ is a feasible solution for* (MBFP*) *and moreover,* $V[(\text{MBFP}^*), \lambda] = \mathit{cap}(S)$.

Combining Theorem 1 and Lemma 2, we get the main result of this paper.

Theorem 2. *Strong duality holds between* (MBFP) *and* (MBFP*)*, i.e., there exists a feasible solution x for* (MBFP) *and a feasible solution λ for* (MBFP*) *in which* $V[(\text{MBFP}),x] = V[(\text{MBFP}^*),\lambda]$.

Acknowledgement

This work is supported by DFG project SK58/7-1 and by the DFG research center MATHEON.

References

1. Anderson, E.J., Nash, P., Philpott, A.B.: A class of continuous network flow problems. Mathematics of Operations Research 7, 501–514 (1982)
2. Anderson, E.J., Philpott, A.B.: Optimisation of flows in networks over time. In: Kelly, F.P. (ed.) Probability, Statistics and Optimisation, ch. 27, pp. 369–382. Wiley, New York (1994)
3. Apostol, T.M.: Mathematical Analysis, 2nd edn. Addison-Wesley, Reading (1974)
4. Fleischer, L.K., Tardos, E.: Efficient continuous-time dynamic network flow algorithms. Operations Research Letters 23, 71–80 (1998)
5. Ford, L.R., Fulkerson, D.R.: Constructing maximal dynamic flows from static flows. Operations Research 6, 419–433 (1958)
6. Ford, L.R., Fulkerson, D.R.: Flows in Networks. Princeton University Press, Princeton (1962)
7. Koch, R., Nasrabadi, E., Skutella, M.: Continuous and discrete flows over time: A general model based on measure theory. Mathematical Methods of Operations Research (2010) (to appear)
8. Philpott, A.B.: Continuous-time flows in networks. Mathematics of Operations Research 15, 640–661 (1990)

Modeling the Gateway Location Problem for Multicommodity Flow Rerouting

Maurizio Bruglieri[1], Paola Cappanera[2], Alberto Colorni[1], and Maddalena Nonato[3]

[1] INDACO, Via Durando 38a, Milano, Italy
{maurizio.bruglieri,alberto.colorni}@polimi.it
[2] DSI, Via S. Marta 3, Firenze, Italy
paola.cappanera@unifi.it
[3] ENDIF, Via L. Saragat 1, Ferrara, Italy
maddalena.nonato@unife.it

Abstract. This paper introduces a new problem, involving the optimal location of a limited number of gateways on the nodes of an uncapacitated network. A multicommodity flow, where each commodity is of single-origin-single-destination type, moves on the network according to its linear objective function c. Gateways are used by the network administrator to reroute flows by obliging each commodity to detour from its c-optimal path and pass by its assigned gateway. Gateways are located and assigned by the administrator so that the resulting c-optimal flows on the detours minimize the administrator's objective function r. Gateways thus provide the administrator with a mechanism for indirect flow control, so that the flow value according to r is improved with respect to the unregulated scenario. To the authors knowledge, this is a new combinatorial optimization problem, that we call the *gateway location problem for multicommodity flow rerouting*. We present three alternative formulations and discuss pros and cons of each. Interesting applications arise in the field of hazardous material transportation. The discussion is supported by computational results on realistic instances from this field.

1 Introduction

When routing flows on a network, different decision makers may have different perspectives on the evaluation of the routing, due to the different criteria according to which the routing is judged. We are concerned with the situation where two classes of decision makers, which for simplicity we identify as the network administrator and a set of network users, operate at different decision levels. The administrator operates indirectly by issueing directives which network users must comply to, when routing their flows on the network infrastructure. The administrator, in this way, implicitly defines the set of the feasible solutions on which the network users will optimize their objective function, say c. In turn, the effect of a specific directive can be evaluated from the administrator point of view only a posteriori, by computing the value of the administrator's objective function, say r, on the users selected c-optimal routes.

Several examples arising in the field of transportation fit into this picture, depending on the kind of directives the administrator may impose: just to mention a few, in road pricing problems, the administrator sets tolls on some arcs, wishing to maximize total income,

J. Pahl, T. Reiners, and S. Voß (Eds.): INOC 2011, LNCS 6701, pp. 262–276, 2011.

while network users select their itineraries wishing to minimize a generalized cost, which depends on both tolls and travel costs [13]. Other applications of toll pricing aim at reducing traffic congestion in certain areas so to achieve system optimal flows [1]. In network design problems, the administrator alters the network infrastructure by setting arcs capacity or modifying traffic signals, while users reroute their commodity in response, each pursuing his own selfish objective [10]. Optimal parking site capacities and fares can be set by predicting user behaviors, as in [11]. Finally, in hazardous material transportation, the administrator may forbid transit on some arcs, in the attempt to reduce total risk. This, in turn, depends on the minimum cost itineraries on the current subnetwork [6]. Toll-setting policies have also been investigated to the same purpose in [14].

In this paper, we propose a new tool for the administrator indirect control over user flows, as an alternative to modifying the network topology or the arc costs and capacities. It consists of imposing check points along the routes, so called *gateways*, and stating the rule that each commodity, on its route from origin to destination, must pass through a specific gateway. Which path to follow along this way is up to the user, and it will be the optimal one according to the user objective function. In turn, the administrator will select the location of each check point and which one has to be assigned to which user, so that user response will optimize the administrator's objective function. We call such problem the *Gateway Location Problem* for multicommodity flow (GLP).

The notion of *gateway path* was first introduced by Lombard and Church in [3] in the context of corridor location, to generate good and spatially different paths, alternative to shortest origin–destination path. A gateway path is defined as a path from origin to destination, obliged to go through a prespecified node, i.e., the gateway. The authors make a smart use of the Dijkstra algorithm to efficiently enumerate the set of all the shortest gateway paths for a single commodity. In this paper, we build upon this concept and generalize it in the context of multicommodity flow. Here, gateways are still used to reroute the flow and steer it away from the original path, hoping not to deteriorate its cost too much. However, in our problem, gateways have first to be located and then to be assigned to commodities, having several commodities sharing the same gateway, thus yielding a global constraints. Moreover, solution quality is evaluated according to a function potentially conflicting with the cost function which drives the computation of the shortest gateway paths.

As far as we know, GLP is a new challenging combinatorial optimization problem which has never been formulated or studied before. We discuss how GLP can be formalized providing three different mathematical models, and highlighting pros and cons of each. In particular, the first one is a new arc based bilevel programming model for which we present a Mixed Integer Linear Programming reformulation. A second contribution of this paper concerns the effectiveness of GLP as an indirect tool for flow control. Computational results computed on realistic instances show that this methodology can provide network administrators with an effective tool for indirect control of the network usage that takes into account users behavior, while not excessively penalizing the network users objective function.

The rest of the paper is organized as follows. In Sect. 2 a formal description of the problem is provided and the three models are presented. In Sect. 3 we discuss pros and cons of each model, while in Sect. 4 computational results on realistic instances are reported, supporting our claim that gateways can be an effective tool for indirect flow control. Conclusions and on going works follow in Sect. 5.

2 Formalizing the Gateway Location Problem

Let us introduce some mathematical notations required to formalize the problem by either Mixed-Integer Linear Programming or Bilevel Integer Linear Programming models. The following notations will be used throughout the paper. Let:

- $V = \{1 \dots, n\}$ be the set of commodities, each associated with its origin-destination pair, $\langle o(v), d(v) \rangle, \forall v \in V$. Let $O = \{o(v),\ v \in V\}$ be the set of origins and, likewise, let $D = \{d(v),\ v \in V\}$ denote the destination set. Let d_v be the demand of commodity v, $\forall v \in V$. In the following, the terms *commodity* and *vehicle* will be used interchangeably.
- N^{GTW} be the set of the m available locations, each representing a candidate site where potentially installing one of the k gateways, with $k < n$ and $k << m$.
- $G = (N, A)$ be a weighted directed graph, such that: $N^{GTW} \cup O \cup D \subseteq N$, that is, the node set includes the candidate gateway locations as well as all the origins and the destinations of the vehicles. $A \subseteq N \times N$, and for each arc $(i, j) \in A$ a positive coefficient c_{ij} and a non-negative coefficient r_{ij} are defined, i.e., c is the vector of the coefficients of the network users objective function, while r is the one of the network administrator $\mathscr{A}$.
- ρ_v^c (ρ_v^r) be the c-optimal (r-optimal) path from $o(v)$ to $d(v)$ for each $v \in V$.
- $\{y_h,\ h \in N^{GTW}\}$ be a set of binary variables, with $y_h = 1$ if a gateway is located at node h and 0 otherwise.
- $\{z_h^v,\ h \in N^{GTW},\ v = 1, \dots, n\}$ be a set of binary variables, with $z_h^v = 1$ if the gateway assigned to vehicle v is the one located at node $h \in N^{GTW}$.

In some applications the objective functions are commodity dependent, i.e., $c_{ij}^v \neq c_{ij}^{v'}$ for $v \neq v'$ or $r_{ij}^v \neq r_{ij}^{v'}$ for $v \neq v'$. However, since our discussion can be fully restated and easily adapted to such a case, hereafter we will omit the commodity index in order to keep notation simple. Moreover, when casting the model into the framework of hazmat transportation, c represents the cost function as perceived by the users, such as distance, travel time, or monetary cost of travel, while r is a function modeling the risk per flow unit when traversing that arc. To follow such intuition, in our further discussion we will refer to c as the *cost* function and to r as the *risk* function.

In the following, three mathematical programming models are presented, each formalizing the problem at a different detail of the decision variables.

The first one is an arc based formulation, where the routes followed by the vehicles for a given gateway assignment are the solution of a minimum cost uncapacitated multicommodity flow problem. In order to take into account such dependency, an outer decision problem is added, where gateways are located and assigned according to risk minimization, thus yielding a bilevel integer linear programming model. We will show how this model can be reformulated as a single level Mixed Integer Linear Programming (MILP) model, by exploiting duality. No preprocessing is required for this formulation to be solved.

The second formulation deals explicitly with decision variables associated with paths. Recall that a "gateway path" is any path from an origin $o(v)$ to the corresponding destination $d(v)$ having at least one potential location for a gateway as an internal node. Each path can be evaluated according to c and to r, so its cost and its risk are known. In particular, shortest gateway paths are constrained to be c optimal. All gateway paths referring to the same gateway belong to one cluster. Only k clusters are allowed. For each commodity one gateway path has to be selected, such that the total risk is minimized.

Finally, if we consider the risk associated with the gateway path of minimum cost for a given commodity as the risk of assigning that gateway to that commodity, then the problem reduces to selecting the k gateways of minimum total risk, thus yielding a sort of k median problem. This last formulation is based on the complete enumeration of the shortest gateway path for each commodity and for each candidate gateway location node, therefore it requires a preprocessing phase whose not negligible computational time adds up to the overall running time.

2.1 A Bilevel Multicommodity Flow Model

This model exploits the potentials of modeling decision variables at their finest granularity, explicitly representing arc flows. Any gateway path of commodity v by gateway h is made of two subpaths, from $o(v)$ to h and from h to $d(v)$. Since each commodity routes its flow according to c, it will select the shortest gateway path, made by the shortest path from $o(v)$ to h and the shortest path from h to $d(v)$. Each subpath is modeled by a separate family of flow variables, namely: flow variables $\overline{x}_{ij}^{v}$, $\forall v \in V$, $\forall (i,j) \in A$: $j \neq d(v)$ are associated with the path from $o(v)$ to the assigned gateway. Likewise, flow variables $\underline{x}_{ij}^{v}$, $\forall v \in V$, $\forall (i,j) \in A : i \neq o(v)$ model the path from the assigned gateway to destination $d(v)$. As usual, $\delta_i^+(\overline{x})$ stands for $\sum_{(i,j)\in FS(i)} \overline{x}_{ij}$ and $\delta_i^-(\overline{x})$ for $\sum_{(j,i)\in BS(i)} \overline{x}_{ji}$; the same notation holds for flow variables $\underline{x}_{ij}$.

The model must capture the hierarchical relationship between the routing part and the location-assignment decisions. Indeed, the network administrator locates gateways at selected candidate sites and assigns them to commodities, in order to optimize his objective function r over the c-optimal flows along the shortest gateway paths. This interaction is captured by the following bilevel multicommodity uncapacitated min cost flow.

$$P^{BL.MCF} : \min \sum_{v=1\ldots n} d_v \sum_{(i,j)\in A} r_{ij}(\overline{\xi}_{ij}^{v} + \underline{\xi}_{ij}^{v}) \quad \text{subject to:}$$

$$\sum_{h=1,\ldots,m} z_h^v = 1 \qquad \forall v \in V \tag{1}$$

$$y_h \geq z_h^v \qquad \forall h \in N^{GTW}, \forall v \in V \tag{2}$$

$$\sum_{h=1,\ldots,m} y_h = k \tag{3}$$

$$z_h^v \in \{0,1\} \qquad \forall h \in N^{GTW}, \forall v \in V \tag{4}$$

$$y_h \in \{0,1\} \qquad \forall h \in N^{GTW} \tag{5}$$

where $\overline{\xi}^{v}_{ij}, \underline{\xi}^{v}_{ij} \in argmin\, P^{SP}$: $\min \sum_{v \in V} \sum_{(i,j) \in A} c_{ij}(\overline{x}^{v}_{ij} + \underline{x}^{v}_{ij})$ subject to:

$$\delta^{+}_{o(v)}(\overline{x}^{v}) = 1 \qquad \forall v \in V \qquad (6)$$

$$\delta^{-}_{h}(\overline{x}^{v}) - \delta^{+}_{h}(\overline{x}^{v}) = z^{v}_{h} \qquad \forall h \in N^{GTW}, \forall v \in V \qquad (7)$$

$$\delta^{-}_{i}(\overline{x}^{v}) - \delta^{+}_{i}(\overline{x}^{v}) = 0 \qquad \forall i \neq o(v), d(v), i \notin N^{GTW}, \forall v \in V \qquad (8)$$

$$\overline{x}^{v}_{ij} \geq 0 \qquad \forall (i,j) \in A, \forall v \in V \qquad (9)$$

$$\delta^{-}_{d(v)}(\underline{x}^{v}) = 1 \qquad \forall v \in V \qquad (10)$$

$$\delta^{+}_{h}(\underline{x}^{v}) - \delta^{-}_{h}(\underline{x}^{v}) = z^{v}_{h} \qquad \forall h \in N^{GTW}, \forall v \in V \qquad (11)$$

$$\delta^{+}_{i}(\underline{x}^{v}) - \delta^{-}_{i}(\underline{x}^{v}) = 0 \qquad \forall i \neq o(v), d(v), i \notin N^{GTW}, \forall v \in V \qquad (12)$$

$$\underline{x}^{v}_{ij} \geq 0 \qquad \forall (i,j) \in A, \forall v \in V \qquad (13)$$

Variables z^{v}_{h} are decision variables at the outer level, while they act as right hand side coefficients of the flow balance constraints at the inner level. Cardinality constraints (3) impose that exactly k gateways are installed at that many locations, while semi-assignment constraints (1) assign to each commodity one open gateway. Constraints (2) link the two choices, so that a gateway must be open in order to be assigned. Furthermore, note that due to the lack of capacity constraints, the inner problem is separable and it corresponds to the solution of $2n$ shortest path problem. Denote them respectively as $SP^{v}(\overline{x})$ (constraints (6–9)) and $SP^{v}(\underline{x})$ (constraints (10–13)). Going into such details allows to take advantage of the tractability of the shortest path problem. The unimodularity of the flow balance constraint matrix allows us to reformulate the problem as a one level optimization problem by exploiting linear programming duality.

Reformulating the Problem as a One Level Optimization Problem. Bilevel programming is usually hard to tackle, see [2], however it can become more tractable if the objective function of the inner problem can be restated in terms of a set of constraints expressing its optimality conditions, taking advantage of linear duality, as here and in [12]. In particular, let us introduce $\pi^{+}_{o(v)v}$, π^{+}_{hv} and π^{+}_{iv} as the dual variables associated with flow balance constraints (6–8), and $\pi^{-}_{d(v)v}$, π^{-}_{hv} and π^{-}_{iv} as those associated with (10–12), respectively. Dual feasibility constraints of the two shortest path problems $SP^{v}(\overline{x})$ and $SP^{v}(\underline{x})$ can now be stated as the usual Bellman's conditions, in (14–15) and (16–17), respectively:

$$\pi^{+}_{jv} - \pi^{+}_{iv} \leq c_{ij} \qquad \forall (i,j) \in A,\ \forall v \in V \qquad (14)$$

$$\pi^{+}_{o(v)v} = 0 \qquad \forall v \in V \qquad (15)$$

$$\pi^{-}_{iv} - \pi^{-}_{jv} \leq c_{ij} \qquad \forall (i,j) \in A,\ \forall v \in V \qquad (16)$$

$$\pi^{-}_{d(v)v} = 0 \qquad \forall v \in V \qquad (17)$$

Optimality is reached when the feasible solutions of the primal and the dual problem have the same objective function value, as stated below.

$$\sum_{(i,j)\in A} c_{ij}\overline{x}_{ij}^{v} = \sum_{h\in N^{GTW}} \pi_{hv}^{+} z_{h}^{v} \qquad \forall v \in V \qquad (18)$$

$$\sum_{(i,j)\in A} c_{ij}\underline{x}_{ij}^{v} = \sum_{h\in N^{GTW}} \pi_{hv}^{-} z_{h}^{v} \qquad \forall v \in V \qquad (19)$$

Let us introduce a new set of variables $\left\{\omega_h^{+v}, \omega_h^{-v} \geq 0\, \forall h \in N^{GTW}, \forall v \in V\right\}$ to linearize optimality conditions (18–19).

$$\sum_{(ij)\in A} c_{ij}\overline{x}_{ij}^{v} = \sum_{h\in N^{GTW}} \omega_h^{+v} \qquad \forall v \in V \qquad (20)$$

$$\sum_{(ij)\in A} c_{ij}\underline{x}_{ij}^{v} = \sum_{h\in N^{GTW}} \omega_h^{-v} \qquad \forall v \in V \qquad (21)$$

$$\omega_h^{+v} \leq L_h^{+v} z_h^{v} \qquad \forall h \in N^{GTW}\ \forall v \in V \qquad (22)$$

$$\omega_h^{+v} \leq \pi_{hv}^{+} \qquad \forall h \in N^{GTW}\ \forall v \in V \qquad (23)$$

$$\omega_h^{-v} \leq L_h^{-v} z_h^{v} \qquad \forall h \in N^{GTW}\ \forall v \in V \qquad (24)$$

$$\omega_h^{-v} \leq \pi_{hv}^{-} \qquad \forall h \in N^{GTW}\ \forall v \in V \qquad (25)$$

Thus ω_h^{+v} equals $\pi_{hv}^{+} z_h^{v}$ as it is set to 0 by (22) if gateway h is not assigned to commodity v ($z_h^v = 0$), and it is bounded from above by the dual variable π_{hv}^{+} in (23) and set equal to the path cost in (20) if $z_h^v = 1$. Likewise for ω_h^{-v}. L_h^{+v} (L_h^{-v}) is the cost of any path from $o(v)$ to h (from h to $d(v)$). The one level MILP reformulation is:

$$P^{1L.MCF} : \min \sum_{v\in V} d_v \sum_{(i,j)\in A} r_{ij}(\overline{x}_{ij}^{v} + \underline{x}_{ij}^{v}) \quad \text{s.t. (1–5), (6–17), (20–25)} \qquad (26)$$

2.2 A Path Based Model

An intermediate level of decision involves variables associated with paths. In particular, this model potentially considers the set of all elementary gateway paths for each commodity. Let P_h^v be the set of gateway paths associated with commodity $v \in V$ and gateway h, so that $P^v = \cup_{h\in N^{GTW}} P_h^v$ is the set of all gateway paths for commodity v. Note that if we consider a gateway path as the pair of the two subpath from $o(v)$ to h and from h to $d(v)$, then two topologically identical paths $p, p' \in P^v$ are seen as different entities if they refer to different gateways. Therefore, selecting a path for a given commodity means to assign a specific gateway to that commodity.

Each path p is characterized by its cost c_p and its risk r_p. As in the arc based model, we assume both the cost and the risk functions to be decomposable on the arcs of the path. For each commodity v, let us introduce a binary variable $w_p\, \forall p \in P^v$. A path based model for GLP can be formulated as follows:

$$P^{path} : \min \sum_{v\in V} d_v \sum_{p\in P^v} w_p r_p$$

$$\sum_{p \in P^v} w_p = 1 \qquad \forall v \in V \tag{27}$$

$$\sum_{p \in P_h^v} w_p \leq y_h \qquad \forall h \in N^{GTW}, \forall v \in V \tag{28}$$

$$\sum_{h=1,\ldots,m} y_h = k \tag{29}$$

$$w_p \in \{0,1\} \qquad \forall p \in P^v, \forall v \in V \tag{30}$$

$$y_h \in \{0,1\} \qquad \forall h \in N^{GTW} \tag{31}$$

$$\pi_{jv}^+ - \pi_{iv}^+ \leq c_{ij} \qquad \forall v \in V \;\; \forall (i,j) \in A : j \neq d(v) \tag{32}$$

$$\pi_{iv}^- - \pi_{jv}^- \leq c_{ij} \qquad \forall v \in V \;\; \forall (i,j) \in A : i \neq o(v) \tag{33}$$

$$\pi_{o(v)v}^+ = 0 \qquad \forall v \in V \tag{34}$$

$$\pi_{d(v)v}^- = 0 \qquad \forall v \in V \tag{35}$$

$$\sum_{p \in P_h^v} w_p c_p \leq \pi_{hv}^+ + \pi_{hv}^- \qquad \forall h \in N^{GTW}, \forall v \in V \tag{36}$$

Only a gateway path which labels an internal node $h \in N^{GTW}$ as its gateway can activate, when selected, the gateway variable y_h by constraints (28). Indeed, constraints (27) not only assign one path to each commodity, but also one gateway, since selecting $p \in P_h^v$ is to assign gateway h to v. Constraints (32–36) model users behavior, since feasible node potentials at a gateway provide a tight upper bound to the cost of the shortest gateway path. Note that, the left hand side of constraint (36) is zero when gateway h is not selected. The choice of the shortest gateway path is thus enforced, but the model leaves room for formalizing alternative path selection rules.

2.3 A k–Median Like Model

The highest decision level consists of directly assigning gateways to commodities. At this level there is no possibility of selecting paths, since for each pair $h \in N^{GTW}$, $v \in V$ the shortest gateway path is the unique alternative. This fact correctly models users behavior, but leaves no extra room for variants. A preprocessing phase is required to carry out the exhaustive computation of each shortest gateway path for each commodity, and evaluate it according to r. Let ρ_h^v denote the shortest gateway path by gateway h for commodity v, and let $r_{\rho_h^v}$ be its risk, computed according to $r_{\rho_h^v} = \sum_{(i,j)\in A} r_{ij}(\overline{x}_{ij}^v + \underline{x}_{ij}^v)$ such that $\overline{x}_{ij}^v$ and $\underline{x}_{ij}^v$ are the optimal solution to $SP^v(\overline{x})$ and $SP^v(\underline{x})$, respectively. Computing ρ_h^v for each gateway and for each commodity is polynomial in the size of the graph G, but such preprocessing has to be considered as part of the solution process and it grows with the number of potential gateway locations.

Call $A = \left[a_h^v\right] \in R^{n\times m}$, with $a_h^v = d_v \cdot r_{\rho_h^v}$, the matrix reporting for each pair h, v the risk value of the shortest gateway path weighted by the demand of the associated commodity. The notion of gateway path is no longer explicitly present in the resulting model, where the problem is to select k columns of A such that is minimum the sum of the minimum element in each row. Formally:

$$P^{km}: \min \sum_{v=1,\ldots,n} \sum_{h=1,\ldots,m} a_h^v z_h^v$$

$$\sum_{h=1,\ldots,m} z_h^v = 1 \qquad \forall v = 1,\ldots,n \tag{37}$$

$$y_h \geq z_h^v \qquad \forall v = 1,\ldots,n, \forall h = 1,\ldots,m \tag{38}$$

$$\sum_{h=1,\ldots,m} y_h = k \tag{39}$$

$$z_h^v \in \{0,1\} \qquad \forall v = 1,\ldots,n,\ \forall h = 1,\ldots,m \tag{40}$$

$$y_h \in \{0,1\} \qquad \forall h = 1,\ldots,m \tag{41}$$

This structure highlights common points between GLP and a well known NP-Hard problem in location science, *k–Median*. Many solution approaches have been proposed to it, heuristic [5] and exact [4], most of which are tailored to the case of facilities and clients being the same set. Advantages and disadvantages of such a compact formulation are discussed in Sect. 3.

3 Comparing the Three Models

Each of three models here presented provides a valid formulation for GLP. However, the different granularity of their decision variables, i.e., arcs, paths and gateways, makes each of them either more or less adapt to be generalized in order to handle specific side constraints, requires specific preprocessing phases, may impact on their robustness w.r.t. the uniqueness of the shortest gateway path, or makes the administrator more or less aware of users costs. Such issues are discussed hereafter.

Let $c_{\rho_v^c}$ and $r_{\rho_v^c}$ be the cost and the risk of ρ_v^c, respectively. Furthermore, recall that $C^* = \{\rho_v^c, v \in V\}$ would be the users itineraries in the unregulated scenario, i.e., no rules enforced by the administrator. A gateway path deviation from ρ_v^c is an interesting alternative from the administrator point of view, only provided that its risk improves upon $r_{\rho_v^c}$. In this sense, $r(C^*) = \sum_{v \in V} r_{\rho_v^c}$ should be an upper bound to $r(G^*)$, the optimal solution value of GLP. Our models can be extended to ensure this property. For each commodity, three cases have to be distinguished, depending on how many shortest gateway paths have a risk below the risk value of the shortest path. i) All the m shortest gateway paths have a risk higher than or equal to the risk of the shortest path ρ_v^c, i.e., $r_{\rho_h^v} \geq r_{\rho_v^c}$ $\forall h \in N^{GTW}$. The unregulated scenario can not be improved by enforcing a detour by any gateway in N^{GTW}: commodity v should be exempted and its vehicle allowed to travel along its shortest path. ii) k' gateways, $k \leq k' < m$, are such that the risk of their shortest gateway path is no lower than $r_{\rho_v^c}$: exemption may improve the risk depending on which gateways are opened, so exempting that commodity must be a decision variable within the model. iii) only $k' < k$ gateways are such that the risk of their shortest gateway path is no lower than $r_{\rho_v^c}$: any selection of k gateways in N^{GTW} will improve upon the risk of the unregulated scenario.

During the preprocessing phase of the *k-median* model cases i) is spotted and recovered by exempting the commodity and removing it from the input. However, exemption must be encoded into all models as a possible choice to guarantee $r(C^*) \geq r(G^*)$ in all cases. Our models can be extended accordingly, as described hereafter, in order to allow the authority to assign to commodity v its shortest path if necessary. Let us introduce into each model a new set of boolean variables $\gamma_v \forall v \in V$ such that $\gamma_v = 1$ if commodity v is exempted, and 0 otherwise.

Modeling exemption in the arc based model: exempting commodity v in the arc based model means to allow arc flow variables to take values mapping the shortest origin-destination path. However, $o(v)$ and $d(v)$ are not connected nor in the subgraph induced by $\left\{\overline{x}_{ij}^v\right\}$ neither in the one induced by $\left\{\underline{x}_{ij}^v\right\}$. Therefore, a variable $\overline{x}_{id(v)}^v$ for each arc $\in BS(d(v))$ must be introduced so that a flow unit from $o(v)$ is able to reach $d(v)$ in case of exemption, by setting the incoming flow into $d(v)$ in terms of the $\overline{x}_{id(v)}^v$ equal to γ_v. Likewise, incoming flow into $d(v)$ in terms of the $\underline{x}_{id(v)}^v$ is set equal to $1 - \gamma_v$ so that, in case of exemption, the subpath modeled by $\left\{\overline{x}_{ij}^v\right\}$ connects $o(v)$ to $d(v)$ while the second one implodes. Accordingly, the right and side of constraint (1) becomes $1 - \gamma_v$ since no gateway must be assigned to commodity v when the vehicle travels along ρ_v^c. Dual feasibility constraints and optimality conditions can be accordingly modified.

Modeling exemption in the path based model: the second model based on path selection can be extended to encompass exemption by adding the term $\sum_{v \in V} \gamma_v r_{\rho_v^c}$ to the objective function. As before, the rhs of constraint (27) becomes $1 - \gamma_v$.

Modeling exemption in the *k-median* like model: again, the term $\sum_{v \in V} \gamma_v r_{\rho_v^c}$ is added to the objective function and the right and side of constraints (37) becomes $1 - \gamma_v$. This extension can be seen as adding a column to matrix A whose selection does not increase the gateway counter.

By encompassing commodity exemption into the models, the optimal solution will never increase the risk w.r.t. the unregulated scenario, and will never increase the user cost without a gain in risk. All our three models prove to be easy to extend and such versions will be used in the experimental campaign. In the following other kinds of generalizations are discussed.

Imposing to detour by the gateways may concentrate the flow of several commodities within a restricted area, in the neighborhood of the selected gateways. This fact may be an issue in the specific framework of hazmat transportation. Knapsack like constraints can be added to each model to impose a capacity on gateways. In particular, constraints (2), (28), and (38) can be restated to take capacity into account, as in classical Capacitated Facility Location. However, the path based model provides additional flexibility in formalizing the capacity issue, as it allows to model capacity restrictions on any part of the network besides gateways: Since the knapsack constraints can be stated directly on the path selection variables, the impact of a set of paths all insisting on the same region of the network can be controlled. Moreover, since any gateway path is potentially represented in the path based model, it is possible to select a gateway path with a higher cost than ρ_h^v that allows to comply with the capacity restrictions at a lower total cost. This can be achieved by relaxing the optimality constraint (36) on the path cost by multiplying the term $\pi_{hv}^+ + \pi_{hv}^-$ by a factor over 1. We believe that this feature is a strength of the path based model which can compensate the drawback of requiring either a path generator engine or a preprocessing phase. In such a phase, users would enumerate all paths they would accept as feasible, possibly resulting from a negotiation process in cooperation with the authority. The user preferences within each set of feasible paths can be enforced as in [8] and replace optimality, cost binding, constraints (36).

The arc based model is sensitive to adding further constraints on flow variables, that tend to destroy the integrality property of the shortest path problems, which is required for the transformation into a one level problem. However, the strength of this model relies on the fact that it does not require any preprocessing phase to be solved, but only the instance description in terms of network topology, the cost and risk functions and the travel demand.

Finally, the k-median model does not allow to choose among alternative paths once a gateway is assigned to a commodity. In this sense, it is the most rigid model. While this formulation allows to tackle GLP by any solution approach developed to k–median, such a compact formulation completely hides the cost structure to the decision maker, and forbids to formalize within the optimization process any tuning between the two objective functions.

A weak point of bilevel programming models is stability [2]. While our solutions proved a posteriori to be stable, as discussed in Sect. 4, the issue should be addressed. The preprocessing phase of the k-median model can take care of computing, among the shortest gateway paths of each pair h, v, which is the one of maximum risk (for non-cooperative users behavior), and use that value as a_h^v to compute matrix A. While it is a longest path problem (maximum risk) with resource constraints (not more costly than $c_{\rho_v^c}$) it can be easily solved, since it is sufficient to explore the acyclic subgraph induced by 0-reduced cost arcs. Therefore, the k-median model yields a correct risk value even in case of several shortest gateway paths.

The path based model, as it models paths explicitly, does not suffer from instability. Concerning the arc based model, a perturbation of the cost coefficients, supposedly strictly positive, of the kind $c_{ij} - \alpha r_{ij}$, with $0 < \alpha << 1$, can make the optimal path unique and model a non-cooperative users attitude towards authority. The parameter α requires careful calibration, which goes beyond the scope of the paper.

As how to avoid to penalize excessively network users, the three models provide different solutions. As for the arc based model, an additional constraint for each commodity, setting an upper bound ub on the cost of each user's itinerary, can be added into the inner problem of the bilevel formulation. Note that such a constraint, whose coefficients are the same as the coefficients of the user's objective function, is either redundant or makes the inner problem infeasible. Therefore, it is still possible to reformulate the extended bilevel model as a single level MILP problem. The threshold ub may be yielded by negotiations between the user and the network administrator or simply be a fixed percentage over 100 of $c_{\rho_v^c}$. The path based model is even more flexible, since only gateway paths whose cost is within a threshold of $c_{\rho_v^c}$ may be allowed, thus guaranteeing a maximum deterioration of users objective function. On the other hand, the k-median model is completely blind to users cost, although an ad hoc constraint can be added, bounding the maximum total cost. Preprocessing should then retrieve information also on $c_{\rho_v^c}$.

4 Computational Results

This section is devoted to the results of an experimental campaign, aimed at verifying the effectiveness of the gateway approach. We tested the multicommodity flow model and the k-median like model on instances derived from published studies in the field of hazardous material transportation. In this sector, ruling the transport in urban and suburban area is still an open problem [9]. Basically, we built our instances on the same data set described in [7], i.e.: an indirected graph with $|N| = 105$ nodes and $|A| = 134$ arcs, being an abstraction of the road network of Ravenna (Italy), as well as a cost and a risk function, not collinear, defined on the arcs. Travel demand is also taken from [7], i.e., $|V| = 35$ commodities with their origin-destination pairs and shipment requests. On this network, we generated 130 instances of the GLP as follows: 5 different sizes for the candidate set N^{GTW} have been considered and expressed as a percentage of $|N|$ (10%, 20%, 30%, 50%, and 100% of the network nodes); for each percentage below 100%, N^{GTW} has been generated three times, and named hereafter as 0, 1 and 2, by independent random sampling with uniform distribution, and without any relationship of inclusion among samples of different percentages. For each one of these thirteen combinations, k varies in $1, \ldots, 10$, thus yielding a total of 130 instances. We solved both models using Cplex 12.1 and performed all testing on a AMD Athlon (tm) 64x2 Dual Core Processor 4200+ (CPU MHz 2211.186). Both models yielded the same results but differed in computing time. The k-median model solved the biggest instances in negligible time, including preprocessing, outperforming the arc based model (requiring just some seconds to solve the whole test bed). It should be mentioned, however, that efficiency is not the target of this campaign and we did not perform any fine tuning on the solver parameters to speed up the computation of the arc based model.

Computational experience reported in this section gives evidence of the following findings: (i) the use of gateways identifies a good strategy in mitigating the transportation risk, and few open gateways are sufficient to achieve a considerable risk reduction; (ii) the risk mitigation occurs at the expenses of an increase in transportation costs: such an increase is however acceptable; (iii) the choice of the candidate gateways represents a crucial issue in mitigating risk; (iv) the solutions obtained are stable from the risk point of view.

For each run a significant quality index is the percentage of risk reduction obtained at a value of k with respect to the risk value obtained at $k = 1$. Given an instance and a percentage of candidate gateways, this index clearly does not decrease as k increases. The value of k beyond which the risk reduction can be considered negligible (i.e., below 0.5%) is said *stable*. The stability analysis w.r.t. k shows that very good levels of risk mitigation are achieved with a small number k of open gateways. Specifically, for the test bed used here, the value of k at which stability occurs is equal to four. In Table 1, for each value of k, the percentage of runs for which the risk stability occurs at k, is given. This motivates our choice to further analyze the case $k = 4$.

Let $G^\star$, $R^\star$ and $C^\star$ denote respectively the optimal solution of the gateway location problem (i.e. the gateway path set), the minimum risk path set (over-regulated scenario) and the minimum cost path set (unregulated scenario); let then $c(P)$ and $r(P)$ denote respectively the total cost and the total risk of path set P. The quantity $(r(G^\star) - r(R^\star))/(r(C^\star) - r(R^\star)) \cdot 100$ yields an aggregate index of solution quality

Table 1. Stability Analysis w.r.t. k

k	1	2	3	4	5	6	7	8	9	10
Frequency	0.00	8.33	41.67	50.00	0.00	0.00	0.00	0.00	0.00	0.00

Table 2. % Cost Reduction w.r.t. Over Regulated Scenario

	% CostRed		
k	Inst 0	Inst 1	Inst 2
3	-22.97	-24.51	-23.51
4	-9.41	-20.49	-16.45
5	-9.67	-20.84	-14.76

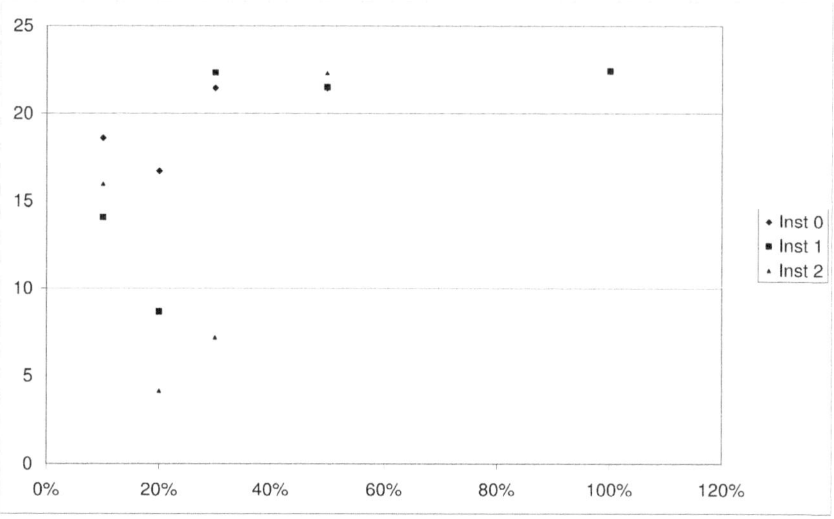

Fig. 1. Percentage of Risk Reduction w.r.t. Gateways Percentage - $k = 4$

since it measures the increase in risk with respect to the over-regulated scenario: the more the index is close to zero, the more the risk of the solution of GLP is close to the risk of the minimum risk solution. For a percentage of candidate gateways fixed to 30 and for $k = 4$, such index is equal respectively to 13.15, 13.14 and 16.62 for instances 0, 1 and 2. It is important to note that these values, already low in itself, are in addition very close to those obtained when all nodes of the network are considered as possible gateways; indeed, in this case, the average index is equal to 13.14 when $k = 4$ and it decreases only to 12.91 when k reaches its upper bound (i.e. 10).

Figure 2 plots for an intermediate percentage of candidate gateways (30%) and for each instance (namely, 0, 1 and 2) the risk reduction when k varies in the range $\{1,\ldots,10\}$. The maximum reduction of risk achievable on those runs is also reported

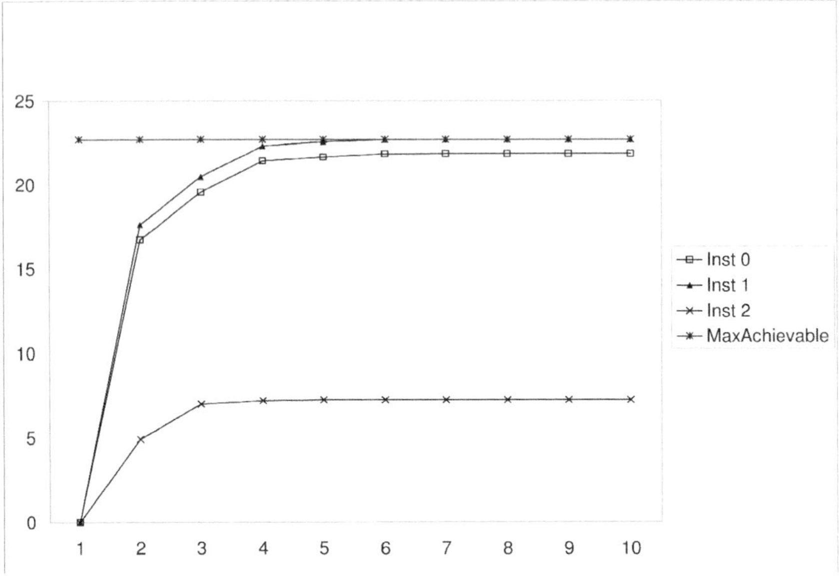

Fig. 2. Percentage of Risk Reduction w.r.t. k - 30% of Candidate Gateways

(it corresponds to the best risk reduction obtained when all nodes are candidate gateways). On the test bed instances, for the 14.29% of vehicles the path of minimum risk is the same as the path of minimum cost, thus generating an exemption; the risk reduction reported in this section is computed over the remaining 85.71% of vehicles. The figure shows that for two of the three instances (namely, 0 and 1) the candidate gateways allow to get very good risk reduction whereas for instance 2 the risk reduction does not grow over 7.24 even if a quite high number of gateways is open. This allows us to conclude that it is more profitable to choose the appropriate pool of candidate gateways rather than increasing its size. Recall that, in the current experiments, the candidate gateways are randomly sampled with uniform probability among all the nodes. This represents the starting phase of the validation of the gateway based approach. Indeed, the selection of the proper set of nodes that can be successfully used in order to mitigate the risk thus identifies a promising line of research which is currently under investigation. To analyze more in depth the relationships between the percentage of candidate gateways and the risk reduction, we report in Figure 1 the risk reduction obtained for $k = 4$ when the percentage of candidate gateways varies in $\{10\%, 20\%, 30\%, 50\%, 100\%\}$, separately for each instance.

As observed previously, there is a tradeoff between risk reduction and cost increase. The cost deterioration of our solution versus the unregulated scenario, and the cost save of our solution versus the over regulated scenario should be both discussed. The cost deterioration is the average computed on all the vehicles of relative increase of the gateway path cost w.r.t. the minimum path cost. Specifically, the cost deterioration is equal respectively to 43.3, 28.66 and 35.51 for the three instances when the 30% of the gateways are selected as candidate and $k = 4$. We believe that a cost deterioration as high as 35.82 on average can be considered acceptable. That being said, the extension of models to the case where an upper bound is given on cost deterioration is surely an interesting line of research we pursue. On the contrary, the cost save is given in Table 2: analogously to the cost deterioration, it is computed as the average relative decrease of the gateway path cost w.r.t. the cost of the minimum risk path.

Finally, we report that the solutions obtained with the gateway method are stable from the risk point of view: we verified experimentally that, fixed the gateway opening and fixed the assignment of gateways to vehicles, for each vehicle the maximum risk path among the minimum cost gateway paths coincides exactly with the path returned by our models.

5 Conclusions and Ongoing Work

In this work we have investigated a new method for indirect control of user flows by part of the network administrator. It consists of imposing check points along the routes, so called gateways, by obliging each user to detour from his original optimal path and pass by the assigned gateway. Gateways are located and assigned by the administrator so that the resulting user optimal flows on the detours minimize the administrators objective function. To manage this method we have proposed three different models: a bilevel multicommodity flow model, a path based model and a k-median like model. Such models present pros and cons concerning to robustness w.r.t. the risk of the shortest gateway path, exemption treatment, preprocessing phase necessity, possibility to be generalized in order to handle specific side constraints or the tradeoff risk-cost.

We have tested the multicommodity flow model and the k-median like model on realistic instances in the field of hazardous material transportation. The experimental campaign shows that the use of gateways is an effective strategy for mitigating the transport risk and good levels of risk reduction are already obtained with a low percentage of candidate gateways (30%) when they are chosen appropriately. In fact, experiments show that different choices of equal-sized candidate gateways sets may produce very different risk mitigation levels. In any case a small number of open gateways (4 for the test bed) is already sufficient to obtain the maximum reduction of risk achievable from a specific candidate gateways choice.

On going works focus on the criteria for generating suitable sets of candidate gateways, on the capacitated extensions of the models, on the risk-cost tradeoff, and on developing appropriate metaheuristics to tackle bigger instances.

References

1. Bergendorff, P., Hearn, D.W., Ramana, M.V.: Congestion toll pricing of traffic networks. In: Pardalos, P.M., et al. (eds.) Network Optimization. LNEMS, vol. 450, pp. 51–71. Springer, Heidelberg (1997)
2. Colson, B., Marcotte, P., Savard, G.: An overview of bilevel optimization. Annals of Operations Research 153(1), 235–256 (2007)
3. Lombard, K., Church, R.L.: The Gateway Shortest Path Problem: Generating Alternative Routes for a Corridor Routing Problem. Geographical Systems 1, 25–45 (1993)

4. Avella, P., Sassano, A., Vasilev, I.: Computational study of large-scale p-median problems. Mathematical Programming 109, 89–114 (2007)
5. Resende, M.G.C., Werneck, R.F.: A hybrid heuristic for the p-median problem. Journal of Heuristics 10, 59–88 (2004)
6. Kara, B.Y., Verter, V.: Designing a Road Network for Hazardous Materials Transportation. Transportation Science 38(2), 188–196 (2004)
7. Erkut, E., Gzara, F.: Solving the hazmat transport network design problem. Computers & Operations Research 35, 2234–2247 (2008)
8. Verter, V., Kara, B.Y.: A path-based approach for hazmat transport network design. Management Science 54(1), 29–40 (2008)
9. Bruglieri, M., Maja, R., Marchionni, G., Rainoldi, G.: Safety in hazardous material road transportation: state of the art and emerging problems. In: Bersani, C., et al. (eds.) Advanced Technologies and Methodologies for Risk Management in the Global Transport of Dangerous Goods. IOS Press, Amsterdam (2008)
10. Migdalas, A.: Bilevel programming in traffic planning: Models, methods and challenge. Journal of Global Optimization 7(4), 381–405 (1995)
11. Garcia, R., Marin, A.: Parking Capacity and Pricing in Park'n Ride Trips: A Continuous Equilibrium Network Design Problem. Annals of Operations Research 116, 153–178 (2002)
12. Cappanera, P., Scaparra, M.P.: Optimal Allocation of Protective Resources in Shortest-Path Networks. Transportation Science 45(1), 64–80 (2011)
13. Labbé, M., Marcotte, P., Savard, G.: A bilevel model of taxation and its application to optimal highway pricing. Management Science 44, 1608–1622 (1998)
14. Marcotte, P., Mercier, A., Savard, G., Verter, V.: Toll policies for mitigating hazardous materials transport risk. Transportation Science 43(2), 228–243 (2009)

Affine Decision Rules for Tractable Approximations to Robust Capacity Planning in Telecommunications

Adam Ouorou

Orange Labs, 38-40 rue du General Leclerc, 92794 Issy-les-Moulineaux Cedex 9
adam.ouorou@orange-ftgroup.com

Abstract. We consider the capacity assignment problem in telecommunications where the demand is uncertain. In our model, given an amount of traffic $\bar{\mu}$, we seek for an optimal link capacity assignment that, given an uncertainty set containing possible demand realizations, limits the loss of traffic to $\bar{\mu}$ in any realization of the demand. To obtain tractable approximations to this problem, we use the so-called *Affinely Adjustable Robust Counterpart* (AARC) concept proposed by Ben-Tal et al. where the adjustable variables are restricted to depend affinely on the uncertain data. Borrowing ingredients from earlier works and the AARC approach, we propose some tractable approximations to this problem.

1 Introduction

Let $\mathscr{G}$ be a network with nodes set $\mathscr{V}$ and links set $\mathscr{A}$. We denote by $\mathscr{K}$ the set of demands, characterized by $K = |\mathscr{K}|$ origin-destination (OD) pairs. Let $X = \{x \in \mathbb{R}^n_+ : x^L \leq x \leq x^U\}$ be the set of feasible capacities, where x^L and x^U are minimum and maximum capacity vectors. For a given $x \in X$, let $\Gamma(x)$ be the set of all possible path flows respecting the capacity x. In this paper, the set $\mathscr{P}_k$ of available paths that can be used to route the demand k is given in explicit form. For a given path $p \in \mathscr{P}_k$, δ^a_{kp} denotes a binary parameter which equals 1 if link a is used by p and 0 otherwise. Let N_p be the total number of available paths. The set $\Gamma(x)$ can be defined as

$$\Gamma(x) = \left\{ z \in \mathbb{R}^{N_p} : z \geq 0, \sum_{k \in \mathscr{K}} \sum_{p \in \mathscr{P}_k} \delta^a_{kp} z_{kp} \leq x_a, \ a \in \mathscr{A} \right\}$$

We propose a simple representation of the uncertain demands. The demand is supposed to lie somewhere in an uncertainty set. In [11], a general polyhedron uncertainty set was considered including the so-called hose model. For the applications we have in mind we consider the uncertainty model already used in [14,2]. Indeed, practitioners can provide a range around a nominal value ξ^n_k for each individual demand, so that the demand could be set as $\xi^n_k + \xi_k \xi^v_k$ to capture this information. The quantity $\xi_k \in [-1,1]$ is a random variable, and ξ^v_k the standard deviation (or any other measure of dispersion). A natural uncertainty set is defined by the constraint

$$||\xi||_\infty = \max_{k \in \mathscr{K}} |\xi_k| \leq \kappa$$

J. Pahl, T. Reiners, and S. Voß (Eds.): INOC 2011, LNCS 6701, pp. 277–282, 2011.

which limits the perturbations (i.e $|\xi_k| \leq \kappa$ for each k). As it is highly improbable that all demands be simultaneously at their highest value $\xi_k^n + \kappa \xi_k^v$, it is reasonable to constrain the uncertainty set further, by adding the global constraint

$$\|\xi\|_1 = \sum_{k \in \mathscr{K}} |\xi_k| \leq \tau.$$

The demand set is then the collection of vectors of $\mathbb{R}^K$ with components $\xi_k^n + \xi_k \xi_k^v$, $k \in \mathscr{K}$, where ξ belongs to the uncertainty set

$$\Xi = \{\xi : \xi \in [-1,1]^K, \ \|\xi\|_\infty \leq \kappa, \ \|\xi\|_1 \leq \tau\}. \tag{1}$$

The parameters κ and τ are safety factors to be set by the decision-maker.

A $\xi \in \Xi$ and a set of path flows $z \in \Gamma(x)$ induce a mismatch cost $f(\xi, z)$ which is 0 if

$$\sum_{p \in \mathscr{P}_k} z_{kp} \geq \xi_k^n + \xi_k \xi_k^v \quad \text{for any} \quad k \in \mathscr{K},$$

i.e if x is sufficient to accomodate the requirements $\xi_k^n + \xi_k \xi_k^v$, $k \in \mathscr{K}$. Whenever ξ is fixed, it is of course convenient to choose z so as to minimize this mismatch. For any $x \in X$ and $\xi \in \Xi$, let $h(x,\xi)$ be the function defined by

$$h(x,\xi) = \min_{z \in \Gamma(x)} f(\xi, z).$$

We consider the total amount of unserved requests as mismatch cost, i.e.,

$$f(\xi, z) = \sum_{k \in \mathscr{K}} \max\{0, \ \xi_k^n + \xi_k \xi_k^v - \sum_{p \in \mathscr{K}} z_{kp}\}.$$

we consider the following model

$$\begin{aligned} \min_x \ & \sum_{a \in \mathscr{A}} c_a x_a \\ \text{s.t.} \ & h(x,\xi) \leq \bar{\mu}, \ \xi \in \Xi, \\ & x \in X, \end{aligned} \tag{2}$$

where $\bar{\mu} \geq 0$ is a given parameter.

2 Affine Decision Rules

The function $h(x,\xi)$ is obtained by solving the following problem

$$h(x,\xi) = \min\{g(s,z) : (s,z) \in \Psi(x,\xi)\}$$

where $g(s,z) = \sum_{k \in \mathscr{K}} s_k$ and,

$$\Psi(x,\xi) = \left\{ (s,z) \in \mathbb{R}^{|\mathscr{K}|} \times \Gamma(x) : \begin{cases} s_k \geq \xi_k^n + \xi_k \xi_k^v - \sum_{p \in \mathscr{P}_k} z_{kp}, \ k \in \mathscr{K}, \\ s_k \geq 0, \ k \in \mathscr{K} \end{cases} \right\},$$

The problem (2) can then be rewritten as

$$\begin{aligned}
\min_x \ & \sum_{a\in\mathscr{A}} c_a x_a \\
\text{s.t.} \ & \min_{s,z}\{g(s,z):\ (s,z)\in\Psi(x,\xi)\}\le\bar{\mu},\ \ \xi\in\Xi, \\
& x\in X,
\end{aligned} \tag{3}$$

We make a first approximation in contenting ourselves with a solution $(s(\xi),z(\xi))$ in $\Psi(x,\xi)$ that respect the upper bound $\bar{\mu}$ instead of the minimizer itself. We come to the following *upper* approximation to (3)

$$\begin{aligned}
\min_x \ & \sum_{a\in\mathscr{A}} c_a x_a \\
\text{s.t.} \ & \sum_{k\in\mathscr{K}} s_k(\xi)\le\bar{\mu},\ \ \xi\in\Xi, \\
& \sum_{k\in\mathscr{K}}\sum_{p\in\mathscr{P}_k}\delta^a_{kp}z_{kp}(\xi)\le x_a,\quad a\in\mathscr{A},\ \xi\in\Xi, \\
& s_k(\xi)\ge\xi^n_k+\xi_k\xi^v_k-\sum_{p\in\mathscr{P}_k} z_{kp}(\xi),\ \ k\in\mathscr{K},\ \ \xi\in\Xi, \\
& z_{kp}(\xi)\ge 0,\quad k\in\mathscr{K},\ p\in\mathscr{P}_k,\ \ \xi\in\Xi, \\
& s_k(\xi)\ge 0,\ \ k\in\mathscr{K},\ \ \xi\in\Xi, \\
& x\in X.
\end{aligned} \tag{4}$$

The capacity x, which must be computed at the first stage, in full uncertainty, they are *non-adjustable* in the parlance of [5], see also [7]. Second the *adjustable* variables, which can be computed after the uncertain parameters are known; these are the path flows which can be tuned themselves to the observed demand. To obtain a tractable approximation for problem (2), we first impose as in [14,2], the dependence of path flows on the random data ξ to follow the *affine decision rule* in the context of robust optimization, that is $z(\xi)$ is an affine function of ξ:

$$z(\xi)=z_0+Z\xi$$

where the new variables are now $z_0\in\mathbb{R}^{N_p}$ with components $z_{0,kp}$, and the $N_p\times|\mathscr{K}|$ matrix with elements $z_{kp,k'}$. Hence,

$$z_{kp}=z_{0,kp}+\sum_{k'\in\mathscr{K}} z_{kp,k'}\xi_{k'}. \tag{5}$$

Likewise, we assume that each $s_k(\xi)$ follows the affine decision rule:

$$s_k(\xi)=s_{k,0}+\sum_{k'\in\mathscr{K}} s_{k,k'}\xi_{k'}. \tag{6}$$

These auxiliary variables $s_k(\xi)$ represent piecewise linear functions $\max\{0,\ \xi^n_k+\xi_k\xi^v_k-\sum_{p\in\mathscr{P}_k} z_{kp}(\xi)\}$. As pointed out in [6], working with variables that represent piecewise linear functions of the data lead to complicated robust counterpart. The affine recourse has been used with efficiency in the context of telecommunications problems [14,2,18]. This paper is a short version of [15].

3 Tractable Approximations

Replacing the variables with their affine decision rules, we come to the following approximation of (4).

$$\min \sum_{a\in\mathscr{A}} c_a x_a \tag{7a}$$

$$\text{s.t.} \sum_{k\in\mathscr{K}} (s_{k,0} + \sum_{k'\in\mathscr{K}} s_{k,k'}\xi_{k'}) \leq \bar{\mu},\ \xi \in \Xi, \tag{7b}$$

$$\sum_{k\in\mathscr{K}} \sum_{p\in\mathscr{P}_k} \delta^a_{kp} z_{0,kp} + \sum_{k\in\mathscr{K}} \sum_{p\in\mathscr{P}_k} \sum_{k'\in\mathscr{K}} \delta^a_{kp}\xi_{k'} z_{kp,k'} \leq x_a,\ a \in \mathscr{A},\ \xi \in \Xi, \tag{7c}$$

$$s_{k,0} + \sum_{k'\in\mathscr{K}} s_{k,k'}\xi_{k'} + \sum_{p\in\mathscr{P}_k} z_{0,kp} + \sum_{p\in\mathscr{P}_k}\sum_{k'\in\mathscr{K}} z_{kp,k'}\xi_{k'} \geq \xi^n_k + \xi_k\xi^v_k,\ k \in \mathscr{K},\ \xi \in \Xi \tag{7d}$$

$$z_{0,kp} + \sum_{k'\in\mathscr{K}} z_{kp,k'}\xi_{k'} \geq 0,\ k \in \mathscr{K},\ p \in \mathscr{P}_k,\ \xi \in \Xi, \tag{7e}$$

$$s_{k,0} + \sum_{k'\in\mathscr{K}} s_{k,k'}\xi_{k'} \geq 0,\ k \in \mathscr{K},\ \xi \in \Xi, \tag{7f}$$

$$z_{0,kp} \in \mathbb{R},\ k \in \mathscr{K},\ p \in \mathscr{P}_k,\ \xi \in \Xi, \tag{7g}$$

$$z_{kp,k'} \in \mathbb{R},\ k \in \mathscr{K},\ p \in \mathscr{P}_k,\ k' \in \mathscr{K},\ \xi \in \Xi, \tag{7h}$$

$$s_{k,0} \in \mathbb{R},\ k \in \mathscr{K},\ \xi \in \Xi, \tag{7i}$$

$$s_{k,k'} \in \mathbb{R},\ k \in \mathscr{K},\ k' \in \mathscr{K},\ \xi \in \Xi, \tag{7j}$$

$$x \in X. \tag{7k}$$

From Theorem 2, Propositions 1 and 4 of [1] , we can express the above problem with a finite number of linear constraints. For instance, constraints (7b) are replaced with

$$\begin{aligned}
&\sum_{k\in\mathscr{K}} s_{k,0} + \tau\gamma + \kappa \sum_{k\in\mathscr{K}} \alpha_k \leq \bar{\mu},\\
&\gamma + \alpha_k - \sum_{k'\in\mathscr{K}} s_{k',k} \geq 0, \quad k \in \mathscr{K},\\
&\gamma + \alpha_k + \sum_{k'\in\mathscr{K}} s_{k',k} \geq 0, \quad k \in \mathscr{K},\\
&\gamma \geq 0,\ \alpha_k \geq 0, \quad k \in \mathscr{K}.
\end{aligned}$$

Proceeding in the same way for all the constraints, results in the first tractable approximation (Model 1). The second and third approximations (Models 3 and 2) are obtained by fixing $z_{kp,k'} = 0$ and $s_{k,k'} = 0$ in (5) and (6) respectively, see [15] for more details. A fourth model is obtained as the capacity planning problem where the demand is at their possible maximum value (Model 4). All these models results in LP and MIP which can be solved using CPLEX. Our preliminary experiments on test problems based on data from SNDLib (Table 1) show however that CPLEX with barrier option fails on some instances (on nobel-us and planar) due to lack of memory. We refer to [15] fo the details of these experiments to respect the limitation imposed on the lenght of the present paper. Table 1 gives some results obtained on the test problems considered with only one path per each origin-destination pair. These experiments show that the overspend from the solutions obtained with the worst case scenario model wich is up to 20.23% but it may be more important, up to 35% on some instances which are not presented

Table 1. Numerical results

Test problems	$\lvert\mathscr{V}\rvert$	$\lvert\mathscr{A}\rvert$	$\lvert\mathscr{K}\rvert$	Model 1 CPU time	Model 1 Cost	Model 2 CPU time	Model 2 Overs. (%)	Model 3 CPU time	Model 3 Overs. (%)	Model 4 CPU time	Model 4 Overs. (%)
pdh	11	34	24	0.94	6.52E+08	0.2	6.72	0.6	0	0.0	6.72
di-yuan	11	42	22	0.73	5.65E+06	0.19	3.44	0.46	1.1	0.0	3.44
polska	12	18	66	6.69	5.84E+06	5.19	15.53	1.81	12.06	0.0	18.93
nobel-us	14	21	91	25.83	1.07E+08	18.36	17.85	5.4	9.66	0.01	20.23
planar30	30	75	92	29.21	7.63E+07	18.87	13.06	4.4	7.39	0.0	14.22

here. This worst case scenario model is likely to be used in general when faced with demand uncertainty. We also observed that allowing path flows to be adjustable is also important even in the case of single routing: the overexpenditure maybe important up to 28%.

References

1. Babonneau, F., Vial, J.-P., Apparigliato, R.: Robust Optimization for environmental and energy planning. In: Handbook on Uncertainty and Environmental Decision Making. International Series in Operations Research and Management Science. Springer, Heidelberg (2009)
2. Babonneau, F., Klopfenstein, O., Ouorou, A., Vial, J.-P.: Robust capacity expansion solutions for telecommunications networks with uncertain demands. Technical paper, Orange Labs R&D (2009)
3. Ben-Ameur, W., Kerivin, H.: New economical virtual private networks. Communication of the ACM 46(6), 69–73 (2003)
4. Ben-Ameur, W., Kerivin, H.: Routing of uncertain traffic demands. Optimization and Engineering 6(3), 283–313 (2005)
5. Ben-Tal, A., Goryashko, A., Guslitzer, E., Nemirovski, A.: Adjustable robust solutions of uncertain linear programs. Mathematical Programming 99(2), 351–376 (2004)
6. Ben-Tal, A., Golany, B., Nemirovski, A., Vial, J.-P.: Retailer-Supplier Flexible Commitments Contracts: A Robust Optimization Approach. Manufacturing & Service Operations Management 7(3), 248–271 (2005)
7. Ben-Tal, A., El Ghaoui, L., Nemirovski, A.: Robust Optimization. Princeton University Press, Princeton (2009)
8. Bertsimas, D., Sim, M.: The price of robustness. Operations Research 52(1), 35–53 (2004)
9. Hijazi, H.: Mixed-integer nonlinear optimization approaches for network design in telecommunications. Ph.D Thesis, Université de Marseille (2010)
10. Klopfenstein, O.: Optimisation robuste des réseaux de télécommunications. Ph.D Thesis, UTC (2008)
11. Lemaréchal, C., Ouorou, A., Petrou, G.: Robust network design in telecommunications under polytope demand uncertainty. European Journal on Operational Research 206, 634–641 (2010)
12. Minoux, M.: On 2-stage robust LP with RHS uncertainty: Complexity results and applications (2009)
13. Minoux, M.: Robust network optimization under polyhedral demand uncertainty is NP-hard. Discrete Applied Mathematics 158(5), 597–603 (2010)

14. Ouorou, A., Vial, J.-P.: A model for robust capacity planning for telecommunications under demand uncertainty. In: Proceedings of Design of Reliable Communications Networks, IEEE Xplore (2007), doi: 10.1109/DRCN.2007.4762287
15. Ouorou, A.: Tractable approximations to a robust capacity assignment model in telecommunications under demand uncertainty (2010) (submitted)
16. Pascali, F.: Chance Constrained Network Design. PhD thesis, Doctorate School in Mathematics for Economic Decisions (2009), `http://etd.adm.unipi.it/theses/available/etd-11262009-005543`
17. Petrou, G., Lemaréchal, C., Ouorou, A.: An approach to robust network design in telecommunications. RAIRO-Operations Research 41(4), 411–426 (2007)
18. Poss, M., Raack, C.: Affine recourse for the robust network design problem: between static and dynamic routing. ZIB-Report 11-03 (February 2011), available on OptimizationOnline

Optimal Download Time in a Cloud-Assisted Peer-to-Peer Video on Demand Service

Pablo Rodríguez-Bocca and Claudia Rostagnol

Instituto de Computación, Facultad de Ingeniería, Universidad de la República, Montevideo, Uruguay
prbocca@fing.edu.uy, crostag@fing.edu.uy

Abstract. In this paper we present a mathematical model to optimize the average download time in a Peer-to-peer *Video on Demand* system, where a set of resources are available in the cloud to assist the service. First, we propose a simple optimization model based on a Markov chain. Then, we provide some numerical results based on simulations and optimizations using a GRASP method on a real scenario.

1 Introduction

Nowadays, lots of applications used to share contents over the Internet are based on the BitTorrent protocol [3,5]. One of such applications is the GoalBit Video Platform [1], currently used to share live content over a P2P network. We are working on adding *Video on Demand* (VoD) support to GoalBit, using the standard BitTorrent protocol. In GoalBit, as in BitTorrent, end-users are called *peers*. They are classified in two groups: if the peer is downloading a content it is a *downloader*; when a peer finishes downloading it becames a *seeder*. There is also an entity or node named *tracker*, which knows all the peers that are sharing a content (seeding or downloading). GoalBit introduces a new type of node to the P2P network named *super-peer*. This kind of node has better bandwidth than a normal peer, and usually is in the network for very long periods of time (very stable peers). The super-peers are intended to store and upload the contents to normal peers (with a very short life in the system). In the current GoalBit protocol specification, super-peers are nodes managed by the operator of the platform and they are hosted in the cloud. We are thinking about the possibility of promoting peers to super-peers in future specifications. For more details about GoalBit specification please refer to [1].

The work in [3] provides an analysis of the estimated average time to download a file (*content*) on a BitTorrent network, assuming that the behavior of peers can be modeled by a Markov chain. This work is generalized in [5], extending the model for several concurrent contents, assuming that most BitTorrent users download several files at the same time. Our goal is to distribute video files over the GoalBit platform depending on the storage capacity and videos' popularity, minimizing the average download time for end-users. To achieve that, we extend those models, so when a video becomes very popular and lots of users want to watch it, we generate more copies for this video in the cloud (if we have enough resources), in order to satisfy the users demand and do not

J. Pahl, T. Reiners, and S. Voß (Eds.): INOC 2011, LNCS 6701, pp. 283–288, 2011.

increase the average download time of the system. When the video becames less popular, we can remove some copies to free space and resources. All this process is made automatically and dynamically based on the model that we defined. In Section 2 we introduce our combinatorial optimization problem based on a fluid model that extends the previous work [5]. In Section 3 we present a GRASP [4] metaheuristic solution for that problem. Finally, in Section 4, we show the performance of the solution in a real scenario, and we present general conclusions of our work.

2 Video-on-Demand Fluid Model

To understand the behavior of peers in a P2P system like GoalBit we should analyze the evolution, scalability, and sharing efficiency of peers. We extend the stochastic fluid model presented in [5], providing insightful results for performance issues and the downloading average time for several contents downloaded simultaneously. Since each peer can download more than one content at time t, peers are grouped in classes: $\{C^1, C^2, \ldots, C^K\}$, such that a peer is in the class C^i if it is downloading i contents at the same time. The data and variables of the model are shown in Figure 1.

K	available videos
s_j	size (in `kbits`) of video j
$x^i_j(t)$	*downloaders* in class C^i downloading video j at time t
$y^i_j(t)$	*seeders* in class C^i seeding video j at time t
$z^i_j(t)$	*super-peers* in class C^i seeding video j at time t
λ^i_j	arrival rate for peers in class C^i requesting video j (where $\sum_i \lambda^i_j$ is the j-th video popularity)
γ	departure rate of *seeders*
c	total download bandwidth for each peer (in `kbps`)
μ	total upload bandwidth for each peer (in `kbps`)
ρ	total upload bandwidth for each *super-peer* (in `kbps`)
η	video sharing effectiveness between peers ($\eta \in [0,1]$)

Fig. 1. Data and variables of the fluid model

To simplify the model representation we assume the following (as in [5]):

1. Resources are used equitably among the contents that are downloaded or served simultaneously. If the peer belongs to class C^i, each video that it downloads will have the i-th part of the peer's bandwidth. Since videos have different sizes, we divide the bandwidth (in `kbps`) by the video size in order to know the download rate (in files per second) for video j. Therefore, if the peer is in class C^i, the file portion downloaded per second for content j is $c^i_j = \frac{c}{is_j}$. The same is applied to $\mu^i_j = \frac{\mu}{is_j}$ and $\rho^i_j = \frac{\rho}{is_j}$ $\forall i, j \in \{1 \ldots K\}$.
2. Peers in class C^i, that at time t are downloading video j, receive from all other ***downloaders*** an amount of content proportional to the upload bandwidth μ^i_j and their population $x^i_j(t)$: $\frac{\mu^i_j x^i_j(t)}{\sum_k \mu^k_j x^k_j(t)} \sum_k \eta \mu^k_j x^k_j(t) = \eta \mu^i_j x^i_j(t)$.

3. Peers in class C^i, that at time t are downloading video j, receive from all the ***seeders*** an amount of content proportional to the download bandwidth c^i_j and their population $x^i_j(t)$: $\frac{c^i_j x^i_j(t)}{\sum_k c^k_j x^k_j(t)} \sum_k \mu^k_j y^k_j(t)$.
4. Peers in class C^i, that at time t are downloading video j, receive from all the ***super-peers*** an amount of content proportional to the download bandwidth c^i_j and their population $x^i_j(t)$: $\frac{c^i_j x^i_j(t)}{\sum_k c^k_j x^k_j(t)} \sum_k \rho^k_j z^k_j(t)$.

Supposing that we can model the problem as a Markovian chain, we want to know the peers' behavior (how x^i_j and y^i_j vary as a function of time). Modeling the behavior as a simple fluid model we get the following equation $\forall i, j \in \{1 \ldots K\}$:

$$\frac{dx^i_j}{dt} = \lambda^i_j - \eta \mu^i_j x^i_j(t) - \frac{c^i_j x^i_j(t)}{\sum_k c^k_j x^k_j(t)} \sum_k \mu^k_j y^k_j(t) - \frac{c^i_j x^i_j(t)}{\sum_k c^k_j x^k_j(t)} \sum_k \rho^k_j z^k_j(t) \tag{1}$$

$$\frac{dy^i_j}{dt} = \eta \mu^i_j x^i_j(t) + \frac{c^i_j x^i_j(t)}{\sum_k c^k_j x^k_j(t)} \sum_k \mu^k_j y^k_j(t) + \frac{c^i_j x^i_j(t)}{\sum_k c^k_j x^k_j(t)} \sum_k \rho^k_j z^k_j(t) - \gamma y^i_j(t) \tag{2}$$

Assuming that the system will reach its *steady state*, where the number of peers is stable ($\frac{dx^i_j}{dt} = \frac{dy^i_j}{dt} = 0 \; \forall i, j \in \{1 \ldots K\}$), we can calculate the steady state value for $x^i_j(t)$ and $y^i_j(t)$:

$$\overline{x^i_j} = \max\{\frac{\lambda^i_j i s_j}{c}, \frac{i\lambda^i_j}{\gamma\mu\eta} \frac{\gamma s_j \sum_k \lambda^k_j - \mu \sum_k \frac{\lambda^k_j}{k} - \gamma\rho \sum_k \frac{z^k_j}{k}}{\sum_k \lambda^k_j}\} \qquad \overline{y^i_j} = \frac{\lambda^i_j}{\gamma} \tag{3}$$

Equations (1) and (2) assume that the bandwidth constraint is in the upload capacity of the system, Equation (3) generalizes this, considering also that the bandwidth constraint can be at the download capacity of peers.

Based on this model, we want to minimize the average download time of the system, making an efficient use of resources, trying to find an optimal distribution of files in super-peers nodes. The average download time for any *downloader* at steady state can be computed applying *Little's law*: $T^i_j = \frac{\overline{x^i_j}}{\lambda^i_j}$. Then, our problem can now be written as a *Combinatorial Optimization Problem* (COP), where we have to add new data:

- E^p_j indicates if the *super-peer* p has a copy of video j (p can seed video j). E^p_j is either 0 or 1: 1 if p has video j, 0 otherwise.
- S^p is the storage capacity of *super-peer* p (in `kbits`).

The optimization problem is shown in Figure 2.

The problem constraint (1) indicates that no *super-peers* can store more videos than its storage capacity. Additionally, in constraint (2) we define that each video must have at least one replica (each video must be stored in at least 2 *super-peers*). Also, the number of video replicas is limited by the peers' download capacities and the seeders' upload capacities for this video, as described in constraint (3). Finally, z^i_j can be computed from E^p_j as the number of super-peers that hosts content j and other $i-1$ contents. Constraints (4)-(7) specify this relationship between z^i_j and E^p_j using some auxiliary variables ($z^{i,p}_j$ and $u^{i,p}$).

$$\min_{E_j^p} \sum_{j=1}^{K} \sum_{i=1}^{K} \lambda_j^i T_j^i \quad = \quad \min_{E_j^p} \sum_{j=1}^{K} \sum_{i=1}^{K} \bar{x}_j^i$$

$$s.t.$$

$$(1) \sum_j E_j^p s_j \leq S^p \; \forall p \qquad (2) \sum_p E_j^p \geq 2 \; \forall j \qquad (3) \sum_k \frac{z_j^k}{k} \leq (\frac{c}{\mu} - \eta) \sum_k \frac{x_j^k}{k} - \sum_k \frac{y_j^k}{k} \; \forall j \qquad (4)\; z_j^i = \sum_p z_j^{i,p} \; \forall i,j$$

$$(5)\; u^{i,p} = \left| \sum_l E_l^p - i \right| \; \forall i,p \qquad (6)\; z_j^{i,p} \geq E_j^p - u^{i,p} \; \forall i,j,p \qquad (7)\; E_j^p \in \{0,1\}, z_j^i \in \{0,P\}, z_j^{i,p} \in \{0,1\}, u^{i,p} \in \mathbb{R}^+ \; \forall i,j,p$$

Fig. 2. Combinatorial Optimization Problem

3 Model Optimization Based on GRASP

Considering that the number of feasible solutions of the problem increases a lot with the size of the problem' instance, we will use a metaheuristic approach in order to solve it (we did not a complexity analysis of the problem at this time). GRASP [4] is a well-known metaheuristic that we have been successfully using to solve other similar hard COPs [2]. It is an iterative process which operates in two phases. In the *Construction Phase* an initial feasible solution is built, whose neighborhood is then explored in the *Local Search Phase*. In Figure 3 we present a GRASP customization to solve our problem. During *construction phase* we must distribute the video files in the super-peers taking into account the constraints of the problem. First, we sort the files depending on their sizes, starting by the largest one. Then, we select the 20% larger files (not copied yet) to create the *Restricted Candidate List* (RCL), and choose one of them randomly to be put in at least two super-peers (selecting super-peers with more storage capacity first). The pseudo-code for this construction phase is shown in Figure 3(a). To improve the solution constructed in the first phase, a local search is applied as second phase. The improvement can be done in 2 ways, applying only one per iteration, selected randomly (both without breaking the problem constraints): (a) inserting a new copy of video *k* in

```
Procedure RandomGreedy
Input: data_0
x ← emptySolution(data_0)
RCL = biggerFiles(data_0, x, 20%)
i = randomSelect(RCL)
while i > 0 do
    sp_1 ← fstBestSP(data_0, x, i)
    x ← x ∪ Copy(i, sp_1)
    if checkRestrictions()
        sp_2 ← sndBestSP(data_0, x, i)
        if checkRestrictions()
            x ← x ∪ Copy(i, sp_2)
        else
            x ← x ∪ Remove(i, sp_2)
        end if
    else
        x ← x ∪ Remove(i, sp_1)
    end if

    RCL = biggerFiles(data_0, x, 20%)
    i = randomSelect(RCL)
end while
return x
```

(a) Construction phase

```
Procedure LocalSearch
Input: x
x* ← clone(x)
i ← 0
while i ≤ i_max do
    x_temp ← randomChange(x)
    if evaluate(x) < evaluate(x*)
        x* ← x_temp
        i = 0
    else
        i++
    end if
        x ← x_temp
end while
return x*
```

(b) Local Search phase

Fig. 3. Pseudo-cide for GRASP phases

the super-peer sp; (b) swaping two videos k_1 and k_2 from two super-peers sp_A y sp_B. The pseudo-code for this local search phase is shown in Figure 3(b).

4 Numerical Results and Discussion

We implement the COP in Matlab, and we calibrate the GRASP algorithm with generated instances. In this paper we present an application into a real scenario in order to show its potential. Using real information got from a local Internet Video-on-Demand Service Provider we construct a real scenario. From the log information of this service, we obtained the popularity and the size of the available video content. Specifically, our scenario has more than 700 videos (K), with an average size of 23 MB (s), 4 super-peers (P) with 100 GB of storage (S) and an upload rate of 80 Mbps (ρ), a peer download rate of 8 Mbps (c), and 4 Mbps of upload (μ), using an effectiveness of 50% ($\eta = 0.5$), having a seeders departure rate of 1 every 10 seconds ($\gamma = 0.1$).

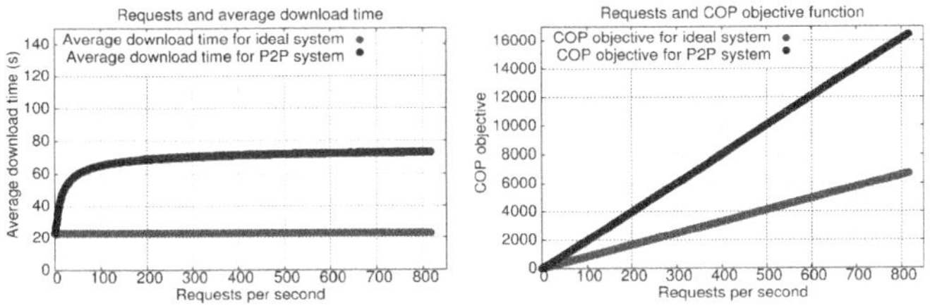

(a) Average download time (b) COP objective function

Fig. 4. Comparing ideal system and P2P system

A lower boundary of the average download time can be computed as the time needed in a system with free upload capacity (i.e. when the download time is determined by the peers download rate). In this ideal scenario, the number of downloaders is $\overline{x_{ideal}}_j^i = \frac{\lambda_j^i s_j}{c}$ and $T_{ideal}{}_j^i = \frac{\overline{x_{ideal}}_j^i}{\lambda_j^i}$. With data provided above, we computed an average ideal time of 23.1 seconds.

In order to determine the scalability of the service, we stress the system keeping the popularity proportional with the real data (i.e. multiplying the real λ by an incremental factor). These results are shown on Figure 4(a), where we can see that our P2P system is a bit far from the ideal system (3 times worse), but it is very scalable since the performance is stable regarding to the increment of requests. With 163 requests per second the average download time is 67 seconds, while with 816 requests the average time is 73 seconds. In Figure 4(b) we show the evolution of the COP objective function for the ideal system and for the P2P system. Notice that to reach the same level of service (the same average download time) in a client-server system we should increase the number of servers (or super-peers) proportionally with the end-user requests, while in the P2P system we have a natural scalability with the growing resources offered by the users (downloaders and seeders).

Therefore, we can conclude that our P2P solution is a good and very scalable approach. Although in this scenario we have a solution 3 times worse, it is better than a client-server option where we should increase the number of servers depending on the number of client requests. We also expect that the efficiency will be more evident in largest deployments.

Currently, we are working to include this model inside the GoalBit platform to test it in a real production scenario.

References

1. Bertinat, M.E., Vera, D.D., Padula, D., Robledo, F., Rodríguez-Bocca, P., Romero, P., Rubino, G.: Goalbit: The first free and open source peer-to-peer streaming network. In: Proceedings of the 5th International IFIP/ACM Latin American Conference on Networking (LANC 2009), pp. 83–93. ACM, New York (September 2009)
2. Martínez, M., Morón, A., Robledo, F., Rodríguez-Bocca, P., Cancela, H., Rubino, G.: A GRASP algorithm using RNN for solving dynamics in a P2P live video streaming network. In: 8th International Conference on Hybrid Intelligent Systems (HIS 2008), Barcelona, Spain, September 10-12 (2008), `http://his2008.lsi.upc.edu/`
3. Qiu, D., Srikant, R.: Modeling and performance analysis of bittorrent-like peer-to-peer networks. In: Proceedings of SIGCOMM 2004, pp. 367–378. ACM, New York (September 2004)
4. Resende, M.G.C., Ribeiro, C.C.: Greedy Randomized Adaptive Search Procedures. In: Glover, F., Kochenberger, G. (eds.) Handbook of Methaheuristics. Kluwer Academic Publishers, Dordrecht (2003)
5. Tian, Y., Wu, D., Ng, K.W.: Analyzing multiple file downloading in bittorrent. In: Proceedings of ICPP 2006, pp. 297–306. IEEE, Los Alamitos (August 2006)

The Maximum Flow Problem with Conflict and Forcing Conditions

Ulrich Pferschy and Joachim Schauer

University of Graz, Department of Statistics and Operations Research,
Universitaetsstr. 15, A-8010 Graz, Austria
{pferschy,joachim.schauer}@uni-graz.at

Abstract. We study the maximum flow problem subject to binary disjunctive constraints in a directed graph: A *negative disjunctive constraint* states that a certain pair of arcs in a digraph cannot be simultaneously used for sending flow in a feasible solution. In contrast to this, *positive disjunctive constraints* force that for certain pairs of arcs at least one arc has to carry flow in a feasible solution.

Negative (positive) disjunctive constraints can be represented by a *conflict (forcing) graph* whose vertices correspond to the arcs of the underlying graph, and whose edges encode the constraints.

We show that the maximum flow problem is strongly $\mathcal{NP}$-hard, even if the conflict graph contains only isolated edges and the network consists only of disjoint paths. For forcing graphs the problem can be solved efficiently if fractional flow values are allowed. If flow values are required to be integral we provide the sharp line between polynomially solvable and strongly $\mathcal{NP}$-hard instances.

1 Introduction

We study two variants of the maximum flow problem (MF) on directed graphs where *binary disjunctive constraints* are given on certain pairs of arcs.

- A negative disjunctive constraint expresses an incompatibility or a conflict between the two arcs in a pair. From each conflicting pair, at most one arc can carry flow in a feasible solution.
- A positive disjunctive constraint enforces that in a feasible solution flow has to be sent over at least one arc from the underlying pair.

We will represent these binary disjunctive constraints by means of an undirected constraint graph: Every vertex of the constraint graph corresponds to an arc of the original digraph, and every edge corresponds to a binary constraint. In the case of negative disjunctive constraints this constraint graph will be called *conflict graph*, and in the case of positive disjunctive constraints this graph will be called *forcing graph*.

Conflict graphs were considered for many other combinatorial optimization problems in the literature from a complexity and an approximation point of

J. Pahl, T. Reiners, and S. Voß (Eds.): INOC 2011, LNCS 6701, pp. 289–294, 2011.

view, in particular with respect to special conflict graph classes, e.g. for knapsack problems [6], bin packing [5], scheduling [2], and minimum spanning trees [3].

In this paper we consider the *maximum flow problem with conflict graph* (*MFCG*) and the *maximum flow problem with forcing graph* (*MFFG*). Given a standard *maximum flow problem* (*MF*) on a directed graph $G = (N, A)$ with n nodes and m arcs, integer capacities u_{ij} for each arc $a = (i, j)$, a designated source node s and sink node t, we use flow variables x_{ij} (or x_a) describing the flow on each arc $a = (i, j)$. Adding to this standard formulation the negative disjunctive structure defined by a conflict graph $H = (A, E)$ with vertices corresponding to the arcs of G gives $MFCG$:

$$(MFCG) \qquad (a, \bar{a}) \in E \implies (x_a = 0 \vee x_{\bar{a}} = 0)$$

$MFFG$ adds to the standard problem MF the following constraints induced by a forcing graph $H = (A, E)$:

$$(MFFG) \qquad (a, \bar{a}) \in E \implies (x_a + x_{\bar{a}} > 0)$$

Note that we allow cyclic flows through some parts of the network to fulfill the forcing condition as long as the flow conservation constraints are met.

We will also use the *maximum flow problem with lower bounds* (*MFLB*), where for certain arcs (i, j) the feasible flow has to reach a given amount l_{ij}:

$$0 \le l_{ij} \le x_{ij} \le u_{ij} \tag{1}$$

The standard definition of s-t cut $(S, \bar{S})$ with capacity $c(S, \bar{S}) = \sum_{(i,j) \in (S, \bar{S})} u_{ij}$ will be used. For use as a conflict (resp. forcing) graph we introduce a 2-*ladder* which is an undirected graph whose components are paths of length one, i.e. isolated edges connecting pairs of vertices.

In this paper we will fully characterize the complexities of $MFCG$ and $MFFG$. $MFCG$ is already strongly $\mathcal{NP}$-hard if the digraph G consists only of disjoint paths from s to t and the conflict graph is a 2-ladder.

$MFFG$ restricted to integer flow values is solvable in polynomial time on a class of elementary instances consisting of disjoint paths from s to t, but adding just one particular arc makes this class strongly $\mathcal{NP}$-hard. Allowing arbitrary nonnegative flow values in $MFFG$, there are two possible solution scenarios: Either we have an optimal solution value equal to the maximum unconstrained flow f_{MF} or we can fulfill the forcing conditions by diminished the flow by some arbitrarily small $\varepsilon > 0$ yielding an optimal solution value of $f_{MF} - \varepsilon$. It is polynomially decidable which of the two cases occurs.

2 $MFCG$ Is Strongly $\mathcal{NP}$-Hard

We derive a strongly $\mathcal{NP}$-hardness result for $MFCG$ even for networks G consisting only of disjoints paths from s to t. Obviously, the maximum flow value f_{MF} can be computed trivially on such instances by taking the minimum of the capacities on each path and summing up over all paths. It seems hard to imagine an even simpler non-trivial flow network. For the conflict graph it suffices to use a 2-ladder which is the simplest meaningful disjunctive constraint structure.

Given an undirected graph Γ with vertex set V the strongly $\mathcal{NP}$-complete *independent set problem* (IS) asks for a subset of vertices $V' \subseteq V$ of cardinality at least K such that no two vertices in V' are adjacent. Let $N(j) \subseteq V$ denote the neighborhood of vertex j in V.

For an instance of IS we construct the digraph G_{MFCG} for our flow network as depicted in Figure 1: Introduce two special vertices s and t and for each vertex $j \in V$ a directed path P_j of length $|N(j)|$ from s to t. Denote the $|N(j)|$ arcs of this path as e_{ji} in arbitrary order where $i \in N(j)$. Obviously, each edge (i, j) in Γ implies in G_{MFCG} the occurrence of an arc e_{ji} in P_j and of e_{ij} in P_i. Define a 2-ladder conflict graph G_{DIS} whose isolated edges have as vertices exactly these arcs e_{ij} and e_{ji} of G_{MFCG} induced by an edge $(i, j) \in \Gamma$. All arc capacities are set to one. Let I be the instance of $MFCG$ defined by G_{MFCG} and G_{DIS}.

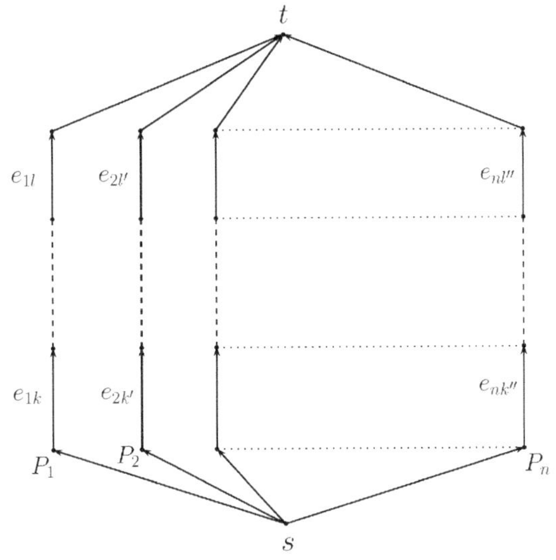

Fig. 1. The digraph G_{MFCG} induced by the independent set problem on a graph Γ

Theorem 1. *$MFCG$ is strongly $\mathcal{NP}$-hard, even if the conflict graph is a 2-ladder and the network consists only of disjoint paths.*

Proof. We show that the following equivalence holds:

$$\exists \text{ a flow } f \text{ in } I \text{ with value } \geq K \Longleftrightarrow \exists \text{ IS } V' \text{ in } \Gamma \text{ with cardinality } \geq K.$$

"$\Longleftarrow$": For every vertex $j \in V'$ send one unit of flow over the path P_j, for all other vertices no flow is sent over P_j. Clearly, this gives a flow with value $|V'| \geq K$. If

$j \in V'$ by definition of (IS) no vertex $i \in N(j)$ can be in V'. Therefore, if there is an arc e_{ji} in P_j, no flow will be sent over arc e_{ij} of P_i. Hence, the conditions of the conflict graph G_{DIS} are satisfied and the described flow is feasible for I.

"$\Longrightarrow$": If f includes some flow over path P_j add vertex j to the independent set of Γ. Since all arc capacities are one at least K paths must contribute to flow f and hence the constructed vertex set has cardinality $\geq K$. Since f fulfills the conditions of the conflict graph G_{DIS} a flow over path P_j forbids a flow over any path P_i for $i \in N(j)$. Hence, the resulting vertex set is independent. □

Note that this reduction preserves the solution values and can easily be extended to the optimization version of the two problems.

The $\mathcal{NP}$-hardness of $MFCG$ can also be deduced from the known $\mathcal{NP}$-hardness of the shortest path problem on a more general graph with forbidden pairs of edges. In fact, it follows from this problem that even checking the feasibility of the problem is $\mathcal{NP}$-complete and hence no polynomial approximation algorithm can exist for $MFCG$ (excluding the trivial zero flow). However, our construction identifies the most elementary network where $\mathcal{NP}$-hardness holds.

3 *MFFG* with Integer Flow Values

Assuming that all flow values are integral the positive disjunctive constraint (1) becomes $x_a + x_{\bar{a}} \geq 1$. Let I be an instance defined by a digraph $G = (N, A)$ consisting of disjoint paths between a source node s and a sink node t and let $H = (A, E)$ be an arbitrary forcing graph for G. Trivially sending as much flow as possible over each path solves the general maximum flow problem and also fulfills all positive disjunctive constraints (at least one unit of flow is routed over every arc) thus giving an optimal solution for I. Adding to such an instance just one new arc which destroys the disjoint paths structure makes the problem already strongly $\mathcal{NP}$-hard, even if the forcing graph is again a 2-ladder.

In our construction we reduce the *vertex cover problem* on an undirected graph Γ with vertex set V to an instance I of $MFFG$ defined as follows: The digraph G_{MFFG} is constructed in the same way as G_{MFCG} in Section 2 (recall Figure 1) with one additional node v that is joined only to s by the new arc (v, s) with capacity k. v becomes the new source node of the network. The forcing graph G_{DIS} is identical to the conflict graph of Section 2.

Vertex cover asks for a subset of vertices $V' \subseteq V$ such that for each edge (i, j) of Γ at least one of the two vertices i and j is in V'. Similar to the reduction of the independent set problem in Section 2 the vertex cover is now reproduced by the forcing graph G_{DIS} as in the proof of Theorem 1. Since vertex cover is a minimization problem we set the capacity k of the new arc (v, s) in a first step equal to one and test if the instance I is feasible. If not, we augment the capacity k by one and iteratively solve I until a feasible solution is found for the first time. The corresponding capacity $k_{\min}$ is reported as the value of the minimum vertex cover in Γ.

Theorem 2. *MFFG is strongly $\mathcal{NP}$-hard, even for a 2-ladder as forcing graph.*

Every approximation algorithm would identify a feasible solution of $MFFG$, if it exists. Hence, we could use such an algorithm to identify the optimal solution of the vertex cover problem by searching iteratively for the smallest value of k where such a feasible solution (and hence the smallest vertex cover) exists. Therefore, there is no polynomial approximation algorithm for $MFFG$.

4 *MFFG* with Arbitrary Flow Values

There are two possible solution scenarios for $MFFG$ with arbitrary nonnegative flow values. Either the maximum flow with value f_{MF} on the digraph $G = (N, A)$ can be rerouted such that the conditions imposed by the forcing graph $H = (A, E)$ are fulfilled or the required diversion of tiny amounts of flows to previously "empty" arcs causes a marginal decrease of the overall flow value yielding an optimal solution value of $f_{MF} - \varepsilon$ for some arbitrarily small $\varepsilon > 0$. We will show that it is polynomially decidable which of the two cases occurs.

First of all let us point out that it is always possible to construct a feasible solution for $MFFG$ with value $f_{MF} - \varepsilon$ for arbitrarily small $\varepsilon > 0$. This can be achieved by applying basic results for the *maximum flow problem with lower bounds* ($MFLB$), see [1, Sec. 6.7]. The details of the construction involving the associated circulation problem are omitted in this short paper.

Now we show how to decide whether also a solution with value f_{MF} exists. In [7] and [4] it is described how to detect all minimum s-t cuts in a digraph. Since there may be exponentially many of them, they cannot be all listed explicitly in polynomial time, but it can be decided efficiently if a given arc belongs to some minimum cut. For our purpose it is important that a directed acyclic graph $DAG_{s,t}$ (a so-called Picard-Queyranne Directed Acyclic Graph) is generated from a maximum flow in the original digraph G and that $DAG_{s,t}$ contains all minimum s-t cuts in its structure. Every node of $DAG_{s,t}$ corresponds to a subset of nodes of G. Moreover, the nodes of $DAG_{s,t}$ induce a partition of N.

By the construction of $DAG_{s,t}$ the node containing t has in-degree 0, the node containing s has out-degree 0. A *closure* of $DAG_{s,t}$ is a subset $\mathcal{C}$ of the nodes of $DAG_{s,t}$ where for every node $A \in \mathcal{C}$ if there is an arc from node A to some node B in $DAG_{s,t}$ then also $B \in \mathcal{C}$. It was shown that all the nodes in G induced by a closure of $DAG_{s,t}$ containing s and not t correspond to the set S of a minimal s-t cut $(S, \bar{S})$ in G.

Therefore we first compute the optimal flow value f_{MF}. Then we consider the corresponding Picard-Queyranne Graph $DAG_{s,t}$. Since all capacities in G are integer, any cut that is not minimal must have a value $\geq (f_{MF} + 1)$. Now consider all violated positive disjunctive constraints, i.e. all edges $E' \subseteq E$ in the forcing graph H where neither of the two arcs of A incident to such an edge in E' carries any flow in f_{MF}. We will distinguish the following two cases:

Case 1. $\forall\ e \in E'$ joining two arcs of G: For at least one of these arcs, say $a = (i, j)$, both vertices i and j are in the same subset of nodes of G corresponding to a node of $DAG_{s,t}$.

Then introducing a lower bound of ε for the capacity of all such arcs (i,j) gives an instance of $MFLB$ which fulfills the positive disjunctive constraints and keeps the same solution value f_{MF}. This follows from the property of $DAG_{s,t}$: Since i and j belong to the same node of $DAG_{s,t}$ the capacities u_{ij} did not contribute to any minimal cut. Therefore, the arcs (i,j) only contributed to cuts with value $\geq (f_{MF}+1)$. Moreover, only on these arcs lower bounds $l_{ij} = \varepsilon$ were introduced. Now considering the cut capacities $u(S,\bar{S})$ of the new problem $MFLB$, every cut containing these modified arcs still has a value $\geq (f_{MF}+1) - \varepsilon|A|$. By choosing ε sufficiently small, a feasible solution with value f_{MF} is derived.

Case 2. $\exists\ e \in E'$ joining two arcs $a = (i,j)$ and $a' = (i',j')$ of G such that i and j as well as i' and j' are in different subsets of nodes of G induced by the nodes of $DAG_{s,t}$.

By the properties of $DAG_{s,t}$ there exist minimum cuts $(S,\bar{S})$, resp. $(S',\bar{S}')$, for each of the two arcs a, resp. a', with $j \in S$ and $i \in \bar{S}$, resp. $j' \in S'$ and $i' \in \bar{S}'$, since both (i,j) and (i',j') carry no flow in f_{MF}. But then the problem can be similarly transformed to an $MFLB$ instance where the new lower bounds $l_{ij} = \varepsilon$, resp. $l_{i'j'} = \varepsilon$, do contribute to the minimum cut $u(S,\bar{S})$, resp. $u(S',\bar{S}')$, and decrease its value thus reducing the optimal flow value. It also follows from the structure of these minimum cuts that no feasible solution to $MFFG$ with value f_{MF} can exist.

Theorem 3. *For $MFFG$ with arbitrary flow values it can be decided in polynomial time if the optimal flow value corresponds to the value f_{MF} of the relaxed problem MF or if it equals $f_{MF} - \varepsilon$ for some arbitrarily small $\varepsilon > 0$.* □

References

1. Ahuja, R., Magnanti, T., Orlin, J.: Network Flows: Theory, Algorithms, and Applications. Prentice Hall, Englewood Cliffs (1993)
2. Bodlaender, H., Jansen, K.: On the complexity of scheduling incompatible jobs with unit-times. In: Borzyszkowski, A.M., Sokolowski, S. (eds.) MFCS 1993. LNCS, vol. 711, pp. 291–300. Springer, Heidelberg (1993)
3. Darmann, A., Pferschy, U., Schauer, J., Woeginger, G.: Paths, trees and matchings under disjunctive constraints. Discrete Applied Mathematics (to appear, 2011)
4. Gusfield, D., Naor, D.: Efficient algorithms for generalized cut trees. In: SODA 1990: Proceedings of the First Annual ACM-SIAM Symposium on Discrete Algorithms, pp. 422–433. SIAM, Philadelphia (1990)
5. Jansen, K.: An approximation scheme for bin packing with conflicts. Journal of Combinatorial Optimization 3, 363–377 (1999)
6. Pferschy, U., Schauer, J.: The knapsack problem with conflict graphs. Journal of Graph Algorithms and Applications 13(2), 233–249 (2009)
7. Picard, J.C., Queyranne, M.: On the structure of all minimum cuts in a network and applications. Mathematical Programming Studies 13, 8–16 (1980)

Algebraic Methods for Stochastic Minimum Cut and Maximum Flow Problems

Katherine C. Hastings and Douglas R. Shier

Department of Mathematical Sciences
Clemson University
Clemson, SC 29634-0975
{hasting,shierd}@clemson.edu

Abstract. We present an algebraic approach for computing the distribution of the capacity of a minimum s-t cut in a network, in which the arc capacities have known (discrete) probability distributions. Algorithms are developed to determine the exact distribution as well as upper and lower bounding distributions on the capacity of a minimum cut. This approach then provides exact and bounding distributions on the maximum flow in such stochastic networks. We also obtain bounds on the expected capacity of a minimum cut (and the expected maximum flow value).

1 Introduction

Network flow models find application in numerous problems arising in transportation, distribution, and communication networks. Additional applications occur, for example, in scheduling [1], open pit mining [15], and voting [17]. Consequently, a variety of algorithms have been proposed for efficient solution of deterministic network flow (and minimum cut) problems [1].

In practice, however, links capacities change over time, and so a more realistic depiction is as a *stochastic maximum flow* model, in which network arcs assume different values (capacity, cost) according to known probability distributions. Although such stochastic flow problems are NP-hard, solution methods have been proposed by a number of authors, for example by applying intelligent state space partitioning [3,7,8,13] or by Monte Carlo simulation [2,9]. By contrast, we study such problems in a distinctly different way: by focusing on the distribution of the capacity of a minimum cut and by employing an algebraic approach. This allows us to obtain both exact and approximate distributions for the capacity of a minimum cut and thus the maximum flow value. We also obtain bounds on the expected minimum cut capacity and expected maximum flow value in stochastic flow networks. This work generalizes previous research [6,14] that treats the important but special case in which arcs have two states, either working or failed.

J. Pahl, T. Reiners, and S. Voß (Eds.): INOC 2011, LNCS 6701, pp. 295–308, 2011.

1.1 Formulation

Let $G = (N, A)$ be a directed network with node set N and arc set A, with distinguished source node s and sink node t. Throughout, we denote $n = |N|$ and $m = |A|$. In deterministic networks, each arc $a \in A$ would have an associated capacity c_a and the capacity of any s-t cut K is given by $c(K) = \sum\{c_a : a \in K\}$. In a stochastic network, however, the capacity of each arc $a \in A$ is modeled by the random variable Y_a. Here we assume that the arc random variable Y_a follows a discrete probability distribution. Specifically, arc a assumes a finite number of states $1, 2, \dots, \sigma_a$; in state k arc a has capacity $c_{a,k}$ with probability $p_{a,k} = \Pr[Y_a = c_{a,k}]$.

When information about the state of each arc in G is available, the *state* of the entire network is given by the *state vector* $\mathbf{x} = [x_1, x_2, \dots, x_m]$, where $x_a \in \{1, 2, \dots, \sigma_a\}$. We assume that the arc capacity random variables are independent, so the probability that the network is in state $\mathbf{x}$ is given by $P(\mathbf{x}) = \prod_{a=1}^{m} p_{a,x_a}$. Since the maximum flow and minimum cut depend on the network state, we focus here on the distribution of the capacity $C = C(\mathbf{x})$ of a minimum s-t cut in G.

A straightforward approach to calculating the distribution of C is by *state space enumeration*, which involves enumerating all possible states $\mathbf{x}$ of the given network. The probability that the minimum cut has capacity w is then

$$\Pr[C = w] = \sum\{P(\mathbf{x}) : C(\mathbf{x}) = w\}.$$

Since there are an exponential number of states, this is not a practical method. We discuss next an algebraic approach to this problem.

2 Algebraic Operations and Properties

2.1 The ⊕ and ⊗ Operations

For simplicity (but without loss of generality), we assume now that the arc capacity levels are nonnegative integers. Thus we can associate with each $a \in A$ an arc capacity polynomial that contains all information about the states and associated capacities of the arc:

$$f_a = f_a(z) = \sum_{k=1}^{\sigma_a} x_{a,k} z^{c_{a,k}}.$$

This generating function (in the symbolic variable z) has as exponents the capacity in state k and as coefficients the Boolean variable $x_{a,k}$ indicating whether arc a is in state k. Of course, $\Pr[x_{a,k} = 1] = p_{a,k}$. We would like to combine these arc polynomials (generating functions) in order to obtain a corresponding polynomial for the capacity of a minimum cut:

$$f_C = f_C(z) = \sum_w \phi_w z^w,$$

where ϕ_w is a Boolean expression enumerating all states in which the minimum capacity cut has capacity w.

The goal of the algebraic approach is to algebraically combine the given arc polynomials f_a using certain algebraic operations $\oplus$ and $\otimes$ to determine the desired polynomial f_C. These operations (defined originally for stochastic shortest path problems [4,11]) are best motivated by first considering series and parallel arcs. Their interpretation will be based on how series and parallel arcs combine for shortest paths.

Suppose arcs a and b have the *arc length* polynomials $g_a = \sum_{k=1}^{\sigma_a} x_{a,k} z^{l_{a,k}}$ and $g_b = \sum_{k=1}^{\sigma_b} x_{b,k} z^{l_{b,k}}$. Consider the case in which arcs $a = (i,k)$ and $b = (k,j)$ are series arcs; that is, arc a is the only arc entering node k and arc b is the only one leaving node k. We wish to obtain the polynomial g_d resulting from removing node k and replacing series arcs a and b by an arc $d = (i,j)$. If arc a is in state q and arc b is in state r, then the length of the i-j path is given by $l_{a,q} + l_{b,r}$ and the corresponding state indicator variable is $x_{a,q} x_{b,r}$; this produces the monomial $x_{a,q} x_{b,r} z^{l_{a,q}+l_{b,r}}$. Consequently we define

$$g_a \otimes g_b = g_d = \sum_{q=1}^{\sigma_a} \sum_{r=1}^{\sigma_b} x_{a,q} x_{b,r} z^{l_{a,q}+l_{b,r}}.$$

Now suppose arcs a and b are parallel arcs (joining the same two nodes i and j) and we wish to replace them by a single arc d joining i and j. If arc a is in state q and arc b is in state r, then the length of the shortest i-j path is given by $\min\{l_{a,q}, l_{b,r}\}$, producing the monomial $x_{a,q} x_{b,r} z^{\min\{l_{a,q}, l_{b,r}\}}$. Consequently we define

$$g_a \oplus g_b = g_d = \sum_{q=1}^{\sigma_a} \sum_{r=1}^{\sigma_b} x_{a,q} x_{b,r} z^{\min(l_{a,q}, l_{b,r})}.$$

Using these operations, the shortest s-t path length polynomial [4,11] can be expressed as

$$\oplus \sum_{P \in \mathcal{P}_{st}} \left(\otimes \prod_{a \in P} g_a \right), \tag{1}$$

where $\mathcal{P}_{st}$ is the set of all (simple) s-t paths. These operations take into account that whereas arcs operate independently of one another, paths may share common arcs and thus exhibit dependent behavior.

The analogous way to combine series arcs and parallel arcs for *arc capacity* polynomials f_a and f_b is (dually) via $f_a \oplus f_b$ for series arcs (select the smaller capacity arc for the cutset) and $f_a \otimes f_b$ for parallel arcs (select both arcs for the cutset). Moreover, the desired minimum cut polynomial f_C can be expressed as

$$f_C = \oplus \sum_{K \in \mathcal{K}_{st}} \left(\otimes \prod_{a \in K} f_a \right), \tag{2}$$

where $\mathcal{K}_{st}$ is the set of all s-t cutsets.

2.2 Properties of $\oplus$ and $\otimes$

Useful properties of $\oplus$ and $\otimes$ have been established by Altenhöfer [4] and Hastings [11].

Theorem 1. *For arc polynomials f, g, h, the following properties hold:*

1. $f \oplus g = g \oplus f, \quad f \otimes g = g \otimes f$
2. $f \oplus (g \oplus h) = (f \oplus g) \oplus h, \quad f \otimes (g \otimes h) = (f \otimes g) \otimes h$
3. $f \oplus (f \otimes g) = f$
4. $f \otimes (g \oplus h) = (f \otimes g) \oplus (f \otimes h)$
5. $f \oplus f = f$

Notice that the other distributive law (analogous to Property 1.4 above) is not listed, since in fact it does not hold. On the other hand we can demonstrate that a corresponding majorization result holds instead.

Definition 1. *Let X and Y be (real) random variables. We say that X is* stochastically less than *Y if $\Pr[X > t] \le \Pr[Y > t]$ for all t. This is denoted by $X \le^{st} Y$.*

Suppose X and Y have generating function polynomials f_X and f_Y respectively. If $X \le^{st} Y$ holds we write $f_X \preceq f_Y$. This relation is a partial order, satisfying the following properties.

Theorem 2. *For arc polynomials f, g, h, the following properties hold:*

1. $f \oplus g \preceq f$
2. $f \preceq f \otimes g$
3. $f \preceq f \otimes (f \oplus g)$
4. $f \oplus (g \otimes h) \preceq (f \oplus g) \otimes (f \oplus h)$
5. $f \preceq g \Rightarrow f \oplus h \preceq g \oplus h$
6. $f \preceq g \Rightarrow f \otimes h \preceq g \otimes h$

The validity of these relations follows from properties of the (ordinary) sum and minimum of random variables.

3 Exact Calculation

Using these algebraic operations, equation (2) expresses f_C in terms of the s-t cutsets. These cutsets can be enumerated in pseudopolynomial time [16].

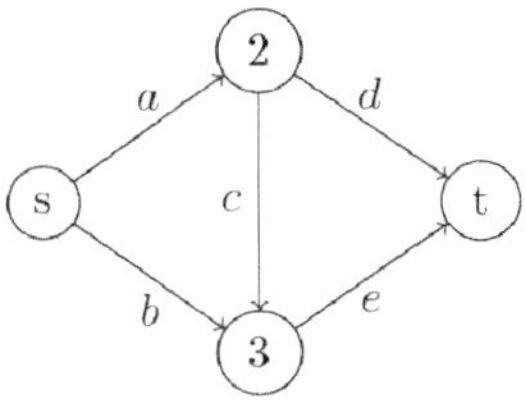

Arc	State 1	State 2
a	5	6
b	1	5
c	0	3
d	2	6
e	2	6

Fig. 1. Bridge network

Example 1. Consider the bridge network shown in Figure 1 in which each arc has two states and the capacities listed. The s-t cutsets are $\{a,b\}$, $\{a,e\}$, $\{b,c,d\}$, and $\{d,e\}$. The arc polynomials are given by

$$f_a = x_{a,1}z^5 + x_{a,2}z^6, \quad f_b = x_{b,1}z^1 + x_{b,2}z^5, \quad f_c = x_{c,1}z^0 + x_{c,2}z^3,$$
$$f_d = x_{d,1}z^2 + x_{d,2}z^6, \quad f_e = x_{e,1}z^2 + x_{e,2}z^6.$$

Using the cutset enumeration formula (2)

$$f_C = (f_a \otimes f_b) \oplus (f_a \otimes f_e) \oplus (f_b \otimes f_c \otimes f_d) \oplus (f_d \otimes f_e)$$

For example, $f_a \otimes f_b = (x_{a,1}z^5 + x_{a,2}z^6) \otimes (x_{b,1}z^1 + x_{b,2}z^5) = x_{a,1}x_{b,1}z^6 + x_{a,2}x_{b,1}z^7 + x_{a,1}x_{b,2}z^{10} + x_{a,2}x_{b,2}z^{11}$ and $f_a \otimes f_e = (x_{a,1}z^5 + x_{a,2}z^6) \otimes (x_{e,1}z^2 + x_{e,2}z^6) = x_{a,1}x_{e,1}z^7 + x_{a,2}x_{e,1}z^8 + x_{a,1}x_{e,2}z^{11} + x_{a,2}x_{e,2}z^{12}$. Since the two terms $f_a \otimes f_b$ and $f_a \otimes f_e$ share dependencies, the application of operation $\oplus$ gives a simplified (correct) expression, using the Boolean identities $x_{a,1}x_{a,1} = x_{a,1}$, $x_{a,2}x_{a,2} = x_{a,2}$, $x_{a,1}x_{a,2} = 0$, and $x_{e,1} + x_{e,2} = 1$:

$$(f_a \otimes f_b) \oplus (f_a \otimes f_e) = (x_{a,1}x_{b,1})z^6 + (x_{a,2}x_{b,1} + x_{a,1}x_{b,2}x_{e,1})z^7 + x_{a,2}x_{b,2}x_{e,1}z^8 + x_{a,1}x_{b,2}x_{e,2}z^{10} + x_{a,2}x_{b,2}x_{e,2}z^{11}.$$

Using such symbolic calculations, the entire polynomial f_C can be calculated:

$$\begin{aligned} f_C = {} & x_{b,1}x_{c,1}x_{d,1}z^3 + (x_{c,2}x_{d,1}x_{e,1} + x_{b,2}x_{c,1}x_{d,1}x_{e,1})z^4 + (x_{a,1}x_{b,1}x_{d,2} \\ & + x_{b,1}x_{c,2}x_{d,1}x_{e,2})z^6 + (x_{a,2}x_{b,1}x_{d,2} + x_{a,1}x_{b,2}x_{d,2}x_{e,1} \\ & + x_{b,2}x_{c,1}x_{d,1}x_{e,2})z^7 + (x_{a,2}x_{b,2}x_{d,2}x_{e,1} + x_{b,2}x_{c,2}x_{d,1}x_{e,2})z^8 \\ & + x_{a,1}x_{b,2}x_{d,2}x_{e,2}z^{10} + x_{a,2}x_{b,2}x_{d,2}x_{e,2}z^{11}. \end{aligned}$$

Consider the coefficient $x_{a,2}x_{b,1}x_{d,2} + x_{a,1}x_{b,2}x_{d,2}x_{e,1} + x_{b,2}x_{c,1}x_{d,1}x_{e,2}$ of z^7. Each of these terms specifies a particular set of network states in which the minimum cut capacity is 7. For example $x_{a,2}x_{b,1}x_{d,2}$ indicates that when arc b is in state 1 and arcs a, d are in state 2, a minimum cut in the resulting deterministic network has capacity 7. In this term, the absence of state indicators for arcs c and e indicates that the states of arcs c and e do not affect the capacity of a minimum cut.

Substituting the probabilities $p_{a,k}$ for each $x_{a,k}$ in the expression for f_C yields a polynomial in which the coefficient of z^w is exactly the probability that a minimum cut has capacity w.

4 Bounding Distributions

While the algebraic approach can be used to find the exact distribution f_C of minimum cut capacities, it is possible to provide upper and lower bounding distributions, often with much less effort. One readily computed lower bound is obtained by ignoring the dependencies that may be present in the terms of (2). That is, we immediately replace the Boolean arc indicator $x_{a,k}$ by its expected value $p_{a,k}$ and then carry out the indicated $\oplus$ operations. To indicate that we are now taking the minimum of presumed *independent* random variables, this operation is denoted $\oplus'$.

Example 2. Returning to Example 1, we now express the arc polynomials more succinctly as $f_a = p_a z^5 + q_a z^6$, $f_b = p_b z^1 + q_b z^5$, $f_c = p_c z^0 + q_c z^3$, $f_d = p_d z^2 + q_d z^6$, $f_e = p_e z^2 + q_e z^6$, where $q_r = 1 - p_r$. Then using the modified $\oplus'$ operation

$$\begin{aligned}(f_a \otimes f_b) \oplus' (f_a \otimes f_e) &= (p_a p_b z^6 + q_a p_b z^7 + p_a q_b z^{10} + q_a q_b z^{11}) \oplus' \\ &\quad (p_a p_e z^7 + q_a p_e z^8 + p_a q_e z^{11} + q_a q_e z^{12}) \\ &= p_a p_b z^6 + (q_a p_b + p_a q_b p_e) z^7 + q_a q_b p_e z^8 + p_a q_b q_e z^{10} \\ &\quad + q_a q_b q_e z^{11}.\end{aligned}$$

It will be seen later that the polynomial resulting from using $\oplus'$ instead of $\oplus$ in the cutset enumeration formula (2) will in fact provide a lower bound on the exact cutset distribution, evaluated at the known $p_{a,k}$.

4.1 Modified Path Enumeration

The s-t cutsets and s-t paths are duals of one another, in the sense that each s-t cutset must intersect every s-t path. Consequently, we can obtain an algebraic analogue of the path enumeration expression (1) by interchanging the operations $\otimes$ and $\oplus$:

$$\otimes \prod_{P \in \mathcal{P}_{st}} \left(\oplus \sum_{a \in P} f_a \right).$$

In fact it will be useful to generalize this idea to the dualization of all s-j paths, for $j \in N$, by considering the *Modified Path Enumeration* polynomial:

$$f_{MPE}(j) = \otimes \prod_{P \in \mathcal{P}_{sj}} \left(\oplus \sum_{a \in P} f_a \right). \tag{3}$$

We will show later that this polynomial provides an upper bound on the polynomial $f_C(j)$ based on enumerating all s-j cutsets:

$$f_C(j) = \oplus \sum_{K \in \mathcal{K}_{sj}} \left(\otimes \prod_{a \in K} f_a \right). \tag{4}$$

Example 3. Consider the bridge network from Example 1, in which the s-t paths are $[a, d]$, $[a, c, e]$, and $[b, e]$. Applying (3) and using the properties in Theorems 1–2 gives

$$\begin{aligned} f_{MPE}(t) &= (f_a \oplus f_d) \otimes (f_a \oplus f_c \oplus f_e) \otimes (f_b \oplus f_e) \\ &\succeq [f_a \oplus (f_d \otimes (f_c \oplus f_e))] \otimes (f_b \oplus f_e) \\ &= (f_a \otimes f_b) \oplus (f_b \otimes f_d \otimes (f_c \oplus f_e)) \oplus (f_a \otimes f_e) \oplus (f_d \otimes f_e \otimes (f_c \oplus f_e)) \\ &\succeq (f_a \otimes f_b) \oplus (f_a \otimes f_e) \oplus (f_b \otimes f_c \otimes f_d) \oplus [(f_d \otimes f_e) \oplus (f_b \otimes f_d \otimes f_e)] \\ &= (f_a \otimes f_b) \oplus (f_a \otimes f_e) \oplus (f_b \otimes f_c \otimes f_d) \oplus (f_d \otimes f_e) = f_C(t). \end{aligned}$$

This verifies that for this example the MPE polynomial majorizes the cutset polynomial.

4.2 Acyclic Networks

In order to develop further bounds, it is convenient to restrict our attention to acyclic networks $G = (N, A)$ where the implementation is simpler, in view of the existence of a *topological ordering* of the nodes of G. Henceforth, we assume that nodes are numbered so that if $(i, j) \in A$ then $i < j$. In such networks we can avoid the enumeration of either paths or cutsets, and can obtain an improved upper bound on $f_C(j)$.

```
algorithm Acyclic
   Input: acyclic G = (N, A); arc polynomials f_a; source node s
   begin
      Label(s) = z^∞; Label(j) = z^0 for j ≠ s
      for j ≠ s {in topological order}
         for a = (i, j) ∈ A
            Label(j) := Label(j) ⊗ (Label(i) ⊕ f_a);
         end for
      end for
   end
```

The outputs of the Acyclic Algorithm are designated $f_{AA}(j) = \text{Label}(j)$.

Example 4. We apply the Acyclic Algorithm to the network in Example 1; nodes have already been numbered in topological order with $s = 1$ and $t = 4$.

Initialization: Label(1) $= z^\infty$; Label(2) $=$ Label(3) $=$ Label(4) $= z^0$.
Iteration 1: Label(2) $= z^0 \otimes$ (Label(1) $\oplus f_a) = f_a$.
Iteration 2: Label(3) $=$ (Label(1) $\oplus f_b) \otimes$ (Label(2) $\oplus f_c$)$= f_b \otimes (f_a \oplus f_c)$.
Iteration 3: Label(4) $=$ (Label(2) $\oplus f_d) \otimes$ (Label(3) $\oplus f_e$)$= (f_a \oplus f_d) \otimes ([f_b \otimes (f_a \oplus f_c)] \oplus f_e)$.

Using the distributive properties from Theorems 1-2, we obtain

$$\begin{aligned} f_{AA}(t) &= (f_a \oplus f_d) \otimes ([f_b \otimes (f_a \oplus f_c)] \oplus f_e) \\ &\preceq (f_a \oplus f_d) \otimes (f_b \oplus f_e) \otimes (f_a \oplus f_c \oplus f_e) = f_{MPE}(t). \end{aligned}$$

Moreover,

$$\begin{aligned} f_{AA}(t) &= (f_a \oplus f_d) \otimes ([f_b \otimes (f_a \oplus f_c)] \oplus f_e) \\ &= [(f_a \oplus f_d) \otimes (f_b \otimes (f_a \oplus f_c)] \oplus [(f_a \oplus f_d) \otimes f_e] \\ &= [f_b \otimes (f_a \oplus f_c) \otimes (f_a \oplus f_d)] \oplus [(f_a \otimes f_e) \oplus (f_d \otimes f_e)] \\ &\succeq [f_b \otimes (f_a \oplus (f_c \otimes f_d))] \oplus [(f_a \otimes f_e) \oplus (f_d \otimes f_e)] \\ &= (f_a \otimes f_b) \oplus (f_b \otimes f_c \otimes f_d) \oplus (f_a \otimes f_e) \oplus (f_d \otimes f_e) = f_C(t). \end{aligned}$$

As seen here, the Acyclic Algorithm not only avoids the need to enumerate paths, but it provides an improved upper bound on f_C. The following result shows that this holds in general, and indeed for all nodes j.

Theorem 3. *$f_C(j) \preceq f_{AA}(j) \preceq f_{MPE}(j)$ for all $j \in N$.*

Proof. We begin by showing inductively that $f_{AA}(j) \preceq f_{MPE}(j)$ holds for all $j \in N$. If $n = 2$, then $f_{MPE}(2) = f_{1,2} = f_{AA}(2)$.

Assume $f_{AA}(j) \preceq f_{MPE}(j)$ for any $j \in N$ with $1 \le |N| \le k$. Let $|N| = k+1$ and let T be the set of sink nodes of G. Define $N' = N \backslash T$ and let G' be the graph induced by N'. Because G is acyclic, $|T| \ge 1$ and so $|N'| \le k$. Thus, by induction, $f_{AA}(j) \preceq f_{MPE}(j)$ for all $j \in N'$. Let $t \in T$ and let $B(t) = \{i : (i,t) \in A\} \subseteq N'$ be the set of immediate predecessors of t. Then

$$\begin{aligned} f_{MPE}(t) &= \otimes \prod_{i \in B(t)} \left[\otimes \prod_{P \in \mathcal{P}_i} \left(\oplus \sum_{(j,k) \in P} f_{j,k} \oplus f_{i,t} \right) \right] \\ &\succeq \otimes \prod_{i \in B(t)} \left[\left(\otimes \prod_{P \in \mathcal{P}_i} \oplus \sum_{(j,k) \in P} f_{j,k} \right) \oplus f_{i,t} \right] \\ &= \otimes \prod_{i \in B(t)} [f_{MPE}(i) \oplus f_{i,t}] \\ &\succeq \otimes \prod_{i \in B(t)} [f_{AA}(i) \oplus f_{i,t}] \\ &= f_{AA}(t), \end{aligned}$$

using the distributive property from Theorem 2 and the induction hypothesis. Therefore, $f_{AA}(j) \preceq f_{MPE}(j)$ for all $j \in N$.

As illustrated in Example 4, we can begin with $f_{AA}(j)$, which is in product-of-sums form, apply both distributive laws, and then expand the polynomial. The result will be a sum-of-products form whose summands represent the s-j cutsets. That is, the resulting polynomial is $f_C(j)$ and so we have $f_{AA}(j) \succeq f_C(j)$. □

5 Bounds on Expected Capacity

We present here various bounds on the expected value of the minimum cut capacity, and consequently the maximum flow value, in a two-terminal flow network. The simplest bound on the expected minimum cut capacity (denoted μ_C) is obtained by replacing each arc distribution by its expected capacity and then solving the resulting deterministic s-t flow problem.

Let $\mathcal{K} = \{K_1, K_2, \ldots, K_r\}$ be the s-t cutsets and let $\mathbf{Y} = (Y_1, Y_2, \ldots, Y_m)$ be the vector of arc capacity random variables. The capacity κ_j of K_j is a linear function of $\mathbf{Y}$: $\kappa_j = \kappa_j(\mathbf{Y}) = \sum\{Y_a : a \in K_j\}$. Then the minimum cut capacity $C = \min\{\kappa_1, \kappa_2, \ldots, \kappa_m\}$ is a concave function of $\mathbf{Y}$. By Jensen's inequality [5], $C(E[\mathbf{Y}]) \geq E[C(\mathbf{Y})]$. In other words, replacing each arc distribution by its expected value produces a minimum cut capacity $\overline{\mu}$ that is an upper bound on the true expected (minimum) cut capacity μ_C.

Another easy upper bound is derived from the MPE polynomial (3), using the fact that $E[g \otimes h] = E[g] + E[h]$. Thus

$$\mu_{MPE} = E[f_{MPE}(t)] = \sum_{P \in \mathcal{P}_{st}} E\left[\left(\oplus \sum_{a \in P} f_a\right)\right] = \sum_{P \in \mathcal{P}_{st}} E[\min_{a \in P} f_a]. \quad (5)$$

Since the arc polynomials f_a are assumed to be independent, the calculation of the last expectation can be done using ordinary polynomials, with $p_{a,k}$ substituted for $x_{a,k}$. Since we have the majorization result $f_C(t) \preceq f_{MPE}(t)$, it follows that $E[f_C(t)] \leq E[f_{MPE}(t)] = \mu_{MPE}$ and so μ_{MPE} is an upper bound on the true μ_C.

Using properties of associated random variables, one can show $g \oplus' h \preceq g \oplus h$ holds for the generating functions considered here. As a result, replacing $\oplus$ by $\oplus'$ in (4) produces

$$f'_C(j) = \oplus' \sum_{K \in \mathcal{K}_{sj}} \left(\otimes \prod_{a \in K} f_a\right) \preceq f_C(j), \quad (6)$$

and so $\mu'_C = E[f'_C(t)]$ is a lower bound on $E[f_C(t)] = \mu_C$. Because $\oplus'$ ignores dependencies, ordinary polynomials with coefficients $p_{a,k}$ can be combined to calculate μ'_C.

5.1 Fulkerson Bound

We present a bound analogous to the bound on expected shortest path lengths in acyclic networks introduced by Fulkerson [10]. The basic idea is that instead of replacing arc weights by expected values, we recursively compute an expected value $\mu_F(j)$ at each node j based on the weight distribution for arcs (i, j) and the node expected values $\mu_F(i)$. We can translate this into our algebraic language by appropriately modifying the Acyclic Algorithm in Section 4.2 to compute expected minimum capacities $\mu_F(j)$ at each node j, relative to the s-j cuts. Here we denote by $\{\alpha\}$ the degenerate distribution with all probability concentrated at the single value α.

```
algorithm Fulkerson Bound
   Input: acyclic G = (N, A); arc polynomials f_a; source node s
   begin
      μ_F(s) = ∞; μ_F(j) = 0 for j ≠ s
      for j ≠ s {in topological order}
         for a = (i, j) ∈ A
            μ_F(j) := μ_F(j) + E[{μ_F(i)} ⊕ f_a];
         end if
      end for
   end
```

Example 5. We apply this algorithm to the network in Example 1, where for simplicity we assume that each arc assumes its lower capacity with probability p and its higher capacity with probability $q = 1-p$: $f_a = pz^5+qz^6$, $f_b = pz^1+qz^5$, $f_c = pz^0 + qz^3$, $f_d = pz^2 + qz^6$, $f_e = pz^2 + qz^6$.

Initialization: $\mu_F(1) = \infty$; $\mu_F(2) = \mu_F(3) = \mu_F(4) = 0$.
Iteration 1: $\mu_F(2) = \mu_F(2)+E[\{\mu_F(1)\}\oplus f_a] = 0+E[\{\infty\}\oplus f_a] = E[f_a] = 6-p$.
Iteration 2: $\mu_F(3) = E[\{\mu_F(1)\}\oplus f_b]+E[\{\mu_F(2)\}\oplus f_c] = E[f_b]+E[\{6-p\}\oplus f_c] = (5-4p)+(3-3p) = 8-7p$.
Iteration 3: $\mu_F(4) = E[\{\mu_F(2)\} \oplus f_d] + E[\{\mu_F(3)\} \oplus f_e]$. The results of the implied minimizations depend on the value of p, yielding

$$\mu_F(4) = \begin{cases} 12 - 9p + p^2 & 0 \le p \le \frac{2}{7} \\ 14 - 18p + 8p^2 & \frac{2}{7} \le p \le \frac{6}{7} \\ 14 - 12p + p^2 & \frac{6}{7} \le p \le 1 \end{cases}$$

We can show that the Acyclic Algorithm produces a bound that always improves upon the Fulkerson upper bound.

Theorem 4. *Let $G = (N, A)$ be an acyclic network. Then $E[f_{AA}(j)] \le \mu_F(j)$ for all $j \in N$.*

Proof. We prove this by induction on $n = |N|$. If $n = 2$, then $f_{AA}(2) = f_{1,2}$ which gives $E[f_{AA}(2)] = E[f_{1,2}] = \mu_F(2)$.

Now assume that for $n \le k$ the stated property holds: that is, $E[f_{AA}(j)] \le \mu_F(j)$ for all $j \in N$. Let $|N| = k+1$ and let T be the set of sink nodes of G. Define $N' = N \setminus T$ and let G' be the graph induced by N'. By induction, $E[f_{AA}(i)] \le \mu_F(i)$ for all $i \in N'$. Select $t \in T$ and let $B(t) = \{i : (i,t) \in A\} \subseteq N'$ be the set of immediate predecessors of t. Since

$$f_{AA}(t) = \otimes \prod_{i \in B(t)} (f_{AA}(i) \oplus f_{i,t}),$$

$$E[f_{AA}(t)] = E[\otimes \prod_{i \in B(t)} (f_{AA}(i) \oplus f_{i,t})]$$

$$
\begin{aligned}
&= \sum_{i \in B(t)} E[(f_{AA}(i) \oplus f_{i,t})] \\
&= \sum_{i \in B(t)} E[\min\{f_{AA}(i), f_{i,t}\}] \\
&\leq \sum_{i \in B(t)} \min\{E[f_{AA}(i)], f_{i,t}\} \\
&\leq \sum_{i \in B(t)} \min\{\mu_F(i), f_{i,t}\} \\
&= \sum_{i \in B(t)} (\{\mu_F(i)\} \oplus f_{i,t}) = \mu_F(t).
\end{aligned}
$$

The first inequality follows since the function $\psi(x) = \min\{x, \alpha\}$ is a concave function and the second inequality follows from the inductive hypothesis. This completes the proof. □

Notice that Theorem 3 guarantees $\mu_C \leq \mu_{AA} \leq \mu_{MPE}$ while Theorem 4 shows $\mu_C \leq \mu_{AA} \leq \mu_F$. In general neither μ_{MPE} nor μ_F is uniformly better than the other.

5.2 Numerical Results

We first show here results for the bridge network of Figure 1, where for ease of presentation each arc assumes its lower capacity with probability p and its higher capacity with probability $1-p$. Thus we can plot the true expected value μ_C, as well as various upper and lower bounds on that value, as a function of p.

Throughout this paper we have discussed various methods for obtaining bounds on the true expected value μ_C. Section 4 derived bounds on the *distribution* of the capacity of a minimum s-t cut, which then allows us to compute a bound on μ_C. Additional bounds on this expected value are developed in Section 5. In particular, the upper bound $\overline{\mu}$ on μ_C is calculated by replacing arc distributions by their expected values rather than by computing the expectation of some distribution. This distinguishes $\overline{\mu}$ from the other upper bounds presented because it is not calculated using our algebraic operations.

The only lower bound μ'_C presented is obtained from the relation (6) by taking expected values. It is a comparatively easy bound to compute because it is based on the operation $\oplus'$ rather than on $\oplus$, thus ignoring dependencies among the cutsets. Because $\overline{\mu}$ and μ'_C share the common property of being readily computed we found it appropriate to consider these bounds together. Figure 2 shows these respective upper and lower bounds $\overline{\mu}$ and μ'_C as well as the true μ_C.

In addition to $\overline{\mu}$ we presented three additional upper bounds on μ_C, all of which are based on the algebraic operations $\oplus$ and $\otimes$. Figure 3 shows these upper bounds μ_{AA}, μ_{MPE}, and μ_F as well as the true μ_C. We observe that $\mu_C \leq \mu_{AA} \leq \mu_{MPE}$ holds for $0 \leq p \leq 1$ as implied by Theorem 3, and that $\mu_{AA} \leq \mu_F$ holds for $0 \leq p \leq 1$ as proved in Theorem 4. Notice that in this example the bound from the Acyclic Algorithm is especially accurate at larger

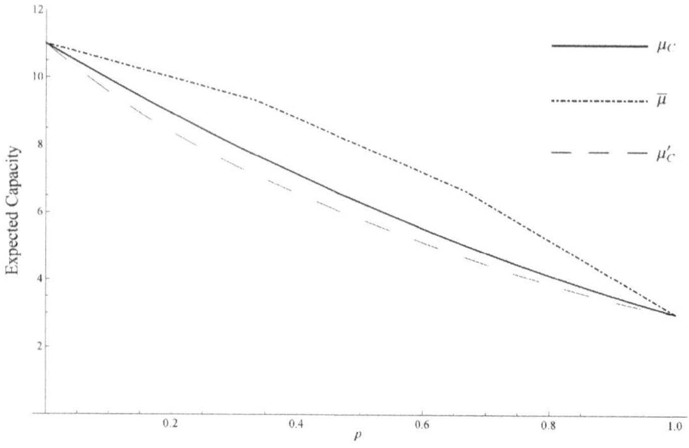

Fig. 2. Estimates of expected capacity of a minimum cut in the bridge network

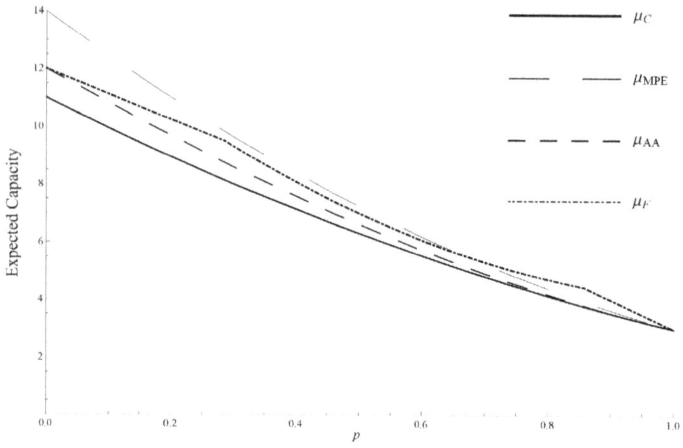

Fig. 3. Estimates of expected capacity of a minimum cut in the bridge network

values of p. Figure 3 also supports our claim that neither μ_{MPE} nor μ_F can be said to be uniformly better than the other.

The methods of this paper have also been applied to the 7-node, 13-arc network of [12, p. 593], which contains 16 s-t paths and 18 s-t cutsets. Results are displayed in Figure 4. The upper bound μ_{MPE} is not shown, as it is substantially dominated by the other bounds. As expected, we see that both μ_{AA} and μ_F provide upper bounds on the true μ_C, with $\mu_{AA} \leq \mu_F$. Here the acyclic bound is quite close to the true expected capacity over the entire range $0 \leq p \leq 1$.

Larger acyclic networks have also been analyzed using this algebraic approach to determine exactly and approximately the expected cut capacity. Further research is ongoing to implement this approach for larger, cyclic networks.

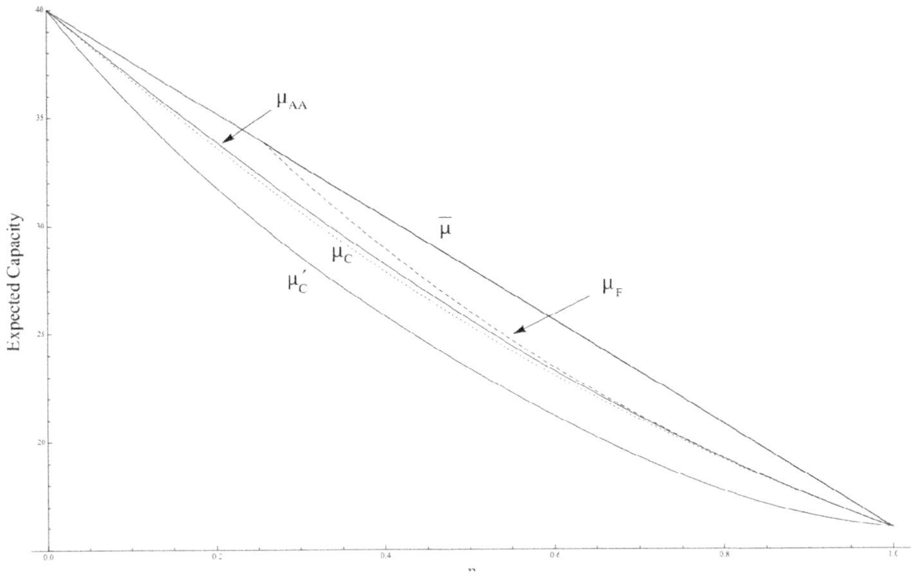

Fig. 4. Estimates of expected capacity of a minimum cut in the second example

References

1. Ahuja, R.K., Magnanti, T.L., Orlin, J.B.: Network Flows: Theory, Algorithms, and Applications. Prentice Hall, Englewood Cliffs (1993)
2. Alexopoulos, C., Fishman, G.S.: Characterizing stochastic flow networks using the Monte Carlo method. Networks 21, 775–798 (1991)
3. Daly, M.S., Alexopoulos, C.: State-space partition techniques for multiterminal flows in stochastic networks. Networks 48, 90–111 (2006)
4. Altenhöfer, M.: An algebraic approach to the stochastic shortest path problem. Department of Mathematical Sciences, Clemson University (June 2007)
5. Billingsley, P.: Probability and Measure, 3rd edn. John Wiley & Sons, New York (1995)
6. Carey, M., Hendrickson, C.: Bounds on expected performance of networks with links subject to failure. Networks 14, 439–456 (1984)
7. Doulliez, P., Jamoulle, E.: Transportation networks with random arc capacities. RAIRO 3, 45–49 (1972)
8. Evans, J.R.: Maximum flows in probabilistic graphs – The discrete case. Networks 6, 161–183 (1976)
9. Fishman, G.S.: The distribution of maximum flow with applications to multi-state reliability systems. Operations Research 35, 607–618 (1987)
10. Fulkerson, D.R.: Expected critical path lengths in PERT networks. Operations Research 10, 808–817 (1962)
11. Hastings, K.C.: Algebraic-based algorithms for determining shortest paths in a stochastic network. Department of Mathematical Sciences, Clemson University (May 2009)
12. Hutson, K.R., Shier, D.R.: Extended dominance and a stochastic shortest path problem. Computers and Operations Research 36, 584–596 (2009)

13. Jarvis, J.P., Shier, D.R.: An improved algorithm for approximating the performance of stochastic flow networks. INFORMS J. Computing 8, 355–360 (1996)
14. Nagamochi, H., Ibaraki, T.: Maximum flows in probabilistic networks. Networks 21, 645–666 (1991)
15. Picard, J.C., Queyranne, M.: Selected applications of minimum cuts in networks. INFOR 20, 394–422 (1982)
16. Provan, J.S., Shier, D.R.: A paradigm for listing (s,t)-cuts in graphs. Algorithmica 15, 351–372 (1996)
17. Pukelsheim, F., Ricca, F., Scozzari, A., Serafini, P., Simeone, B.: Network flow methods for electoral systems. Networks (to appear)

Reliable and Restricted Quickest Path Problems

Stefan Ruzika and Markus Thiemann

University of Kaiserslautern, Paul-Ehrlich-Straße 14, 67663 Kaiserslautern, Germany
{ruzika,thiemann}@mathematik.uni-kl.de

Abstract. In a dynamic network, the quickest path problem asks for a path minimizing the time needed to send a given amount of flow from source to sink along this path. In practical settings, for example in evacuation or transportation planning, the reliability of network arcs depends on the specific scenario of interest. In this circumstance, the question of finding a quickest path among all those having at least a desired path reliability arises. In this article, this *reliable quickest path problem* is solved by transforming it to the *restricted quickest path problem*. In the latter, each arc is associated a nonnegative cost value and the goal is to find a quickest path among those not exceeding a predefined budget with respect to the overall (additive) cost value. For both, the restricted and reliable quickest path problem, pseudopolynomial exact algorithms and fully polynomial-time approximation schemes are proposed.

1 Introduction

In *dynamic* networks, flow units take time to traverse an arc and, there, the *quickest path problem* generalizes the shortest path problem. Given an amount of flow U and two nodes s and t, the goal of the quickest path problem is to find an s-t-path with minimum transmission time, that is the total travel time from s to t of this path plus the number of repetitions to send all U flow units along this path (cf. [4]). The quickest path problem appears in communication networks, transportation networks, and evacuation modeling (see, among others, [5,7,12]). Polynomial time solution algorithms were established for this problem by reducing it to the shortest path problem in a modified network ([4], [14]).

Numerous variants and extensions of the quickest path problem have been considered, including constrained quickest path problems ([3]), robust quickest path problems ([15]), and extensions of the quickest path problem to a stochastic-flow network ([9]).

In practice, operability of arcs in the network may be subject to their reliability, i.e. the probability not to fail. To have a calculable probability of a path's functioning, it is of interest to comprise path reliability in the quickest path problem. Quickest paths with reliabilities have been considered by Xue [16] and Bang et al.[1]. There, a *most reliable quickest path* and a *quickest most reliable path* is sought which is to find a quickest path among the most reliable ones in the first case and a most reliable among the quickest paths in the latter case. In contrast, the reliable quickest path problem considered in this article at hand is understood as finding a quickest path among all paths with at least a desired path reliability predefined by a decision maker. This problem generalizes the most reliable quickest path problem, since it does not require the path to have the best possible reliability.

J. Pahl, T. Reiners, and S. Voß (Eds.): INOC 2011, LNCS 6701, pp. 309–314, 2011.

If usage of an arc is associated with costs and a budget that is not to be exceeded is given, the restricted quickest path problem is of interest. This problem seeks for a quickest path among those paths which obey the budget constraint. It is a generalization of the restricted shortest path problem which has attracted great attention in the literature [8,11]. [2] considered the related *Minimum Cost Quickest Path Problem with Multiple Delay Bounds* where a minimum cost path among those paths not exceeding a given transmission time has to be found.

In [10] k paths in a stochastic flow network are sought to maximize the system reliability, that is the probability that U units of flow can be sent through k s-t-paths satisfying a given time constraint T and within a given budget B.

This article is subsequently organized as follows. The next section introduces the quickest path problem, defines its reliable and restricted variants and depicts the equivalence of the two problems. In Section 3, the restricted quickest path problem is solved with a pseudopolynomial algorithm and approximated polynomially. Identical results are deduced for the reliable quickest path problem. The last section gives a conclusion of the article.

2 Problem Definition

A dynamic network $G = (N, A)$ with node set N and arc set A is equipped with two kinds of parameters: capacities $u_{ij} \in \mathbb{Z}^+$ and travel times $\tau_{ij} \in \mathbb{Z}_0^+$ for all $(i,j) \in A$. The former limits the number of flow units that can enter arc (i,j) in a single time step, the latter is the time needed for a flow unit to traverse arc (i,j). Let $s,t \in N$ denote the source and sink node, respectively. The number of nodes and arcs is denoted n and m, i.e., it is $n = |N|$ and $m = |A|$. Given an initial amount of flow $U \in \mathbb{Z}^+$, the *Quickest Path Problem* seeks for an s-t-path P with minimum *transmission time* $\sigma(P) := \tau(P) + \lceil \frac{U}{u(P)} \rceil$ where $\tau(P) := \sum_{(i,j)\in P} \tau_{ij}$ and $u(P) := \min_{(i,j)\in P} u_{ij}$ are the travel time and capacity of path P, respectively. Let $\mathscr{P}$ refer to the set of all s-t-paths in G.

Assume that for each arc $(i,j) \in A$ a reliability $r_{ij} \in (0,1]$ is given, which describes the probability that this arc is operational. For a path $P \in \mathscr{P}$, $\prod_{(i,j)\in P} r_{ij}$ is called the *path reliability* and expresses the probability of a path's functioning. Let $R \in (0,1]$ denote the *desired minimum path reliability*. The *Reliable Quickest Path Problem* asks for the quickest path that has a path reliability of at least R:

$$\begin{aligned} \min \quad & \sigma(P) \\ \text{s. t.} \quad & P \in \mathscr{P} \\ & \prod_{(i,j)\in P} r_{ij} \geq R. \end{aligned} \tag{1}$$

Note that the constraint (1) is equivalent to $\sum_{(i,j)\in P} \ln(1/r_{ij}) \leq \ln(1/R)$. Since $x \mapsto \ln(1/x)$ is a bijective mapping from $(0,1]$ to $[0,\infty)$, the reliable quickest path problem is equivalent to the *Restricted Quickest Path Problem*, where cost values $c_{ij} \in \mathbb{R}_0^+$ for all arcs $(i,j) \in A$ and a budget $C \in \mathbb{R}_0^+$ are given and the goal is to find a quickest s-t-path with cost not exceeding the budget C:

$$\begin{aligned} \min \quad & \sigma(P) \\ \text{s. t.} \quad & P \in \mathscr{P} \\ & c(P) := \sum_{(i,j)\in P} c_{ij} \le C. \end{aligned}$$

Solutions of the restricted and reliable quickest path problem are referred to as *restricted quickest paths* and *reliable quickest paths*, respectively.

Given $\varepsilon > 0$, an s-t-path Q is a $(1+\varepsilon)$-approximation of the restricted quickest path $P^\star$, if $c(Q) \le C$ and $\sigma(Q) \le (1+\varepsilon)\sigma(P^\star)$. Accordingly, Q is a $(1+\varepsilon)$-approximation of the reliable quickest path $\tilde{P}$, if $\prod_{(i,j)\in Q} r_{ij} \ge R$ and $\sigma(Q) \le (1+\varepsilon)\sigma(\tilde{P})$. A minimization problem is said to admit a fully polynomial-time approximation scheme (FPTAS), if there is a $(1+\varepsilon)$-approximation algorithm with running time polynomial in the input size and in $1/\varepsilon$ for all instances of the problem (cf. [13]).

3 Algorithms

The restricted quickest path problem generalizes the restricted shortest path problem: find an s-t-path P with minimum travel time $\tau(P)$ in the set of all paths with costs $c(P)$ at most C [8]. A restricted quickest path for $U = 1$ obviously defines a restricted shortest path. Since the restricted shortest path problem is known to be NP-hard [6], this applies to the restricted quickest path problem, too. For the restricted shortest path problem on directed acyclic networks, a pseudopolynomial algorithm has been developed by [8] who also proposed an FPTAS. [11] suggested an improved FPTAS for general directed networks with running time $\mathscr{O}(\frac{mn^2}{\varepsilon}\log\frac{n}{\varepsilon})$.

These algorithms for the restricted shortest path problem require the budget C to be a nonnegative integer. The integer restriction can be avoided by adding a super source S to N and an artificial arc (S,s) to A with $\tau_{Ss} = 0$, $u_{Ss} = \infty$ and cost $c_{Ss} = C - \lfloor C \rfloor$. Then, a restricted shortest S-t-path with budget $\lceil C \rceil$ refers to a solution of the corresponding restricted s-t-path problem with budget C.

[8]'s exact algorithms for the restricted shortest path problem on directed acyclic networks are only described for travel times and costs both being integral. However, examining Algorithm B in [8] reveals that this algorithm only operates on the integer valued travel times. Thus, the arc costs are not required to be integral and, hence, the algorithm is capable of computing exact solutions for restricted shortest path problems as considered in this article. The algorithm has computational complexity $\mathscr{O}(mB)$ where B is an upper bound on the optimal value of the restricted shortest path problem (e.g. the sum of all arc travel times). The generalization of this algorithm for arbitrary directed networks runs in $\mathscr{O}(nmB)$ [11].

For $k \ge 0$ let $G(k)$ be the network with arc set $A(k) := \{(i,j) \in A : u_{ij} \ge k\}$. For a restricted quickest path problem, the corresponding restricted shortest path problem in $G(k)$ for $k \ge 0$ is defined on the same cost values and budget constraint.

The following lemma describes a relation between a restricted quickest path and a restricted shortest path.

Lemma 1. *Let Q be a solution of the restricted quickest path problem. Then, Q solves the corresponding restricted shortest path problem in $G(u(Q))$.*

Proof. Let P be a restricted shortest path in $G(u(Q))$. Then, $\tau(P) \leq \tau(Q), u(P) \geq u(Q)$, $c(P) \leq C$, and $c(Q) \leq C$. Since Q is a restricted quickest path, it is

$$\tau(Q) + \left\lceil \frac{U}{u(Q)} \right\rceil \leq \tau(P) + \left\lceil \frac{U}{u(P)} \right\rceil .$$

It follows that $\tau(Q) \leq \tau(P)$ and, hence, Q is a restricted shortest path in $G(u(Q))$. □

Theorem 1. *Let $u_1, \ldots, u_l$ be the distinct capacities in G. For each u_j, $j = 1, \ldots, l$, let P_j be a restricted shortest path in $G(u_j)$. Let*

$$P_k \in \operatorname*{argmin}_{j=1,\ldots,l} \sigma(P_j).$$

Then, P_k solves the restricted quickest path problem.

Proof. Let Q be a solution of the restricted quickest path problem. Let $u_{j_0} = u(Q)$ and consider the network $G(u_{j_0})$. According to Lemma 1, Q is a restricted shortest path in $G(u_{j_0})$. Therefore, it is $\tau(P_{j_0}) = \tau(Q)$ and $u(P_{j_0}) \geq u(Q)$. Thus,

$$\sigma(P_k) \leq \tau(P_{j_0}) + \left\lceil \frac{U}{u(P_{j_0})} \right\rceil \leq \tau(Q) + \left\lceil \frac{U}{u(Q)} \right\rceil$$

and, hence, P_k is a restricted quickest path. □

Corollary 1. *The restricted quickest path problem can be solved in $\mathcal{O}(nm^2B)$.*

Proof. A restricted quickest path can be found by solving l restricted shortest path problems, see Theorem 1. Since $l \leq m$, the computational complexity follows directly from that of the restricted shortest path problem. □

For $\varepsilon > 0$, a $(1+\varepsilon)$-approximation of the restricted shortest path is a path obeying the budget constraints and having a travel time within a factor $(1+\varepsilon)$ of the optimal travel time.

Theorem 2. *Let $\varepsilon > 0$ and let $u_1, \ldots, u_l$ be the distinct capacities in G. For each u_j, $j = 1, \ldots, l$ let P_j be a $(1+\varepsilon)$-approximation of the restricted shortest path in $G(u_j)$. Let*

$$P_k \in \operatorname*{argmin}_{j=1,\ldots,l} \sigma(P_j).$$

Then, P_k is a $(1+\varepsilon)$ approximation of the restricted quickest path.

Proof. Let Q be a solution of the restricted quickest path problem and let $u_{j_0} = u(Q)$. Then, Q is a restricted shortest path in $G(u_{j_0})$ due to Lemma 1 and from definition it is $\tau(P_{j_0}) \leq (1+\varepsilon)\tau(Q)$. With $u(P_{j_0}) \geq u(Q)$ it follows that

$$\tau(P_{j_0}) + \left\lceil \frac{U}{u(P_{j_0})} \right\rceil \leq (1+\varepsilon)\tau(Q) + \left\lceil \frac{U}{u(Q)} \right\rceil \leq (1+\varepsilon)\left(\tau(Q) + \left\lceil \frac{U}{u(Q)} \right\rceil\right). \quad \square$$

Corollary 2. *The restricted quickest path problem admits an FPTAS running in $\mathcal{O}(\frac{m^2n^2}{\varepsilon} \log \frac{n}{\varepsilon})$.*

Proof. Using Theorem 2, at most m $(1+\varepsilon)$-approximations of restricted shortest paths have to be computed, each of which takes $\mathscr{O}(\frac{mn^2}{\varepsilon}\log\frac{n}{\varepsilon})$ [11]. □

As shown in Section 2, the reliable and restricted quickest path problems are equivalent. Thus, the reliable quickest path problem is also NP-hard. Further, the results on pseudopolynomial and approximation algorithms arise from the corresponding results for the restricted quickest path problem.

Corollary 3. *The reliable quickest path problem can be solved in* $\mathscr{O}(nm^2B)$. *Moreover, the reliable quickest path problem admits an FPTAS running in* $\mathscr{O}(\frac{m^2n^2}{\varepsilon}\log\frac{n}{\varepsilon})$.

4 Conclusion

Two variants of the quickest path problem are investigated. The reliable quickest path problem is of interest in dynamic networks where arcs may have a probability of failure. The goal of this problem is to find a quickest path among those having at least a desired path reliability predefined by a decision maker. With a parameter transformation, it is shown that the reliable quickest path problem is equivalent to the restricted quickest path problem, where cost values are given for all arcs and the goal is to find a quickest path among those not exceeding a predefined budget. A pseudopolynomial exact algorithm and an FPTAS are proposed for both problems. Since the problems are NP-hard, the presented algorithms are the best achievable.

Acknowledgement

This work has been partially supported by BMBF, Project REPKA, FKZ 13N9961 (TU KL).

References

1. Bang, Y.C., Choo, H., Mun, Y.: Reliability problem on all pairs quickest paths. In: Sloot, P.M.A., Abramson, D., Bogdanov, A.V., Gorbachev, Y.E., Dongarra, J., Zomaya, A.Y. (eds.) ICCS 2003. LNCS, vol. 2660, pp. 518–523. Springer, Heidelberg (2003)
2. Bang, Y.C., Hong, I., Lee, S., Ahn, B.: On algorithms for minimum-cost quickest paths with multiple delay-bounds. In: Laganá, A., Gavrilova, M.L., Kumar, V., Mun, Y., Tan, C.J.K., Gervasi, O. (eds.) ICCSA 2004. LNCS, vol. 3043, pp. 1125–1133. Springer, Heidelberg (2004)
3. Chen, Y.: Finding the k quickest simple paths in a network. Inform. Process. Lett. 50(2), 89–92 (1994)
4. Chen, Y., Chin, Y.: The quickest path problem. Comput. Oper. Res. 17(2), 153–161 (1990)
5. Clímaco, J., Pascoal, M., Craveirinha, J., Captivo, M.: Internet packet routing: Application of a K-quickest path algorithm. European J. Oper. Res. 181(3), 1045–1054 (2007)
6. Garey, M., Johnson, D.: Computers and Intractability. A Guide to the Theory of NP-Completeness. A Series of Books in the Mathematical Sciences. W.H. Freeman, San Francisco (1979)

7. Hamacher, H., Tjandra, S.: Mathematical modelling of evacuation problems–a state of the art. In: Schreckenberger, M., Sharma, S. (eds.) Pedestrian and Evacuation Dynamics, pp. 227–266. Springer, Berlin (2002)
8. Hassin, R.: Approximation schemes for the restricted shortest path problem. Math. Oper. Res. 17(1), 36–42 (1992)
9. Lin, Y.: Extend the quickest path problem to the system reliability evaluation for a stochastic-flow network. Comput. Oper. Res. 30(4), 567–575 (2003)
10. Lin, Y.: System reliability for quickest path problems under time threshold and budget. Comput. Math. Appl. 60, 2326–2332 (2010)
11. Lorenz, D., Raz, D.: A simple efficient approximation scheme for the restricted shortest path problem. Oper. Res. Lett. 28(5), 213–219 (2001)
12. Moore, M.: On the fastest route for convoy-type traffic in flowrate-constrained networks. Transport. Sci. 10(2), 113–124 (1976)
13. Papadimitriou, C., Steiglitz, K.: Combinatorial Optimization: Algorithms and Complexity. Prentice-Hall, Englewood Cliffs (1982)
14. Rosen, J., Sun, S., Xue, G.: Algorithms for the quickest path problem and the enumeration of quickest paths. Comput. Oper. Res. 18(6), 579–584 (1991)
15. Ruzika, S., Thiemann, M.: Min-max quickest path problems. Tech. rep., University of Kaiserslautern, Department of Mathematics, Report in Wirtschaftsmathematik Nr. 130 (2010)
16. Xue, G.: End-to-end data paths: quickest or most reliable? IEEE Commun. Lett. 2(6), 156–158 (1998)

Modeling and Optimization of Production and Distribution of Drinking Water at VMW

Derek Verleye[1] and El-Houssaine Aghezzaf[2]

[1] Department of Industrial Management, Faculty of Engineering, Ghent University, Building 903, Technologiepark, 9052 Zwijnaarde
derek.verleye@ugent.be
[2] Department of Industrial Management, Faculty of Engineering, Ghent University, Building 903, Technologiepark, 9052 Zwijnaarde
elhoussaine.aghezzaf@ugent.be

Abstract. We develop and discuss an operational planning model aiming at minimizing production and distribution costs in large drinking water networks containing buffers with free inflow. Modeling drinking water networks is very challenging due of the presence of complex hydraulic constraints, such as friction losses and pump curves. Non-linear, non-convex constraints result from the relationships between pressure and flow in power terms. Also, binary variables are needed to model the possibility of free inflow or re-injection of water at reservoirs. The resulting model is thus a non-convex Mixed-Integer Non-Linear Program (MINLP). A discrete-time setting is proposed to solve the problem over a finite horizon made of several intervals. A commercial solver, *BONMIN*, suited for convex MINLP models is used to heuristically solve the problem. We are able to find a good solution for a small part of an existing network operated by the Vlaamse Maatschappij voor Watervoorziening (VMW), a major drinking water company in Flanders.

1 Introduction

After an optimal design of the drinking water network of water production and distribution systems, further economic efficiency can be reached through optimal management of operational activities such as production scheduling and pump switching. Most drinking water companies neglect these savings by operating their networks based upon experience. In this regard, an operational support model that minimizes production and distribution costs over a finite horizon is extremely useful. Distribution and production in drinking water networks can be modeled as a minimum cost flow problem [1,2] with many side constraints. The arcs represent pipelines, whereas nodes are used to represent junctions, buffers etc. Conservation of flow and bounds on the flow rate are imposed as restrictions. This model is expanded with more complicated constraints resulting from the network hydraulics.

Early research papers on this subject handle energy-efficient operation of pumps [8,9], possibly in combination with guaranteed supply by optimal management of water inventory in buffers [6]. Nonlinearity of the objective function is an immediate consequence of the 'power term', which is a function of two main variables of the system:

J. Pahl, T. Reiners, and S. Voß (Eds.): INOC 2011, LNCS 6701, pp. 315–326, 2011.

pump head and flow rate. In [7], an approach to linearize this term is proposed. The resulting MILP for a hydro scheduling commitment was hereby successfully solved. The optimization of large-scale drinking water network with multiple production centers is a less studied topic [5,4]. The authors present a complete model for the drinking water supply in Berlin, using sophisticated functions to describe the friction losses in pipes as well as accurate mathematical models for pump curves. Their formulation results in a MINLP model, which is very difficult to solve for such a large network. Hence, the model is approximated by a practical useful NLP (nonlinear problem) formulation. These authors state however, that more complicated MINLP models will become practical in the future with improvements in computing technology and mathematical algorithms.

In [10], the first steps in developing such a model are taken. This paper summarizes the key aspects of this work. A specific component is free inflow and re-injection of water at reservoirs, requiring additional binary variables in the model. The relationship between flow and head in pressure losses and the power term lead to nonlinear, non-convex constraints. The resulting model is tested on a small part of the existing network of VMW (Vlaamse Maatschappij voor Watervoorziening), the largest drinking water company in Flanders. The commercial solver *BONMIN* [3] is used to generate a suboptimal solution over a horizon of one day, divided in several discrete periods, in a reasonable computation time.

2 Description of the Model

In this section, the structure as well as the different components in the supply network are defined. The different hydraulic constraints will be mathematically formulated for each of these components. Note that we only consider a reduced supply network in the province West Flanders (Flanders, Belgium) which is showed in Fig. 1. The supply network is the pipeline structure wherein large volumes of fresh water are sent from the production station to the buffers, delivery points and junctions with the distribution network. The latter one consist of the pipes with small diameters through which water is distributed to the end user. In this report, demand in the distribution network is clustered in so called demand parameters in the supply network. The supply network is modeled as a graph $G = [K, L]$, where K denotes the set of nodes and the set L contains all the pipes in the network. Nodes will be indicated by the superscript i, whereas pipes connecting nodes i and j are indicated by superscript ij. In what follows, the time division in discrete periods is explained. An overview of the most important symbols is given. Next, the constraints in different components in the considered network during each of these periods, are described. Afterwards, the goal function is defined.

2.1 Discrete Time Setting

As an optimization over a period of at least one day is desired, a day is divided into discrete periods during each of which the state of the network is assumed to be constant. Because a calculation of the network state at every hour is very time-consuming, a division of a day as in Fig. 2 is proposed. This division is based upon a typical hourly

Fig. 1. Test Configuration with Junctions (V), Reservoirs (R), Water Towers (W) and Production Centers (WPC)

demand pattern and the different day and night tariffs for electricity. We will denote an interval by $t \in [1,T]$, where T represents the number of periods. The length of each period is denoted by τ_t.

2.2 System Components and Restrictions

Throughout this section restrictions will be written per period, for each $t \in 1..T$. Furthermore, the following notation will be used:

- K_B: set of buffers
- L_G: set of pipes wherein water flows gravitationally out of a buffer
- L_L: set of pipes wherein no pump is active
- L_R: set of pipes wherein a raw water pump is active
- L_P: set of pipes wherein a pure water pump is active
- H: the piezometric head as the sum of the geometric height h and the manometric water pressure p / γ (m)
- Q: flow through a pipe (m^3/ h)
- B: fluid level in tank at the end of a period (m)
- V: volume in tank at the end of a period (m^3)
- M: mean head of water in tank (m)
- ΔH: pump head (m)
- P: power delivered by pump (W)

The main decision variables are $Q_{ij}, ij \in L_R$, the amount of water produced in the water production centers and ΔH, the head delivered by each pump. Knowledge of the

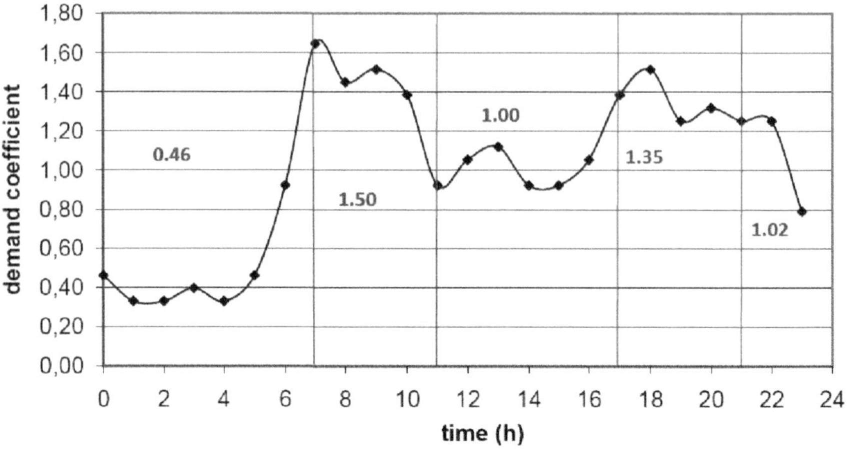

Fig. 2. Discrete Time Intervals Based Upon Demand Coefficients and Electricity Tariffs

value of these variables will allow operators to optimally control the network. All other variables are thus dependent on these two decision variables.

Node Modeling. We distinguish following nodes:

1. Junctions and delivery points
2. Buffers
3. Basins

Junctions and delivery points. In this paper, a junction is defined as the place where the material or diameter of a pipe changes, where the pipe splits up in two or more sections or where a nonzero demand parameter v (in m^3/ h) is assigned to a point in the network.

Apart from conservation of mass, limits to the allowed pressure are also imposed. As a general rule, no negative pressures are allowed and the manometric pressure cannot exceed 10 bar. As these limits are calculated in the junctions, we assume that the height of the actual pipeline is in between that of the endpoints. Otherwise stated, the pipelines are monotone ascending or descending in between two junctions such that the pressure limits are respected everywhere.

Delivery points are situated at the border of two different drinking water networks. If one of the parties has difficulties of delivering water without capacity expansion or the use of boosters and the other party is able to easily deliver extra water in this point, a long-term contract can be signed. A delivery parameter, that can be either positive or negative, is assigned to these nodes.

Buffers. Buffers play a major role in drinking water networks. Apart from their storage capacity, they are of a major economical importance. At night, when the energy tariff is low, water can be stored in the reservoir using pumps. The next day this water can

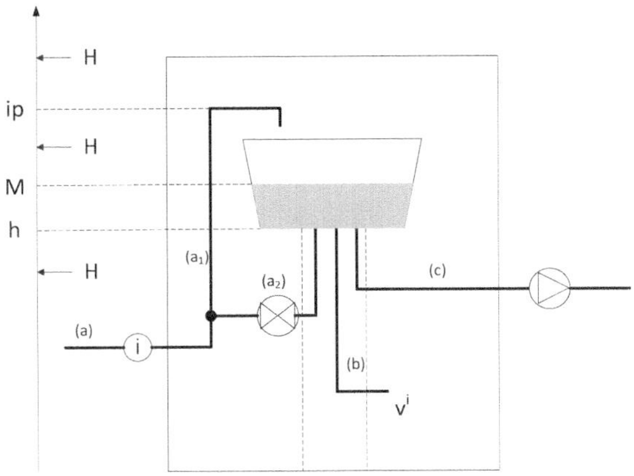

Fig. 3. Model of a Buffer. Depending on the variable value of the head (H) at the entrance i, water can flow either in or out.

flow gravitationally back into the network to meet part of the demand. In this way, high energy costs due to excessive pumping during the day are avoided.

In buffers, we distinguish both pure water tanks and water towers. The latter one can be interpreted as an elevated tank on an underneath construction, thus falling under the same category.

The most general representation of a buffer is given in Fig. 3. Despite its complexity, it is modeled as a single conceptual node. At the entrance (connection (a)) water flows in or out at a variable head. At the exit (connection (c)) water is pumped out at the pressure exercised by the fluid level of the tank. Water storage and distribution take place at the tank itself.

The conservation of mass is expressed as:

$$\sum_{k:(k,i)\in L} Q_t^{ki} - \sum_{j:(i,j)\in L_L \cup L_R} Q_t^{ij} = I_t^{i+} - I_t^{i-} \tag{1}$$

Here, I^{i+} and I^{i-} represent in- and outflow at the buffer, respectively. The necessity of these two additional variables will soon be explained.

Other relationships that take place at the buffer will be explained using Fig. 3. At (a), water in the connected pipes will flow either in or out of the buffer, depending on the head H. Since water can flow freely into the tank, actual inflow only takes place through connection (a1) in periods where the head is bigger than the height of the inflow point ip. Under some circumstances, the mean head of the water M in the tank is higher than the head at the entrance i. As a consequence, water will now be re-injected into the pipe network through connection (a2). This is expressed by the following relations:

$$H_t^i \geq ip^i \Rightarrow I_t^{i+} \geq 0$$
$$H_t^i \leq M_t^i \Rightarrow I_t^{i-} \geq 0$$

By introducing some binary variables, we translate these relations into constraints.

If the head at the entrance of the buffer is in between these values, neither in- nor outflow will occur. In this case, the operator is nevertheless able to open the bypass at (a2) in order to allow water to flow into the tank. Doing this has some unwanted qualitative effects and forces us to add extra binary variables, so we decide not to include this bypass in our model.

At (c), boosters or fresh water pumps transporting water out of the buffer as well as pipes in which water flows out gravitationally, induce an outflow U^{i-}:

$$\sum_{j:(i,j)\in L_P \cup L_G} Q_t^{ij} = U_t^{i-} \tag{2}$$

An important variable is the tank volume at the end of a period, V^i, as it links two subsequent periods through the filling rate:

$$V_t^i = V_{t-1}^i + (I_t^{i+} - I_t^{i-} - U_t^{i-} - v_t^i)\,\tau_t \tag{3}$$

Notice that a buffer may directly be feeding the distribution network with demand v^i through (b).

Necessary initial and terminal conditions are:

$$V_0^i = V^i(0) \tag{4}$$

$$V_T^i \geq V^i(0) \tag{5}$$

where $V^i(0)$ is the tank volume present at the start of the simulation. The second constraint prevents the optimal configuration from depleting all buffers in the last period.

The fluid level in the tank at the end of period t is defined as

$$B_t^i = \frac{V_t^i}{A^i} \tag{6}$$

The cross section of the tank, A^i, is approximated as a constant circle over the entire height, despite the fact that the tank is possibly conic. The fluid level is restricted by a lower bound due to quality considerations and an upper bound equal to the total height of the tank.

To obtain an approximation of the mean head of the water in the tank during a period, the mean fluid level is augmented with the geometric height of the tank floor:

$$M_t^i = h^i + \frac{B_t^i + B_{t-1}^i}{2} \tag{7}$$

Basins. In basins, raw water is captured at surface water treatments. In an optimization over one period, they act as an infinite source of water. Therefore, no restrictions take place at this node. By subtracting the outflow, the volume at the end of the optimization is generated in the output in order to know how much reserve is left.

Pipe Modeling. Friction losses in pipes are calculated by the formula of Darcy-Weisbach for turbulent flow:

$$\Delta p = \lambda \frac{l}{d} \rho \frac{v^2}{2}$$

where l denotes the length of the pipe and d is the inner diameter. The (dimensionless) friction coefficient λ depends on the value of the Reynolds number. Here, we work with the (simplified) law of Prandtl-Kármán for hydraulically rough pipes:

$$\lambda = (2\log \frac{k}{3.71\,d})^{-2}$$

where k (mm) represents the roughness of the pipe and is dependent of the material. Substituting $v = \frac{Q}{A}$, $A = \frac{\pi d^2}{4}$, the friction loss equation leads to:

$$H_t^i - H_t^j = \frac{8\,l^{ij}}{3600^2\,g\,\pi^2\,(d^{ij})^5}\;\lambda^{ij}\,Q_t^{ij}|Q_t^{ij}|\,,\;\forall (i,j) \in L_L \tag{8}$$

for general pipes, and to

$$M_t^i - H_t^j = \frac{8\,l^{ij}}{3600^2\,g\,\pi^2\,(d^{ij})^5}\;\lambda^{ij}\,(Q_t^{ij})^2\,,\;\forall (i,j) \in L_G \tag{9}$$

for water flowing gravitationally out of a buffer.

The formulas (8) and (9) will lead to an underestimation of the friction losses for small values of the flow (up to 20 m^3/ h). In the supply network, such small values for the flow are not common. This justifies the approximation. For a more precise calculation of the friction losses, see Burgschweiger et al., 2000.

The pressure losses for a specific pipe, calculated according to (8), are displayed in Fig. 4. This constraint is clearly non-convex.

A pipe that is out of use, must be excluded from the model. Hereto we add a binary control parameter x for every pipe in the network.

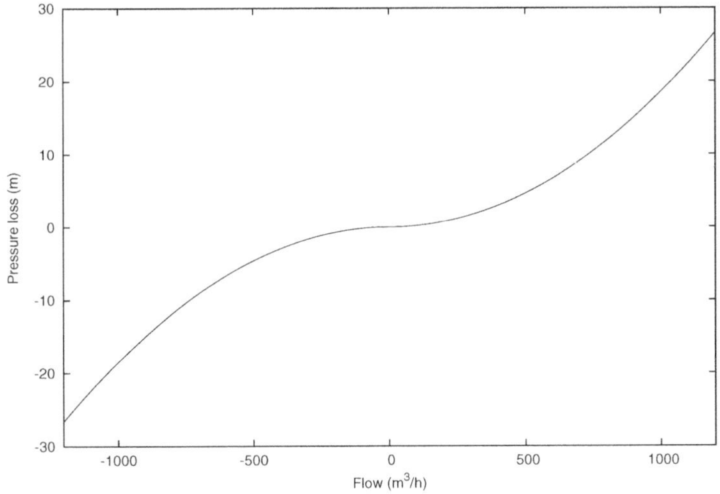

Fig. 4. Pressure losses in a pipe with l = 5 km, d = 500 mm, k = 0,5

Pump Modeling. We distinguish:

1. delivery pumps and boosters
2. raw water pumps

Delivery pumps and boosters. Delivery pumps add pressure to distribute the water from the fresh water basements at the production centers throughout the network. Boosters are placed on well chosen places in the network where extra pressure is needed due to friction losses. In the pipes where the pump is active, we find the following equations:

$$H_t^i - H_t^j = \frac{8\, l^{ij}}{3600^2\, g\, \pi^2\, (d^{ij})^5}\ \lambda^{ij}\, (Q_t^{ij})^2 - \Delta H_t^{ij}\ ,\ \forall (i,j) \in L_P : i \in K \backslash K_B \qquad (10)$$

$$M_t^i - H_t^j = \frac{8\, l^{ij}}{3600^2\, g\, \pi^2\, (d^{ij})^5}\ \lambda^{ij}\, (Q_t^{ij})^2 - \Delta H_t^{ij}\ ,\ \forall (i,j) \in L_P : i \in K_B \qquad (11)$$

where ΔH^{ij} is the head added by the pump. This head depends on the flow that goes through the pump and can be derived from the pump curve. The operating point is the point where this curve and the system curve — the pressure loss curve augmented with the static hight difference — meet. Figure 5 shows this principle for a delivery pump which is currently in use by VMW in West Flanders.

To correctly model this curve, an extensive study is needed. However, such a study does not fit in the scope of this research. Since most of the pumps in the network are variable speed pumps, we undertake a different approach. The working region of a variable speed pump is much larger. Moreover, the flow is within a wide range of values

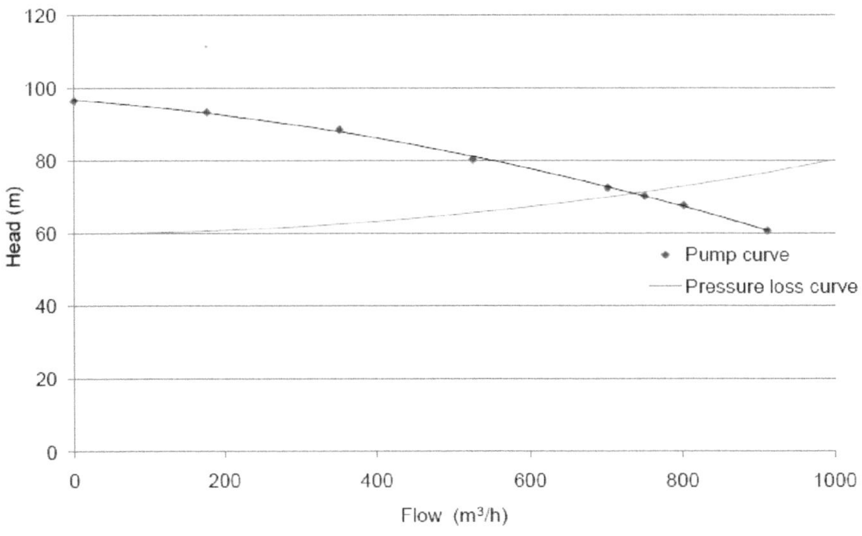

Fig. 5. Operating Ooint of a Pump at a Given Pressure Loss and Static Height Difference

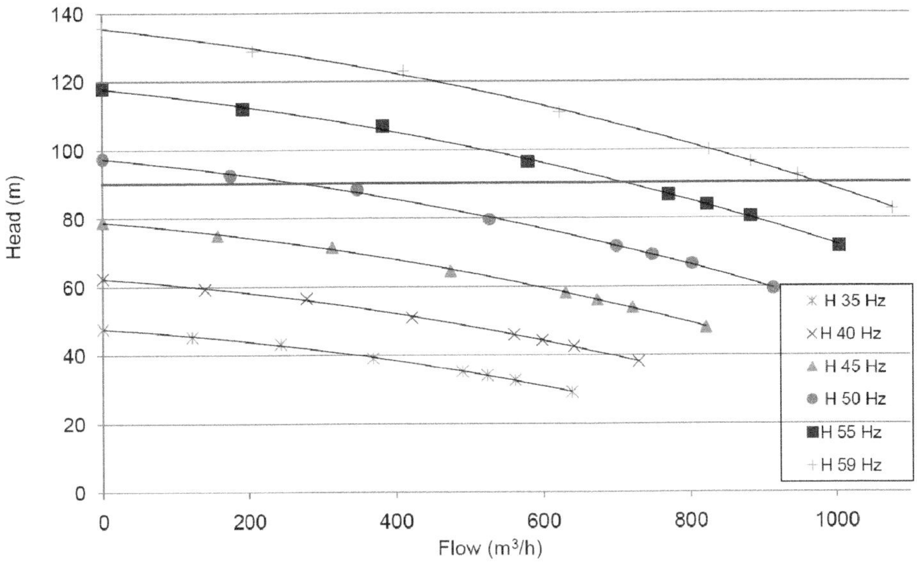

Fig. 6. Pump Curve at Different Frequencies, Fixed Pressure of 90 m

for a fixed value of the pressure, as can be seen from Fig. 6. That is why we force the outlet pressure of this pumps to be fixed at a value u^{ij}, giving:

$$\Delta H_t^{ij} = x^{ij}(u^{ij} - H_t^i)\ , \quad \forall i \in K \backslash K_B \tag{12}$$

$$\Delta H_t^{ij} = x^{ij}(u^{ij} - M_t^i)\ , \quad \forall i \in K_B \tag{13}$$

x is, again, a binary control parameter that takes the value 0 if the pump is not working. The values for u are divided in a winter and a summer regime and are chosen with the aid of the experienced operators at VMW.

For pumps with fixed speed, we opt for a linear approximation of the pump curve. Departing from three fixed points on the pump curve, namely $[q(1), \Delta h(1)]$, $[q(2), \Delta h(2)]$, $[q(3), \Delta h(3)]$, with $q(1) < q(2) < q(3)$ and $h(1) > h(2) > h(3)$, we note:

$$U_t^{ij}(1) \leq W_t^{ij}(1),\ U_t^{ij}(2) \leq W_t^{ij}(1) + W_t^{ij}(2),\ U_t^{ij}(3) \leq W_t^{ij}(2) \tag{14}$$

$$U_t^{ij}(1) + U_t^{ij}(2) + U_t^{ij}(3) + W_t^{ij}(3) = 1 \tag{15}$$

$$W_t^{ij}(1) + W_t^{ij}(2) = 1 \tag{16}$$

$$Q_t^{ij} = x^{ij}[U_t^{ij}(1)\, q^{ij}(1) + U_t^{ij}(2)\, q^{ij}(2) + U_t^{ij}(3)\, q^{ij}(3)] \tag{17}$$

$$\Delta H_t^{ij} = x^{ij}[U_t^{ij}(1)\, \Delta h^{ij}(1) + U_t^{ij}(2)\, \Delta h^{ij}(2) + U_t^{ij}(3)\, \Delta h^{ij}(3)] \tag{18}$$

In this formulation, the variables U are real, whereas W are binary. $W(3)$ is needed to allow the solution $Q = \Delta H = 0$.

The power is given by

$$P_t^{ij} = \frac{2,73\,\Delta H_t^{ij}\, Q_t^{ij}}{\eta_t^{ij}} \tag{19}$$

for all pure water pumps. Here, η denotes pump efficiency and is dependent on the pump's operating point. For simplicity we state $\eta_t^{ij} = \eta^{ij} = 0,65$ which is a rather safe assumption.

Finally, the flow through the pump has to lie within the allowable range:

$$x^{ij}\, Q^{ij}(min) \leq Q_t^{ij} \leq x^{ij}\, Q^{ij}(max) \tag{20}$$

Raw water pumps. Those pumps which push water from the basins all the way through the treatment at the production center are not explicitly modeled. The reason hereto is the fact that the total pressure needed through to get water through these treatment steps is almost constant. We thus consider the cost of the energy delivered by these pumps as part of the production cost and thus linearly dependent with the production flow. The only restrictions are the treatment capacities and raw water contracts as an upper bound to the total produced fresh water.

2.3 Goal Function

As stated before, the goal function consists of two terms: a production and an energy cost. The production cost, kp (in €/ m^3), is time-independent and consists of the costs of chemicals, taxes and the electricity cost. Not only the energy for lighting and treatment is considered, but also for operating the raw water pumps at a fixed pressure. The energy cost of the other pumps is calculated with ke, the time-dependent energy tariff (in €/ kWh). The total cost function that is to be minimized can thus be written as:

$$\text{Min} \sum_{t=1}^{T} \left(\sum_{(i,j)\in L_P} P_t^{ij} \frac{ke_t}{1000} + \sum_{(i,j)\in L_R} Q_t^{ij}\, kp^{ij} \right) \tau_t \tag{21}$$

3 Testing and Results

The resulting MINLP model is non-convex due to the pressure loss constraint. Therefore, a global optimal solution is not guaranteed. As a programming language, *AMPL* is used and the problem is solved using the open-source solver *BONMIN*. As this solver is suited for convex problems, one can expect a heuristic approach of the optimal solution.

Firstly, a configuration is made on which the model will be tested. An optimization is done over a horizon of two periods. Once the working of the model is validated, the existing network in West Flanders is optimized. An optimization is executed over a period of 24 hours. While optimizing over 3 periods does not pose any difficulties, no solution is found from 4 periods onwards. The reason hereto are the fixed pressure restrictions at variable speed pumps. During periods of high demand, friction losses take on high values due to the large pumped flows. In some cases, the fixed pressure may be insufficient to compensate for these losses. Therefore, restrictions 12 and 13 are dropped. This will allow free combinations of flow-head values without considering the actual operating area of the pump.

The test configuration (see also Fig. 1) consists of 2 production centers, 20 junctions, 1 delivery point and 5 buffers, from which 4 are water towers and one is a reservoir. All these nodes are connected by 28 pipes, 2 delivery pumps and 2 boosters. In an acceptable calculation time, we are able to find an optimal solution for a relatively small network. On a HP Pavilion dv6700, 1.83 GHz Processor with 3 GB RAM we found a solution within a time of 885 s. Despite the 'free' behavior of the variable speed pumps, flow-head values were within decent ranges. Nevertheless, the pump curves should be mathematically formulated in a future stadium. Not only will this prevent solutions which are infeasible in practice, the bounds on these two main variables will certainly allow a faster convergence to a (sub)optimal solution.

Some important results can be derived from the optimal solution. All clean water tanks maintain a maximum fluid level during all periods, except the final one. This can be explained through the gain in energy efficiency from the smaller head that delivery pumps behind those reservoirs have to add. During periods of low energy costs, all water towers maintain a minimal fluid level. During periods of low energy cost, these towers are filled back again. These facts lead to the graph in Fig. 7, which displays the total pumping power together with the energy tariff.

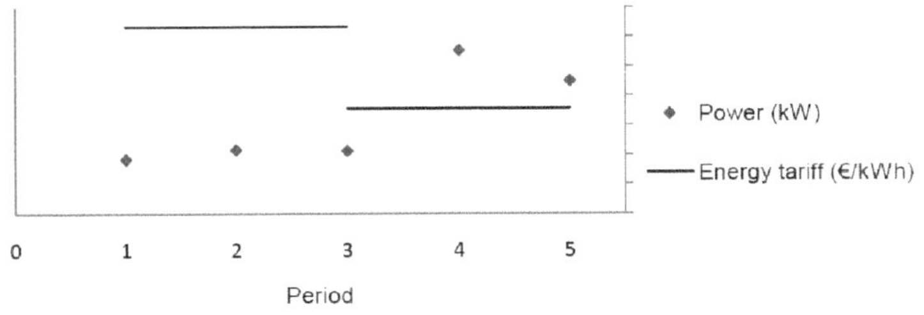

Fig. 7. Pump Power Versus Energy Tariff

4 Conclusion

A model for operative planning in a drinking water network over a finite horizon was constructed. An important component is the basin with free inflow and outflow possibilities, as it is the major link between subsequent periods. The result is a non-convex MINLP model that is based on several simplifications, such as an empirical approximation of the friction coefficient. To solve the problem, we relied on the open-source solver *BONMIN* [3], which was able to generate a solution for a part of the existing network within several minutes. The optimal configuration gives good results concerning the time-dependent electricity tariffs, mostly due to the efficient filling and depleting of the basins.

An important step for further improvement is a detailed modeling of the characteristic pump curves. At the same time, the pressure losses should be calculated with more precision. Additional components such as valves can be modeled as well. Once the model accurately describes the complete network hydraulics, attempts shall be undertaken to minimize the calculation time. This can be accomplished by smart reformulations of the complicating constraints. Additionally, the use of other exact and heuristic methods will be tested.

References

1. Ahuja, R.K., Magnanti, T.L., Orlin, J.B.: Network Flows: Theory, Algorithms, and Applications. Prentice-Hall, Englewood Cliffs (1993)
2. Bertsekas, D.P.: Network Optimization: Continuous and Discrete Models. Optimization and Computation Series, Athena Scientific, Post Office Box 391 Belmont, Massachusetts (1998)
3. Bonami, P., Lee, J.: BONMIN Users' Manual (August 2007)
4. Burgschweiger, J., Gnädig, B., Steinbach, M.C.: Optimization Models for Operative Planning in Drinking Water Networks. Optim. Eng. 10, 43–73 (2008)
5. Burgschweiger, J., Gnädig, B., Steinbach, M.C.: Nonlinear programming techniques for operative planning in large drinking water networks. Open. Appl. Math. J. 3, 14–28 (2009)
6. Crawley, P.D., Dandy, G.C.: Optimal operation of multiple-reservoir system. J. Water Resour. Plan. Manage. 119(1), 1–17 (1993)
7. D'Ambrosio, C.: Application-oriented Mixed Integer Non-Linear Programming. Phd thesis, University of Bologna (2009)
8. Pezeshk, S., Helweg, O.J., Oliver, K.E.: Optimal operation of ground-water supply distribution systems. J. Water Resour. Plan. Manage. 120(5), 573–586 (1994)
9. Ulanicki, B., Rance, J.P., Davis, D., Chen, S.: Computer-aided optimal pump selection for water distribution networks. J. Water Resour. Plan. Manage. 119(5), 542–562 (1993)
10. Verleye, D.: Modellering en optimalisatie van waterproductie en -verdeling bij de Vlaamse Maatschappij voor Watervoorziening. Master thesis, University of Ghent (2010)

On the Hazmat Transport Network Design Problem

Edoardo Amaldi[1], Maurizio Bruglieri[2], and Bernard Fortz[3]

[1] Dipartimento di Elettronica e Informazione, Politecnico di Milano, Piazza Leonardo da Vinci 32, 20133 Milano, Italy
amaldi@elet.polimi.it
[2] INDACO, Politecnico di Milano, via Durando 38/a, 20158 Milano, Italy
maurizio.bruglieri@polimi.it
[3] Département d'Informatique, Faculté des Sciences, Université libre de Bruxelles, CP212, Boulevard du Triomphe, 1050 Bruxelles, Belgium
bernard.fortz@ulb.ac.be

Abstract. We consider the problem of designing a network for hazardous material transportation where the government can decide which roads have to be forbidden to hazmats and the carriers choose the routes on the network. We assume that the government is interested in minimizing the overall risk of the shipments whereas the carriers minimize the route costs. In spite of the rich literature on hazmat transportation and of the relevance of this network design problem, it has received little attention and only quite recently. In this work we prove that the version of the hazmat transport network design problem where a subset of arcs can be forbidden is NP-hard even when a single commodity has to be shipped. We propose a bilevel integer programming formulation that guarantees solution stability and we derive a (single-level) mixed integer linear programming formulation that can be solved in reasonable time with a state-of-the-art solver.

1 Introduction

In this work we consider the following *Hazmat Transport Network Design Problem* (HTNDP). A given set of hazmat shipments has to be sent over a road transportation network in order to transport a given amount of hazardous materials from specific origin points to specific destination points. The government can decide which roads have to be forbidden to hazmats and the carriers can choose the routes on the resulting network. In spite of the rich literature on hazmat transportation (see for instance [1,2,3,5]) and of the relevance of this network design problem, it has received little attention and only recently, being studied just in [9,6,7,10]. In [9] the problem is formulated as a bilevel mixed integer program where the leader problem models the government decisions, while the inner problem corresponds to the carrier route selection. Since the bilevel program may be ill-posed (this happens when the inner problem has multiple optimal solutions with different risk values), a heuristic method able to find stable solutions is presented in [7]. The easier case where the network to be designed is constrained to be a tree is considered in [6] (here the problem becomes single level since in a tree there is a unique path between the origin and the destination of each commodity). Finally, in [10] the authors describe a single level path-based formulation, where a set of acceptable (economically viable) paths is considered for each carrier. The proposed framework

J. Pahl, T. Reiners, and S. Voß (Eds.): INOC 2011, LNCS 6701, pp. 327–338, 2011.

can be used to identify road-closure decision based on the risk-cost trade-off between the regulator and the carriers.

In this paper we consider a generalization of the HTNDP where a subset of roads can be forbidden by the government. Such a generalization is motivated by the fact that in Europe the ADR2007 [12] is regulating the hazmat transportation in tunnels, whereas the hazmat transportation outside tunnels remains unconstrained. Thus in this case just the tunnel arcs of a road network can be forbidden. We prove that this generalization of the hazmat transport network design problem is NP-hard even when a single commodity has to be shipped. We propose a bilevel integer programming formulation that guarantees stability and we derive a single-level mixed integer programming formulation that is more compact than the one in [9] and can be solved more efficiently with state-of-the-art solvers such as CPLEX. Our solution method is compared with the exact approach of [9] and the heuristic of [7] on the benchmark real-world instances introduced in [7].

2 Problem Description

Suppose a transport road network is represented by a directed graph $G = (N, A)$, where N is the set of nodes corresponding to road intersections, and A is the set of arcs corresponding to road segments of the network. Let $T \subseteq A$ be the set of arcs that can be forbidden by the government. K hazardous commodities have to be transported from their (single) origin to their (single) destination. Let (o_k, d_k) denote the origin-destination pair of commodity $k = 1,...,K$ and φ^k the corresponding transport demand (amount or number of shipments). Let c_{ij}^k and r_{ij}^k the cost and risk associated to the transport of a unit flow of commodity k on arc $(i, j) \in A$, respectively. The HTNDP consists in finding a subnetwork, i.e., a subgraph of G, $G' = (N', A')$, with $A \setminus A' \subseteq T$, such that the total transport risk is minimized when each commodity moves in such subnetwork from its origin to its destination according to a minimum cost path.

A feasible solution of the HTNDP is called *stable* if the subnetwork does not admit for any commodity multiple minimum cost paths having different risk values.

Note that in general the solution of the HTNDP does not coincide with the union over all the commodities of the minimum risk path for each commodity. Consider for instance the directed graph depicted in Figure 1 with the given costs and risks, where 4 hazardous commodities with the same transport demands (e.g. $\varphi^k = 1$ for $k = 1,...,4$) have to be shipped from the origins to the destinations indicated. In this case if we take as subnetwork for the hazmat transport the union of the minimum risk paths of every commodity, depicted in Figure 2 case a), we obtain that each commodity follows the minimum cost path indicated in the second column of Table 1, with total risk 45. Notice that in this way some commodity follows a path different from its minimum risk path since can travel also some arcs that belong to the minimum risk path of some other commodity. Whereas if we consider the subnetwork depicted in Figure 2 case b) we obtain that each commodity follows the minimum cost path indicated in the second column of Table 2, with total risk 22. It is possible to prove that the latter is the optimal solution of the HTNDP for the instance considered. Moreover it is easy to verify that both solutions are stable. Notice that the solution a) of Figure 2 becomes unstable if the cost of arc (3,6) is 18 rather than 1, because for instance for the fourth commodity there

Table 1. Minimum cost paths in network a) of Figure 2

Commodities	Minimum cost paths	Cost	Risk
(1,7)	1,3,6,7	3	5
(2,5)	2,3,6,7,5	13	15
(8,5)	8,7,5	9	11
(3,5)	3,6,7,5	3	14

Table 2. Minimum cost paths in network b) of Figure 2

Commodities	Minimum cost paths	Cost	Risk
(1,7)	1,3,6,7	3	5
(2,5)	2,3,4,5	30	3
(8,5)	8,9,5	20	12
(3,5)	3,4,5	20	2

will be two minimum cost paths: 3,6,7 and 3,4,5, both with cost 20, but the first path has risk 14 whereas the second one has risk 2. Like observed in [7] in this case the HTNDP is ill-posed since the government cannot induce the carriers to use the paths that achieve the lowest risk and so the government objective function is not single-valued for this government choice. In such a case it is preferable for the government to find a perturbed network with a small deviation from the best one and on which the risk of the carriers is stable.

3 Computational Complexity

In this section, we prove that the HTNDP is strongly NP-hard by a reduction from 3-SAT. The construction was inspired by a similar proof in [11].

Theorem 1. *The HTNDP is strongly NP-hard even for a single commodity.*

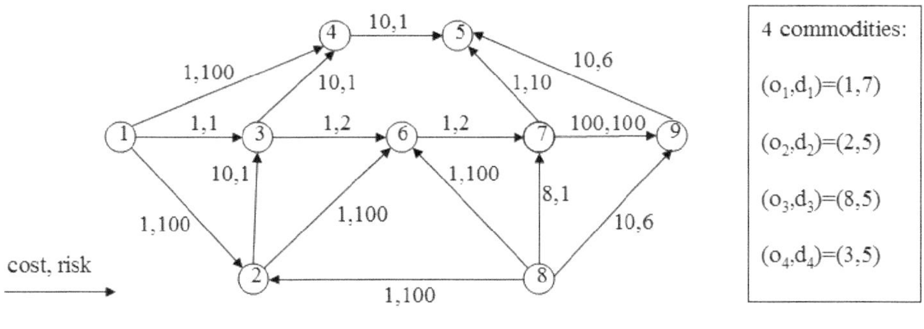

Fig. 1. Instance of the HTNDP with 4 commodities

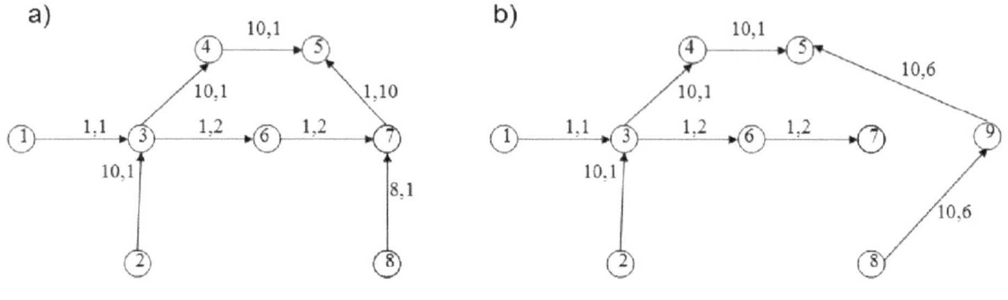

Fig. 2. a) minimum risk path network; b) optimal solution of HTNDP

Proof. Let $x_1,\dots,x_n$ be n Boolean variables and $F = \bigwedge_{i=1}^{m}(l_{i1} \vee l_{i2} \vee l_{i3})$ be a 3-CNF formula consisting of m clauses with literals (variables or their negations) l_{ij}. To reduce 3-SAT to an HTNDP instance with one commodity, we construct for each clause a subnetwork comprising one arc in T (i.e. that can be forbidden) for each literal as shown in Figure 3. More precisely, we create $m+1$ nodes s_i, $i = 1,\dots,m+1$, and for $i = 1,\dots,m$ and $j = 1,2,3$, nodes u_{ij} and v_{ij}. For clause i, we have one arc $(s_i, s_{i+1}) \notin T$ with a cost of 4 and a risk of 1, and arcs $(s_i, u_{ij}) \notin T$, $(u_{ij}, v_{ij}) \in T$ and $(v_{ij}, s_{i+1}) \notin T$ for $j = 1,2,3$, all with cost 1 and risk 0. For $i > 1$, the subnetwork associated to clause i shares node s_i with the subnetwork associated to clause $i-1$.

We have one commodity with origin s_1, destination s_{m+1} and a requested flow of 1, i.e. we want to find a single path from s_1 to s_{m+1}. The idea of the construction is that, if the optimal path goes through an arc (u_{ij}, v_{ij}), then the corresponding literal l_{ij} is TRUE (if $l_{ij} = \overline{x_k}$, then $x_k =$ FALSE).

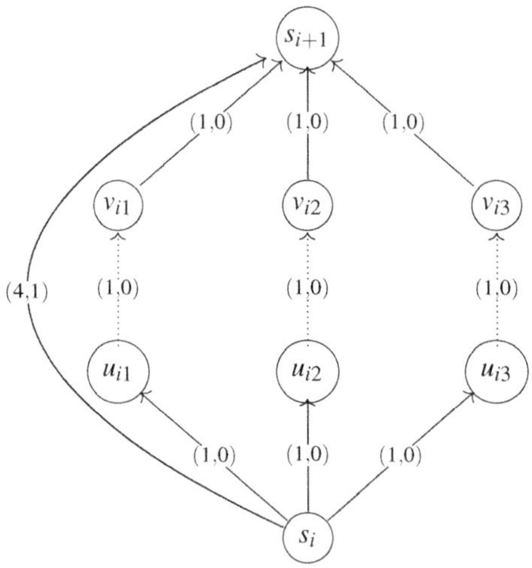

Fig. 3. Subnetwork for clause $(l_{i1} \vee l_{i2} \vee l_{i3})$. The label on an arc a is (c_a, r_a), i.e. the arc a has cost c_a and risk r_a. Dotted arcs can be forbidden.

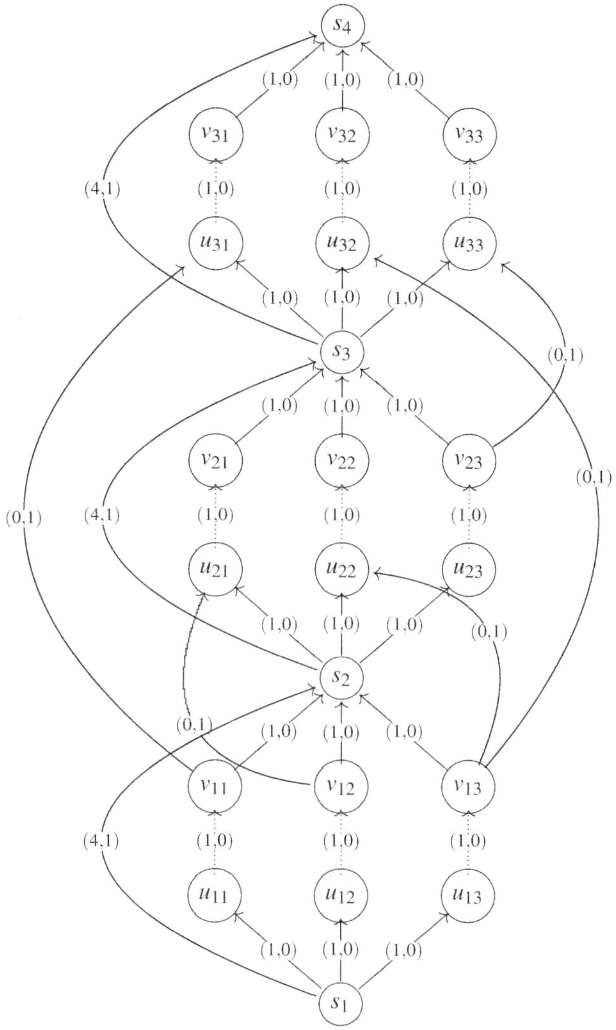

Fig. 4. Network for instance $(x_1 \vee x_2 \vee \overline{x_3}) \wedge (\overline{x_2} \vee x_3 \vee \overline{x_4}) \wedge (\overline{x_1} \vee x_3 \vee x_4)$

If F is satisfiable, we want the optimal path to go through a single arc from T per subnetwork (i.e., one true literal per clause) and simultaneously, we want to make sure that the corresponding assignment of variables is consistent; that is, paths that include a variable and its negation must be ruled out. For that purpose, we assign to every pair of literals corresponding to a variable and its negation an interclause arc between the corresponding arcs, with cost 0 and risk 1 (see Fig. 4). These arcs cannot be forbidden.

We next show that F is satisfiable iff there exists an (optimal) path of risk 0. First suppose there exists an optimal path with risk 0. Clearly, by construction, this path uses exactly one arc u_{ij}, v_{ij} for each clause i. Without loss of generality, we may assume that the only arcs in T that are open are on the path , i.e. l_{ij} literals on the path are set to TRUE. To show that F is satisfiable, it is sufficient to show that a variable and its negation cannot appear together on this path. If it was the case, a less costly path could be constructed by using the interclause arc between the occurrences of the variable and its negation, leading to a positive risk. Hence the path constructed would not be optimal. Now suppose F is satisfiable. If we open the arcs corresponding to literals that are set to TRUE, then interclause arcs cannot belong to a path from s_1 top s_{m+1} as it would mean a variable and its negation are set to TRUE. Since at least one arc must be opened for each subnetwork corresponding to a clause i, the subpath of cost 3 using that arc is less costly than the direct arc (s_i, s_{i+1}). It follows that only arcs with risk 0 belong to a cheapest path and the result follows.

4 A Mixed Integer Linear Programming Formulation

For each arc (i, j) in A and for each k with $1 \leq k \leq K$, we define variable

$$x_{ij}^k = \begin{cases} 1 & \text{if arc } (i,j) \text{ is used by hazardous commodity } k, \\ 0 & \text{otherwise,} \end{cases}$$

and for each arc (i, j) in T, we define

$$y_{ij} = \begin{cases} 1 & \text{if tunnel } (i,j) \text{ is available for hazmat transportation,} \\ 0 & \text{otherwise.} \end{cases}$$

We start with the following bilevel integer programming formulation of HTNDP:

$$(P1) \quad \min \quad \sum_{k=1}^{K} \sum_{(i,j)\in A} r_{ij}^k \varphi^k x_{ij}^k \tag{1}$$

$$\text{s.t.} \quad y_{ij} \in \{0,1\} \qquad \forall (i,j) \in T \tag{2}$$

where $\{x_{ij}^k\}_{1 \leq k \leq K, (i,j) \in A}$ is an optimal solution of the *follower* problem:

$$(P2) \quad \min \quad \sum_{k=1}^{K} \left(\sum_{(i,j)\in A} c_{ij}^k x_{ij}^k - \frac{1}{R} \sum_{(i,j)\in A} r_{ij}^k x_{ij}^k \right) \tag{3}$$

$$\text{s.t.} \quad \sum_{i\in\delta^-(j)} x_{ij}^k - \sum_{l\in\delta^+(j)} x_{jl}^k = \begin{cases} 0, & \text{if } j \neq o_k, d_k \\ -1, & \text{if } j = o_k \\ 1, & \text{if } j = d_k \end{cases} \qquad \forall j \in N, \forall k = 1,\ldots,K \tag{4}$$

$$x_{ij}^k \leq y_{ij} \qquad \forall (i,j) \in T, \forall k = 1,\ldots,K \tag{5}$$

$$x_{ij}^k \in \{0,1\} \qquad \forall (i,j) \in A, \forall k = 1,\ldots,K \tag{6}$$

where constant R is the sum over all the commodities of the maximum risk path values.

This formulation differs from the formulation presented in [9] by the objective function of the follower problem. It is worth pointing out that, unlike in [7] where a heuristic is presented to tackle the stability problem, the objective function (3) allows us to solve the stability problem in an *exact* way. Indeed, when the minimum cost path of a carrier is not unique, our formulation selects, among all of them, one having maximum risk. This amounts to consider the worst-case scenario in terms of risk.

Instead of linearizing the Karush-Kuhn-Tucker (KKT) conditions of the follower problem as in [9], we propose a different way to solve the bilevel problem (1)-(6) to optimality. First, we observe that since the network is uncapacitated, the follower problem decomposes in K subproblems, corresponding to the K commodities, that can be solved separately because they do not share any resource. More precisely, the Integer Linear Programming (ILP) problem to solve for each k, with $1 \leq k \leq K$, is given by the objective function contained in the round brackets of (3) subject to constraints (4), (5) and (6). Since the constraint matrix of every such problem is totally unimodular, we can substitute constraints (6) on the variables with

$$0 \leq x_{ij}^k \leq 1 \quad \forall (i,j) \in A, \forall k = 1, \ldots, K \tag{7}$$

and obtain K Linear Programming (LP) relaxations. According to the weak and strong duality theorems, every such LP problem can be replaced with the primal feasibility constraints, the dual feasibility constraints, and the reverse weak duality inequality. Specifically, the follower problem is replaced with constraints (4), (5), (7) (corresponding to primal feasibility) and the following constraints:

$$\pi_j^k - \pi_i^k \leq c_{ij}^k - \frac{r_{ij}^k}{R} \qquad \forall (i,j) \in A \setminus T \tag{8}$$

$$\pi_j^k - \pi_i^k \leq c_{ij}^k - \frac{r_{ij}^k}{R} + M(1 - y_{ij}) \qquad \forall (i,j) \in T \tag{9}$$

$$\sum_{(i,j)\in A} c_{ij}^k x_{ij}^k - \frac{1}{R} \sum_{(i,j)\in A} r_{ij}^k x_{ij}^k \leq \pi_{d_k}^k - \pi_{o_k}^k. \tag{10}$$

Constraints (8) are the classical dual constraints of the minimum cost path problem (with arc cost $c_{ij}^k - (r_{ij}^k/R)$): they impose that the difference between the potentials associated to the head and the tail of each arc (i,j) does not exceed its cost. Constraints (9) concern the tunnels: they coincide with constraints (8) when the tunnel (i,j) is open whereas they become redundant when the tunnel (i,j) is closed. Finally, constraint (10) imposes that the values of the primal and dual objective functions are equal.

Thus we obtain the following (single-level) mixed integer programming (MILP) formulation of HTNDP:

$$\min \quad \sum_{k=1}^{K} \sum_{(i,j)\in A} r_{ij}^k \varphi^k x_{ij}^k \tag{11}$$

$$\text{s.t.} \quad \sum_{i\in\delta^-(j)} x_{ij}^k - \sum_{l\in\delta^+(j)} x_{jl}^k = \begin{cases} 0, & \text{if } j \neq o_k, d_k \\ -1, & \text{if } j = o_k \\ 1, & \text{if } j = d_k \end{cases} \qquad \forall j \in N, \forall k = 1,\ldots,K \tag{12}$$

$$x_{ij}^k \leq y_{ij} \qquad \forall (i,j) \in T, \forall k = 1,\ldots,K \tag{13}$$

$$\pi_j^k - \pi_i^k \leq c_{ij}^k - \frac{r_{ij}^k}{R} \qquad \forall (i,j) \in A \setminus T, \forall k = 1,\ldots,K \tag{14}$$

$$\pi_j^k - \pi_i^k \leq c_{ij}^k - \frac{r_{ij}^k}{R} + M(1 - y_{ij}) \qquad \forall (i,j) \in T, \forall k = 1,\ldots,K \qquad (15)$$

$$\sum_{(i,j)\in A} (c_{ij}^k - \frac{r_{ij}^k}{R}) x_{ij}^k \leq \pi_{d_k}^k - \pi_{o_k}^k \qquad \forall k = 1,\ldots,K \qquad (16)$$

$$0 \leq x_{ij}^k \leq 1 \qquad \forall (i,j) \in A, \forall k = 1,\ldots,K \qquad (17)$$

$$y_{ij} \in \{0,1\} \qquad \forall (i,j) \in T \qquad (18)$$

Although the constraint matrix is no longer totally unimodular due to constraints (14)-(15), we have the following simple property.

Proposition 1. *The mixed integer programming formulation (11)-(18) of HTNDP admits an optimal solution with variables $x_{ij}^k \in \{0,1\}$, for all $(i,j) \in A$ and k, $1 \leq k \leq K$, even if they are not constrained to be integer.*

Proof. Suppose there exists an optimal solution of (11)–(18) with a fractional variable x_{ij}^k for at least one arc $(i,j) \in A$ and a given commodity $k \in \{1,\ldots,K\}$. Due to the flow conservation constraints (12) such a solution corresponds to the superposition of fractional flows along s different paths P_l, $l = 1,\ldots,s$, from o_k to d_k. Let C_l and R_l be the cost and the risk of these paths P_l, respectively. Since constraints (16) hold and $\pi_{d_k}^k - \pi_{o_k}^k$ is a lower bound on the minimum cost path w.r.t. the perturbed costs $\tilde{c}_{ij}^k = c_{ij}^k - (r_{ij}^k/R)$, all paths P_l have to be of minimal perturbed cost. Moreover, since objective function (11) minimizes the total risk, all paths P_l must have the same risk. Therefore, by sending the whole flow along a single path P_l with $l \in \{1,\ldots,s\}$, we obtain a feasible solution of the same perturbed cost and risk as the original one. This clearly corresponds to an optimal solution of (11)-(18) where all variables x_{ij}^k are binary.

The advantage of our MILP formulation with respect to the ILP formulation proposed in [9] is that it contains a much smaller number of binary variables. Indeed, it includes just $|T|$ binary variables (y_{ij} for all (i,j) in T), whereas by linearizing the KKT conditions as in [9] we need $K|A| + |T|$ binary variables (y_{ij} for all (i,j) in T and x_{ij}^k for all (i,j) in A, with $k = 1,\ldots K$).

5 Computational Experiments

The ILP linearization of the bilevel formulation of the HTNDP described in subsection 4 has been implemented in AMPL [8] and solved with CPLEX 11.00 on a PC Intel Xeon with 2.80 GHz, 512KB L2 cache and 2GB RAM. We tested our solution method on the same set of instances used in [7].

5.1 Instance Testbed

The instances are derived from real data of the city of Ravenna (Italy) used in [6]. The road network is composed of $|N| = 105$ nodes and $|A| = 134$ arcs. Based on the population densities around the arcs and the locations and populations of places of assembly in the city (such as schools, churches, hospitals, factories, etc.) the authors in [6] have calculated an *aggregate risk* measure by using the population in places of assembly within 500 m of the arcs, whereas the arc costs are given by the actual distance. As observed in [7], such a risk measure is not generally colinear to the cost, thus the HTNDP may have unstable solutions.

Using this road network, 50 instances with *aggregate risk* have been derived in [7] by varying the number of commodities $K = 20, 30, 40, 50, 60$ and by generating, for each value of K, 10 instances by randomly choosing K origin and destination pairs from the 105 vertices of the network. The shipment demands φ^k of each commodity $k = 1, \ldots, K$ are generated uniformly in the interval [10,100] and rounded to the nearest integer.

5.2 Comparison with Previous Exact and Heuristic Methods

To compare our solution method with the heuristic approach of [7] on an equal footing, we have added to the ILP model presented in Section 4 the following set of constraints that was also present in [7] $y_{ij} = y_{ji}\ \forall (i, j) \in T$. These constraints state that both arcs (i, j) and (j, i) have to be open to hazmat transit if any one of them is open. Moreover, for the same reason we have chosen $T = A$, i.e., the administrator can forbid any arc of the road network to hazmat transit (provided that no commodity origin and destination pair is disconnected).

The results obtained for the testbed of 50 instances described in the previous subsection are summarized in Table 3. The first column indicates the number K of commodities, the second column identifies the instance generated for each value of K, the third column is the ratio between the total risk of the minimum risk paths (over-regulated scenario), R2, and the total risk, R4, given by the best of the two heuristics of [7] (exactly as in column R2/R4 of Table 3 of [7]). The fourth column reports the same ratio obtained with our exact method rather than with the heuristics (R represents the total risk of our solution). The fifth column corresponds to the improvement in risk achieved with our method, namely Δ risk=100*[(R2/R)/(R2/R4)-1]. The successive three columns are analogous to the three previous ones but they are related to costs. The sixth column reports the ratio between the total cost of minimum cost paths and the total cost of the best heuristic solution of [7] (exactly the values of column C1/C4 of Table 3 of [7]), the seventh column reports the same ratio with our solution rather than the heuristic one and the eighth column is the cost variation Δ cost=100*[(C1/C)/(C1/C4)-1]. Finally, the last two columns report the CPU time in seconds of the best heuristic of [7] and of our exact method, respectively.

Note that we omit the comparison with Kara-Verter's exact method [9] because in [7] the authors show that when checked for stability (i.e., when the worst risk of each minimum cost path is considered) the actual total risk of each Kara-Verter's solution is never better than those they obtain, with a relative average risk difference of 4.47%.

Compared with the results of [7], our method provides solutions with a lower overall risk for almost half of the instances (for 24 of them) with an average relative improvement of 0.56%. In general, the risk improvement produces a cost deterioration and in fact the cost of our solutions is higher in average by a factor of 6.45% w.r.t. the cost of the heuristic solutions found in [7]. Interestingly, there are 6 instances where we improve not only the risk but also the transportation cost, obtaining solutions which

Table 3. Results on Erkut-Gzara's instance testbed

K	Inst.	R2/R4	R2/R	Δ risk	C1/C4	C1/C	Δ cost	CPU E-G	CPU exact
20	1	100	100.00	0.00	50.82	49.49	-2.61	2.12	2.38
20	2	97.79	96.88	**-0.93**	35.88	34.67	-3.37	4.20	953.14
20	3	100	100.00	0.00	37.96	37.26	-1.84	2.38	3.85
20	4	99.37	99.50	0.13	47.93	41.59	-13.23	2.95	40.08
20	5	100	100.00	0.00	71.24	62.85	-11.78	1.99	1.89
20	6	98.89	99.91	1.03	51.15	51.46	**0.61**	3.09	40.68
20	7	100	100.00	0.00	52.55	43.54	-17.15	2.26	2.67
20	8	100	100.00	0.00	46.83	43.77	-6.53	2.40	3.01
20	9	99.6	99.41	**-0.19**	52.65	47.88	-9.06	3.16	61.79
20	10	100	100.00	0.00	65.07	62.21	-4.39	2.49	2.61
30	1	98.57	98.78	0.21	55.79	51.06	-8.48	4.74	58.51
30	2	100	100.00	0.00	49.96	43.81	-12.32	3.70	7.29
30	3	100	100.00	0.00	46.34	41.97	-9.43	3.89	6.28
30	4	98.4	99.19	0.80	61.15	56.23	-8.04	6.12	102.02
30	5	98.9	99.78	0.89	44.30	43.48	-1.86	4.72	132.22
30	6	100	100.00	0.00	45.56	39.52	-13.26	4.12	7.08
30	7	100	100.00	0.00	45.72	37.31	-18.40	3.54	6.72
30	8	100	100.00	0.00	54.82	49.23	-10.19	3.49	7.90
30	9	100	100.00	0.00	46.41	42.14	-9.19	3.59	6.84
30	10	100	100.00	0.00	48.04	48.08	0.08	3.82	7.58
40	1	99.44	99.90	0.47	51.52	45.28	-12.11	9.37	92.27
40	2	98.03	98.24	0.22	50.03	49.58	-0.90	8.88	1119.45
40	3	99.59	99.60	0.01	43.29	40.51	-6.42	7.00	63.19
40	4	89.33	93.18	4.31	44.46	43.61	-1.92	8.57	4024.47
40	5	100	100.00	0.00	48.97	38.50	-21.38	4.87	39.88
40	6	99.31	99.34	0.03	47.99	47.93	-0.12	6.81	2799.95
40	7	100	100.00	0.00	47.00	37.30	-20.65	5.13	10.88
40	8	98.37	99.25	0.89	57.43	51.64	-10.09	6.31	1006.15
40	9	97.19	98.90	1.76	36.40	34.98	-3.90	6.88	1137.27
40	10	100	100.00	0.00	51.04	42.48	-16.77	5.33	13.86
50	1	100	100.00	0.00	50.99	46.99	-7.85	6.63	14.63
50	2	100	100.00	0.00	47.52	44.81	-5.71	6.25	13.80
50	3	99.83	99.72	**-0.11**	48.43	45.01	-7.07	9.56	642.85
50	4	92.4	96.27	4.19	47.00	54.93	**16.87**	11.45	823.92
50	5	99.04	99.38	0.34	43.90	39.18	-10.76	9.41	194.28
50	6	99.13	99.29	0.16	48.64	47.58	-2.17	10.47	5164.05
50	7	99.2	99.28	0.08	41.05	38.15	-7.06	8.56	389.72
50	8	98.21	99.07	0.87	54.60	50.96	-6.67	13.26	426.02
50	9	100	100.00	0.00	55.82	45.03	-19.34	6.53	15.53
50	10	98.79	99.77	0.99	50.14	49.01	-2.25	11.11	360.63
60	1	99.41	99.85	0.44	48.90	45.66	-6.62	16.28	623.99
60	2	100	100.00	0.00	54.67	49.77	-8.97	8.17	12.89
60	3	98.94	99.28	0.35	42.45	45.12	**6.30**	10.80	471.89
60	4	98.53	98.80	0.27	51.34	53.01	**3.25**	13.70	1461.33
60	5	94.03	98.30	4.54	49.74	53.11	**6.77**	16.61	10470.50
60	6	100	100.00	0.00	52.68	47.24	-10.34	8.90	30.41
60	7	99.59	99.57	**-0.02**	56.77	56.47	-0.52	14.08	3001.67
60	8	99.54	99.42	**-0.12**	49.18	48.81	-0.74	11.21	623.85
60	9	93.08	94.29	1.29	56.73	62.52	**10.20**	18.27	5145.38
60	10	99.49	99.87	0.38	58.35	56.22	-3.65	15.57	2011.40
Average				0.56			-6.45	7.29	873.21

dominate the ones of [7]. See the numbers emphasized in bold in the eighth column of Table 3.

The computational experiments show 5 exceptions. For five instances, the total risk of the solution found by the heuristic is lower than the one obtained by solving our exact mixed integer linear formulation. These instances have negative values, emphasized in bold, in the column Δ risk. In one case (K=60, instance 7) this may be due to machine rounding error because the difference is of about 10^{-2}. We have carefully checked the other cases and we believe that either there is an error in the results reported in [7] or the instances we used are slightly different.

Finally, it is worth pointing out that the optimal solutions are found within a reasonable CPU time, in average in 873.21 seconds. The heuristic described in [7] is clearly faster (in average 7.29 sec. on another machine) but does not provide an optimality guarantee and often yields worse quality solutions.

6 Concluding Remarks

We have investigated the problem of designing a network for hazardous material transportation where the government can decide which roads have to be forbidden to hazmats so as to minimize the overall risk and the carriers choose the routes so as to minimize transportation costs. Motivated by the real application of ADR2007 [12], we have considered a generalized version of the problem where just a subset of roads (e.g., the tunnels) can be forbidden by the government.

We have proved that the problem is NP-hard even in the case of a single commodity and we have proposed a mixed integer linear programming formulation to solve the problem to optimality. The advantages of our method with respect to the one in [9] is that it finds stable solutions and requires a lower computational effort. The advantage with respect to the approach in [7] is that although both look for stable solutions, our method solves the problem exactly, whereas [7] solves it heuristically. The computational experiments on a testbed of 50 instances show that for 24 instances we obtain solutions with lower overall risk (with an average relative improvement of 0.56%). Moreover, for 6 instances not only the risk but also the transportation cost is improved.

Our mixed integer linear programming formulation for HTNDP is currently being tested on instances deriving from hazmat transportation in tunnels, according to [12], that is where T is a proper subset of the arc set A. Our approach can also be adapted to deal with the risk-cost trade off, by minimizing the factor by which the minimum cost route of each carrier is increased due to the forbidden arcs. This is an important issue to make the administrator decisions economically acceptable. Future work also include strengthening our mixed integer linear programming formulation in order to speed up solution for large-size instances.

Acknowledgment

The authors would like to thank Dr. F. Gzara (York University, Canada) for kindly providing the instances used in [7] and Prof. R. Maja (Politecnico di Milano, Italy) for useful discussion concerning the hazmat transportation and ADR 2007. Bernard

Fortz was supported by an "Actions de Recherche Concertées" (ARC) projet of the "Communauté française de Belgique" and a FRFC program "Groupe de Programmation Mathémathique".

References

1. Bruglieri, M., Maja, R., Marchionni, G., Rainoldi, G.: Safety in hazardous material road transportation: state of the art and emerging problems. In: Bersani, C., et al. (eds.) Advanced Technologies and Methodologies for Risk Management in the Global Transport of Dangerous Goods, pp. 88–129. IOS Press, Amsterdam (2007)
2. Carotenuto, P., Giordani, S., Ricciardelli, S.: Finding minimum and equitable risk routes for hazmat shipments. Computers & Operations Research 34(5), 1304–1327 (2007)
3. Centrone, G., Pesenti, R., Ukovich, W.: Hazardous Materials Transportation: A Literature Review and an Annotated Bibliography. In: Bersani, C., et al. (eds.) Advanced Technologies and Methodologies for Risk Management in the Global Transport of Dangerous Goods, pp. 33–62. IOS Press, Amsterdam (2007)
4. Colson, B., Marcotte, P., Savard, G.: An overview of Bilevel optimization. Annals of Operations Research 153, 235–256 (2007)
5. Dell'Olmo, P., Gentili, M., Scozzari, A.: On finding dissimilar Pareto-optimal paths. European Journal of Operational Research 162(1), 70–82 (2005)
6. Erkut, E., Alp, O.: Designing a road network for dangerous goods shipments. Computers & Operations Research 34(5), 1241–1242 (2007)
7. Erkut, E., Gzara, F.: Solving the hazmat transport network design problem. Computers & Operations Research 35(7), 2234–2247 (2008)
8. Fourer, R., Gay, D.: The AMPL Book. Duxbury Press, Pacific Grove (2002)
9. Kara, B.Y., Verter, V.: Designing a road network for hazardous materials transportation. Transportation Science 38(2), 188–196 (2004)
10. Verter, V., Kara, B.Y.: A path-based approach for hazmat transport network design. Management Science 54(1), 29–40 (2008)
11. Roch, S., Savard, G., Marcotte, P.: An approximation algorithm for Stackelberg network pricing. Networks 46(1), 57–67 (2005)
12. UNECE, United Nations Economic Commission for the Europe, European Agreement concerning the international carriage of Dangerous goods by Road (ADR), 73-76 (2007), 1.9.5, `http://www.unece.org/trans/danger/publi/adr/adr2007/English/01-0%20E_Part%201.pdf`

Complexity of Inverse Shortest Path Routing

Mikael Call and Kaj Holmberg

Linköping University, Department of Mathematics, SE-581 83 Linköping
{mikael.call,kaj.holmberg}@liu.se

Abstract. The inverse shortest path routing problem is to decide if a set of tentative routing patterns is simultaneously realizable. A routing pattern is defined by its destination and two arc subsets of required shortest path arcs and prohibited non-shortest path arcs. A set of tentative routing patterns is simultaneously realizable if there is a cost vector such that for all routing patterns it holds that all shortest path arcs are in some shortest path and no non-shortest path arc is in any shortest path to the destination of the routing pattern. Our main result is that this problem is NP-complete, contrary to what has been claimed earlier in the literature. Inverse shortest path routing problems naturally arise as a subproblem in bilevel programs where the lower level consists of shortest path problems. Prominent applications that fit into this framework include traffic engineering in IP networks using OSPF or IS-IS and in Stackelberg network pricing games. In this paper we focus on the common subproblem that arises if the bilevel program is linearized and solved by branch-and-cut. Then, it must repeatedly be decided if a set of tentative routing patterns is realizable. In particular, an NP-completeness proof for this problem is given.

1 Introduction

In this paper we are concerned with the complexity of the inverse shortest path routing problem (ISPR), i.e., to decide if a set of tentative routing patterns is simultaneously realizable. Let $G = (N, A)$ be a digraph. For each destination $l \in L \subseteq N$, a (tentative) routing pattern is represented by a shortest path graph (SP-graph), defined by an arc subset pair, $(A_l, \bar{A}_l) \subset A \times A$. The arcs in A_l are required to be shortest path arcs (SP-arcs) and the arcs $\bar{A}_l$ are non-shortest path arcs (non-SP-arcs), i.e., prohibited to be on shortest paths. A family of SP-graphs, $\{(A_l, \bar{A}_l) : l \in L\}$ is realizable if there is a strictly positive cost vector, $w \in \mathbb{R}^A_+$, such that all SP-arcs in all SP-graphs are in some shortest path to their respective destinations and no non-SP-arc is in a shortest path to its destination. Our main result is that this problem is NP-complete contrary to what has been claimed earlier in the literature, e.g. in [5, § 8.3.2].

An important relaxation of ISPR, referred to as compatibility, is to decide if there is a cost vector, $w \in \mathbb{R}^A_+$, such that for each $l \in L$ there is a node potential, $\pi^l \in \mathbb{R}^N$, such that the implied reduced costs are compatible with $(A_l, \bar{A}_l)$, i.e.,

$$w_{ij} + \pi^l_i - \pi^l_j \begin{cases} = 0 \text{ if } (i,j) \in A_l, \\ > 0 \text{ if } (i,j) \in \bar{A}_l, \\ \geq 0 \text{ otherwise.} \end{cases} \tag{1}$$

J. Pahl, T. Reiners, and S. Voß (Eds.): INOC 2011, LNCS 6701, pp. 339–353, 2011.

This problem has received much attention in the OSPF literature, see Chapter 8 in [5] and the references therein. Until now, it was believed that compatibility is equivalent to realizability, we provide a counter example to this below.

The realizability and compatibility problems are significantly different from the ordinary inverse shortest path (ISP) problem in [2]. In particular, in ISPR a partial desired optimal solution is specified via SP-arcs and non-SP-arcs and we must *determine if it can be completed to an optimal solution without implicitly introducing some undesired non-SP arcs*. The introduction of *prohibited arcs is actually what makes the problem hard*. Also, the actual cost vector in ISPR is irrelevant, while many ISPs involve the minimization of a deviation from some ideal cost vector. A counterpart for ISPR would be to find small integral weights, i.e. a small deviation from 0. This related problem is considered in [3].

The primary motivation for studying ISPRs is that they naturally arise as crucial subproblems in bilevel programs where the lower level is a set of shortest path problems. We refer to these problems as bilevel shortest path problems (BSP). In a BSP, a leader decides upon a cost vector that implicitly affects the followers response in their resulting shortest path problems.

To model a BSP, let $G = (N,A)$ be a digraph and $\mathscr{K}$ a set of followers. The demand h_k associated with follower $k \in \mathscr{K}$ should be sent from the origin $o(k)$ to the destination $d(k)$. Three sets of variables are used: the leaders control variables, u, the followers flow variables, x, and the cost variables, w. The cost, w_{ij}, for arc $(i,j) \in A$ depends explicitly on the leaders control variables, u. The exact relation is application dependent and modelled via the set $W(u)$ which is assumed to be strictly positive and integral (to ease the presentation). The follower flow variable, x^k_{ij}, denotes the fraction of the demand, h_k, sent on an arc, $(i,j) \in A$, by follower $k \in \mathscr{K}$. Finally, the feasible combinations of leader decisions and follower flow assignments are modelled via the set Π.

The objective is for the leader to maximize his objective function, $F(u,x)$, while followers minimize their costs by using shortest paths w.r.t. w, i.e. to solve

$$\begin{aligned} &\max F(u,x) \\ &s.t.\ (u,x) \in \Pi, \\ &\quad\ w \in W(u), \\ &\quad\ x^k \in \mathscr{S}_k, \qquad k \in \mathscr{K}, \end{aligned} \tag{2}$$

where

$$\begin{aligned} \mathscr{S}_k = \arg\min &\sum_{(i,j)\in A} w_{ij}\bar{x}_{ij} \\ s.t. &\sum_{j\in\delta^+(i)} \bar{x}_{ij} - \sum_{j\in\delta^-(i)} \bar{x}_{ij} = b^k_i, \qquad && i \in N, \\ &\bar{x}_{ij} \geq 0, && (i,j) \in A, \end{aligned} \tag{3}$$

and

$$b^k_i = \begin{cases} 1 \text{ if } i = o(k), \\ -1 \text{ if } i = d(k), \\ 0 \text{ otherwise.} \end{cases} \tag{4}$$

A large number of practically important optimization problems can be formulated as BSPs. Prominent applications that fit into the framework include OSPF routing problems, traffic engineering in IP networks and Stackelberg network pricing games. These problems have received much attention in the literature, cf. [5, 9, 11, 13] and the references therein.

To arrive at the ISPR subproblem, we adopt the commonly used approach to re-write the bilevel program into a single-level program with bilinear constraints by utilizing the complementary slackness optimality conditions. This is feasible since the leaders optimal decisions guarantee the absence of negative cost cycles in the followers shortest path problems. Dual feasibility is modelled by introducing node potentials and requiring appropriate reduced costs for all arcs.

Observe that it is possible to combine all followers with the same destination and use a common node potential for all of them. Denote the set of destinations by $L = \{i \in N : i = d(k) \text{ for some } k \in \mathscr{K}\}$ and by $\mathscr{K}_l = \{k \in \mathscr{K} : d(k) = l\}$ the followers with destination l. Hence it suffices to use a shortest path indicator variable y_{ij}^l that is set to 1 if $x_{ij}^k > 0$ for some $k \in \mathscr{K}_l$ and 0 otherwise. This yields the integer bilinear model below.

$$
\begin{array}{lll}
\max & F(u,x) & \\
s.t. & (u,x) \in \Pi, & \\
& w \in W(u), & \\
& \sum_{j \in \delta^+(i)} x_{ij}^k - \sum_{j \in \delta^-(i)} x_{ij}^k = b_i^k, & i \in N,\ k \in \mathscr{K}, \\
& 0 \le x_{ij}^k \le y_{ij}^l, & (i,j) \in A,\ k \in \mathscr{K}_l, \\
& w_{ij} + \pi_i^k - \pi_j^k \ge 1 - y_{ij}^l, & (i,j) \in A,\ k \in \mathscr{K}_l, \\
& \left(w_{ij} + \pi_i^k - \pi_j^k\right) y_{ij}^l = 0, & (i,j) \in A,\ k \in \mathscr{K}_l. \\
& y_{ij}^l \in \mathbb{B}, & (i,j) \in A,\ l \in L.
\end{array} \qquad (5)
$$

Note that the shortest path indicator variables y^l induce an acyclic SP-graph to node l via the union of all shortest path intrees.

An issue with BSPs is how to handle the case with multiple shortest paths from some node to some destination. It is commonly assumed that the leader freely selects an optimal solution for the follower that best suits his objective. In model (5) it is required that all arcs with reduced cost 0 actually carry some flow. In the OSPF literature, this requirement is closely related to the ECMP principle where traffic is divided "evenly" on all shortest paths.

When it comes to solving model (5) it is not clear how to handle the bilinear constraints. Linearizing them with big-Ms typically yields weak (sometimes extremely weak) LP-relaxations. Regarding the big-Ms, they may have to be as large as $2^{24} - 1$ in some IP network problems. It is also shown in [3] that it is APX-hard to find an integral cost vector that minimize the maximal arc cost for a set of realizable SP-graphs, i.e. to find suitable big-Ms.

We also need to point out that there are differences between applications, e.g. the actual arc costs in IP routing are insignificant but the complicating issues of arc capacity and congestion come into play. In other problems, such as the toll setting problem in [11], the arc cost corresponds to the operators revenue and is of course one of the most significant aspects. Another factor that typically makes some problems very hard is the number of followers, in some applications there are many followers, i.e. the OD-matrix is dense and sometimes even complete. Clearly, the size of the big-Ms may be an issue in some applications.

The context of the rest of this paper is that the bilinear constraints in model (5) have been linearized and that the resulting model is solved by branch-and-cut. This "requires" that at each node in the enumeration tree it is determined if a partial integral y solution can be completed to a solution. If this is not the case, valid cuts that prohibit the routing conflict induced by y should be augmented to the formulation.

From the above it is clear that a very fundamental issue in BSPs is to decide if the family of SP-graphs induced by a y solution is realizable, i.e. to solve ISPR.

An outline of the paper is as follows. In Section 2 an inadequate formulation of ISPR from the literature is presented along with a new model for ISPR. In Section 3 the main result that ISPR is indeed NP-complete is proved. We conclude in Section 4.

2 The Inverse Shortest Path Routing Problems

Recall that $G = (N, A)$ is a strongly connected digraph and $L \subseteq N$ a set of destinations. An *SP-graph* to destination $l \in L$ is a pair of arc subsets, $(A_l, \bar{A}_l)$ such that an arc $(i, j) \in A_l$ if and only if it is required that (i, j) is in some shortest path to l and $(i, j) \in \bar{A}_l$ if and only if (i, j) is not allowed to be in any shortest path to l. The arcs in A_l and $\bar{A}_l$ are called *SP-arcs* and *non-SP-arcs*, respectively.

Note that the SP-arcs correspond to the shortest path indicator variables y^l_{ij} equal to 1, or equivalently, the reduced costs zero arcs. Non-SP-arcs similarly correspond to y variables equal to 0. When y is fractional, the induced SP-graphs may be non-spanning and disconnected, since the non-binary part of y is ignored.

To formally define realizability the mapping $\mathscr{I}_l : \mathbb{Z}^A_+ \to 2^A$ is introduced. For a cost vector, w, let $\mathscr{I}_l(w)$ be the set of all arcs that are on some shortest path to destination l w.r.t. w. I.e., $\mathscr{I}_l(w)$ is the reduced cost zero arcs w.r.t. w and destination l. Note that an arc set $\mathscr{I}_l(w)$ always induces a spanning SP-graph to destination l since G is strongly connected.

Definition 1. *Let $\mathscr{A}_L = \{(A_l, \bar{A}_l)\}_{l \in L}$ be a family of SP-graphs for the set of destinations, L, and let $w \in \mathbb{Z}^A_+$ be a cost vector.*

- *A family of SP-graphs, $\mathscr{A}_L$, is compatible with w if for each $l \in L$ there exists a feasible node potential, π^l, such that π^l is tight w.r.t. w for all arcs in A_l and π^l is not tight w.r.t. w for any arc in $\bar{A}_l$.*
- *A family of SP-graphs, $\mathscr{A}_L$ is fully compatible with w if for all $l \in L$ it holds that $A_l \subseteq \mathscr{I}_l(w)$ and $\bar{A}_l \cap \mathscr{I}_l(w) = \emptyset$.*

We say that $\mathscr{A}_L$ is compatible *if it is compatible with some w and* realizable *if it is fully compatible with some w.*

2.1 Mathematical Formulations of ISPR Problems

Given a family of SP-graphs, $\mathscr{A}_L = \{(A_l, \bar{A}_l) : l \in L\}$, we consider the ISPR problems to decide if $\mathscr{A}_L$ is compatible and if $\mathscr{A}_L$ is realizable.

The compatibility problem has earlier been considered in the literature, e.g. in [5]. It has in fact (incorrectly, as shown in Example 2 below) been believed that the compatibility and realizability problems are equivalent.

The model below is widely available in the OSPF literature. It is based on arc weights, w, and the introduction of a set of node potentials, π^l, one for each destination, $l \in L$. Then, the reduced costs on SP-arcs are forced to 0 and the reduced costs of non-SP-arcs must be strictly positive.

$$\begin{aligned} w_{ij} + \pi_i^l - \pi_j^l &= 0, && (i,j) \in A_l,\ l \in L, \\ w_{ij} + \pi_i^l - \pi_j^l &\geq 1, && (i,j) \in \bar{A}_l,\ l \in L, \\ w_{ij} + \pi_i^l - \pi_j^l &\geq 0, && (i,j) \in A \setminus \left(A_l \cup \bar{A}_l\right),\ l \in L, \\ w_{ij} &\geq 1, && (i,j) \in A. \end{aligned} \tag{ISPR-C}$$

Similar, essentially equivalent, models can also be found elsewhere, e.g. in [1,4,7,8, 10,12]. It is well known from the literature that a family of SP-graphs, $\mathscr{A}_L$, is compatible if and only if (ISPR-C) is feasible.

Intuitively, $\mathscr{A}_L$ is not compatible if a subset of SP-arcs and non-SP-arcs directly induce a reduced cost routing conflict. Two such conflicts are given in Example 1.

Let us consider the realizability problem. That is, to find a cost vector that induce a spanning family of SP-graphs that comply with all SP- and non-SP-arcs. A compatible SP-graph family is realizable *if it can be completed to a spanning family of SP-graphs*. Clearly, compatibility is a necessary condition for realizability.

Proposition 1. *If a family of SP-graphs, $\mathscr{A}_L$, is realizable then it is compatible.*

However, compatibility does not in general imply that it is possible to complete a family of SP-graphs to spanning ingraphs. To derive a model that can be used to decide if a family of SP-graphs is realizable, the following observation is crucial.

Proposition 2. *A family of SP-graphs, $\mathscr{A}_L$, is realizable if and only if there is a cost vector w and tight node potentials, $\{\pi^l : l \in L\}$, induced by w that are feasible in* (ISPR-C).

This proposition yields the straightforward modelling approach to force the existence of (reverse) arborescences rooted at each destination, $l \in L$, where all reduced costs are 0, since this yields tight node potentials. We can reuse the shortest path indicator variables for this purpose, i.e. y_{ij}^l is 1 if the arc (i,j) is in $\mathscr{I}_l(w)$ and 0 otherwise. The following bilinear integer program can be used to determine if a family of SP-graphs is realizable.

$$\begin{aligned} w_{ij} + \pi_i^l - \pi_j^l + y_{ij}^l &\geq 1 && (i,j) \in A,\ l \in L, \\ \left(w_{ij} + \pi_i^l - \pi_j^l\right) y_{ij}^l &= 0 && (i,j) \in A,\ l \in L, \\ \sum_{(i,j) \in A} y_{ij}^l &\geq 1 && i \in N \setminus \{l\},\ l \in L, \\ y_{ij}^l &= 1 && (i,j) \in A_l,\ l \in L, \\ y_{ij}^l &= 0 && (i,j) \in \bar{A}_l,\ l \in L, \\ w_{ij} &\geq 1 && (i,j) \in A, \\ y_{ij}^l &\in \mathbb{B} && (i,j) \in A,\ l \in L. \end{aligned} \tag{ISPR-R}$$

The correctness of the model is motivated by the following observations. First, since all costs are strictly positive, the y^l-variables can not induce a directed cycle. Because of this, the outdegree constraints imply that the y^l-variables induce an ingraph that contains an arborescence rooted at l for each $l \in L$. Hence, the node potentials are tight and satisfy the resulting (ISPR-C) part of the model.

Theorem 1. *A family of partial SP-graphs, $\mathscr{A}_L$, is realizable if and only if* (ISPR-R) *has a feasible solution.*

The major difference between compatibility and realizability is that compatibility, cf. model (ISPR-C), only take dual feasibility of the shortest path problem into account, but partly neglects primal feasibility and complementary slackness.

2.2 Incompatible and Unrealizable SP-Graphs

In this section we give three examples of (potential) routing conflicts required later in the paper. A comprehensive analysis and numerous examples of routing conflicts are given in [8].

Example 1. The first potential routing conflict in this example is referred to as a valid cycle in [6, 7, 8]. It will be used in the NP-completeness proof in Section 3.

Consider the SP-graph family, $\mathscr{A}_L$, with destinations $a, b \in L$ drawn in Figure 1. Suppose that there are two (start) nodes, s_1 and s_2 and two (end) nodes, e_3 and e_4 that forms the following SP-arc patterns,

$$A_a = \{(s_1, e_3),\ (s_2, e_4)\} \text{ and } A_b = \{(s_1, e_4),\ (s_2, e_3)\}. \tag{6}$$

In all feasible solutions to (ISPR-C) (and (ISPR-R)) it must hold that

$$w_{13} + \pi_1^b - \pi_3^b = w_{23} + \pi_2^a - \pi_3^a = w_{14} + \pi_1^a - \pi_4^a = w_{24} + \pi_2^b - \pi_4^b = 0. \tag{7}$$

This is direct from the surrogate constraint composed of the corresponding reduced cost constraints,

$$\begin{aligned} 0 = {} & w_{13} + \pi_1^b - \pi_3^b + w_{23} + \pi_2^a - \pi_3^a + w_{14} + \pi_1^a - \pi_4^a + w_{24} + \pi_2^b - \pi_4^b \\ & - \left(w_{13} + \pi_1^a - \pi_3^a + w_{23} + \pi_2^b - \pi_3^b + w_{14} + \pi_1^b - \pi_4^b + w_{24} + \pi_2^a - \pi_4^a\right) \\ \leq {} & 0 + 0 + 0 + 0 - (0 + 0 + 0 + 0) = 0. \end{aligned} \tag{8}$$

Hence, the arcs (s_1, e_3) and (s_2, e_4) are induced SP-arcs to destination b and the arcs (s_1, e_4) and (s_2, e_3) are induced SP-arcs to destination a.

A similar, but slightly more complicated example is illustrated in Figure 2. This result is also required for the proof in Section 3. Now, the SP-graph family, $\mathscr{A}_L$, with destinations $a, b, c \in L$ contains the following SP-arc patterns.

$$A_a = \{(s_1, e_3),\ (s_2, e_5)\},\ A_b = \{(s_1, e_4),\ (s_2, e_3)\} \text{ and } A_c = \{(s_1, e_5),\ (s_2, e_4)\}. \tag{9}$$

A similar argument as above shows that all feasible solutions to (ISPR-C) satisfy

$$\begin{aligned} & w_{13} + \pi_1^b - \pi_3^b = w_{14} + \pi_1^c - \pi_4^c = w_{15} + \pi_1^a - \pi_5^a = \\ & w_{23} + \pi_2^a - \pi_3^a = w_{24} + \pi_2^b - \pi_4^b = w_{25} + \pi_2^c - \pi_5^c = 0. \end{aligned} \tag{10}$$

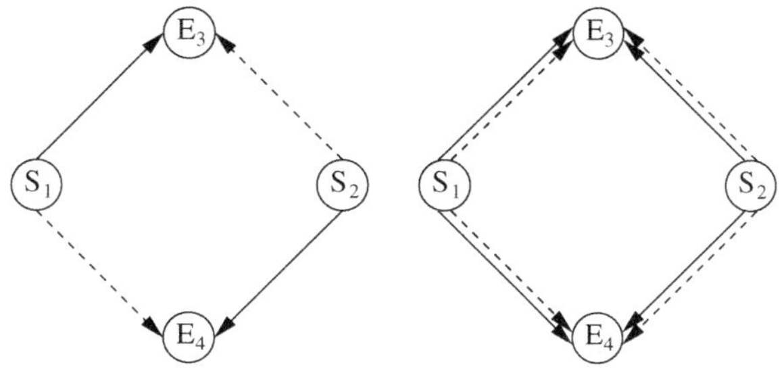

Fig. 1. The example setting is illustrated on the left and the induced SP-arcs are drawn on the right. The SP-arcs to destinations $a, b \in L$, are drawn with solid and dashed arrows, respectively. In the right part, the upper arcs of parallel arc pairs are the original SP-arcs and the lower are the induced arcs.

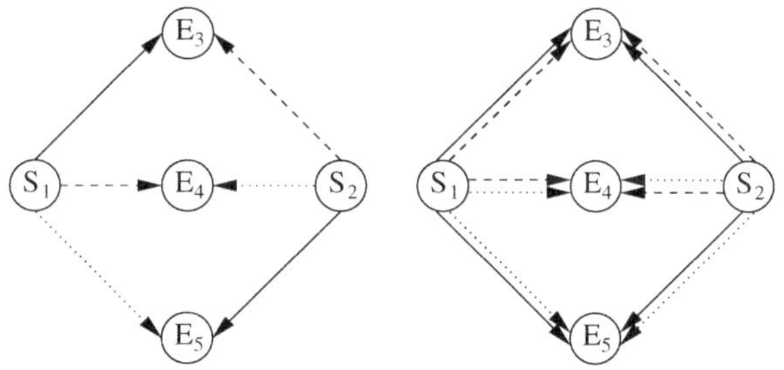

Fig. 2. The example setting is illustrated on the left and the induced SP-arcs are drawn on the right. The SP-arcs to destination a, b and c are drawn with solid, dashed and dotted arrows, respectively. In the right part, the upper arcs of parallel arc pairs are the original SP-arcs and the lower are the induced arcs.

Hence, the arcs (s_1, e_5) and (s_2, e_3) are induced SP-arcs to destination a, the arcs (s_1, e_3) and (s_2, e_4) are induced SP-arcs to destination b and the arcs (s_1, e_4) and (s_2, e_5) are induced SP-arcs to destination c.

We conclude the section with an example of a family of SP-graphs that is compatible but not realizable. Hence, the inadequacy of model (ISPR-C) and the need for model (ISPR-R) is well motivated.

Example 2. Consider the graph in Figure 3 and the SP-graphs with arc sets

$$\begin{array}{ll} A_0 = \{(1,2),\ (2,3),\ (3,s)\}, & \bar{A}_0 = \{(1,a),\ (2,a),\ (2,b),\ (3,b),\ (3,c)\}, \\ A_a = \{(1,a)\}, & \bar{A}_a = \{(1,2)\}, \\ A_b = \{(2,b)\}, & \bar{A}_b = \{(2,3),\ (2,a)\}, \\ A_c = \{(3,c)\}, & \bar{A}_c = \{(3,s),\ (3,b)\}. \end{array} \tag{11}$$

This family of partial SP-graphs is not realizable. Note that the dashed path starting at node 1 must be augmented to a path that ends in node 0. Any augmentation of the path

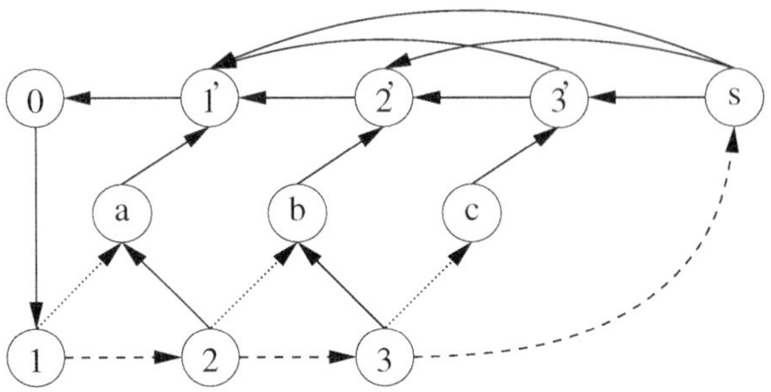

Fig. 3. A graph and four SP-graphs that can not be realized. All dashed arcs are SP-arcs to destination 0. The dotted arcs emanating from node 1, 2 and 3, respectively, are SP-arcs to destinations a, b and c, respectively.

implies that it goes via node 1', 2' or 3'. Therefore, the shortest path subpath optimality property implies that there are two disjoint shortest paths from 1 to 1', 2 to 2' or 3 to 3'. This is however not feasible w.r.t. the non-SP-arcs.

Now consider compatibility. Setting the cost on arcs entering node a, b and c to a large number and the costs on all other arcs to 1 yields a feasible solution to (ISPR-C). Hence, the instance is compatible but not realizable.

Note that the node potentials in the "feasible" solution are not tight. Also, the tight node potentials are not feasible w.r.t. the reduced cost constraints. It is easy to generalize this example to one with arbitrarily many nodes and the same property.

3 Complexity of ISPR Problems

In this section we focus on the complexity of ISPR. From model (ISPR-C) above it is clear that compatibility can be decided in polynomial time by solving a linear (feasibility) program. The complexity of the problem to decide if a family of partial SP-graphs is realizable has been open until now.

To prove that *it is NP-complete to decide if a family of SP-graphs is realizable* a polynomial reduction to the exclusive 1 in 3 Boolean satisfiability (X3SAT) problem will be described. Recall that the X3SAT problem is to decide if a formula (in conjunctive normal form) where each clause contains three literals has a truth assignment that makes *exactly one literal in each clause is true*. Canonical X3SAT instances only contain sorted clauses where all variables are positive and different and no two clauses share more than one variable.

Definition 2. *An X3SAT instance,* $I = (X, \mathscr{C})$*, with the variable set* $X = \{x_1, \ldots, x_n\}$ *and clause collection* $\mathscr{C} = \{C_1, \ldots, C_m\}$ *is canonical if*

- *For each clause* $C = (x_i \vee x_j \vee x_k) \in \mathscr{C}$ *it holds that* $i < j < k$.
- *No pair of variables is included in two or more clauses.*

It is actually no restriction to only consider canonical X3SAT instances. Indeed, if there are clauses $(a \vee b \vee c)$ and $(a \vee b \vee d)$, then $c = d$ in all feasible truth assignments and one variable and (at least) one clause can be dropped. Since X3SAT is NP-complete, applying this argument recursively yields the following result.

Theorem 2. *It is NP-complete to decide if a canonical X3SAT instance is satisfiable.*

The reduction in our proof requires the use of a next operator. Informally, it yields the modulo-wise next variable in a clause.

Definition 3. *If $C = (x_i \vee x_j \vee x_k) \in \mathscr{C}$ is a clause in a canonical X3SAT instance, $I = (X, \mathscr{C})$, then the next operator, $n : X \times \mathscr{C} \to X$, is defined by*

$$n(x,C) = \begin{cases} x_j \text{ if } x = x_i \\ x_k \text{ if } x = x_j \\ x_i \text{ if } x = x_k. \end{cases} \tag{12}$$

Example 3. The X3SAT instance with variable set $\{a,b,c,d,e,f,g\}$ and clause collection $\{C_1,C_2,C_3,C_4\} = \{a \vee b \vee c,\ a \vee d \vee e,\ a \vee f \vee g,\ b \vee d \vee f\}$ is in canonical form. It has three feasible assignments (set $\{b,e,g\}$ to true, $\{c,d,g\}$ to true, or $\{c,d,g\}$ to true). Note that the instance remains to be in canonical form if the clause $(c \vee e \vee f)$ is added, but not if the clause $(b \vee e \vee f)$ is added since C_4 already contains both b and f. The next operator takes the following values for clause C_1: $n(a,C_1) = b$, $n(b,C_1) = c$ and $n(c,C_1) = a$.

Before a formal reduction to X3SAT from realizability is given we will briefly describe the idea behind the graph that is used in the realizability problem. For each clause, *ABC* say, a node *ABC* is created with the purpose to force at least one variable in the clause to be true. Then, three auxiliary nodes *AB*, *AC* and *BC* are created such that it is guarantee that at most one of the variables in each pair (hence, also in the clause) is true. To accomplish these tasks, potential potential routing conflicts of the kind in Example 1 are created with the SP-arcs and the non-SP-arcs. The actual truth assignment is determined by arcs from an auxiliary starting node to auxiliary variable nodes. The graph obtained from clause *ABC* is given in Figure 4.

We may now formally describe how to construct a realizability instance from a canonical X3SAT instance that is feasible if and only if the X3SAT instance is. Given a canonical X3SAT instance, $I = (X, \mathscr{C})$, the graph $G(I) = (N,A)$ and the family of SP-graphs, $\mathscr{A}(I)$, is created as follows. For each variable $x \in X$, create three nodes in G: x^+, x^- and x. For each clause $C \in \mathscr{C}$, create four nodes C_{ij}, C_{ik}, C_{jk} and C. Also add an auxiliary starting node, S.

To determine the set of arcs consider each variable $x \in X$ and add the arcs

$$\begin{array}{l} (S,x^+),\ (x^+,x),\ (S,x^-), (x^-,x), \\ (x^+,y),\ (x^-,y),\ (x,y), \\ (C,x^-)(C_{ij},x^+),\ (C_{ik},x^+),\ (C_{jk},x^+), \end{array} \tag{13}$$

where $y \neq x$ is also a variable and $C = (x_i \vee x_j \vee x_k) \in \mathscr{C}$ a clause that contains x.

It remains to construct the family of SP-graphs. For each variable, $x \in X$, form an SP-graph to the node $l = x$ with SP-arcs, A_l, non-SP-arcs $\bar{A}_l$ and unrestricted arcs $U_l = A \setminus (A_l \cup \bar{A}_l)$ determined according to the rules below.

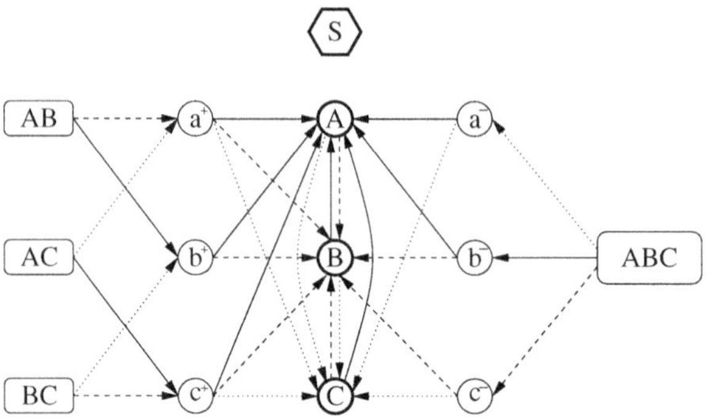

Fig. 4. The SP-graphs associated with variables A,B and C in a realizability instance obtained from a canonical X3SAT instance with clause ABC. The solid, dashed and dotted arcs are required to be SP-arcs for destination A, B and C, respectively. Note that appropriate destination assignments to arc (S,A^-), (S,B^-) and (S,C^-) can be used to create the situation in Figure 2 in Example 1. Similarly, arc (S,A^+), (S,B^+) and (S,C^+) can be used to create the situation in Figure 1 in Example 1.

1. Add the arcs (x^+,x) and (x^-,x) to A_l as SP-arcs. Also add the arc (y,x) to A_l for each variable $y \neq x$.
2. Add the arcs (S,x^+) and (S,x^-) to U_l.
3. Add all arcs emanating from S except (S,x^+) and (S,x^-) to $\bar{A}_l$ as non-SP-arcs.
4. For each clause C, let $y = n(x,C)$, then add the arcs (C,y^-) and (y^-,x) to A_l.

For each clause $C = (x_i \vee x_j \vee x_k)$ add arcs as SP-arcs to the associated SP-graphs according to the following rules.

5. Add (C_{ij},x_j^+) and (x_j^+,x_i) to A_l if $l = x_i$ and (C_{ij},x_i^+) and (x_i^+,x_j) to A_l if $l = x_j$.
6. Add (C_{ik},x_k^+) and (x_k^+,x_i) to A_l if $l = x_i$ and (C_{ik},x_i^+) and (x_i^+,x_k) to A_l if $l = x_k$.
7. Add (C_{jk},x_k^+) and (x_k^+,x_j) to A_l if $l = x_j$ and (C_{jk},x_j^+) and (x_j^+,x_k) to A_l if $l = x_k$.
8. Arcs not put in A_l or $\bar{A}_l$ due to one of the rules above is put in U_l.

Note that the SP-arcs for destination $l = x$ induce a tree that spans all nodes in G associated with a variable or clause that is connected to x via some clause. Also, S is not contained in any such tree and all its emanating arcs are in U_l or $\bar{A}_l$.

These rules applied to the single clause $(A \vee B \vee C)$, yields the result in Figure 4.

Theorem 3. *Given a canonical X3SAT instance, $I = (X,\mathscr{C})$, let $G(I) = (N,A)$ and $A_l \cup U_l \cup \bar{A}_l$ be constructed from rules 1-8 above for each variable $x_l \in X$. Denote the induced family of SP-graphs by $\mathscr{A}_L$. Then, the X3SAT instance $I = (X,\mathscr{C})$ is feasible if $\mathscr{A}_L$ is realizable.*

To prove this theorem some lemmas are required.

Lemma 1. *Given an SP-graph family, $\mathscr{A}_L$, and a cost vector, w, that verifies that $\mathscr{A}_L$ is realizable, let $a,b \in L$ be two destinations. If there are nodes $s_1,s_2,e_3,e_4 \in N$ such that the SP-arc sets satisfy $\mathscr{I}_a \supseteq \{(s_1,e_3)\} \cup \{(s_2,e_4)\}$ and $\mathscr{I}_b \supseteq \{(s_1,e_4)\} \cup \{(s_2,e_3)\}$.*

Then, the induced SP-arc sets must also satisfy $\mathscr{I}_a \supseteq \{(s_1,e_4)\} \cup \{(s_2,e_3)\}$ *and* $\mathscr{I}_b \supseteq \{(s_1,e_3)\} \cup \{(s_2,e_4)\}$.

Lemma 2. *Given an SP-graph family,* $\mathscr{A}_L$, *and a cost vector,* w, *that verifies that* $\mathscr{A}_L$ *is realizable, let* $a,b,c \in L$ *be three destinations. If there are nodes* s_1,s_2,e_3,e_4 *and* e_5 *such that the SP-arc sets satisfy* $\mathscr{I}_a \supseteq \{(s_1,e_3)\} \cup \{(s_2,e_5)\}$, $\mathscr{I}_b \supseteq \{(s_1,e_4)\} \cup \{(s_2,e_3)\}$ *and* $\mathscr{I}_c \supseteq \{(s_1,e_5)\} \cup \{(s_2,e_4)\}$. *Then, the induced SP-arc sets must also satisfy* $\mathscr{I}_a \supseteq \{(s_1,e_5)\} \cup \{(s_2,e_3)\}$, $\mathscr{I}_b \supseteq \{(s_1,e_3)\} \cup \{(s_2,e_4)\}$ *and* $\mathscr{I}_c \supseteq \{(s_1,e_4)\} \cup \{(s_2,e_5)\}$.

These lemmas were exemplified along with a proof outline in Example 1 above. They are used to derive properties of realizable instances obtained from canonical X3SAT instances as described above. The properties are summarized in Lemma 3 which is the foundation of the proof of Theorem 3.

Lemma 3. *Let* $\mathscr{A}_L$ *be an SP-graph family induced by a canonical X3SAT instance,* $I = (X,\mathscr{C})$. *Let* w *be a cost vector that verifies that* $\mathscr{A}_L$ *is realizable, i.e.,*

$$A_l \subseteq \mathscr{I}_l(w) \quad \text{and} \quad \bar{A}_l \cap \mathscr{I}_l(w) = \emptyset, \quad \text{for all } l \in L. \tag{14}$$

Then, the following properties of $\mathscr{I}_l(w)$ *are satisfied for all* $l \in L$. *Here, shortest paths, SP-arcs and non-SP-arcs are considered w.r.t.* w, *i.e., SP-arcs are in* $\mathscr{I}_l(w)$ *and non-SP-arcs are not.*

1. *For any clause,* $(A \vee B \vee C)$ *say, at least one of the arcs* (S,A^+), (S,B^+) *and* (S,C^+) *is an SP-arc to destination* A,B *and* C *respectively.*
2. *For any clause,* $(A \vee B \vee C)$ *say, it holds that: (1) at most one of the arcs* (S,A^+) *and* (S,B^+) *is an SP-arc to destination* A *and* B, *respectively, (2) at most one of the arcs* (S,A^+) *and* (S,C^+) *is an SP-arc to destination* A *and* C, *respectively, and (3) at most one of the arcs* (S,B^+) *and* (S,C^+) *is an SP-arc to destination* B *and* C, *respectively.*
3. *For any clause,* $(A \vee B \vee C)$ *say, exactly one of the arcs* (S,A^+), (S,B^+) *and* (S,C^+) *is an SP-arc to destination* A,B *and* C *respectively.*
4. *For any variable,* x *say, exactly one of the arcs* (S,x^-) *and* (S,x^+) *is an SP-arc to destination* x.

Proof. All properties essentially follows from Lemmas 1 and 2. By construction, given a variable x, it holds that an arc, (S,i), emanating from the starting node is a non-SP-arc unless i equals x^+ or x^-. Therefore, at least one of the arcs (S,x^-) and (S,x^+) is an SP-arcs to destination x. Using this, we prove Property 1-4.

1. Consider a clause, $(A \vee B \vee C)$ say. Assume that none of the arcs (S,A^+), (S,B^+) and (S,C^+) is an SP-arc to destination A,B and C, respectively. Then, all the arcs (S,A^-), (S,B^-) and (S,C^-) must be SP-arcs to destination A,B and C, respectively. Recall that also (ABC,B^-), (ABC,C^-) and (ABC,A^-) are SP-arcs to destination A,B and C, respectively, by construction. This yields that the requirements in Lemma 2 are satisfied with start nodes $s_1 = S$ and $s_2 = ABC$ and end nodes $e_3 = A^-, e_4 = B^-$ and $e_5 = C^-$. Therefore, (for instance) the arc (S,A^-) is also an SP-arc to destination C which yields a contradiction.

2. Consider a clause, $(A \vee B \vee C)$ say. Assume that both of the arcs (S, A^+) and (S, B^+) are SP-arcs to destination A and B, respectively. Recall that also (AB, B^+) and (AB, A^+) are SP-arcs to destination A and B, respectively, by construction. This yields that the requirements in Lemma 1 are satisfied with start nodes $s_1 = S$ and $s_2 = AB$ and end nodes $e_3 = A^+$ and $e_4 = B^+$. Therefore, (for instance) the arc (S, A^+) is also an SP-arc to destination B which yields a contradiction. The cases AC and BC are proved analogously.
3. Consider a clause, $(A \vee B \vee C)$ say. Combining the three constraints in 2 yields that at most one of the arcs $(S, A^+), (S, B^+)$ and (S, C^+) is an SP-arc to destination A, B and C respectively. Since Property 1 states that at least one of the arc is an SP-arc to the respective destination, exactly one SP-arc must be so.
4. Consider a variable x and a clause $C = (x \vee y \vee z)$. At least one of the arcs (S, x^-) and (S, x^+) is an SP-arc to destination x. Assume that both are. This yields that (S, y^-) and (S, z^-) are SP-arcs to destinations y and z, respectively, from Property 2 with xy and xz. Since (S, x^-) was also assumed to be an SP-arc to destination x the same situation as in the proof of Property 1 occurs and Lemma 2 yields a contradiction.

Proof (Theorem 3). Given a realizability certificate construct the assignment from the SP-arcs emanating from the starting node as follows. If (S, x^+) is an SP-arc, then set x to true, otherwise, set x to false. It now follows from Lemma 3 that exactly one variable in each clause is true and the assignment is feasible.

To complete our NP-completeness proof it is required to construct a realizability certificate for a given Boolean assignment that satisfies the X3SAT instance.

Theorem 4. *Given a canonical X3SAT instance, $I = (X, \mathscr{C})$, let $G(I) = (N, A)$ and $A_l \cup U_l \cup \bar{A}_l$ be constructed from rules 1-8 above for each variable $x_l \in X$. Denote the induced family of SP-graphs by $\mathscr{A}_L$. Then, $\mathscr{A}_L$ is realizable if the X3SAT instance $I = (X, \mathscr{C})$ is feasible.*

Proof. It suffices to find a cost vector, $w(\widetilde{X})$, from a given Boolean assignment, $\widetilde{X}$, that verifies the realizability of $\mathscr{A}_L$ in $G(I) = (N, A)$. The rules in Table 1 are used to determine w from $\widetilde{X}$. Here x and y are different variables and $C = (x \vee y \vee z)$ is a clause. Costs on arcs not covered by a rule below are set to 5.

Table 1. Rules to Determine the Cost Vector From a Truth Assignment

(i,j)	if x is true	if x is false
(S, x^+)	1	1
(S, x^-)	1	1
(x^+, x)	1	5
(x^-, x)	5	1
(x, y)	5	5

(i,j)	x is true	y is true	z is true
(C_{xy}, x^+)	3	1	3
(C_{xy}, y^+)	1	3	3
(C_{xz}, x^+)	3	3	1
(C_{xz}, z^+)	1	3	3
(C_{yz}, y^+)	3	3	1
(C_{yz}, z^+)	3	1	3
(C, x^-)	1	2	3
(C, y^-)	3	1	2
(C, z^-)	2	3	1

(i,j)	w_{ij}
(x^+, y)	2
(y^+, x)	2
(x^+, z)	2
(z^+, x)	2
(y^+, z)	2
(z^+, y)	2
(x^-, z)	2
(y^-, x)	2
(z^-, y)	2

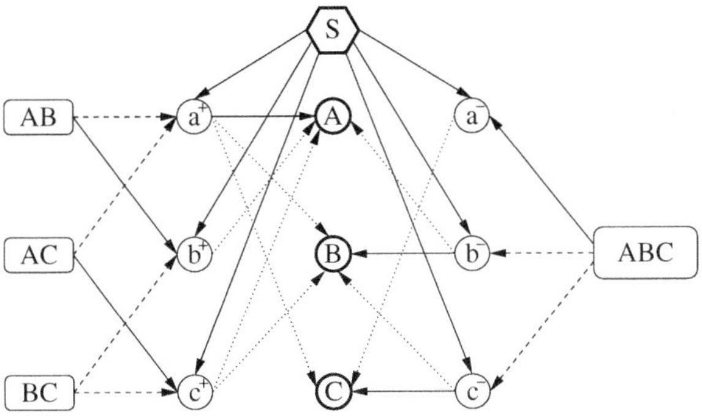

Fig. 5. The part of the graph that involves nodes associated with the clause *ABC*. When *A* is true, the cost of an arc is 1, 2 or 3, if it is solid, dotted or dashed, respectively. An arc that is not drawn have cost 5.

Since the X3SAT instance is canonical, there is no variable pair that is in two clauses. This implies that the rules above are unambiguous. The possible source of ambiguity would be for a clause, $C = (x \vee y \vee z)$, from x^-, y^- or z^- to x, y or z. However, since no other clause can contain two of the variables x, y or z this is of no concern. This "independence" property yields that it essentially suffices to consider one clause in isolation.

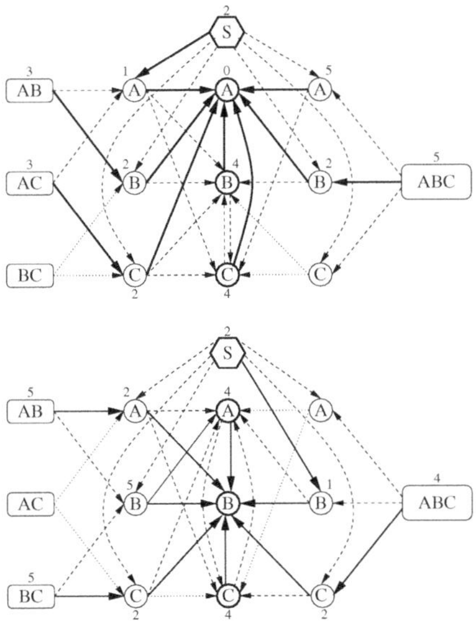

Fig. 6. The parts of the shortest path tree to destinations *A* (top) and *B* (bottom) that involve nodes associated with the clause *ABC* which is assumed to be satisfied by *A*. Solid arcs represent SP-arcs and dotted arcs represent non-SP-arcs. The dotted arcs have not been SP-arcs or non-SP-arcs.

The cost vector obtained from the rules in Table 1 is illustrated in Figure 5 for the part of the graph involving the clause *ABC* when *A* is assigned to true.

From these rules it is straightforward to derive the tight node potential for any destination. Due to the independence there are only three cases to consider for each clause, $C = (x \vee y \vee z)$, depending on which position x has relative to the true variable in C. The result is illustrated for the clause *ABC* and destinations *A* and *B* when *A* is true in Figure 6. From this it is easy to verify that all required SP-arcs are SP-arcs and that no non-SP-arc is an SP-arc. That is, the family of SP-graphs is realizable.

It follows from Theorems 3 and 4 that realizability is NP-complete. Note that realizability is NP-complete also when it is required that all shortest paths are unique.

Theorem 5. *It is NP-complete to decide if a family of SP-graphs is realizable.*

From the proof above it is clear that a stronger statement holds. Namely, it is NP-complete to decide if a family of SP-graphs is realizable even if for each SP-graph the SP-arcs form a rooted, non-spanning tree with maximum depth 2.

4 Conclusion

In this paper we consider inverse shortest path routing problems. In particular, we note that there is a difference between the previously considered compatibility variant of the problem and the more complete variant, here referred to as realizability. We provide a mathematical programming formulation for the realizability problem. Most importantly, it is proved that the realizability problem is NP-complete. This result has significant theoretical consequences for bilevel programs where the lower level is a shortest path problem.

References

1. Ben-Ameur, W., Gourdin, E.: Internet routing and related topology issues. SIAM J. Discrete Math 17, 18–49 (2003)
2. Burton, D., Toint, P.L.: On an Instance of the Inverse Shortest Paths problem. Mathematical Programming 53, 45–61 (1992)
3. Bley, A.: Inapproximability results for the inverse shortest paths problem with integer lengths and unique shortest paths. Networks 50, 29–36 (2007)
4. Bley, A.: Routing and Capacity Optimization for IP Networks. PhD thesis, TU Berlin (2007)
5. Bley, A., Fortz, B., Gourdin, E., Holmberg, K., Klopfenstein, O., Pióro, M., Tomaszewski, A., Ümit, H.: Optimization of OSPF routing in IP networks. In: Koster, A.M.C.A., Muñoz, X. (eds.) Graphs and Algorithms in Communication Networks: Studies in Broadband, Optical, Wireless and Ad Hoc Networks, ch. 8, pp. 199–240. Springer, Heidelberg (2009)
6. Broström, P.: Holmberg. K.: Valid cycles: A source of infeasibility in OSPF routing. Networks 52, 206–215 (2008)
7. Broström, P., Holmberg, K.: Compatible weights and valid cycles in non-spanning OSPF routing patterns. Algorithmic Operations Research 4, 19–35 (2009)
8. Call, M.: Inverse shortest path routing problems in the design of ip networks. Linköping Studies in Science and Technology. Thesis No. 1448 (2010)

9. Dempe, S.: Annotated bibliography on bilevel programming and mathematical programs with equilibrium constraints. Optimization 52, 333–359 (2003)
10. Dzida, M., Zagozdzon, M., Pióro, M.: Optimization of Resilient IP Networks with Shortest Path Routing. In: International Workshop on the Design of Reliable Communication Networks (DRCN), La Rochelle, France (2007)
11. Labbé, M., Marcotte, P., Savard, G.: A bilevel model of taxation and its application to optimal highway pricing. Management Science 44, 1608–1622 (1998)
12. Pioro, M., Medhi, D.: Routing, Flow, and Capacity Design in Communication and Computer Networks. Morgan Kaufmann Publishers, San Francisco (2004)
13. Van Hoesel, S.: An overview of Stackelberg pricing in networks. European Journal of Operational Research 189, 1393–1402 (2008)

The Skill Vehicle Routing Problem

Paola Cappanera[1], Luís Gouveia[2], and Maria Grazia Scutellà[3]

[1] Dipartimento di Sistemi e Informatica, Università degli Studi di Firenze, Firenze, Italy
paola.cappanera@unifi.it
[2] Departamento de Estatística e Investigação Operacional,
Universidade de Lisboa, Lisboa, Portugal
legouveia@fc.ul.pt
[3] Dipartimento di Informatica, Università di Pisa, Pisa, Italy
scut@di.unipi.it

Abstract. Given a (complete) directed network, where a skill level s_j is associated with each node j other than the depot, stating requirements for the associated service call, and given a set of available technicians, each one operating at a certain skill level, we address the problem of defining the tour of each technician, each one starting and ending at the depot, in such a way that each service requirement is fulfilled by exactly one technician, and proper skill level constraints are satisfied. These constraints state that the service requirement at node j can be operated by any technician having a skill level at least s_j. Given travelling costs for the arcs of the network, which are skill level dependent, we want to determine a minimum cost set of tours which satisfy the skill level constraints. This problem, named Skill VRP, originates from a real application context and it specializes, to some extents, the Site Dependent VRP (SDVRP) and, thus, the Periodic VRP. Furthermore, it shows strict relationships with Home Care Scheduling problems. Various ILP formulations are proposed for Skill VRP, which are tested on a large suite of randomly generated instances. The obtained results show that the stated problem may be very difficult to solve exactly. However, some of the proposed models produce LP bounds which are very close to the optimum cost, and which can be determined in an efficient way. Thus, these ILP models constitute a very promising starting point for the solution of Skill VRP, and for the design of efficient cutting plane approaches for real application extensions.

1 The Skill Vehicle Routing Problem

Let $G = (N,A)$ be a (complete) directed network, with $|N| = n$ and $|A| = m$. Let each node j, $j \neq 1$, represent a service requirement, and s_j denote the skill level required by node j for the associated service call. On the other hand, node 1 denotes the depot. Assume to have a set T of available technicians, each one operating at a certain skill level, and s_t denote the skill level of technician t. Assume also that the service requirement at node j can be operated by any technician having a skill level at least s_j, for $j \in N \setminus \{1\}$, and S denote the set of skills given by the union of the skill requirement set at the nodes and the skill set associated with the technicians, i.e. $S = \{s_i : i \in N\} \cup \{s_t : t \in T\}$.

Given non-negative skill dependent travelling costs c^t_{ij} for each $(i,j) \in A$ and technician $t \in T$, we study the problem of defining the tour of the technicians, each one starting and ending at node 1, in such a way that each service requirement is fulfilled

J. Pahl, T. Reiners, and S. Voß (Eds.): INOC 2011, LNCS 6701, pp. 354–364, 2011.

by exactly one technician, and the skill level constraints are satisfied. Hereafter this problem will be referred to as Skill VRP.

The Skill VRP, for which mathematical models have never been proposed before to the best of our knowledge, originates from a real application concerning the dispatch of technicians to customer service requests in after-sales service management, referred to as *field service* ([1], [7]). Furthermore it specializes, to some extents, the Site Dependent VRP (SDVRP) [4], [2] and, thus, the Periodic VRP [4], [3]. In fact, by modelling the technicians in terms of vehicles, Skill VRP is the special case of the SDVRP where a suitable ordering among the types of vehicles is imposed (in fact any service call j can be operated by technicians having a skill level at least s_j), whereas in the general SDVRP no hierarchy among the vehicles does exist. See also [6] for an example of Tabu Search approach for SDVRP with time windows. Furthermore, the Skill VRP shows strong resemblance with Home Care Scheduling problems (HCS), where coordinators have to assign the care of every client to an operator, considering his workload, the particular skills that the operator should have in order to satisfy the clients needs and the travel distance of the operator for reaching the clients home. See [5] for an example of Integer Linear Programming (ILP) model developed to support human resource short term planning in home care.

In this paper various ILP models and related valid inequalities are proposed for formulating and solving the Skill VRP. The proposed models are based on multicommodity flow variables and constraints, on increasing disaggregation levels which allow one to strengthen the associated LP bounds. Furthermore, the models strongly exploit the skill information at the nodes, which is the key aspect differentiating the proposed models from more classical VRP models. As shown in the computational section, a deep skill information consideration revealed to be crucial in determining good LP bounds. The models have been tested on a large suite of randomly generated instances. The obtained results show that the Skill VRP may be very difficult to solve exactly on reasonable sized networks. However, some of the proposed models (the more disaggregated ones) produce LP bounds which are very close to the optimum cost. The time required to compute these LP bounds, however, may be still too large. Therefore, some families of valid inequalities have been proposed to enhance some less disaggregated ILP models. The obtained results are very good. In fact, the proposed models, when equipped with those valid inequalities, produce LP bounds near to the optimum at a reasonable computational cost. The proposed models thus appear to be a very promising starting point for the solution of the Skill VRP. Further, they appear to be a reasonable basis for the design of efficient cutting plane approaches, based on the proposed formulation and valid inequalities, to address real application extensions, which take into account additional constraints in addition to the investigated skill constraints.

2 ILP Models for the Skill VRP

Let us introduce the design variables which are common to all models. For each $(i, j) \in A$ and $t \in T$ such that $s_t \geq \max\{s_i, s_j\}$, let:

$$x_{ij}^t = \begin{cases} 1 \text{ if } (i, j) \text{ belongs to the tour of technician } t \\ 0 \text{ otherwise.} \end{cases}$$

Observe that we imposed $s_t \geq \max\{s_i, s_j\}$ in any feasible solution that uses arc (i, j) for technician t according to the skill level constraints. This property is peculiar to the particular structure of the Skill VRP. Next we will present two models for the Skill VRP, for increasing disaggregation levels of auxiliary multicommodity flow variables and constraints.

2.1 The Aggregated Formulation

As common in some flow based VRP formulations, let us introduce a flow variable y_{ij} for each $(i, j) \in A$ in addition to the design variables x^t_{ij}, in order to break subtours. Using these variables we can state the following model, referred to as Agg:

$$\min \sum_{(i,j)\in A} \sum_{t:s_t \geq \max\{s_i,s_j\}} c^t_{ij} x^t_{ij} \tag{1}$$

$$\sum_{i\in N} \sum_{t:s_t \geq \max\{s_i,s_j\}} x^t_{ij} = 1 \qquad j = 2,..,n \tag{2}$$

$$\sum_{i\in N:s_t \geq s_i} x^t_{ij} = \sum_{i\in N:s_t \geq s_i} x^t_{ji} \qquad j = 2,..,n, t : s_t \geq s_j \tag{3}$$

$$\sum_{j\in N} y_{1j} = n - 1 \tag{4}$$

$$\sum_{i\in N} y_{ij} - \sum_{i\in N} y_{ji} = 1 \qquad j \in N \setminus \{1\} \tag{5}$$

$$y_{ij} \leq (n-1) \sum_{t:s_t \geq \max\{s_i,s_j\}} x^t_{ij} \qquad (i, j) \in A \tag{6}$$

$$y_{ij} \geq 0 \qquad (i, j) \in A$$

$$x^t_{ij} \in \{0, 1\} \qquad (i, j) \in A, t : s_t \geq \max\{s_i, s_j\}$$

Constraints (2) and (3) model the tour of each technician t by taking into account the skill constraints. The flow constraints (4) and (5) and the successive linking constraints (6) allow one to break subtours, i.e. tours which are disconnected from the depot. Also in this case the constraints do incorporate the skill requirements at the nodes. The objective function (1), to be minimized, gives the cost of the tours.

2.2 The Destination Disaggregated Model

In order to enhance the linear programming (LP) bound of the aggregated formulation, let us disaggregate the auxiliary flow variables y_{ij} by destination. Also this way of strengthening the LP relaxation of single commodity flow models is well-known. Consider a model where one unit of flow is sent from the depot to each node $k \in V \setminus \{1\}$ by using disaggregated flow variables y^k_{ij} with $j \neq 1$ and $i \neq k$, that specify the amount of flow traversing arc (i, j) that is destined to that node. The resulting model, referred to as *WeakDestDis*, is the following:

$$\min \sum_{(i,j)\in A} \sum_{t:s_t \geq \max\{s_i,s_j\}} c^t_{ij} x^t_{ij}$$

$$\sum_{i \in N} \sum_{t:s_t \geq \max\{s_i,s_j\}} x^t_{ij} = 1 \qquad j = 2,..,n$$

$$\sum_{i \in N: s_t \geq s_i} x^t_{ij} = \sum_{i \in N: s_t \geq s_i} x^t_{ji} \qquad j = 2,..,n, t : s_t \geq s_j$$

$$\sum_{\substack{(i,j) \in A: \\ i \neq k, j \neq 1}} y^k_{ij} - \sum_{\substack{(j,i) \in A: \\ i \neq 1, j \neq k}} y^k_{ji} = \begin{cases} 1 & \text{if } i = 1 \\ 0 & \text{if } i \neq 1, k \\ -1 & \text{if } i = k \end{cases} \qquad k = 2, \ldots, n \tag{7}$$

$$y^k_{ij} \leq \sum_{t:s_t \geq \max\{s_i,s_j\}} x^t_{ij} \qquad (i,j) \in A,\ k \in V : k \neq 1, i, j \neq 1 \tag{8}$$

$$x^t_{ij} \in \{0,1\} \qquad (i,j) \in A, t : s_t \geq \max\{s_i,s_j\}$$

$$y^k_{ij} \geq 0 \qquad (i,j) \in A : j \neq 1, k \neq 1, i$$

Constraints (7) are the classical flow conservation constraints, stating that one unit of flow must be pushed to each destination node k. Since this unit of flow is sent from the depot, we can reinterpret the flow variables as indicating whether arc (i,j) is in the path to node k. This immediately justifies the linking constraints (8).

However, here we are able to specialize this technique to the particular structure of the Skill VRP. Constraints (8) can be replaced by the following constraints, that tighten the LP relaxation of the problem:

$$y^k_{ij} \leq \sum_{t:s_t \geq \max\{s_i,s_j,s_k\}} x^t_{ij} \quad (i,j) \in A,\ k \in V : k \neq 1, i,\ j \neq 1. \tag{9}$$

The resulting model is referred to as *StrongDestDis*. The computational results reported in Section 4 show that the LP bounds computed via *StrongDestDis* are substantially better than the ones obtained by *WeakDestDis*, by proving that an accurate modelling of the skill information at the nodes is a crucial aspect for determining good LP bounds.

2.3 Some Valid Inequalities

The LP relaxation of *StrongDestDis* shows two critical issues that suggested some families of valid inequalities in order to enhance the LP bound. Specifically, by looking at the x^t_{ij} variables in the LP solution, and specifically at the subgraphs induced by the positive x^t_{ij}, for each t, we observed the following facts:

1. the LP solution contains two-cycles (i.e. cycle of type $(i - j - i)$) or composition of two-cycles in the subgraphs related to technicians t not used in the solution (a technician t is not used when $x^t_{1j} = 0$ for all nodes j other than the depot)
2. the LP solution contains two-cycles (or again composition of two-cycles) also in the subgraphs of technicians used, i.e. connected to the depot.

In order to eliminate the two-cycles of the first type, the following valid inequalities have been added to the LP relaxation of *StrongDestDis*:

$$\sum_{(i,j):i,j \neq 1, s_t >= \max\{s_i,s_j\}} x^t_{ij} \leq n \sum_{j \in N \setminus \{1\}: s_t \geq s_j} x^t_{1j} \quad t \in T \tag{10}$$

In addition, the model has been enhanced via the following families of valid inequalities:

$$x^t_{ik} + x^t_{ki} \leq \sum_{j \in N \setminus \{1\}: s_t \geq s_j} x^t_{1j} \quad t \in T, \quad i,k \neq 1 : s_t >= \max\{s_i, s_k\} \tag{11}$$

$$x^t_{ik} \leq \sum_{(j,i): j \neq k, s_t \geq s_j} x^t_{ji} \quad t \in T, \quad i,k \neq 1 : s_t >= \max\{s_i, s_k\}. \tag{12}$$

Both (11) and (12) aim at eliminating the two-cycles of the second type. Hereafter, these valid inequalities will be referred to as TC (which stands for Two-Cycles) and EC (which stands for Enhanced Connectivity), respectively.

As reported in Section 4, TC and EC do not behave equivalently when added to *StrongDestDis*. Both of them, instead, are implied by the stronger model in Section 3, which is obtained by further disaggregating *StrongDestDis*. As shown by the computational results in Section 4, this further disaggregation step allowed us to generate LP bounds which are very close to the optimum values, but this has been achieved at the expenses of a significant computational effort. This has motivated the proposal of (11) and (12) as a tool to enhance the more tractable model *StrongDestDis*, as better discussed in Section 4.

StrongDestDis equipped with all the families of valid inequalities discussed above is referred to as *TightStrongDestDis*.

3 The Destination-Technician Disaggregated Model

In order to strengthen the LP bound, a further disaggregation has been performed by considering a disaggregation by technician in addition to the disaggregation by destination. This double disaggregation is innovative, in our opinion, in modelling issues for VRP.

Consider a model which extends *StrongDestDis* via disaggregated flow variables y^{kt}_{ij}, with $j \neq 1$, $i \neq k$ and $t : s_t \geq \max\{s_i, s_j, s_k\}$, where y^{kt}_{ij} denotes the unit of flow which is sent from the depot to node $k \in V \setminus \{1\}$ via technician t, if technician t serves k, and which holds zero otherwise. The model uses additional variables y^t_k, $k = 2, .., n, t : s_t \geq s_k$, where y^t_k denotes whether technician t transports one unit of flow from the depot to node k (equivalently, whether arc (i, j) is in the path from the depot to node k, and this path belongs to the tour of technician t).

The resulting model, referred to as *StrongDestTechDis*, is the following:

$$\min \sum_{(i,j) \in A} \sum_{t: s_t \geq \max\{s_i, s_j\}} c^t_{ij} x^t_{ij}$$

$$\sum_{i \in N} \sum_{t: s_t \geq \max\{s_i, s_j\}} x^t_{ij} = 1 \qquad j = 2, .., n$$

$$\sum_{i \in N: s_t \geq s_i} x^t_{ij} = \sum_{i \in N: s_t \geq s_i} x^t_{ji} \qquad j = 2, .., n, t : s_t \geq s_j$$

$$\sum_{\substack{(i,j)\in A:\\ i\neq k, j\neq 1,\\ s_t\geq s_i,s_j}} y_{ij}^{kt} - \sum_{\substack{(j,i)\in A:\\ i\neq 1, j\neq k,\\ s_t\geq s_i,s_j}} y_{ji}^{kt} = \begin{cases} y_k^t & \text{if } i=1 \\ 0 & \text{if } i\neq 1,k \\ -y_k^t & \text{if } i=k \end{cases} \quad k=2,\ldots,n, t: s_t\geq s_k \tag{13}$$

$$\sum_{t:s_t\geq s_k} y_k^t = 1 \quad k=2,..,n \tag{14}$$

$$y_{ij}^{kt}\leq x_{ij}^t \quad (i,j)\in A,\ k\in V: k\neq 1, i,\ j\neq 1,\ t: s_t\geq \max\{s_i,s_j,s_k\} \tag{15}$$

$$x_{ij}^t\in\{0,1\} \quad (i,j)\in A, t: s_t\geq \max\{s_i,s_j\}$$

$$y_k^t\in\{0,1\} \quad k=2,..,n, t: s_t\geq s_k$$

$$y_{ij}^{kt}\geq 0 \quad (i,j)\in A: j\neq 1, k\neq 1, i,\ t: s_t\geq \max\{s_i,s_j,s_k\}$$

Constraints (13) are the flow conservation constraints related to the disaggregated flow variables y_{ij}^{kt}. They state that, if $y_k^t=1$, i.e. technician t serves node k, then one unit of flow must be pushed from the depot to node k. Constraints (14) guarantee that each node k is served by exactly one technician, while constraints (15) are the corresponding linking constraints. The computational results presented in the next section will provide evidence that the LP bounds produced by *StrongDestTechDis* are very tigth, and this is essentially due to the fact that the LP solution contains very few two-cycles (or, again, compositions of two-cycles) for technicians used, i.e. connected to the depot.

In particular, it is easy to show that *StrongDestTechDis* implies the following *cut constraints*:

$$\sum_{i\in N\setminus S, j\in S: s_t\geq \max\{s_i,s_j\}} x_{ij}^t \geq y_k^t \quad S\subseteq N\setminus\{1\}, k\in S, t\in T: s_t\geq s_k. \tag{16}$$

Moreover, the *cut constraints* imply *EC*, as shown by the following lemma:

Lemma 1. *Cut constraints, and so model StrongDestTechDis, imply EC.*

Proof. Consider *EC* for given $i,k\neq 1$ and $t\in T$ such that $s_t >= \max\{s_i,s_k\}$, that is:

$$x_{ik}^t \leq \sum_{(j,i): j\neq k, s_t\geq s_j} x_{ji}^t. \tag{17}$$

Add $\sum_{(j,k): j\neq i, s_t\geq s_j} x_{jk}^t$ to both sides of (17). Since $\sum_{(j,k): s_t\geq \max\{s_j,s_k\}} x_{jk}^t \geq y_k^t$ is implied by model *StrongDestTechDis*, then we get the *cut constraint* for $S=\{i,k\}$. □

The *cut constraints* imply also *TC*, as shown below:

Lemma 2. *Cut constraints, and so model StrongDestTechDis, imply TC.*

Proof. Consider *EC* for given $i,k\neq 1$ and $t\in T$ such that $s_t >= \max\{s_i,s_k\}$, that is:

$$x_{ik}^t \leq \sum_{(j,i): j\neq k, s_t\geq s_j} x_{ji}^t. \tag{18}$$

Recall that *EC* is implied by the *cut constraints* by Lemma 1. Add x^t_{ki} to both sides of (18). Since $\sum_{j \in N\setminus\{1\}: s_t \geq s_j} x^t_{1j} \geq \sum_{(j,i): s_t \geq s_j} x^t_{ji}$, we derive

$$x^t_{ik} + x^t_{ki} \leq \sum_{j \in N\setminus\{1\}: s_t \geq s_j} x^t_{1j}, \tag{19}$$

which is the corresponding *TC* constraint. □

Thus model *StrongDestTechDis* appears to be very strong, and this is confirmed by the computational results in Section 4. Clearly, its main drawback is in its size, which causes high computational times. Anyway, we will show next that model *StrongDest-Dis*, when equipped with the *TC* and the *EC* valid inequalities, allows one to obtain LP bounds which are comparable with those produced by the stronger disaggregated model, at a reasonable computational time.

4 The Computational Experience

We present here the results of a wide computational experience, on a large suite of randomly generated Skill VRP instances, aimed at comparing the previously introduced ILP models both in terms of quality of the returned LP bounds and also in terms of the required computational times.

4.1 The Instances

We used a test set for VRP which was properly extended with the data relative to the skills. Specifically, the original instances are characterized by the number of nodes (either 20 or 40) and by the type (4 different choices). There are two types of Euclidean instances referred to as *tc* and *te* according to the position of the depot in the grid. In the *tc* instances the depot is located on a corner of the grid, while in the *te* instances the depot is located in the centre of the grid. In addition, there are two types of random instances: *tra* denotes asymmetric instances and *trs* denotes the symmetric ones. 5 instances are available for each given number of nodes and type, thus summing up to a total of 40 ($2 \times 4 \times 5$) instances.

In order to generate the Skill VRP instances we used instance number one (over the five available) for each type and for each given number of nodes, thus summing up to 8 original instances over 40.

In the instances there is a one-to-one correspondence between skills and technicians, i.e. we have a technician for each skill. We set both the number of technicians and the number of skills to 3 where skill 1 denotes the lowest skill and skill 3 is the highest (w.l.o.g. the depot is assumed to have skill 1).

For each original instance we generated 18 Skill VRP instances according to the pattern used to assign a skill to the nodes. The patterns used are the following:

50-10-40 (for a total of 6 permutations)
50-20-30 (for a total of 6 permutations)
40-40-20 (for a total of 3 permutations)
30-30-40 (for a total of 3 permutations)

where the numbers indicate the percentage of nodes requiring a specific skill.

In the Skill VRP instances the costs depend on both the arc and the technician, differently from what happens in the original VRP instances, where the costs depend on the arc only. In particular, we assume that the cost of an arc (i,j) is increasing with the technician skills, i.e., the more a technician is qualified higher the cost. The formula used to generate the costs is the following:

$$c_{ij}^{t} = \begin{cases} c_{ij} & \text{if } t = 1 \\ c_{ij} \cdot \delta \cdot (t-1) & \text{if } t > 1, \end{cases}$$

where c_{ij} represents the cost of arc (i,j) in the original instances. Observe that these costs satisfy the triangular property for each technician when the original costs do. Trivially if $c_{ij} \leq c_{ih} + c_{hj}$, the inequality $c_{ij} + \delta \cdot k \leq c_{ih} + c_{hj} + 2\delta \cdot k$ holds true. Under the hypothesis of costs not satisfying the triangular property, the proposed models do not assure that, for a given technician, the schedule is composed of a single tour: it might be convenient coming back to the depot more than once.

4.2 Computational Results

Both the instances characterized by 20 nodes that the ones having 40 nodes have been tested. In Tables 2 and 3, we give a comparison of the models *Aggregated*, *WeakDestDist*, *StrongDestDis*, *TightStrongDestDis* and *StrongDestTechDis* on the larger instances, characterized by 40 nodes, in terms of percentage gap of the LP bound with respect to the optimum value (computed via the model *Aggregated*) and of average time required to solve the continuous relaxation. For the aggregated model also the computational time required to get the optimum is reported.

It appears that computing the integer solution is computational expensive (see the average computational times reported under *AggIPTime* in Table 3). However, by enhancing the disaggregation level the proposed models tend to produce very good LP bounds (see the average gaps in Table 2). In particular, *StrongDestTechDis* provides an average gap going from the minimum 0.03 (*te* family) to the maximum 0.75 (*tra* family). Clearly, the required computational times becomes high when the disaggregation level increases, reaching an average time of 1033,51 sec. for the *trs* instances.

On the other hand, the computational time is not an issue for the instances with 20 nodes. Therefore, time is not in Table 1, where a comparison among the LP bounds produced by a subset of the proposed models is shown.

Finally, Table 4 reports average gaps obtained when the *StrongDestDis* model is equipped respectively with the Enhanced Connectivity valid inequalities (*StrongDestDis+EC*), with the Two-Cycles valid inequalities (*StrongDestDis+TC*) and with all the families of valid inequalities proposed in Section 2.3.

5 Conclusions

The problem of defining a set of tour, each one operated by a skilled technician, so as to fulfill service requirements asking for a particular skill, has been defined and studied. Three models, characterized by an increasing level of variables disaggregation have

Table 1. Model Comparison in Terms of Optimality Gap - 20 Node Instances

Instance	Agg	StrongDestDis
tc20-1-10-2-8.dat	13.62	0.86
tc20-1-10-8-2.dat	18.50	2.74
tc20-1-2-10-8.dat	13.36	0.86
tc20-1-8-10-2.dat	18.59	3.21
tc20-1-2-8-10.dat	12.66	0.00
tc20-1-8-2-10.dat	12.40	0.00
tc20-1-10-6-4.dat	17.58	1.11
tc20-1-10-4-6.dat	15.46	1.10
tc20-1-4-10-6.dat	14.39	1.07
tc20-1-6-10-4.dat	16.70	1.36
tc20-1-4-6-10.dat	12.08	0.00
tc20-1-6-4-10.dat	10.41	0.00
tc20-1-8-8-4.dat	16.42	2.89
tc20-1-4-8-8.dat	12.08	0.00
tc20-1-8-4-8.dat	13.55	0.64
tc20-1-8-6-6.dat	14.73	0.43
tc20-1-6-8-6.dat	14.65	1.07
tc20-1-6-6-8.dat	13.58	0.43
Avg tc20-1	14.49	0.99
te20-1-10-2-8.dat	18.24	0.00
te20-1-10-8-2.dat	20.75	2.11
te20-1-2-10-8.dat	14.97	0.60
te20-1-8-10-2.dat	19.61	0.60
te20-1-2-8-10.dat	13.82	0.00
te20-1-8-2-10.dat	15.24	0.00
te20-1-10-6-4.dat	20.45	2.55
te20-1-10-4-6.dat	19.08	1.78
te20-1-4-10-6.dat	17.27	0.00
te20-1-6-10-4.dat	17.62	2.21
te20-1-4-6-10.dat	14.62	0.00
te20-1-6-4-10.dat	15.20	0.00
te20-1-8-8-4.dat	18.98	0.80
te20-1-4-8-8.dat	14.76	0.20
te20-1-8-4-8.dat	15.16	0.40
te20-1-8-6-6.dat	17.45	1.21
te20-1-6-8-6.dat	15.91	0.80
te20-1-6-6-8.dat	14.96	0.20
Avg te20-1	16.89	0.75
tra20-1-10-2-8.dat	2.29	0.00
tra20-1-10-8-2.dat	11.48	6.42
tra20-1-2-10-8.dat	0.36	0.00
tra20-1-8-10-2.dat	7.78	4.13
tra20-1-2-8-10.dat	0.33	0.00
tra20-1-8-2-10.dat	1.31	0.19
tra20-1-10-6-4.dat	6.18	2.70
tra20-1-10-4-6.dat	2.63	0.87
tra20-1-4-10-6.dat	5.54	2.95
tra20-1-6-10-4.dat	6.23	3.72
tra20-1-4-6-10.dat	2.29	0.00
tra20-1-6-4-10.dat	0.36	0.00
tra20-1-8-8-4.dat	7.88	3.38
tra20-1-4-8-8.dat	0.33	0.00
tra20-1-8-4-8.dat	0.33	0.00
tra20-1-8-6-6.dat	5.50	3.46
tra20-1-6-8-6.dat	4.18	0.00
tra20-1-6-6-8.dat	0.77	0.00
Avg tra20-1	3.65	1.55
trs20-1-10-2-8.dat	14.59	2.39
trs20-1-10-8-2.dat	20.00	5.61
trs20-1-2-10-8.dat	12.64	3.50
trs20-1-8-10-2.dat	18.33	4.97
trs20-1-2-8-10.dat	10.48	1.11
trs20-1-8-2-10.dat	13.47	0.48
trs20-1-10-6-4.dat	19.16	7.21
trs20-1-10-4-6.dat	18.25	5.25
trs20-1-4-10-6.dat	15.57	4.30
trs20-1-6-10-4.dat	16.27	2.38
trs20-1-4-6-10.dat	9.36	0.00
trs20-1-6-4-10.dat	11.15	0.96
trs20-1-8-8-4.dat	18.63	3.98
trs20-1-4-8-8.dat	11.50	0.48
trs20-1-8-4-8.dat	14.20	0.96
trs20-1-8-6-6.dat	15.54	1.19
trs20-1-6-8-6.dat	15.64	3.14
trs20-1-6-6-8.dat	12.07	1.11
Avg trs20-1	14.83	2.72

Table 2. Model Comparison in Terms of Optimality Gap - 40 Node Instances

Instance	AggIPTime	Agg	WeakDestDis	StrongDestDis	TightStrongDestDis	StrongTechDestDis
tc40-1-20-4-16.dat	740.01	16.67	7.56	2.24	0.21	0.00
tc40-1-20-16-4.dat	138099.46	20.81	12.61	3.55	1.69	0.56
tc40-1-4-20-16.dat	292.92	12.58	3.92	0.33	0.00	0.00
tc40-1-16-20-4.dat	66387.62	20.00	11.48	3.43	1.74	0.25
tc40-1-4-16-20.dat	269.38	11.76	3.08	0.14	0.00	0.00
tc40-1-16-4-20.dat	153.52	13.83	5.60	0.28	0.00	0.00
tc40-1-20-12-8.dat	7154.01	18.52	9.80	3.34	0.84	0.00
tc40-1-20-8-12.dat	4623.10	18.61	9.52	1.75	0.21	0.00
tc40-1-8-20-12.dat	1376.12	15.43	6.72	0.95	0.00	0.00
tc40-1-12-20-8.dat	21599.58	16.09	8.12	2.42	0.00	0.00
tc40-1-8-12-20.dat	269.24	11.74	3.08	0.00	0.00	0.00
tc40-1-12-8-20.dat	293.61	14.38	5.32	0.28	0.00	0.00
tc40-1-16-16-8.dat	8792.80	18.20	8.96	2.52	0.23	0.00
tc40-1-8-16-16.dat	382.51	13.62	4.20	0.28	0.00	0.00
tc40-1-16-8-16.dat	595.59	15.67	7.00	0.77	0.00	0.00
tc40-1-16-12-12.dat	2747.81	16.69	8.40	1.61	0.00	0.00
tc40-1-12-16-12.dat	1680.03	16.34	7.56	1.37	0.00	0.00
tc40-1-12-12-16.dat	536.05	12.91	5.32	0.28	0.00	0.00
Avg tc40-1	14221.85	15.77	7.13	1.42	0.27	0.04
te40-1-20-4-16.dat	199.26	6.33	4.42	1.67	0.00	0.00
te40-1-20-16-4.dat	137139.71	9.08	7.25	5.11	0.89	0.38
te40-1-4-20-16.dat	41.51	2.39	1.07	0.49	0.00	0.00
te40-1-16-20-4.dat	167422.29	10.40	7.65	5.78	0.66	0.24
te40-1-4-16-20.dat	6.35	1.30	0.49	0.39	0.00	0.00
te40-1-16-4-20.dat	63.62	2.66	1.02	0.29	0.00	0.00
te40-1-20-12-8.dat	9104.19	8.66	6.01	3.10	0.28	0.00
te40-1-20-8-12.dat	1163.74	6.91	4.24	2.03	0.00	0.00
te40-1-8-20-12.dat	232.59	4.18	2.92	1.02	0.00	0.00
te40-1-12-20-8.dat	7744.17	7.31	6.18	3.14	0.23	0.00
te40-1-8-12-20.dat	23.42	3.05	1.10	0.58	0.00	0.00
te40-1-12-8-20.dat	18.04	2.85	1.39	1.16	0.00	0.00
te40-1-16-16-8.dat	59379.08	7.39	6.15	4.69	0.51	0.00
te40-1-8-16-16.dat	25.76	2.48	1.19	0.49	0.00	0.00
te40-1-16-8-16.dat	712.57	7.28	5.12	4.22	0.00	0.00
te40-1-16-12-12.dat	1948.18	6.01	4.22	1.77	0.00	0.00
te40-1-12-16-12.dat	917.58	5.00	2.92	1.96	0.00	0.00
te40-1-12-12-16.dat	177.64	4.95	3.90	2.98	0.00	0.00
Avg te40-1	21462.21	5.46	3.73	2.27	0.14	0.03
tra40-1-20-4-16.dat	16.56	5.67	3.27	1.98	1.13	0.85
tra40-1-20-16-4.dat	6071.55	9.35	8.55	5.14	1.42	1.09
tra40-1-4-20-16.dat	8.73	3.44	3.09	2.47	0.18	0.18
tra40-1-16-20-4.dat	1930.30	7.05	6.65	4.42	1.03	1.00
tra40-1-4-16-20.dat	31.45	2.46	1.67	1.20	0.07	0.00
tra40-1-16-4-20.dat	321.51	5.94	4.29	3.70	0.71	0.41
tra40-1-20-12-8.dat	2513.81	9.64	7.73	5.39	2.37	1.86
tra40-1-20-8-12.dat	991.68	7.75	6.35	6.09	1.53	0.49
tra40-1-8-20-12.dat	844.34	6.65	5.29	3.57	1.24	0.83
tra40-1-12-20-8.dat	1122.82	5.78	5.75	5.03	1.41	1.09
tra40-1-8-12-20.dat	19.58	2.58	1.83	1.44	0.00	0.00
tra40-1-12-8-20.dat	5.04	1.70	1.05	0.61	0.00	0.00
tra40-1-16-16-8.dat	2419.07	9.21	8.97	5.89	1.93	1.63
tra40-1-8-16-16.dat	55.96	3.83	2.44	1.98	0.00	0.00
tra40-1-16-8-16.dat	339.80	5.02	4.92	3.43	0.64	0.12
tra40-1-16-12-12.dat	771.73	6.46	6.11	3.33	2.00	1.33
tra40-1-12-16-12.dat	886.28	7.66	6.60	5.45	2.59	2.50
tra40-1-12-12-16.dat	104.89	6.74	4.78	3.53	0.59	0.13
Avg tra40-1	1025.28	5.94	4.96	3.59	1.05	0.75
trs40-1-20-4-16.dat	1042.82	23.00	13.33	6.39	0.98	0.00
trs40-1-20-16-4.dat	24210.90	24.86	17.45	8.11	1.81	0.78
trs40-1-4-20-16.dat	92.81	14.22	5.87	2.08	0.00	0.00
trs40-1-16-20-4.dat	2993.72	19.42	11.67	3.33	0.94	0.15
trs40-1-4-16-20.dat	260.99	12.73	5.57	1.77	0.00	0.00
trs40-1-16-4-20.dat	1514.70	19.17	12.71	4.40	1.56	0.81
trs40-1-20-12-8.dat	12724.45	24.03	15.42	7.39	1.40	0.60
trs40-1-20-8-12.dat	3304.62	23.93	14.54	6.90	0.99	0.72
trs40-1-8-20-12.dat	1438.01	18.77	11.45	5.91	0.32	0.25
trs40-1-12-20-8.dat	9238.33	21.86	14.68	6.24	1.30	1.18
trs40-1-8-12-20.dat	100.96	13.71	5.87	1.28	0.00	0.00
trs40-1-12-8-20.dat	87.79	12.82	5.62	0.89	0.00	0.00
trs40-1-16-16-8.dat	1607.80	20.63	11.96	4.96	0.51	0.00
trs40-1-8-16-16.dat	491.16	17.55	8.31	3.73	0.72	0.20
trs40-1-16-8-16.dat	156.85	18.03	10.27	3.15	0.00	0.00
trs40-1-16-12-12.dat	3191.11	20.67	13.05	6.53	1.63	0.31
trs40-1-12-16-12.dat	5044.69	19.95	12.22	6.15	0.90	0.64
trs40-1-12-12-16.dat	494.40	18.95	10.76	3.89	0.00	0.00
Avg trs40-1	3777.56	19.13	11.15	4.62	0.73	0.31

Table 3. Comparison in Terms of Average Time

Instance	AggIP	Agg	Disagg	WeakDestDis	Strong DestDis	TightStrong DestDis	StrongTech DestDis
tc40-1	14221.85	0.24	0.94	214.84	324.91	209.27	538.71
te40-1	21462.21	0.32	0.98	162.48	149.87	114.14	454.97
tra40-1	1025.28	0.35	0.72	142.71	112.28	45.86	232.92
trs40-1	3777.56	0.27	1.01	199.88	442.46	206.13	1033.51

Table 4. Comparison in Terms of Gap

Instance	StrongDestDis+EC	StrongDestDis+TC	TightStrongDestDis
tc40-1	0.61	0.56	0.27
te40-1	2.00	0.18	0.14
tra40-1	3.33	1.17	1.05
trs40-1	1.73	2.37	0.73

been formulated and tested on a large suite of randomly generated instances. Computational results have shown that introducing a greater level of disaggregation permits to obtain very good LP bounds, unfortunately at the cost of greater computational times required to solve the models. A model with an intermediate level of disaggregation equipped with a set of valid inequalities which have proven to be computationally effective, seems thus to represent a good trade-off between quality of the LP bound and computational burden. Such a model thus represents a promising starting point to develop exact as well heuristic approaches to solve the problem to optimality.

References

1. Agnihothri, S.R., Mishra, A.K.: Cross-training decisions in field services with three job types and server-job mismatch. Decision Sciences 35(2), 239–257 (2004)
2. Baldacci, R., Bartolini, E., Mingozzi, A., Roberti, R.: An exact solution framework for a broad class of vehicle routing problems. Computational Managagement Science 7, 229–268 (2010)
3. Baldacci, R., Bartolini, E., Mingozzi, A., Valletta, A.: An exact algorithm for the period routing problem. Operations Research (to appear, 2011)
4. Baldacci, R., Toth, P., Vigo, D.: Exact algorithms for routing problems under vehicle capacity constraints. Annals of Operations Research 175, 213–245 (2009)
5. Borsani, V., Matta, A., Beschi, G., Sommaruga, F.: A home care scheduling model for human resources. In: IEEE International Conference on Service Systems and Service Management, pp. 449–454 (2006)
6. Cordeau, J.F., Laporte, G.: A tabu search algorithm for the site dependent vehicle routing problem with time windows. INFOR 39(3), 292–298 (2001)
7. Rapaccini, M., Sistemi, A., Visintin, F.: A simulation-based DSS for field service delivery optimization. In: Proceedings of the International Workshop on Modelling and Applied Simulation, Campora San Giovanni, Amantea (CS), Italy, pp. 116–122 (2008)

The Biobjective Inventory Routing Problem – Problem Solution and Decision Support

Martin Josef Geiger[1] and Marc Sevaux[1,2]

[1] Helmut Schmidt University,
Logistics Management Department, Hamburg, Germany
m.j.geiger@hsu-hh.de
[2] Université de Bretagne-Sud – Lab-STICC,
Lorient, France
marc.sevaux@univ-ubs.fr

Abstract. The article presents a study on a biobjective generalization of the inventory routing problem, an important optimization problem found in logistical networks in which aspects of inventory management and vehicle routing intersect. A compact solution representation and a heuristic optimization approach are presented, and experimental investigations involving novel benchmark instances are carried out. As the investigated problem is computationally very demanding, only a small subset of solutions can be computed within reasonable time. Decision support is nevertheless obtained by means of a set of reference points, which guide the search towards the Pareto-front.

1 Introduction

Many logistical activities are commonly concerned with moving required items through a graph, often referred to as a "supply chain" or "supply network". Naturally, numerous aspects influence the way in which such processes are organized, including macro- and microeconomical conditions, the objectives of the players along the supply chain, their interrelations, and their technical abilities of communicating and thus exchanging the required information. Consequently, different implementations of coordination and replenishment strategies can be found in practice. While some supply chains are organized in a rather loose fashion, with sporadic interactions among the players, others consists of permanent, long-term business relations with frequent deliveries.

A more recent and prominent example of how to organize the logistical activities in a supply chain is the Vendor Managed Inventory-strategy. In this approach, the vendor/ supplier takes considerable control over the inventories held at the downstream entity, *i.e.* the buyer/ retailer, by determining the level of inventories and the corresponding replenishment cycles/ patterns. In this sense, responsibility for adequate inventory levels is shifted upstream, while inventories as such are still held downstream. From an operations research (OR) point of view, such problems combine decision variables describing shipping quantities and dates, as well as variables determining the routing of the vehicles used for transporting the goods. The obtained models therefore lie in the intersection of two classical OR problems, namely the vehicle routing problem and

J. Pahl, T. Reiners, and S. Voß (Eds.): INOC 2011, LNCS 6701, pp. 365–378, 2011.

inventory management problems, and have led to what is commonly referred to as *inventory routing problems* (IRP) [3, 18].

Inventory routing problems have drawn a considerable interest in past years. While vehicle routing and inventory management are challenging issues on their own, both aspect combined together significantly increase the difficulty of the problem at hand. Contrary to most vehicle routing problems, dynamically developing inventories require multi-period models in which deliveries are made over a certain time horizon, avoiding stockouts, if possible. On the other hand, it may be beneficial to serve customers which are close to each other together, reducing the traveled distances of the vehicles. Cost minimization in the IRP thus combines the three components of (i) costs from routing, (ii) held inventory, and (iii) stockouts. Obviously, the third component is comparably difficult to measure, and may in practice be replaced by a criterion measuring the service-level, leading to the introduction of an additional side constraint or to a multi-objective model formulation. Evaluation of solutions for the IRP nevertheless remains a difficult issue, which is why some dedicated research is going in this direction [22].

Besides these general problem characteristics, uncertainties about future demand, deliveries, production rates, etc. are often present, leading to dynamic, stochastic models [6].

Due to the difficulty of the obtained formulations, many solution approaches reintroduce some simplifying assumptions. For example in [5], the amount of goods delivered to the customers is chosen such that a pre-defined maximum inventory level is reached again, while the work of [9] concentrates on the optimization of the delivery volume only. Other simplifications address the routing, e. g. by direct deliveries to the customers [4, 14, 17]. Alternative approaches allow the backlogging of parts, thus treating the problem of stockouts in an particular way [1]. In the case of deterministic demand, cyclic formulations of the IRP lead to the construction of solutions with constant replenishment intervals [13, 19, 20].

Not surprisingly, solution approaches often employ (meta-) heuristics [10]. Prominent examples include Iterated Local Search, Variable Neighborhood Search, Greedy Randomized Adaptive Search [8], Memetic Algorithms [7] and decomposition approaches making use of Lagrangian relaxations [23]. Common to all ideas is that they have been successfully adopted to the particular chosen problem.

Then however, and in contrast to other domains from operations research/ management science, comparably little work has been done with respect to the proposition of a common ground for experiments, such as the proposition of benchmark instances that would allow for a detailed comparison of the solution approaches. One of the few examples is found in the "Praxair"-case, an international industrial gases company facing an IRP [8]. Besides, a description of how to compute instances is given in [2]. From this perspective, the definition and publication of benchmark data allowing a future comparison appears to be a beneficial contribution.

Besides, and with with respect to practical side constraints of such problems, rule-based solution approaches might provide an interesting field of research [15]. Contrary to most (meta-) heuristics, which are based on the local search paradigm, rule-based decision making approaches explicitly state why certain decision are made in particular situations. In addition to providing a solution methodology, they also contribute to the explanation of the choices made by an decision support system, thus allowing an interpretation of the system's behavior. Moreover, human experts may directly adopt the decision making rules, an aspect that also contributes to the acceptance of such (semi-) automated optimization/ approximation approaches.

This article contributes to two of the above mentioned issues of the Inventory Routing Problem. After the description of the IRP tackled in this paper in the following Section 2, we present a compact solution representation and a heuristic solution approach in Section 3. A set of benchmark instances is introduced in Section 4, and experimental investigations of the solution approach on the proposed data are carried out and reported. Conclusions are presented in Section 5.

2 Description of the Investigated IRP

We consider an IRP in which a given set of n customers needs to be delivered with goods from a single depot. The problem-immanent vehicle routing problem (VRP) aspects of the ones of the classical capacitated VRP. This means that we assume a unlimited number of available trucks, each of which has a limited capacity C for the delivered goods.

Decision variables of the problem are on the one hand delivery quantities q_{it} for each customer i, $i = 1, \ldots, n$, and each period t, $t = 1, \ldots, T$. On the other hand, a VRP must be solved for each period t of the planning horizon T, combining the delivery quantities q_{it} into tours/ routes for the involved fleet of vehicles. We assume $q_{it} \geq 0 \; \forall i, t$, thus forbidding the pickup of goods at all times. Following the definition of the classical capacitated VRP, we do not permit split-deliveries, and therefore $q_{it} \leq C \; \forall i, t$ where C is the truck capacity.

At each customer i, a demand d_{it} is to be satisfied for each period t. Inventory levels L_{it} at the customers are limited to a maximum amount of Q_i, *i.e.* $L_{it} \leq Q_i \; \forall i, t$. An incoming material flow ϕ_{it}^{+} and an outgoing material flow ϕ_{it}^{-} links the inventory levels at the customers over the time horizon: $L_{it+1} = L_{it} + \phi_{it}^{+} - \phi_{it}^{-} \; \forall i, t = 1, \ldots, T-1$.

The incoming material flow is given by $\phi_{it}^{+} = q_{it}$ iff $q_{it} \leq Q_i - L_{it-1}$, and $\phi_{it}^{+} = Q_i - L_{it-1}$ otherwise. This implies that the maximum inventory levels are never exceeded, independent from the choice of the shipping quantities q_{it}. On the other hand, delivery quantities q_{it} with $q_{it} > Q_i - L_{it-1}$ can be excluded when solving the problem. Obviously, as all inventory levels $L_{it} \geq 0$, any delivery quantity $q_{it} > Q_i$ will not play a role also.

The outgoing material flow assumes $\phi_{it}^{-} = d_{it}$ iff $d_{it} \leq L_{it-1} + \phi_{it}^{+}$, and $\phi_{it}^{-} = L_{it-1} + \phi_{it}^{+}$ otherwise. The latter case is also referred to as a stockout-situation, and the stockout-level can be computed as $d_{it} - L_{it-1} - q_{it}$. As we however assume d_{it} to be known in advance, we can easily avoid such situations by shipping enough goods in advance or just in time.

Two objective functions are considered, leading to a biobjective formulation. First, we minimize the total inventory as given in expression (1). Besides, the minimization of the total distances traveled by the vehicles for shipping the quantities in each period are of interest. While the solution of this second objective as such presents an $\mathcal{NP}$-hard problem, we denote this second objective with expression (2).

$$\min \sum_{i=1}^{n} \sum_{t=1}^{T} L_{it} \tag{1}$$

$$\min \sum_{t=1}^{T} \text{VRP}_t(q_{1t}, \ldots, q_{nt}) \tag{2}$$

The two objectives are clearly in conflict to each other. While large shipping quantities allow for a minimization of the routes, small quantities lead to low inventory levels over time. In between, we can expect numerous compromise solutions with intermediate values.

Without any additional information or artificial assumptions about the tradeoff between the two functions, no sensible aggregation into an overall evaluation is possible. Consequently the solution of the problem at hand lies in the identification of all optimal solutions in the sense of Pareto-optimality, which constitute the Pareto-set P.

Treating the IRP as a biobjective optimization problem has several advantages. First, no cost functions need to be found for the aggregation of inventory levels and routing distance. This is particularly important as some components of such underlying cost functions might vary over time, with fuel prices as a prominent example. Besides, eliciting sensible statements about tradeoffs between objectives from a decision maker can be a complicated and time consuming process. Second, it allows the analysis of the problem on a more tactical planning level. In such a setting, the effect of a reduced routing on inventory levels can be studied. This might be especially interesting for companies who want to investigate the possibilities and consequences of reducing emissions of logistical activities, such as CO_2, over a longer time horizon.

3 Solution Representation and Heuristic Search

The solution approach of this article is based on the decomposition of the problem into two phases. First, quantity decisions of when and how much to ship to each customer are made. This is done such that only customers running out of stock are served, *i.e.* we set $q_{it} > 0$ iff $d_{it} > L_{it-1}$ $\forall i, t$. Second, the resulting capacitated VRPs are solved for each period t by means of an appropriate solution approach.

Solving the IRP in this sequential manner has several advantages. Most importantly, we are able to introduce a compact encoding for the first phase, which allows an intuitive illustration of how the results of the procedure are obtained. We believe this to be an important aspect especially for decision makers from the industry. Besides, searching the encoding is rather easy by means of local search/ metaheuristics.

3.1 Solution Encoding

Alternatives are encoded by a n-dimensional vector $\pi = (\pi_1, \ldots, \pi_n)$ of integers. Each element π_i of the vector corresponds to a particular customer i, and encodes for how many periods the customer is served in a row (covered). In the following, we refer to these values π_i as a customer delivery *'frequency'*. Precisely, if $d_{it} > L_{it-1}$, then q_{it} may be computed as given in expression (3).

$$q_{it} = \min \left\{ \sum_{l=t}^{t-1+\pi_i} d_{il} - L_{it-1}, Q_i - L_{it-1}, C \right\} \tag{3}$$

A day-to-day delivery policy is obtained for values of $\pi_i = 1$. On the other hand, large values π_i ship up to the maximum of the customer/ vehicle capacity, and intermediate values lead to compromise quantities in between the two extremes. Obviously, such an encoding may easily interpreted by practitioners, thus providing a practical benefit.

On the basis of the delivery quantities, and as mentioned above, VRPs are obtained for each period which do not differ from what is known in the scientific literature.

3.2 Heuristic Search

Throughout the optimization runs, an archive $\widehat{P}$ of nondominated solutions is kept that presents a heuristic approximation to the true Pareto set P. By means of local search, and starting from some initial alternatives, we then aim to get closer to P, updating $\widehat{P}$ such that dominated solutions are removed, and newly discovered nondominated ones are added.

In theory, any local search strategy that modifies the frequency values of vector π might be used for finding better solutions. Clearly, values $\pi_i < 1$ are not possible, and exceptional large values do not make sense either due to the upper bounds on the delivery quantities as given in expression (3).

Construction Procedure. First, we generate solutions for which the delivery frequency assumes identical values for all customers, starting with 1 and increasing the frequency in steps of 1, up to the point where the alternative cannot be added to $\widehat{P}$ any more. Then, we construct alternatives for which the frequency values are randomly drawn between two values: j and $j+1$, thus mixing the frequency values of the first phase and therefore computing some solutions in between the ones with purely identical frequencies. The so obtained values of π are then used to determine the delivery quantities as described above.

The vehicle routing problems are then solved using two alternative algorithms. On the one hand, a classical savings heuristic [12] is employed, which allows for a comparable fast construction of the required routes/ tours. On the other hand, the more advanced record-to-record travel algorithm [16] is used. Having two alternative approaches is particularly interesting with respect to the practical use of our system. While the first algorithm will allow for a fast estimation of the routing costs, improved results are possible by means of the second approach, however at the cost of more time-consuming computations.

Improvement Procedure. The improvement procedure put forward in this article is based on modifying the values within π, and re-solving the subsequently modified VRPs using the algorithms mentioned above. For any element of $\widehat{P}$, we compute the set of neighboring solutions by modifying the values π_i by ± 1 while avoiding values of $\pi_i < 1$. The maximum number of neighboring solutions therefore is $2n$. Search for improved solution naturally terminates in case of not being able to improve any element of $\widehat{P}$. In this sense, the approach implements the concept of a multi-point hillclimber for Pareto-optimization.

Contrary to searching all elements of $\widehat{P}$, a representative smaller subset of $\widehat{P}$ can be taken for the improvement procedure. This reduces the computational effort to a considerable extent. One possibility is to assume a set of reference points, and to select the elements in $\widehat{P}$ that minimize the distance to these reference points. As a general guideline, the reference points should be chosen such that the solutions at the extreme ends of the two objective functions (1) and (2) are present in the subset. A favorable implementation of the distance function lies in the use of the weighted Chebyshef distance metric, as it allows for a selection of convex-dominated alternatives and thus provides some theoretical advantages over other approaches.

4 Experimental Investigation

4.1 Proposition of Benchmark Data

Using the geographical information of the 14 benchmark instances given in chapter 11 of [11], new data sets for inventory routing have been proposed. While the classical VRP commonly lacks multiple-period demand data, we filled this gap by proposing three demand scenarios, each of which contains a total of $T = 240$ demand periods for all customers.

Scenario a: The average demand of each customer is constant over time. Actual (integer) demand values for each period however are drawn from an interval of $\pm 25\%$ around this average.

Scenario b: We assume an increasing average demand, doubling from the initial value at $t = 1$ to $t = 240$. Again, the actual demand is drawn from an interval of $\pm 25\%$ around this average.

Scenario c: In this scenario, the average demand doubles from $t = 1$ to $t = 120$, and goes back to its initial value in $t = 240$, following the shape of a sinus curve. Identical to the above presented cases a deviation of $\pm 25\%$ around these values is allowed.

The resulting set of 42 benchmark instances have been made available to the scientific community under `http://logistik.hsu-hh.de/IRP` [21].

4.2 Implementation of a Decision Support System

A test program has been developed and is used to test new and innovative strategies. Fig. 1 shows a typical screen shot of our solver. The upper part on the left gives the name of the instance and the vehicle capacity. Then below, the decision maker is able to display the different alternatives computed by the software.

For the current alternative, and the current period, a text window gives the inventory level, the number of vehicle used and information on the tours. The box in the bottom

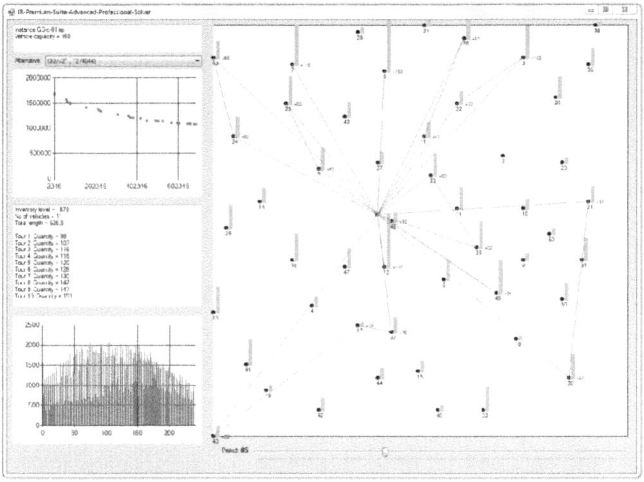

Fig. 1. Screen shot of the Inventory Routing Solver

left part represents the evolution of the total inventory over all periods. The large window on the right presents the current alternative and period routing. Green bars are the stock level at the customer location at the end of the period.

4.3 Results

Initial Solutions. Fig. 2 presents typical output of the frequency policies. The top left black dot is the day-to-day delivery policy, which clearly minimizes the inventory cost but has a routing cost which is important. Black dot on the bottom right represents the other alternative which apply the order-up-to-level policy. A large cloud of small

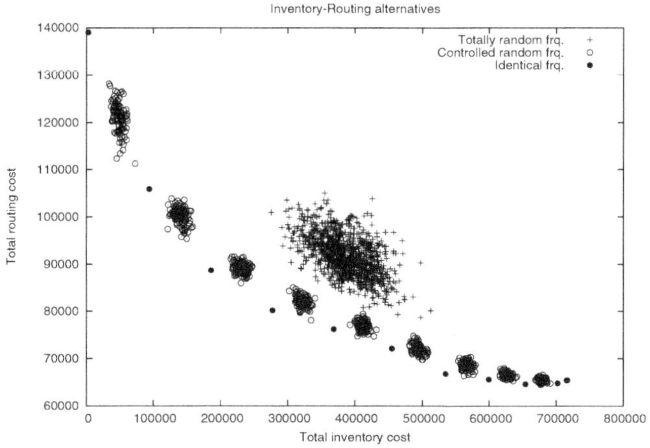

Fig. 2. First approximation using identical and randomized frequencies

crosses in the center results from a totally random frequency strategy. The black dots represent the solutions when all customers are served with the same frequency. To fill the gaps, a controlled random frequency (random frequencies but between two consecutive frequency values) is used and produces the results represented by white circles.

Using the work done in [16], the routing cost can be improved. Fig. 3 shows the previous approximate Pareto-front resulting from the left figure with an improved routing. With low frequencies, the routing cost can be reduced greatly.

Results of Local Search. Running the multi-point hillclimber introduced in Section 3.2 on all elements of $\widehat{P}$ turns out to be not feasible for most of the proposed benchmark instances. On the one hand, the computations are too time consuming: Even when only making use of the savings approach for the vehicle routing part, $T = 240$ periods need to be solved. On the other hand, $\widehat{P}$ is quickly populated with several thousands of solutions, and the computational effort for checking the neighborhood of all these solutions grows too large.

Fig. 4 shows the obtained set of local optima for the smallest instance, scenario a, `GS-01-a.irp`. While the results seem to form a line/ Pareto-front, they are, indeed, 2485 discrete alternatives.

4.4 The Decision Maker's Point of View

Of course, from a theoretical point of view, having the previous results are good, even excellent. Producing 2485 alternatives ensures us to have a better coverage of the Pareto front and hence not miss any of the good solutions. But, from the decision maker point of view, this is also a very important drawback. The decision maker would wish to choose one alternative among a few, not among too many.

We have then oriented our research towards the selection of representative alternatives. This is a very complicated task and can lead to errors or bad solutions. It is the reason why this specific study is still a prospective research part and will be improve in the future.

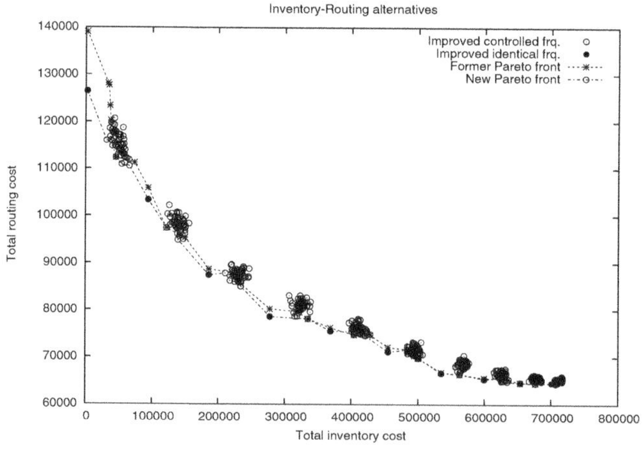

Fig. 3. Improvements by means of the record-to-record travel algorithm

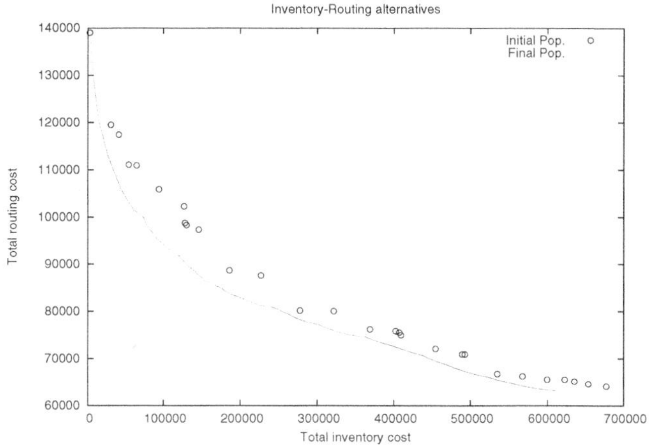

Fig. 4. Results for instance `GS-01-a.irp`. Initial population and final set of local optima after applying local search.

The Global Strategy. The global strategy that we have implemented is the following. Among the initial population, we select a subset of solutions that, in our opinion, is a representative subset of the alternatives at the current stage. The number of solutions in this subset is a parameter that the decision maker will be able to adjust. From this subset, we start the local search on each of the alternatives and improve the solutions. We keep in an archive all the non-dominated solutions produced during this second phase. The local search is run until no more improvements are found.

With this new subset of alternatives, we start again the procedure by selecting a subset of alternatives and by again running the local search over these solutions. This methods converges rather quickly (in the number of steps) but is still time consuming. Nevertheless, it comes closer to an acceptable running time for a decision maker and moreover to a good final number of alternatives for the decision maker. This will be explained with the help of Fig. 5 in the sequel.

The subset of alternatives that we select is part of the critical step. We use the reference point technique that seems to be the most appropriate in that kind of situations. To make the technique consistent, the two objectives have been normalized. Fig. 5 depicts a typical situation. The complete set of non-dominated solutions is represented by the white circles and the two extreme solutions by black squares (the two solutions surrounded by circles at the top left corner and the bottom right corner). The two extreme solutions plus the "ideal point" (the solution circled in the bottom left corner of the figure) having the best value in terms of routing cost and inventory cost compose the three initial reference points.

Then, depending on the number of final alternatives that the decision maker wants, we split the two axes (represented by the dotted line between circles) in equal proportions. In the case presented in Fig. 5, the two axes have been split in 5, generating four additional references points for the the vertical axe and four additional reference points for the horizontal axe. These new points are represented on the figure by the black

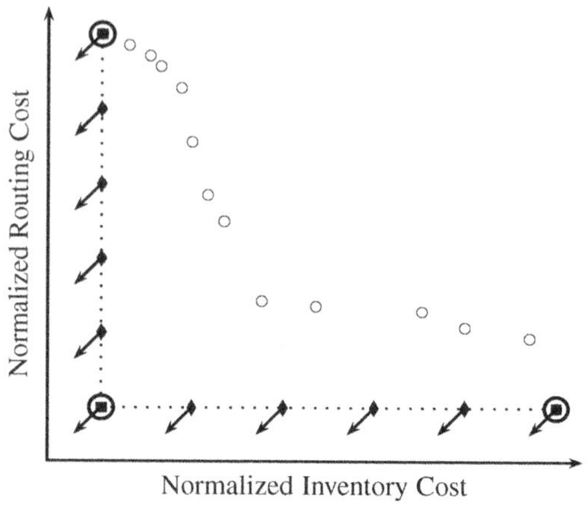

Fig. 5. The reference set point selection and directions of search

diamond. With each of the 11 reference points, we will select in the complete solution set, the solution that minimizes the Chebyshef distance metric. This will give us at most 11 solutions that will be used in the sequel for improvement as mentioned in Section 3.2. The direction of search is indicated on the figure with the arrows.

Further Experiments. To see how the search evolves, the different steps for a single run over instance `GS-01-a.irp` are reported int Table 1. The first column are the different steps identification (step 0 stands for the initial solutions). The second column

Table 1. Evolution of the search for instance `GS-01-a.irp`

Steps	# evaluations	Subset size	Cumulative CPU (s)
0	235	28	34.79
1	1233	105	221.29
2	2179	150	387.58
3	3022	173	535.42
4	3962	198	699.27
5	4799	200	844.52
6	5733	221	1006.00
7	6564	241	1152.02
8	7320	243	1276.47
9	8149	258	1424.31
10	9074	261	1592.00
11	9765	275	1702.35
12	10355	280	1795.94
13	10944	275	1883.19
14	11532	272	1977.23
15	11819	268	2035.96
16	12105	274	2094.83
17	12390	272	2153.83
18	12574	269	2205.63
19	12660	272	2232.77

reports the cumulative number of evaluation done during the search. The third column represents the size of the subset before the local search improvement phase. And the last column reports the cumulative CPU time (in seconds) of the different steps.

We can see that at step 0, *i.e.* we have generated 28 solutions. Among these solutions, a maximum subset of 11 solutions are selected and improved by the local search phase producing a subset of 105 non-dominated solutions for the next step. Again, from the 105 solutions, we selected 11 representatives and improve them with the local search producing 150 solutions. At the end, after 19 steps, the set of non-dominated solutions comprises 272 solutions from which we can select only 11 to present to the decision maker.

As one can see, the cumulative CPU time is still large but remain acceptable since this problem is solved in a strategic phase for the company. As mentioned before, the number of reference points selected by the decision maker is the parameter that can be adapted. It clearly influences the running time of the algorithm as shown in Table 2.

Table 2. Evolution of running time with the number of reference points for instance `GS-01-a.irp`

# RP	# Steps	# evaluations	Subset size	Cumulative CPU (s)
3	14	2885	113	360.30
5	19	5429	159	802.59
11	19	12660	272	2232.77

Another good result from the method is that the quality does not depend of the number of alternatives we are going to present to the decision maker. On the same instance `GS-01-a.irp`, we draw the final approximate Pareto sets obtained at the end of the

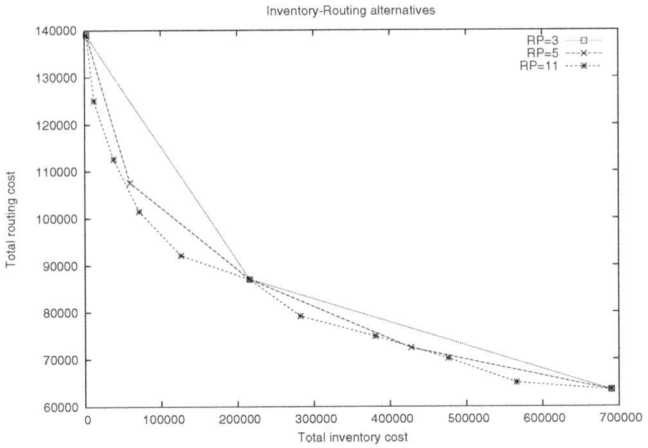

Fig. 6. Comparison of the final alternatives for instance `GS-01-a.irp`

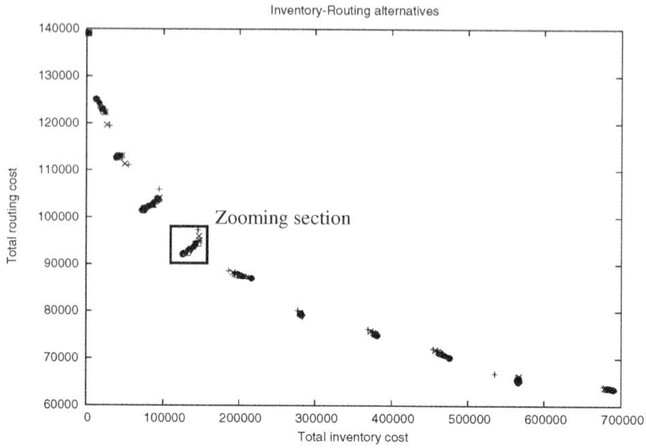

Fig. 7. Complete display of the subsets over the search for instance GS-01-a.irp and the zooming section

search with different reference set sizes. The three solutions proposed when the reference set size is 3 (RF=3) on the figure are included in the two other sets. Other solutions of R=5 are not dominated by the solutions proposed when RF=11. We also expect to obtain these results with the complete set of instances.

Finally, it is interesting to see the evolution of the search over the different steps. The complete figure showing the total set of generated solutions during the search is not worth to display since too many solutions are close to each other making the figure overloaded. Instead, we display only the subsets of solutions that we select over the search. For the same instance GS-01-a.irp, they are represented on Fig. 7. Because the figure contains different alternatives, it is necessary to have a closer look. We decided to zoom a specific section showed in Fig. 7.

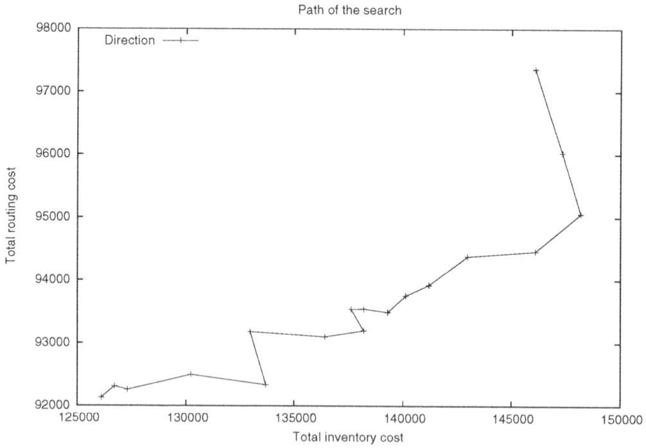

Fig. 8. Path of the local search for instance GS-01-a.irp

The result is shown in Fig. 8. On purpose, we have drawn the different solutions with a line, showing the path followed by the search over the different steps. Again, the conclusion drawn from this figure is that the solutions generated over time are not dominated by the ones of the previous iterations (otherwise they would have been eliminated from the search). Moreover, the improvement is done on both objectives, following more or less the direction imposed by the weighted Chebyshef distance measure.

5 Conclusions

In this paper, we have presented a practical bi-objective framework for solving the inventory routing problem. We have generated a set of new instances that are available on the internet. Preliminary experiments indicate the good general impression of the solving method and need to be confirmed by further experiments. Nevertheless, with the reference set strategy, we have been able to overcome the long computational running times, and at the same time the too numerous solutions that need to be presented to the decision maker at the end.

References

1. Abdelmaguid, T.F., Dessouky, M.M., Ordóñez, F.: Heuristic approaches for the inventory-routing problem with backlogging. Computers & Industrial Engineering 56(4), 1519–1534 (2009)
2. Archetti, C., Bertazzi, L., Laporte, G., Speranza, M.G.: A branch-and-cut algorithm for a vendor managed inventory routing problem. Transportation Science 41(3), 382–391 (2007)
3. Baita, F., Ukovich, W., Pesenti, R., Favaretto, D.: Dynamic routing-and-inventory problems: A review. Transportation Research-A 32(8), 585–598 (1998)
4. Barnes-Schuster, D., Bassok, Y.: Direct shipping and the dynamic single-depot/multi-retailer inventory system. European Journal of Operational Research 101(3), 509–518 (1997)
5. Bertazzi, L., Paletta, G., Speranza, M.G.: Deterministic order-up-to level policies in an inventory routing problem. Transportation Science 36(1), 119–132 (2002)
6. Bertazzi, L., Savelsbergh, M., Speranza, M.G.: Inventory routing. In: Golden, B., Raghavan, S., Wasil, E. (eds.) The Vehicle Routing Problem: Latest Advances and New Challenges. Operations Research/Computer Science Interfaces, vol. 43, pp. 49–72. Springer Science+Business Media, LLC, Heidelberg (2008)
7. Boudia, M., Prins, C.: A memetic algorithm with dynamic population management for an integrated production-distribution problem. European Journal of Operational Research 195(3), 703–715 (2009)
8. Campbell, A.M., Savelsbergh, M.W.P.: A decomposition approach for the inventory-routing problem. Transportation Science 38(4), 488–502 (2004)
9. Campbell, A.M., Savelsbergh, M.W.P.: Delivery volume optimization. Transportation Science 38(2), 210–223 (2004)
10. Campbell, A.M., Savelsbergh, M.W.P.: Efficient insertion heuristics for vehicle routing and scheduling problems. Transportation Science 38(3), 369–378 (2004)
11. Christofides, N., Mingozzi, A., Toth, P., Sandi, C.: Combinatorial optimization. John Wiley, Chichester (1979)
12. Clarke, G., Wright, J.W.: Scheduling of vehicles from a central depot to a number of delivery points. Operations Research 12(4), 568–581 (1964)

13. Day, J.M., Wright, P.D., Schoenherr, T., Venkataramanan, M., Gaudette, K.: Improving routing and scheduling decisions at a distributor of industrial glass. Omega 37(1), 227–237 (2009)
14. Kleywegt, A.J., Nori, V.S., Savelsbergh, M.W.P.: The stochastic inventory routing problem with direct deliveries. Transportation Science 36(1), 94–118 (2002)
15. Lee, K.K.: Fuzzy rule generation for adaptive scheduling in a dynamic manufacturing environment. Applied Soft Computing 8(4), 1295–1304 (2008)
16. Li, F., Golden, B., Wasil, E.: A record-to-record travel algorithm for solving the heterogeneous fleet vehicle routing problem. Computers & Operations Research 34(9), 2734–2742 (2007)
17. Li, J.A., Wu, Y., Lai, K.K., Liu, K.: Replenishment routing problems between a single supplier and multiple retailers with direct deliveries. European Journal of Operational Research 190(2), 412–420 (2008)
18. Moin, N.H., Salhi, S.: Inventory routing problems: a logistical overview. Journal of the Operational Research Society 58(9), 1185–1194 (2007)
19. Raa, B.: New models and algorithms for the cyclic inventory routing problem. 4OR 6(1), 97–100 (2008)
20. Raa, B., Aghezzaf, E.H.: A practical solution approach for the cyclic inventory routing problem. European Journal of Operational Research 192(2), 429–441 (2009)
21. Sevaux, M., Geiger, M.J.: Inventory routing and on-line inventory routing file format. Research Report RR-11-01-01, Helmut-Schmidt-University, University of the Federal Armed Forces Hamburg, Hamburg, Germany (January 2011)
22. Song, J.H., Salvelsbergh, M.: Performance measurement for inventory routing. Transportation Science 41(1), 44–54 (2007)
23. Yu, Y., Chen, H., Chu, F.: A new model and hybrid approach for larger scale inventory routing problems. European Journal of Operational Research 189(3), 1022–1040 (2008)

Problem Transformations for Vehicle Routing and Scheduling in the European Union

Asvin Goel

Zaragoza Logistics Center, Spain
agoel@zlc.edu.es
Applied Telematics/e-Business Group, Department of Computer Science,
University of Leipzig, Germany
asvin.goel@uni-leipzig.de

Abstract. In the European Union working hours of truck drivers must comply with regulation (EC) No 561/2006 which entered into force in April 2007. The regulation has a significant impact on travel times including the driving time and the time required for compulsory breaks and rest periods. Recently, several approaches for solving vehicle routing and scheduling problems have been proposed in which European Union legislation must be complied with. All of these approaches restrict the application domain by constraining the maximum amounts of driving time and working time by the "weekly" limits of 56 hours and 60 hours imposed by regulation (EC) No 561/2006. In this paper it is shown that the amounts of driving time and working time a driver may accumulate within a period of six days can be significantly higher due to inconsistent definitions in the regulation. Problem transformation rules are presented which can be embedded in the previously developed approaches to exploit these inconsistencies.

1 Introduction

The vehicle routing problem is the problem of determining a set of vehicle routes to service a set of customers at minimal costs. There are many applications of the vehicle routing problem and a multitude of variants of the classical vehicle routing problem have been proposed considering various problem characteristics found in real-life applications. Many of these variants are discussed in [5]. The scope of this paper is an extension of the vehicle routing problem in which each customer must be visited within a given time window and working hours of truck drivers must comply with applicable legislation. In the European Union regulation (EC) No 561/2006 regulates working hours of truck drivers. The combined vehicle routing and truck driver scheduling problem in the European Union has been introduced by [1] and studied by [3] and [4].

The most prominent rules of regulation (EC) No 561/2006 are that a truck driver may not drive for more than four and a half hours without taking a break period of at least 45 minutes, and that a driver may not drive for more than 9 hours without taking a daily rest period of at least 11 hours. The regulation

J. Pahl, T. Reiners, and S. Voß (Eds.): INOC 2011, LNCS 6701, pp. 379–384, 2011.

defines a *week* as the period of time between 00.00 on Monday and 24.00 on Sunday and a *weekly rest period* as a period of rest of at least 45 hours. Some of the constraints of the regulation are imposed on the time between two weekly rest periods and some on the time during a week. A driver may drive for up to 56 hours and work for up to 60 hours during a week. At most 144 hours (six days) may elapse between the end of a weekly rest period and the start of the next weekly rest period. Twice a week, a driver may drive up to 10 hours without taking a daily rest period. Three times in between two weekly rest periods a driver may take a reduced daily rest period of at least 9 hours. It most be noted that several other constraints are imposed by the regulation. For brevity and w.l.o.g. these other constraints will not be considered in the remainder of this paper. For a comprehensive discussion of the entire set of rules imposed by regulation (EC) No 561/2006 the reader is referred to [2].

If we only consider a planning horizon of six days which does not include the night from Sunday to Monday, the inconsistent definition regarding weeks and the time between two weekly rest periods has no ramification on the feasibility check of a vehicle route. The heuristics for the combined vehicle routing and truck driver scheduling problem presented by [1], [3] and [4] as well as the exact truck driver scheduling method presented by [2] take advantage of this and make such a restriction.

In the general case, however, the planning horizon may begin at any time of the week. Let us for example consider a planning horizon of six days starting at 0.00 on Thursday. The driver may have up to 7 daily driving periods and 6 daily rest periods within the planning horizon. Without making use of the options to extend the daily driving time or reduce the daily rest, the driver can drive up to $7 \cdot 9 = 63$ hours (the total amount of break required is $7 \cdot \frac{3}{4} = 5\frac{1}{4}$ hours and the total amount of rest is $6 \cdot 11 = 66$ hours). As the planning horizon ranges across two weeks, the driver may have four extended daily driving times of 10 hours: two in the first week and two in the second week. Thus, a total of $4 \cdot 10 + 3 \cdot 9 = 67$ hours of driving can be accumulated (the total amount of break required is $4 \cdot (2 \cdot \frac{3}{4}) + 3 \cdot \frac{3}{4} = 8\frac{1}{4}$ hours and the minimum amount of rest required is $3 \cdot 9 + 3 \cdot 11 = 60$ hours). The "weekly" driving limit of 56 hours imposed by the regulation can therefore be exceeded by almost 20 percent. Similarly, the accumulated amount of working time within six days can exceed the "weekly" working limit of 60 hours significantly. In the next section we see how we can check feasibility of a vehicle route with a planning horizon starting at any time of the week.

2 Problem Transformation

The approaches for combined vehicle routing and truck driver scheduling presented by [1] and studied by [3] and [4] consist of a heuristic framework to determine potential vehicle routes and a feasibility check determining whether all customers in the route can be visited within the given time windows and without violating the regulation. Let us denote with $n_1, n_2, \ldots, n_\lambda$ the locations of a vehicle route and with $\delta_{\mu,\mu+1}$ the driving time required for moving from a node n_μ to a node $n_{\mu+1}$. At each location n_μ some stationary work of duration w_μ shall be conducted. This work shall begin within a given time window denoted by $[t_\mu^{\min}, t_\mu^{\max}]$. Let s_0 denote the state of the driver at the beginning of the planning horizon of six days. With each period of driving, conducting other work, taking a break or rest, or waiting idle, the state of the driver is changed. The *European Union truck driver scheduling problem (EU-TDSP)* is the problem of determining whether state s_0 can be successively changed into a state s with the following characteristics

1. the driver has visited λ locations
2. at the μth location the driver conducted some stationary work of duration w_μ
3. the stationary work at location n_μ started within time window $[t_\mu^{\min}, t_\mu^{\max}]$
4. the total amount of driving between location n_μ and $n_{\mu+1}$ is $\delta_{\mu,\mu+1}$
5. the driver complies with regulation (EC) No 561/2006

A mathematical formulation of the EU-TDSP including all the constraints of the regulation (EC) No 561/2006 is given by [2] who also presented an exact approach that can be used if

$$\sum_{\mu=1}^{\mu<\lambda} \delta_{\mu,\mu+1} \leq 56 \text{ and } \sum_{\mu=1}^{\mu<\lambda} \delta_{\mu,\mu+1} + \sum_{\mu=1}^{\mu\leq\lambda} w_\mu \leq 60.$$

The same assumption has been made by [1], [3], and [4]. All of these approaches determine a set of labels for each location in the tour representing different driver states. If this assumption can not be made, the approaches may determine labels representing driver states violating the regulation. However, we can transform the problem representation in such a way that these approaches can still be used. For this we have to make sure that the accumulated amount of driving time and the accumulated amount of working time are not exceeded in both weeks that may belong to the planning horizon.

Let λ' be the index for which

$$\sum_{\mu=1}^{\mu<\lambda'} \delta_{\mu,\mu+1} \leq 56 \text{ and } \sum_{\mu=1}^{\mu<\lambda'+1} \delta_{\mu,\mu+1} > 56.$$

Then, the accumulated amount of driving time exceeds 56 hours on the trip from location $n_{\lambda'}$ to location $n_{\lambda'+1}$. We have to make sure that the accumulated amount of driving time in the first week of the planning horizon does not exceed 56 hours. This can be achieved by inserting a virtual location n' between locations $n_{\lambda'}$ and $n_{\lambda'+1}$. The time window of this location begins at the beginning of the second week and ends at the end of the planning horizon. The working time at location n' is set to zero, the driving time from $n_{\lambda'}$ to n' is set to $56 - \sum_{\mu=1}^{\mu<\lambda'} \delta_{\mu,\mu+1}$, and the driving time from n' to $n_{\lambda'+1}$ is set to $\sum_{\mu=1}^{\mu<\lambda'+1} \delta_{\mu,\mu+1} - 56$. By this, we ensure that the accumulated amount of driving in the first week cannot exceed 56 hours if all time window restrictions are satisfied.

Analogously, let λ'' be the index for which

$$\sum_{\mu=\lambda''}^{\mu<\lambda} \delta_{\mu,\mu+1} > 56 \text{ and } \sum_{\mu=\lambda''+1}^{\mu<\lambda} \delta_{\mu,\mu+1} \leq 56.$$

Then, the accumulated amount of driving time (when counting backwards) exceeds 56 hours on the trip from location $n_{\lambda'}$ to location $n_{\lambda'+1}$. We have to make sure that the accumulated amount of driving time in the second week of the planning horizon does not exceed 56 hours. This can be achieved by inserting a virtual location n'' between locations $n_{\lambda''}$ and $n_{\lambda''+1}$. The time window of this location starts at the beginning of the planning horizon and ends at the end of the first week. The working time at location n'' is set to zero, the driving time from $n_{\lambda''}$ to n'' is set to $\sum_{\mu=\lambda''}^{\mu<\lambda} \delta_{\mu,\mu+1} - 56$, and the driving time from n'' to $n_{\lambda''+1}$ is set to $56 - \sum_{\mu=\lambda''+1}^{\mu<\lambda'} \delta_{\mu,\mu+1}$.

By inserting the virtual locations n' and n'' in the route of a vehicle we make sure that the accumulated amount of driving time does not exceed the limit imposed for either week of the planning horizon. Similarly we can make sure that accumulated amount of working time does not exceed the limit imposed for either week of the planning horizon.

Let λ''' be the index for which

$$\sum_{\mu=1}^{\mu<\lambda'''} \delta_{\mu,\mu+1} + \sum_{\mu=1}^{\mu\leq\lambda'''} w_\mu \leq 60 \text{ and } \sum_{\mu=1}^{\mu<\lambda'''+1} \delta_{\mu,\mu+1} + \sum_{\mu=1}^{\mu\leq\lambda'''+1} w_\mu > 60.$$

If $\sum_{\mu=1}^{\mu<\lambda'''} \delta_{\mu,\mu+1} + \sum_{\mu=1}^{\mu\leq\lambda'''} w_\mu + \delta_{\lambda''',\lambda'''+1} > 60$ then the accumulated amount of working time exceeds 60 hours on the trip from location $n_{\lambda'''}$ to $n_{\lambda'''+1}$ and we insert a virtual location n''' between these locations. The time window of this location begins at the beginning of the second week and ends at the end of the planning horizon. The working time at location n''' is set to zero, the driving time from $n_{\lambda'''}$ to n''' is set to $60 - (\sum_{\mu=1}^{\mu<\lambda'''} \delta_{\mu,\mu+1} + \sum_{\mu=1}^{\mu\leq\lambda'''} w_\mu)$, and the driving time from n' to $n_{\lambda'+1}$ is set to $\sum_{\mu=1}^{\mu<\lambda'''} \delta_{\mu,\mu+1} + \sum_{\mu=1}^{\mu\leq\lambda'''} w_\mu + \delta_{\lambda''',\lambda'''+1} - 60$. If $\sum_{\mu=1}^{\mu<\lambda'''} \delta_{\mu,\mu+1} + \sum_{\mu=1}^{\mu\leq\lambda'''} w_\mu + \delta_{\lambda''',\lambda'''+1} \leq 60$ then the limit of 60 hours is reached during the stationary work conducted at location $n_{\lambda'''+1}$. Instead of inserting a virtual location n''' between locations $n_{\lambda'''}$ and $n_{\lambda'''+1}$, we increase (if necessary) the beginning of the time window of location $n_{\lambda'''+1}$ to the beginning of the second week reduced by $60 - (\sum_{\mu=1}^{\mu<\lambda'''} \delta_{\mu,\mu+1} + \sum_{\mu=1}^{\mu\leq\lambda'''} w_\mu + \delta_{\lambda''',\lambda'''+1})$.

Analogously, let λ'''' be the index for which

$$\sum_{\mu=\lambda''''}^{\mu<\lambda} \delta_{\mu,\mu+1} + \sum_{\mu=\lambda''''}^{\mu\leq\lambda} w_\mu > 60 \text{ and } \sum_{\mu=\lambda''''+1}^{\mu<\lambda} \delta_{\mu,\mu+1} + \sum_{\mu=\lambda''''+1}^{\mu\leq\lambda} w_\mu \leq 60.$$

If $\sum_{\mu=\lambda''''+1}^{\mu<\lambda} \delta_{\mu,\mu+1} + \sum_{\mu=\lambda''''+1}^{\mu\leq\lambda} w_\mu + \delta_{\lambda'''',\lambda''''+1} > 60$ then we insert a virtual location n'''' between locations $n_{\lambda''''}$ and $n_{\lambda''''+1}$. The time window of this location starts at the beginning of the planning horizon and ends at the end of the first week. The working time at location n'''' is set to zero, the driving time from $n_{\lambda''''}$ to n'''' is set to $\sum_{\mu=\lambda''''+1}^{\mu<\lambda} \delta_{\mu,\mu+1} + \sum_{\mu=\lambda''''+1}^{\mu\leq\lambda} w_\mu + \delta_{\lambda'''',\lambda''''+1} - 60$, and the driving time from n'''' to $n_{\lambda''''+1}$ is set to $60 - (\sum_{\mu=\lambda''''+1}^{\mu<\lambda} \delta_{\mu,\mu+1} + \sum_{\mu=\lambda''''+1}^{\mu\leq\lambda} w_\mu)$. If $\sum_{\mu=\lambda''''+1}^{\mu<\lambda} \delta_{\mu,\mu+1} + \sum_{\mu=\lambda''''+1}^{\mu\leq\lambda} w_\mu + \delta_{\lambda'''',\lambda''''+1} \leq 60$ then the limit of 60 hours is reached (when counting backwards) during the stationary work conducted at location $n_{\lambda''''}$. Instead of inserting a virtual location n'''' between locations $n_{\lambda''''}$ and $n_{\lambda''''+1}$, we decrease (if necessary) the end of the time window of location $n_{\lambda''''}$ to the end of the first week reduced by $w_{\lambda''''} - (60 - (\sum_{\mu=\lambda''''+1}^{\mu<\lambda} \delta_{\mu,\mu+1} + \sum_{\mu=\lambda''''+1}^{\mu\leq\lambda} w_\mu + \delta_{\lambda'''',\lambda''''+1}))$.

By making these modifications to the problem representation before invoking the feasibility check presented by [1], [2], [3], or [4] we can make sure that the accumulated amounts of driving time and working time do not exceed the limit imposed for either week of the planning horizon even if conditions (1) and (2) are not satisfied. By this the inconsistencies in the time frames used by the regulation can be exploited without the need to modify the existing methods for checking feasibility of a vehicle route. [3] and [4] also presented variants of their methods which make use of the provision of the regulation that a driver may drive up to 10 hours twice a week without a daily rest period. If the planning horizon ranges across two weeks the driver may drive up to 10 hours four times within the planning horizon. To fully consider this provision of the regulation, the number of extended daily driving times can be increased to four and the approaches need to verify that only two extended daily driving times are are used in each week. If necessary, the "weekly" limits can be adjusted to consider previous activities conducted by the driver and the resulting impact on the maximum amount of driving and working time within the planning horizon.

3 Conclusions

In the European Union a truck driver may only accumulate 56 hours of driving time and 60 hours of working time within a week. This paper shows that, due to inconsistent definitions of the regulation, the amount of driving and working time within a period of six days can be significantly higher if the planning horizon ranges across two weeks. Although driving on weekends is restricted in some member states, there are various exemptions from weekend driving bans. The significant increase in accumulated driving and working time may promote driving on weekends.

Recently, several approaches for combined vehicle routing and truck driver scheduling in the European Union have been proposed. These approaches only consider a planning horizon starting on Monday or later and ending on Sunday of the same week or earlier. This paper shows that the feasibility check of these approaches can also be used for a planning horizon starting on any day of the week. This only requires relatively simple modifications of the problem representation.

References

1. Goel, A.: Vehicle scheduling and routing with drivers' working hours. Transportation Science 43(1), 17–26 (2009)
2. Goel, A.: Truck Driver Scheduling in the European Union. Transportation Science 44(4), 429–441 (2010)
3. Kok, A.L., Meyer, C.M., Kopfer, H., Schutten, J.M.J.: A Dynamic Programming Heuristic for the Vehicle Routing Problem with Time Windows and European Community Social Legislation. Transportation Science 44(4), 442–454 (2010)
4. Prescott-Gagnon, E., Desaulniers, G., Drexl, M., Rousseau, L.M.: European Driver Rules in Vehicle Routing with Time Windows. Transportation Science 44(4), 455–473 (2010)
5. Toth, P., Vigo, D.: The Vehicle Routing Problem, Philadelphia. SIAM Monographs on Discrete Mathematics and Applications (2002)

New Models for and Numerical Tests of the Hamiltonian p-Median Problem

Stefan Gollowitzer[1], Dilson Lucas Pereira[2], and Adam Wojciechowski[3]

[1] Department of Statistics and Operations Research, University of Vienna, Austria
stefan.gollowitzer@univie.ac.at
[2] Universidade Federal de Minas Gerais, Brasil
dilsonlucas@gmail.com
[3] Chalmers University of Technology, Gothenburg, Sweden
wojcadam@chalmers.se

Abstract. The Hamiltonian p-median problem (HpMP) was introduced by [Branco90]. It is closely related to two well-known problems, namely the Travelling Salesman problem (TSP) and the Vehicle Routing problem (VRP). The HpMP is to find exactly p node-disjoint cycles of minimum edge cost, such that each node of the graph is contained in exactly one cycle. We present three new models for the HpMP problem which differ with regard to the constraints that enforce a maximum number of cycles. We demonstrate that one of the models (SEC) is dominated by another model (PCON) with regard to the LP relaxation. Further, we introduce a class of symmetry breaking constraints. We present results regarding the quality of the lower bounds provided by the respective LP relaxations for two of the models, and provide computational results that demonstrate the computational efficiency.

1 Introduction

We consider the Hamiltonian p-median problem (HpMP) on an undirected graph. It is a generalization of a well-known problem, namely the Travelling Salesman problem (TSP). This problem has been studied in numerous variants, with single and multiple salesmen, time-dependent, capacitated and many more. Considering the HpMP as a generalization of the TSP it is the following: For p salesmen without a given start position, find a tour for each of them such that each city is visited by exactly one salesman and the distance of all tours is minimal.

The HpMP is also connected to a generalized Vehicle Routing problem (VRP) which allows several depots and in which we assume distinct depots for all vehicles in a fleet of size p, and require every customer to be visited by exactly one vehicle. We obtain the HpMP by considering the problem of simultaneously locating p depots and considering the sum of costs for routing the vehicles from all depots.

In this article, we consider the HpMP on an undirected graph and assume a minimum tour length of 3 nodes.

Definition 1 (HpMP). *Given an undirected graph $G = (V,E)$, find exactly p node-disjoint cycles of minimum edge cost, such that each node is contained in exactly one cycle and each cycle contains at least 3 nodes.*

J. Pahl, T. Reiners, and S. Voß (Eds.): INOC 2011, LNCS 6701, pp. 385–394, 2011.

As this problem generalizes an NP-hard problem, it is itself NP-hard in the general case. Furthermore, [**?**] shows that the HpMP is NP-hard for each p.

A relaxation of the HpMP is the problem of partitioning the nodes into (any number of) disjoint cycles such that the total distance is minimized. This is a relaxation of the TSP obtained by removing the subtour elimination constraints and is known as the 2-matching problem. We obtain a description of the convex hull of the 2-matching problem by adding the 2-matching inequalities (see for instance [NW99, p. 276]) and the problem can hence be solved in polynomial time. In order to obtain exactly p cycles, we have to introduce constraints that enforce this.

There are several possible applications of the HpMP. Consider, for instance, the assignment of p guards to n objects and assume that each guard cycles among the assigned protection objects. A similar problem is the assignment of p maintenance/inspection vehicles that need to maintain/inspect n machines. [Glaab98] present an application where p lasers are used to visualize a set of cuts to be performed on leather. In all of these applications, however, it might be more adequate to minimize (or put a bound on) the longest cycle. This is an extension of the HpMP which we will consider in future research.

Literature: The HpMP belongs to a class of problems denoted location routing problems (LRPs, see [nagy2007] for a survey). These problems combine the problem of facility location with the problem of finding an optimal routing. An early work on a LRP closely related to the HpMP is a paper by [Laporte83]. The article contains, among others, a mixed integer linear program (MILP) for the problem of finding at most p cycles of minimum cost.

The HpMP was first introduced by [Branco90]. The authors consider the HpMP on a directed graph and present two models and several heuristics for the problem. The first model (P1) is a partitioning problem where the nodes are partitioned into p subsets. For each set of each partition the TSP solution is assumed to be given as data. The second model (P2) is more similar to the models presented here. One variable is assigned to each combination of edge and cycle, and vertex and cycle of the graph. The number of variables is therefore $|E|p + |V|p$. The constraints that guarantee a maximum number of cycles (7) depend on the variables and the model can therefore not be formulated as a MILP. [Glaab00] introduce a model where the variables corresponding to the nodes are dropped and constraints that guarantee the maximum number of cycles based on partitions are suggested. Finally, [Zohrehbandian07] presents a model of the HpMP as an extended VRP where a virtual depot is added for each cycle. The number of variables is $2|E|p + |V|$. The models for the HpMP presented in our paper have one variable for each edge, and for each combination of node and cycle, hence the number of variables is $|E| + p|V|$.

A related problem to the HpMP is the subject of [Bauer02]. Here the problem is to find a minimum cost cycle with a constraint on the number of edges denoted the cardinality constrained circuit problem. The authors present mixed integer programming formulations based on connectivity, discuss a number of facet defining inequalities and devise separation algorithms for them.

Finally, a topic which is relevant to the HpMP is the question of how many nodes in a given graph can be covered by p node disjoint cycles, and is the subject of a publication by [Wong03]. Given a graph with $n \geq 3p$ nodes and minimum degree $d \geq 2p$, the authors prove that the graph contains p node disjoint cycles that cover at least $\min\{2d,n\}$ nodes. In the article the authors also present counter examples that show that this is a minimum requirement on the node degree. This implies that we have to require a node degree of at least $\max(n/2,2p)$ to ensure that a feasible solution of the HpMP exists.

2 MILP Models

We present three new MILP models for the HpMP that differ with respect to the constraints that enforce a maximum number of cycles. The first model, SEC, uses an extension of the subtour eliminating constraints of the TSP. The second model, PCON, uses a class of connectivity constraints based on node partitions. The connectivity constraints in PCON dominate the constraints of SEC. The third model NCON uses a class of node based connectivity constraints. We demonstrate that neither PCON nor NCON is dominant. Constraints similar to (SEC) and (PCON) were both introduced in [Glaab00]. The constraints (NCON) are a generalization of the connectivity based approach used in [Bauer02].

2.1 Notation

For a graph $G=(V,E)$ the set of edges is given as $E=\{e=ij \mid i,j\in V, i<j\}$. We write the set of adjacent edges of node $i\in V$ as $\delta(i) := \{e=(j,k)\in E \mid i=j \vee i=k\}$. Furthermore, we define an index set for an integer k as $I_k := \{1,\dots,k\}$.

Let $C\subseteq E$ such that $|C|=|V|$ denote a subset of edges defining at least $k+1$ vertex-disjoint cycles covering every node of G, and $\mathbf{C}_{k+1}$ denote the set of all subsets C as just described. By P_C we denote the partition of the nodes in V into subsets containing exactly the nodes of one cycle in C respectively.

The set of partitions of the node set V into $k+1$ sets of size greater than or equal to 3 is denoted by

$$\mathbf{P}^3_{k+1} := \left\{ \{S_1,\dots,S_{k+1}\} \mid \bigcup_{i=1}^{k+1} S_i = V,\ |S_i|\geq 3,\ S_i\cap S_j=\emptyset\ \forall i\neq j,\ i,j\in I_{k+1} \right\}.$$

The set of edges between the sets of such a partition $P=\{S_1,\dots,S_{k+1}\}\in\mathbf{P}^3_{k+1}$ is given by

$$E_P := \{ij \mid i\in S_v,\ j\in S_w,\ S_v\neq S_w,\ S_v,S_w\in P\}.$$

We will use the following variables:

$$x_{ij} = \begin{cases} 1, & \text{if edge } ij \text{ is in the solution,} \\ 0, & \text{otherwise,} \end{cases} \qquad ij\in E,$$

$$z_i^m = \begin{cases} 1, & \text{if node } i \text{ is in cycle } m, \\ 0, & \text{otherwise,} \end{cases} \qquad i\in V,\ m\in I_k.$$

2.2 Subtour Elimination Constraints

Subtour elimination constraints have proven successful for solving problems of a number of classes, especially the TSP. Thus, using them for the subtour partitioning problem suggests itself. Formulation *SEC* is based on an adapted variant of the classical SECs.

$$\text{(SEC)} \quad \min \sum_{e \in E} c_e x_e$$

$$\sum_{m \in I_k} z_i^m = 1, \qquad i \in V, \tag{1a}$$

$$\sum_{i \in V} z_i^m \geq 3, \qquad m \in I_k, \tag{1b}$$

$$\sum_{e \in \delta(i)} x_e = 2, \qquad i \in V, \tag{1c}$$

$$\sum_{e \in C} x_e \leq |V| - 2, \quad C \in \mathbf{C}_{k+1}, \tag{SEC}$$

$$z_k^m + x_e \leq 1 + z_l^m, \quad e = ij \in E,\ \{k,l\} \in \{\{i,j\},\{j,i\}\},\ m \in I_k, \tag{1d}$$

$$x_e \in \{0,1\}, \quad e \in E, \tag{1e}$$

$$z_i^m \in \{0,1\}, \quad i \in V,\ m \in I_k. \tag{1f}$$

Equalities (1a) ensure that each node is in exactly one cycle. Inequalities (1b) ensure that each cycle contains at least three nodes. Constraints (1c) ensure that every node has 2 active incident edges. The subtour elimination constraints (SEC) eliminate solutions with more than k cycles. They limit the number of active edges in each subset of edges forming more than k cycles. Inequalities (1d) link variables x and z and thereby force a pair of nodes with an active edge between them to be in the same cycle.

2.3 Formulations Based on Connectivity Concepts

Another successful concept in the realms of network design problems (and others) is connectivity. We propose two different connectivity concepts for the HpMP, based on partitions and node variables respectively.

Partition Based Connectivity. We can replace constraints (SEC) from the first proposed formulation by the following inequalities, based on partitions of the node set of the underlying graph,

$$\sum_{e \in E_P} x_e \geq 2, \quad P \in \mathbf{P}_{k+1}^3. \tag{PCON}$$

We denote this formulation by $PCON$. The partition based connectivity constraints (PCON) make sure that for every partition of the node set into $k+1$ subsets, at least 2 edges between these sets are active, thereby ensuring that an optimal solution contains no more than k tours.

Denote by $\upsilon_{LP}(P)$ the optimal solution value of the linear programming relaxation of formulation P. Then we have the following

Lemma 1. *Constraints* (PCON) *dominate the subtour elimination constraints* (SEC), *i.e.*

$$\upsilon_{LP}(PCON) \geq \upsilon_{LP}(SEC)$$

Proof. By summing the constraints (1c) for all nodes we obtain

$$\sum_{e \in E} x_e = |V|. \tag{2}$$

Using equality (2), we can rewrite constraints (PCON) as follows:

$$|V| - 2 \geq \sum_{e \notin E_P} x_e \geq \sum_{e \in C} x_e \quad C \in \mathbf{C}_{k+1},\ P = P_C \tag{3}$$

The first inequality is equivalent to constraints (PCON). The second inequality holds because the edges not in E_{P_C} are a superset of those forming the cycles C. □

An example for which the second relation in (3) is strict is shown in Figure 1. In this example $k = 3$ and dashed edges denote corresponding variables values of 0.5. The partition constraint for

$$P = \{\{1,2,3\},\{4,\ldots,7\},\{8,\ldots,11\},\{12,\ldots,15\}\}$$

is violated but all constraints (SEC) are satisfied.

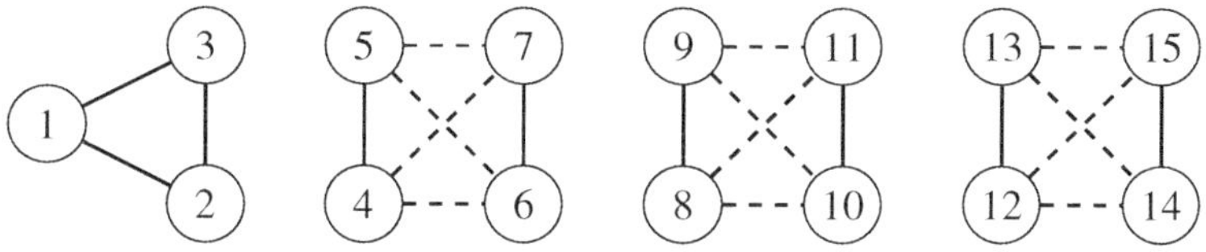

Fig. 1. Example for the proof of Lemma 1

Generalized Partition-Based Subtour Elimination Constraints We can generalize constraints (PCON) for partitions into more than $k+1$ sets:

$$\sum_{ij \in E_P} x_{ij} \geq 1 + u, \quad P \in \mathbf{P}^3_{k+u}. \tag{4}$$

This set of constraints contains inequalities (PCON) as special cases and therefore dominates them.

Node Based Connectivity. We can replace inequalities (PCON) by constraints that ensure connectivity between each two nodes that are in the same cycle:

$$\sum_{e \in \delta(S)} x_e \geq 2(z_i^m + z_j^m - 1), \quad S \subset V : 3 \leq |S| \leq |V| - 3, i \notin S, j \in S. \tag{NCON}$$

We refer to the formulation obtained by replacing (PCON) with (NCON) as *NCON*.

Lemma 2. *Formulations PCON and NCON are not comparable with respect to their LP lower bounds.*

Proof. Consider the graph $K_{15} = (V,E)$, i.e. a complete graph with $V = \{1,\dots,15\}$ and let $k = 3$. A feasible solution for the LP relaxation of NCON is the following: The active edges form cycles as in Figure 2a. The node variables have the following values: $z_i^1 = 1$ for $i \in \{1,\dots,3\}$, $z_i^2 = z_i^3 = 0.5$, for $i \in \{4,\dots,15\}$ and $z_i^m = 0$ otherwise. However, constraint (PCON) is violated for the partition $\{\{1,\dots,3\},\{4,\dots,7\},\{8,\dots,11\},\{12,\dots,15\}\}$.

Consider now the solution depicted in Figure 2b. Dashed edges denote corresponding variable values x_e of 0.5 and node variables have the values $z_i^1 = z_i^2 = 0.5$, for $i \in \{1,\dots,7\}$, $z_i^3 = 1$, for $i \in \{8,\dots,15\}$ and $z_i^m = 0$ otherwise. No partition of the nodes into more than 3 subsets can be found such that the respective constraint (PCON) is violated. However, constraint (NCON) is violated for $S = \{8,9,10,11\}$ and $i \in \{12,13,14,15\}$. □

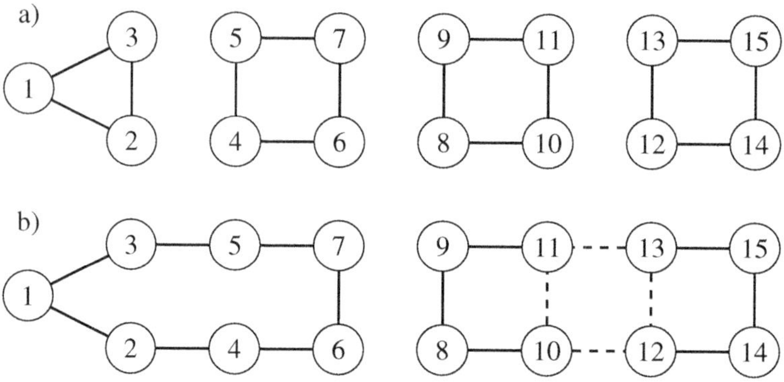

Fig. 2. Illustration of the examples used in the proof of Lemma 2

2.4 Symmetry Breaking and Valid Inequalities

By assigning nodes to numbered cycles our models allow a lot of symmetries. Especially in a branch-and-bound scheme this could lead to extremely long running times, as for each node up to k branches are needed. We eliminate symmetries using the following strategy. All cycles are indexed by the lowest index of all contained nodes.

$$z_i^m \leq \sum_{j=1}^{i-1} z_j^{m-1} \quad i = 3,\dots,n-3,\ m = 2,\dots,k. \tag{5}$$

Symmetry breaking constraints (5) state that, if for a given node i there is no node with a lower index in the cycle indexed by $m-1$, node i cannot be in cycle m. Accordingly, for a graph with node set $V = \{1,\dots,n\}$ node 1 will always be in cycle 1. Node 2 will either be in cycle 1 or 2. If node 2 is in cycle 1, node 3 should be in cycle 1 or 2 and so on. As a consequence we can fix a number of node variables z to zero.

$$z_i^m = 0 \quad i \in I_{k-1},\ m = i+1,\dots,k. \tag{6}$$

We indicate adding constraints (5) and (6) to a model M by writing M^+.

3 Computational Study

In this section we compare the models NCON, PCON, and the one comprising all constraints from both of these models. We denote this model by PNCON. We will investigate the importance of the symmetry breaking constraints (5) and (6). The models are compared with respect to running time, quality of the lower bound provided by the respective LP relaxation and the number of branch-and-bound nodes needed to solve the problems.

3.1 Branch-and-Cut Framework

We used the Branch-and-Bound framework provided by IBM CPLEX (version 12.2) and IBM Concert Technology. In the following we describe the separation of constraints (PCON) and (NCON) and a primal heuristic.

Separation of Inequalities (PCON). We compute the connected components of the support graph, i.e. the graph containing edges for which the corresponding variable is greater or equal to some ε in the current solution. If there are more then k connected components, there is a violated partitioning constraint for the partition into the node sets of the connected components. In this case, the constraint added states that the sum of the edges between the components should be greater than or equal to 2.

Note that this separation procedure is not exact. An exact separation would require setting $\varepsilon = 1$ and at this point we do not know how to choose the partition optimally such that the obtained partition based subtour elimination constraint is the most violated one.

Separation of Inequalities (NCON). Given k, the connectivity constraints state that the sum of the values of the edges in the cut $(S:\bar{S})$ has to be greater or equal $2(z_i^k + z_j^k - 1)$ for all pairs i and j such that $i \in S$ and $j \notin S$. Thus we calculate the minimum cut (=maximum flow) between the nodes in each such pair on the support graph. For each $m \in I_k$ we check whether constraint (NCON) is violated and, if that's the case, add it to the model.

Primal Heuristic. We use a simple primal heuristic to generate a starting solution before solving the model. It consists of the following steps:

1. Compute a minimum spanning tree T of G.
2. Remove the most expensive edges from T, such that k connected components are left.
3. For each component C with less than 3 vertices search the components with more than 3 vertices for the vertex closest to any of the nodes in C and add this vertex to C.
4. Run the nearest neighbour heuristic for the Travelling Salesman problem on the nodes of each component.
5. Number the cycles according to the rules specified in Section 2.4.

3.2 Instances

For our computational experiments we use randomly generated instances. The node were chosen randomly on a square of $[0,100]^2$ and as weight for each edge we used the Euclidean distance of the adjacent nodes. We generated 5 instances for each value of $|V|$ in $\{20,40,60,80\}$ respectively.

3.3 Results

All experiments were performed on a 2.67 GHz machine with 4 GB RAM. Each run was performed on a single processor.

LP bounds. We first test the strength of the proposed models by comparing the lower bounds provided by their LP relaxations. We test 6 settings. Models PCON, NCON and PNCON, with and without symmetry breaking constraints (5) and (6) respectively. We test on all 20 test instances with $k \in \{3,4,5\}$ and the 2 larger groups ($|V| \in \{60,80\}$) with $k = 10$. We compute LP gaps in Table 1 as $(OPT - v_{LP}(.))/OPT$.

The symmetry breaking constraints have a marginal effect on the LP gaps. For some test instances NPCON yields a better LP bound than NPCON$^+$. This is due to the fact that the symmetry breaking constraints affect which inequalities from set (PCON) are found to be violated.

Symmetry Breaking. To show the importance of the symmetry breaking constraints we compare the runtimes of the 3 model setups with and without symmetry breaking constraints. We test this only on the two smaller instance groups ($|V| \in \{20,40\}$) with $k \in \{3,4,5\}$. The reason is that the model (NCON) without symmetry breaking constraints was already unable to solve some instances in the group with $|V| = 40$ within the allotted time. The results of the tests are reported in Table 2.

Adding the symmetry breaking constraints improves the performance of NCON and PNCON. For PCON, we observe an improvement on the smaller test instances but a worsening on the larger test instances. This behaviour is not surprising as constraints (PCON) only involve variables linked to edges but not to cycle indices whereas constraints (NCON) share the symmetry in the cycle index with variables z.

Table 1. Results showing the mean LP gap in percent for NCON, PCON and NPCON with and without symmetry breaking constraints. * indicates that, for this group of test problems, some of the 5 instances were not solved within the allotted 1 h of cpu time. The LP bound for these instances was computed using the best feasible solution found (in all cases by PCON$^+$).

		NCON		PCON		NPCON	
$\|V\|$	k	sym.	no sym.	sym.	no sym.	sym.	no sym.
20	3	3.5	4.1	3.7	3.9	3.5	3.9
20	4	3.4	3.9	3.4	3.7	3.2	3.7
20	5	3.6	4.3	3.5	4.1	3.3	4.1
40	3	3.7	4.4	3.2	3.2	3.3	3.2
40	4	3.7	4.1	3.6	3.6	3.4	3.6
40	5	2.8	3.5	2.8	2.8	2.5	2.8
60	3	4.4	5.0	3.9	3.9	3.9	3.9
60	4	4.2	4.7	3.8	3.8	3.7	3.8
60	5	3.6	3.9	3.1	3.1	2.9	3.1
60	10*	4.0	4.1	3.3	3.3	3.5	3.3
80	3*	2.7	2.9	2.5	2.5	2.3	2.5
80	4*	3.3	3.4	2.9	2.9	3.0	2.9
80	5*	3.0	3.2	3.2	3.2	2.9	3.2
80	10*	5.9	6.0	6.0	6.0	5.9	6.0

Table 2. Results showing the mean and worst case cpu-times for the NCON, PCON and the NPCON with and without symetry breaking constraints. *1 out of 5 instances was not solved within the allotted 1 h cpu-time. †2 out of 5 instances were not solved within the allotted 1 h cpu-time.

		NCON				PCON				NPCON			
		sym.		no sym.		sym.		no sym.		sym.		no sym.	
$\|V\|$	k	mean	w.c.	mean	w.c.	mean	w.c.	mean	w.c.	mean	w.c.	mean	w.c.
20	3	**0.50**	1.00	2.72	7.91	**0.43**	0.58	1.03	3.27	**0.60**	1.44	2.09	8.32
20	4	**0.79**	2.44	5.73	11.92	**0.70**	1.93	1.70	6.04	**0.87**	2.43	3.28	12.39
20	5	**1.58**	3.25	14.27	22.29	**1.23**	2.93	5.84	13.41	**1.34**	3.05	7.76	19.57
40	3	**11.51**	39.12	860.47*	3600.13	30.03	123.85	**18.21**	73.52	**84.42**	412.71	145.25	680.38
40	4	**37.04**	119.44	1428.26*	3600.13	52.32	200.52	**45.12**	102.28	**130.58**	531.04	134.55	546.32
40	5	**92.69**	360.17	1507.31†	3600.17	45.23	181.98	**27.82**	88.13	111.65	527.65	**102.03**	454.08

Overall Performance. We compare the runtime and number of branch-and-bound nodes needed by the 4 approaches PCON$^+$, PCON, NCON$^+$ and PNCON$^+$ to solve the instance/k combinations from Section 3.3. The results are reported in Table 3. We note that PCON and PCON$^+$ are the best models with respect to runtime. NCON$^+$ and NPCON$^+$ run out of memory on several of the larger instances. NCON$^+$, however, uses fewer B&B nodes in the solution of the problems. We also conclude that complexity is stronger correlated with the number of nodes $|V|$ than with the number of subtours k.

Table 3. The results show the mean cpu-time and number of B&B nodes used to solve the test instance within each group. *indicates that some instances within the group were not solve within the allotted 1h cpu-time. †incicates that CPLEX ran out of memory on some instances within the group, these results were not included into the mean.

		NCON$^+$		PCON$^+$		PCON		PNCON$^+$	
$\|V\|$	k	time	nodes	time	nodes	time	nodes	time	nodes
20	3	0.50	**20**	**0.43**	29	1.03	72	0.60	21
20	4	0.79	18	**0.70**	26	1.70	66	0.87	**17**
20	5	1.58	22	**1.23**	31	5.84	182	1.34	**19**
40	3	**11.51**	30	30.03	450	18.21	431	84.42	299
40	4	**37.04**	**94**	52.32	604	45.12	500	130.58	314
40	5	92.69	**161**	45.23	351	**27.82**	303	111.65	221
60	3	796.45	**576**	**318.81**	1772	413.45	2017	1237.06†	856
60	4	2281.16*†	1417	**587.86**	1813	857.21	2460	2144.96*	788
60	5	2130.10*	782	758.78	1481	**502.32**	1143	1838.75*	665
60	10	1056.03*	171	**867.69***	469	2216.53*	3256	1103.20*	177
80	3	2244.99*†	502	**1830.88***	4493	1973.30*	4449	2012.02†	542
80	4	3602.40*†	443	2375.51*	3278	**2159.20***	3182	3601.64*	428
80	5	3601.46*	361	2368.62*	2430	**2368.12***	3115	3586.56*†	447
80	10	3076.29*	128	2230.13*	377	**2136.47***	1091	2762.51*	122

4 Conclusions

The approaches we tested have proven to be efficient for medium sized instances. To solve larger instances further enhancements seem necessary. We believe that efficient branching rules based on the structure of the problem could improve model NCON+ considerably. A better separation of constraints (PCON) should improve the runtimes and lower bounds for model PCON and PCON+. Another open question is whether other objectives like the length of the longest cycle influence the difficulty of the problem. In future work we also intend to perform computational and theoretical comparisons of the models presented here with models in the literature.

Acknowledgements

We would like to thank Prof. Gilbert Laporte for pointing out this very interesting problem, Prof. Luis Gouveia for hosting a fruitful Winter School and Prof. Martine Labbé and Ivana Ljubić for hints and discussions regarding symmetry breaking.

Further we would like to thank the referees for their valuable comments, suggestions and hints on existing literature. They have helped us to improve the paper significantly.

References

[Bauer02] Bauer, P., Linderoth, J., Savelsbergh, M.: A branch and cut approach to the cardinality constrained circuit problem. Mathematical Programming 91(2), 307–348 (2002)

[Branco90] Branco, I., Coelho, J.: The Hamiltonian p-median problem. European Journal of Operational Research 47(1), 86–95 (1990)

[Cerdeira86] Cerdeira, J.: The hamiltonian k-median problem for any given k is NP-complete. Tech. rep., Centro de Estatistica e Aplicacoes, nota no. 14/86 (1986)

[Wong03] Egawa, Y., Hagita, M., Kawarabayashi, K.I., Wang, H.: Covering vertices of a graph by k disjoint cycles. Discrete Mathematics 270(1-3), 115–125 (2003)

[Glaab98] Glaab, H., Löfflad, S., Pott, A.: Rundreiseprobleme beim halbautomatischen Lederzuschnitt. In: Proceedings of Operations Research 1997, pp. 86–101. Springer, Heidelberg (1998)

[Glaab00] Glaab, H., Pott, A.: The Hamiltonian p-median problem. The Electronic Journal of Combinatorics 7(R42), 2 (2000)

[Laporte83] Laporte, G., Nobert, Y., Pelletier, P.: Hamiltonian location problems. European Journal of Operational Research 12(1), 82–89 (1983)

[nagy2007] Nagy, G., Salhi, S.: Location-routing: Issues, models and methods. European Journal of Operational Research 177(2), 649–672 (2007)

[NW99] Nemhauser, G., Wolsey, L.: Integer and combinatorial optimization. Wiley, New York (1999)

[Zohrehbandian07] Zohrehbandian, M.: A New Formulation of the Hamiltonian p-Median Problem. Applied Mathematical Sciences 1(8), 355–361 (2007)

Solving Variants of the Vehicle Routing Problem with a Simple Parallel Iterated Tabu Search

Mirko Maischberger[1] and Jean-François Cordeau[2]

[1] Global Optimization Laboratory and DSI,
Università degli Studi di Firenze, Via di S. Marta 3, 50139 Firenze, Italy
[2] Canada Research Chair in Logistics and Transportation and CIRRELT,
HEC Montráal, 3000 chemin de la Côte-Sainte-Catherine, Montréal, Canada H3T 2A7

Abstract. We introduce a parallel iterated tabu search heuristic for solving eight different variants of the vehicle routing problem. Through extensive computational results we show that the proposed heuristic is both general and competitive with specific heuristics designed for each problem type.

1 Introduction

We present a simple parallel iterated tabu search heuristic (ITS/N) for several variants of the Vehicle Routing Problem: the Capacitated VRP (CVRP), the Periodic VRP (PVRP), the Multi-Depot VRP (MDVRP), and the Site-Dependent VRP (SDVRP), all with or without time window constraints. We build on the previous work of [7] and [9,10], and on the Iterated Local Search (ILS) framework [15], adding some improvements and an original parallel evolution of ILS to be used on multi-core computers and clusters.

The VRP is a hard combinatorial optimization covered by an extensive literature of both exact and heuristic methods [27,12,14]. It can be modelled on a complete directed graph $G = (V, A)$, where $V = \{v_0, v_1, \dots, v_n\}$ is a vertex set and $A = \{(v_i, v_j) : v_i, v_j \in V, i \neq j\}$ is the arc set with associated non-negative travel costs c_{ij}. Vertex v_0 is the depot while vertices $v_1, \dots, v_n$ represent customers with a non-negative demand q_i and a non-negative service duration d_i that must be satisfied by a fleet of m vehicles. Vehicles must respect a maximum capacity Q_k and a route duration limit D_k. The solution consists in the determination of m vehicle routes of minimum total cost such that each route starts and ends at the depot and each customer is visited by exactly one vehicle.

In the PVRP, we consider a planning horizon of t days during which each customer i requires f_i visits. These visits must follow an allowable combination of visit days as specified by the set C_i. For example, on a weekly planning horizon a customer i with $f_i = 2$ and $C_i = \{\{Mon, Wed\}, \{Tue, Thu\}, \{Wed, Fri\}\}$ must be visited either on Monday and Wednesday, on Tuesday and Thursday, or on Wednesday and Friday. The PVRP consists in assigning an allowable visit combination to each customer and in designing vehicle routes for each day as in the VRP. For a review of the PVRP see [11].

As shown in [7,8], the MDVRP and SDVRP can be seen as special cases of PVRP. In the VRPTW, PVRPTW, MDVRPTW, and SDVRPTW variants, service at customer i must start within a time window $[e_i, l_i]$.

Recent successful heuristics for the VRP were introduced by [22], [25] and [19] (denoted by NB09 in the following). The Adaptive Large Neighbourhood Search (ALNS)

J. Pahl, T. Reiners, and S. Voß (Eds.): INOC 2011, LNCS 6701, pp. 395–400, 2011.

of [22] has been applied not only to the VRP but also to the MDVRP, the SDVRP, the VRPTW, and the Open VRP. Heuristics for the PVRP were introduced, e.g., by [2,6,7,1,13], the best of which being that of [13] (denoted by HDH). Effective heuristics for the MDVRP were designed by [4,26,7,22] while [18,5,8,22] introduced heuristics for the SDVRP. For both the MDVRP and the SDVRP, the best results were obtained by the ALNS heuristic of [22].

The VRPTW is the most studied variant of the VRP and a very large number of heuristics have been proposed to solve the problem; see, e.g., [17] and [20] (denoted by NBD). For a survey of heuristics for the VRPTW, we refer to [3]. The PVRPTW, MD-VRPTW and SDVRPTW have received less attention than their counterparts without time windows. The PVRPTW and MDVRPTW were first addressed with a tabu search heuristic by [9,10] (denoted by CLM). An improved VNS heuristic for the PVRPTW was recently introduced by [21] (denoted by PR08) while [24,23] introduced a VNS (denoted by PHDR) for the MDVRPTW. To the best of our knowledge, only [9,10] addressed the SDVRPTW.

2 Short Description of the Iterated Tabu Search Heuristic

A common *Iterated Tabu Search* (ITS) heuristic is presented that solves all eight problem variants considered. The ITS is based on the ILS framework [15] and consists in an alternation between TS phases and perturbations.

In our implementation of ITS, since the heuristic allows intermediate infeasible solutions, the classical framework is slightly adjusted to check whether the solution is feasible before it can replace the best solution s^*. Moreover, when the improved solution is not accepted the search returns to s^*. We stop the algorithm when the number of iterations λ (neighborhood explorations) exceeds a limit η.

In the *perturbation mechanism* (inspired by [22]) a cluster is constructed by choosing a seed customer at random and by identifying the π closest customers to the seed, where π is randomly selected in the interval $[0, \lceil\sqrt{n}\rceil]$. These $\pi + 1$ customers are removed from their routes, assigned a new visit combination randomly chosen in the set C_i, and reinserted in the solution so as to minimize the cost.

The algorithm described in [7,8,9,10] is used as the improving strategy, during route manipulations, and in the construction of the initial solutions. The only notable difference is that the search parameters controlling the intensity of the diversification and the penalty on repeated moves that lead to infeasibility are randomized in the interval $[0,1)$, and the tabu list tenure is sampled in $[0,\sqrt{nmt})$ at each ITS iteration. Moreover the stopping criterion has been replaced by a maximum number of iterations without improvement equal to $\sqrt{(\eta-\lambda)\pi}$.

The most important difference between the *tabu search* (TS) heuristic used here and the one introduced by [7] is the use of an *intra-route optimization* mechanism. During the course of the search, we sequentially remove and reinsert each customer every 200 iterations.

The ITS uses an *acceptance criterion* to accept non improving solutions with decreasing probability from the start to the end of the ITS algorithm. This mechanism ensures diversification in the initial phase of the execution and intensification around the best solution towards the end. Whenever an improving solution is found during a call to the TS heuristic, it replaces the current best solution s^*. However, the working solution $\tilde{s}$ returned by the TS heuristic at the end of an improvement phase corresponds to the last solution visited by the search and not necessarily to the best one seen during this improvement phase. This solution $\tilde{s}$ is accepted with probability $1-(\lambda/\eta)^2$.

The sequential algorithm described in the previous paragraphs (ITS/1) is extended to a parallel version with some modifications. In the *parallel algorithm* (ITS/N) each process generates a different starting solution and proceeds with ITS iterations. To further improve differentiation, every time a TS phase starts, each process chooses the search parameters independently using random distributions.

At the end of each improvement phase each process $p \in \{1,2,\ldots,N\}$ decides whether to accept the working solution $\tilde{s}$ or to revert to the j-th best solution, with $j = \lfloor\sqrt{p}\rfloor$. The working solution is accepted with probability $1-(\lambda/\eta)^2$. This mechanism is used to balance the exploration of the best solutions found during the search.

At the beginning the algorithm can broadly explore different parts of the space, while near the end all processes will eventually try to improve the best solutions.

At the end of each parallel execution of TS, with 10% probability, a crossover operation is performed which is similar to the perturbation, but uses current visit combination information from another randomly chosen solution.

3 Computational Results

The algorithm was coded in C++ using GCC v4.x and the COIN-OR METSlib framework [16]. The sequential algorithm was run on an Intel Xeon CPU X7350 at 2.93GHz, while the parallel runs were made on the Cottos cluster of the RQCHP, a Linux cluster with 128 nodes, each node equipped with dual Xeon E5472 processors running at 3GHz interconnected using Infiniband.

We solved the relevant instances available at `http://neumann.hec.ca/chairedistributique/data` and report results averaged over instance groups. Table 1 reports results for CVRP and VRPTW over 10 runs for the ITS/1 algorithm. Regarding the CVRP we report the average and minimum gaps using 10^5 and 10^6 iterations, while for the VRPTW only the best solution is shown. Tables 2–3 report the results for the multi-level problems. In these tables 10^5 and 10^6 refer to ITS/1 with a different number of iterations. Columns /8 and /64 refer to the ITS/8 and ITS/64 using 125 000 iterations for each core. Experiments with less than 64 cores were run ten times whereas those involving 64 cores were run five times. It is worth noting that the overall number of iterations when using 8 cores is the same as for the sequential algorithm with $\eta = 10^6$ or for the minimum over 10 runs with $\eta = 10^5$, so a direct comparison is possible. The results show that the parallel algorithm is usually better than the sequential algorithm which, in turn, is better than 10 independent runs of the same algorithm using 1/10 of the iterations. The *Our best* column reports the best average gap obtained overall considering two more independent runs using 32 and 64 cores and 10^6 iterations per core.

In conclusion we proposed simple modifications to the unified algorithm described in [7,9,10] for solving many VRP variants. The modifications consist in: *i)* embedding

Table 1. CVRP and VRPTW results: comparison with respect to best known solutions (% gaps and absolute values)

CVRP	NB09		ALNS 50K		ITS		
	avg.	best	avg.	best	Avg. 10^5	Avg. 10^6	Best 10^6
CMT	0.03	0.00	0.31	0.11	0.60	0.27	0.07
GWKC	0.12	0.02	1.02	0.49	1.30	0.79	0.41

VRPTW Group	NBD avg.	NBD best	ALNS avg.	ALNS best	Best 10^6
R1	11.92	11.92	12.03	11.92	12.00
	1210.34	1210.34	1215.16	1212.39	1209.19
R2	2.73	2.73	2.75	2.73	2.73
	951.71	951.03	965.94	957.72	951.17
C1	10.00	10.00	10.00	10.00	10.11
	828.38	828.38	828.38	828.38	849.56
C2	3.00	3.00	3.00	3.00	3.00
	589.86	589.86	589.86	589.86	589.86
RC1	11.50	11.50	11.60	11.50	11.50
	1384.30	1384.16	1385.56	1385.78	1385.90
RC2	3.25	3.25	3.25	3.25	3.25
	1119.43	1119.24	1135.46	1123.49	1120.53

Table 2. PVRP, MDVRP, and SDVRP: Average gaps with respect to best known solutions

PVRP	Average gaps (%)						Minimum gaps (%)				
	HDH		ITS				HDH	ITS			
	10^7	10^9	10^5	10^6	/8	/64		10^5	10^6	/8	Our Best
a	1.39	0.33	0.54	0.23	0.11	-0.08	0.03	0.15	-0.02	-0.12	-0.21
b	1.53	0.44	1.40	0.59	0.58	0.18	0.00	0.81	0.20	0.21	-0.17
ab	1.42	0.36	0.75	0.31	0.22	-0.02	0.02	0.31	0.04	-0.04	-0.20
t (s)	185		91		144	213					

MDVRP	Average gaps (%)						Minimum gaps (%)				
	ALNS		ITS				ALNS	ITS			
	25*K*	50*K*	10^5	10^6	/8	/64		10^5	10^6	/8	Our Best
a	0.54	0.35	0.39	0.14	0.11	0.00	0.00	0.12	0.01	0.07	-0.03
b	0.47	0.34	0.85	0.33	0.25	-0.02	0.00	0.29	0.05	0.20	-0.14
ab	0.52	0.34	0.53	0.20	0.15	-0.01	0.00	0.18	0.02	0.11	-0.07
t (s)	118	237	110		149	197					

SDVRP	Average gaps (%)						Minimum gaps (%)				
	ALNS		ITS				ALNS	ITS			
	25*K*	50*K*	10^5	10^6	/8	/64		10^5	10^6	/8	Our Best
a	0.69	0.52	0.88	0.20	0.19	-0.05	0.00	0.13	-0.06	-0.09	-0.16
b	1.01	0.80	1.99	0.79	0.72	0.19	0.00	0.92	0.25	0.24	-0.13
ab	0.80	0.62	1.26	0.40	0.37	0.03	0.00	0.40	0.05	0.02	-0.15
t (s)	81	162	154		238	348					

Table 3. PVRPTW, MDVRPTW, and SDVRPTW: Average gaps with respect to best known solutions. When using the forward time slack, [21] report only best solutions found, not averages.

PVRPTW	Average gaps (%)					Minimum gaps (%)				
	PR08	ITS				PR08	ITS			
	RVNS/2	10^5	10^6	/8	/64		10^5	10^6	/8	Our Best
a	-	1.02	0.03	0.08	-0.45	0.00	0.32	-0.49	-0.44	-1.01
b	-	1.25	-0.10	-0.01	-0.80	0.00	0.23	-1.05	-0.67	-1.50
ab	-	1.13	-0.04	0.03	-0.62	0.00	0.28	-0.75	-0.55	-1.25
t (s)		293		456	679					

MDVRPTW	Average gaps (%)					Minimum gaps (%)				
	PHDR	ITS				PHDR	ITS			
	10^8	10^5	10^6	/8	/64		10^5	10^6	/8	Our Best
a	0.82	1.45	0.66	0.58	0.14	0.00	0.60	0.28	0.12	-0.10
b	1.40	1.99	0.62	0.56	0.06	0.00	0.82	-0.03	0.08	-0.20
ab	1.11	1.72	0.64	0.57	0.10	0.00	0.71	0.12	0.10	-0.15
t (s)	149	180		249	394					

SDVRPTW	Average gaps (%)						Minimum gaps (%)				
	CLM		ITS				CLM	ITS			
	10^5	10^6	10^5	10^6	/8	/64		10^5	10^6	/8	Our Best
a	1.44	0.49	0.63	-0.25	-0.23	-0.72	0.00	-0.28	-0.72	-0.76	-1.05
b	1.62	0.52	0.57	-0.40	-0.54	-0.90	0.00	-0.48	-1.00	-1.01	-1.33
ab	1.53	0.50	0.60	-0.33	-0.39	-0.81	0.00	-0.38	-0.86	-0.89	-1.19
t (s)	13		160		272	336					

the existing algorithm into an ILS framework achieving significant improvement over the traditional approach; *ii)* a straightforward parallel computing extension that further improves the results. On the 260 benchmark instances tested we found 58 new best solutions using the sequential algorithm and 90 overall.

References

1. Alegre, J., Laguna, M., Pacheco, J.: Optimizing the periodic pick-up of raw materials for a manufacturer of auto parts. European Journal of Operational Research 179, 736–746 (2007)
2. Beltrami, E.J., Bodin, L.D.: Networks and vehicle routing for municipal waste collection. Networks 4, 65–94 (1974)
3. Braysy, O., Gendreau, M.: Vehicle routing problem with time windows, part II: Metaheuristics. Transportation Science 39, 119–139 (2005)
4. Chao, I.M., Golden, B.L., Wasil, E.A.: A new heuristic for the multi-depot vehicle routing problem that improves upon best-known solutions. Am. J. Math.-S 13, 371–406 (1993)
5. Chao, I.M., Golden, B., Wasil, E.: A computational study of a new heuristic for the site-dependent vehicle routing problem. INFOR 37, 319–336 (1999)

6. Chao, I.M., Golden, B.L., Wasil, E.: An improved heuristic for the period vehicle routing problem. Networks 26, 25–44 (1995)
7. Cordeau, J.F., Gendreau, M., Laporte, G.: A tabu search heuristic for periodic and multi-depot vehicle routing problems. Networks 30, 105–119 (1997)
8. Cordeau, J.F., Laporte, G.: A tabu search algorithm for the site dependent vehicle routing problem with time windows. INFOR 39, 292–298 (2001)
9. Cordeau, J.F., Laporte, G., Mercier, A.: A unified tabu search heuristic for vehicle routing problems with time windows. J. Oper. Res. Soc. 52, 928–936 (2001)
10. Cordeau, J.F., Laporte, G., Mercier, A.: Improved tabu search algorithm for the handling of route duration constraints in vehicle routing problems with time windows. J. Oper. Res. Soc. 55, 542–546 (2004)
11. Francis, P., Smilowitz, K., Tzur, M.: The period vehicle routing problem and its extensions. In: Golden, B., Raghavan, S., Wasil, E. (eds.) The Vehicle Routing Problem: Latest Advances and New Challenges, pp. 73–102. Springer, Berlin (2008)
12. Golden, B., Raghavan, S., Wasil, E. (eds.): The Vehicle Routing Problem: Latest Advances and New Challenges, vol. 43. Springer, Heidelberg (2008)
13. Hemmelmayr, V.C., Doerner, K.F., Hartl, R.F.: A variable neighborhood search heuristic for periodic routing problems. European Journal of Operational Research 195, 791–802 (2009)
14. Laporte, G.: Fifty years of vehicle routing. Transport Science 43, 408–416 (2009)
15. Lourenço, H., Martin, O., Stülze, T.: Iterated local search. In: Handbook of Metaheuristics, ch. 11, pp. 321–353. Kluwer Academic Publisher, Dordrecht (2003)
16. Maischberger, M.: Routing of vehicles. From point to point trips to fleet routing. Ph.D. thesis, Università degli Studi di Firenze (2010)
17. Mester, D., Bräysy, O.: Active guided evolution strategies for large-scale vehicle routing problems with time windows. Computers & Operations Research 32, 1593–1614 (2005)
18. Nag, B., Golden, B., Assad, A.: Vehicle routing with site dependencies. In: Golden, B.L., Assad, A.A. (eds.) Vehicle Routing: Methods and Studies, pp. 149–159. North-Holland, Amsterdam (1988)
19. Nagata, Y., Bräysy, O.: Edge assembly-based memetic algorithm for the capacitated vehicle routing problem. Networks 54, 205–215 (2009)
20. Nagata, Y., Bräysy, O., Dullaert, W.: A penalty-based edge assembly memetic algorithm for the vehicle routing problem with time windows. Computers & Operations Research 37, 724–737 (2010)
21. Pirkwieser, S., Raidl, G.R.: A variable neighborhood search for the periodic vehicle routing problem with time windows. In: Prodhon, C. (ed.) Proceedings of the 9th EU/MEeting on Metaheuristics for Logistics and Vehicle Routing (2008)
22. Pisinger, D., Ropke, S.: A general heuristic for vehicle routing problems. Computers & Operations Research 34, 2403–2435 (2007)
23. Polacek, M., Benkner, S., Doerner, K., Hartl, R.F.: A cooperative and adaptive variable neighborhood search for the multi depot vehicle routing problem with time windows. BuR Business Research Journal 1 (2008)
24. Polacek, M., Hartl, R.F., Doerner, K., Reimann, M.: A variable neighborhood search for the multi depot vehicle routing problem with time windows. Journal of Heuristics 10, 613–627 (2004)
25. Prins, C.: A GRASP $\times$ evolutionary local search hybrid for the vehicle routing problem. In: Bio-inspired Algorithms for the Vehicle Routing Problem, pp. 35–53. Springer, Berlin (2009)
26. Renaud, J., Laporte, G., Boctor, F.F.: A tabu search heuristic for the multi-depot vehicle routing problem. Computers & Operations Research 23, 229–235 (1996)
27. Toth, P., Vigo, D. (eds.): The Vehicle Routing Problem. Discrete Mathematics and Applications. SIAM, Philadelphia (2002)

The Multi-Commodity One-to-One Pickup-and-Delivery Traveling Salesman Problem: A Matheuristic

I. Rodríguez-Martín and Juan José Salazar-González

DEIOC, Universidad de La Laguna, Spain
{irguez,jjsalaza}@ull.es

Abstract. This paper addresses an extension of the TSP where a vehicle with a limited capacity must transport certain commodities from their origins to their destinations. Each commodity has a weight, and the objective is to find a minimum length Hamiltonian tour satisfying all the transportation requests without ever violating the capacity constraint. We propose for this problem a heuristic approach that combines mathematical programming and metaheuristic techniques. The method is able to improve the best known solutions for a set of instances from the literature in a reasonable amount of computation time.

1 Introduction

The *Multi-Commodity One-to-One Pickup-and-Delivery Traveling Salesman Problem* (m-PDTSP) is a routing problem that generalizes the classical *Traveling Salesman Problem* (TSP). We are given a set of locations and the non necessarily symmetric travel distances among them. One specific location is considered to be the depot of a capacitated vehicle, while the other locations correspond to customers. The depot is denoted by 0, and the customers by $i \in \{1,...,n\}$. The capacity of the vehicle is denoted by Q. There is a set of m different commodities or products, and each of them must be transported from a given pickup customer (origin) to a given delivery customer (destination). The origins and destinations of different commodities are not necessarily disjoint. Each product $k \in \{1,...,m\}$ has a known weight q_k. All locations must be visited by the vehicle, even if some are neither origin nor destination of a commodity. The m-PDTSP consists of finding a minimum length Hamiltonian route for the vehicle that satisfies all the transportation requirements without ever exceeding the vehicle capacity. This problem is of interest when designing a capacitated network with a ring topology to satisfy a set of customer-to-customer requests.

The m-PDTSP is NP-hard since it can be seen as the capacitated version of the *Asymmetric Traveling Salesman Problem with Precedence Constraints*, also known as the *Sequential Ordering Problem* [1,6]. It was introduced by Hernández-Pérez and Salazar-González [9], who presented two different mathematical formulations and described an exact branch-and-cut algorithm, and it is related to many other problems in the literature such as the *Traveling Salesman Problem with Pickup and Delivery* [11] and the *One-commodity Pickup-and-Delivery Traveling Salesman Problem* [7,8]. For a complete survey on pickup and delivery problems see [2] and [12,13].

In this paper we present a heuristic method that provides good feasible solutions for the m-PDTSP. The method combines mathematical programming techniques with

J. Pahl, T. Reiners, and S. Voß (Eds.): INOC 2011, LNCS 6701, pp. 401–405, 2011.

typical metaheuristic ingredients, such as neighborhood definitions, local searches and diversification mechanisms, and so it can be considered a *matheuristic* [10]. The computational experiments compare the heuristic results with those given by the exact branch-and-cut algorithm described in [9].

2 The Heuristic Algorithm

The algorithm we propose for solving the m-PDTSP is a MIP-based greedy randomized adaptive search procedure (GRASP). GRASP is a multi-start metaheuristic algorithm, where each iteration consists of two phases: constructing a feasible solution and improving it. Both phases are repeated until a stopping criterion is satisfied. In our case, the stopping criterion is a time limit.

To generate an initial solution we apply a randomized greedy heuristic in the spirit of the *nearest neighborhood heuristic* for the TSP. Starting from the depot, it extends a partial route randomly choosing at each iteration one of the three nearest customers such that the extended route is feasible, i.e., it is a simple path where the destination of a commodity is not visited before its origin, and the load of the vehicle is below its capacity limit. If a feasible solution is not found, the procedure restarts from the depot.

The improvement phase starts from a feasible solution given by the constructive algorithm, and tries to improve it using a modification of the branch-and-cut algorithm described in [9]. The modification consists of *fixing variables* to get an easier-to-solve model. More precisely, a given percentage of binary variables, corresponding to arcs of the initial solution, are fixed to 1. Then, the relaxed MIP model is solved using the branch-and-cut approach in [9] with a time limit. The process that combines fixing variables and solving the corresponding MIP model is repeated until there is no further improvement or a time limit is reached.

Note that fixing some variables to predefined values (*hard fixing*) and solving the relaxed model serve to explore the neighborhood of a given feasible solution. This is also the idea of the Local Branching strategy described in [4]. The difference is that in Local Branching, the neighborhoods are defined by introducing linear inequalities (called *local branching cuts*) in the mathematical model. These cuts produce a variable *soft fixing* effect, i.e., they limit the distance between the current solution and the solution of the new model. Soft fixing does not reduce the number of variables in the model, and it does increase the number of constraints. According to our preliminary experiments on the m-PDTSP, the use of local branching cuts resulted in MIP models as hard to solve as the original one, even for medium-sized instances. Thus we opted by a variable hard fixing strategy related to the Relaxation Induced Neighborhood Search (RINS) methodology [3].

Moreover, each solution found, either by the construction algorithm or by the branch-and-cut algorithm, is submitted to edge-exchange procedures such as the standard TSP 2-opt and 3-opt. These exchange procedures have been modified in order to keep the feasibility of the solutions in the spirit of [5].

The whole scheme of the heuristic is outlined in Algorithm 1.

Algorithm 1. MIP-based GRASP for the m-PDTSP

```
repeat
  s_0 ← GenerateInitialSol()
  s_0 ← 2-and-3-opt(s_0)
  s_best ← s_0
  s_local_best ← s_0
  while there is solution improvement and LS_timeLimit not reached do
    Fix_variables(s_local_best)
    s' ← solveMIP()
    s' ← 2-and-3-opt(s')
    if s' is better than s_local_best then
      s_local_best ← s'
    end if
  end while
  if s_local_best is better than s_best then
    s_best ← s_local_best
  end if
until timeLimit reached
return s_best
```

3 Computational Results

The algorithm was implemented in C++ and the program was run on a personal computer with Intel Core 2 CPU at 2.4 GHz under Windows XP, using CPLEX 12.1 as MIP solver. Computational experiments were carried on the subset of hardest instances used in [9]. These are randomly generated instances with node coordinates in $[-500,500] \times [-500,500]$, $q_k \in [1,5]$, and Euclidean distances. Class 2 groups instances with $n = 25$, $m = 15$ and $Q \in \{15,20,25,30\}$, and Class 3 instances with $n = 30$, $m = 15$ and $Q \in \{5,10,15,20\}$. There are 10 instances for each combination of n, m and Q. In order to potentially know the optimal solution for these instances, the exact branch-and-cut method in [9] was executed with a time limit of 7200 seconds on each of them. If the time limit was reached, there is no guarantee to have the optimal solution.

As for the heuristic parameters, we set the total time limit to 180 seconds for each run, and the time limit for the improvement or local search phase and the branch-and-cut procedure to 120 seconds. The choice of the percentage of variables to fix is an important issue, since the resulting model is very sensitive to this parameter. A large figure results in an easy to solve but probably infeasible model. If the figure is too low, the relaxed model may be hard to solve. According to our computational experience, fixing a number of variables equal to 20% of the number of nodes proved effective for the instances in our test bed. Finally, a CPLEX parameter was modified in order to make more emphasis in getting feasibility than optimality.

We ran the heuristic ten times over each instance. The results are summarized in Table 1. Column headings stand for: number of nodes (n), number of commodities (m), vehicle capacity (Q), instance name (#), percentage deviation between the best and the average heuristic value and the value reported in [9] (*best*/*B&C* and *aver*/*B&C*, respectively), and finally, number of times, over 10, that the heuristic solution equals or improves the branch-and-cut solution for each instance (*#improv*). A bold figure in Column # indicates that the optimal solution value is known for that instance; an asterisk indicates that the branch-and-cut procedure in [9] was unable to find any feasible solution within the time limit of 2 hours. This is the case for two instances, one in Class 2 and another in Class 3. We do not report any deviation percentage for those instances, since there is not a value to compare with. Note that the heuristic is able to find a feasible solution for those two instances the ten times it is run on each of them.

Negative figures in Columns *best*/*B&C* and *aver*/*B&C* mean that the heuristic algorithm outperforms the branch-and-cut method, and therefore, new best known solutions have been found for the corresponding instances. A zero deviation percentage means that the heuristic result equals the branch-and-cut result. For all instances, the best heuristic value either coincides with the optimal value, when it is known, or it is better than the result reported in [9]. Even taken in average, the heuristic value is equal or better than the branch-and-cut value for 49 out of the 80 instances in the tables, and

Table 1. Heuristic results

			Class 2 instances								Class 3 instances			
n	m	Q	#	best/B&C	aver/B&C	# improv	n	m	Q	#	best/B&C	aver/B&C	# improv
25	15	15	0	-2.74	-1.56	8	30	15	5	**0**	0.00	0.00	10
			1	0.00	0.00	10				**1**	0.00	0.03	9
			2	-3.91	-3.47	10				**2**	0.00	5.27	2
			3	-15.58	-11.31	10				3	-1.66	1.53	3
			4	-2.89	-2.38	9				**4**	0.24	5.58	0
			5*	-	-	10				**5**	0.00	5.31	3
			6	0.00	0.05	7				**6**	0.00	2.07	1
			7	0.00	0.00	8				**7**	0.00	9.10	1
			8	-21.91	-19.11	10				**8**	0.00	1.20	2
			9	-26.95	-24.73	10				9	-1.79	0.65	5
25	15	20	**0**	0.00	0.92	8	30	15	10	0	-1.77	2.74	3
			1	0.00	0.00	10				1	-48.89	-46.60	10
			2	-1.73	-0.65	9				2	-34.27	-30.61	10
			3	-7.83	-6.97	10				3	-5.46	-1.03	8
			4	0.00	0.00	10				4	-13.26	-9.27	10
			5	-11.45	-6.60	10				5	-43.82	-40.60	10
			6	0.00	0.00	10				6	-44.76	-39.41	10
			7	0.00	2.65	7				**7**	0.00	1.02	7
			8	-12.09	-11.01	10				8	-34.46	-30.59	10
			9	0.00	2.27	4				9	-37.56	-32.47	10
25	15	25	**0**	0.00	0.00	10	30	15	15	**0**	0.00	4.56	4
			1	0.00	0.00	10				1	-24.37	-19.79	10
			2	0.00	0.00	10				2*	-	-	10
			3	0.00	0.14	9				**3**	0.00	1.09	7
			4	0.00	0.00	10				4	-1.08	0.43	5
			5	0.00	0.00	10				**5**	0.00	1.64	4
			6	0.00	0.00	10				6	-4.19	-0.99	7
			7	0.00	0.00	10				**7**	0.00	0.00	10
			8	0.00	0.40	7				**8**	0.00	1.73	6
			9	0.00	0.08	9				9	-32.44	-27.89	10
25	15	30	**0**	0.00	0.00	10	30	15	20	**0**	0.00	2.01	5
			1	0.00	0.00	10				**1**	0.00	1.32	8
			2	0.00	0.00	10				**2**	0.00	4.02	2
			3	0.00	0.00	10				**3**	0.00	0.00	10
			4	0.00	0.00	10				**4**	0.00	1.20	2
			5	0.00	0.00	10				**5**	0.00	0.64	4
			6	0.00	0.00	10				**6**	0.00	1.04	6
			7	0.00	0.00	10				**7**	0.00	0.00	10
			8	0.00	0.00	10				**8**	0.00	0.00	10
			9	0.00	0.00	10				**9**	0.00	3.12	3

it substantially improves the branch-and-cut result many times (deviation below -20%), while when it is worse, the deviation error is never higher than 10%.

References

1. Ascheuer, N., Jünger, M., Reinelt, G.: A branch & cut algorithm for the Asymmetric Traveling Salesman Problem with Precedence Constraints. Computational Optimization and Applications 17, 61–84 (2000)
2. Berbeglia, G., Cordeau, J.-F., Gribkovskaia, I., Laporte, G.: Static Pickup and Delivery Problems: A Classification Scheme and Survey. TOP 15, 1–31 (2007)
3. Danna, E., Rothberg, E., Le Pape, C.: Exploring relaxation induced neighborhoods to improve MIP solutions. Mathematical Programming A 102, 71–90 (2005)
4. Fischetti, M., Lodi, A.: Local Branching. Mathematical Programming B 98, 23–47 (2003)
5. Gambardella, L.M., Dorigo, M.: An Ant Colony System Hybridized with a New Local Search for the Sequential Ordering Problem. INFORMS Journal on Computing 13, 237–255 (2000)
6. Gouveia, L., Pesneau, P.: On extended formulations for the Precedence Constrained Asymmetric Traveling Salesman Problem. Networks 48, 77–89 (2006)
7. Hernández-Pérez, H., Salazar-González, J.J.: A branch-and-cut algorithm for a Traveling Salesman Problem with Pickup and Delivery. Discrete Applied Mathematics 145, 126–139 (2004)
8. Hernández-Pérez, H., Salazar-González, J.J.: Heuristics for the One-Commodity Pickup-and-Delivery Traveling Salesman Problem. Transportation Science 38, 245–255 (2004)
9. Hernández-Pérez, H., Salazar-González, J.J.: The multi-commodity one-to-one Pickup-and-Delivery Traveling Salesman Problem. Europoean Journal of Operational Research 196, 987–995 (2009)
10. Maniezzo, V., Stützle, T., Voß, S. (eds.): Matheuristics. Hybridizing Metaheuristics and Mathematical Programming, Annals of Information Systems. Springer, Heidelberg (2009)
11. Mosheiov, G.: The traveling salesman problem with pickup and delivery. European Journal of Operational Research 79, 299–310 (1994)
12. Parragh, S.N., Doerner, K.F., Hartl, R.F.: A survey on pickup and delivery problems. Part I: Transportation between customers and depot. Journal für Betriebswirtschaft 58, 21–51 (2008)
13. Parragh, S.N., Doerner, K.F., Hartl, R.F.: A survey on pickup and delivery problems. Part II: Transportation between pickup and delivery locations. Journal für Betriebswirtschaft 58, 81–117 (2008)

An Adaptive Large Neighborhood Search Heuristic for a Snow Plowing Problem with Synchronized Routes

M. Angélica Salazar-Aguilar[1], André Langevin[2], and Gilbert Laporte[3]

[1] CIRRELT, HEC Montréal, Canada H3T 2A7
angelica.salazar@cirrelt.ca
[2] CIRRELT and Département de mathématiques et de génie industriel, École Polytechnique de Montréal
andre.langevin@polymtl.ca
[3] CIRRELT and Canada Research Chair in Distribution Management, HEC Montréal
gilbert.laporte@cirrelt.ca

Abstract. A synchronized arc routing problem (SyARP) for snow plowing operations is introduced. Given a network of streets and a fleet of snow plowing vehicles, the SyARP consists of determining a set of routes such that all streets, some of which have multiple lanes, are plowed by using synchronized vehicles, and the ending time of the longest route is minimized. A mathematical formulation and an adaptive large neighborhood search heuristic are proposed. The performance of the proposed solution procedure is evaluated over a large set of instances. Computational results reveal that the proposed procedure yields good quality feasible solutions.

1 Introduction

The problems encountered by winter road maintenance planners are complex and site specific because of the variation of climatic conditions, demographics, economics, and technology [1,6,7]. The cost of snow plowing operations can often be reduced through the application of appropiate operational research techniques [2,3]. We introduce a synchronized arc routing problem (SyARP) arising in snow plowing operations. We first define the problem, and we then present a model, an adaptive large neighborhood heuristic and illustrative computational results. This work is preliminary. The definitive version will be submitted for publication later.

2 Problem Description

Given a network of streets and a fleet of snow plowing vehicles based at a depot, the SyARP consists of finding a set of routes such that all streets, some of which have multiple lanes, are plowed by using synchronized vehicles, and the ending time of the longest route is minimized. The street segments have one or two directions, and for each direction the number of lanes goes from 1 to 3. All lanes belonging to the same segment (in the same direction) must be plowed simultaneously. This means that the number of

J. Pahl, T. Reiners, and S. Voß (Eds.): INOC 2011, LNCS 6701, pp. 406–411, 2011.

vehicles that service a given segment (i, j) is equal to the number of lanes that go from i to j. Deadheading is allowed, so that any vehicle can traverse any arc without providing service to it.

2.1 Mathematical Model

Let $G = (V, A)$ be a directed multi-graph where $V = \{0, 1, ..., n\}$ is the vertex set and $A = \{(i, j) : i, j \in V \text{ and } i \neq j\}$ is the arc set. Vertex 0 is the depot. An additional vertex $0'$ is used to represent an artificial depot. Artificial arcs $(0', 0)$ and $(0, 0')$ are used to start and end the routes. Let R be the set of available vehicles. The maximal number of arcs included in any route is given by e and we define K as $\{1, ..., e\}$. Let n_{ij} be the number of lanes on arc (i, j). Each arc $(i, j) \in A$ has two associated times called t_{ij} and t'_{ij} for service and traversal. We define the decision variables

$$x_{ijr}^{k} = \begin{cases} 1 & \text{if arc } (i,j) \text{ is traversed by vehicle } r \text{ and appears} \\ & \text{in the } k\text{-th position of the route while deadheading} \\ 0 & \text{otherwise.} \end{cases}$$

$$y_{ijr}^{k} = \begin{cases} 1 & \text{if arc } (i,j) \text{ is serviced by vehicle } r \text{ and appears} \\ & \text{in the } k\text{-th position of the route} \\ 0 & \text{otherwise.} \end{cases}$$

Note that variables $y_{0'0r}^{1}$ and $y_{00'r}^{k}$ should be equal to zero given that the artificial arcs do not need service.

t_{ijr}^{k} is the starting time of service or traversal of arc (i, j) by vehicle r and this arc appears in the k-th position of the route.

w_{ijr}^{k} is the waiting time of vehicle r after service or traversal of arc (i, j) when it appears in the k-th position of the route.

The objective is to minimize the makespan:

$$\text{Minimize} \quad z = \max_{r \in R, k \in K} \left\{ t_{00'r}^{k} \right\} \tag{1}$$

subject to:

$$x_{0'0r}^{1} + y_{0'0r}^{1} = 1 \qquad r \in R \tag{2}$$

$$\sum_{(0,j) \in A} x_{0jr}^{2} + y_{0jr}^{2} = 1 \qquad r \in R \tag{3}$$

$$\sum_{k \in K} x_{00'r}^{k} + y_{00'r}^{k} = 1 \qquad r \in R \tag{4}$$

$$\sum_{k \in K} \sum_{r \in R} y_{ijr}^{k} = n_{ij} \qquad (i, j) \in A \tag{5}$$

$$(t_{ijr}^{k} + t_{ij}) y_{ijr}^{k} + (t_{ijr}^{k} + t'_{ij}) x_{ijr}^{k} + w_{ijr}^{k} = \sum_{(j,h) \in A \cup (0,0')} t_{jhr}^{k+1} (x_{jhr}^{k+1} + y_{jhr}^{k+1}) \qquad r \in R, (i, j) \in A; \; k \in K \setminus \{1, e\} \tag{6}$$

$$\sum_{(i,j) \in A \cup \{(0',0),(0,0')\}} x_{ijr}^{k} + y_{ijr}^{k} \leq 1 \qquad r \in R;$$

$$k \in K \quad (7)$$

$$y_{ijr}^{k}\left(t_{ijr}^{k} - \sum_{q\in R}\sum_{c\in K\setminus\{1\}} t_{ijq}^{c} y_{ijq}^{c}\right)/n_{ij} = 0 \qquad (i,j) \in A, r \in R;$$

$$k \in K\setminus\{1\} \quad (8)$$

$$\sum_{(i,j)\in A} (x_{ijr}^{k} + y_{ijr}^{k}) \leq \sum_{(j,h)\in A\cup(0,0')} (x_{jhr}^{k+1} + y_{jhr}^{k+1}) \qquad r \in R, j \in V;$$

$$k \in K\setminus\{1,e\} \quad (9)$$

$$x_{ijr}^{k} \in \{0,1\} \qquad r \in R, k \in K, (i,j) \in A;$$

$$y_{ijr}^{k} \in \{0,1\} \qquad r \in R, k \in K, (i,j) \in A;$$

$$t_{ijr}^{k} \geq 0 \qquad r \in R, k \in K, (i,j) \in A;$$

$$w_{ijr}^{k} \geq 0 \qquad r \in R, k \in K, (i,j) \in A.$$

Objective (1) is the makespan. Constraints (2)-(4) guarantee that all routes start and end at the depot. Constraints (5) state that all lanes of arc (i,j) should be serviced. Constraints (6) ensure time consistency of service and traversal times. Constrains (7) guarantee that at most one arc appears in each position of a route. Constraints (8) ensure that each vehicle servicing an arc (i,j) starts at the same time as all vehicles that service the same arc. Constraints (9) ensure that each vertex is connected to some vertex; note that when a vertex j is reached in the k-th position of route r, the outgoing arc (j,h) should be in the $(k+1)$-th position of the route.

3 Summary of the Adaptive Large Neighborhood Search Heuristic

The proposed solution procedure consists of a generation of initial solutions method which provides initial solutions to the ALNS. The adaptive large neighborhood search (ALNS) heuristic was proposed in [4] and extends the LNS heuristic proposed by [5].

3.1 Initial Solutions Generator

In the SyARP, one of the hardest constraints is the synchronization of routes. In our initial solutions we tackled this constraint by sending fleets of vehicles in such a way that each fleet has as many vehicles as the number of lanes of the arcs that they will service. Therefore, given the maximum number of lanes $\bar{l}$ and the total of available vehicles $|R|$, a classification of the arcs according to the number of lanes and different fleet distributions is carried out at the beginning of this procedure. For instance, suppose that we have an instance with 12 available vehicles and arcs having one, two or three lanes. This means that we need at least 6 vehicles to service the different type of arcs. In this case, we would send at least three fleets, one fleet with one vehicle, another one with two vehicles, and the last one with three vehicles. The rest of the vehicles would be distributed in fleets with different sizes and all combinations of fleets which require 12 vehicles would be considered to generate multiple initial solutions. In this example, with 12 available vehicles and maximum three lanes, we have seven possible combinations of fleets: (1,1,3), (1,4,1), (2,2,2), (3,3,1), (4,1,2), (5,2,1), and (7,1,1). The fifth combination (4,1,2) means that four fleets of one vehicle each are needed to service the arcs with one lane, one fleet of two vehicles is needed to service the arcs with two lanes, and two fleets of three vehicles are needed to service the arcs with three lanes. Therefore, our solution procedure provides routes for each fleet in the given distribution of vehicles. Each of these solutions is an input to the improvement phase which we now describe.

3.2 Improvement Phase

Given a initial solution I, the improvement phase of our ALNS heuristic is applied until a stopping criterion is reached. All multiple initial solutions are successively used as input. At the end of the procedure we report the best solution. During the iterative process, each destroy/repair operator d has a given weight ρ_d which is adjusted dynamically during the search. At the first iteration all operators d have the same weight ρ_d. In the following iterations, these weights are adjusted dynamically, based on the past performance of the operator. The selection of an operator d is done by following a *roulette wheel principle*. An operator is applied until 3 consecutive non-improving solutions are found. A non-improving solution can be accepted if its value does not exceed that of the incumbent solution by more than 10%. For each operator d, at most 50 non-improving solutions can be accepted. We have developed 5 destroy/repair operators. For each of these, the insertion of any sequence of arcs is allowed if and only if the fleet size associated with a given arc is at least equal to the number of lanes.

N1: Best Arcs Sequence Removal-Best Insertion. The goal of this destroy/repair operator is to reduce the ending time of the longest route. A good removal and insertion of arcs reduces directly the makespan. Therefore, a sequence of arcs is removed from the longest route and is inserted in another route. The insertion point is after the arc that has the shortest path from the initial arc of the sequence.

N2: Random Arcs Sequence Removal-Insertion. The goal of this destroy/repair operator is to diversify the search. A sequence of arcs is removed from the longest route and is inserted in another route. The insertion point is randomly chosen.

N3: Best Interchange of Arc Status (Serviced-Traversed). This destroy/repair operator attempts to intensify the search process. An arc (i, j) serviced by the longest route is randomly chosen. If another route uses this arc just for traversal, the statuses of this arc in these routes are interchanged. The routes are reoptimized by computing shortest paths in the links affected by the change in arc (i, j).

N4: Worst Interchange of Arc Status (Traversed-Serviced). This destroy/repair operator attempts to diversify the search. The route with the shortest ending time is chosen and an arc (i, j) which is just traversed by the route is randomly selected. Another route servicing the arc (i, j) is identified and an interchange of statuses in the arc (i, j) is carried out in both routes.

N5: First Traversed-First Serviced. This destroy/repair operator attempts to intensify the search process. All routes in the current solution are reoptimized. Recall that each arc can be traversed several times by the same route but it should receive service just once. Therefore, this procedure is applied to each route p_r in such a way that the arcs that are both serviced and traversed by p_r are reordered to attempt a makespan reduction. Thus, each arc that is both serviced and traversed by p_r will be forced to be serviced the first time that this arc is used by route p_r.

4 Preliminary Computational Results

The algorithm was coded in C++ and compiled on a 2191.72 MHz AMD Opteron(tm) Processor 248 with 1GB of RAM under the Linux operating system. Three instance sets with 42, 180, and 300 vertices were generated on the same grid shape. The minimum number of arcs for each set was 113, 499, and 795, respectively. Fifteen instances were generated for each set and all arcs have one, two or three lanes. The number of available vehicles for each set was 12, 18, and 25, respectively. The maximum number of iterations in the improvement phase was set to 1500. Because this problem has never addressed in the literature, no comparative data and no competing heuristic exist. We have evaluated the performance of our ALNS heuristic by using different combinations of our destroy/repair operators. For each instance, we have identified the best found solution over the different combinations of neighborhoods used in our ALNS heuristic. Using these values, we then computed the percentage of gap associated to each combination of neighborhoods. Tables 1, 2, and 3 display these results for various neighborhood choices ("N" is the set of all neighborhoods). For instances with 42 and 180 vertices (Tables 1 and 2), the ALNS with all proposed destroy/repair operators found the best solution most of the time. In contrast, Table 3 shows that for largest instances the best solutions were found with "N1", "N2", and "N3" only.

Table 1. % of Gap with respect to the best found solution, instances set (42,12)

Gap(%)	**N1**	**(N1,N2)**	**N\(N4,N5)**	**N\(N5)**	**N**
Min	3.26	3.26	0.00	0.00	0.00
Ave	11.81	11.45	5.08	4.82	0.68
Max	18.89	18.89	12.13	10.56	7.07

Table 2. % of Gap with respect to the best found solution, instances set (180,18)

Gap(%)	**N1**	**(N1,N2)**	**N\(N4,N5)**	**N\(N5)**	**N**
Min	3.00	3.68	0.00	0.00	0.00
Ave	11.26	11.45	1.95	1.54	1.01
Max	21.30	21.33	9.34	7.19	4.15

Table 3. % of Gap with respect to the best found solution, instances set (300,25)

Gap(%)	**N1**	**(N1,N2)**	**N\(N4,N5)**	**N\(N5)**	**N**
Min	1.46	1.46	0.00	0.00	0.00
Ave	9.66	9.97	1.40	1.78	2.58
Max	21.75	25.70	5.71	5.56	7.82

5 Conclusions

We have introduced a synchronized arc routing problem for snow plowing operations. A mixed integer non-linear problem has been proposed. A solution procedure based on adaptive large neighborhood search metaheuristic was developed. Five destroy/repair operators were developed. The performance of the proposed ALNS procedure was evaluated over large instances. The best set of operators depend on the instance size.

Acknowledgement

This work was supported by CONACYT under the program of postdoctoral training abroad 2009-2010 and the SMM-FSK 2010.

References

1. Norrman, J., Eriksson, M., Lindqvist, S.: Relationship between road slipperiness, traffic accident risk and winter road maintenance activity. Climate Research 15, 185–193 (2005)
2. Perrier, N., Langevin, A., Campbell, J.F.: A survey of models and algorithms for winter road maintenance. Part I: System design for spreading and plowing. Computers & Operations Research 33, 209–238 (2006)
3. Perrier, N., Langevin, A., Campbell, J.F.: A survey of models and algorithms for winter road maintenance. Part IV: Vehicle routing and fleet sizing for plowing and snow disposal. Computers & Operations Research 34, 258–297 (2007)
4. Ropke, S., Pisinger, D.: An adaptive large neighbourhood search heuristic for the pickup and delivery problem with time windows. Transportation Science 40, 455–472 (2006)
5. Shaw, P.: Using constraint programming and local search methods to solve vehicle routing problems. In: Maher, M.J., Puget, J.-F. (eds.) CP 1998. LNCS, vol. 1520, pp. 417–431. Springer, Heidelberg (1998)
6. Usman, T., Fu, L., Miranda-Moreno, L.F.: Quantifying safety benefit of winter road maintenance: Accident frequency modeling. Accident Analysis and Prevention 42, 1878–1887 (2010)
7. Venäläinen, A., Kangas, M.: Estimation of winter road maintenance costs using climate data. Meteorological Applications 10, 69–73 (2003)

A Novel Column Generation Algorithm for the Vehicle Routing Problem with Cross-Docking

Fernando Afonso Santos, Geraldo Robson Mateus, and Alexandre Salles da Cunha

Federal University of Minas Gerais, Computer Science Department, Belo Horizonte – Brazil
{fsantos,mateus,acunha}@dcc.ufmg.br

Abstract. In this paper we present a novel column generation (CG) formulation and a branch-and-price (BP) algorithm for the Vehicle Routing Problem with Cross-Docking (VRPCD). Our BP algorithm is compared with a previous algorithm to solve the VRPCD and the computational results show that our approach dominates the other in terms of the quality of lower and upper bounds and also can evaluate optimal solutions faster.

1 Introduction

Supply chain management (SCM) is a set of approaches used to efficiently integrate suppliers, manufacturers, warehouses and stores in order to minimize system-wide costs while satisfying service level requirements [17]. A lot of tools have been proposed to help the SCM in the last few years aiming to reduce production and inventory costs for different kinds of manufacturing business.

Among problems solved by SCM tools, the vehicle routing problem (VRP) is one of the most important, since the problem of transporting goods from suppliers to customers can arise in different parts of supply chains and implies elevated costs. Furthermore, VRP can be integrated with other supply chain problems, increasing complexity of the solutions. The inventory routing problem (IRP) is an example of such integration and concerns assigning routes for vehicles as well as warehouses to store undelivered goods within a time interval (varying from weeks to months). Different authors deal with IRP, we refer to [5], [13] and [16] for details.

In the last few years, several studies were conducted in different areas of SCM aiming to improve its efficiency and effectiveness. A novel distribution strategy were proposed for helping in the transportation of goods without inventory: the Cross-Docking (CD) warehouse [1]. Items delivered to a CD warehouse by inbound vehicles are immediately sorted out, reorganized based on customer demands and loaded into outbound vehicles for delivery without the items being actually held in inventory at the warehouse. If any item is held in storage, it is usually for a brief period of time that is generally less than 24 hours [19]. The problem of assigning routes for vehicles and organizing loads at CD is called vehicle routing problem with Cross-Docking (VRPCD).

A formal definition of VRPCD is given using a directed graph $G = (V, A)$ where $V = \{\{0\}, S, C\}$, $S = \{1, \ldots, n\}$ is a set of suppliers, $C = \{1', \ldots, n'\}$ is a set of customers and 0 denotes the CD. Assume that $A = A_S \cup A_C$ $(A_S \cap A_C = \emptyset)$, where $A_S = \{(i, j) : i, j \in \{0, 1, \ldots, n\}\}$ denotes the set of arcs connecting the suppliers as well as the CD

J. Pahl, T. Reiners, and S. Voß (Eds.): INOC 2011, LNCS 6701, pp. 412–425, 2011.

and $A_C = \{(i,j) : i,j \in \{0,1',\ldots,n'\}\}$ denotes the set of arcs connecting the customers and CD. Let us assign costs $\{c_{ij} \geq 0 : (i,j) \in A\}$ to the arcs of G. Consider that $P = \{p_i \leftarrow (i,i',q_i) : i = 1,\ldots,n\}$ denotes a set of requests, each one representing a given (unsplitable) load $q_i > 0$ to be shipped from a supplier i to a customer i'. Consider as well that a set of K homogeneous trucks (or vehicles) of capacity Q are available to guarantee that all requests will be satisfied (collected and delivered). Finally, assume that before delivering goods to customers, each vehicle must necessarily stop at CD, in order to allow loads to change trucks and eventually make their delivery easier.

The VRPCD consists in finding routes for all K vehicles to visit once suppliers and customers in order to collect and delivery items, respectively, without exceeding the vehicle capacity. Each route must start and end at CD, furthermore, items can change from vehicles at CD to facilitate its delivery, but whenever a load q_i moves from/to vehicle k to/from another vehicle k', a cost $c_i \geq 0$ is incurred at CD. Thus, the objective of VRPCD is to find routes such that the total cost (given by the sum of the arcs traversed by the vehicles and the costs of changing loads at CD) is minimized. In Figure 1, we depict a set of routes for vehicles k_1 and k_2, where requests p_1, p_2 changed from vehicle k_1 to k_2 and request p_4 from k_2 to k_1.

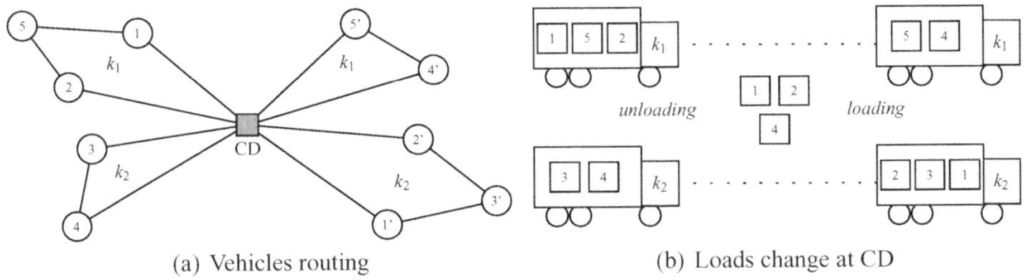

Fig. 1. An example of a VRPCD solution

The costs incurred to change items at CD define the VRPCD structure. Suppose we assign costs $c_i = 0$, $\forall p_i \in P$ to change items at CD. This problem can be reduced to a 2-VRP that consists in finding $2K$ minimum cost routes for visiting suppliers and customers, disregarding costs of changing items at CD. On the other hand, if such costs are assigned as $c_i > U_c$, $\forall p_i \in P$, where U_c is a threshold to ensure that no item will be changed at CD in the optimal solution, the VRPCD is reduced to a vehicle routing problem with pickups and deliveries (VRPPD). According to [2], in VRPPD each vertex either has a pickup or a delivery request but not both. Solutions for VRPPD concerns collecting items using a vehicle k at suppliers and delivering them to the respective customers using the same vehicle. In this paper we addressed the VRPCD with costs $0 < c_i < U_c$, since solutions for 2-VRP and VRPPD are well known in the literature.

In the case where the routing costs are disregarded ($c_{ij} = 0$, $\forall (i,j) \in A$), the VRPCD is reduced to the truck scheduling problem. In this problem we deal only with the scheduling of loads at CD, aiming to organize items from inbound trucks into outbound trucks in order to optimize Cross-Docking operations (cost/time). Different works addressed the truck scheduling problem, see [3], [4] and [19] for details.

The main challenge of VRPCD is to integrate routing and scheduling decisions in order to minimize some objective function. A slightly different variant of the proposed VRPCD was introduced by [10]. The authors addressed the VRPCD with time windows constraints at vertices, but they neglected costs of changing items at CD. A mathematical formulation to solve the problem was presented and a heuristic algorithm based on Tabu search was implemented in order to approximate results for larger instances. The results were reported using random generated instances with 10, 30 and 50 vertices and the results achieved 5% from optimal values in the worst case.

Another work on VRPCD was presented by [18] with similar characteristics from that introduced in [10], however including operational constraints from a real world problem obtained from a logistic company. The authors also presented a mathematical formulation and a tabu search heuristic to solve the problem. Experimental results were conducted based on real data instances with up to 200 pairs of nodes and the solutions from heuristic were within 5% from optimal values.

An exact algorithm for VRPCD was introduced in [15]. The authors proposed a CG formulation to evaluate tighter lower bounds and implemented a branch-and-price algorithm to solve VRPCD instances derived from [18]. They used integer decision variables λ and γ to denote respectively routes for suppliers and customers, indexed for K vehicles in order to catch loads changing at CD. Even though this formulation suffers with a large symmetry, their computational results dominates those obtained using a Linear Programming based Branch-and-bound method that relies on a previous network flow formulation solved with CPLEX MIP package.

The aim of this paper is to present a novel exact algorithm to solve the VRPCD. For this purpose we introduce a newer CG formulation that does not suffer with symmetrical solutions and also a branch-and-price algorithm based on such formulation. The remaining of the paper is organized as follows. In Section 2 we introduce the proposed CG formulation for VRPCD, while in Section 3 we present a mathematical formulation and algorithms used to solve the CG subproblem. The Branch-and-price algorithm is described in Section 4 and the computational results are reported in Section 5. We conclude this work on Section 6 where we also provide further research directions.

2 Column Generation Formulation

We propose a mathematical formulation for VRPCD based on Dantzig-Wolfe decomposition using a directed graph $G' = (V, A')$ where

$$A' = (A \setminus (A_{S0} \cup A_{0C})) \cup A_{SC}$$

$$A_{S0} = \{(i,0) \forall i \in S\}, \quad A_{0C} = \{(0,j) \forall j \in C\}, \quad A_{SC} = \{(i,j) : i \in S, j \in C\}.$$

We also assign costs $c_{ij} = c_{i0} + c_{0j}$, $\forall (i,j) \in A_{SC}$. Note that arcs from suppliers to depot and from depot to customers in A are replaced to arcs connecting suppliers to customers in A', and for this reason routes in G' must leave CD to collect items at suppliers and visit customers immediately latter, returning to CD at end of route. We model change of items at CD by adding a cost c_i if a given vehicle (route) visits a customer i' without visiting the respective supplier i of a request $p_i \in P$.

Let R be the set of all feasible routes in G' visiting a vertex $i \in V \setminus \{0\}$ at most once and do not exceeding the capacity limit of vehicles. For each route $r \in R$ let c_r denote its cost, given by the sum of its arcs. Assume we are given parameters a_{ir} indicating if route $r \in R$ visits or not vertex $i \in V$ (with value 1 or 0, respectively) and $b_{ir} = \max(0, a_{i'r} - a_{ir})$ that assumes value 1 if and only if route r visits customer i' and do not visit the respective supplier i of a given request $p_i \in P$. The CG formulation is accomplished by using decision variables λ_r indicating if a route r is used or not in the solution, with value 1 or 0, respectively, and decision variables τ_i to control the changing of items at CD, assuming value 1 if vertex i' of request p_i is visited on route r but vertex i is not, and 0 otherwise. An integer programming formulation for VRPCD based on CG follows.

$$\min \sum_{r \in R} c_r \lambda_r + \sum_{p_i \in P} c_i \tau_i \tag{1}$$

$$\sum_{r \in R} \lambda_r = K \tag{2}$$

$$\sum_{r \in R} a_{ir} \lambda_r = 1 \qquad \forall i \in V \setminus \{0\} \tag{3}$$

$$\tau_i - \sum_{r \in R} b_{ir} \lambda_r \geq 0 \qquad \forall p_i \in P \tag{4}$$

$$\lambda \in \mathbb{B}^{|R|}, \tau \in \mathbb{B}^{|P|} \tag{5}$$

The objective function (1) minimizes the sum of routing costs plus load/unload costs at CD. Convexity constraint (2) assures that all K vehicles are used in the routing solution, while equalities (3) impose that each vertex must be visited exactly once. Inequalities (4) control load/unload operations at CD, enforcing $\tau_i = 1$ if and only if some route visits a customer without visiting the respective supplier. Finally, constraints (5) define decision variables bounds.

Assume we relax the integrality of variables λ and τ and replace the set of all feasible routes R for a restricted set $\hat{R}$, obtaining a Restricted Linear Master Problem (RLMP). Assigning dual variables $\alpha \in \mathbb{R}$, $\{\theta_i \in \mathbb{R} : i = \{1, \ldots, n, 1', \ldots, n'\}\}$ and $\{\chi_i \in \mathbb{R}_+ : i = \{p_i, \ldots, p_n\}\}$ respectively to constraints (2), (3) and (4), we obtain a dual formulation of RLMP given by (6)–(8):

$$\max \qquad K\alpha + \sum_{i \in V \setminus \{0\}} \theta_i \tag{6}$$

$$\alpha + \sum_{i \in V \setminus \{0\}} a_{ir} \theta_i - \sum_{p_i \in P} b_{ir} \chi_i \leq c_r \qquad \forall r \in \hat{R} \tag{7}$$

$$\chi_i \leq c_i \qquad \forall p_i \in P \tag{8}$$

Suppose w is the optimal value obtained by solving RLMP for a given set $\hat{R}$. If we take the optimal dual variables values and rewrite inequality (7) in terms of network arcs, we can identify negative reduced cost routes looking for $r \in R \setminus \hat{R}$ satisfying

$$\sum_{(i,j) \in r} c_{ij} - \sum_{i \in V} a_{ir} \theta_i + \sum_{p_i \in P} b_{ir} \chi_i < \alpha \tag{9}$$

The CG subproblem consists of minimizing the LHS of inequality (9). Suppose r^* is the minimum cost route obtained by solving the above subproblem. If r^* satisfies (9), we do $\hat{R} \leftarrow \hat{R} \cup r^*$ and start a new iteration of CG. Otherwise, CG stops and $w = w_{CG}$ is an lower bound for VRPCD.

3 Column Generation Subproblem

The CG subproblem described in Section 2 is a slight variant of the Elementary Shortest Path Problem with Resource Constraints (ESPPRC). Such a variant occurs due to changing costs incurred when a route (elementary path) visits a customer vertex i' without visiting the respective supplier i of a given request p_i. ESPPRC appears as subproblem in many CG based algorithms, in particular, vehicle routing problems. Dynamic Programming (DP) is the most common algorithm to solve ESPPRC. Feillet et al. [6] were the first to introduce a DP algorithm for ESPPRC, while the better results in the literature are reported by Salani's algorithm [14]. However, DP algorithms present some shortcomings when routes are long and the resources are not tighter, slowing convergence to optimal solutions.

Besides of DP algorithms shortcomings, our CG subproblem demands new dominance rules to prevent DP algorithm to discard feasible solutions as well as to ensure that all feasible solutions are evaluated. After a first experience implementing a DP algorithm to our subproblem we decided to change the algorithm paradigm due to the large computing time to solve even small size instances.

Aiming to overcome DP shortcomings, [11] introduced a branch-and-cut (BC) algorithm to solve the Elementary Shortest Path Problem with Capacity Constraint (ESPPCC), a particular case of ESPPRC with a single resource, that arises as a CG subproblem of the capacitated vehicle routing problem (CVRP). The results of BC algorithm outperform those obtained by DP algorithm to ESPPCC, specially for instances where capacity constraints are not tight.

After analysing the arguments above we decided to implement a BC algorithm to solve our CG subproblem. We also implemented a heuristic algorithm based on GRASP (Greedy Randomized Adaptive Search Procedure) metaheuristic to solve the subproblem, in order to obtain negative reduced cost routes faster. In the next subsections we present the implementation issues related to the BC and the heuristic algorithms. In what follows, we modify graph G' by adding vertex $0'$, an artificial copy of the depot, in order to look for paths instead of routes. We also embed the optimal values of dual variables θ on arcs costs of A' obtaining $c'_{ij} = c_{ij} - \theta_j, \forall (i,j) \in A'$ to facilitate algorithms implementations.

3.1 Branch-and-Cut Algorithm

We introduce a mathematical formulation to solve the CG subproblem using decision variables x_{ij} to control whether an arc $(i,j) \in A'$ is selected or not in the solution, assuming respectively value 1 or 0. Decision variables y_i indicate if a vertex $i \in V$ is visited or not, with value 1 or 0, respectively. We also introduce a decision variable t_i assuming value 1 if, for a given request p_i, the customer i' is visited and its respective supplier i is not, 0 otherwise. The integer programming formulation is given by (10) – (19) in the following.

$$\min \sum_{(i,j)\in A} c_{ij}x_{ij} \quad + \quad \sum_{p_i\in P} c_i t_i \tag{10}$$

$$\sum_{j\in S} x_{0j} = 1 \tag{11}$$

$$\sum_{i\in C} x_{i0'} = 1 \tag{12}$$

$$\sum_{j\in \Gamma_i^+} x_{ij} = -y_i \qquad \forall i \in V \setminus \{0,0'\} \tag{13}$$

$$\sum_{h\in \Gamma_i^-} x_{hi} = -y_i \qquad \forall i \in V \setminus \{0,0'\} \tag{14}$$

$$\sum_{i\in W,\, j\in V-W} x_{ij} \geq y_j \qquad \forall 0 \in W \subset V, \forall j \in V \setminus W \tag{15}$$

$$t_i \geq y_i - y_{i'} \qquad \forall p_i \in P \tag{16}$$

$$\sum_{i\in S} y_i q_i \leq Q \tag{17}$$

$$\sum_{i'\in C} y_{i'} q_{i'} \leq Q \tag{18}$$

$$x \in \mathbb{B}^{|A|}, y \in \mathbb{B}^{|V|}, t \in \mathbb{B}^{|P|} \tag{19}$$

The objective function (10) minimizes the sum of arcs costs plus changing costs. Equalities (11) and (12) ensure that exactly one arc is used to leave and arrive the depot and its artificial copy, respectively. From (13) and (14) we have the couple constraints to assure together that if any arc of a given vertex is used such a vertex must be visited. Directed cutset constraints (15) enforce connectivity and subtour elimination on solution. Inequalities (16) controls variables t_i associated with changing costs, while (17) and (18) are capacity constraints used respectively to prevent a vehicle from collecting (or delivering) more loads than its capacity. Decision variables space is given by constraints (19).

The above model holds an exponential number of constraints due to (15). Our BC algorithm concerns solving model (10)-(19) disregarding constraints (15) and adding violated constraints by demand, until no such constraints are found. For BC implementation we use callbacks routines of CPLEX optimization software (release 12) keeping all default parameters, except pre-processing and heuristic routines to prevent improper feasible solutions, since we deal only with a subset of constraints.

To identify violated constraints, we use CPLEX cut callback routine to take values of decision variables y_i^* and x_{ij}^* and build a graph $G_o = (V_o, A_o)$, where

$$V_o = \{0\} \cup \{i : y_i^* > 0\} \text{ and } A_o = \{(i,j) : x_{ij}^* > 0\}.$$

Then, we solve $|V_o| - 1$ max-flow (min-cut) problems on G_o from vertex $s = 0$ to $t = \{i : i \in V_o \setminus \{0\}\}$, assigning arcs capacities $C_{ij} = x_{ij}^*, \forall (i,j) \in A_o$.

Let $f(0,i)$ be the max-flow from vertex 0 to i on G_o and f_{ij} be the flow passing through arc $(i,j) \in A_o$ in the max-flow solution. Suppose $\exists i \in V_o : f(0,i) < y_i^*$, exist a set W leading to a violated constraint of (15). If we build a graph $G_r = (V_o, A_r)$, where $A_r = \{(i,j) : C_{ij} - f_{ij} > 0,\ \forall (i,j) \in A_o\}$, set W is composed by vertices $i \in V_o$ for which exist a path from 0 to i on G_r. We add only the most violated constraint calculating $z = \arg\min_{i \in V \setminus W} (y_i^* - f_i)$ generating constraint

$$\sum_{i \in W,\, j \in V - W} x_{ij} \geq y_z.$$

3.2 Heuristic Algorithm

GRASP metaheuristic [7] consists basically in embedding random characteristics within greedy procedures, aiming to achieve good quality local minima. A construction phase builds feasible solutions using greediness plus randomization characteristics, while local search procedures are used to improve such solution. Solutions are built at construction phase using a Restricted Candidate List (RCL) and a parameter $0 \leq \alpha \leq 1$ to control the greediness to choose its elements. After that, local search procedures are applied to the solution in order to achieve a local minimum. This procedure is repeated until a stopping criterion is achieved and the best solutions are stored along the iterations.

In the construction phase, the algorithm generates a feasible solution starting by vertex 0 and filling up RCL with the best adjacent vertices, according to the parameter α. A candidate from RCL is randomly selected and a new iteration starts. Suppose vertex i is selected to be adjacent of 0, we include arc $(0,i)$ in the solution and start a new iteration looking for the best adjacents of vertex i to fill the RCL. These steps are repeated until select vertex $0'$ for the path.

Before to explain the implementation issues about the algorithm, let us first introduce an important notation used along this section. The subset $\Gamma_i^+ \subset V$ denotes a set of feasible vertices for which $i \in V$ can visit (feasible adjacent vertices of i) in a given iteration. Similarly, $\Gamma_i^- \subset V$ is the set of vertices able to visit i. Therefore, $j \in \Gamma_i^+$ if, and only if, $(i,j) \in A'$, j was not visited yet and the load q_j does not exceed the residual capacity of vehicle.

The algorithm fills the RCL based on visiting costs estimative for adjacent vertices. Suppose i is the last vertex of a given path, the estimated visiting cost $a_j, \forall j \in \Gamma_i^+$ is calculated as

$$a_j = \begin{cases} c_{ij} + \min_{k \in \Gamma_j^+}(c_{jk}); & \text{if } j \in S \\ c_{ij} + \min_{k \in \Gamma_j^+}(c_{jk}) + b_j c_j; & \text{if } j \in C \\ c_{i0'}; & \text{if } j = 0' \end{cases}$$

Minimum $\lfloor a_j \rfloor = \min_{j \in \Gamma_i^+}(a_j)$ and maximum $\lceil a_j \rceil = \max_{j \in \Gamma_i^+}(a_j)$ estimative costs are calculated and the RLC is composed by adjacents of i with values within $\lfloor a_j \rfloor \leq a_j \leq \lfloor a_j \rfloor + \alpha \times (\lceil a_j \rceil - \lfloor a_j \rfloor)$. Note that, assigning $\alpha = 0$ we perform a greedy construction phase, while $\alpha = 1$ makes the search totally random. The algorithm chooses one element of RCL randomly and continues until reaches vertex $0'$.

Local search movements are applied to each solution from the construction phase leading to its local minimum. The local search implemented in our algorithm is a Variable Neghborhood Decent (VND) [12]. Instead of using a single neighborhood, VND uses H not necessarily related neighborhoods, applying to a given solution a larger number of movements. Suppose $f(x)$ is a cost function for a given solution x and $N_1, N_2, \ldots, N_H$ are neighborhoods of x, Algorithm 1 describes the VND local search.

Algorithm 1. VND Local Search

Require: Initial solution x_0
Ensure: Local minimum solution x_o^*
  $x \leftarrow x_0$; $k \leftarrow 1$;
  while $h \leq H$ **do**
    if ($\exists s \in N_h(x) : f(s) < f(x)$) **then**
      $x \leftarrow s$; $k \leftarrow 1$; **break;**
    end if
  end while
  return x

In our heuristic we define $H = 3$ neighborhoods to be used within VND, which are described as N_1, N_2 and N_3 in the following. Let V_r be a set of vertices visited by a given route r. N_1 is an insertion neighborhood composed by routes r' on which $V_{r'} = V_r \cup \{i\} : i \in V \setminus V_r$. Solutions $r' \in N_2$ consists in removing one vertex of solution r for which $V_{r'} = V_r \setminus \{i\} : i \in V_r$ and finally N_3 is a swap neighborhood where vertices visiting order of a route r is changed. The movements are performed in the respective neighborhood only if they are feasible using the best improving strategy.

To help intensification of solutions obtained with GRASP heuristic we implemented a path-relinking algorithm [8]. Path-relinking uses solutions from a set of good quality solutions, named elite set, to guide the search of other solutions along solutions space. For our GRASP implementation, path-relinking consists in using an initial solution i obtained from each iteration of GRASP (construction phase + VND local search) and applying movements according to the neighborhoods N_1,N_2 and N_3 to convert it on a solution e, chosen from elite set. Different strategies are available to build the set of elite solutions, as well as to choose a solution from this set as guiding, but in our implementation we store in elite set only the overall minimum cost solution. Suppose $I \subset R$ is the set of intermediate solution obtained by converting i into e using path-relinking movements. If $\exists i^* \in I : f(i^*) < f(i)$, we apply VND local search on i^* to ensure it achieves its local minimum.

4 Branch-and-Price Algorithm

Our Branch-and-price (BP) algorithm for VRPCD first evaluates the linear relaxation value (w_{CG}) of model (1)–(5) using the CG approach described in Section 2. For this purpose we start $\hat{R}$ with K randomly generated routes, one for each vehicle, to compose a feasible solution for VRPCD. Each iteration of CG begins looking for negative reduced cost routes using the heuristic algorithm. If such routes are found, all of them are added to $\hat{R}$ and a new iteration starts. In the case of the heuristic algorithm is not able to find negative reduced cost routes in a given iteration, the BC algorithm is called to evaluate the optimal route r^*. If r^* has negative reduced cost, it is included in $\hat{R}$ and the above steps are repeated, otherwise we achieved w_{CG}.

Along CG steps the BP algorithm also evaluates a subestimate for w_{CG}. Such estimative is described in details in [9] and is calculated as $w_{CG}^{-} = w_{CG} - K f(r^*)$, where $f(r^*)$ is a cost function for the optimal route r^* calculated with BC. Although w_{CG} is a tighter lower bound for VRPCD, w_{CG}^{-} can be useful to keep a valid lower bound for VRPCD even if we stop the CG early.

After CG evaluates w_{CG}, if all variables λ present integer values (variables τ are also integer due to constraints (4)), w_{CG} is the optimal solution for VRPCD and BP algorithm stops. Otherwise, we must resort to branching. In our BP implementation we branch on arc variables x_{ij} obtained from RLMP as

$$x_{ij} = \sum_{r \in \hat{R}} a_{ijr} \lambda_r$$

where parameter a_{ijr} indicates whether a route r pass through the arc $(i,j) \in A'$ or not, assuming value 1 or 0, respectively. To select the branching variable we evaluate the violation level v_{ij} of arcs as

$$v_{ij} = \min(|x_{ij} - 0.5|)$$

and calculate the greater violation value $v_{ij}^{*} = \min(\{v_{ij} : (i,j) \in A'\})$. We compose the set of candidate arcs for branching $B = \{(i,j) \in A' : v_{ij} = v_{ij}^{*}\}$. If $|B| = 1$ we branch on the most violated arc, creating two nodes on BP tree by fixing $x_{ij} = 1$ and $x_{ij} = 0$. Otherwise, if $|B| > 1$ we decide for which arc to branch using the dual information. Such decision is made adding on RLMP the constraint

$$\sum_{r \in R} a_{ijr} \lambda_r = v_{ij}^{*} \qquad \forall (i,j) \in B$$

and branching on that arc with the greater absolute dual value. For node selection on BP tree, we use the best bound strategy.

Aiming to improve upper bound values, we implemented a column generation heuristic (CGH) for VRPCD. Let $\hat{R}^{*}$ be the set of routes priced out on RLMP to achieve w_{CG} for a given node. We replace set R on model (1)–(5) to $\hat{R}^{*}$ and solve the resulting integer programming problem to obtain an upper bound u_{CG} for the problem.

5 Computational Results

We conducted computational experiments to evaluate the performance of the proposed algorithms BC, GRASP and BP. Our test set is built upon VRPCD instances introduced in [18], where real world data belonging to a Danish Logistics Consultant were used to build five instances with $|P| = 200$ requests: 200a, 200b, 200c, 200d and 200e. Since such instances are out of reach for exact solution approaches, we extracted for each instance in [18] smaller subsets of requests with $|P| \in \{10, 15, 20, 25, 30\}$. Arc costs on our test set are based on Euclidean distance while costs to change items at CD are fixed as $c_i = 30, \forall p_i \in P$, except if we state otherwise on the text. These instances are available to download at `http://www.dcc.ufmg.br/~fsantos/instances/`.

The algorithms discussed here were coded in C++, using ILOG CPLEX release 12.1 as the LP and MIP solver. Computational experiments were conducted with a Intel Core 2 Quad machine running at 2.2 GHz, with 4 Gbytes of RAM memory, under Linux Operating System. A time limit of four hours was imposed on the execution of the algorithms.

Table 1. GRASP performance according to the number of iterations

	1000 it		10000 it		20000 it		30000 it		optimal	
instance	cost	time	cost	time	cost	time	cost	time	cost	time
30*a*	208.93	0.42	201.68	4.27	199.96	8.71	199.28	13.23	193.23	221.3
30*b*	90.19	0.30	73.95	3.39	72.81	7.03	70.91	10.87	61.73	1488.1
30*c*	10.16	0.26	7.62	3.01	7.40	6.08	7.14	9.02	-70.77	360.6
30*d*	240.93	0.36	204.22	4.00	188.12	8.19	182.01	12.57	172.12	147.0
30*e*	-328.25	0.29	-340.76	3.52	-343.70	8.02	-345.22	13.31	-355.92	2068.4
average	44.39	0.32	29.34	3.63	24.91	7.60	22.82	11.8	0.07	857.1

We started evaluating the performance of BC and heuristic algorithms, in particular we evaluated the heuristic concerning its parameters: number of iterations (it) and α. In order to obtain negative arcs costs we introduced dual costs on vertices obtained from a CG iteration before to achieve w_{CG}. We report in Table 1 the average results of 50 runs of the heuristic algorithm solving instances with 30 requests (60 vertices). We show for each instance the minimum route cost and the execution time attained by the heuristic algorithm for the number of iterations varying from 1000 to 30000, with parameter α fixed in 0.5. The two last columns of Table 1 show respectively the optimal route value and the execution time to achieve it with the BC algorithm.

Results from Table 1 show that the quality of solutions improves along the iterations while the execution time increases. Note that, near optimal solutions can be obtained faster using the heuristic algorithm, furthermore for instances 30*a*, 30*b*, 30*d* and 30*e* the optimal value was achieved in up to 10 runs of the heuristic when the number of iterations is 10000 or higher. For BP implementation we decided to fix the number of iterations in 10000 because the improvements are not attractive for higher number of iterations due to the increasing on execution time.

In Figure 2 we depict the quality of solutions in terms of parameter α when the heuristic runs over 10000 iterations. On axis x, α starts with value 0 and ends with 1 with step 0.1, while axis y presents the distance of solutions from the optimal value in percentual figures.

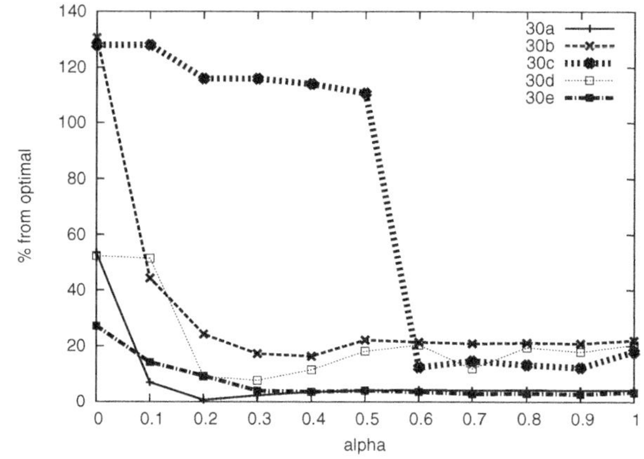

Fig. 2. Evaluation of parameter α on heuristic behavior

The parameter α controls the greediness to build solutions in the construction phase of GRASP heuristics. According to Figure 2, building solutions randomly ($\alpha = 1$) is the worst strategy. On the other hand, building such solutions according to a greedy strategy ($\alpha = 0$) presents results within 20% from optimal values. However, the algorithm performs better with α on the interval $[0.2, 0.6]$ and for this reason on BP implementation it must choose randomly a value for α within this interval before each iteration.

We conducted the next experiment aiming to compare our BP algorithm with that presented in [15]. We labeled both algorithms as NBP (Novel Branch-and-Price) and PBP (Previous Branch-and-Price proposed in [15]) in order to facilitate explanations. For all instances of our test set we evaluated the LP bound w_{CG} and the execution time t_w for the root node and also the best lower (BLB) and upper (BUB) bounds achieved within the time limit (4 hours), besides of the associated percentual duality gap, defined as $\frac{BUB-BLP}{BUB}$. The results reported on table 2 present on columns $3-7$ values obtained using NBP while in columns $8-12$ are the values from PBP. A superscript symbol '$*$' implies w_{CG} achieved with integer variables leading to the optimal solution for VRPCD, while '$-$' indicates that the lower bound w_{CG}^{-} is used instead of w_{CG} because it could not be evaluated within the time limit.

Table 2. Computational Results for the proposed BP the BP proposed in [15]

		NBP					PBP proposed in [15]				
$\|P\|$	instance	w_{CG}	t_w	BLB	BUB	GAP	w_{CG}	t_w	BLB	BUB	GAP
10	a	1783.1*	18s	1783.1	1783.1	0.0%	1678.0	0.9s	1783.1	1783.1	0.0%
	b	1739.0	14s	1741.4	1741.4	0.0%	1637.1	0.7s	1741.4	1741.4	0.0%
	c	1978.6*	10s	1978.6	1978.6	0.0%	1888.1	1s	1978.6	1978.6	0.0%
	d	1766.8	24s	1770.2	1770.2	0.0%	1677.4	0.8s	1770.2	1770.2	0.0%
	e	2192.3*	12s	2192.3	2192.3	0.0%	2079.0	1.1s	2192.3	2192.3	0.0%
15	a	2587.7*	43s	2587.7	2587.7	0.0%	2436.1	1.2s	2560.4	2587.7	1.0%
	b	2671.2*	49s	2671.2	2671.2	0.0%	2495.6	1.1s	2662.3	2671.2	0.3%
	c	2943.3	44s	2948.3	2948.3	0.0%	2852.2	1.3s	2948.3	2948.3	0.0%
	d	2553.5	29s	2568.2	2568.2	0.0%	2432.2	1s	2545.7	2568.2	0.8%
	e	2959.1	101s	2964.5	2964.5	0.0%	2809.0	1.4s	2964.5	2964.5	0.0%
20	a	3170.5	102s	3173.3	3178.9	0.1%	2989.3	4.2s	3082.0	3236.6	4.7%
	b	3397.8*	1274s	3397.8	3397.8	0.0%	3185.5	1.5s	3312.4	3518.0	5.8%
	c	3724.3	196s	3727.2	3773.7	1.2%	3568.1	1.8s	3670.5	3801.2	3.4%
	d	3105.5	119s	3111.7	3126.5	0.4%	2936.7	3.4s	3021.6	3234.6	6.5%
	e	3685.1*	1410s	3685.1	3685.1	0.0%	3508.5	2s	3644.9	3685.1	1.0%
25	a	3972.3	1201s	3975.8	4012.2	0.9%	3736.0	10s	3797.6	4243.2	10.5%
	b	4326.5	12655s	4326.5	4398.5	1.6%	4076.2	5s	4138.0	4642.3	10.8%
	c	4525.8	7317s	4525.8	4605.6	1.7%	4258.5	4	4386.3	4612.4	4.9%
	d	3888.6	1812s	3890.5	3949.8	1.5%	3675.0	12s	3725.9	4258.4	12.5%
	e	4480.2	10501s	4480.2	4519.3	0.8%	4244.0	4s	4322.3	4649.7	7%
30	a	4360.1	8708s	4360.1	4456.9	2.1%	4114.1	28s	4166.9	4894.6	14.8%
	b	4801.5⁻	14400s	4801.5	5173.1	7.1%	4768.4	21s	4834.9	5337.5	9.4%
	c	4455.8⁻	14400s	4455.8	5241.2	14.9%	4771.1	12s	4827.8	5417.9	10.8%
	d	4354.5	13785s	4354.5	4393.7	0.8%	4099.1	30s	4127.0	4977.6	17.0%
	e	4671.7⁻	14400s	4671.7	5301.1	11.8%	4929.8	12s	4990.4	5580.7	10.5%

For all instances on which NBP can evaluate w_{CG} the values are tighter than those evaluated with PBP, however the execution time for NBP grows almost one order of magnitude if the number of requests increases 5 units. In average NBP spends 99% of execution time on CG subproblem, in particular 95% is due to BC algorithm. If we drop the heuristic algorithm from NBP the execution time increases in average 500% and if we solve the CG subproblem using Dynamic Programming NBP is not able to achieve w_{CG} even for problems with 20 requests within 4 hours. These results emphasize how much difficult is our CG subproblem. Moreover, w_{CG} are worth the time because from those 22 instances on which NBP is able to evaluate w_{CG} within the time limit, 7 are optimal and in average w_{CG} is 1% from the optimal value for the others. For instances $30b$, $30c$ and $30e$ a single BC execution can spend hours and for this reason NBP could not evaluate w_{CG} within the time limit, and we report $w_{\overline{CG}}$. On the other hand, PBP evaluates loosen w_{CG} values faster for all instances. This occurs because PBP evaluates two CG subproblems, one for suppliers and other for customers, dealing with small subgraphs for each pricing. However this approach implies symmetry on solutions and for this reason w_{CG} are not as tighter as those evaluated with NBP.

Analysing overall solutions for VRPCD the results show that NBP performs better than PBP because it finds feasible solutions with lower cost for all instances. For the most of instances, such solutions are evaluated using the CGH described in Section 4. A similar heuristic is also used to evaluate feasible solutions for PBP, but the quality of routes priced out for NBP is greater because it acomplishes suppliers and customers together and for this reason achieves better feasible solutions on CGH. Considering only those instances solved at optimality by both algorithms, NBP also dominates PBP in terms of time need to achieve the optimal solution, in average 684 seconds for NBP against 3678 for PBP. Finally, the average duality gap attained by NBP for instances with $10, 15, 20, 25$ and 30 requests are respectively $0.0\%, 0.0\%, 0.34\%, 1.3\%$ and 7.34% while PBP achieves for the same instances gaps $0.0\%, 2.5\%, 4.28\%, 9.14\%$ and 12.5%.

We finished our computational experiments illustrating how NBP and PBP algorithms perform dealing with VRPCD for different costs to change loads at CD. We assigned changing costs as $c_i = \{10, 30, 50\}, \forall p_i \in P$ and evaluated w_{CG} for instances with 20 requests with both algorithms. Results of such comparisons are in Table 3. For each instance we report the optimal value followed by the lower bound implied by w_{CG} and the upper bound (U_{CG}) calculated using the CGH for $c_i = \{10, 30, 50\}$. Values from PBP are shown within brackets.

Table 3. Comparing lower and upper bounds from NBP and PBP (within brackets) for different costs to load/unload items at CD

	$c_i = 10$			$c_i = 30$			$c_i = 50$		
instance	optimal	w_{CG}	U_{CG}	optimal	w_{CG}	U_{CG}	optimal	w_{CG}	U_{CG}
$20a$	3068.2	3059.3 (2989.3)	3069.3 (3081.1)	3178.9	3170.5 (2989.3)	3216.3 (3216.5)	3232.5	3220.5 (2989.3)	3258.9 (3406.1)
$20b$	3297.8	3277.3 (3185.5)	3352.7 (3311.9)	3397.8	3397.8 (3185.5)	3397.8 (3442.6)	3444.6	3439.5 (3185.5)	3444.6 (3539.9)
$20c$	3692.4	3646.3 (3568.1)	3715.8 (3701.0)	3766.1	3724.3 (3568.1)	3786.9 (3779.5)	3805.1	3764.2 (3568.1)	3814.4 (3847.3)
$20d$	3045.9	3030.2 (2936.7)	3086.3 (3069.6)	3124.2	3105.5 (2936.7)	3124.2 (3236.6)	3124.2	3105.5 (2936.7)	3124.2 (3417.4)
$20e$	3592.4	3585.5 (3508.5)	3592.4 (3595.7)	3685.1	3685.1 (3508.5)	3685.1 (3688.3)	3765.1	3755.6 (3508.5)	3765.1 (3807.8)

These results illustrates clearly that PBP is plagued by a deep symmetry problem. For each instance, the lower bound w_{CG} evaluated with PBP keeps constant regardless the value assigned for c_i. Moreover, the upper bound U_{CG} increases according to c_i values enforcing a large duality gap. Concerning NBP, values for both w_{CG} and U_{CG} are closer from the optimal values for all instances, since the pricing problem evaluates routes for suppliers and customers together.

6 Concluding Remarks

We presented in this work an exact solution approach for VRPCD, a recent problem that integrates routing and scheduling decisions in supply chains. Aiming to improve previous results reported in [15] we presented a novel column generation algorithm to fix shortcomings from the previous algorithm. The resulting column generation subproblem of our algorithm implies a huge computational complexity and for this reason we implemented a branch-and-cut algorithm to obtain optimal solutions and a heuristic algorithm based on GRASP metaheuristic to approximate such solutions faster. We evaluated the performance of our algorithms using a real based data set introduced in [18] and the overall results outperformed those obtained from the previous algorithm concerning both lower and upper bounds.

In the current step of research we are evaluating another column generation formulation to prevent symmetrical solutions as well as to facilitate the resulting subproblem. As further research we plan to investigate state space relaxation approaches toward to solve the subproblem faster, since it spends up to 99% of the execution time for the larger instances.

Acknowledgements

Alexandre Salles da Cunha is partially funded by FAPEMIG PRONEX APQ-01201-09 and CNPq grants 477863/2010-8 and 302276/2009-2.

References

1. Apte, M.U., Viswanathan, S.: Effective cross docking for improving distribution efficiencies. International Journal of Logistics: Research and Applications 3, 291–302 (2000)
2. Berbeglia, G., Cordeau, J.F., Gribkovskaia, I., Laporte, G.: Static pickup and delivery problems: a classification scheme and survey. TOP 15, 1–31 (2007)
3. Boysen, N.: Truck scheduling at zero-inventory cross docking terminals. Computers & Operations Research 37, 32–41 (2010)
4. Chen, F., Lee, C.Y.: Minimizing the makespan in a two-machine cross-docking flow shop problem. European Journal of Operational Research 193, 59–72 (2009)
5. Federgruen, A., Simchi-Levi, D.: Analysis of Vehicle Routing and Inventory-Routing Problems. In: Handbook in operations research and management science. Elsevier, Amsterdam (1995)
6. Feillet, D., Dejax, P., Gendreau, M., Gueguen, C.: An exact algorithm for the elementary shortest path problem with resource constraints: Application to some vehicle routing problems. Networks 44, 216–229 (2004)

7. Feo, T.A., Resende, M.G.C.: Greedy randomized adaptive search procedures. Journal of Global Optimization 6, 109–133 (1995)
8. Glover, F.: A Template for Scatter Search and Path Relinking. In: Hao, J.-K., Lutton, E., Ronald, E., Schoenauer, M., Snyers, D. (eds.) AE 1997. LNCS, vol. 1363, pp. 13–54. Springer, Heidelberg (1998)
9. Huisman, D., Jams, R., Peeters, M., Wagelmans, A.: Combining Column Generation and Lagrangian Relaxation (chapter 9). In: Column Generation, pp. 247–270. Springer, Heidelberg (2005)
10. Lee, Y.H., Jung, J.W., Lee, K.M.: Vehicle routing scheduling for cross-docking in the supply chain. Computers and Industrial Engineering 51(2), 247–256 (2006)
11. Jepsen, M.K., Petersen, B., Spoorendonk, S.: A Branch-and-Cut Algorithm for the Elementary Shortest Path Problem with a Capacity Constraint Technical Report 08/01, DIKU (2008)
12. Mladenovic, N., Hansen, P.: A variable neighborhood search. Computers & Operations Research 24, 1097–1100 (1997)
13. Moin, N.H., Salhi, S.: Inventory routing problems: a logistical overview. Journal of the Operational Research Society 58, 1185–1194 (2006)
14. Salani, M.: Branch-and-price algorithms for vehicle routing problems. Ph.D. thesis, Universita degli studi di Milano, Italy (2005)
15. Santos, F.A., Mateus, G.R., Cunha, A.S.: A Branch-and-price algorithm for a Vehicle Routing Problem with Cross-Docking VI LAGOS - Latin American Algolrithms, Graphs and Optimization Symposium (to appear, 2011)
16. Sarmiento, A.M., Nagi, R.: A review of integrated analysis of production-distribution systems. IIE Transactions 31, 1061–1074 (1999)
17. Simchi-Levi, D., Kaminsky, P., Simchi-Levi, E.: Designing and managing the supply chain: concepts, strategies and case studies. Irwin/McGraw-Hill, New York (2003)
18. Wen, M., Larsen, J., Clausen, J., Cordeau, J.-F., Laporte, G.: Vehicle routing with cross-docking. Journal of Operational Research Society 60, 1708–1718 (2008)
19. Yu, W., Egbelu, P.J.: Scheduling of inbound and outbound trucks in cross docking systems with temporary storage. European Journal of Operational Research 184(1), 377–396 (2008)

Impacts of Imprecise Demand Forecasts in Network Capacity Control: An Online Analysis

Jörn Schönberger and Herbert Kopfer

University of Bremen, Wilhelm-Herbst-Straße 5, 28359 Bremen, Germany
{jsb,kopfer}@uni-bremen.de

Abstract. Revenue Management (RM) provides the opportunity to generate additional income by segmenting the overall market and by defining adequate prices for products from a given product portfolio. Typical application fields for RM are: service and leisure industries but also production-to-order-applications. In the last years RM-techniques have penetrated into 'non-traditional' sectors. Among them is the freight transport sector. Air cargo represents the most prominent example but also in maritime contexts, RM-tools are used. However, the application circumstances in road-based freight transportation are quite different compared to the aforementioned applications. The road network is close-meshed so that the number of possible services in a given network is extremely large. A further challenge is that the requests under consideration are subject of high uncertainty. On the one hand, the announced weight and volume often differs from the actual dimensions of an accepted request to be served by a given network. On the other hand, a lot of dispatchers grant ad-hoc discounts on the prices defined within the RM system. Thus, inconsistencies between calculated and actual dispatching-relevant data occur. This contribution addresses the revealing and understanding of impacts of imprecise demand forecast on the performance of the RM-tool capacity control.

1 Introduction

Revenue Management (RM) provides the opportunity to generate additional income by segmenting the overall market and by defining adequate prices for products from a given product portfolio. Typical application fields for RM are: service and leisure industries but also production-to-order-applications. In the last years RM-techniques have penetrated into 'non-traditional' sectors. Among them is the freight transport sector. Air cargo represents the most prominent example but also in maritime contexts, RM-tools are used.

Recent studies (e.g. [13]) suggests the application of RM in road-based freight transportation with the objective to earn additional revenues from performing transport services in a road-based network. Two basic ideas are followed. At first it is tried to filter the incoming stream of requests and to reject unprofitable requests (*capacity control*). At second it is tried to increase the number of sold requests by segmenting the corresponding market (*price segmentation*).

Capacity control subsumes efforts to scan an incoming stream of consecutively arriving requests for the utilization of scarce resources with the goal to filter the most

J. Pahl, T. Reiners, and S. Voß (Eds.): INOC 2011, LNCS 6701, pp. 426–438, 2011.

profitable requests out from the proposed stream ([10]). Basic components of capacity control systems are the control of utilized resource capacities and the prediction of expected future profits (revenues).

In road-based freight transportation (road haulage), freight forwarding companies as well as freight carriers work together to fulfill the transport demand of customers. A freight forwarder receives the customers' demand and organizes a transportation chain through a given transportation network. Often, a forwarding company has no own transport equipment or is not able to use own equipment to realize the complete transport chain from the customer-specified pickup location to the desired delivery point. In this case, the freight forwarder organizes parts or the overall transport chain by hiring third parties that fulfill the physical movement of goods. The third parties are involved into the transshipment facilities and the transport operations. The latter operations are fulfilled by freight carriers who are engaged by the freight forwarder to move customers' goods with their equipment (trucks, trailers, vans, ...). Here, the demand is expressed as a request that can only be fulfilled in complete or not at all.

There are several impediments preventing the transfer of RM-tools from the aforementioned successfully served application areas. This contribution reveals the specific challenges and shortcomings but also addresses proposals to handle these challenges. We address two challenges (i) the management of the extreme large portfolio of possible transport services in a network and (ii) the impreciseness of demand forecasts which are typically caused by a volatile market (number of expected requests) but also on organizational deficiencies like ad-hoc discounts granted by dispatchers to customers. For coping with (i) we use the concept of so-called "flexible products" for the control of road-transport capacities. The resulting mathematical decision models are then evaluated in computational simulation experiments in order to prove the following research hypothesis: *In early stages of a booking period, imprecise demand forecasts have small impacts but in later stages of the selling interval, forecast errors have severe impacts on earned revenues.*

We start our report with the description and analysis of the decision situation (Section 2). Afterwards, we present capacity control approaches for the investigated dynamic decision problem (Section 3). A comprehensive simulation study incorporating the proposed capacity control approaches is reported, the observed results are discussed and an answer to the motivating research hypothesis is derived (Section 4).

2 Decision Situation

We survey literature related to our investigation subject (Subsection 2.1). An informal description of the analyzed scenarion is given (Subsection 2.2) and a decision model is proposed (Subsection 2.3).

2.1 Literature

There is a very high number of scientific publications dealing with operational transport process planning in road-based freight transportation. These contributions are classified into three classes according to the investigated subproblem.

The most often referred subproblem class in operational freight transportation comprises the *vehicle routing problem* and its variants ([7]), in which least cost routes and schedules are compiled for the available trucks. The *externalization* of subsets of requests (subcontracting) is also investigated [11]. Often, these two subproblems of operations transport process planning are considered as an integrated problem [12,15]. The third referred subproblem is the request acceptance problem. Here, the dispatcher has to select a subset of requests from a given request portfolio. Chosen requests are compiled into routes or externalized and fulfilled but unchosen requests are rejected.

Static and dynamic request acceptance situations are distinguished. In a static situation, the overall requests portfolio is known at the time when the acceptance decision is made. Typically, variants (extensions) of the vehicle routing problem like the selective vehicle routing problem [24,1], the profitable salesman problem [4] or pickup and delivery selection problems [22] are investigated. If the portfolio is not known completely at the acceptance decision time, the acceptance problem is dynamic. Whenever additional requests are released, it has to be decided again, which subset of the additional requests is accepted and which requests are rejected. Dynamic acceptance problems belong to the class of capacity control problems [23] which are an important decision problem class in operational revenue management [3]. The basic challenge in capacity control is to filter an incoming stream of requests so that a scarce resource is used in the most profitable way by the selected requests. *Network capacity control* considers several linked resources simultaneously [9].

The major challenge in capacity control is that a once made acceptance decision cannot be revised later on. However, if flexible products are offered, it is possible to alter the request fulfillment process after a request has been definitively accepted ([14,6]). In road based freight transportation, a product is flexible, if there are two or more routes or services to fulfill a request. This dispatching option enables a network service provider to increase its profitability [16].

Capacity control problems are typically represented as dynamic programming (DP) models. However, the tree structure to store and evaluate all future system states in the DP is too large to be efficiently handled. Thus, so called *deterministic linear programs (DLP)* are used [10] as decision models. In a DLP, the evaluation of the future states is simplified. Only real-valued expected values (e.g. the expected number of upcoming requests or the expected sum of expected revenues collectible from the upcoming requests) are considered to represent future system states.

Capacity control strategies represent rules that describe under which circumstances a request is accepted [17]. Two forms of strategies are discussed in the scientific literature. At first, quotas are determined for products specifying the maximally allowed number of accepted requests [18]. A proposed request is accepted without individual inspection as long as the quota is not exceeded. At second, in a bid-price approach [9], each request is individually checked for profitability at its arrival time. However, the determination of the required margin costs (least prices) for requests is a very challenging problem. Several approximation concepts are proposed [1,9].

The application of capacity control techniques for road-based freight transport is reported only in very few contributions. The consulting study of [13] concludes that road-based freight carriage companies would like to apply capacity control techniques.

Their primary goal is to reduce the portion of empty trips among the necessary operations. [20] report a first outline of a capacity control system applied to a single vehicle freight transportation planning scenario. The achieved revenues and capacity utilizations after the completion of the selling are compared. Within this report we perform an online evaluation of the network capacity control system introduced for road-based transportation in [19] in order to explain the observed results.

In the following we give a description of the problem and of the capacity control system introduced in [19].

2.2 Dynamic Decision Problem

A freight carrier is confronted with an incoming stream of consecutively arriving request proposals originated by freight forwarders searching for transport resources. Each request expresses the indivisible demand for the movement of goods of a certain quantity through the network $\mathscr{N} := (\mathscr{V}, \mathscr{A})$ formed by the set of pickup and delivery points $\mathscr{V}$ and connections between these nodes collected in the set $\mathscr{A}$.

The definition of $\mathscr{N}$ is based on planned transportation services necessary to fulfill demand specified in longer term contracts. A regular visit of the nodes in $\mathscr{V}$ by a truck of the carrier is necessary for loading and/or unloading goods according to the previously mentioned contracts. To enable the fulfillment of the long term contracts the carrier operates regular transport services. A service S is an ordered sequence $S = (s_1, s_2, \ldots, s_{n(s)})$ of nodes taken from $\mathscr{V}$. It is fulfilled by a truck that offers a given capacity CAP_S. This truck visits the nodes $s_1, s_2, \ldots, s_{n(s)}$ in the defined sequence. The services are collected in the set $\mathscr{S}$.

The arcs $a(S,1) := (s_1, s_2)$, $a(S,2) := (s_2, s_3)$, ..., $a(S, n(s)-1) := (s_{n(s)-1}, s_{n(s)})$ form the arc set $\mathscr{A}$. Each arc (s_i, s_j) has been indexed with the donating service S in order to have several copies of (s_i, s_j) offered by different services in the set $\mathscr{A}$. An arc $a \in \mathscr{A}$ is referred to as **resource** a and its capacity $C(a)$ is initialized by $C(a) := CAP_S$. For a given resource $a \in \mathscr{A}$ $s(a)$ refers to the service that provides a. All resources are stored in $\mathscr{A}$.

Products are derived from the specified services running in $\mathscr{N}$. A product p is an ordered pair of two nodes $p := (u, v) \in \mathscr{V} \times \mathscr{V}$ so that there is at least one service $s^* \in \mathscr{S}$ that is able to pick up some goods at u and moves them without transshipment to v where the goods are unloaded. All derived products are collected in the set $\mathscr{P}$. The binary indicator $SPR(s, p, a)$ is set to 1 if and only if service $s \in \mathscr{S}$ offers product $p \in \mathscr{P}$ and, doing so, uses the resource $a \in \mathscr{A}$.

A **request** r expresses the demand for a specific product $p := P(r) \in \mathscr{P}$. The carrier expects a profit $REV(P(r))$ from the execution of request r.

At time t^k (k=0,1,...,K_{max}) several spot-market requests are proposed to the carrier by one or more forwarding companies. These requests are proposed in addition to the regular demand described in the long-term contracts. The carrier is free to decide if it wants to take the fulfillment responsibility for a spot-market request. However, the carrier has to decide immediately about the acceptance of the recently proposed requests. We assume that the acceptance of a spot-market request does not lead to considerable additional accountable costs. Thus, the carriers profit is lifted by the collected revenue $REV(P(r))$ associated with the spot-market request r. The carriers' goal is now to filter the most profitable requests from the stream of consecutively arriving requests considering the already made irreversible decisions as well as the scarce capacity of the available resources. Therefore it decides about the acceptance/rejection of the currently appeared requests guided by a capacity control strategy. The carrier relies its decisions on calculated bid-prices of the available resources or on previously fixed product- and service-specific quotas. Thus, the carriers' decision problem can be interpreted as a dynamic bid-price and/or quota determination problem.

At the decision time t^k, the carrier knows the achievable revenues $REV(P(r))$ associated with a request r, the number $Y^k(P(r))$ of requests for product $P(r)$ already accepted not later than t^{k-1}, the expected demand-to-come $DTC^k(P(r))$ for product $P(r)$ in the remaining booking phase $[t^k, t^{K_{max}}]$ as well as the remaining capacity of the available resources. It must be found out if r is accepted or if the potentially scarce capacity of the resources will be assigned to later arriving requests, which are more profitable but whose appearance is unsure (which results in the rejection of the currently waiting request r). Although the carrier is unable to revise the acceptance decision of a once accepted or rejected request it may revise the assignment of an already accepted request r^* from a service s' to another service s^* if the product $P(r^*)$ is flexible, e.g. if two or even more services offer product $P(r)$ and provide free capacity.

2.3 Model-Based Determination of Quotas and Bid-Prices

We represent the previously outlined dynamic decision problem as an online decision model $M = (M^1, M^2, \ldots, M^{K_{max}})$ [21]. Whenever additional requests arrive at time t^k we adjust the stored bid-prices $BP(a)$ of all resources $a \in \mathscr{A}$ and the service-specific quotas $Q(p,s)$ ($p \in \mathscr{P}$, $s \in \mathscr{S}$) for the products to the current workload and to the expected future demand.

In order to define the decision model instance M^k at time t^k we introduce the two decision variable families y_p^k ($p \in \mathscr{P}$) and $z_{p,s}^k$ ($p \in \mathscr{P}$, $s \in \mathscr{S}$) to code the revenue maximal quotas for each product p (the y-variables) and the service-specific quotas (the z-variables).

$$\sum_{p \in \mathscr{P}} REV(p) \cdot y_p^k \rightarrow \max \tag{1}$$

$$\sum_{p \in \mathscr{P}} SPR(s(a), p, a) \cdot z_{p,s(a)}^k \leq C(a) \; \forall a \in \mathscr{A} \tag{2}$$

$$\sum_{s \in \mathscr{S}} z_{p,s}^k = y_p^k \; \forall p \in \mathscr{P} \tag{3}$$

$$y_p^k \geq Y^k(p) \; \forall p \in \mathscr{P} \tag{4}$$

$$y_p^k \leq Y^k(p) + DTC^k(p) \; \forall p \in \mathscr{P} \tag{5}$$

$$y_p^k, z_{p,s}^k \geq 0 \; \forall p \in \mathscr{P}, s \in \mathscr{S} \tag{6}$$

The total sum of revenues is maximized (1) by finding adequate quotas y_p^k for each product $p \in \mathscr{P}$. [5] propose the linear constraints (2) - (6) to encode the collections of feasible quotas y_p^k as well as $z_{p,s}^k$. Constraint (2) ensures that the capacity of the considered resources is not exceeded. All accepted requests must be distributed among the available services that can fulfill the considered request (3) and it is guaranteed for each already accepted request that enough capacity is reserved for each product p so that there is at least one service for fulfilling the already accepted requests associated with product p (4). We do not reserve capacity for more requests than expected to appear during the remaining booking phase $[t^k, t^{K_{max}}]$ (5). [9] explains why it is unnecessary to enforce the integer-property of the quota decision variables in the model.

Model (1)-(6) is a DLP. Information about the uncertainty and value of the upcoming spot-market requests are compressed in the expectation values $DTC^k(p)$ $(p \in \mathscr{P})$.

We (re-)solve the instance M^k of this program at every time point t^k $(k = 1, \ldots, K^{max})$ to update the so far used quotas $Q(p,s)$ with respect to the meanwhile arrived requests and to the still available resources. After the solution of (1) - (6) has been determined, we update the quotas $Q(p,s)$ by setting $Q(p,s) := z^k_{p,s}$ $(\forall p \in \mathscr{P}, \forall s \in \mathscr{S})$.

If we use a simplex algorithm to solve the DLP (1) - (6) we are able to get shadow prices of all resources $a \in \mathscr{A}$ from the final simplex tableau. For a given resource a its shadow price $sp^k(a)$ equals the entry in the final simplex tableau that is found in the objective function row in the column of the slack variable belonging to the restriction of a's capacity in the constraint family (2). The shadow price $sp^k(a)$ of a resource a represents opportunity costs for the utilization of one unit of resource a. Hence, the shadow price of resource a represents the least necessary revenue that must be earned if one unit of this capacity is used in order to increase the overall sum of earned revenues. Consequently, the shadow price of a resource a represents a reasonable bid price for resource a at time t^k. We update $BP(a)$ at time t^k by setting $BP(a) := sp^k(a) \forall a \in \mathscr{A}$.

In conclusion, the solving of the DLP (1)-(6) enables an update of service-specific product quotas as well as resource specific bid-prices. The updated values of $Q(p,s)$ and of $BP(a)$ are calculated under consideration of all knowledge acquired until time t^k and consideration of the forecast for the remaining booking period from t^k to $t^{K_{max}}$.

The definition of artificial benchmark instances that enable a parameterizable evaluation of the dynamic decision situation is given in [19].

3 Capacity Control System

We start with an outline of the proposed capacity control system (Subsection 3.1). Next, a decision model for the tentative assignment of already accepted requests is proposed in Subsection 3.2. The definition of control policies completes the description of the capacity control system (Subsection 3.3).

3.1 Outline of the System

The procedure *capacity_control*() whose pseudo-code is shown in Fig. 1 is used to process the online model $(M^1, \ldots, M^{K_{max}})$. At first, the network $\mathscr{N}$ is determined, the decision times are fixed and the capacity control strategy is specified (a). Next, the iteration counter k is initialized (b) and the set of already accepted requests is inaugurated (c). For every decision time t^k the loop (d)-(o) is executed. The current time is fetched first (e). Then a tentative resource allocation is made for the already accepted requests in order to determine the still available capacities of the resources in $\mathscr{A}$ (f). It is continued

(a) **procedure** capacity_control($\mathcal{N}$,$\{t^1,t^2,\ldots,t^{K_{max}}\}$, *strategy*);
(b) $k := 1$;
(c) $ACC := \emptyset$;
(d) **while**($k \leq K^{max}$) **do**;
(e) *current_time* := t^k;
(f) make_tentative_assignment(ACC, $\mathcal{N}$);
(g) M^k:=define_parameter_update_model(ACC);
(h) (Q,BP):=update_control_parameters(M^k);
(i) REQ^k:=fetch_waiting_requests();
(j) **if** ($strategy == QUOTA$) **then** process_by_quota(REQ^k,$\mathcal{N}$, Q);
(k) **if** ($strategy == BP$) **then** process_by_bp(REQ^k,$\mathcal{N}$, BP);
(l) propagate_decisions();
(m) update(ACC);
(n) $k := k+1$;
(o) **wend**;

Fig. 1. Pseudo-code of the capacity-control system

with the definition of the current instance of the parameter update model (1)-(6) in line (g). This model is solved (h) and the quotas as well as resource bid prices are updated. Now, the currently waiting spot-market requests are collected in the set REQ^k (i). If a quota-based capacity control strategy is used then the waiting requests are processed calling the decision making procedure *process_by_quota*() (j) but if a bid-price-based capacity control strategy is incorporated then the processing of the fetched requests is done by calling the procedure *process_by_bp*() (k). After all requests have been processed the acceptance decisions are send back to the customers (l). The set ACC of already accepted requests is updated (m) and the iteration counter is increased by 1 (n).

3.2 Resource (Re-)Allocation

When the acceptance decision is made for a request then it is ensured that enough capacity is available and a tentative resource allocation is made. Since the demand forecast and/or the expected revenue differ from the actually appearing demand it is necessary to evaluate and re-work the assignment regularly in order to ensure an optimal resource utilization. Therefore, the procedure make_tentative_assignment(ACC, $\mathcal{N}$) is called. Its general function is described in this subsection.

At time t^k we have to assign each accepted request $r \in ACC$ to an appropriate service $s \in S$ operating in the considered network $\mathcal{N}$. For coding the necessary allocation decisions we introduce the family x^k_{rs} ($r \in ACC, s \in \mathcal{S}$) of binary decision variables. We define x^k_{rs} to be 1 if and only if request r is assigned to service s at time t^k. In a preprocessing step, we determine the value of the indicators $\rho(r,s)$. If service s offers the product of request r we set $\rho(r,s)$ to 1 otherwise we set $\rho(r,s)$ to 0 (*operability*).

It is necessary to assign each request to exactly one service. In order to be able to allocate resources belonging to service s for request r it is necessary to limit the entirety of selectable services. Only those services offering product $P(r)$ become potential candidate services for request r. The capacity of the transport resources $a \in \mathcal{A}$ are limited and only $CAP(a)$ capacity units can be moved at the same time by a resource.

$$\sum_{s \in \mathcal{S}} x^k_{rs} = 1 \; \forall r \in ACC \tag{7}$$

$$x^k_{rs} \leq \rho(r,s) \; \forall r \in ACC, s \in \mathcal{S} \tag{8}$$

$$\sum_{s \in \mathcal{S}} \sum_{r \in \mathcal{R}} SPR(S(a), P(r), a) \cdot x_{rs} \leq C(a) \forall a \in \mathcal{A} \tag{9}$$

$$x^k_{rs} \in \{0,1\} \; \forall r \in ACC, s \in \mathcal{S} \tag{10}$$

Every assignment scheme that respects the constraints (7)-(10) is a feasible one. A partition of the current portfolio of accepted requests is achieved by incorporating constraint (7). The consideration of constraint (8) ensures that request r is executable by service s. Constraint (9) must be satisfied in order to respect the limited capacities of the resources.

$$\sum_{r \in ACC} \sum_{s \in \mathcal{S}} BP(r,s) x^k_{rs} \rightarrow \min \tag{11}$$

Among the available allocations represented by the feasible assignment schemes we are looking for the most beneficiary one. In our investigation, we evaluate an allocation by its profit. Therefore, we determine the assignment specific costs $BP(r,s)$ that account if resources of service s are allocated for the fulfillment of request r. We approximate $BP(r,s)$ by summing up the current bid-prices of resources belonging to service s and being deployed in the fulfillment of r. The overall costs for an assignment scheme are achieved by summing up the $BP(r,s)$-values for each selected assignment. We select the assignment scheme that minimizes this sum (11) while fulfilling (7)-(10).

3.3 Control Policies

This subsection outlines the procedure that processes the additionally arrived requests and that determines the required acceptance decisions.

Quota-Based Control (QUOTA). By solving the DLP (1)-(6) we determine service- and product-specific contingents $Q(p,s)$ (quotas). In order to gain the maximal possible revenues from the stream of spot-market requests the following acceptance rule is applied: an additionally arrived request for product p using service s is accepted as long as the contingent $Q(p,s)$ is not completely exhausted.

At first, the procedure *process_by_quota*() determine an arbitrarily generated order of the requests contained in REQ^k. These requests are then processed one after another. In order to decide whether request r is accepted, the procedure *process_by_quota* first determines the product $P(r)$ of request r. Next, it checks if there are still service(s) available that can fulfill r. For that purpose, it tests consecutively for all available services s, if $Q(P(r),s) > 0$. Let s^* be the first service whose quota is still available. If such a service does not exist, r is rejected. Otherwise, r is accepted and the quota is updated by $Q(P(r),s^*) := Q(P(r),s^*) - 1$. Request r is tentatively assigned to service s^*.

Bid-Price Control (BP). Also the procedure *process_by_bp* first sequentializes the waiting requests and then processes the requests one after another. According to commonly used acceptance policies for flexible products [16], the carrier accepts r if and only if there is at least one service s^* available that can be used to fulfill r and that has a bid-price $BP(r,s)$ below the revenues $REV(r)$ associated with the fulfillment of r. Hence, r is accepted if and only if $BP(r,s^*) \leq REV(r)$ and if $CAP(s^*) > 0$ for at least one service s^*. Request r is then tentatively assigned to the service $s^\star$ that leads to the highest positive contribution margin $REV(r) - BP(r,s^*)$.

4 Computational Experiments

We report on computational simulation experiments in which the capacity control system developed in Section 3 is evaluated. The booking processes for the demand scenarios introduced in [19] are simulated. The metrics used to represent the simulation results are introduced in Subsection 4.1 and the observed results are presented and discussed in Subsection 4.2.

4.1 Metrics for Evaluation

We execute a comprehensive evaluation of the two capacity control policies $QUOTA$ and BP introduced in Subsection 3.3. To facilitate such an evaluation, we define different performance indicators in order to represent the policy behavior during the normalized booking period $[0,1]$ with the acceptance decision times $0, 0.1, \ldots, 1.0$. In order to evaluate the performance of the two sophisticated acceptance strategies BP and QUOTA, we compare them with two benchmark strategies. The first benchmark strategy is $FCFS$ (first-come / first-serve) and for the second benchmark strategy BP^{exact} we assume that an exact forecast of the future demand is available for the BP-strategy. For the reason of simplicity, we restrict the presentation of the observed results to some meaningful configurations. Actually, we compare $FCFS$ and BP^{exact} with $QUOTA$ and BP in the event that $\alpha = 0.6$ and $\beta = 0.2$.

Firstly, we fetch the sum of revenues $rev_t(i,\phi,\alpha,\beta)$ gained up to time t for the i-th realization (draw) of a demand pattern and with the applied control policy $\phi \in \{QUOTA, BP\}$ and the forecast of quality (α,β). For each simulation experiment configuration (ϕ,α,β) we calculate the average $rev_t(\phi,\alpha,\beta)$ over all demand pattern draws. Let $rev^{max}(\alpha,\beta)$ be the maximum of the set $\{rev_t(\phi,\alpha,\beta) \mid t \in \{0, 0.1, \ldots, 1.0\},\ \phi \in \{QUOTA, BP\}\}$. The normalized sum of gained revenues for the configuration (ϕ,α,β) is then defined as $r_t(\phi,\alpha,\beta) := \frac{rev_t(\phi,\alpha,\beta)}{rev^{max}(\alpha,\beta)}$. These values vary between 0 and 1.

The capacity utilization rate $cap_t(\phi,\alpha,\beta)$ represents the average of the capacity utilization observed for all demand pattern i if the control policy ϕ is applied. Similarly to the normalized sum of gained revenues we calculate the normalized capacity utilization rate $c_t(\phi,\alpha,\beta)$.

A third performance indicator is the normalized bid price $bp_t(\phi,\alpha,\beta)$ for the utilization of one capacity unit. Finally, the forth and last observed performance indicator is the normalized acceptance rate $a_t(\phi, \alpha, \beta)$ that expresses the portion of requests accepted at time t among the requests released at this time. We observe the development of performance indicator values during the booking phases' progress (*online analysis*).

4.2 Results

The so far presented and analyzed simulation results generated by the application of the previously described capacity control system [19] are the outcome of *offline evaluations* of QUOTA and BP. The indicator variable values determine the control policy performance after the complete booking period has been processed. In order to find out the reasons for the quite different final indicator variable results we perform *online evaluations* throughout the booking phase. Now, we fetch the current indicator values at the time points $t = 0, 0.1, \ldots, 1.0$ during the booking phase. The development of the performance indicators during this period can then be captured.

All four observed acceptance strategies are able to collect revenues throughout the overall booking period from $t = 0$ to $t = 1.0$. However, they exhibit a different behavior with respect to the overall sum of collected revenues and the part of the booking period where the revenues are collected (left picture in Fig. 2). The highest sum of revenues is collected by the BP-strategy if an exact demand forecast is exploited and the lowest sum is earned by FCFS, which leads to a portion of 70% of the revenue sum observed for BP^{exact}. For the first half of the booking period ($0 \leq t \leq 0.5$), $FCFS$ collects more revenues than all other strategies but during the second half of the booking period $FCFS$ is outperformed by all other strategies so that finally, $FCFS$ captures the least sum of revenues. During the interval from 0.5 to 0.75 $QUOTA$ as well as BP outperform even BP^{exact}. $QUOTA$ exhibits the best performance. However, during the last quarter of the booking period, $QUOTA$ as well as BP are finally outperformed by BP^{exact}. The $QUOTA$-strategy reaches 81% of the revenues of BP^{exact} but BP approximates BP^{exact} by 88%. In conclusion, imprecise demand forecasts interfere with the achieved sum of revenues but the "more intelligent strategies" outperform the greedy $FCFS$-strategy significantly. Finally, the BP acceptance strategy is working better than the $QUOTA$-strategy. This verifies the initially stated research hypothesis.

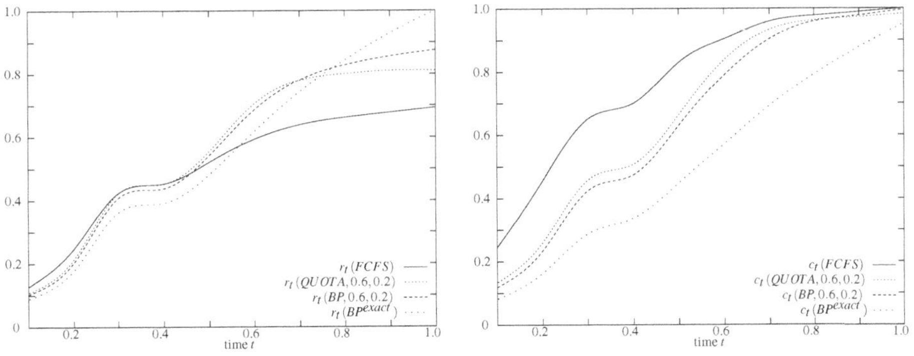

Fig. 2. Development of the normalized sum of gained revenues $r_t(\phi, 0.6, 0.2)$ (left picture), of the normalized capacity utilization degree $c_t(\phi, 0.6, 0.2)$ (right picture)

A first attempt is made to explain the quite different behavior of the acceptance strategies. We compare the development of the capacity utilization during the progress of the booking period (right picture in Fig. 2). At first, we see that BP^{exact} leads to a significantly lower capacity utilization rate than $FCFS$, BP and $QUOTA$. At the end of the booking period the three last-mentioned strategies achieve a normalized capacity utilization rate above 95% but BP^{exact} does not climb over 95%. $FCFS$, BP as well as $QUOTA$ demonstrate a significantly greedier behavior and they reserve more capacity in the first three quarters of the booking period. Vice-versa, after three quarters of the booking periods length is over, these three acceptance strategies have already exhausted 95% of the normalized capacity. At the same time, BP^{exact} has only spent 80% of the capacity and has enough remaining capacity for incorporating profitable requests.

The inspection of the acceptance rate development during the booking period supports the theory that the earning of additional revenues is prevented by missing capacity at the end of the booking period. In the left picture of Fig. 3 we have compiled the acceptance rates of the four acceptance strategies over the length of the booking period. In the event that the demand forecast is close to be perfect (BP^{exact}) the acceptance rate remains stable throughout the booking period. In average, 50% of the incoming requests are identified to be profitable. If one of the three other strategies is applied then a drastic reduction of the acceptance rate is detected. While FCFS accepts alle incoming requests at the entry of the booking period this rate declines down to 10% at the end of the booking period. The two 'intelligent' strategies BP and $QUOTA$ demonstrate another development trail. At the beginning, they selects around 70% of the requests for acceptance but at the end of the booking period they do not allow more than 18% of the proposed requests to be accepted.

We have calculated the bid-prices also for the situation in which we use the $QUOTA$ acceptance strategy. The right picture in Fig. 3 reveals the impacts of demand forecast impreciseness. An underestimation of the future demand leads to very low bid prices of the remaining capacity. This observation is true for both reservation strategies BP as well as $QUOTA$. In the event that the expected demand is underestimated then the bid-prices decrease from time $t = 0.6$ because the small number of expected upcoming

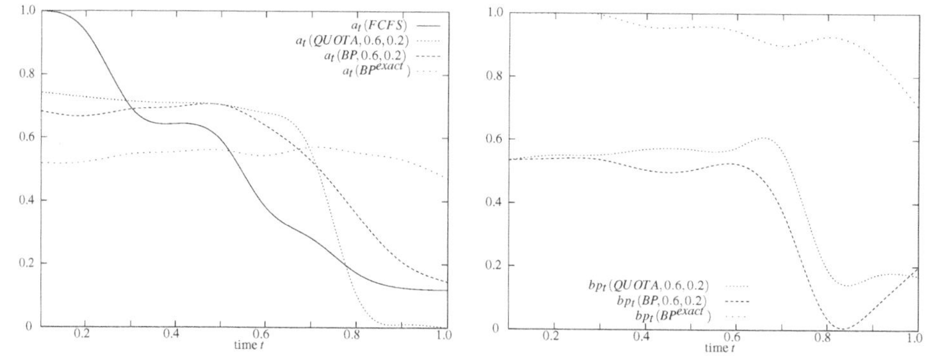

Fig. 3. Development of the normalized acceptance of a capacity unit $a_t(\phi, 0.6, 0.2)$ (left picture) and of the normalized average bid-price rate $bp_t(\phi, 0.6, 0.2)$ (right picture)

requests is tried to be exploited at maximal intensity. It becomes more and more unlikely that the remaining capacity can be exhausted with the still expected additional requests. Thus, the capacity is not identified anymore as scarce so that the corresponding bid-price collapses. Immediately after the failure of the demand forecast becomes obvious (because further unexpected requests arrive) scarceness of the resources is expected and the bid-prices of the capacity increase again. However, since the remaining capacity is too low caused by the too aggressive acceptance of requests at the beginning, no significant additional revenues can be realized.

5 Conclusions

Within the reported research we have proposed a capacity control system for supporting the sales and capacity management of scheduled services in road-based freight transportation. A quota-based capacity control strategy as well as a bid-price-founded capacity control strategy have been proposed. Within comprehensive computational simulation experiments their general applicability has been proven. We were able to proof our research hypothesis that impacts of demand forecast errors will become effective during the second half of the booking period.

Future research efforts are currently in progress with the goal to find out the impacts of wrong demand volume forecasts on revenues. In additional we want to integrate the service generation with the acceptance strategies in order to exploit the maximal revenue increase potentials in road-based freight transportation. Another research trail is the development of countermeasures to be applied online during the selling interval as soon as a wrong demand forecast is revealed.

Acknowledgement

This research was supported by the German Research Foundation (DFG) as part of the Collaborative Research Centre 637 *Autonomous Cooperating Logistic Processes* (Subproject B7).

References

1. Aras, N., Aksen, D., Tekin, M.: Selective multi-depot vehicle routing problem with pricing. Transportation Research Part C: Emerging Technologies (2010) (to appear)
2. Becker, M., Wenning, B.-L., Görg, C., Gehrke, J.D., Lorenz, M., Herzog, O.: Agent-based and discrete event simulation of autonomous logistic process. In: Borutzky, W., Orsoni, A., Zobel, R. (eds.) 20th European Conference on Modelling and Simulation, pp. 566–571 (2006)
3. Corsten, H., Gössinger, R.: Kapazitätssteuerung im revenue management. Zeitschrift für Betriebswirtschaftslehre 67, 31–52 (2005)
4. Feillet, D., Dejax, P., Gendreau, M.: Traveling salesman problems with profits. Transportation Science 39, 188–205 (2005)
5. Gallego, G., Iyengar, G., Phillips, R., Dubey, A.: Managing flexible products on a network. Technical report, Ithaca, United States (2004)

6. Gallego, G., Phillips, R.: Revenue management of flexible products. Manufacturing&Service Operations Management 6, 321–337 (2004)
7. Golden, B., Raghavan, S., Wasil, E.: The Vehicle Routing Problem. Springer, Heidelberg (2008)
8. Kimms, A., Müller-Bungart, M.: Simulation of stochastic demand data streams for network revenue management problems. OR Spectrum 29, 5–20 (2007)
9. Klein, R.: Network capacity control using self-adjusting bid-prices. OR Spectrum 29, 39–60 (2007)
10. Klein, R., Steinhardt, C.: Revenue Management. Springer, Heidelberg (2009)
11. Kopfer, H., Wang, X.: Combining vehicle routing with forwarding - extension of the vehicle routing problem by different types of sub-contraction. Journal of the Korean Institute of Industrial Engineers 35, 1–14 (2009)
12. Krajewska, M., Kopfer, H.: Transportation planning in freight forwarding companies - tabu search algorithm for the integrated operational transportation planning problem. European Journal of Operational Research (EJOR) 197, 741–751 (2009)
13. Kunkel, M.: Yield-Management-Studie 2009, Steria Mummert Consulting AG (2010)
14. Müller-Bungart, M.: Revenue Management with Flexible Products. Springer, Heidelberg (2007)
15. Pankratz, G.: Speditionelle Transportdisposition. DUV (2002)
16. Petrick, A., Steinhard, C., Gönsch, J., Klein, R.: Using flexible products to cope with demand uncertainty in revenue management. OR Spectrum (2010a) (to appear)
17. Petrick, A., Gönsch, J., Steinhard, C., Klein, R.: Dynamic control mechanisms for revenue management with flexible products. Computers&Operations Research, 2027–2039 (2010b)
18. Rehkopf, S.: Revenue Management-Konzepte zur Auftragsannahme bei kundenindividueller Produktion. DUV (2006)
19. Schönberger, J., Kopfer, H.: Approaching the application borders of network capacity control in road haulage. In: Hülsmann, M., Scholz-Reiter, B., Windt, K. (eds.) Autonomous Cooperation and Control in Logistics. Springer, Heidelberg (2011)
20. Schönberger, J., Kopfer, H.: Kapazitätssteuerung in der Transportlogistik. Ein Ansatz zur Ertrags-maximierenden Ressourcen-Allokation im Straßengüterverkehr. In: Schumann, M., Kolbe, L.M., Breitner, M.H., Frerichs, A. (Hrsg.) Tagungsband der Multikonferenz Wirtschaftsinformatik 2010 (MKWI 2010), Universitätsverlag Göttingen, pp. S.383–S.384 (2010)
21. Schönberger, J., Kopfer, H.: Online decision making and automatic decision model adaptation. Computers & Operations Research 36, 1740–1750 (2009)
22. Schönberger, J.: Operational Freight Carrier Planning. Springer, Heidelberg (2005)
23. Talluri, K., Ryzin, G.V.: The theory and practice of revenue management. Springer, Heidelberg (2005)
24. Valle, C., da Cunha, A.S., Mateus, G., Martinez, L.: Exact algorithms for a selective vehicle routing problem where the longest route is minimized. Electronic Notes in Discrete Mathematics 35, 133–138 (2009)

A Branch-and-Price Algorithm for the Risk-Equity Constrained Routing Problem

Nora Touati-Moungla[1], Pietro Belotti[2], Vincent Jost[3], and Leo Liberti[4]

[1] École polytechnique, Laboratoire d'informatique (LIX), 91128 Palaiseau Cedex, France
touati@lix.polytechnique.fr
[2] Department of Mathematical Sciences of Clemson University
pbelott@clemson.edu
[3] École polytechnique, Laboratoire d'informatique (LIX), 91128 Palaiseau Cedex, France
vincent.jost@lix.polytechnique.fr
[4] École polytechnique, Laboratoire d'informatique (LIX), 91128 Palaiseau Cedex, France
liberti@lix.polytechnique.fr

Abstract. We study a multi-criteria variant of the problem of routing hazardous material on a geographical area subdivided in regions. The two objective functions are given by a generally defined routing cost and a *risk equity* equal to the maximum, over each region, of the risk *perceived* within a region. This is a multicommodity flow problem where integer variables are used to define the number of trucks used for the routing. This problem admits a straightforward path formulation, for which a branch-and-price problem where, for each node of the branch-and-bound tree, column generation is used to obtain a lower bound.

1 Introduction

The transportation of hazardous materials (*hazmat* from now on) has received a large interest in recent years, this results from the increase in public awareness of the dangers of hazmats and the enormous amount of hazmats being transported [3]. The main target of this problem is to select routes from a given origin-destination pair of nodes such that the risk for the surrounding population and the environment minimum, without producing excessive economic costs. When solving such a problem by minimizing both cost and the total risk, typically several vehicles share the same (short) routes which results in high risks associated to regions surrounding these paths whereas other regions are less affected. In this case, one may wish to distribute the risk in an equitable way over the population and the environment. The computation of routes with a fairly distributed risk consists in generating dissimilar origin-destination paths, i.e paths which relatively don't impact the same zones. We classify solution approaches in two sets, *resolution-equity-based methods* and *model-equity-based methods*.

In *resolution-equity-based methods*, equity constraints are taken into account in the resolution process. These methods are based on a dissimilarity index which permits to indicate when two paths are considered as dissimilar. The iterative penalty method [10] consists of computing iteratively a shortest path and penalize its arcs by increasing their weights for discouraging the selection of the same arc set in the generated paths set in the next iteration. The Gateway shortest-paths method [13] consists of generating

J. Pahl, T. Reiners, and S. Voß (Eds.): INOC 2011, LNCS 6701, pp. 439–449, 2011.

dissimilar paths by forcing at each time a new path to go through a different node (called the gateway node), the dissimilarity index is defined as the absolute difference between areas under the paths (areas between paths and the abscissa axis). The minimax method [12] consists of selecting k origin-destination shortest-paths and select among them iteratively a subset of dissimilar paths by means of a dissimilar index defined as the length of common parts between the paths. The p-dispersion method [1] generates an initial set U of paths and determines a maximal dissimilar subset S, i.e., the one with the maximum minimum dissimilarity among its paths, the dissimilarity index is the length of common parts or the common impact zones between the paths. The efficiency of these methods is based on the dissimilarity index and the initial set of paths [3].

Model-equity-based methods consist of taking into account equity constraints in the model formulation. In [8,9], the authors propose an equity shortest path model that minimizes the total risk of travel, while the difference between the risks imposed on any two arbitrary zones does not exceed a given threshold, the authors solve the lagrangean relaxation of the problem and a gap-closing procedure is presented. In [3] is proposed a multi-commodity flow model for routing of hazmat, where each commodity is considered as one hazmat type. The objective function is formulated as the sum of the economical cost and the cost related to the consequences of an incident for each material (commodity). To deal with risk equity, the costs are defined as functions of the flow traversing the arcs, this imposes an increase of the arc's cost and risk when the number of vehicles transporting a given material increases on the arc.

Our problem is similar to that proposed in [3]. We consider the problem where a set of given quantities of hazmats has to be routed over a transportation network from specific origin points to specific destination points. Our goal is the minimization of the total routing cost *and* the maximization of the risk equity, the latter broadly defined as the risk shared by a set of regions that compose the geographical area under consideration. Thus our focus is a multi-criteria optimization problem which we describe more in detail below. The originality of our work is the integration of the objective of minimization of the maximum of risk imposed on all regions during the transportation activity into the multi-commodity flow model which can be solved using a Branch-and-Price algorithm.

This paper is organized as follows. In Section 2, the problem is described and an optimization model is given. In Section 3 a path formulation of the problem is given and a column generation procedure is described. We present in Section 5 a branch-and-price procedure, in Section 5 we present some preliminary experiments and we close the paper in Section 6.

2 Description of the Problem

Let the transportation network be represented as a directed graph $G = (N,A)$, with N being the set of n nodes and A the set of m arcs. Let C be the set of commodities, given as a set of point-to-point demands to transport a certain amount of hazmats. For any commodity $c \in C$, let s^c and t^c be respectively the source node and the destination node, and let D^c be the amount of hazmats to be shipped, by means of a set of trucks of given capacity F_c, from s^c to t^c. We for now assume that each commodity is associated with a unique type of hazmat.

We assume that the risk is computed on each arc of the network and is proportional to the flow traversing such an arc. We consider a set Q of regions, each given as subsets N_q of nodes for each $q \in Q$ of the transportation network, and we define r_{ij}^{cq} as the risk imposed on region $q \in Q$ when the arc $(i, j) \in A$ is used for the transportation of one unit of hazmats of type c. We remark that we employ a notion of *spread risk*, in that an accidental event on arc (i, j) within region $q \in Q$ may strongly affect another region $q' \in Q$.

2.1 Multiple Objective Functions

The problem of transporting hazmat is multi-objective in nature: one usually wants to minimize two (or more) objectives, namely the total cost of transportation, computed as a function of the amount of hazmat transported throughout the network and the trucks used for the transportation, and the *distributed risk*, which can be defined as a measure of risk that is shared among different regions. More specifically, for a given solution each region $q \in Q$ will be affected by a risk which is dependent on the transportation patterns in all other regions, and which can be summarized by a quantity ω_q. The second objective will then be $\max_{q \in Q} \omega_q$, and has to be minimized.

2.2 An Optimization Model

We introduce a flow variable f_{ij}^c defining the portion of commodity c being transported on arc (i, j). These variables are subject to flow conservation constraints

$$\sum_{j \in \delta^+(i)} f_{ij}^c - \sum_{j \in \delta^-(i)} f_{ji}^c = b_i^c \qquad \forall i \in N, c \in C$$

where $\delta^-(i)$ and $\delta^+(i)$ are the forward and backward star of i, i.e.,

$$\delta^-(i) = \{j \in N : (j,i) \in A\}, \qquad \delta^+(i) = \{j \in N : (i, j) \in A\},$$

and

$$b_i^c = \begin{cases} 1 & \text{if } i = s^c \\ -1 & \text{if } i = t^c \\ 0 & \text{otherwise.} \end{cases}$$

Also, y_{ij}^c defines the number of trucks to be used on arc (i, j) for commodity c. The link between variables f and y is given by the constraint

$$D_c f_{ij}^c \leq F_c y_{ij}^c \qquad \forall (i, j) \in A, c \in C.$$

The first objective is a function of both f and y variables and is to be minimized: $\sum_{c \in C} \sum_{(i,j) \in A} (\alpha_{ij}^c f_{ij}^c + \beta_{ij}^c y_{ij}^c)$, with α and β suitable cost coefficients which we assume nonnegative. We define the risk ω_q imposed on a region $q \in Q$ as a linear combination of the flow variables:

$$\omega_q := \sum_{c \in C} \sum_{(i,j) \in A} r_{ij}^{cq} f_{ij}^c$$

and add a new variable $z := \max_{q \in Q} \omega_q$, which therefore is subject to the constraints

$$z \geq \sum_{c \in C} \sum_{(i,j) \in A} r_{ij}^{cq} f_{ij}^{c} \qquad \forall q \in Q.$$

The y variables represent trucks that transport hazmat from each source to each destination, and are therefore subject to flow conservation constraints. We write such constraints here for each commodity and for all intermediate nodes of each commodity, as the source and destination flow balance is redundant here (i.e., it is strictly dependent on the flow variables f):

$$\sum_{j \in \delta^{+}(i)} y_{ij}^{c} - \sum_{j \in \delta^{-}(i)} y_{ji}^{c} = 0 \qquad \forall i \in N \setminus \{s^c, t^c\}, \forall c \in C$$

The optimization model is therefore as follows:

$$\begin{aligned}
\min \quad & \textstyle\sum_{c \in C} \sum_{(i,j) \in A} (\alpha_{ij}^{c} f_{ij}^{c} + \beta_{ij}^{c} y_{ij}^{c}) && (1)\\
\min \quad & z && (2)\\
s.t. \quad & \textstyle\sum_{j \in \delta^{+}(i)} f_{ij}^{c} - \sum_{j \in \delta^{-}(i)} f_{ji}^{c} = b_i^c, \ \forall i \in N, c \in C && (3)\\
& \textstyle\sum_{j \in \delta^{+}(i)} y_{ij}^{c} - \sum_{j \in \delta^{-}(i)} y_{ji}^{c} = 0 \quad \forall i \in N \setminus \{s^c, t^c\}, \forall c \in C && (4)\\
& D_c f_{ij}^{c} \leq F_c y_{ij}^{c} \qquad \forall (i,j) \in A, c \in C && (5)\\
& z \geq \textstyle\sum_{c \in C} \sum_{(i,j) \in A} r_{ij}^{cq} f_{ij}^{c} \qquad \forall q \in Q && (6)\\
& f_{ij}^{c} \in [0,1] \qquad \forall (i,j) \in A, c \in C && (7)\\
& y_{ij}^{c} \in \mathbb{Z} \qquad \forall (i,j) \in A, c \in C. && (8)
\end{aligned}$$

Notice that constraints (4) and (5) guarantee that a sufficient number of trucks is allocated for each commodity regardless of the flow of hazmat. The path formulation described below is unable to provide such a guarantee and will therefore have to be modified.

3 Column Generation Formulation

The above *arc-flow* formulation is polynomial in $|N|$, $|A|$, $|Q|$, and $|C|$, but its size can make it impractical to solve real-world instances of our problem. A common approach is to use a *path-flow* formulation [7]. In these formulations, for each commodity c a variable is associated with every *path* from s^c to t^c. We denote by $\mathscr{P}^c$ the set of paths from s^c to t^c for a commodity $c \in C$ and by $\mathscr{P}_{ij}^c$ the set of paths in $\mathscr{P}^c$ containing arc $(i,j) \in A$. A new path variable f_p, $\forall p \in \mathscr{P}^c, \forall c \in C$, represents the portion of commodity transported on path p.

As for the flow of hazmat, in this formulation the number of trucks, previously denoted by variables y_{ij}^c, might be dependent on path variables f_p. They are by definition the number of trucks to be used on arc $(i,j) \in A$ for commodity $c \in C$. In practice, each truck drives on the whole path p, hence there should be a variable y_p that counts the number of trucks and that is related to variable f_p as follows:

$$F_c y_p \geq D_c f_p \qquad \forall p \in \mathscr{P}^c, c \in C. \tag{9}$$

This constraint substitutes the flow conservation constraint (4) and the "capacity" constraint (5). Let us write this path formulation for completeness:

$$\begin{aligned}
&\min \textstyle\sum_{c\in C}(\sum_{p\in \mathscr{P}^c} \alpha_p f_p + \sum_{(i,j)\in A} \beta_{ij} y^c_{ij}) && && (10)\\
&\min && z && (11)\\
&s.t. && \textstyle\sum_{p\in \mathscr{P}^c} f_p \geq 1 \quad && \forall c \in C && (12)\\
& && F_c y_p - D_c f_p \geq 0 && \forall p \in \mathscr{P}^c, c \in C && (13)\\
& && z - \textstyle\sum_{c\in C} \sum_{p\in \mathscr{P}^c} r^q_p f_p \geq 0 && \forall q \in Q && (14)\\
& && f_p \geq 0 && \forall p \in \mathscr{P}^c, c \in C && (15)\\
& && y^c_p \in \mathbb{Z} && \forall p \in \mathscr{P}^c, c \in C, && (16)
\end{aligned}$$

where $\alpha_p = \sum_{(i,j)\in p} \alpha^c_{ij}$ are cost coefficients on the path $p \in \mathscr{P}^c$ and $r^q_p = \sum_{(i,j)\in p} r^{cq}_{ij}$ is the risk imposed on region $q \in Q$ when the path $p \in \mathscr{P}^c$ is used for the transportation of one unit of hazmats of type c. Constraint (12) is the path-flow counterpart of the flow conservation constraint (3) and requires that, regardless of the set of paths used, each commodity is fully routed. Constraints (13) and (14) are straightforward extensions of (5) and (6) respectively, given that the flow of commodity $c \in C$ on arc $(i,j) \in A$ is equal to $\sum_{p\in \mathscr{P}^c_{ij}} f_p$.

When restricting to a single-objective optimization problem, this model is an integer multicommodity flow problem. Regardless of considering only one objective, problem (11)-(16) contains $V = \sum_{c\in C} |\mathscr{P}^c|$ variables, which can be exponential in $|N|$. Therefore, solving it by introducing all path variables is in general impractical using the usual combinatorial optimization methods.

Column generation algorithms are very well suited for solving this kind of problems [4]. They use a relatively small initial set of columns to solve a problem, and iteratively introduce a new column when necessary to improve the objective function. Specifically, given a set of columns with negative reduced cost (among those that haven't been considered yet), one can introduce one or more such variables and apply a primal simplex method to resolve the amended problem. The problem with an initially small subset of columns is called the *restricted master problem*, while the problem of finding a variable (column) with negative reduced cost is called the *pricing problem*.

Constraint (13) introduces a major issue in the problem. In principle, introducing y variables indexed on paths rather than arcs and commodities allows to further reduce the number of columns, as we only need $1 + 2\sum_{c\in C} |\mathscr{P}^c|$ variables. However, analogously to columns, we do not want to have exponentially many rows (there are exponentially many paths). The above constraint could be *dynamically* generated, hence instead of column generation we would need *row-column* generation. One huge problem here is that to dynamically generate paths one needs to know all dual variables σ_p, for each $p \in \mathscr{P}^c$ and for all $c \in C$, to solve a pricing problem, and most of these dual variables are *not* available since we didn't generate all of them.

One possible way to deal with this is to use *surrogate constraints*: rather than impose all such constraints or generate them dynamically, we consider a *cover* of such set of constraints and impose conic combinations thereof (see for instance [14]). More specifically, for each $(i,j) \in A$, consider all constraints (9) summed up for all paths containing (i,j). We obtain

$$F_c \sum_{p \in \mathscr{P}_{ij}^c} y_p \geq D_c \sum_{p \in \mathscr{P}_{ij}^c} f_p \qquad \forall (i,j) \in A, c \in C. \tag{17}$$

The problem has now $|C|(1+m)+|Q|$ rows and $1+2\sum_{c\in C}|\mathscr{P}^c|$ variables. Since we relax all of the path constraints (9), the model (10)-(16) constitutes a relaxation of (1)-(8). Column generation can be applied safely now, although it has to be applied to both f and y variables, and it converges to a *dual feasible* solution which gives a lower bound but not necessarily an optimal solution of the continuous relaxation of (1)-(8).

Suppose an integer solution is found as the optimal solution of the LP relaxation (solved with column generation). If at least one of the constraints (13) is violated, we are stuck with a solution that has no physical value but that cannot be proven primal infeasible unless a constraint is added. What we can do is therefore to create a second branching rule which discriminates between integer feasible solutions and eliminates the integer (but infeasible) solution just found. We will detail this procedure later in this paper, and instead provide insight on how to generate variables.

3.1 Handling One Objective Only

We consider from now on a continuous relaxation of (11)-(16) amended by the surrogate constraints:

$$\begin{aligned}
\min \quad & z && && (18)\\
s.t. \quad & \textstyle\sum_{p\in\mathscr{P}^c} f_p \geq 1 && \forall c \in C && (19)\\
& F_c \textstyle\sum_{p\in\mathscr{P}_{ij}^c} y_p \geq D_c \sum_{p\in\mathscr{P}_{ij}^c} f_p && \forall (i,j) \in A, c \in C && (20)\\
& z - \textstyle\sum_{c\in C}\sum_{p\in\mathscr{P}^c} r_p^q f_p \geq 0 && \forall q \in Q && (21)\\
& f_p \geq 0 && \forall p \in \mathscr{P}^c, c \in C && (22)
\end{aligned}$$

We associate the dual variables vector $\mu \in \mathbb{R}_+^{|C|}$ with constraints (19), $\sigma \in \mathbb{R}_+^{m|C|}$ with constraints (20), and $\lambda \in \mathbb{R}_+^{|Q|}$ with constraints (21). We first analyze this problem considering the single objective (11). Let us define the subset of paths $\bar{\mathscr{P}}^c \subset \mathscr{P}^c, \forall c \in C$. The restricted master problem (RMP from now on) of (11)-(16), generated on a restricted subset of variables $f_p, p \in \bar{\mathscr{P}}^c, c \in C$, is as follows:

$$\begin{aligned}
\min \quad & z && && (23)\\
s.t. \quad & \textstyle\sum_{p\in\bar{\mathscr{P}}^c} f_p \geq 1 && \forall c \in C && (24)\\
& F_c \textstyle\sum_{p\in\mathscr{P}_{ij}^c} y_p - D_c \sum_{p\in\bar{\mathscr{P}}_{ij}^c} f_p \geq 0 && \forall c \in C, (i,j) \in A && (25)\\
& z - \textstyle\sum_{c\in C}\sum_{p\in\bar{\mathscr{P}}^c} r_p^q f_p \geq 0 && \forall q \in Q && (26)\\
& f_p \geq 0 && \forall p \in \mathscr{P}^c, c \in C. && (27)
\end{aligned}$$

It is barely worth noting here that (23)-(27) is a restriction of the continuous relaxation of (11)-(16), which therefore provides neither a lower nor an upper bound. Only by applying column generation to (23)-(27), i.e., by iteratively amending columns with negative reduced cost, can we find a lower bound of (11)-(16).

The reduced cost of variables f_p, for each $c \in C, p \in \mathscr{P}^c$, is as follows:

$$\begin{aligned} h(f_p) &= -\mu_c + D_c \sum_{(i,j)\in p} \sigma^c_{ij} + \sum_{q\in Q} r^q_p \lambda_q \\ &= -\mu_c + D_c \sum_{(i,j)\in p} \sigma^c_{ij} + \sum_{q\in Q} \sum_{(i,j)\in p} r^{cq}_{ij} \lambda_q. \end{aligned} \tag{28}$$

Suppose an optimal primal solution $(\bar{f}, \bar{y}, \bar{z})$ and an optimal dual solution $(\bar{\mu}, \bar{\sigma}, \bar{\lambda})$ is given. At each iteration of the column generation algorithm, we look for a negative reduced cost variable by solving the problem:

$$\min_{c\in C, p\in \mathscr{P}^c} h(f_p),$$

which provides the column with most negative reduced cost. The *pricing problem* consists of finding the path p that minimizes (28), and is equivalent to solving a *shortest path* problem on a graph G where each arc $(i,j) \in A$ has weight $w_{ij} = D_c \bar{\sigma}^c_{ij} + \sum_{q\in Q} r^{cq}_{ij} \bar{\lambda}_q$. The path must have an origin-destination pair among those defined by the commodities in C. Suppose that, for the shortest path obtained, $-\bar{\mu}_c + D_c L_c + \sum_{q\in Q} r^q_p \bar{\lambda}_q < 0$. Then variable f_p has a negative reduced cost and can be introduced in the model.

One may also look for a negative reduced cost variable for *each* commodity, and add at most $|C|$ such variables. Although this usually speeds up convergence in terms of number of iterations, adding many column every time slows the primal simplex used to obtain a new solution. We obtain $|C|$ origin-destination shortest path problems, therefore the pricing problem becomes $|C|$ times slower — this is negligible given that most of the CPU time is usually spent on the primal simplex.

Notice that y variables do not need to be generated for the risk-objective problem: they only appear in the surrogate constraint, which makes them completely useless given that their value can be decided from an optimal f. This only happens if we consider the second objective function, while the first does contain those variables and would force us to generate them as well. Actually, no f variable is needed either as long as the y variables are only contained in the capacity constraint. The next subsection should shed light on this and introduce another use for y variables.

3.2 Risk on Trucks

Another consideration is on risk equity associated to trucks: is the risk (especially the perceived one) only related to the real quantity, or portion, of hazmat transported, or is it also related on the trucks? If both quantity of hazmat and number of trucks should be considered, then the risk equity constraint would change. In this case, we could probably use a parameter s^{cq}_{ij} with an analogous meaning to that of parameter r, i.e., the influence of one truck driving through (i,j), transporting commodity $c \in C$, on region q, and modify (26) as follows:

$$z \geq \sum_{c\in C} \sum_{p\in \mathscr{P}^c} (r^q_p f_p + s^q_p y_p) \qquad \forall q \in Q.$$

where, similarly to r, we define $s^q_p := \sum_{(i,j)\in p} s^{cq}_{ij}$. This provides a motivation for the generation of both f and y variables. In fact, now the procedure to generate y variable can be defined as one that aims at finding a path p such that the reduced cost of the corresponding y_p is minimum:

$$\min_{c\in C, p\in \mathscr{P}^c} h(y_p) = \min\{-D_c \sum_{(i,j)\in p} \sigma_{ij}^c + \sum_{q\in Q} s_p^q \lambda_q : p \in \mathscr{P}\}$$

which provides a more difficult problem given that now the shortest path has to be found on a network with possibly both positive and negative weights.

4 Branch-and-Price for Single Objective Problems

In order to find an optimal integer solution to problem (10)-(16), the column generation approach outlined above must be coupled with a branch-and-bound algorithm. This class of algorithms, better known as *branch-and-price*, solve each branch-and-bound node by applying column generation on each lower bounding (continuous) subproblem [2]. For the single objective routing problem, we outline below an implementation of a branch-and-price, which we have implemented in ABACUS.

If only integer variables y_{ij}^c are not dynamically generated (but this no longer seems to be the case), the branching rule is rather simple: consider an optimal solution $(\bar{f}, \bar{y}, \bar{z})$ obtained after column generation at a branch-and-bound node. If, for all $c \in C$ and $(i,j) \in A$, we have $\bar{y}_{ij}^c \in \mathbb{Z}$, then the node can be fathomed as the solution is integer feasible. Otherwise, we select an arc $(i,j) \in A$ and a commodity $c \in C$ such that $\bar{y}_{ij}^c \notin \mathbb{Z}$ and generate two new branch-and-bound nodes with the amended constraints $y_{ij}^c \leq \left\lfloor \bar{y}_{ij}^c \right\rfloor$ and $y_{ij}^c \geq \left\lceil \bar{y}_{ij}^c \right\rceil$, respectively.

If we use y_p variables instead, we need to take special care in branching rules: given that these variables are generated, the branching rules have dual variables that need to be taken into account in the pricing problem. Furthermore, simple branching rules would not work and the branch-and-bound algorithm would not converge: the branching rule $y_p \leq k$, with $k \in \mathbb{Z}$, does not impose anything on the pricing problem, which might generate another variable that uses the same path as p with reduced cost. Another issue is making sure that the pricing problem remains a shortest path problem. One common branching rule for these cases is that used by Barnhart et al. [2].

5 Preliminary Experiments and Perspectives

In a first time, we test the efficiency of our model. We implement the formulation (11)-(16) in AMPL (A Modeling Language for Mathematical Programming) [6]. We report a sampling of our computational experiences with the model. We consider an instance with $N = 31$, $|C| = 3$ and $|Q| = 16$ (figure 1). We focus on risk equity objective function (11). Figure 2 present the solution obtained.

Throw our experimentations, we remarked that improving the equity of the risks imposed results in increased in the total risks imposed. When distributing the risk in an equitable way, routes can be longer, this increases both the total risk and the economic costs. We present on Table 1 the solution values generated by the weights (γ, δ), where

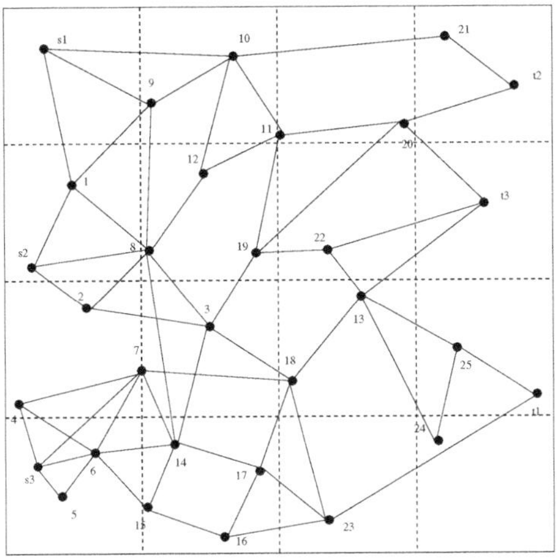

Fig. 1. Transportation risk model: network used in sample problem

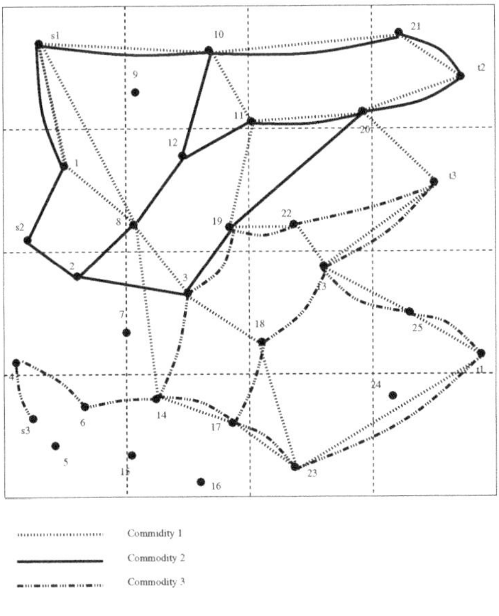

Fig. 2. Transportation risk model: solution of a sample problem

γ is the weight on the equity objectives and δ is weight on the total risk objectives (the objective function became: $\gamma z + \delta(\sum_{c \in C,(i,j) \in A, q \in Q} r_{ij}^{cq} f_{ij}^{c})$).

The tradeoffs among risk and the equity of the risk imposed are complex and the number of options are extremely large. Distributing the risk in an equitable way can result in an increase in the total risk and economic cost. In this case, it seems to be not realistic to consider the model with only one objective function. In a branch-and-price algorithm, we can consider two possibilities for considering more than one objective function, (1) the weighting method can be applied, and (2) the total risk objective

Table 1. Weighting approach

(γ,δ)	$(0,1)$	$(0.3,0.7)$	$(0.5,0.5)$	$(0.7,0.8)$	$(1,0)$
z	18	16.0	13.8	11.7	10.3
total risk	124	124.7	126.5	129.3	152.6

function can be taken into account during the column generation algorithm, where the pricing problem will compute Pareto optimal paths considering both the reduced cost and the total risk generated by the path.

6 Conclusion

The transportation of hazmats is an important optimization problem in the field of sustainable development and in particular the equitable distribution of risks is of high interest. Within this study, we formalize this transportation problem as the minimization of two objectives (risk equity and economic cost) and show that a third objective function (total risk) has to be taken into account. Note that, for the moment an actual implementation has to prove in the future what is the effectiveness of the algorithm, which additional accelerating techniques of column generation can be used for solving large instances and how can we take into account many objective functions.

References

1. Akgun, V., Erkut, E., Batta, R.: On finding dissimilar paths. European Journal of Operational Research 121(2), 232–246 (2000)
2. Barnhart, C., Hane, C.A., Vance, P.H.: Using Branch-and-Price-and-Cut to Solve Origin-Destination Integer Multicommodity Flow Problems. Oper. Res. 48(2), 318–326 (2000)
3. Caramia, M., Dell'Olmo, P.: Multiobjective management in freight logistics: increasing capacity, service level and safety with optimization algorithms. In: Hazardous Material Transportation Problems, ch. 4, pp. 65–101. Springer-Verlag London Ltd, Heidelberg (2008)
4. Desrosiers, J., Lübbecke, M.E.: A Primer in Column Generation, Column Generation, pp. 1–32. Springer, US (2006)
5. Erkut, E., Tjandra, S., Verter, V.: Transportation. In: Barnhart, C., Laporte, G. (eds.) Hazardous Materials Transportation. Handbooks in Operations Research and Management Science, vol. 14, ch. 9, pp. 539–621. Elsevier, Amsterdam (2007)
6. Fourer, R., Gay, D.M., Kernighan, B.W.: AMPL A modelling Language for Mathematical Programming, 2nd edn. Duxbury Press Brooks Cole Publishing Co. (2003)
7. Bertsekas, D.P., Gendron, B., Tsai, W.K.: Implementation of an Optimal Multicommodity Network Flow Algorithm Based on Gradient Projection and a Path Flow Formulation, Massachusetts Institute of Technology, LIDS-P-1364, pp. 1-66 (1984)
8. Gopalan, R., Batta, R., Karwan, M.H.: The equity constrained shortest path problem. Computers and Operations Research 17, 297–307 (1990)
9. Gopalan, R., Kolluri, K.S., Batta, R., Karwan, M.H.: Modeling equity of risk in the transportation of hazardous materials. Operations Research 38(6), 961–973 (1990)
10. Johnson, P.E., Joy, D.S., Clarke, D.B., Jacobi, J.M.: HIGWAY 3.01, An enhanced highway routing model: Program, description, methodology and revised user's manual, Oak Ridge National Laboratory, ORLN/TM-12124, Oak Ridge, TN (1992)

11. Jünger, M., Thienel, S.: The ABACUS system for branch-and-cut-and-price algorithms in integer programming and combinatorial optimization. Software: Practice and Experience 30(11), 1325–1352 (2000)
12. Kuby, M., Zhongyi, X., Xiaodong, X.: A minimax method for finding the k best differentiated paths. Geographical Analysis 29(4), 298–313 (1997)
13. Lombard, K., Church, R.L.: The gateway shortest path problem: Generating alternative routes for a corridor location problem. Geographical Systems 1, 25–45 (1993)
14. Orlowski, S., Raack, C., Koster, A.M.C.A., Baier, G., Engel, T., Belotti, P.: Branch-and-cut techniques for solving realistic two-layer network design problems. In: Graphs and Algorithms in Communication Networks, ch. 3, pp. 95–117. Springer, Heidelberg (2009)

A Matheuristic for the Dial-a-Ride Problem

Roberto Wolfler Calvo[1] and Nora Touati-Moungla[2]

[1] 99, Avenue J.-B. Clement, 93430 Villetaneuse, France
Roberto.Wolfler@lipn.univ-paris13.fr
[2] École polytechnique, Laboratoire d'informatique (LIX), 91128 Palaiseau Cedex France
touati@lix.polytechnique.fr

Abstract. The Dial-a-Ride is a transport system on demand. A fleet of vehicles, without fixed routes and schedules, carries people from their pickup points to their delivery points, during a pre-specified time interval. It can be modeled as an $\mathcal{NP}$-hard routing and scheduling problem, with a suitable mixed integer programming formulation. Exact approaches to this problem are too limited to tackle real-life instances: time dependent network, requests received on-line, different objective functions. In this paper we propose an algorithm to solve the off-line Dial-a-Ride Problem (DARP), based on a Granular Tabu Search method. This algorithm was fast and effective, when tested on instances created ad hoc using the Milan network.

1 Introduction

The Dial-a-Ride (DAR) system concerns the management of a fleet of vehicles in order to satisfy transport demands. The customer requests the service by calling a central unit and specifying: the pick-up point, the delivery point (respectively, *origin* and *destination*), the number of passengers and some limitations on the service time (e. g., the earliest departure time). Such transportation system is called *on demand*: the routes and schedules of the vehicles change dynamically on the basis of the current requests of the users. By better exploiting vehicle capacity, they offer the comfort and flexibility of private cars and taxies at a lower cost. DAR is suited to service sparsely populated areas, weak demand periods or special classes of passengers with specific requirements (elderly, disabled). Several models of the DAR service have been proposed in the literature: with or without time windows, with a fixed or unlimited fleet of vehicles, and so on. In the "static" DAR, the customer asks for service in advance and the vehicles are routed before the system starts to operate; in the "dynamic" DAR, the customer can call during the service time and the routes are updated on-line. Different objective functions have been taken into account: minimization of the number of vehicles used or the total travel time, maximization of the number of customers served or the level of service provided to the user. This paper addresses the static DARP with time windows and a fixed fleet of vehicles. Each request corresponds to a single passenger, and the objective function maximizes firstly the number of customers served, then it minimizes the number of vehicles used and finally it maximizes the level of service provided on average to the customers. The DARP is $\mathcal{NP}$*-hard in the strong sense*, as it generalizes the *Pickup and Delivery Problem with Time Windows* (PDPTW) [6]. Nevertheless, exact

J. Pahl, T. Reiners, and S. Voß (Eds.): INOC 2011, LNCS 6701, pp. 450–463, 2011.

algorithms for the multi-vehicle case [1,6] have been developed. However, heuristics and matheuristics are needed [3,8,9,13], while exact methods are useless, for dealing with the dynamic problem with time dependent network.

This paper describes a fast and effective matheuristic for improving the feasible initial solution obtained at the end of the algorithm described in [16] to solve the off-line DARP. The algorithm is based on a Granular Tabu Search where the proposed procedure for obtaining the granular graph is different from the one proposed in [14], since it is based on the reduced costs matrix obtained by solving a simpler, but useful, subproblem. The algorithm transforms the original graph of nodes (i.e. pick-up and delivery nodes for each customer) into a graph of customers, which is smaller than the original one. A value is assigned to each arc of the auxiliary graph, measuring the *distance* between each pair of customers. Then, a classical assignment problem is defined and solved on this graph. The solution obtained gives several information on the instance. The most useful one is the reduced costs matrix which shows how close customers are to each other (a *short* edge has a reduced cost equal zero). The set of arcs is therefore ordered by increasing value of the reduced costs. The proposed Granular Tabu Search exploits this information to guide the local search.

The paper is organized as follows. The problem is described in the next section together with its mixed linear programming formulation. In Section 3 the general Granular Tabu Search approach is described. Section 4 explains the problems addressed for applying the Granular Tabu Search to the DARP and how they have been solved. The whole algorithm used is described in Section 5. Computational results and conclusions close the paper.

2 The Problem

Let $R = \{1...n\}$ be a set of requests (customers). For each request i two nodes (i^+ and i^-) are defined: a load q_i must be taken from i^+ to i^-. Let $N^+ = \{i^+|i \in R\}$ be the set of pick up nodes and $N^- = \{i^-|i \in R\}$ the set of delivery nodes. A positive amount $q_{i^+} = q_i$ is associated with the pick up node, a negative amount $q_{i^-} = -q_i$ with the delivery node. A time window is also associated with each node (i.e. pick up node $[e_{i^+}, l_{i^+}]$ and delivery node $[e_{i^-}, l_{i^-}]$). The fleet of vehicles is denoted as V; all vehicles have the same capacity Q and time window $[e_0, l_0]$. Let $G = (N,A)$ be a directed graph, whose set of vertices is defined as $N = N^+ \cup N^- \cup \{0\}$, where node 0 is the depot. The set of arcs A is defined as $A = \{(i,j) : i,j \in N, i \neq j\}$ with a distance $d_{i,j}$ or a travel time $t_{i,j}$ assigned to each arc $(i,j) \in A$. Another set $E = \{(i,j) : i,j \in N^+ \cup N^-, i \neq j\}$ represents the subset of arcs whose extremes are customer nodes. The problem consists in finding a set of routes starting and ending at the depot, such that an objective function is optimized. The pick up node of each customer must be visited before the delivery node in the same route. Capacity and time window constraints must also be respected. The variables $x^v_{i,j}$ are equal to 1 if vehicle v uses arc $(i,j) \in A$ and equal to 0 otherwise; p_i represents the departure time from node $i \in N^+ \cup N^-$; y_i is the load of the vehicle leaving node i.

$$\max z(P) = \max \left(\alpha_1 \sum_{i \in N^+} \sum_{v \in V} \sum_{j \in N} x^v_{i,j} - \alpha_2 \sum_{v \in V} \sum_{j \in N^+} x^v_{0,j} - \alpha_3 \sum_{i \in N^-} S_i \right) \quad (1)$$

subject to

$$\sum_{v\in V}\sum_{j\in N} x^v_{i,j} \le 1 \qquad \forall i \in N^+ \tag{2}$$

$$\sum_{j\in N} x^v_{i,j} - \sum_{j\in N} x^v_{j,i} = 0 \qquad \forall v \in V, \quad \forall i \in N^+ \cup N^- \tag{3}$$

$$\sum_{j\in N} x^v_{i^+,j} - \sum_{j\in N} x^v_{i^-,j} = 0 \qquad \forall v \in V, \quad \forall (i^+,i^-) \in N^+ \cup N^- \tag{4}$$

$$x^v_{i,j}(y_i + q_j) \le y_j \qquad \forall v \in V, \quad \forall (i,j) \in E \tag{5}$$

$$q_i \le y_i \le Q \qquad \forall i \in N^+ \tag{6}$$

$$0 \le y_i \le Q - q_i \qquad \forall i \in N^- \tag{7}$$

$$x^v_{i,j}(p_i + t_{i,j}) \le p_j \qquad \forall v \in V, \quad \forall (i,j) \in E \tag{8}$$

$$e_i \le p_i \le l_i \qquad \forall i \in N \tag{9}$$

$$p_{i^+} + t_{i^+,i^-} \le p_{i^-} \qquad \forall i = (i^+,i^-) \in N^+ \cup N^- \tag{10}$$

$$\sum_{v\in V}\sum_{j\in N^+} x^v_{0,j} \le |V| \tag{11}$$

$$x^v_{i,j} \in \{0,1\} \qquad \forall v \in V, \quad \forall (i,j) \in A \tag{12}$$

Expression (1) represents the three-level objective function addressed in this paper. The first term is the number of serviced customers. The second term represents the minimization of the number of used vehicles. The third term represents the level of service (indicated in the following with S) over the set of served customers. The three terms are suitably weighted with $\alpha_1, \alpha_2, \alpha_3$ in order to guarantee that the first term is the main objective, the second term is the second one and the last term is the third. A reasonable measure of the quality of service perceived by a customer i is introduced as the ratio between the service time offered by the DAR system for the trip and the minimum time the customer would need to go from the origin to the destination (i.e. the value associated to the arc (i^+,i^-)). More precisely, the quality of service for customer i is defined as $S_i = \frac{(a_{i^-} - e_{i^+})}{t_{i^+i^-}}$, where a_{i^-} is the arrival time in node i^-, $(a_{i^-} - e_{i^+})$ denotes the total time needed to reach the destination using the DAR transportation system and $t_{i^+i^-}$ denotes the minimum time needed to go from the origin to the destination directly. Obviously $S_i \ge 1\ \forall i$ and the higher its value, the lower the service quality for customer i. Moreover, it is important to stress that the maximum number of vehicles is fixed and that even finding a feasible solution is itself a $NP-complete$ problem.

The first three groups of constraints ensure that each customer is served by at most one vehicle. Constraints (2) make sure that at most one vehicle exits from each origin node i^+. Constraints (3) impose that the number of vehicles entering and exiting each node be the same. Finally constraints (4) establish that the same vehicle, if any, visits the pickup and the delivery node. Constraints (5), (6) and (7) ensure the feasibility of the loads. The number of passengers in a given vehicle varies according to the number of people boarding it or getting off it. The vehicle capacity cannot be exceeded. The last three classes of constraints impose the feasibility of the schedule. Constraints (8) represent the compatibility requirements between routes and schedules. Constraints (9) ensure that the begining of service time takes place during the time window: when the vehicle arrives at node i before e_i the driver must wait; it is unfeasible to arrive at node i after l_i. Constraints (10) imply that for each trip the delivery node is visited after the pickup node. Finally, constraint (11) imposes a limit on the number of used vehicles (i.e. fleet size). Constraints (8) and constraints (5) can be linearized respectively as $M(1-x^v_{ij}) \geq p_i + t_{ij} - p_j$ and $O(1-x^v_{ij}) \geq y_i + q_j - y_j$. The former equations are a generalization of the classical *TSP subtour* elimination constraints proposed by Miller, Tucker and Zemlin [11].

3 The Granular Search Approach

Tabu Search (TS) is a memory-based search method introduced by Glover [10]. It is an iterative improvement procedure that starts from any initial solution and attempts to determine a better solution. Generally, TS is characterized by its ability to avoid local optima and prevent cycling by using flexible memory of search history. The Granular Tabu Search (GTS) [14] is a TS, but with a particular focus on the computational time. The method uses a drastically restricted neighborhood obtained from that standard by removing the moves that involve only elements which are not likely to belong to high-quality feasible solutions. This neighborhood is called a *granular neighborhood*. This method was applied in [14] to the Vehicle Routing Problem (VRP). The authors initially verify that good solutions rarely contain long edges. Thus, the proposed algorithm reduces the graph used by the TS in a simple and straightforward way. Given the best solution found so far, the edges longer than a certain threshold are removed from the graph. Therefore, the local search explores a smaller, but promising neighborhood. From time to time, when no improving solutions are found, the threshold is increased until the original graph is restored and whenever the best solution found so far is improved, the graph is reduced again.

This approach needs to be reviewed when the problem to solve is a routing and scheduling problem. This family of problems needs a different definition of the distance among customers, taking into account the spatial and the temporal dimension. Indeed, two customers (i and j) close together in terms of travel time, but far away in terms of time windows (i.e. the distance $e_i - e_j$ is wide and greater than $l_i - e_i$) will rarely be adjacent in the optimal solution. Therefore, a useful definition of the distance between customers for routing and scheduling problems must consider the time windows.

4 Applying Granular Search to DARP

In the DARP each customer is identified with a pair of nodes: the pick-up node and the delivery node. It means that in a two dimensional plane the spatial distance between customers must be measured in terms of distance between directed segments. In addition, the time window associated with each costumer makes even harder the definition of the distance (spacial and temporal). Thus, for applying a GTS algorithm guided by the reduced cost to the DARP, a further improvement is necessary. The purpose is to deal with a graph of customers instead of a graph of nodes, by associating a cost function to each arc of this smaller graph. This function measures the distance between each pair of customers so that solving an assignment problem for creating clusters of customers is useful on this graph. The solution of the assignment problem can be easily transformed into an initial feasible solution of the addressed problem, as described in [16]. Then, the GTS improves this initial solution. Therefore, we explain in the next how the distance between pairs of customers is defined and how the auxiliary graph is obtained (i.e. a method to deal with customers without neglecting their own initial characteristics: a pick-up node and a delivery node for each customer and a time window associated to each node).

• **The average departure time** $\bar{p}_{i,j}$: The value $\bar{p}_{i,j}$ is the average departure time from the node j^- in a four nodes (i.e. two customers) sequence starting as earliest as possible from i^+. It measures the spatial and temporal distance between customer i and customer j. The value $\bar{p}_{i,j}$ sums up most of the information necessary to evaluate the feasibility and the cost of any possible path to go directly from customer i to customer j, starting with node i^+: $\{i^+,i^-,j^+,j^-\}$, $\{i^+,j^+,i^-,j^-\}$ and $\{i^+,j^+,j^-,i^-\}$, as shown in Figure 1. To compute $\bar{p}_{i,j}$ constraints(8) and (9) are used. They impose that for any pair of consecutive nodes i, j in a sequence, the departure time from j is $p_j = \max\{p_i + t_{i,j}, e_j\}$. When the sequence is made by only two nodes, the equation becomes $p_j = \max\{e_i + t_{i,j}, e_j\} = e_i + t_{i,j} + \tilde{w}_{i,j}$, where $\tilde{w}_{i,j} = \max\{0, e_j - e_i - t_{i,j}\}$ is the *a priori* waiting time at node j.

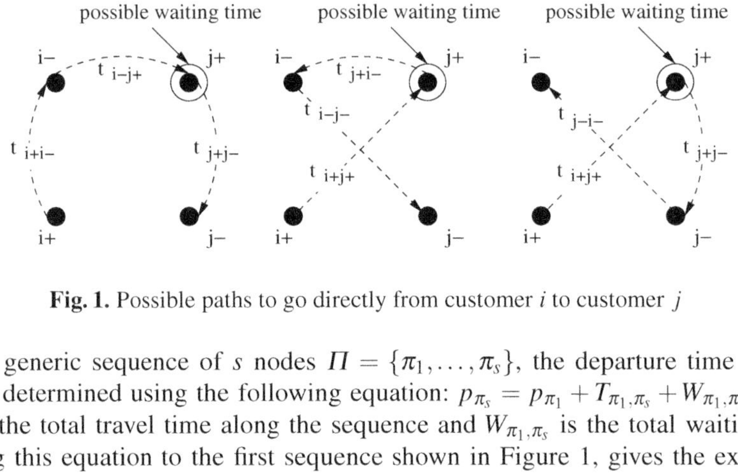

Fig. 1. Possible paths to go directly from customer i to customer j

For a generic sequence of s nodes $\Pi = \{\pi_1, \ldots, \pi_s\}$, the departure time in node s can be determined using the following equation: $p_{\pi_s} = p_{\pi_1} + T_{\pi_1,\pi_s} + W_{\pi_1,\pi_s}$, where T_{π_1,π_s} is the total travel time along the sequence and W_{π_1,π_s} is the total waiting time. Applying this equation to the first sequence shown in Figure 1, gives the expression $p^1_{j^-} = e_{i^+} + t_{i^+,i^-} + t_{i^-,j^+} + t_{j^+,j^-} + w_{j^+}$, and similar expressions can be written for the other two sequences by obtaining the value of $p^2_{j^-}$ and of $p^3_{i^-}$. It is important to remark that there is only one possible waiting time, i.e. w_{j^+} in node j^+, since the sequence starts in i^+ and the waiting time before the delivery nodes does not exist: $e_{i^-} = e_{i^+} + t_{i^+i^-}$. If $e_{i^-} > e_{i^+} + t_{i^+i^-}$, the customer would wait before getting down, which makes no sense; if $e_{i^-} < e_{i^+} + t_{i^+i^-}$, the time window can be reduced [5], since part of it will be never used. So $p_{i^-} = a_{i^-}$, where a_{i^-} is the arrival time in node i. Then, the coefficient $\bar{p}_{i,j}$ representing the average departure time from node j^-, can be defined as $\bar{p}_{i,j} = \frac{\sum_{r=1}^{3} p^r k_r}{\sum_{r=1}^{3} k_r}$, where $k_r = 1$ if $p^r \neq \infty$, $k_r = 0$ otherwise. Of course, when $\sum_{r=1}^{3} k_r = 0$ the procedure sets $\bar{p}_{i,j} = \infty$: there are no feasible ways to go from customer i to customer j. The value $\bar{p}_{j,i}$ is defined in the same way and in general $\bar{p}_{i,j} \neq \bar{p}_{j,i}$.

• **The auxiliary graph and the assignment problem:** The auxiliary graph, $\hat{G} = (\hat{R}, \hat{A})$, is simpler and smaller than the initial one, since it has a node for each customer without any time window associated and the value $\bar{p}_{i,j}$ associated to each arc (i, j). This smaller graph is used for building clusters of customers. The method chosen for creating the clusters is based on the idea of defining and solving an assignment problem on this simpler graph. In order to make the solution of the assignment problem useful, it would be desirable to enlarge the auxiliary graph with the minimum number of vehicles (i.e. minimum number of clusters) necessary to satisfy the requests. Since this number is unknown we settle for a lower bound. When the fleet includes homogeneous vehicles and all the vehicles are based on a common depot the lower bound of the fleet size can be calculated. It is necessary to identify the maximal set of customers which cannot be loaded by the same vehicle, since going from customer i to customer j violates some time windows. This value, indicated with m, is calculated as the maximal clique on the incompatible graph (see for example [16]). Then, the set of nodes of the graph is defined as $\hat{R} = R \cup \{n+1, ..., n+m\}$, while $\hat{A} = \{(i, j) : i, j \in \hat{R}, i \neq j\}$ is the set of arcs. The cost function $\bar{p}_{i,j}$ is assigned to each arc, if i and j are both customers. When i and j are both vehicles $\bar{p}_{ij} = \infty$. The values given to $\bar{p}_{ij}$, when $i \in V$ and $j \in N$ or viceversa is $\bar{p}_{ij} = t_{ij} + w_j$.

It is easy to show that the assignment problem is a useful relaxation of the DARP. Indeed, the constraints involving the time windows, namely constraints (8) and (9), can be relaxed, since they are used in the computation of $\bar{p}_{ij}$. Moreover, the minimization of the objective function takes care of the right part of constraints (9) (i.e. $p_i \leq l_i$). Constraints (4) and constraints (10) are irrelevant to this graph $\hat{G}$, hence removed. Constraints (2), (3), (5), (6) and (7) have to be taken into account. The last three constraints, ensuring the feasibility of the loads, are relaxed and the feasibility of the solution obtained is verified at a later stage. Constraints (2) and (3), are replaced by the classical assignment constraints and the problem to solve becomes a standard assignment problem whose size is $n + m$.

The solution of the assignment problem is a set of clusters of sequenced customers: some of these clusters contain the depot (i.e. the vehicle) while the rest of them, called *subtours*, are composed by customers only. The solution of the assignment problem can be easily transformed in an initial feasible solution of the addressed problem. It is enough to make the paths starting and ending from the depot feasible, and to insert in these main routes the customers in the subtours (see [16]). In addition, the assignment problem gives the reduced cost value for each arc (i, j) calculated as $\bar{c}_{i,j} = \bar{p}_{i,j} - u_i - v_j$ where u_i and v_j are the dual variables of the assignment problem. The reduced costs matrix is useful for knowing how much it costs to replace an edge, actually in the solution, with another one outside. There are several possible solutions with similar costs, because there are at least $n+m-1$ arcs not used in the solution but with a reduced cost value equal 0. In other words the solution of the assignment problem proposes how the customers in the same cluster may be served by the same vehicle. Nevertheless, there are pairs of customers that in the current solution are placed in different clusters, but that could be served by the same vehicle without increasing the solution value.

5 The Proposed Granular Tabu Search

The main components of a GTS are: the neighborhood, the method to obtain the granular neighborhood, the diversification/intensification procedure, the stopping criteria and the aspiration criteria. The algorithm stops after a predefined number of iterations or when a maximum number of iterations without improvement is reached. The other components are described in the following sections.

• **The neighborhood structure:** The neighborhood $N(s)$ of the current solution s consists in all the solutions that can be obtained applying a single move to s. This space includes only feasible solutions and the possible moves are: (i) move a request from a route into another one, (ii) insert an unserved request in a route and (iii) remove a request from a route (clearly it becomes unserved). More complex transformations can be achieved through sequences of the described simple moves: for example rerouting a customer in the same route is obtained combining the move *remove* with the move *insert*. By considering the moves just defined, the neighborhood can be described using the graph $\hat{G} = (\hat{R}, \hat{A})$, but with the attention of removing from $\hat{A}$ arcs connecting two requests served by the same vehicle.

Table 1. A Small Example with Four Customers and Two Vehicles

request	*pick-up node* i^+	*delivery node* i^-	e_{i^+}	l_{i^+}	e_{i^-}	l_{i^-}
1	a	b	7000	8000	7484	8553
2	c	d	7400	8400	8116	9474
3	e	f	7500	8500	8012	8121
4	f	e	22000	23000	22447	23416

Table 1 reports the data (nodes and time windows) of an example used in the following, constituted by four customers $(1,2,3,4)$. A possible feasible solution is the following: requests 1 and 2 served by vehicle (A) and requests 3 and 4 served by vehicle (B). The graph $\hat{G} = (\hat{R}, \hat{A})$ associated to this solution is reported in Figure 2, without the copy nodes of the depot.

Each arc of this graph identifies two new different, partially routed, clusters of customers. For example arc $(3,2)$ suggests $\{d,4,d\},\{d,1,3,2,d\}$ and $\{d,1,d\},\{d,3,2,4,d\}$. In a similar way arc $(2,3)$ suggests $\{d,4,d\},\{d,1,2,3,d\}$ and $\{d,1,d\},\{d,2,3,4,d)\}$. Let's notice that the difference is the order of visiting the pickup nodes of clients 2 and 3. Arc $(3,2)$ states node 3^+ before node 2^+, while arc $(2,3)$ imposes 2^+ before 3^+. The arc under consideration suggests the position of the pickup node of the moved client, then the algorithm looks for the best position for inserting the delivery node. The local search algorithm explores all the possible feasible positions for inserting the delivery node and in the worst case the complexity is $O(n)$. In Figure 3 the initial solution (i.e. the complete sequence of pick-up and delivery nodes) is reported and Figure 4 shows an example

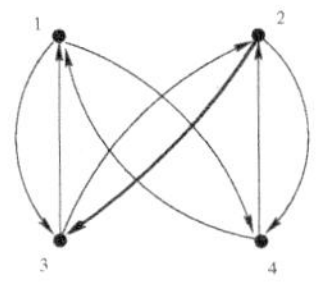

Fig. 2. The graph $\hat{G} = (\hat{R}, \hat{A})$ associated to the solution $\{d, 1, 2, d\}$, $\{d, 3, 4, d\}$, without the copy nodes of the depot

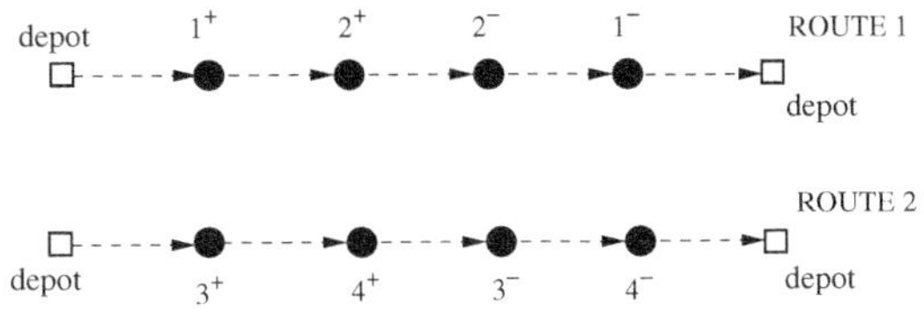

Fig. 3. The Complete Initial Solution

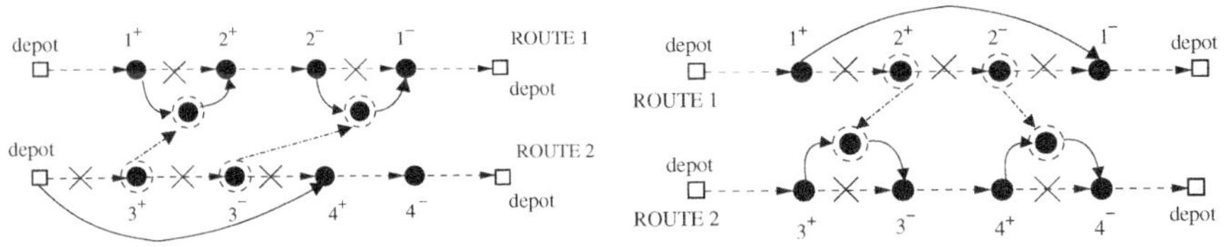

(a) move node 3 into the route of node 2 (b) move node 2 into the route of node 3

Fig. 4. Two possible feasible solutions induced by arc $(3,2)$, node 3^+ always before node 2^+

of two solutions generated by arc $(3,2)$. Each iteration of the Tabu Search optimization process requires the evaluation of the whole neighborhood: any possible feasible move obtained by using the arcs in the graph $\hat{G} = (\hat{R}, \hat{A})$.

• **The granular neighborhood:** The local search on a granular neighborhood considers only moves which are generated by arcs belonging to the granular graph. The way used to reduce the graph is important, as suggested in [14]; for building an effective GTS any edge extracted from the granular graph should completely identify a move. This is the case for the proposed GTS. The information used for reducing the neighborhood are obtained by the reduced costs matrix given by the solution of the assignment problem. An arc having a reduced cost value equal zero means that two customers are close to each other and, if they are located in two different vehicles, this suggests a potentially interesting move. The threshold starting value is 0, but it can be modified to obtain the diversification and intensification strategies. Figure 5 reports the reduced costs matrix associated with the example presented in Table 1. The solution can be read on this matrix by following the cell with value 0. For example from row of *Bus A* to column of *client 1* and from its row to column of *client 2* and then back to the Depot (i.e. column of *Bus A*). Figure 6 shows how the neighborhood changes when the threshold change.

• **Tabu list and aspiration criteria:** In a TS framework, search is guided by a memory. This memory, called tabu list corresponds to a short term memory [15] and it is used for avoiding cycling and for helping the search process to escape local minima. At each iteration the move executed is declared *tabu* for a given number of iterations (*tabu tenure*) [7], forbidding its reversal. Thus, moves involving only *tabu* arcs can be executed if they lead to a solution whose objective function value is better than the best one found so far: this is called *aspiration criterion*.

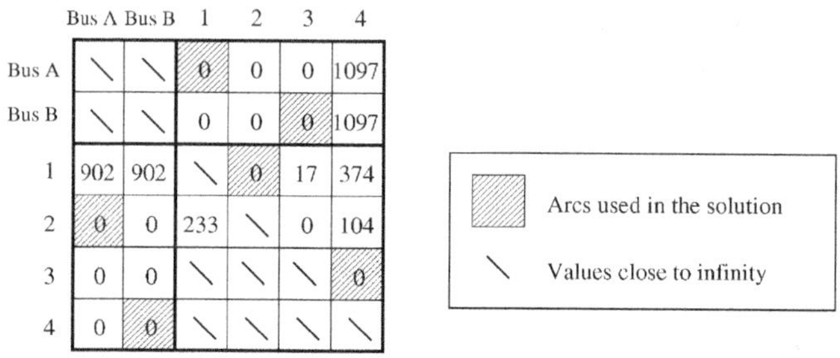

Fig. 5. The reduced cost matrix given by the assignment solution. The number in the cell represents the value of the reduced cost. For example $\bar{c}_{1,3} = c_{1,3} - u_1 - v_3 = 17$.

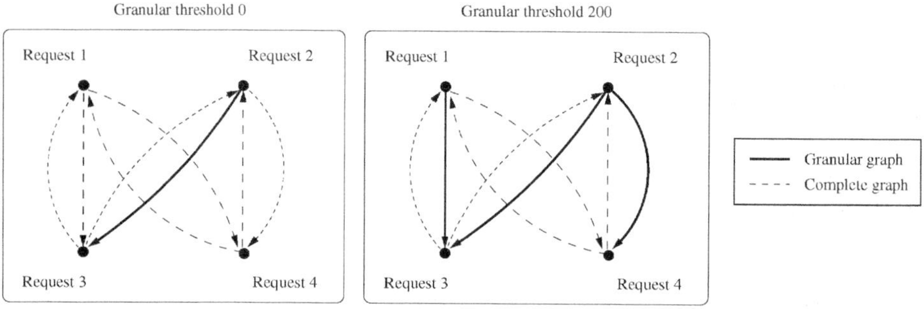

Fig. 6. The granular neighborhood obtained by using two different values for the threshold

The proposed algorithm sets as tabu all the edges involved when moving a customer. For example, Figure 7 shows which arcs become tabu after the removal of request j from its route: they are marked with bold arrows. The number of arcs inserted in the tabu list depends on the position of customer j. The arcs directly connecting nodes to the depot or joining the origin to the destination of the same request are not considered: they are useless for an univocal identification of the implemented move.

• **Diversification and intensification:** *Intensification* and *diversification* strategies are used to improve the effectiveness of the Tabu Search method [7]. *Intensification* tries to focus the search on promising portions of the solutions space, while *diversification* moves the algorithm to another unexplored region (i.e. trying to make up for the local search drawbacks). Three diversification strategies have been used in this work:

Tabu tenure dynamic variation: The *tabu tenure* value (Tt) can be constant or can vary according to several strategies [7]. In the proposed algorithm this variation is based on the objective function evolution. If improving moves have been executed for a consecutive pre-defined number of iterations (Nit), the search process is probably exploring an interesting portion of the solutions space, thus intensification is required and the *tabu tenure* value is reduced. On the contrary, if the objective function value has not been improved for the same number of iterations (Nit), the search process may have reached a local minimum, thus diversification is required and the *tabu tenure*

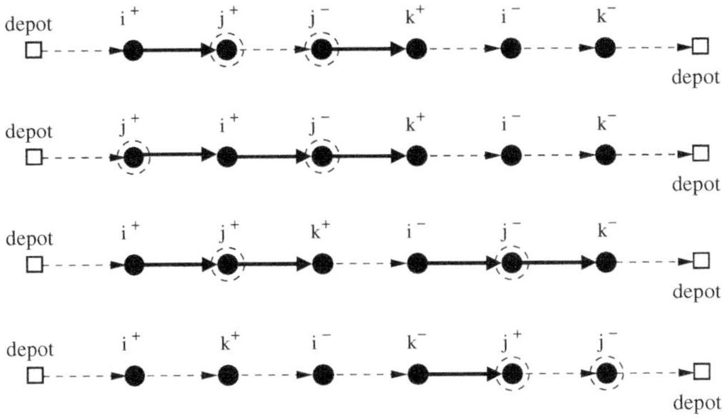

Fig. 7. Arcs Saved Into the *Tabu List*

value is increased. The function used to increase or to decrease Tt is the following: $Tt = Tt \pm \varepsilon Tt$. The maximal value and the minimal value for Tt are bounded by, respectively, $Tt = Tt + \vartheta Tt$ and $Tt = Tt - \vartheta Tt$.

Frequency-based penalization: This technique uses a long term memory for recording the number of times an arc appears in the incumbent solution.

Let s be the current solution and each solution $\bar{s} \in N(s)$ such that $f(\bar{s}) > f(s)$ is penalized by a factor $p(s) = \lambda\rho\sqrt{n \cdot m} f(\bar{s})$. ρ is the mean value of the number of times each considered arc has been added to the current solution, λ is a parameter used to control the intensity of the diversification and $\sqrt{n \cdot m}$ is a scaling factor required to adjust the penalties with respect to the problem size. This strategy has been proposed by Taillard [12] and successfully used in many other *tabu search* algorithms applied to the *vehicle routing problem*, for example in [4].

Granularity threshold variation: The granularization process, by reducing the neighborhood, naturally intensifies the search, since few moves are evaluated during each iteration. However several diversification strategies can also be defined by varying the granularity threshold (thd) and thus by dynamically changing the neighborhood structure. The step, identified in the following with δ and used to increase or to decrease the threshold, is calculated using the following formula $\delta = \gamma\left(\max_{(ij)\in\hat{A}} \bar{c}_{i,j} - \min_{(ij)\in\hat{A}} \bar{c}_{i,j}\right)$. At the beginning $thd = 0$ and it is increased (i.e. $thd = thd + \delta$) when at least one of the following conditions is verified: (a) all the feasible solutions in the current neighborhood are tabu or (b) the algorithm is unable to improve the best solution found so far for a fixed number of iterations (FIt). The threshold is set equal 0 again whenever one of the following conditions is verified: (1) the algorithm improves the *best solution* found so far or (2) the algorithm improves the *current solution*. Three versions of the algorithm have been obtained by combining the methods for changing the threshold (i.e. (a),(b),(1) and (2)). **The Granular Tabu Search Fixed (GTSF)** uses conditions (a) and (1), **the Granular Tabu Search Variable Current (GTSVC)** uses conditions (a), (b) and (2) and **the Granular Tabu Search Variable Optimal (GTSVO)** uses conditions (a), (b) and (1). These three versions produce different quality solutions within different times, as shown in the next section.

6 Computational Study and Conclusions

The algorithms have been tested, by considering *non time-dependent* and *time-dependent* networks. Moreover, the use of different diversification strategies based on the granularity threshold variation gives three different versions of the algorithm (GTSF, GTSVC and GTSVO) tested and compared both among them and with a TS algorithm. The TS is obtained by setting $thd = \infty$.

• **Instances:** To the best of authors knowledge, there are no benchmark instances for the DARP with time dependent network and a fleet of vehicles of fixed size. Therefore, the data set has been created by using the complete network of Milan (about 17000 arcs and 7000 nodes). The set of instances has been generated by choosing randomly a pair of nodes for each customer. Each customer requires either the latest arrival time (l_{i^-}) or the earliest departure time (e_{i^+}) and that value is generated randomly. The time windows construction is basically based on the maximum ride time as described in [16]. Almost all the tests have been performed using instances consisting of 100, 250 and 500 requests with the following characteristics:

- *requests temporal clustered*: all the earliest departure times required by the customers are randomly generated within an interval two hours wide ($8.30 - 10.30$). The same for customers asking the latest arrival time, but within a different interval (i.e. $13.30 - 15.30$). The service starts at 7.00 and the nodes (spatial dimension) are generated by considering the whole network.
- *requests spatial clustered*: the nodes have been randomly generated from two different areas, smaller than the whole network, each of them with 2 km of radius. The pick-up nodes are selected in one of them, and the delivery nodes in the other one. The earliest departure times (latest arrival time) are generated by using the whole day (i.e. 10 hours). This set of instances are inspired by the class $C1$ and $C2$ of the well known solomon instances for the VRPTW.
- *requests without clusters*: the requests are randomly generated using the whole network and within a time interval of 10 hours. Nevertheless, using the same data, two sets have been created using two different values for the maximum ride time ($maxS$).
  - The first set has been obtained using a large value for $maxS$ (i.e. 2.6 for trip shorter equal than 10 minutes, 1.9 for trip longer equal than 40 minutes and the following hyperbole function $maxS = 1.5 + \frac{1000}{t_{i^+i^-}}$ in between). This corresponds to wide time windows. This set of instances are inspired by the class $R2$ and $RC2$ of the well known solomon instances for the VRPTW.
  - The second set has a smaller value for $maxS$ (i.e. 2.1 for trip shorter equal than 10 minutes, 1.4 for trip longer equal than 40 minutes and the following hyperbole function $maxS = 1.0 + \frac{900}{t_{i^+i^-}}$ in between). This corresponds to tight time windows. This set of instances are inspired by the class $R1$ and $RC1$ of the well known solomon instances for the VRPTW.

• **Implementation, parameters and algorithms compared:**A fleet size suitable for each size of instances has been determined through some preliminary tests: the algorithm was allowed to use as many vehicles as it needed for serving on average at least 90% of the customers. In this way the required fleet size has been estimated to be equal to 6, 12 and 20 vehicles to serve 100, 250 and 500 requests, respectively. The algorithm uses always the whole fleet of vehicles, therefore the number of vehicles used is reported in tables only when it is smaller than the whole fleet size (i.e. Table 7, column *veh.*). The speed over the *non time-dependent* network used is 25 km/h. *Time-dependent* network has been described considering 3 time intervals whose data are shown in Table 2.

Table 2. Time Intervals

Start time	*End time*	*Speed (km/h)*
7.00	10.00	20
10.00	12.00	25
12.00	17.00	28

Table 3. Parameters Used

		tabu tenure			
Parameter	λ	*Tt*	*Nit*	ε	ϑ
Value	0.015	$7.5\log_{10} n$	3	20%	70%
	granular threshold			***objective function***		
Parameter	*thd*	*Flt*	γ	α_1	α_2	α_3
Value	0	15	10%	100	10	1

The parameters used are set to the values reported in Table 3. They have been selected after a preliminary testing phase, in order to provide the best average solution values.

• **Commented results:** The results are reported in Table 4. The columns in the tables are as follows: the first one lists the algorithm (or the procedure) used, (n) the number of clients, (*uns.*) the percentage of unserved clients and (*avg.* S) the average level of service. The computational time is reported in the last column. Moreover, in Tables from 6 to 9, (*iter.*) lists the number of iterations. The results reported in the Tables are obtained as average over five different instances for each type of problem.

The aim of this first test is to evaluate the performance and the quality of results produced by the different versions of the optimization algorithm: (GTSF, GTSVC and GTSVO) and to compare them with the TS. The requests have been generated without clusters and the time windows are wide. Table 4 shows the results obtained after 2500 tabu search iterations with a non time-dependent network. Every optimization algorithm succeeds in serving much more requests than the initial solution does, while providing a better level of service to the customers. The number of used vehicles is equal to the fleet size, since there is always at least one unserved request. TS and GTSVO produce better solutions as expected, but are very slow. GTSF and GTSVC generate very good solutions, within shorter computing time. In particular, the GTSVC version of the algorithm is the best compromise between efficiency and effectiveness.

• **Impact of clusters:** As a consequence of the previous results, the algorithms compared are: GTSF, GTSVO and GTSVC without the TS. The network is no-time dependent. A slightly different stopping condition is adopted: instead of using a fixed number of iterations (2500 or 1000), the algorithms stop when unable to improve the best solution found so far for a fixed number of iterations (150). In this way the trend of the objective function is taken into account.

Tables 6, 7, 8 and 9 report the results. The instances with temporal clusters (Table 6) are more difficult, since the time windows distance between nodes is shorter than the travel time and therefore there are few feasible solutions. Nevertheless, the optimization phase is able to strongly improve the initial solution. In the case of requests spatial

Table 4. Optimization Algorithms Analysis

Algorithm	*n*	*uns.*	*avg* S	*time (s)*
INITIAL SOL.	100	3.4%	0.36769	0.097
GTSF	100	0.4%	0.19034	40.616
GTSVC	100	0.2%	0.16928	63.956
GTSVO	100	0.2%	0.16875	652.762
TS	100	0.2%	0.16862	1092.071
INITIAL SOL.	250	4.0%	0.45250	0.922
GTSF	250	1.0%	0.23460	107.678
GTSVC	250	0.4%	0.19336	218.585
GTSVO	250	0.3%	0.16494	2648.395
TS	250	0.4%	0.18535	6469.919

Table 5. Optimization Algorithms Analysis: Results with a Time-Dependent Network

Algorithm	*n*	*uns.*	*avg.* S	*time (s)*
INITIAL SOL.	100	2.6%	0.38346	4.328
GTSF	100	0.6%	0.18165	197.341
GTSVC	100	0.4%	0.17711	1209.111
GTSVO	100	0.4%	0.16084	10986.608
INITIAL SOL.	250	4.9%	0.44249	22.450
GTSF	250	1.6%	0.24035	366.703
GTSVC	250	1.5%	0.20855	4798.447
GTSVO	250	1.6%	0.18429	22522.980

Table 6. Results on Instances Temporal Clustered with 100, 250 and 500 Customers

Algorithm	*n*	*uns.*	*avg.* S	*iter.*	*time (s)*
INIT SOL.	100	27.0%	0.48121	0	0.100
GTSF	100	18.8%	0.39537	342.4	5.219
GTSVC	100	17.4%	0.38231	355.2	7.162
GTSVO	100	16.2%	0.36564	469.8	42.662
INIT SOL.	250	34.1%	0.51788	0	0.922
GTSF	250	24.0%	0.41472	466.0	28.528
GTSVC	250	20.2%	0.42068	877.4	82.672
GTSVO	250	17.3%	0.39979	1147.0	405.025
INIT SOL.	500	41.1%	0.53928	0	4.741
GTSF	500	30.5%	0.44663	717.2	196.662
GTSVC	500	24.8%	0.42300	1589.2	609.969
GTSVO	500	20.0%	0.42835	2289.8	2529.436

Table 7. Results on Instances Spatial Clustered with 100, 250 and 500 Customers

Algorithm	*n*	*uns.*	*avg.* S	*veh.*	*iter.*	*time (s)*
INIT SOL.	100	0.2%	0.46747	5.6	0	0.100
GTSF	100	0.0%	0.17882	5.4	359.4	20.359
GTSVC	100	0.0%	0.14047	5.4	549.6	40.013
GTSVO	100	0.0%	0.12231	5.4	694.8	315.740
INIT SOL.	250	0.0%	0.54006	10.6	0	0.922
GTSF	250	0.0%	0.20251	10.6	568.4	136.797
GTSVC	250	0.0%	0.14714	10.6	1266.0	511.131
GTSVO	250	0.0%	0.11795	10.6	1762.0	4557.998
INIT SOL.	500	0.0%	0.56200	17.6	0	4.741
GTSF	500	0.0%	0.20671	17.4	941.0	888.591
GTSVC	500	0.0%	0.15758	17.4	1757.0	2759.460
GTSVO	500	0.0%	0.12234	17.4	3792.0	39411.360

Table 8. Results Obtained with Requests: Randomly Generated, wide time Windows and 100, 250 and 500 Clients

Algorithm	*n*	*uns.*	*avg.* S	*iter.*	*time (s)*
INIT SOL.	100	2.0%	0.47015	0	0.100
GTSF	100	1.2%	0.18717	413.2	15.091
GTSVC	100	1.0%	0.16606	389.2	18.897
GTSVO	100	0.6%	0.15969	586.4	127.956
INIT SOL.	250	4.4%	0.55046	0	0.922
GTSF	250	0.7%	0.25489	623.0	65.400
GTSVC	250	0.5%	0.23167	651.8	109.688
GTSVO	250	0.3%	0.17508	1339.6	1434.552
INIT SOL.	500	13.3%	0.56418	0	4.741
GTSF	500	4.8%	0.36405	900.2	383.906
GTSVC	500	1.6%	0.31724	1761.0	1129.446
GTSVO	500	1.3%	0.26867	2358.0	4249.736

Table 9. Results Obtained with Requests: Randomly Generated, Narrow Time Windows and 100, 250 and 500 Clients

Algorithm	*n*	*uns.*	*avg.* S	*iter.*	*time (s)*
INIT SOL.	100	5.4%	0.23281	0	0.100
GTSF	100	4.0%	0.20780	296.4	14.869
GTSVC	100	4.0%	0.19339	340.0	22.241
GTSVO	100	3.4%	0.21670	295.6	56.909
INIT SOL.	250	4.9%	0.38327	0	0.922
GTSF	250	3.0%	0.26821	440.2	38.756
GTSVC	250	2.2%	0.25879	522.0	68.681
GTSVO	250	2.2%	0.25013	641.4	397.784
INIT SOL.	500	10.3%	0.48664	0	4.741
GTSF	500	6.1%	0.38374	715.0	260.585
GTSVC	500	5.0%	0.33455	980.4	547.415
GTSVO	500	4.3%	0.33708	1057.4	1678.794

clustered (Table 7) the algorithms are able to reduce the number of used vehicles. This is expected, since all the requests are already satisfied in the initial solution which allows the algorithms in trying to reduce the number of used vehicles and in improving S. The number of iterations reported in Table 6 and Table 7 are of the same order of magnitude, but the computational time is different. This is due to the number of feasible solutions contained in each neighborhood: the more they are, the longer is each iteration. Finally Table 8 and Table 9 report the results obtained when solving the instances with wide time windows and instances with tight time windows. The quality of the obtained initial solutions is similar, but after the optimization phase the number of served clients is different. Instances with tight time windows are harder to solve as expected, since it is easier to insert customers when the problem has wide time windows.

References

1. Ropke, S., Cordeau, J.-F.: Branch-and-Cut-and-Price for the Pickup and Delivery Problem with Time Windows. Transportation Science 43, 267–286 (2009)
2. Ropke, S., Cordeau, J.-F., Laporte, G.: Models and a Branch-and-Cut Algorithm for Pickup and Delivery Problems with Time Windows. Networks 49, 258–272 (2007)
3. Cordeau, J.F., Laporte, G.: A tabu search heuristic for the static multi-vehicle dial-a-ride problem. Transportation Research Part B 37, 579–594 (2003)
4. Cordeau, J.F., Laporte, G.: Tabu Search Heuristic for the Vehicle Routing Problem. Metaheuristic Optimization via Memory and Evolution Operations Research/Computer Science Interfaces Series, vol. 30, Part II, pp. 145–163 (2005)
5. Desrosiers, J., Dumas, Y., Solomon, M.M., Soumis, F.: Time Constrained Routing and Scheduling, Network Routing. In: Handbooks in Operations Research and Management Science, vol. 8, pp. 35–139. North-Holland, Amsterdam (1995)
6. Dumas, Y., Desrosiers, J., Soumis, F.: The Pickup and Delivery Problem with Time Windows. EJOR 54, 7–22 (1991)
7. Gendreau, M.: An Introduction to Tabu Search. In: Handbook of Metaheuristics. International Series in Operations Research & Management Science, vol. 57, pp. 37–54 (2003)
8. Gendreau, M., Potvin, J.Y.: Dynamic Vehicle Routing and Dispatching. In: Crainic, T.G., Laporte, G. (eds.) Fleet Management and Logistics, pp. 115–126. Kluwer, Boston (1998)
9. Ghiani, G., Guerriero, F., Laporte, G., Musmanno, R.: Real-time vehicle routing: Solution concepts, algorithms and parallel computing strategies. European Journal of Operational Research 151, 1–11 (2003)
10. Glover, F.: Tabu Search. Part I - ORSA Journal on Computing 1, 190–206 (1990)
11. Miller, C.E., Tucker, A.W., Zemlin, R.A.: Integer Programming Formulations and Traveling Salesman Problems. ACM 7, 326–329 (1960)
12. Taillard, É.D.: Parallel iterative search methods for vehicle routing problem. Networks 23, 661–673 (1993)
13. Toth, P., Vigo, D.: Heuristic Algorithms for the Handicapped Persons Transportation Problem. Transportation Science 31(1), 60–71 (1997)
14. Toth, P., Vigo, D.: The Granular Tabu Search and its Application to the Vehicle-Routing Problem. INFORMS Journal of Computing 15, 333–346 (2003)
15. Glover, F., Laguna, M.: Tabu Search. Kluwer Academic Publisher, Norwell (1997)
16. Wolfler Calvo, R., Colorni, A.: An effective and fast heuristic for the Dial-a-Ride Problem. A Quarterly Journal of Operations Research (4OR) 5(1), 1–13 (2007)

A MILP-Based Heuristic for Energy-Aware Traffic Engineering with Shortest Path Routing

Edoardo Amaldi, Antonio Capone, Luca G. Gianoli, and Luca Mascetti

DEI, Politecnico di Milano, Piazza L. da Vinci 32, Italy
last_name@elet.polimi.it

Abstract. Internet energy consumption is rapidly becoming an issue due to the exponential traffic growth and the rapid expansion of communication infrastructures worldwide. We address the problem of energy-aware intra-domain traffic engineering in networks operated with a shortest path routing protocol. We consider the problem of switching off (putting in sleeping mode) network elements (links and routers) and of adjusting the link weights so as to minimize the energy consumption as well as maximizing a measure of effectiveness of the routing weight configuration. We propose a three-phase MILP-based heuristic for tackling this multi-objective problem with priority (first minimize the energy consumption and then the overall cost of link utilization), which exploits the IGP-WO heuristic proposed for optimizing the link weights so as to minimize the total cost of link utilization. For comparison purposes, we also developed a greedy randomized search procedure with path-relinking. The computational results for four real network topologies and different types of traffic matrices show that it is possible to switch off a substantial number of core nodes during low and moderate traffic periods, while guaranteeing the same point-to-point service quality and moderately increasing the network total cost of link utilization.

1 Introduction

Data reported in [9] show that in 2007 Internet had been responsible for 5.5% of the total energy consumption in the world and that the annual increment rate can be estimated around 20-25%. For these reasons, the issues of energy saving in IP networks and of power awareness in network design have recently become of great interest in the scientific community and have attracted the interest of device manufacturers and Internet Service Providers (ISP).

Energy management in the Internet exploits the fact that networks are designed and dimensioned to serve the estimated peak traffic demand. Usually, during network operation, traffic load varies remarkably over time and even during peak hours it is usually well below network capacity. Unfortunately, current network device architectures and transmission technologies make their power consumption almost independent of the traffic load. As a result, networks consume energy as if they were always fully loaded.

We consider the most widely used Internal Gateway Protocol (IGP) in IP networks, namely the Open Shortest Path First (OSPF) protocol. Traffic demands are routed from origin to destination along the shortest paths computed with respect to the weights assigned to the links. If the equal cost multi-path (ECMP) rule is considered, the packets

J. Pahl, T. Reiners, and S. Voß (Eds.): INOC 2011, LNCS 6701, pp. 464–477, 2011.

are evenly split at nodes where more outgoing links belong to shortest paths to the destination. Link weights are managed by network operators, who may modify them in order to optimize routing and reduce network congestion.

Let the directed graph $G = (V,A)$ represent the network topology, where V is the set of nodes and A the set of links. We distinguish two types of nodes: *edge* nodes and *core* nodes. Edge nodes can be both source and destination of traffic demands, while core nodes play only the role of transit routers. Let D denote the traffic matrix, where d_{ij} is the traffic demand for each pair of edge nodes i and j, and $d_{ij} = 0$ for all other pairs of nodes.

We consider the following extension of the IGP weight optimization problem with ECMP rule for intra-domain Traffic Engineering, that we refer to as *Energy-aware Traffic Engineering* (E-TE).

E-TE: Given a directed graph $G = (V,A)$, representing the topology of an IP network composed by routers and links with capacities on the links, and a traffic matrix D, decide which network elements (routers and links) to switch off and determine the link weights so as to minimize the total network energy consumption (primary objective) and a measure of effectiveness of the routing weight configuration (secondary objective), while guaranteeing that all the traffic demands are routed and the maximum utilization constraint is satisfied for each link.

According to the distinction between primary and secondary objectives, we first look for a sub-network with minimum total energy consumption and then minimize the total cost of link utilization proposed in [6] on the sub-network corresponding to the active elements. The unnecessary routers and links can then be excluded from the shortest path trees, by assigning a very large value to the corresponding link weights. Note that by switching off a node or a link we do not necessarily mean to turn it completely off but to put it in sleeping mode.

After a summary of related work in Section 2, a mixed integer programming formulation is presented in Section 3. Since it is very challenging even for small size instances, in Section 4 we propose a three-phase heuristic for tackling the E-TE problem combining greedy procedures with the iterative solution of a relaxed MILP-formulation for a sub-network with increasing traffic matrices. Building on previous work on intra-domain traffic engineering, we use the IGP-WO algorithm proposed by Fortz and Thorup [6] that, given a network topology and a traffic matrix, aims at a set of link weights that minimizes the total cost of link utilization. For comparison purposes we have also implemented a greedy randomized procedure (GRASP) with path-relinking. In Section 5 we report an discuss the computational results obtained on four real network topologies with different types of traffic matrices. Finally, Section 6 contains some concluding remarks.

2 Related Work

Since the seminal work by Gupta and Singh [8], the research community has started developing technologies for manufacturing energy efficient network devices, methodologies for power aware network design, and energy management strategies for reducing energy wastes of networks in operation.

To the best of our knowledge, there are only a few recent works on energy-aware traffic engineering.

The approach proposed in [16] aims at switching off the line-cards (network links) guaranteeing QoS constraints (maximum utilization and maximum path length constraints) in a scenario where an hybrid MPLS/OSPF scheme is adopted. The approach is based on a MIP formulation where the traffic demands are routed through a set of k-shortest path previously calculated.

In [4] the authors describe some heuristics that, given a traffic matrix and a fully powered network, are able to switch off nodes and links while respecting traffic constraints. In [15] some Energy-Aware Traffic Engineering (EATe) techniques are presented for optimizing links and routers power consumption, by considering their rate-dependant energy profiles. Assuming that the energy consumption of the network elements depends on the different rates, EATe algorithms try to switch off the underutilized elements or to reduce their rate, by re-routing part of the traffic in other network portions without increasing the rate of any element. Unlike in our work, the approaches in [4] and [15] follow a flow-based strategy that is suitable for the Internet geographic backbone where label switching protocols are adopted and not for routing domains based on shortest paths.

The energy management algorithm for IP networks called Energy Aware Routing (EAR) algorithm and presented in [5] is able to switch off network elements exploiting a modified version of the OSPF protocol. EAR algorithm selects a subset of routers, Importers Router (IR), that do not calculate their own shortest path tree (SPT) but use that of some neighboring routers, Exporters Routers (ER). In general a small number of active SPTs reduces also the number of links used that can be switched off. There are several important differences between our work and [5]. As in the literature on OSPF traffic engineering [13], we keep the OSPF protocol unchanged and focus on optimizing the link weights, while in the EAR algorithm the weights are assumed to be given and the protocol needs to be modified to implement ERs and IRs. Moreover, we explicitly consider link capacity limitations and minimize the network congestion level in order to guarantee service quality, while in [5] neither traffic load nor network capacity are clearly considered.

The reader is referred to [3] and [10] for summaries of the work done over the past decade on link weights optimization for intra-domain traffic engineering. Different objective functions have been considered (e.g., link utilization cost function minimization [6], residual capacity maximization and load balancing maximization) and different heuristic methods have then been developed (e.g., local search, genetic algorithm, Lagrangian approach). The code of the well-known IGP-WO algorithm [6] is available from the TOTEM toolbox [2]. To the best of our knowledge, the issue of energy-aware link weights optimization has not yet been addressed.

3 Mixed Integer Programming Formulation

Let p_{ij} and p_k be the power consumption of link (i, j) and node k respectively. Let c_{ij} be the capacity of the link (i, j). If the binary decision variables x_{ij} and y_k represent the power status (on/off) of links and routers respectively, a Mixed Integer Linear Programming (MILP) formulation of the part of the E-TE problem involving the the energy consumption minimization is given by:

$$\min \sum_{(i,j)\in A} p_{ij}x_{ij} + \sum_{k\in V} p_k y_k \tag{1}$$

s. t.

$$x_{ij} \leq y_i \qquad (i,j) \in A \tag{2}$$

$$x_{ij} \leq y_j \quad (i,j) \in A \tag{3}$$

$$\sum_{i\in V} f_{it}^t = \sum_{s\in V} d_{st} \quad t \in N \tag{4}$$

$$\sum_{j\in V} f_{vj}^t - \sum_{i\in V} f_{iv}^t = d_{vt} \quad v,t \in V, t \neq v \tag{5}$$

$$\sum_{t\in V} f_{ij}^t \leq x_{ij}\alpha c_{ij} \quad (i,j) \in A \tag{6}$$

$$0 \leq z_i^t - f_{ij}^t \leq (1-u_{ij}^t) \sum_{v\in V} d_{vt} \quad t \in V, (i,j) \in A \tag{7}$$

$$f_{ij}^t \leq u_{ij}^t \sum_{v\in V} d_{vt} \quad t \in V, (i,j) \in A \tag{8}$$

$$0 \leq r_j^t + \omega_{ij} - r_i^t \leq (1-u_{ij}^t)M \quad t \in V, (i,j) \in A \tag{9}$$

$$1 - u_{ij}^t \leq r_j^t + \omega_{ij} - r_i^t \quad t \in V, (i,j) \in A \tag{10}$$

$$u_{ij}^t \leq x_{ij} \quad (i,j) \in A, t \in V \tag{11}$$

$$\omega_{ij} \geq (1-x_{ij})\omega_{max} \tag{12}$$

$$1 \leq \omega_{ij} \leq \omega_{max} \quad (i,j) \in A \tag{13}$$

$$\omega_{ij} \in Z \quad (i,j) \in A \tag{14}$$

$$u_{ij}^t \in \{0,1\} \quad t \in V, (i,j) \in A \tag{15}$$

$$x_{ij}, y_k \in \{0,1\} \quad (i,j) \in A, k \in V \tag{16}$$

$$f_{ij}^t \geq 0 \quad (i,j) \in A, t \in V \tag{17}$$

$$r_i^t, z_i^t \geq 0 \quad i,t \in V, \tag{18}$$

where M is a large enough constant. The objective function (1) aims at minimizing the total energy consumption of the network. Constraints (2)-(3) ensure that if a node is switched off all incident links are turned off. Obviously a node can be switched off only if there are no traffic demands having it as source or destination (edge or core node). Constraints (4)-(5) are the classical flow conservation constraints, where the (real) positive variable f_{ij}^t indicates the amount of flow routed through the link $(i,j) \in A$ destined to node $t \in N$. Constraints (6) are the maximum utilization constraints imposing that the total flow through each link does not exceed the link maximum utilization and forcing the flow to 0 if the link (i,j) is powered off; the parameter α is comprised between 0 and 1. The binary variables $u_{ij}^t = 1$ appearing in Constraints (7)-(9) describe the routing configuration: $u_{ij}^t = 1$ if and only if the link (i,j) belongs to one of the shortest paths from node i to node t. Constraints (7) make sure that if $u_{ij}^t = 1$ then the flow f_{ij}^t destined to node t is equal to the (real) variable z_i^t, which is the common value of the flow assigned to all links outgoing from i and belonging to the shortest paths from i to t. Constraints (8) force $f_{ij}^t = 0$ for all links (i,j) that do not belong to a shortest path to node t. Finally, the shortest path routing Constraints (9)-(15) assure that the routing vector u defines shortest paths consistent with the link weight vector ω and forbid switched off links to belong to a shortest path; moreover the switched off links weights are put equal to the maximum value w_{max}. For each pair of nodes j and t, the (real) variable r_j^t corresponds to the length of the shortest path from node j to node t.

Unfortunately, the above MILP-formulation, which is an extension of the one given in [10] for intra-domain traffic engineering, turns out to be very challenging even for small size networks. For instance, we did not manage to find a feasible integer solution for a small network with 10 nodes and 42 links in 10 hours of computing time.

4 Heuristic Algorithms

4.1 Greedy Algorithm with Dual Weights Initialization (GA-DW)

A first simple approach to tackle the E-TE problem is to adopt a greedy strategy. Given a network topology G and a traffic matrix D, an initial set of links weights with low total cost of link utilization is obtained by applying the IGP-WO algorithm (see below for the initialization). Then we sort the network elements (nodes and links) according to some intuitive criteria, consider them in that order and try to switch off as many of them as possible.

We use three criteria for sorting nodes. In Least-Link (LL), Least-Flow (LF), and Sum-of-Weights (SW), nodes are sorted in non-decreasing order according to, respectively, the degree (number of incident links), the total amount of traffic flowing through them, and the sum of the weights of all the incident (active) links. We consider two criteria for sorting links. In Least-Flow (LF) and Traffic-Engineering (TE), links are sorted in non-decreasing order according to, respectively, the total amount of traffic flowing through them and their weight. Two of these criteria, Least-Link (LL) and Least-Flow (LF), were already used in a different context in [4]. The six combined node-link sorting policies are given in Table 1. The nodes are always considered before the links because of their higher energy consumption.

Table 1. Combinations of sorting criteria, the rows correspond to the link criteria and the columns to the router criteria

	LF	LL	SW
LF	LF-LF	LL-LF	SW-LF
TE	LF-TE	LL-TE	SW-TE

At each step of the greedy procedure, we verify whether the next available active network element according to the sorting order can be turned off. The considered element is actually turned off if the OSPF routing determined on the reduced network by the link weights of the active links, is able to support the traffic matrix (all traffic demands) without exceeding the link maximum utilization limit α comprised between 0 and 1. A run terminates when all the network elements have been tested.

The initial set of link weights is determined by applying 150 iterations of IGP-WO with a maximum weight value of 100. Since the java implementation of IGP-WO in the TOTEM toolbox [2] is computationally heavy, we speed up the procedure by following the suggestion in [3]. The idea is to warm-start IGP-WO with the set of link weights obtained with the procedure described in [14], namely by taking as initial link weights the values of the dual variables of the following linear programming multicommodity flow relaxation:

$$\min \quad \Phi = \sum_{(i,j)\in A} \phi\left(c_{ij}, l_{ij}\right) \tag{19}$$

$s.\ t.$

$$\sum_{j\in V} f^t_{vj} - \sum_{j\in V} f^t_{jv} = \begin{cases} -\sum_{s\in V} d_{st} & if \quad v = t \\ d_{vt} & if \quad v \neq t \end{cases} \qquad (v,t) \in N \tag{20}$$

$$l_{ij} = \sum_{t\in V} f^t_{ij} \quad (i,j) \in A \tag{21}$$

$$\phi\left(c_{ij}, l_{ij}\right) \geq \alpha_z l_{ij} - \beta_z c_{ij} \quad (i,j) \in A, z \in Z \tag{22}$$

$$f^t_{ij} \geq 0 \quad (i,j) \in A, t \in V. \tag{23}$$

The objective function (19) is, as in the IGP-WO algorithm [6], the sum of piecewise linear convex functions $\phi\left(c_{ij}, l_{ij}\right)$ based on the links utilization. Constraints (20) are the classic flow conservation constraints, while Constraints (21) force the total flow on each arc to be equal to the sum of all the flows routed through the arc itself. The dual variables used as link weights are those corresponding to Constraints (21). Constraints (22) define the link utilization cost function. Finally Constraints (23) define the positive flow variables f^t_{ij}, that are equal to the amount of traffic routed through the arc (i, j) and destined to node t.

In the *Greedy Algorithm with Dual Weights Initialization* (GA-DW) the greedy procedure is run six times (once for each combined node-link sorting policy) and the best solution found is returned. Finally, a set of link weights for the resulting sub-network is obtained by executing 150 iterations of IGP-WO so as to reduce the total cost of link utilization.

Note that the set of values of the link weights is determined only twice: at the beginning for the initial network topology and at the end of the algorithm for the resulting sub-network.

4.2 Two-Stage Algorithm with Dual Weights Initialization (TA-DW)

Since the problem of finding a minimum energy sub-network of G with an optimized set of link weights is very challenging, we split it into two stages. The TA-DW procedure includes a switching-off stage and a feasible routing stage.

The *Switching-off Stage*, which receives as input the complete network topology G and the given traffic matrix denoted by D, aims at selecting the set of network elements that could be switched off. This is achieved by solving within a 3% gap the following Mixed Integer Linear Program (MILP) that is a subset of the E-TE formulation (1)-(18):

$$\min \sum_{(i,j)\in A} p_{ij}x_{ij} + \sum_{k\in V} p_k y_k \tag{24}$$

s. t.

$$x_{ij} \le y_i \quad (i,j) \in A \tag{25}$$

$$x_{ij} \le y_j \quad (i,j) \in A \tag{26}$$

$$\sum_{i\in V} f_{it}^t = \sum_{s\in V} d_{st} \quad t \in N \tag{27}$$

$$\sum_{j\in V} f_{vj}^t - \sum_{i\in V} f_{iv}^t = d_{vt} \quad v,t \in V, t \neq v \tag{28}$$

$$\sum_{t\in V} f_{ij}^t \le x_{ij}\alpha c_{ij} \quad (i,j) \in A \tag{29}$$

$$x_{ij}, y_k \in \{0,1\} \quad (i,j) \in A, k \in V \tag{30}$$

$$f_{ij}^t \ge 0 \quad (i,j) \in A, t \in V. \tag{31}$$

The objective function (24) aims at minimizing the network energy consumption. Constraints (25)-(31) are identical to Constraints (2)-(6) and (16)-(17) of the E-TE formulation. Since the node power consumption p_i is in general much larger (about ten times) than the link power consumption p_{ij}, the formulation will give the priority to switching off the nodes. Note that the traffic demands routing is considered fully splittable, see Constraints (31). This formulation falls within the well-known class of capacitated multi-commodity minimum cost flow problems (CMCF) [7].

The *Feasible Routing Stage*, which receives as input the sub-network $G(D)$ determined at the *Switching-Off Stage*, aims at finding a set of link weights that allows to route through the sub-network all the traffic demands according to shortest paths, without exceeding the link maximum utilization α $(0 < \alpha \leq 1)$. The link weights configuration of the second stage is determined by applying 150 iterations of the IGP-WO algorithm with a maximum weights value of 100. As for GA-DW, IGP-WO is warm-started using as input the dual weights computed by solving (19)-(23) for the reduced network $G(D)$ determined at the *Switching-Off Stage*.

Unfortunately, given the sub-network topology $G(D)$ obtained at the first stage and its respective traffic matrix D, there is no guarantee that there exists a set of link weights allowing feasible routing of all traffic demands. This may occur because the first stage considers a fully splittable routing that can be hardly reproduced by the OSPF protocol. In case no feasible OPSF weights set exists (or is found), we slightly increase the original traffic matrix D by multiplying it with a fixed parameter γ, and we repeat the first stage with the increased traffic matrix $D(\gamma)$. γ is equal to 1 at the first iteration, and is increased by 0.1 every time IGP-WO fails to find a feasible set of link weights. If the maximum utilization level is greater than α but smaller than $\alpha + 0.01$, the γ parameter is increased by 0.05 rather than 0.1. This operation clearly leads to a sub-network $G(D(\gamma))$ with more active elements as input of the second stage (the second stage is always run with the original traffic matrix D). The *Feasible Routing Stage* ends when a feasible set of link weights is found.

To check whether some other elements of the resulting sub-network can still be switched off, we apply a last run of the GA-DW algorithm.

4.3 MILP-Based Algorithm

To achieve high quality solutions in a reasonable computing time, we combine the two above algorithms. The idea is to achieve a trade-off between the very low computational load of GA-DW and the robustness of TA-DW.

The resulting *MILP-based Algorithm* (MILP-BA) is composed of the following two stages:

- A complete run of GA-DW with a maximum utilization level of $\alpha - 0.1$.
- A complete run of TA-DW, where the nodes switched-off at the first stage, are forced to remain off. Moreover, the traffic matrix increase criteria are slightly changed, i.e., the γ scaling parameter that is set to 1 at the first iteration, is increased by 0.05 if the maximum utilization level obtained by IGP-WO is comprised between α and $\alpha + 0.1$, or increased by 0.1 if the maximum utilization exceeds $\alpha + 0.1$.

On the one hand, GA-DW allows us to reduce the computational load of TA-DW by substantially reducing the number of variables in the MILP formulation. On the other hand, by decreasing the maximum utilization α by 0.1 in the first run of GA-DW, we tend to avoid that GA-DW forces TA-DW to switch off the wrong network elements.

Table 2. Rocketfuel network topologies

Network	Type	Nodes	Links	$\textbf{Edge}_{node}$	$\textbf{Core}_{node}$	$\%\textbf{Core}_{node}$
Ebone	Backbone	87	322	31	56	64.4
Exodus	Backbone	79	294	38	41	51.9
Sprint	Access	52	168	52	0	0
AT&T	Access	115	296	115	0	0

5 Computational Experiments

5.1 Network Topologies and Traffic Matrices

We have carried out computational tests on four real network topologies provided by the Rocketfuel project [1]. Since our algorithms aim at switching-off both nodes and links, the main focus is on backbone networks that contain edge nodes as well as core nodes and whose core nodes may be switched-off. However, we have also considered access networks that only contain edge nodes, which cannot be switched-off. The characteristics of the four network topologies are summarized in Table 2. The two access networks, Sprint and AT&T, have been used in [16] for testing other energy-aware traffic engineering approaches. Unfortunately, although the network topologies are known, no accurate information is available concerning link capacities and network equipments. For the backbone networks, we assume that all the network links have the same capacity, and we equip each node with routers M10i (power consumption p_i of $86.4W$), and each link with a Gigabit Ethernet line card (power consumption p_{ij} of $7.3W$). Since a router M10i can support at most eight Gigabit Ethernet line cards, the number of routers in each node directly depends on the degree g_i of the node ($\lceil \frac{8}{g_i} \rceil$ routers in each node). For the access networks, we use the capacity values and the equipment configuration kindly provided to us by the authors of [16]. In the case of backbone networks, also the information on the subdivision between edge and core nodes is missing. Since edges nodes can be both source and destination of traffic demands and core nodes play only a role of transit routers, core nodes are the only one that can be powered off. We have randomly selected a set of edge routers for each one of the three network topologies. To avoid the trivial and unrealistic cases where core leaf nodes can be easily powered off, all the leaf nodes are considered as edge nodes. At least one edge node has also been selected for each city.

As to the traffic matrices, for access networks we have used the same traffic matrices as in [16]; the matrices have been obtained by multiplying with different scaling factors the basic matrices computed with the gravity model. For the backbone networks, we have generated the traffic matrices in two different ways:

1. **Constant and Poisson:** generated using the Totem toolbox [2], the maximum load matrices with constant and Poisson traffic distribution that can be supported by the (complete) networks with OSPF *hop-count* routing.

2. **LP-based multicommodity flow matrices:** generated by scaling with a parameter $\beta \in (0,1)$ a maximum supported traffic matrix obtained with a linear programming (LP) formulation. By maximum supported traffic matrix, we mean that all the traffic demands can only be satisfied by switching on all the network links/routers and performing fully splittable routing.

For the sake of simplicity, in our tests we have considered the maximum utilization parameter $\alpha = 1$.

5.2 Results

The computational experiments have been carried out on an Intel Pentium Duo 3.0GHz with 3.5GB of RAM. The results are reported in Tables 3 and 5 for backbone networks and in Table 4 for access networks. The first column indicates the instance considered (network and matrix). Ex, Eb and Spr are the abbreviations for, respectively, Exodus, Ebone and Sprint, while Letters C and P, and numbers (30-40-50 in backbone cases, 7-12-14-21-24-36 in access cases) that follow the networks acronyms correspond to the traffic matrix considered (C for constant matrices, P for Poisson matrices, 30-40-50 for the maximum supported LP matrices scaled by 0.3, 0.4 and 0.5, 7-12-14-21-24-36 for the basic gravity matrices scaled by 7, 12, 14, 21, 24 and 36). Note that by multiplying the Sprint basic matrix by 36 and the AT&T basic matrix by 21, the maximum utilization level obtained performing the OSPF routing is above 90%. The columns $C-E$, L, E_c^{tot} (W), $Cong_{min}$ report respectively, the number of core and edge network nodes, the number of network links, the energy consumption of the complete network and the optimized congestion obtained with the complete network. As measure of the congestion level we use the value of the cost function defined by IGP-WO, namely the total cost of link utilization. The remaining columns correspond to the solutions returned by the algorithms; E_c (W) is the energy consumption, gap is the ratio $(E_c - E_c^b)/E_c^b$, where E_c^b is the bound on the energy consumption. Note that this energy consumption bound value is calculated by considering the energy consumption of the optimum solution of the MILP formulation for the switching-off stage. $Cong$ indicates the solutions congestion level, $Cong_{\%}$ reports the ratio $Cong/Cong_{min}$. N_{off} and L_{off} show respectively the number of nodes and links switched-off, while t and t_{nor} are respectively the total computing time and the computing time normalized w.r.t. the computing time of GA-DW.

For comparison purposes, we have also adapted the general Greedy Randomized Adaptive Search Procedure (see e.g. [11]) to the E-TE problem. At each iteration of the greedy algorithm, the network element to be switched off is randomly selected among the first $k\%$ elements of the ordered list derived from the sorting criterion. This randomized greedy procedure is run with the same sorting criterion for a predefined number of iterations and the best solution obtained is returned as an approximate solution. We have also endowed our GRASP procedure with a path-relinking feature that allows to intensify the search between elite solutions [12]. Unfortunately, in this case path-relinking only slightly improved the solution quality.

In Table 3 we compare the results obtained running GA-DW and the corresponding version GA-RW with random weights initialization on the backbone networks. The results clearly confirm the impact of this type of link weight initialization. GA-DW results are better in nine cases out of ten.

Table 3. Computational results. Comparison between GA-RW (Greedy Algorithm with Random Weights) and GA-DW on the backbone networks.

				GA-RW				GA-DW				Bound
Inst	$C-E$	L	$Cong_{min}$	$E_c\,(W)$	N_{off}	L_{off}	$Cong$	$E_c\,(W)$	N_{off}	L_{off}	$Cong$	$E_c^b\,(W)$
Ex30	41-38	294	155052	5146.1	31	169	381870	5239.8	30	168	354534	4546.2
Ex40	41-38	294	253240	5883.5	25	139	499842	5753.3	26	145	566605	5131.5
Ex50	41-38	294	382600	6620.9	19	109	761000	6418.9	21	113	616497	5536.7
ExC	41-38	294	147639	5130.3	30	183	388132	5058.5	31	181	386235	4537.7
ExP	41-38	294	164764	5440.6	27	176	382164	5346.8	28	177	354508	4653.3
Eb30	56-31	322	111265	5677.9	34	207	306940	5677.9	34	207	300163	5540.4
Eb40	56-31	322	169811	6364.2	28	184	353066	6248.6	29	188	369041	5872.6
Eb50	56-31	322	242613	7324.3	19	159	537422	7136.9	21	161	462120	6327.7
EbC	56-31	322	179798	6616.1	25	185	310032	6313.1	28	191	379940	5865.3
EbP	56-31	322	202439	6537.0	26	184	377949	6320.4	28	190	423698	5865.3

Table 4. Computational results obtained with MILP-BA for the Sprint and AT&T access networks

					MILP-BA				
Inst	$C-E$	L	$E_c^{tot}\,(W)$	$Cong_{min}$	$E_c\,(W)$	L_{off}	$t\,(s)$	$Cong$	$Cong_{\%}$
Spr12	0-52	168	24972	28785	11950 (47.9%)	85 (50.6%)	578	160774	558%
Spr24	0-52	168	24972	59651	13339 (53.4%)	76 (45.2%)	584	476098	798%
Spr36	0-52	168	24972	96214	13795 (55.2%)	73 (43.5%)	1241	412256	428%
AT&T7	0-115	296	43344	38990	30504 (70.4%	82 (27.7%)	1816	215903	553%
AT&T14	0-115	296	43344	77980	31026 (71.6%)	79 (26.7%)	1854	323802	415%
AT&T21	0-115	296	43344	117347	32388 (74.7%)	70 (23.6%)	1990	616849	525%

In Table 5 we report the results obtained running GA-DW, G-GA-DW (the adaptation of GRASP to E-TE), TA-DW and MILP-BA on the backbone networks.

The solutions obtained with G-GA-DW are of slightly better quality than those provided by GA-DW but computing times are much higher. In one case the normalized computing time t_{nor} reaches even the value of 20. This is due to the multi-start strategy where the greedy is repeated 100 times for each one of the sorting policies.

TA-DW is very heavy computationally (its normalized computing time t_{nor} reaches the value of 15 in the worst cases) but the quality of the solutions is much better than those found by G-GA-DW. The gap is small, in average smaller than 5%. However, a computing time of 2 up to 8 hours is needed to solve the Ebone instances with up to 87 nodes (56 core nodes and 322 links).

MILP-BA turns out to achieve a remarkable trade-off. Computing times are much smaller than those of TA-DW (maximum value of t_{nor} of 6, and usually smaller than 2), while the solution quality remains the same (except for Ex40). Computing times are generally less than half-hour, and only about half of the time is used to solve the MILP formulations.

Table 5. Computational results. Comparison between GA-DW and G-GA-DW, TA-DW and MILP-BA on the backbone networks. The bound on the energy consumption obtained by solving the MILP formulation is also reported.

					GA-DW								**G-GA-DW**								**Bound**
$Inst$	$C-E$	L	$E_c^{tot}(W)$	$Cong_{min}$	$E_c(W)$	gap	N_{off}	L_{off}	$t(s)$	t_{nor}	$Cong$	$Cong_{\%}$	$E_c(W)$	gap	N_{off}	L_{off}	$t(s)$	t_{nor}	$Cong$	$Cong_{\%}$	$E_c^b(W)$
Ex30	41-38	294	9058.2	155052	5239.8	15.26%	30	168	1159	1	354534	229%	5124.2	12.71%	31	172	16224	14.00	409715	264%	4546.2
Ex40	41-38	294	9058.2	253240	5753.3	12.12%	26	145	1146	1	566605	224%	5738.7	11.83%	26	147	21152	18.46	604029	239%	5131.5
Ex50	41-38	294	9058.2	382600	6418.9	15.93%	21	113	1252	1	616497	161%	6519.9	17.76%	20	111	25323	20.23	621392	162%	5536.7
ExC	41-38	294	9058.2	147639	5058.5	11.48%	31	181	767	1	386235	262%	4928.3	8.61%	32	187	14632	19.08	331051	224%	4537.7
ExP	41-38	294	9058.2	164764	5346.9	14.91%	28	177	957	1	354508	215%	5152.2	10.72%	30	180	15329	16.02	305352	185%	4653.3
Eb30	56-31	322	10126.6	111265	5677.9	2.48%	34	207	2217	1	300163	270%	5670.6	2.35%	34	208	10121	4.57	334146	300%	5540.4
Eb40	56-31	322	10126.6	169811	6248.6	6.40%	29	188	1897	1	369041	217%	6162.2	4.93%	30	188	13053	6.88	328896	194%	5872.6
Eb50	56-31	322	10126.6	242613	7136.9	12.79%	21	161	1818	1	462120	190%	7107.7	12.33%	21	165	18218	10.02	440984	182%	6327.7
EbC	56-31	322	10126.6	179798	6313.1	7.63%	28	191	1624	1	379940	211%	6298.5	7.39%	28	193	13075	8.05	353153	196%	5865.3
EbP	56-31	322	10126.6	202439	6320.4	7.76%	28	190	1434	1	423698	209%	6226.7	6.16%	29	191	12163	8.48	407699	201%	5865.3

					TA-DW								**MILP-BA**								**Bound**
$Inst$	$C-E$	L	$E_c^{tot}(W)$	$Cong_{min}$	$E_c(W)$	gap	N_{off}	L_{off}	$t(s)$	t_{nor}	$Cong$	$Cong_{\%}$	$E_c(W)$	gap	N_{off}	L_{off}	$t(s)$	t_{nor}	$Cong$	$Cong_{\%}$	$E_c^b(W)$
Ex30	41-38	294	9058.2	155052	4929.5	8.43%	33	175	4732	4.08	492486	318%	4922.2	8.27%	33	176	2189	1.89	465289	300%	4546.2
Ex40	41-38	294	9058.2	253240	5342.0	4.10%	30	154	10509	9.17	666063	263%	5399.2	5.22%	29	158	2024	1.77	892690	352%	5131.5
Ex50	41-38	294	9058.2	382600	6216.9	12.29%	23	117	10020	8.00	663916	174%	6195.0	11.89%	23	120	4983	3.98	652835	171%	5536.7
ExC	41-38	294	9058.2	147639	4682.5	3.19%	34	197	1939	2.53	465046	315%	4704.4	3.67%	34	194	1863	2.43	441493	299%	4537.7
ExP	41-38	294	9058.2	164764	4820.0	3.58%	33	190	4369	4.57	400072	243%	4805.4	3.27%	33	192	1798	1.88	416809	253%	4653.3
Eb30	56-31	322	10126.6	111265	5670.6	2.35%	34	208	9062	4.09	312596	281%	5569.6	0.53%	35	210	2674	1.21	336005	302%	5540.4
Eb40	56-31	322	10126.6	169811	6111.1	4.06%	30	195	29808	15.71	456886	269%	6096.5	3.81%	30	197	2606	1.37	474118	279%	5872.6
Eb50	56-31	322	10126.6	242613	6689.1	5.71%	25	175	26303	14.47	490739	202%	6667.2	5.37%	25	178	3663	2.01	516547	213%	6327.7
EbC	56-31	322	10126.6	179798	6002.8	2.34%	30	198	9701	5.97	432694	241%	5931.0	1.12%	31	198	1868	1.15	503908	280%	5865.3
EbP	56-31	322	10126.6	202439	6096.5	3.94%	30	197	7657	5.34	490676	242%	6010.1	2.47%	31	197	2648	1.85	423935	209%	5865.3

Concerning the congestion level, it is worth pointing out that in general the total cost of link utilization increases reasonably.

Finally, Table 4 contains the results obtained with MILP-BA for the two access networks. Also in this case a large number of the links can be switched-off (for Spr12 up to 50% of the links). The lower percentage of energy saving achievable for AT&T networks (about half compared with that for Sprint networks) is due to the lower link redundancy and the higher number of leafs that characterize the AT&T topology. The largest computing times are of the order of 30 minutes. Note that, although the percentage increase in congestion is much larger than for backbone networks (up to 8 times), its absolute value is still reasonable.

6 Concluding Remarks

We have investigated the relevant and challenging problem of energy-aware IP traffic engineering with shortest path routing. We have proposed an efficient three-phase MILP-based heuristic which aims at minimizing the energy consumption as well as the total cost of link utilization. The computational results for two real network topologies and different types of traffic matrices show that it allows to switch off a substantial number of core nodes during low and moderate traffic periods, while guaranteeing the same point-to-point service quality and reasonably increasing the network total cost of link utilization.

We leave as future work the extension to account for single link failure and uncertainty in the traffic matrices.

References

1. Rocketfuel Project, `www.cs.washington.edu/research/networking/rocketfuel`
2. Totem Projec, `http://totem.run.montefiore.ulg.ac.be`
3. Altın, A., Fortz, B., Thorup, M., Ümit, H.: Intra-domain traffic engineering with shortest path routing protocols. 4OR: A Quarterly Journal of Operations Research 7(4), 301–335 (2009)
4. Chiaraviglio, L., Mellia, M., Neri, F.: Reducing power consumption in backbone networks. In: IEEE International Conference on Communications, ICC 2009, pp. 1–6. IEEE, Los Alamitos (2009)
5. Cianfrani, A., Eramo, V., Listanti, M., Marazza, M., Vittorini, E.: An energy saving routing algorithm for a green OSPF protocol. In: INFOCOM IEEE Conference on Computer Communications Workshops, pp. 1–5. IEEE, Los Alamitos (2010)
6. Fortz, B., Thorup, M.: Internet traffic engineering by optimizing OSPF weights. In: Proceedings of IEEE Nineteenth Annual Joint Conference on the IEEE Computer and Communications Societies, INFOCOM 2000, vol. 2, pp. 519–528. IEEE, Los Alamitos (2000)
7. Ghamlouche, I., Crainic, T., Gendreau, M.: Cycle-based neighbourhoods for fixed-charge capacitated multicommodity network design. Operations Research 51(4), 655–667 (2003)
8. Gupta, M., Singh, S.: Greening of the Internet. In: Proceedings of the 2003 Conference on Applications, Technologies, Architectures, and Protocols for Computer Communications, pp. 19–26. ACM, New York (2003)
9. Koomey, J.: Estimating total power consumption by servers in the US and the world. Final Report (2007)

10. Pióro, M., Medhi, D.: Routing, flow, and capacity design in communication and computer networks. Morgan Kaufmann Publishers, San Francisco (2004)
11. Resende, M., Ribeiro, C.: Greedy randomized adaptive search procedures. In: Glover, F., Kochenberger, G. (eds.) Handbook of Metaheuristics. International Series in Operations Research & Management Science, vol. 57, pp. 219–249. Springer (2003)
12. Resende, M., Ribeiro, C.: GRASP with path-relinking: Recent advances and applications. In: Ibaraki, T., Nonobe, K., Yagiura, M. (eds.) Metaheuristics: Progress as Real Problem Solvers, pp. 29–63. Springer (2005)
13. Sridharan, A., Guerin, R., Diot, C.: Achieving near-optimal traffic engineering solutions for current OSPF/IS-IS networks. IEEE/ACM Transactions on Networking 13(2), 234–247 (2005)
14. Ümit, H., Fortz, B.: Fast heuristic techniques for intra-domain routing metric optimization. In: Proc. of the 3rd International Network Optimization Conference (INOC 2007), Spa, Belgium, vol. 6 (2007)
15. Vasić, N., Kostić, D.: Energy-aware traffic engineering. In: Proc. of the 1st International Conference on Energy-Efficient Computing and Networking, pp. 169–178. ACM, New York (2010)
16. Zhang, M., Yi, C., Liu, B., Zhang, B.: GreenTE: Power-Aware Traffic Engineering. In: IEEE International Conference on Network Protocols (2010)

Designing AC Power Grids Using Integer Linear Programming

Arie M.C.A. Koster and Stephan Lemkens

Lehrstuhl II für Mathematik, RWTH Aachen University, 52056 Aachen, Germany
{koster,lemkens}@math2.rwth-aachen.de

Abstract. Recent developments have drawn focus towards the efficient calculation of flows in AC power grids, which are difficult to solve systems of nonlinear equations. The common linearization approach leads to the well known and often used DC formulation, which has some major drawbacks. To overcome these drawbacks we revisit an alternative linearization of the AC power flow. Work on this model has already been done in the 1990s but was intractable at that time. In view of recent developments in the field of integer programming, we show that this model is computationally tractable.

1 Introduction

A power grid is a transmission network transporting electrical energy from power plants to some substations near urban or industrial centers. To reduce the loss of energy, different high voltages (at least 110kV) and alternating currents (AC) are used to transport the power. The so called AC power flow consists of two single power flows called active and reactive flow.

The network consists of different subnetworks each with its own level of voltage, whereby a single network contains nodes with a specific demand of active and reactive power which are connected through lines. The nodes may represent substations which lead to different networks of a lower voltage level or groups of customers. The lines may represent underground power cables or overhead power lines.

In this paper, we consider the design of AC power grids, including the placement of supply equipment (called generators). The potential topology is modelled as an undirected graph $N = (V, E)$, where the set V denotes the demand nodes and E the set of all possible lines between the nodes. The design problem is to find the minimum cost network which fulfills all demands. Given a selection of the lines, we need to calculate the power flow in the network. In an AC network, we have an active and a reactive power flow which periodically reverse their direction. The computation of these bidirectional power flows involves complex numbers and nonlinear functions.

The most common way to handle the nonlinearities is to use the so-called *DC model* which provides linear approximations. Although the DC modelling of the power flow has proven to be very fast, its major drawback is that information about the reactive flows is lost. As engineers depend on them, they have shifted their focus towards metaheuristics like genetic algorithms to solve power flow problems [7]. In view of recent developments in integer linear programming, we revisit a less known linearization which approximates both the active and the reactive power flow.

J. Pahl, T. Reiners, and S. Voß (Eds.): INOC 2011, LNCS 6701, pp. 478–483, 2011.

This paper gives an overview of our model and some preliminary computational studies which show some promising results towards future work. We hope that by using more powerful tools like the generation of valid inequalities, our model can provide an alternative to the often used DC model.

2 The General Model

Given the potential topology $N = (V, E)$, let the set A consist of both arcs (v, w) and (w, v) for all $\{v, w\} \in E$. Additionally, we have a set $\mathscr{G}$ of possible generators with different construction and operating costs c_g and maximum power feed Ψ_g for each $g \in \mathscr{G}$. All these generators operate on the same voltage level $\mathbf{U}$. Generators can be installed at a subset $S \subseteq V$ to fulfill the power requirements.

For every potential line $e \in E$, let c_e the operating and construction costs, $\hat{c}_e$ a cost factor for the active power losses, R_e the line resistance, and X_e the reactance.

For every line, we compute the conductance $G_e = R_e/(R_e^2 + X_e^2)$ and susceptance $B_e = -X_e/(R_e^2 + X_e^2)$.

Finally let P_v and Q_v denote the active and reactive power demand of node $v \in V$.

At each node $v \in V$, we have to calculate the voltage $|U_v| \cdot e^{i\vartheta_v}$, where i denotes the imaginary unit. $|U_v|$ is called the voltage magnitude and ϑ_v the voltage angle. Therefore, we introduce continuous variables U_v and ϑ_v for each node, which are bounded by $U_{\min}$ and $U_{\max}$ and accordingly $\vartheta_{\min}$ and $\vartheta_{\max}$. In addition, let $P_{\text{gen}\,v}$ and $Q_{\text{gen}\,v}$ denote the active and reactive power feed at node v. For each $e \in E$ variables $x_e \in \{0, 1\}$ denote whether or not the line is constructed. Variables $y_{vg} \in \mathbb{Z}_0^+$ denote how many generators of type g are installed at node v.

Let $P(a)$ and $Q(a)$ be functions (depending on U_v and ϑ_v, defined below) which model the active and reactive flow on arc $a \in A$ and $f(P(e))$ a function wich represents the power losses on line $e \in E$. We consider the following nonlinear model describing the optimal network design:

$$\min \sum_{e \in E} (\hat{c}_e \cdot f(P(e)) + c_e) \cdot x_e + \sum_{v \in V} \sum_{g \in \mathscr{G}} c_g \cdot y_{vg}$$

$$\sum_{(v,w) \in A} P((v,w)) \cdot x_{\{v,w\}} = P_v - P_{\text{gen}\,v} \qquad \forall v \in V \tag{1a}$$

$$\sum_{(v,w) \in A} Q((v,w)) \cdot x_{\{v,w\}} = Q_v - Q_{\text{gen}\,v} \qquad \forall v \in V \tag{1b}$$

$$\sum_{g \in \mathscr{G}} \Psi_g \cdot y_{vg} \geq P_{\text{gen}\,v} + Q_{\text{gen}\,v} \qquad \forall v \in S \tag{1c}$$

$$x \in X$$

Here the constraints (1a) and (1b) ensure that the active and reactive power demands are fulfilled at each node. Additionally, they conserve the flow in the network. Constraint (1c) guarantees generators are needed at a node before it can feed their power into the network. Further desired properties regarding the network topology can be modelled by a set X, e.g., two-connectivity.

If a node feeds into the network, it will have a fixed voltage magnitude $\mathbf{U}$. As the variable y_{vg} is integer we need to introduce a binary variable z_v which is set to one iff $\sum_{g\in\mathscr{G}} y_{vg} \geq 1$. For every possible feeding node $v \in S$ this leads to the constraints

$$\begin{aligned} U_v + (U_{\min} - \mathbf{U}) \cdot z_v &\geq U_{\min}, \qquad & z_v - \sum_{g\in\mathscr{G}} y_{vg} &\leq 0, \\ U_v + (U_{\max} - \mathbf{U}) \cdot z_v &\leq U_{\max}, \qquad & -M \cdot z_v + \sum_{g\in\mathscr{G}} y_{vg} &\leq 0, \end{aligned} \tag{1d}$$

with M sufficiently large. The constraints on the left side guarantee that a node with at least one generator has a voltage magnitude of $\mathbf{U}$. The other constraints force z_v to be one iff $\sum_{g\in\mathscr{G}} y_{vg} \geq 1$.

The nonlinear functions for the AC power flow on arc $(k,j) \in A$ are [1]:

$$P((k,j)) = U_k^2 G_{kj} - U_k U_j G_{kj} \cos(\vartheta_k - \vartheta_j) - U_k U_j B_{kj} \sin(\vartheta_k - \vartheta_j), \tag{2a}$$

$$Q((k,j)) = -U_k^2 B_{kj} + U_k U_j B_{kj} \cos(\vartheta_k - \vartheta_j) - U_k U_j G_{kj} \sin(\vartheta_k - \vartheta_j). \tag{2b}$$

The difference between the two active flows on a line is called the active power loss. For a line $e = \{k,j\}$ with corresponding arcs $a_+ = (k,j)$ and $a_- = (j,k)$ it is:

$$f(P(e)) = P(a_+) + P(a_-) = G_e \cdot \left(U_k^2 + U_j^2 - 2U_k U_j \cos(\vartheta_k - \vartheta_j)\right), \tag{3}$$

Note that these nonlinear functions are multiplied with a binary variable. Although recent developments in the field of mixed integer nonlinear programming (MINLP) have yield some promising results, the resulting MINLP model is very difficult to handle, cf. [2].

3 Linear Approximations of the Power Flow Functions

As one main difficulty lies in the choice of f, P and Q, we now discuss two linear choices for these functions.

3.1 The DC Power Flow

The most common linearization of (2a) and (2b) is the so called *DC power flow model*. To linearize the model, we assume $\cos(\vartheta_k - \vartheta_j) \approx 1$, $\sin(\vartheta_k - \vartheta_j) \approx \vartheta_k - \vartheta_j$, $R_{kj} \ll X_{kj}$, and $|U_k| = |U_j| = U_0$ with U_0 a fitting constant. From $R_{kj} \ll X_{kj}$ it follows that $G_{kj} \approx 0$ and we get the DC power flow equations

$$P((k,j)) = -U_0^2 B_{kj} \cdot (\vartheta_k - \vartheta_j), \qquad Q((k,j)) = 0.$$

Notice, for $U_0 = 1$ this reduces to the well-known $P((k,j)) = (\vartheta_k - \vartheta_j)/X_{kj}$.

The above assumptions guarantee that we get a symmetric flow, meaning that $P_{kj} = -P_{jk}$ holds. Therefore, we have no active power losses ($f = 0$). This linearization is used in a variety of different integer linear programs concerning power grid problems, cf. [3,4,6].

One major advantage of this linear model is, its interpretation as DC power flows and not just only as an approximation of the AC power flows. The drawback, however, is the missing insight on the reactive flows Q. We like to stress the fact that the above assumptions can only be made, if the underlying network fulfills some specific properties. The checking whether these conditions hold is, however, often forgotten. We refer to [9] for a discussion of the importance of these conditions.

3.2 Approximation of the AC Power Flow

The following approach was first introduced by Moser [8] and improved by Braun [5], but to our knowledge no further work has been conducted since then. For each arc $a = (k, j) \in A$, we introduce new variables $\Delta U_a := U_k - U_j$ and $\Delta \vartheta_a := \vartheta_k - \vartheta_j$.

Like in the DC linearisation we assume $\cos(\Delta \vartheta_{kj}) \approx 1$, $\sin(\Delta \vartheta_{kj}) \approx \Delta \vartheta_{kj}$, and $|U_k| = |U_j| = U_0$, but without setting $\Delta U_{kj} = 0$ and $G_{kj} = 0$. This leads to the following linearisation of the AC power flow equations:

$$\begin{aligned} P((k,j)) &= |U_k|^2 G_{kj} - |U_k||U_j| G_{kj} \cos(\Delta \vartheta_{kj}) - |U_k||U_j| B_{kj} \sin(\Delta \vartheta_{kj}) \\ &\approx |U_k| G_{kj}(|U_k| - |U_j|) - |U_k||U_j| B_{kj} \Delta \vartheta_{kj} \\ &= U_0 G_{kj} \Delta U_{kj} - U_0^2 B_{kj} \Delta \vartheta_{kj} \end{aligned} \tag{4a}$$

$$\begin{aligned} Q((k,j)) &= -|U_k|^2 B_{kj} + |U_k||U_j| B_{kj} \cos(\Delta \vartheta_{kj}) - |U_k||U_j| G_{kj} \sin(\Delta \vartheta_{kj}) \\ &\approx |U_k| B_{kj}(-|U_k| + |U_j|) - |U_k||U_j| G_{kj} \Delta \vartheta_{kj} \\ &= -U_0 B_{kj} \Delta U_{kj} - U_0^2 G_{kj} \Delta \vartheta_{kj} \end{aligned} \tag{4b}$$

This approximation leads to a symmetric power flow as well (and therefore $f = 0$), but it allows the computation of a reactive flow.

4 Linearization of the Network Design Model

In this section, we construct a mixed integer linear program to determine the optimal network design with respect to the AC power flow. Even by using the linear power flow equations (4a) and (4b), we have to overcome the fact that the constraints (1a) and (1b) are nonlinear. Therefore, we substitute these constraints by

$$\sum_{(v,w) \in A} P((v,w)) = P_v - P_{\text{gen}\,v} \quad \forall v \in V \tag{5a}$$

$$\sum_{(v,w) \in A} Q((v,w)) = Q_v - Q_{\text{gen}\,v} \quad \forall v \in V. \tag{5b}$$

To guarantee that $P((v,w)) = Q((v,w)) = 0$ if $x_{\{v,w\}} = 0$, we force $\Delta U_{(v,w)} = \Delta \vartheta_{(v,w)} = 0$ iff $x_{\{v,w\}} = 0$. For this, we introduce for each $a = (v,w)$ complementary variables $\Delta \tilde{U}_a$ and $\Delta \tilde{\vartheta}_a$ satisfying

$$\Delta \tilde{U}_a = U_v - U_w - \Delta U_a \quad \text{and} \quad \Delta \tilde{\vartheta}_a = \vartheta_v - \vartheta_w - \Delta \vartheta_a. \tag{5c}$$

The requirement now holds iff for each $a = (v,w)$ and $e = \{v,w\}$ the constraints

$$-\Delta U_{\max} \cdot x_e \leq \Delta U_a \leq \Delta U_{\max} \cdot x_e,$$
$$-\Delta \vartheta_{\max} \cdot x_e \leq \Delta \vartheta_a \leq \Delta \vartheta_{\max} \cdot x_e,$$
$$-\Delta U_{\max} \cdot (1 - x_e) \leq \Delta \tilde{U}_a \leq \Delta U_{\max} \cdot (1 - x_e),$$
$$-\Delta \vartheta_{\max} \cdot (1 - x_e) \leq \Delta \tilde{\vartheta}_a \leq \Delta \vartheta_{\max} \cdot (1 - x_e), \tag{5d}$$

are satisfied, with appropriate values for $\Delta U_{\max}$ and $\Delta \vartheta_{\max}$, computed from the given bounds for U and ϑ.

5 Computational Experience and Further Remarks

Note that in the above model there are no explicit bounds on the power flows given. However, the bounds on the voltage drop imply bounds on the maximal flow per line. To reduce the running time, we focused our studies on a medium sized 110 kV network with 28 nodes and 67 possible lines. We used the above model with both approximations for the power flow equations.

The DC model has proven to be very fast, as the problem was solved within seconds. As the DC model operates under the assumption of a constant voltage magnitude, the bounds on the voltage magnitudes are meaningless. Therefore, the optimal solution of the DC model was the minimum cost circuit of the network.

The overall performance of the linearized AC model was acceptable (but significantly slower than the DC model) and the optimal designs produced have the expected shape, meaning a couple of intersected circuits.

For both solutions, we calculated nonlinear power flows for the computed topologies using Newton's method. We observed that the calculated AC power flow approximation yields a good starting solution for the method. The error margin of the approximation is small, e.g., less than 6% for the voltage magnitude. Furthermore, the nonlinear solution fulfills the bounds of the voltage drop. In contrast, the nonlinear power flow solution of the topology given by the DC model does not satisfy these bounds, and therefore the topology is not valid for the problem setting.

In the future we like to derive more information from the very fast DC model and use it for the AC approximation. In addition, much work still has to be done to enhance the performance of the AC model. We hope that by deriving valid inequalities the performance of these problems can be signficantly improved.

Our next focus lies on the calculation of an approximation of the active power loss. Although our linearization forces the power loss to be zero, we can use a different approach to successfully approximate the power loss. This model is way more complex than the above, but surprisingly it seems to outperform it significantly regarding computation time.

Acknowledgement

This work was supported by a RWTH Seed Funds Project 2010, funded by the Excellence Initiative of the German federal and state governments.

References

1. Andersson, G.: Modelling and Analysis of Electric Power Systems, ETH Zürich (September 2008), http://www.eeh.ee.ethz.ch/uploads/tx_ethstudies/modelling_hs08_script_02.pdf
2. Bautista, G., Anjos, M.F., Vannelli, A.: Formulation of Oligopolistic Competition in AC Power Networks: An NLP Approach. IEEE Transactions on Power Systems 22(1) (2007)
3. Bienstock, D., Mattia, S.: Using mixed-integer programming to solve power grid blackout problems. Discrete Optimization 4(1), 115–141 (2007)
4. Bienstock, D., Verma, A.: The N-k Problem in Power Grids: New Models, Formulations and Numerical Experiments. SIAM J. on Optimization 20(5), 2352–2380 (2010)
5. Braun, A.: Anlagen- und Strukturoptimierung von 110-kV-Netzen. Ph.D. thesis, RWTH Aachen University (2001)
6. Marshall, A., Boffey, T., Green, J., Hague, H.: Optimal design of electricity distribution networks. IEE Proceedings on Generation, Transmission and Distribution, C 138, 69–77 (1991)
7. Maurer, H.C.G.: Integrierte Grundsatz- und Ausbauplanung für Hochspannungsnetze. Ph.D. thesis, RWTH Aachen University (2004)
8. Moser, A.: Langfristig optimale Struktur und Betriebsmittelwahl für 110-kV-Überlandnetze. Ph.D. thesis, RWTH Aachen University (1995)
9. Purchala, K., Meeus, L., Dommelen, D.V., Belmans, R.: Usefulness of DC Power Flow for Active Power Flow Analysis. In: Power Engineering Society General Meeting, 2005, vol. 1, pp. 454–459. IEEE, Los Alamitos (2005)

Energy Saving in Fixed Wireless Broadband Networks*

David Coudert[1], Napoleão Nepomuceno[2], and Issam Tahiri[1]

[1] MASCOTTE, INRIA, I3S (CNRS/UNS), France
firstname.lastname@inria.fr
[2] IMADA, Syddansk Universitet, Denmark
napoleao@imada.sdu.dk

Abstract. In this paper, we present a mathematical formulation for saving energy in fixed broadband wireless networks by selectively turning off idle communication devices in low-demand scenarios. This problem relies on a fixed-charge capacitated network design (FCCND), which is very hard to optimize. We then propose heuristic algorithms to produce feasible solutions in a short time.

1 Introduction

Fixed broadband wireless communications is a sector of the communication industry that holds great promise for delivering private high-speed data connections [1]. Such network comprises remote locations, each of them served by a radio base station (RBS), connected by means of high-capacity microwave radio links. A bidirectional link connecting two RBSs requires a dedicated pair of outdoor units (ODUs), each one directly coupled to a high-directional antenna.

Commonly, in this context, the network is built in a robust fashion to guarantee fault protection and to support the extremely bursty traffic behaviors. As a drawback, since ODUs consume substantial power whenever the link is up, it brings forth important energy waste to provide extra resources which could be used only in critical situations. Therefore, the traffic fluctuation over the time offers an opportunity to energy savings by handling traffic efficiently and turning off devices used to keep microwave radio links whose capacities are underused.

In this work, we consider the problem of deciding both the network's configuration and flows that minimize the total energy expenditure. Particularly, by configuration, we mean the choice of which communication devices we need to keep on to successfully meet the traffic requirements. This problem relies on a fixed-charge capacitated network design (FCCND), which is very hard to optimize [6]. Among others, [4] and [7] tackled similar problems on different networks. We present an exact formulation for this problem and propose heuristics that may be employed to produce good feasible solutions in a short time.

2 Problem Modeling and Linear Formulation

The network topology is modeled by a digraph $H = (V,E)$ where every node $v \in V$ represents a base station and every arc $vw \in E$ represents a radio link. Every link has a

* This work has been partially supported by project APRF RAISOM (PACA & FEDER), ANR DIMAGREEN, and Villum Kann Rasmussen foundation.

J. Pahl, T. Reiners, and S. Voß (Eds.): INOC 2011, LNCS 6701, pp. 484–489, 2011.

capacity c_{vw} and can be either active (while consuming energy) or not. Traffic demands are defined by $|D|$ pairs (s^d, t^d), with $s^d, t^d \in V$ and by an average volume per demand h^d. We assume that H is symmetric since in the type of studied networks, radio links are usually symmetric. This implies that for every node $v \in V$, the entering neighborhood is the same as the leaving neighborhood, i.e. $\delta^+(v) = \delta^-(v) = \delta(v)$. Also the cost of an active link is constant and equal to CL (as shown in [5]). Another assumption that we consider in this work, and which is not always true, is the possibility of routing the traffic of the same demand d through different paths from s^d to t^d (multi-routing).

The problem can be formulated as a mixed integer program (MIP). We define two types of variables: to represent the state of link vw we consider a binary decision variable, being equal to 1 if the link is active and 0 otherwise. Since symmetric links must be in the same state, and in order to reduce the total number of binary variables, we use a single variable u_{vw} with $v < w$ (assuming some ordering of the nodes) for the pair of symmetric links vw and wv. We also employ a variable x^d_{vw} to indicate the volume fraction of the demand d which is routed through the link vw. In the MIP formulation, Eq. (1) is the objective function, Eq. (2) are the capacity constraints on the links, and Eq. (3) are the flow conservation constraints.

$$\min \sum_{vw \in E} CL \cdot u_{vw} \tag{1}$$

$$s.t. \sum_{d=1}^{D} x^d_{vw} \le c_{vw} u_{vw} \qquad \forall vw \in E \tag{2}$$

$$\sum_{w \in \delta(v)} x^d_{wv} - \sum_{w \in \delta(v)} x^d_{vw} = \begin{cases} -h^d, & \text{if } v = s^d, \\ h^d, & \text{if } v = t^d, \\ 0, & \text{otherwise} \end{cases} \qquad \forall v \in V, \forall d = 1, \ldots, |D| \tag{3}$$

$$x^d_{vw} \in [0, h^d] \qquad \forall vw \in E, \forall d = 1, \ldots, |D| \tag{4}$$

$$u_{vw} \in \{0, 1\} \qquad \forall vw \in E \tag{5}$$

3 Hybrid Algorithm

The model cited in the previous section is a mixed integer linear program. Even though it can be handled by a solver like "CPLEX", this may take a very long time on relatively large networks (containing more than a hundred nodes). The number of variables and constraints can be huge and not even fit in memory. For such networks, we built a hybrid solution by combining a heuristic based on simulated annealing and a linear program with real variables (Multi Commodity Flow - MCF). The former would be the master and on every iteration it chooses the links to turn on/off. The latter, which is the slave, will only find out whether there is a feasible solution with this configuration or not (see Figure 1). Actually the linear program will have the same formulation except that the link state variables u_{vw} will now be constant. Therefore there will be no more integer variables in the program, which makes it faster to solve.

At each iteration of the simulated annealing process, a new network configuration is generated from the current solution by switching the state (Up Down or Down Up)

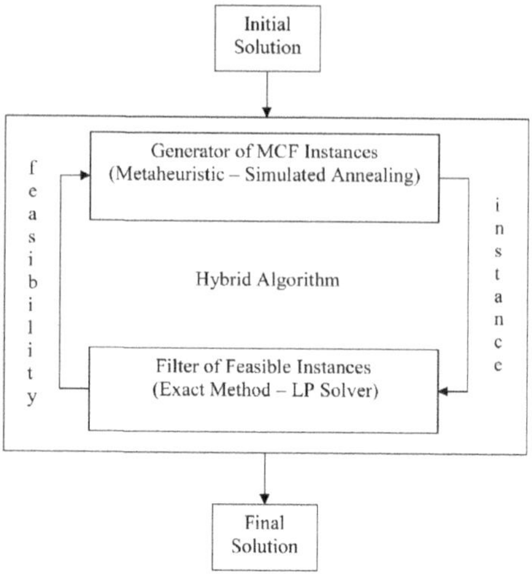

Fig. 1. The framework of the hybrid algorithm

of a couple of symmetric radio links. This couple is randomly chosen. Then, we apply the filter to check if this new configuration leads to a feasible solution. In the affirmative case, we attribute the total power utilization as energy score value of this solution. Otherwise, we discard this infeasible solution. We then follow the original description of the simulated annealing process [9] , where the algorithm replaces the current solution by the new solution with a probability that depends on the difference between the corresponding energy score values and a global parameter T (called the temperature).

4 Sparse Cuts-Based Algorithm

As an alternative approach, we propose a heuristic to construct a good configuration. It is also hybrid since the LP solver is still responsible for testing the feasibility of a configuration. This method is based on the sparse cuts of the graph. Roughly speaking, a cut is considered sparse if it has a high average load per link. Since this notion is mostly studied in the context of undirected graphs, we will consider for this heuristic the underlying undirected graph of our initial symmetric digraph.

Let us define the sparsity more formally. Let $G = (V,E)$ be the undirected graph. Let D be the set of demands (pairs of vertices). Let c be the capacity function that given a subset of edges returns the sum of capacities of all edges in this subset. Let also h be the function that for a given subset of demands returns the sum of volumes of demands in this subset. For any cut separating S (a subset of V) from $\bar{S}$, let $\alpha(S)$ denote the sparsity of this cut, with: $\alpha(S) = c(E(S,\bar{S}))/h((S*\bar{S})\cap D)$. If we consider a uniform version in which all edges have capacity 1 and to every pair of vertices corresponds a demand with the same volume 1, then the sparsity is $\alpha(S) = |E(S,\bar{S})|/|S|*|\bar{S}|$.

Algorithm 1. Maximize the number of shutdown links

Require: $G_0 = (V, E)$ is the initial graph
Require: c is a capacity function on the edges
Require: $cutsList$ is the list of precomputed sparse cuts
Require: D is the set of demands
Ensure: $G' = (V, E')$ with $E' \subseteq E$ is a subgraph ensuring the routing of all demands.

```
nonPotentialLinks ⇐ ∅
E' ⇐ E
while nonPotentialLinks ≠ E do
  SORT(cutsList) in descending order of average load per link in up state
  potentialList ⇐ ∅
  for i from 1 to SIZE(cutsList) do
    potentialList ⇐ potentialList|LINKS(cutsList[i])
  potentialList ⇐ potentialList|(E − (potentialList ∪ nonPotentialLinks))
  for i from SIZE(potentialList) downto 1 do {in LIFO order}
    if ROUTING_LP((V, E' − potentialList[i]), c, D) is feasible then
      E' ⇐ E' − potentialList[i]
      nonPotentialLinks ⇐ nonPotentialLinks ∪ potentialList[i]
      break
    else
      nonPotentialLinks ⇐ nonPotentialLinks ∪ potentialList[i]
```

As said before, our heuristic is based on sparse cuts; those which have a small sparsity. Finding the minimum-sparsity cut, also called the sparsest cut, is a well-known combinatorial optimization problem in the literature. Although it is NP-complete, many work targeting improved approximations have been carried out [11,10,3,8], and recently an approximation that relies on a semi definite program (SDP) relaxation to achieve a $\sqrt{\log n}$ approximation factor has been proposed [2]. In order to obtain many different sparse cuts, we run the approximation many times while adding each time a small randomness in the objective function of the SDP.

The motivation behind using sparse cuts is the fact that the sparser the cut is, the more loaded its edges tend to be, and the less efficient deleting one of them would be. So every link will have an estimated load equal to the average load per link of the sparsest cut covering this edge. The heuristic will try to remove the links in ascending order of estimated load. At each iteration, the feasibility of routings will be checked with a linear program and, after removing a link, the sparsity of the cuts will be updated. An efficient way to implement this heuristic is described in Algorithm 1 which performs $\Omega(m)$ calls to the LP solver (m being the number of edges).

5 Simulation Results

We will follow the same scenario for the different approaches previously explained: the exact formulation using the linear program, the hybrid algorithm, and the sparse cuts based algorithm. The simulations have all been executed on SNDlib topologies which represent backbones (France with 45 links and Norway with 51 links). The main reason for this choice is that we were not able to find instances of topologies for fixed wireless

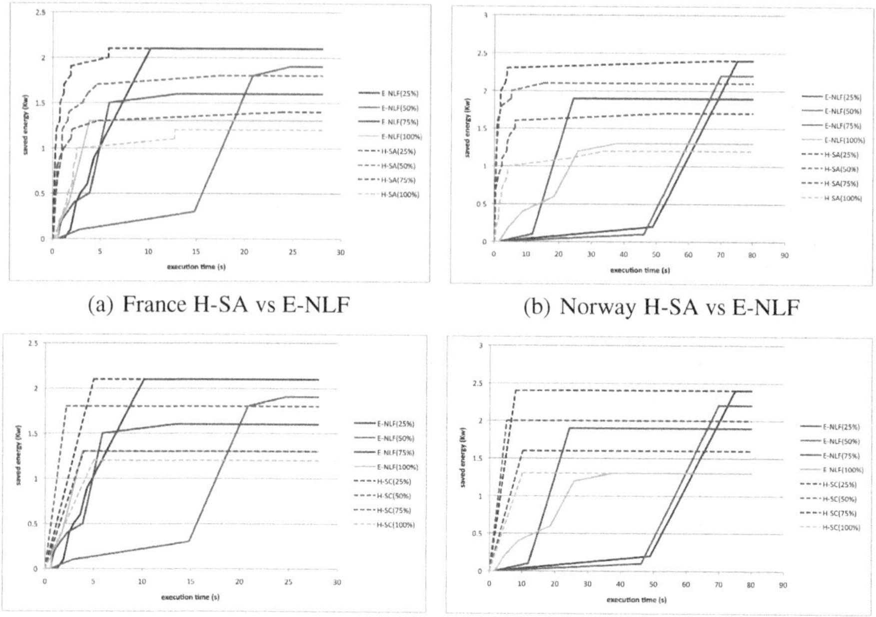

(a) France H-SA vs E-NLF

(b) Norway H-SA vs E-NLF

(c) France H-SC vs E-NLF

(d) Norway H-SC vs E-NLF

Fig. 2. The saved power as a function of the execution time

networks in the litterature. Link capacities and energy consumption were set for a typical fixed wireless network scenario. For the traffic matrix, we consider a uniform traffic matrix. The demand volume for every topology will take four values. Supposing that P is the maximum weight that we can achieve for that topology in an all-to-all scheme, these values would be $\{\ 100\%P, 75\%P, 50\%P, 25\%P\}$.

All the simulations have been executed on the same kind of machine equipped with dual core processors operating at 3GHz and 2GB of RAM. As MIP solver, we used CPLEX 12.1 to which MIP emphasis was set to "feasibility over optimality". In order to solve SDPs, we use the CSDP software. For notation, we use E-NLF for the exact (node-link) formulation. The two heuristics are labeled H-SA for the simulated-annealing-based heuristic and H-SC for the sparse cuts based heuristic.

Since our objective is to show that some heuristics can yield good solutions when the decision making is constrained by a limited execution time, we are not interested in the best solutions found by those heuristics. Hence, in Figure 2 we have reported the behavior of the algorithms before the Exact algorithm finds the optimal solution or near optimal (for the topology of Norway, in some cases, there was a slight improvement of the objective function after a thousand of seconds).

6 Conclusion

We first gave a mathematical formulation to the problem of minimizing energy consumption in fixed broadband wireless networks. Then we compared the behavior of two heuristic algorithms with that of an LP solver. We found out that heuristics would give a better solution if the execution time is very limited, especially when the size of the network increases. Thanks to the sparse cuts based algorithm we were able to verify that these cuts are an important argument to consider while designing heuristics for the studied problem. Therefore, one of our perspectives is to find an efficient way of choosing k cuts out of the sparsest in a given graph.

References

1. Anderson, H.: Fixed Broadband Wireless System Design. John Wiley & Sons, Chichester (2003)
2. Arora, S., Rao, S., Vazirani, U.: Geometry, flows, and graph-partitioning algorithms. Commun. ACM 51, 96–105 (2008)
3. Chawla, S., Gupta, A., Racke, H.: Embeddings of negative-type metrics and an improved approximation to generalized sparsest cut. In: 16th Annual ACM-SIAM Symposium on Discrete Algorithms (SODA), pp. 102–111 (2005)
4. Chiaraviglio, L., Mellia, M., Neri, F.: Reducing power consumption in backbone networks. In: IEEE International Conference on Communications, pp. 1–6 (2009)
5. Coudert, D., Nepomuceno, N., Rivano, H.: Power-efficient radio configuration in fixed broadband wireless networks. Computer Communications 33(8), 898–906 (2010)
6. Giroire, F., Mazauric, D., Moulierac, J., Onfroy, B.: Minimizing routing energy consumption: from theoretical to practical results. In: International Conference on Green Computing and Communications (GreenCom 2010), Hangzhou, China, p. 8 (2010)
7. Idzikowski, F., Orlowski, S., Raack, C., Woesner, H., Wolisz, A.: Saving energy in ip-over-wdm networks by switching off line cards in low-demand scenarios. In: Proceedings of the 14th Conference on Optical Network Design and Modeling, ONDM 2010, pp. 42–47 (2010)
8. Khandekar, R., Rao, S., Vazirani, U.: Graph partitioning using single commodity flows. In: STOC 2006: Proceedings of the Thirty-Eighth Annual ACM Symposium on Theory of Computing, pp. 385–390. ACM, New York (2006)
9. Kirkpatrick, S., Gelatt, C.D., Vecchi, M.P.: Optimization by simulated annealing. Science 220(4598), 671–680 (1983)
10. Leighton, T., Rao, S.: Multicommodity max-flow min-cut theorems and their use in designing approximation algorithms. J. ACM 46(6), 787–832 (1999)
11. Shmoys, D.B.: Cut problems and their application to divide-and-conquer (1996)

MIP Modeling of Incremental Connected Facility Location*

Ashwin Arulselvan[1], Andreas Bley[1], Stefan Gollowitzer[2], Ivana Ljubić[3], and Olaf Maurer[1]

[1] Technische Universität Berlin, MATHEON, Straße des 17. Juni 136, 10623 Berlin, Germany
{arulsel,maurer,bley}@math.tu-berlin.de

[2] Department of Statistics and Operations Research, University of Vienna, Austria
stefan.gollowitzer@univie.ac.at

[3] Institute of Computer Graphics and Algorithms, Vienna University of Technology, Austria
ivana@ads.tuwien.ac.at

Abstract. We consider the incremental connected facility location problem, in which we are given a set of potential facilities, a set of interconnection nodes, a set of customers with demands, and a planning horizon. For each time period, we have to select a set of facilities to open, a set of customers to be served, the assignment of these customers to the open facilities, and a network that connects the open facilities. Once a customer is served, it must also be served in subsequent periods. Furthermore, in each time period the total demand of all customers served must be at least equal to a given minimum coverage requirement for that period. The objective is to maximize the net present value of the network, which is given by the discounted revenues of serving the customers and by the discounted investments and maintenance costs for the facilities and the network. We study different MIP models for this problem, discuss some valid inequalities to strengthen these formulations, and present a branch and cut algorithm for finding its solution. Finally, we report (preliminary) computational results of our implementation of this algorithm.

1 Introduction

Practical Problem. The problem under consideration is an optimal design of a network topology in the context of a multi-period planning of local access networks. In this setting, a telecommunication company wants to increase the speed of broadband connections by combining fiber optic technology with existing copper connections, i.e., by means of the *Fiber-to-the-Curb* (FTTC) technology. Street segments along which fiber optic cables can be installed, determine the *core network*. Potential optical and *existing copper cables* intersect at locations where potential multiplexor devices need to be installed. Between a multiplexor

* This work was supported by the DFG Research Center MATHEON – Mathematics for key technologies.

J. Pahl, T. Reiners, and S. Voß (Eds.): INOC 2011, LNCS 6701, pp. 490–502, 2011.

and an end-customer, the existing copper connection is used. The existing copper paths are pre-processed building an *assignment network* whose edges are assignment links between potential multiplexor locations and end-customers. To build an FTTC network, one has to decide on which locations to install multiplexor devices so that each end-customer is assigned to a multiplexor, and each multiplexor is connected to the *central office* by a fiber optic path.

Due to the huge investment needed to build an FTTC network, the deployment is done in several stages. The company takes the strategic decision of fixing a minimal percentage of *customer demands* that should be served at each of the stages. Thereby, demand of a customer is defined as the number of end-subscribers (e.g., offices and/or households) behind the customer's address. The coverage of customer demands need to be increased over time.

We define the *incremental connected facility location problem*, denoted as *incremental ConFL*, as follows: We are given three disjoint sets of nodes: a set of facilities F, a set of customers R, and a set of Steiner nodes M. We denote $S = F \cup M$ and $V = S \cup R$. The potential connections among the nodes in S build the *core network* and are given as the undirected edge set E_S. The corresponding directed arc set is $A_S = \{(i,j),(j,i) \mid ij \in E_S\}$. The possible connections between the facilities F and the customers R are given by the edges $E_R \subseteq F \times R$, which define the directed arc set $A_R = \{(i,j) \in F \times R \mid ij \in E_R\}$. Note that it is sufficient to consider only arcs directed from facilities to customers here. We let $A = A_S \cup A_R$ and $E = E_S \cup E_R$. The considered planning horizon is given as a discrete set of (not necessarily equally long) time periods $T = \{1, \ldots, \mathcal{T}\}$, $\mathcal{T} > 1$. In addition, we are given fixed costs for edges $c : E \rightarrow \mathbb{R}_+$ and facilities $g : F \rightarrow \mathbb{R}_+$ for opening the edge or facility for the first time, and maintenance costs for edges $m : E \rightarrow \mathbb{R}_+$ and facilities $m_f : F \rightarrow \mathbb{R}_+$ that arise for each period an edge or a facility is actually used. The pre-period revenue for serving the customers is given by $p : R \rightarrow \mathbb{R}_+$. Finally, we are given customer demands $d : R \rightarrow \mathbb{Z}_+$ and a minimum coverage requirement D^t for each time period $t \in T$.

We seek for a schedule that, for each time period, describes which subset of facilities to use, which set of customers to serve by these facilities, how to assign the served customers to the open facilities, and how to build the core network in order to connect the open facilities. In each time step, the total demand of the served customers must satisfy the minimum coverage requirement and the chosen edges in E_S must form a network connecting the open facilities. Furthermore, a customer must be served in all periods after it has been served for the first time. The goal is to maximize the net present value of the network.

Related Multi-Period Optimization Problems: Facility location problem over time is a well-studied problem. A recent survey is given in [13]. In a recent work, [2] consider a multi-period *incremental facility location problem*, where the coverage of customer demand needs to be increased over time. The authors combine subgradient optimization and a Lagrangian approach and generate feasible solutions with a Lagrangian based heuristic.

There has been intense research on multi-period network design problems since publication of the seminal articles by [6], [7] and [17]. Optimization methods have been used for designing networks for telecommunication, transportation [16], distribution of gas or water [14] and many others.

Most of the literature on applications in the telecommunications sector consider capacitated problems. Recent contributions are, e.g., [5,11]. Much less literature is available on the Connected Facility Location problem.

Single-Period Connected Facility Location: Early work on ConFL mainly includes approximation algorithms. The problem can be approximated within a constant ratio and the currently best-known approximation ratio is provided by [8]. [12] describes a hybrid heuristic combining Variable Neighborhood Search with a reactive tabu search method. The author compares it with an exact branch and cut approach. In [15], a Greedy Randomized Adaptive Search Procedure (GRASP) for the unrooted ConFL problem is presented. The authors also provide a transformation that enables solving ConFL as the Steiner arborescence problem. [3] develop a dual-based local search (DLS) heuristic for a generalization of the ConFL problem. The presented DLS heuristic computes lower and upper bound using a dual-ascent and then improves the solution with a local search procedure. [9] study MIP formulations for ConFL, both theoretically and computationally. The authors provide a complete hierarchy of ten MIP formulations with respect to the quality of their LP bounds.

The remainder of this paper is organized as follows. In Section 2 we present integer programming formulations for the incremental ConFL problem and discuss a class of valid inequalities that may be used to strengthen these formulations. Section 3 provides a description of the separation subroutines that we implemented in order to solve these models. In Section 4 we describe the benchmark data sets, details of our implementation of the branch and cut algorithm, and (preliminary) results of our computational experiments.

2 MIP Modeling

In this section we present two alternative integer programming formulations for the incremental connected facility location problem.

We assume that one of the facilities, denoted as root r is open and used in all time periods. This node corresponds to the central office with an uplink to the backbone network of the area corresponding to the respective instance.

In order to model the connectivity constraints among the open facilities, it is sufficient to ensure that all other open facilities in F are connected to the root r [9]. For notational simplicity, we let F denote the set of all facilities except r throughout the remainder of this paper. Furthermore, we denote $\delta^-(W) := \{(i,j) \in A \mid j \in W,\, i \notin W\} \forall W \subset V$ and $F(j) := \{i \in F \mid (i,j) \in A_R\} \forall j \in R$.

In order to describe which customers and facilities are served and used at each time period, we introduce binary variables $y_j^t \in \{0,1\} \forall j \in R$ and $t \in T$ and $z_i^t \in \{0,1\} \forall i \in F$ and for all $t \in T$. These variables are interpreted as $y_j^t = 1$ if customer j is served in time period t, and 0 otherwise, and $z_i^t = 1$ if facility i is used in time period t, and 0 otherwise. The assignment of the served customers to the open facilities and the network connecting the open facilities to the root node are modeled together by the arc variables $x_{ij}^t \in \{0,1\}$ for all directed arcs $(i,j) \in A$ and for all time periods $t \in T$, which are interpreted as $x_{ij}^t = 1$ if arc (i,j) is used in time period t, and 0 otherwise. To describe the initial opening of facilities and edges, we also introduce the facility variables $\tilde{z}_i^t \in \{0,1\} \forall i \in F$ and all $t \in T$ and the aggregated edge variables $\tilde{x}_e^t \in \{0,1\} \forall e \in E \forall t \in T$, which are interpreted as $\tilde{z}_i^t = 1$ if facility i is opened for the first time in time period t,and $=$ otherwise and $\tilde{x}_e^t = 1$ if edge e is opened for the first time in time period t, and 0 otherwise. Observe that variables $\tilde{x}_e^t$ are associated to edges instead to arcs of the core network for the following reason: In the general case, a facility $i \in F$ may be opened in period $t \in T$, and closed in period $t+k \in T$ $(k > 0)$. Consequently, an arc that was oriented like (i,j) in period t, may be used in the opposite direction in period $t+k$. Since the edge opening costs need to be payed only once, we have to leave the direction of set-up variables $\tilde{x}$ unspecified.

With these variables and notations, the objective function of the incremental ConFL problem can be formulated as follows:

$$\mathbf{f}(x,y,z) = \sum_{t=1}^{T} (1+\alpha)^{-t} \left[\sum_{j \in R} p_j y_j^t - \sum_{e \in E} c_e \tilde{x}_e^t - \sum_{(i,j) \in A} m_{ij} x_{ij}^t \quad - \sum_{i \in F} g_i \tilde{z}_i^t - \sum_{i \in F} m_i z_i^t \right]$$

For each period $t \in T$, the objective function comprises the collected profit for customers served in period t decreased by the investment (maintenance) costs that need to be paid for each edge and facility that are opened (used) in this period. The following mixed integer programming formulation models the incremental ConFL:

$$(CUT_F): \quad \max \mathbf{f}(x,y,z)$$

$$\sum_{i \in F(j)} x_{ij}^t = y_j^t \qquad \forall j \in R,\ t \in T \qquad (1)$$

$$x_{ij}^t \le z_i^t, \qquad \forall (i,j) \in A_R,\ t \in T \qquad (2)$$

$$x_{ij}^t + x_{ji}^t \le \sum_{k=1}^{t} \tilde{x}_e^k \qquad \forall (i,j) = e \in E,\ t \in T \qquad (3)$$

$$z_i^t \le \sum_{k=1}^{t} \tilde{z}_i^k \qquad \forall i \in F,\ t \in T \qquad (4)$$

$$\sum_{j \in R} d_j y_j^t \ge D^t \qquad \forall t \in T \qquad (5)$$

$$y_j^t \ge y_j^{t-1} \qquad \forall j \in R,\ t \in T \qquad (6)$$

$$\sum_{(u,v) \in \delta^-(W)} x_{uv}^t \ge z_j^t \qquad \forall W \subseteq S \backslash \{r\},\ j \in W \cap F \neq \emptyset,\ t \in T \qquad (7)$$

$$x_{kl}^t, y_j^t, z_i^t \in \{0,1\} \qquad \forall (k,l) \in A,\ j \in R,\ i \in F,\ t \in T \qquad (8)$$

$$\tilde{x}_e^t, \tilde{z}_i^t \in \{0,1\} \qquad \forall e \in E,\ i \in F,\ t \in T \qquad (9)$$

Constraints (1) model the fact that a customer is served only if there is a facility connected to it. Constraints (2) enforce that a facility is open if it is used to serve a customer. Inequalities (3) and (4) ensure that we *open* edges and facilities as soon as they are *used*. Constraint set (5) expresses the minimum demand coverage requirement for each time period. Inequalities (6) enforce the continuance of service for each customer (i.e., if customer j was served in period $t \in T$, it also need to remain served in all consecutive periods). Finally, the exponentially large constraint set (7) ensures that, in each time period, all open facilities are connected to the root node. The inequalities in constraint set (7) enforce that for every subset $W \subseteq S$ that includes a facility j and does not include the root node r, at least one of the arcs in the set of all incoming arcs in W must be *used* if facility j is open. These inequalities correspond to the directed cutset inequalities in the Steiner tree formulation [9,12].

Instead of enforcing at least one arc in each directed cut that separates a chosen facility from the root node, as done by constraints (7), we may model the connectivity constraints by enforcing at least one arc in every directed cut that separates a chosen customer from the root node. This leads to the following alternative formulation for the incremental ConFL problem:

$$
\begin{aligned}
(CUT_R): \quad & \max \ \mathbf{f}(x,y,z) \\
& (x,y,z) \text{ satisfies } (1)-(6) \\
& \sum_{(u,v)\in\delta^-(W)} x^t_{uv} \geq y^t_j \qquad \forall W \subseteq V\backslash\{r\},\ j \in W \cap R,\ t \in T \qquad (10)
\end{aligned}
$$

2.1 Valid Inequalities

In this section we provide two new families of valid inequalities that can strengthen the previous two models. The third group of constraints presented here are several degree-inequalities that were very useful throughout our computations.

Cover Inequalities: The minimum coverage constraints (5) imply a set of *cover inequalities* that can be defined for each single period $t \in T$. We call a subset of facilities $I^t \subset F$ a *cover* if its complement, $\bar{I}^t = F \setminus I^t$, cannot serve enough customers to satisfy the minimal demand requirements for the time period t. We denote by $COV^t \subseteq 2^F$ the family of all covers for period t. We call an inclusion-wise minimal such facility set I^t a *minimal cover*. In other words, I^t is a minimal cover if $\bar{I}^t$ cannot satisfy the minimum demand requirement of period t even if all the facilities in $\bar{I}^t$ are open, but for any $i \in I^t$ the facility set $J = \bar{I}^t \cup \{i\}$ would allow to serve enough customers to meet the minimum coverage constraint. In such a case, obviously at least one facility from I^t needs to be opened. Consequently, the following set of *cover inequalities* are valid for all solutions of (CUT_F) and (CUT_R):

$$
\sum_{i\in I^t} z^t_i \geq 1 \quad \forall t \in T,\ I^t \in COV^t \qquad (11)
$$

It is easy to verify that the cover inequality for any non-minimal cover I^t is dominated by the cover inequality for any minimal cover $I^t_{\min} \subseteq I^t$. Furthermore, any non-minimal cover I^t can be easily turned into a minimal cover by iteratively removing all those facilities, whose removal still results in a cover.

It is also not difficult to construct examples where the addition of cover inequalities (11) strengthens the LP relaxations of (CUT_F) and (CUT_R). These inequalities are similar to the cover inequalities studied for knapsack constraints.

The separation of cover inequalities is a modification of the knapsack problem, and hence it is an NP-hard problem. Our separation algorithm for cover inequalities is described in Section 3.2.

Cut-Set-Cover Inequalities: The set of cover inequalities (11) also implies the following exponentially large family of cut-set inequalities, that we will refer to as *cut-set-cover inequalities*:

$$\sum_{uv \in \delta^-(W)} x^t_{uv} \geq 1 \quad \forall t \in T,\ I^t \in COV^t,\ W \subseteq S \setminus \{r\}, I^t \subseteq W \tag{12}$$

These inequalities state that, in each period $t \in T$, we have to establish a path between the root and at least one of the facilities from the set I^t. Once the corresponding covers I^t become known, the separation of these new inequalities can be done in polynomial time by means of a maximum flow algorithm, see Section 3.2.

Again, it is not difficult to show that the addition of the cut-set-cover inequalities (12) strengthens the LP relaxations of (CUT_F) and (CUT_R).

In-Arc Inequalities: The requirement that, in each time period, the root node is connected to any open facility, implies the following *in-arc inequalities*:

$$z^t_i \leq \sum_{(j,i) \in \delta^-(i)} x^t_{ji} \qquad \forall i \in F,\ t \in T \tag{13}$$

$$x^t_{ik} \leq \sum_{(j,i) \in \delta^-(i): j \neq k} x^t_{ji} \qquad \forall (i,k) \in A_S,\ i \neq r,\ t \in T \tag{14}$$

Inequalities (13) imply that there is at least one arc entering any chosen facility. Inequalities (13) ensure that there is at least one arc entering any facility or Steiner node if there is an arc leaving that node.

Note that these inequalities are implied by the cut inequalities (7) or (10), but not vice versa. However, there is only a polynomial number of inequalities of type (13) and (14), which makes these inequalities very useful in practical computations [9,10].

Furthermore, we add the inequalities

$$\sum_{(j,i) \in \delta^-(i)} x^t_{ji} \leq 1 \qquad \forall i \neq r,\ t \in T \tag{15}$$

to the LP relaxations of (CUT_F) and (CUT_R). The inequalities ensure that the indegree of every node except the root node is at most 1. These inequalities may cut off feasible solutions but as there are no capacity constraints associated with the facilities and edges, there always *exists an optimal* solution of incremental ConFL that satisfies these inequalities. Adding these inequalities to the formulations substantially reduced the solution times in our experiments.

3 Separation Algorithms

In this section we explain separation algorithms for the cover inequalities and the three groups of cut-set inequalities described above.

3.1 Separation of Cut-Set Inequalities

We now present the separation routine to generate cut inequalities of type (7). Let $\hat{x}^t$ and $\hat{z}^t$ by the values of the arc variables and of the facility variables of the current optimal LP solution. In order to find a violated inequality of type (7), we compute for each time period $t \in T$ and each facility node $j \in F$ a minimum r-j-cut in the digraph $G(S, A_S)$ with arc capacities $\hat{x}^t$, solving the corresponding maximum flow problem. Let $\Gamma(r, j)$ be the set of arcs in the minimum cut obtained from this maximum flow computation. If the corresponding maximum flow value is less than $\hat{z}_i^t$, the corresponding cut inequality

$$\sum_{(u,v) \in \Gamma(r,j)} x_{uv}^t \geq z_j^t \tag{16}$$

is violated and we add this inequality to the current formulation.

The separation of the customer based cutset inequalities (10) is carried out analogously. We now consider the entire digraph $G(V, A)$ with capacities $\hat{x}^t$ given by the LP solution's arc variable values and solve the maximum flow problem with the root node r as the source and the customer node j as the sink for each customer $j \in R$ and each time period. Again, let $\Gamma(r, j)$ be the arcs of the corresponding minimum cut. If the maximum flow value is less than y_i^t, we add the violated cut

$$\sum_{(u,v) \in \Gamma(r,j)} x_{uv}^t \geq y_i^t \; . \tag{17}$$

3.2 Separation of Cover and Cut-Set-Cover Inequalities

Let $t \in T$ and let $\hat{z}^t$ be the values of the facility variables in the current LP solution. In order to find a cover I^t for which the corresponding cover inequality (11) is violated, we introduce variables $\alpha_i \in \{0, 1\}$ for all $i \in F$ indicating which facilities are contained in I^t and $\beta_j \in \{0, 1\}$ for all $j \in R$ indicating which customers can be served by any of the facilities not in I^t. Clearly, a cover I^t that maximizes the violation of inequality (11) corresponds to an optimal solution of the following integer program:

$$\min \sum_{i \in F} \hat{z}_i^t \alpha_i \tag{18}$$

$$\sum_{j \in R} d_j \beta_j \leq D^t - \epsilon \tag{19}$$

$$\beta_j \geq 1 - \alpha_i \qquad \forall (i,j) \in A_R \tag{20}$$

$$\alpha_i, \beta_j \in \{0,1\} \qquad \forall i \in F, j \in R \tag{21}$$

Inequalities (20) guarantee that all clients that have at least one neighboring facility not in I^t are served, while constraint (19) ensures that the total demand of all served clients is strictly less than the demand required to meet the coverage constraint. Together, these constraints ensure that, for any integer solution of (18) - (21), the set of facilities i with $\alpha_i = 1$ forms a cover. Note that the objective value of a solution of (18) - (21) is equal to the left hand side of the corresponding cover inequality for the current LP solution. Finding a violated cover inequality thus is equivalent to finding a time period $t \in T$ and a solution of (18) - (21) with objective value strictly less than 1. In our implementation, we solve this integer program for all $t \in T$.

To separate the cut-set-cover inequalities for a given cover I^t, we create an artificial sink node l and connect the nodes in I^t to l. We then compute a maximum r-l flow in the graph $G(S \cup \{l\}, A_S \cup I^t \times \{l\})$ with capacities $\hat{x}^t$ for the arcs in A_S and capacity 1 for the artificial arcs in $I^t \times \{l\}$. If the maximum flow value is less than 1, we add the violated cut-set-cover inequality

$$\sum_{(u,v) \in \Gamma(r,l)} x_{uv}^t \geq 1 \tag{22}$$

where $\Gamma(r,l)$ is the arc set of a corresponding minimum cut.

4 Experiments

Benchmark Instances: In [9], a set of instances for connected facility location was generated by combining a set of benchmark instances for the Uncapacitated Facility location (UFL) problem from the UflLib [1] with instances of the Steiner tree problem (STP) from the OR-library [4]. The ConFL input graphs are generated in the following way: first f nodes of the STP instance are selected as potential facility locations (where f denotes the number of facilities in the corresponding UFL instance), and the node with index 1 is selected as the root. The number of facilities, the number of customers, opening costs and assignment costs are provided in UFL files. STP files provide edge-costs and additional Steiner nodes.

We consider a set of 32 instances obtained by combining four UFL instances `mp1`,`mp2` and `mq1`,`mq2` (of the size 200×200 and 300×300, respectively) with eight STP instances {`c`,`d`}`n`, for $n \in \{5, 10, 15, 20\}$. These instances define the core networks with between 500 and 1000 nodes and with up to 25,000 edges.

We extend these instances to include demands and time periods. We generate demands uniformly between 20 and 40 for each customer and we consider a time horizon $T = 5$. In the test instances generated in [9], the facility set F and customers R induce a complete bipartite graph. We desire a more sparse setting for our demand satisfaction and the cover set inequalities. Therefore, we only considered the connections of the first 20 closest facilities for each customer. Such obtained instances contain up to 1300 nodes and 45,000 edges. Finally, the minimum coverage required for time period t is defined as

$$D^t = \frac{\sum_{j \in R} d_j}{1.25(T-t)} \text{ for } t \in \{0, 1, 2, 3, 4\} \text{ and } T = 5.$$

The experiments were performed on an Intel Core2 Quad 2.66 Ghz systems with 2GB RAM. Each run was carried out on a single processor.

4.1 Branch and Cut Implementation

To test the effectiveness of the presented formulations and inequalities, we implemented a branch and cut algorithm using CPLEX 12.2 and Python API, a commercial integer programming solver with a branch and cut framework.

The integer linear programs initially contain all variables and the constraints (1) – (6). The cut inequalities (7) and (10), the cover inequalities (11), and the cut-set-cover inequalities (12) are applied in a standard cutting plane approach, iteratively adding those inequalities that are violated by the current fractional solution.

We add all indegree constraints (15) to the initial LP formulation. We generate a cut pool with all the in-arc inequalities (13) and (14), which are added at the root node if they are violated. We then call the maximum flow separation routine that generates the inequalities (7). This separation consists of randomly selecting 50 terminals at every time period and generating the violated cuts. We restrict the number of calls to the separation routine at every node by 10, to enable branching and avoid multiple calls to the separation routines. In addition to the above, the separation routine is called at node depth of multiples of 10 and at every occasion an incumbent is rejected. The intuition behind this scheme is that it would provide us with a balance between the time spent in generating the cuts and branching, as branching helps us reducing the search space (due to the priority strategies described below). The enhanced cuts and customer cuts are combined in the same separation routine. Each test run was limited to 2000 CPU seconds and the optimality gap at this point of time is reported in the results.

Branching: The assignment variables x_{ij}^t, when branched (set to 0 or 1), does not affect the search space as much as the facility variables z_i^t. So we give them the highest priority in the branching. This was also observed in [12], but unlike the connected facility location problem, in our incremental version of the problem, we also have uncertainty in determining the set of customers to be served at each time period. So, we provide them with the next highest priority in branching.

Separation Routine: We observed that the cuts generated by the maximum flow algorithm when the root is treated as source tend to generate cuts that are closer to the root node and there will be edges repeated in the various minimum cuts generated for various terminals. In order to avoid this, we treat the root as the sink and the facilities as the source. This was appropriately captured in the primal heuristic and the in-arc inequalities as well. We also perform nested cuts, wherein we resolve maximum flow for the same facility by setting the capacity of the edges in current minimum cutset to 1. The cover (11) and cut-set-cover inequalities (12) rely on solving an integer program at every call of the separation routine, which is run for every time period. The integer program terminates if the elapsed running time is over 100 seconds or if the objective value drops below 1. We use this exact separation to test the impact of these inequalities on the lower bound and in the event they are useful they will be replaced with heuristic methods similar to the techniques used to generate cover inequalities for knapsack constraints.

Primal Heuristics: We also implemented and tested a naive primal heuristic. After our initial runs we decided to turn off the CPLEX heuristics as this was leading to poor performance. The primal heuristic rounds up all the z variables that indicate the usage of a facility as well as the y variables, which indicate the service to a customer. We run a minimum cost flow algorithm with a linear cost estimator with the open facilities (rounded up values) as sinks and the root node as source to generate our Steiner tree.

4.2 Results

Our preliminary computational study has shown that CUT_R formulation is not competitive against the CUT_F model, due to the size of the support graph and the large number of cut-set inequalities that need to be separated. This is also consistent with the results obtained by [9] for the single-period ConFL.

Therefore, in our computational study, we compared the performance of the following two branch and cut settings:

- CUT_F formulation,
- CUT_F+ formulation extended by cover inequalities (11) and cut-set-cover inequalities (12).

For each of the two settings, we report on the following values given in Table 1: the overall percentage gap obtained after the time limit of 2000 seconds calculated as Gap $= (UB - LB)/LB$, where UB is the best obtained upper bound, and LB is the global lower bound; the number of all constraints separated throughout the execution of the algorithm, denoted by "Cuts"; the number of branch and bound nodes, denoted by "B&B".

Comparing the number of inserted cuts by the two approaches, we observe that the inclusion of coverage-related cuts (i.e., (11) and (12)) reduces the overall

Table 1. Comparison of two branch and cut settings: plain CUT_F model vs. CUT_F extended by cover and cut-set-cover inequalities

		CUT_F				CUT_F + (11) + (12)			
Instance	best LB	best UB	Gap[%]	cuts	B&B	best UB	Gap[%]	cuts	B&B
c10-mp1	164,136	166,691	1.53	2103	170	166,670	1.52	695	104
c10-mp2	160,278	168,347	4.79	1781	99	168,258	4.74	297	20
c10-mq1	346,866	386,409	10.23	457	11	386,054	10.15	161	0
c10-mq2	348,929	387,501	9.95	559	14	387,088	9.86	136	0
c15-mp1	165,004	166,900	1.14	1539	107	166,985	1.19	212	15
c15-mp2	161,333	168,487	4.25	1508	56	168,525	4.27	299	16
c15-mq1	352,583	386,520	8.78	567	15	386,326	8.73	140	0
c15-mq2	348,640	387,614	10.05	422	10	387,452	10.02	138	0
c20-mp1	155,919	167,227	6.76	334	6	167,184	6.74	121	0
c20-mp2	154,157	168,658	8.60	360	3	168,656	8.60	137	0
c20-mq1	349,075	386,640	9.72	298	0	386,540	9.69	55	0
c20-mq2	348,628	387,792	10.10	311	0	387,681	10.07	60	0
c5-mp1	162,521	166,466	2.37	1927	190	166,320	2.28	590	75
c5-mp2	158,230	168,042	5.84	1630	45	167,892	5.76	347	19
c5-mq1	346,924	386,236	10.18	869	15	385,491	10.00	133	0
c5-mq2	348,453	387,330	10.04	817	15	386,744	9.90	143	0
d10-mp1	164,160	166,945	1.67	2008	45	166,706	1.53	126	10
d10-mp2	155,885	168,381	7.42	1902	21	168,167	7.30	291	15
d10-mq1	342,278	386,234	11.38	508	6	385,844	11.29	121	0
d10-mq2	348,389	387,584	10.11	642	9	387,062	9.99	110	0
d15-mp1	158,402	167,103	5.21	813	15	167,005	5.15	176	6
d15-mp2	158,835	168,398	5.68	1103	15	168,468	5.72	193	1
d15-mq1	346,494	386,579	10.37	407	0	386,258	10.29	95	0
d15-mq2	346,129	387,683	10.72	415	6	387,458	10.67	118	0
d20-mp1	155,821	167,168	6.79	256	0	167,208	6.81	139	0
d20-mp2	154,141	168,661	8.61	291	0	168,675	8.62	175	0
d20-mq1	348,738	386,621	9.80	180	0	386,580	9.79	62	0
d20-mq2	348,365	387,801	10.17	135	0	387,772	10.16	16	0
d5-mp1	163,074	166,680	2.16	2666	50	166,216	1.89	1387	225
d5-mp2	163,182	167,967	2.85	2495	104	167,634	2.66	1172	165
d5-mq1	346,120	386,141	10.36	1230	15	385,396	10.19	185	0
d5-mp1	344,089	387,304	11.16	1541	15	386,517	10.98	166	0

number of cuts generated within a given time limit. This can easily be explained by the large separation times needed to solve the integer program (18)-(21). Despite the reduced number of separated inequalities, in 27 out of 32 instances we obtained reduced duality gaps when the coverage-related inequalities were used. This indicates the strength of the coverage-related cuts, but also the trade-off between their strength and their separation time.

We also observe that due to the branching and separation strategies that we choose, there is no direct correlation between the usage of coverage-related constraints and the number of branch and bound nodes.

5 Conclusions

In this work we introduce a new combinatorial optimization problem that models the design of fiber-to-the-curb networks over time. The problem is a multi-period version of the connected facility location problem that has been intensively studied in the literature in the last decade. Besides two mixed integer programming models, we also introduce two new families of valid inequalities derived from the incremental coverage constraints over time. We provide separation algorithms needed to detect the new coverage-related inequalities within a cutting plane framework. The problem is then solved by means of a branch and cut algorithm that makes use of the cut-set inequalities and the new coverage-related constraints. In the (preliminary) computational study we show that the new inequalities are useful for small and/or sparser instances, where the obtained duality gaps can be significantly reduced. For larger instances, it turns out that the there is a trade-off between the separation time of the coverage-related family of inequalities and the obtained improvement of the quality of lower bounds.

In a future work we intend to investigate the performance of the branch and cut algorithm on a larger set of benchmark instances. We also want to study the influence of the minimum coverage rate D^t to the quality of lower bounds of the proposed models. Further problem-related inequalities will be derived as well. One of the problems addressed by our computational results is the computational inefficiency of the integer program needed to separate the coverage-related inequalities. To overcome this problem, one needs to develop more efficient exact or heuristic approaches for the separation. Finally, it will be also interesting to compare decomposition based approaches (e.g., Lagrangian or Benders decomposition) with the proposed branch and cut framework.

References

1. UflLib, `http://www.mpi-xinf.mpg.de/departments/d1/projects/benchmarks/UflLib/`
2. Albareda-Sambola, M., Fernández, E., Hinojosa, Y., Puerto, J.: The multi-period incremental service facility location problem. Computers & Operations Research 36, 1356–1375 (2009)
3. Bardossy, M.G., Raghavan, S.: Dual-Based Local Search for the Connected Facility Location and Related Problems. INFORMS Journal on Computing 22(4), 584–602 (2010)
4. Beasley, J.E.: OR-library, `http://people.brunel.ac.uk/~mastjjb/jeb/orlib/steininfo.html`
5. Bienstock, D., Raskina, O., Saniee, I., Wang, Q.: Combined network design and multiperiod pricing: Modeling, solution techniques, and computation. Operations Research 54(2), 261–276 (2006)
6. Christofides, N., Brooker, P.: Optimal expansion of an existing network. Mathematical Programming 6, 197–211 (1974)
7. Doulliez, P.J., Rao, M.R.: Optimal network capacity planning: A shortest-path scheme. Operations Research 23(4), 810–818 (1975)
8. Eisenbrand, F., Grandoni, F., Rothvoß, T., Schäfer, G.: Connected facility location via random facility sampling and core detouring. Journal of Computer and System Sciences 76, 709–726 (2010)
9. Gollowitzer, S., Ljubić, I.: MIP models for connected facility location: A theoretical and computational study. Computers & Operations Research 38(2), 435–449 (2011)
10. Koch, T., Martin, A.: Solving Steiner tree problems in graphs to optimality. Networks 32, 207–232 (1998)
11. Lardeux, B., Nace, D.: Multi-period network design of optical transmission networks. In: 12th IEEE Symposium on Computers and Communications, ISCC 2007, pp. 935–940 (2007)
12. Ljubić, I.: A hybrid VNS for connected facility location. In: Bartz-Beielstein, T., Blesa Aguilera, M.J., Blum, C., Naujoks, B., Roli, A., Rudolph, G., Sampels, M. (eds.) HCI/ICCV 2007. LNCS, vol. 4771, pp. 157–169. Springer, Heidelberg (2007)
13. Owen, S.H., Daskin, M.S.: Strategic facility location: A review. European Journal of Operational Research 111(3), 423–447 (1998)

14. Suhl, U.H., Hilbert, H.: A branch-and-cut algorithm for solving generalized multi-period Steiner problems in graphs. Networks 31(4), 273–282 (1998)
15. Tomazic, A., Ljubić, I.: A GRASP algorithm for the connected facility location problem. In: Proceedings of 2008 International Symposium on Applications and the Internet (SAINT), pp. 257–260 (2008)
16. Ukkusuri, S.V., Patil, G.: Multi-period transportation network design under demand uncertainty. Transportation Research Part B: Methodological 43(6), 625–642 (2009)
17. Zadeh, N.: On building minimum cost communication networks over time. Networks 4(1), 19–34 (1974)

A Computational Study of the Pseudo-Boolean Approach to the p-Median Problem Applied to Cell Formation

Boris Goldengorin[1] and Dmitry Krushinsky[2]

[1] Department of Applied Mathematics and Informatics,
Nizhny Novgorod branch of Higher School of Economics, Russia
bgoldengorin@hse.ru

[2] Department of Operations, University of Groningen, The Netherlands
d.krushinsky@rug.nl

Abstract. In this study we show by means of computational experiments that a pseudo-Boolean approach leads to a very compact presentation of p-Median problem instances which might be solved to optimality by a general purpose solver like CPLEX, Xpress, etc. Together with p-Median benchmark instances from OR and some other libraries we are able to solve to optimality many benchmark instances from cell formation in group technology which were tackled in the past only by means of different types of heuristics. Finally, we show that this approach is flexible to take into account many other practically motivated constraints in cell formation.

1 Introduction

The p-Median problem (PMP) is a well-known NP-hard problem which was originally defined by Hakimi [19] and involves the location of p facilities on a network in such a manner that the total weighted distance of serving all demands is minimized. Being a classical problem in combinatorial optimization, the PMP has been widely studied in literature and applied in cluster analysis, quantitative psychology, marketing, telecommunications industry [10], sales force territories design [23], political districting [5], optimal diversity management [9], cell formation in group technology [33], vehicle routing [21], and topological design of computer communication networks [25].

The basic PMP model that has remained almost unchanged during recent 30 years is the so called ReVelle and Swain [27] integer LP formulation (in fact, a Boolean LP formulation). The PMP has since been the subject of considerable research involving the development of adjusted model formats (Rosing *et al.* [29], Cornuejols *et al.* [13], Church [11,12]), and recently by AlBdaiwi *et al.* [2], and Elloumi [15], as well as the development of advanced solution approaches, e.g. Beltran *et al.* [7], Avella *et al.* [4]. For a comprehensive review of the PMP we address the reader to Reese [26], Mladenovich *et al.* [22] and ReVelle *et al.* [28].

In this paper we consider an application of PMP to the industrial engineering problem of cell formation (CF) in group technology. Cell formation suggests decomposition of a manufacturing system into several subsystems such that these subsystems, *manufacturing cells*, are as independent as possible. This ensures that machines processing

J. Pahl, T. Reiners, and S. Voß (Eds.): INOC 2011, LNCS 6701, pp. 503–516, 2011.

the same parts are placed closer to each other and time spent by parts on travelling from one machine to another is substantially reduced. Moreover, smaller systems (cells) are easier to manage (e.g. scheduling). PMP was applied to cell formation by a number of authors, e.g. [33]. However, due to NP-hardness of the PMP all of them used heuristics to solve the resulting PMP instances.

The purpose of this paper is two-fold. First, we show by means of numerical experiments that a pseudo-Boolean approach allows a very compact representation of the PMP instance data and can be used to derive an efficient Mixed-Integer Linear Programming (MILP) formulation. Second, we show that the PMP can be used as a flexible framework for cell formation, as in CF applications PMP can be very efficiently solved to optimality and any real-world constraints can be included. Our experiments show that even the largest CF instances used in literature can be solved within a second and the quality of solutions outperforms even the most contemporary heuristics.

In the next section we describe an efficient MILP formulation of the PMP based on a pseudo-Boolean approach and in Section 3 we conclude that all known reductions are contained in our model. Section 4 reports our computational study with benchmark instances. In Section 5 we discuss the possibilities induced by our model in CF applications and Section 6 summarizes this paper and provides future research directions.

2 The Mixed Boolean Pseudo-Boolean Model (MBpBM)

Recall that given sets $I = \{1,2,\dots,m\}$ of sites in which plants can be located, $J = \{1,2,\dots,n\}$ of clients, a non-negative matrix $C = [c_{ij}]$ of costs of serving each $j \in J$ from each $i \in I$, the number p of plants to be opened, and assuming a unit demand at each client site, the p-Median Problem (PMP) is one of finding a set $S \subseteq I$ with $|S| = p$, such that the total serving cost is minimized:

$$f_C(S) = \sum_{j \in J} \min\{c_{i,j} | i \in S\} \tag{1}$$

The *Combinatorial Formulation of PMP* is to find a subset $S^\star$ such that

$$S^\star \in \arg\min\{f_C(S) : \emptyset \subset S \subseteq I, \quad |S| = p\} \ . \tag{2}$$

The objective function $f_C(S)$ of the PMP (1) can be reformulated in terms of a pseudo-Boolean polynomial, pBp, (see Hammer [20] and Beresnev [6]) in the following way. Consider a vector $\mathbf{y} = (y_1,\dots,y_m)$ of Boolean variables such that $y_i = 0$ iff i-th location is opened (i.e. iff $i \in S$). For some client j a corresponding column of C contains the costs of serving this client from any location. Clearly, the demand of client j cannot be satisfied cheaper than $c_{\pi_{1j},j}$, where π_{1j} is the index of the smallest entry in j-th column of the costs matrix. This minimum value is attained only if location π_{1j} is opened and $y_{\pi_{1j}} = 0$. Otherwise, the cheapest way of satisfying client j is to use the second smallest entry in j-th column. In this case the cost is $c_{\pi_{2j},j} = c_{\pi_{1j},j} + y_{\pi_{1j}}(c_{\pi_{2j},j} - c_{\pi_{1j},j})$. If the location corresponding to the second smallest entry is also closed, the minimum costs of serving client j is $c_{\pi_{3j},j} = c_{\pi_{1j},j} + y_{\pi_{1j}}(c_{\pi_{2j},j} - c_{\pi_{1j},j}) + y_{\pi_{1j}}y_{\pi_{2j}}(c_{\pi_{3j},j} - c_{\pi_{2j},j})$. This intuition can be extended further and the costs of serving client j can be expressed as:

$$c_{\pi_{1j},j} + \sum_{k=1}^{m-1} (c_{\pi_{k+1,j},j} - c_{\pi_{kj},j}) \prod_{r=1}^{k} y_{\pi_{rj}} \qquad (3)$$

This representation naturally induces two objects related to j-th column: a permutation $\Pi^j = (\pi_{1j}, \ldots, \pi_{mj})$ that sorts the entries from the corresponding column of the costs matrix in a nondecreasing order, and the vector of differences $\Delta^j = (\delta_{0j}, \ldots, \delta_{m-1,j})$ defined as follows:

$$\begin{aligned} \delta_{0j} &= c_{\pi_{1j},j} \ , \\ \delta_{rj} &= c_{\pi_{r+1,j},j} - c_{\pi_{rj},j} \text{ for } 1 \le r \le m-1 \ , \end{aligned} \qquad (4)$$

By extending the above reasoning to all clients and defining a permutation matrix $\Pi = (\Pi^1, \ldots, \Pi^n)$ and a differences matrix $\Delta = (\Delta^1, \ldots, \Delta^n)$ the total cost function (1) can be represented by the following polynomial:

$$\mathscr{B}_{C,\Pi}(\mathbf{y}) = \sum_{j=1}^{n} \left\{ \delta_{0j} + \sum_{k=1}^{m-1} \delta_{kj} \prod_{r=1}^{k} y_{\pi_{rj}} \right\} \ . \qquad (5)$$

The expressions $\alpha_S \prod_{i \in S} y_i$ and $\prod_{i \in S} y_i$ are called a *monomial* and a *term*, respectively. In this paper monomials with the same term are called *similar monomials*. We say that a pseudo-Boolean polynomial is in *the reduced form* if for any two of its monomials the corresponding terms differ. In other words, the algebraic summation of similar monomials is called *reduction*.

AlBdaiwi et al. [2] show that the total cost function (5) is identical for all possible permutation matrices Π, hence we can remove it from notations without any confusion.

The Hammer-Beresnev polynomial $\mathscr{B}_C(\mathbf{y})$ contains less than $m \cdot n$ monomials and their number can be further reduced by using that for any feasible solution $\mathbf{y}$ to a PMP instance holds $\sum_{i=1}^{m} y_i = m - p$. This implies that any monomial in the pBp expressed as a constant multiplied by more than $m - p$ variables necessarily evaluates to zero. This is formalized in Theorem 1 (AlBdaiwi *et al.* [2]).

Theorem 1. *For any PMP instance C with $p \le m$ the following assertions hold:*

1. *The degree of a truncated Hammer-Beresnev polynomial $\mathscr{B}_{C,p}(\mathbf{y})$ is at most $m - p$;*
2. *Each column of the costs matrix C can be p-truncated by setting all p largest entries in a column to the value of the smallest entry among these p largest.*

The above theorem allows to substitute $\mathscr{B}_C(\mathbf{y})$ (5) by $\mathscr{B}_{C,p}(\mathbf{y})$ defined as

$$\mathscr{B}_{C,p}(\mathbf{y}) = \sum_{j=1}^{n} \left\{ \delta_{0j} + \sum_{k=1}^{m-p} \delta_{kj} \prod_{r=1}^{k} y_{\pi_{rj}} \right\} \ . \qquad (6)$$

We can reformulate (2) in terms of Hammer-Beresnev polynomials as the *pseudo-Boolean formulation of PMP*:

$$\mathbf{y}^{\star} \in \arg\min\{\mathscr{B}_{C,p}(\mathbf{y}) : \mathbf{y} \in \{0,1\}^m, \sum_{i=1}^{m} y_i = m - p\} \ . \qquad (7)$$

Let us denote by $|B|$ the number of monomials in $\mathscr{B}_{C,p}(\mathbf{y})$, by $T_r \in \{1,\dots,m\}$ a set of indices of variables in the r-th monomial of the pBp and by α_r coefficients of the monomials (e.g. $\alpha_0 = \sum_{j=1}^{n} \delta_{0j}$). Now the truncated reduced Hammer-Beresnev polynomial can be expressed as

$$\mathscr{B}_{C,p}(\mathbf{y}) = \alpha_0 + \sum_{r=1}^{m} \alpha_r y_r + \sum_{r=m+1}^{|B|} \alpha_r \prod_{i\in T_r} y_i \tag{8}$$

and by introducing nonnegative variables z_r $(r = m+1,\dots,|B|)$ we have linearised it (see e.g., Wolsey[32]) in order to obtain a linear objective function

$$f(\mathbf{y},\mathbf{z}) = \alpha_0 + \sum_{r=1}^{m} \alpha_r y_r + \sum_{r=m+1}^{|B|} \alpha_r z_r \ . \tag{9}$$

By introducing for each variable $z_r = \prod_{i\in T_r} y_i$ the constraints

$$z_r \geq \sum_{i\in T_r} y_i - |T_r| + 1 \quad , \quad z_r \geq 0 \tag{10}$$

we obtained our Mixed Boolean pseudo-Boolean Model (MBpBM):

$$\alpha_0 + \sum_{r=1}^{m} \alpha_r y_r + \sum_{r=m+1}^{|B|} \alpha_r z_r \longrightarrow \min \tag{11}$$

$$\text{s.t.} \sum_{i=1}^{m} y_i = m - p \ , \tag{12}$$

$$\sum_{i\in T_r} y_i - |T_r| + 1 \leq z_r,\ r = m+1,\dots,|B| \ , \tag{13}$$

$$z_i \geq 0, \quad i = m+1,\dots,|B| \ , \tag{14}$$

$$y_i \in \{0,1\},\ i = 1,\dots,m \tag{15}$$

Example 1. Consider a PMP instance from [15] with $m = 4$, $n = 5$, $p = 2$ and

$$C = \begin{bmatrix} 1 & 6 & 5 & 3 & 4 \\ 2 & 1 & 2 & 3 & 5 \\ 1 & 2 & 3 & 3 & 3 \\ 4 & 3 & 1 & 8 & 2 \end{bmatrix} \ . \tag{16}$$

A possible ordering and differences matrices for this problem are given by

$$\Pi = \begin{bmatrix} 1 & 2 & 4 & 1 & 4 \\ 3 & 3 & 2 & 2 & 3 \\ 2 & 4 & 3 & 3 & 1 \\ 4 & 1 & 1 & 4 & 2 \end{bmatrix} \text{ and } \Delta = \begin{bmatrix} 1 & 1 & 1 & 3 & 2 \\ 0 & 1 & 1 & 0 & 1 \\ 1 & 1 & 1 & 0 & 1 \\ 2 & 3 & 2 & 5 & 1 \end{bmatrix} \ . \tag{17}$$

The Hammer-Beresnev polynomial representing the total cost function for this instance in the form (5) is

$$\begin{aligned} \mathscr{B}_C(\mathbf{y}) = {} & [1 + 0y_1 + 1y_1y_3 + 2y_1y_2y_3] + \\ & [1 + 1y_2 + 1y_2y_3 + 3y_2y_3y_4] + \\ & [1 + 1y_4 + 1y_2y_4 + 2y_2y_3y_4] + \\ & [3 + 0y_1 + 0y_1y_2 + 5y_1y_2y_3] + \\ & [2 + 1y_4 + 1y_3y_4 + 1y_1y_3y_4] \ . \end{aligned} \tag{18}$$

After reduction of similar monomials and truncation we obtain the following pseudo-Boolean representation of the instance:

$$\begin{aligned} \mathscr{B}_C(\mathbf{y}) = &\ 8 + 1y_2 + 2y_4 + 1y_1y_3 + 1y_2y_3 + 1y_2y_4 + 1y_3y_4 \longrightarrow \min \\ \text{s.t. } &\ y_1 + y_2 + y_3 + y_4 = m - p = 2, \quad \mathbf{y} \in \{0,1\}^m \ . \end{aligned} \tag{19}$$

After introduction of new variables $z_5 = y_1y_3$, $z_6 = y_2y_3$, $z_7 = y_2y_4$, $z_8 = y_3y_4$ the MBpBM is:

$$8 + y_2 + 2y_4 + z_5 + z_6 + z_7 + z_8 \longrightarrow \min \tag{20}$$

$$\text{s.t. } \sum_{i=1}^{4} y_i = m - p = 2 \tag{21}$$

$$z_5 + 1 \geq y_1 + y_3 \tag{22}$$

$$z_6 + 1 \geq y_2 + y_3 \tag{23}$$

$$z_7 + 1 \geq y_2 + y_4 \tag{24}$$

$$z_8 + 1 \geq y_3 + y_4 \tag{25}$$

$$z_i \geq 0,\ i = 5, \ldots, 8 \tag{26}$$

$$y_i \in \{0,1\},\ i = 1, \ldots, 4 \ . \tag{27}$$

This MBpBM (20)–(27) has the same decision variables y_i as the pseudo-Boolean formulation (7), but its objective function is a linear in y_i and z_i. Note that our model contains 7 coefficients in the objective function, 5 constraints, 4 Boolean and 4 non-negative variables, while for the best Elloumi's model NFexr [15] these numbers are 10, 11, 4 and 7, correspondingly.

In the following Lemma 1 we explain how to reduce the number of Boolean variables y_i involved in the restrictions (13).

Lemma 1. *Let $\emptyset \neq T_r \subset T_q$ be a pair of embedded sets of Boolean variables y_i. Thus, two following systems of inequalities are equivalent:*

$$\begin{array}{ccc} \sum\limits_{i \in T_r} y_i - |T_r| + 1 \leq z_r & & \sum\limits_{i \in T_r} y_i - |T_r| + 1 \leq z_r \\ \sum\limits_{i \in T_q} y_i - |T_q| + 1 \leq z_q & \textit{and} & z_r + \sum\limits_{i \in T_q \setminus T_r} y_i - |T_q \setminus T_r| \leq z_q \\ z_r \geq 0, \quad z_q \geq 0 & & z_r \geq 0, \quad z_q \geq 0 \end{array} \tag{28}$$

Based on a compact representation of a PMP instance within pseudo-Boolean formulation (7) one may conclude that this formulation has extracted only essential information to represent the PMP. It means that we are in a position to check whether our MBpBM is an optimal model within the class of Mixed Boolean LP models. If we were able to show that the matrix of all linear constraints induced by non-linear monomials contains the smallest number of non-zero entries, then taking into account that the objective function has the smallest number of non-zero coefficients and linear constraints induced by non-linear terms one may conclude that our MBpBM is the optimal one [17]. Unfortunately, in general this is not the case. It is not difficult to show that the problem of finding a constraint matrix with the smallest number of non-zero entries is at least as hard as the *classic set covering problem* (see e.g. Garey and Johnson [16]). This means that to find an optimal model within the class of Mixed Boolean LP models is an NP-hard problem, even if the number of linear constraints is a linear function on the number of all decision variables. However, if only numbers of variables, constraints and terms in the objective function are taken into account, MBpBM is an optimal one [17].

Despite the efficiency of MBpBM it can be further reduced based on a decomposition of the whole search space into subspaces induced by terms in a pBp. By using upper and lower bounds on the cost of optimal solutions one may prove that some subspaces do not contain optimal solutions (see Goldengorin *et al.* [18]). Suppose, we know from some heuristic a (global) upper bound f^{UB} on the cost of optimal solutions. Let us now consider some term $\mathscr{T}$ and the set of indices of its variables T. One can also compute a lower bound $f^{LB}_{\mathscr{T}}$ over a subspace of solutions for which $\mathscr{T}$ is nonzero, i.e. all locations from T are closed. In case $f^{LB}_{\mathscr{T}} > f^{UB}$ one can conclude that for every optimal solution $\mathscr{T}$ evaluates to 0 and a constraint for the corresponding z-variable can be modified respectively. Moreover, all terms containing $\mathscr{T}$ also evaluate to 0 and the corresponding z-variables and constraints can be dropped. We call the model with this reduction MBpBMb and the lower bound that we used is given by:

$$f^{LB}_{\mathscr{T}} = f_C(\overline{T}) + \min_{k_i \in \overline{T}} \sum_{i=1}^{|\overline{T}|-p} [f_C(\overline{T} \setminus \{k_i\}) - f_C(\overline{T})], \tag{29}$$

where $f_C(.)$ – cost function of the PMP and $\overline{T}$ denotes the complement of T, i.e. $\overline{T} = \{1,\ldots,m\} \setminus T$.

3 Reduction Tests for the p-Median Problem

Looking at PMP models described in the literature it can be noticed that all improvements over the classical formulation by ReVelle and Swain [27] are done by application of various reduction tests to the instances of the problem. These tests can be classified into the following two broad groups: pure and optimizational.

The first group includes all kinds of tests exploiting structural redundancies in the input data. For example, presence of equal entries in a column of the costs matrix within a MILP model was first used by Cornuejols [13] and is present in most of the recent models, including Elloumi's ([15]), Church's [11,12] and our MBpBM (11)-(15). Another pure reduction excludes from the formulation largest $p-1$ entries of each column of the costs matrix. It was used by Avella *et al.* ([3], reduction test $FIX1$) and Church [11], and in our model is done by truncation of the pBp. This reduction has a universal nature in a sense that it allows truncation of exactly $p-1$ entries from each column of the cost matrix, irrespectively of the particular instance data. The next structural peculiarity that can be exploited for strengthening the formulation stems from the order in which potential locations are listed if being sorted by increasing distance from a client. If for two clients these orders are equal, they may be considered as one aggregated client. In our model this rule is applied by reducing similar monomials, while in Elloumi's NF (reduction rules R2 and R3 in [15]) and Church's COBRA [11] this is done by "substitution of equivalent variables" (as it is called in [11]). Finally, we would like to

Table 1. Effect of reduction tests for selected benchmark instances

instance	m	p	$\|C\|$	$\|B\|$	**red.**(%)	instance	m	p	$\|C\|$	$\|B\|$	**red.**(%)
pmed26	600	5	360000	25440	**92.93**	rw100	100	10	10000	5683	**43.17**
pmed27	600	10	360000	24117	**93.30**		100	20	10000	5045	**49.55**
pmed28	600	60	360000	19142	**94.68**		100	30	10000	4404	**55.96**
pmed29	600	120	360000	17724	**95.08**		100	40	10000	3771	**62.29**
pmed30	600	200	360000	16920	**95.30**		100	50	10000	3138	**68.62**
pmed31	700	5	490000	25940	**94.71**	rw500	500	10	250000	154622	**38.15**
pmed32	700	10	490000	26384	**94.62**		500	50	250000	141991	**43.20**
pmed33	700	70	490000	21030	**95.71**		500	100	250000	126281	**49.49**
pmed34	700	140	490000	18684	**96.19**		500	150	250000	110487	**55.81**
pmed35	800	5	640000	27788	**95.66**		500	250	250000	78946	**68.42**
pmed36	800	10	640000	28579	**95.53**	rw1000	1000	10	1000000	625052	**37.49**
pmed37	800	80	640000	23324	**96.36**		1000	50	1000000	599698	**40.03**
pmed38	900	5	810000	29230	**96.39**		1000	100	1000000	568197	**43.18**
pmed39	900	10	810000	27638	**96.59**		1000	300	1000000	441946	**55.81**
pmed40	900	90	810000	24165	**97.02**		1000	500	1000000	315682	**68.43**

mention Elloumi's reduction rule R1. The essence of this rule is that some z-variables can be expressed in terms of y-variables in a linear way. Our model implies R1 as we do not introduce any new variables for linear monomials. Thus, our model incorporates all known pure reductions by excluding monomials with zero coefficients, truncation and reduction of similar monomials. The effect of pure reductions can be illustrated by Table 1 where reduction tests were applied to several selected benchmark instances that are widely used for testing PMP-related algorithms. The first three columns contain the title of an instance, numbers of potential locations m and medians p, correspondingly. The next two columns indicate the number of entries in a costs matrix $|C|$ and the number of non-zero coefficients in the pBp $|B|$. The last column displays the achieved reduction (based on truncation and reducing similar monomials) that we computed as $red. = (|C| - |B|)/|C| \times 100\%$.

The second group of reduction tests includes optimizational approaches that suggest (pre-)solving the problem. These are reductions based on comparison of upper and (restricted) lower bounds on the optimal solution. Such a reduction was used by Avella *et al.* ([3], reduction test $FIX2$) and is implemented in our formulation MBpBMb leading to even more compact model and, as will be shown in the next section, reduced computing times.

The presented analysis can be summarized as follows: MBpBM in a natural way incorporates all available in literature pure reductions and can be subjected to optimizational problem size reduction techniques.

4 Computational Results for Benchmark Instances

In order to show the applicability of our compact MBpBM formulation, a number of computational experiments were held. We used benchmark instances from two of the

Table 2. Comparison of computing times for our and Elloumi's NF formulations, and Avella et al.'s B&C&P algorithm (15 largest OR-library instances)

instance	m	p	MBpBM	MBpBMb	Elloumi	Avella et al.
pmed26	600	5	163.84	**111.81**	180.31	187
pmed27	600	10	27.59	**21.31**	43.73	47
pmed28	600	60	2.48	**2.13**	3.61	
pmed29	600	120	1.78	**1.31**	2.91	
pmed30	600	200	1.50	**0.78**	4.81	
pmed31	700	5	153.22	**57.05**	90.95	106
pmed32	700	10	**33.13**	43.39	37.64	65
pmed33	700	70	3.09	**2.69**	4.73	
pmed34	700	140	3.72	**1.97**	7.11	
pmed35	800	5	**70.30**	154.41	514.72	189
pmed36	800	10	2256.83	4252.13	6737.25	**453**
pmed37	800	80	3.91	**3.08**	7.00	
pmed38	900	5	1328.34	2041.28	**307.00**	320
pmed39	900	10	572.81	444.08	473.95	**271**
pmed40	900	90	5.39	**4.02**	8.42	

Table 3. Comparison of computing times for our and Elloumi's NF formulations (Resende and Werneck random instances)

instance	m	p	MBpBM	MBpBMb	Elloumi
rw100	100	10	678.91	**452.52**	845.30
	100	20	4.00	**2.22**	5.25
	100	30	0.09	**0.03**	0.13
	100	40	0.08	**0.02**	0.14
	100	50	0.06	**0.02**	0.13
rw500	500	10	–	–	–
	500	100	–	–	–
	500	150	2.97	**1.22**	12.27
	500	200	2.25	**0.28**	4.11
	500	250	1.77	**0.13**	4.36
rw1000	1000	10	–	–	–
	1000	200	–	–	–
	1000	300	118.91	**13.40**	234.99
	1000	400	11.49	**1.16**	21.81
	1000	500	9.08	**0.77**	28.47

– Not solved within 1 hour.

most widely used libraries: J. Beasley's OR-library [24] and randomly generated RW instances by Resende and Werneck (see e.g. Elloumi [15]).

The experiments were conducted on a PC with Intel 2.33 GHz 1.95 GB and Xpress-MP as an MILP solver. Tables 2 and 3 summarize the computational results obtained for the 15 largest OR instances and RW instances, correspondingly. The first three columns contain the name of an instance, the number $m = |I| = |J|$ of nodes and the number p of medians. The next three columns are related to the running times (in seconds) of our models: MBpBM and its modification with reduction based on bounds. The next column reflects computing times for Elloumi's NF that we implemented and tested within the mentioned environment to ensure consistent comparison. The last column displays times reported by Avella et al. [4] for their Branch-and-Cut-and-Price (B&C&P) algorithm (Intel 1.8 GHz 1 GB).

Computational experiments with OR and RW instances can be summarized as follows. Our basic MBpBM formulation outperforms Elloumi's NF and Avella et al.'s B&C&P in most of the tested cases. The reduction based on bounds MBpBMb outperformed other considered models for all but five ORlib instances (in two of these cases our MBpBM was faster).

We also mention an instance from TSP library [30] fl1400 with $p = 400$ unsolved in Avella *et al.* [4] and solved to optimality by MBpBM in 598.5 sec. Note that Beltran's *et al.* [7] advanced semi-Lagrangean approach based on Proximal-Analytic Center Cutting Plane Method has not solved fl1400 with $p = 400$ to optimality and returns an approximation within 0.11% in 678 sec.

5 Application to Cell Formation

The p-Median Problem (PMP) was applied to cell formation in group technology by a number of researchers (see [33], [14] and references within). However, to the best of our knowledge, in all CF related papers PMP (as well as any other model based on graph partitioning or MILP) is solved by some heuristic method. At the same time, for the p-Median problem there exist efficient formulations (like MBpBM or the one in [15]) that allow solving large instances to optimality.

Recall that for a directed weighted graph $G = (V, A, C)$ with $|V|$ vertices, set of arcs $(i, j) \in A \subseteq V \times V$ and weights (distances, dissimilarities, etc.) $C = \{c_{ij} : (i, j) \in A\}$, the PMP consists of determining p nodes (the *median nodes*, $1 \le p \le |V|$) such that the sum of weights of arcs joining any other node and one of these p nodes is minimized. In terms of cell formation, vertices represent machines and weights c_{ij} represent dissimilarities between machines i and j. These dissimilarities can be derived from the sets of parts that are being processed by either of the machines (e.g. if two machines process almost the same set of parts they have small dissimilarity and are likely to be in the same cell) or from any other desired characteristics (e.g. workers skill matrix, operational sequences, etc.). Moreover, usually there is no need to invent a dissimilarity measure as it can be derived from one of the available similarity measures using an expression $d(i, j) = c - s(i, j)$, where $d(.,.)/s(.,.)$ is a dis/similarity measure and c – some constant large enough to keep all dissimilarities non-negative. As can be seen from the literature, several similarity measures were proposed and the particular choice can influence results of cell formation. For our experiments we have chosen one of the most widely used – Wei and Kern's "commonality score" [31], and derived our dissimilarity measure as

$$d(i, j) = r \cdot (r - 1) - \sum_{k=1}^{r} \Gamma(a_{ik}, a_{jk}) \tag{30}$$

where

$$\Gamma(a_{ik},a_{jk}) = \begin{cases} (r-1), & if\ a_{ik} = a_{jk} = 1 \\ 1, & if\ a_{ik} = a_{jk} = 0 \\ 0, & if\ a_{ik} \neq a_{jk} \end{cases} \tag{31}$$

where r - number of parts, a_{ij} - entries of the $m \times r$ *machine-part incidence matrix* (i.e. $a_{ij} = 1$ if part j needs machine i and $a_{ij} = 0$, otherwise).

Thus, if applied to cell formation, the p-Median problem means finding p machines that are best representatives (centres) of p manufacturing cells, i.e. the sum over all cells of dissimilarities between such a centre and all other machines within the cell is minimized. Once p central machines are found, the cells can be generated by assigning each other machine to the central one such that their dissimilarity is minimum. Note that the desired number of cells p is part of the input for the model and must be known beforehand. Otherwise, it is possible to solve the problem for several numbers of cells and pick the best solution.

Example 2. Let the instance of the cell formation problem be defined by the machine-part incidence matrix:

$$\begin{array}{c} \\ 1 \\ 2 \\ 3 \\ 4 \end{array} \begin{array}{c} 1\,2\,3\,4\,5 \\ \begin{bmatrix} 0\,1\,0\,1\,1 \\ 1\,0\,1\,0\,0 \\ 0\,1\,0\,1\,0 \\ 1\,0\,1\,0\,0 \end{bmatrix} \end{array} \tag{32}$$

with 4 machines and 5 parts. One can construct the machine-machine dissimilarity matrix C by applying the defined above dissimilarity measure (30):

$$C = \begin{bmatrix} 6 & 20 & 10 & 20 \\ 20 & 9 & 19 & 9 \\ 10 & 19 & 9 & 19 \\ 20 & 9 & 19 & 9 \end{bmatrix} \tag{33}$$

For example, the top left entry a_{11} is obtained in the following way:

$$\begin{aligned} a_{11} &= r(r-1) - \sum_{k=1}^{r} \Gamma(a_{1k},a_{1k}) = \\ &= 5(5-1) - \Gamma(0,0) - \Gamma(1,1) - \Gamma(0,0) - \Gamma(1,1) - \Gamma(1,1) = \\ &= 20 - 1 - 4 - 1 - 4 - 4 = 6 \end{aligned} \tag{34}$$

If one is interested in having two manufacturing cells then the number of medians p in should be set to 2 and the MBpBM formulation of the given instance of cell formation is as follows:

$$f(\mathbf{y},\mathbf{z}) = 33 + 4y_1 + 1y_3 + 20z_5 + 20z_6 \longrightarrow \min \tag{35}$$

$$y_1 + y_2 + y_3 + y_4 = 2 \tag{36}$$

$$z_5 \geq y_1 + y_3 - 1 \tag{37}$$

$$z_6 \geq y_2 + y_4 - 1 \tag{38}$$

$$z_i \geq 0,\ i = 5,6 \tag{39}$$

$$y_i \in \{0,1\},\ i = 1,\dots,4 \tag{40}$$

Its solution $y = (0,0,1,1)^T$, $z = (0,0)^T$ leads to the following cells:

$$\begin{array}{c} \\ 1 \\ 3 \\ 2 \\ 4 \end{array} \begin{array}{c} 2\,4\,5\;1\,3 \\ \left[\begin{array}{ccc|cc} 1 & 1 & 1 & 0 & 0 \\ 1 & 1 & 0 & 0 & 0 \\ \hline 0 & 0 & 0 & 1 & 1 \\ 0 & 0 & 0 & 1 & 1 \end{array}\right] \end{array} \quad (41)$$

We conducted numerical experiments in order to compare this approach with other contemporary ones. The aim of our numerical experiments was twofold. First, we would like to show that the model based on PMP produces high-quality cells and in many cases outperforms other contemporary heuristics, thus making their use questionable. Second, by showing that computation times are negligibly small, we argue the use of heuristics for solving PMP itself. In order to compare solutions quality two following measures were used:

$$GCI = 1 - \frac{e}{o} \times 100\% \quad \text{and} \quad \eta = \frac{1}{2}\left[\frac{\alpha}{\alpha+v} + \frac{\beta}{\beta+e}\right] \times 100\%, \quad (42)$$

where *exceptional elements* are those nonzero entries of the block-diagonalized machine-parts incidence matrix that lie outside the blocks, m – the number of machines, r – the number of parts, o – the total number of ones in the machine-part incidence matrix, e – the number of exceptional elements, v – the number of zeroes in diagonal blocks, $\alpha = o - e$ and $\beta = mr - o - v$.

We compared our experimental results with those reported in four recent papers. The main focus was made on the largest 24 instances we could find in literature on CF with $m \times r$ ranging from 8×20 to 50×150. The first paper [33] uses a p-Median approach but solves PMP by a heuristic procedure. It uses Wei and Kern's [31] similarity measure and GCI as a quality measure. We were not able to derive the value of η because solutions are not provided in that paper. The second paper [34] applies the ART1 neural network to cell formation, thus using a completely different approach. The authors used η-measure to estimate solution quality and included solutions (block-diagonalized matrices) in their paper, thus making it possible for us to compute GCI. The third paper [1] demonstrates an application of a decision-making technique to the cell formation problem. Authors report values of η and we derived values of GCI from their solutions.

The most recent paper considered in our computational experiments is [8]. It uses a model that is very similar to the p-Median problem but differs in the following detail: a restriction specifying the number of cells is replaced by a constraint ensuring that each cell has at least two machines. We implemented the models for machine cell formation and part assignment from [8] in Xpress to perform the experiments. Like in the previous cases we used only machine-part incidence matrices as an input and Wei and Kern's (dis)similarity measure. Taking into account that the model from [8] automatically defines the best number of cells, we had to solve our PMP based model for all possible values of p.

The results of the experiments can be summarized as follows. With regard to the solutions quality our model outperforms all its considered counterparts both in GCI

and η measures. Even though there were scarce instances for which our approach was slightly dominated (this can be explained by the fact that the PMP does not explicitly optimize the considered quality measures), both average and best improvement of quality is noticeable. The difference between quality of our solutions and the ones reported in [33], [34], [1] and obtained by the model from [8] both in terms of GCI and η can be summarised as follows:

$$\begin{array}{lccc} & \textit{worst} & \textit{average} & \textit{best} \\ \Delta GCI & -1.90 & 2.86 & 15.20 \\ \Delta\eta & -0.55 & 3.82 & 17.41 \end{array} \tag{43}$$

Concerning the computing times, we would like to mention that each of the considered instances was solved within 1 second on a standard PC using Xpress. Even if some other approach can do faster, the difference is negligible.

5.1 Additional Constraints

Above we considered a PMP approach to the CF problem in its simplest form with only the machine-part relations taken into account. Yet, in real manufacturing systems there exist additional factors that must be considered to generate "reasonable" cells. Clearly, there are three places in our model where additional factors can be incorporated:

- dissimilarity coefficients;
- objective function (structure);
- constraints;

The easiest way of introducing additional factors into the model is via the dissimilarity coefficients. For example, dissimilarity between a pair of machines can be made dependent not only on the number of parts that need these machines but also on the weight, volume, processing time of these parts or their operational sequences. Moreover, workers able to operate these machines can be taken into account. Thus, suitable choice of dissimilarity coefficients allows to account for capacity, workload and skills issues without changing the structure of the model. At the same time, a variety of restrictions can be inserted either by changing the structure of the objective function (e.g. by adding terms penalising assignment of some machines to the same cell) or by adding new constraints. It is easy to understand that the only requirement on new constraints is their linearity. The fact that almost any real-world constraints can be either expressed or approximated in a linear form makes the PMP-based approach quite flexible. Taking into account that our model for PMP is very compact, any additional constraints are welcome.

6 Summary and Future Research Directions

The paper presented a new approach to formulation of models for the p-Median problem. We started with a pseudo-Boolean formulation with just one constraint on the number of medians. We then reduced the size of the objective function by truncation and reducing similar monomials, and linearised all non-linear terms in it with additional variables and constraints. The resulting model, a Mixed Boolean pseudo-Boolean Model, incorporates all known reductions and has the smallest number of constraints related to non-negative variables. As we have shown, the matrix of constraints would be as sparse as possible if we were able to solve a generalization of the classic set covering problem defined on the set of all terms involved in the pseudo-Boolean formulation of PMP. Unfortunately, this set covering problem is NP-hard [16]. Anyway, if we evaluate the size of a model by the number of non-zero coefficients in the objective function and corresponding constraints, our MBpBM is the smallest one (within mixed Boolean LP models) and instance specific! Note, however, that a smaller formulation does not guarantee smaller solution time (due to NP-hardness of the problem).

The MBpBM can also be applied to cell formation problems, leading to an improved solutions quality compared to the most contemporary approaches. As well, computing times for the largest CF instances are within 1 second and thus are competitive with any heuristic. In the numerical experiments we considered the simplest possible approach to cell formation aimed at block-diagonalising the machine-part incidence matrix without taking into account additional real-world factors. There are two reasons for this. First, we wanted to demonstrate that even a computationally intractable model of cell formation (at least in its simplest form) can be solved to optimality, and this possibility, to the best of our knowledge, was overlooked in literature. Second, this choice was partially governed by available recent papers in the field with which we wanted to compare our results. At the same time, we showed that several types of constraints can be incorporated into the PMP-based CF model thus making it more realistic and allowing to use all the available information about the manufacturing system.

To summarize, in this paper we have shown that our model extends the possibilities of solving p-Median problem instances to optimality by means of general-purpose MILP software, e.g. Xpress.

References

1. Ahi, A., Aryanezhad, M.B., Ashtiani, B., Makui, A.: A novel approach to determine cell formation, intracellular machine layout and cell layout in the CMS problem based on TOPSIS method. Comput. Oper. Res. 36(5), 1478–1496 (2009)
2. AlBdaiwi, B.F., Ghosh, D., Goldengorin, B.: Data aggregation for p-median problems. J. Comb. Optim. 21(3), 348–363 (2011)
3. Avella, P., Sforza, A.: Logical reduction tests for the p-median problem. Ann. Oper. Res. 86, 105–115 (1999)
4. Avella, P., Sassano, A., Vasil'ev, I.: Computational study of large-scale p-median problems. Math. Prog., Ser. A 109, 89–114 (2007)
5. Belenky, A.S. (ed.): Mathematical modeling of voting systems and elections: Theory and Applications. Math. Comput. Model. 48(9-10), 1295–1676 (2008)
6. Beresnev, V.L.: On a problem of mathematical standardization theory (in Russian). Upravliajemyje Sistemy 11, 43–54 (1973)
7. Beltran, C., Tadonki, C., Vial, J.-P.H.: Solving the p-median problem with a semi-Lagrangian relaxation. Comput. Optim. Appl. 35, 239–260 (2006)
8. Bhatnagar, R., Saddikuti, V.: Models for cellular manufacturing systems design: matching processing requirements and operator capabilities. J. Opl. Res. Soc. 61, 827–839 (2010)
9. Briant, O., Naddef, D.: The optimal diversity management problem. Oper. Res. 52, 515–526 (2004)

10. Brusco, M.J., Köhn, H.-F.: Optimal partitioning of a data set based on the p-median problem. Psychometrika 73(1), 89–105 (2008)
11. Church, R.L.: COBRA: a new formulation of the classic p-median location problem. Ann. Oper. Res. 122, 103–120 (2003)
12. Church, R.L.: BEAMR: An exact and approximate model for the p-median problem. Comput. Oper. Res. 35, 417–426 (2008)
13. Cornuejols, G., Nemhauser, G., Wolsey, L.A.: A canonical representation of simple plant location problems and its applications. SIAM J. Matrix Anal. A (SIMAX) 1(3), 261–272 (1980)
14. Deutsch, S.J., Freeman, S.F., Helander, M.: Manufacturing cell formation using an improved p-median model. Comput. Ind. Eng. 34(1), 135–146 (1998)
15. Elloumi, S.: A tighter formulation of the p-median problem. J. Comb. Optim. 19, 69–83 (2010)
16. Garey, M.R., Johnson, D.S.: Computers and Intractability. Freeman, New York (1979)
17. Goldengorin, B., Krushinsky, D.: Complexity evaluation of benchmark instances for the p-median problem, Math. Comput. Model. 53, 1719–1736 (2011)
18. Goldengorin, B., Tijssen, G.A., Ghosh, D., Sierksma, G.: Solving the simple plant location problems using a data correcting approach. J. Global Optim. 25, 377–406 (2003)
19. Hakimi, S.L.: Optimum locations of switching centers and the absolute centers and medians of a graph. Oper. Res. 12, 450–459 (1964)
20. Hammer, P.L.: Plant location – a pseudo-Boolean approach. Israel J. Technol. 6, 330–332 (1968)
21. Koskosidis, Y.A., Powell, W.B.: Clustering algorithms for consolidation of customer orders into vehicle shipments. Transp. Res. 26B, 365–379 (1992)
22. Mladenovic, N., Brimberg, J., Hansen, P., Moreno-Peréz, J.A.: The p-median problem: A survey of metaheuristic approaches. Eur. J. Oper. Res. 179, 927–939 (2007)
23. Mulvey, J.M., Beck, M.P.: Solving capacitated clustering problems. Eur. J. Oper. Res. 18, 339–348 (1984)
24. OR-Library, `http://people.brunel.ac.uk/~mastjjb/jeb/orlib/pmedinfo.html`
25. Pirkul, H.: Efficient algorithms for the capacitated concentrator location problem. Comput. Oper. Res. 14(3), 197–208 (1987)
26. Reese, J.: Solution Methods for the p-Median Problem: An Annotated Bibliography. Networks 48, 125–142 (2006)
27. ReVelle, C.S., Swain, R.: Central facilities location. Geogr. Anal. 2, 30–42 (1970)
28. ReVelle, C.S., Eiselt, H.A., Daskin, M.S.: A bibliography for some fundamental problem categories in discrete location science. Eur. J. Oper. Res. 184, 817–848 (2008)
29. Rosing, K.E., ReVelle, C.S., Rosing-Vogelaar, H.: The p-Median and its Linear Programming Relaxation: An Approach to Large Problems. J. Oper. Res. Soc. 30, 815–822 (1979)
30. TSP–library, `http://www.iwr.uni-heidelberg.de/groups/comopt/software/TSPLIB95/`
31. Wei, J.C., Kern, G.M.: Commonality analysis, a linear cell clustering algorithm for group technology. Int. J. Prod. Res. 27(12), 2053–2062 (1989)
32. Wolsey, L.: Mixed integer programming. In: Wah, B. (ed.) Wiley Encyclopedia of Computer Science and Engineering, John Wiley & Sons, Inc., Chichester (2008)
33. Won, Y., Lee, K.C.: Modified p-median approach for efficient GT cell formation. Comput. Ind. Eng. 46, 495–510 (2004)
34. Yang, M.-S., Yang, J.-H.: Machine-part cell formation in group technology using a modified ART1 method. Eur. J. Oper. Res. 188(1), 140–152 (2008)

Cache Location in Tree Networks: Preliminary Results

Bauguion Pierre[1], Ben Ameur Walid[2], and Gourdin Eric[1]

[1] Orange Labs, France Telecom R&D,
38-40 bd du Général Leclerc, 92794 Issy-les-Moulineaux Cedex 9
{pierre.bauguion,eric.gourdin}@orange-ftgroup.com
[2] Institut Telecom, Telecom SudParis, UMR CNRS 5157,
9 rue Charles Fourier, 91011 Evry, France
walid.benameur@it-sudparis.eu

Abstract. One popular approach to overcome the expected congestions due to the spectacular development of various multimedia applications consists in installing transparent caches at strategically chosen places inside telecommunication networks. The problem of locating caches is a difficult optimization problem, closely related to the p-median problem. In the case where the network is a tree, some cache location problems have already been investigated. In this paper, we propose to refine these models by taking into account a dynamic effect due to cache replacement policies. In our model, only the most popular contents are stored in the caches. The hierarchical effect of several successive caches is also captured by the model. A Mixed Integer Programming model and a Dynamic Programming algorithm are proposed and compared on a preliminary set of numerical experiments.

1 Introduction

One of the major trends in modern telecommunications is the spectacular development of a wide range of multimedia services. As a direct consequence, one can expect a huge increase in terms of bandwidth consumption and storage resources. Maintaining a reasonable level of QoS (Quality of Service) is probably one of the main challenges for the telecommunication network operators and various service providers.

Many actors in the Internet community are considering various potential solutions to cope with the problem of traffic increase. The term CDN, for Content Delivery Network [12], seems to federate most of the current activities related to efficient content distribution. One key feature of a CDN is an equipment called *cache*. A cache deployed in some transit node can intercept (i) a request made by clients for certain content, and (ii) the content itself when it is forwarded back from a central server. The cache will store some of the content (ii) according to a certain *replacement policy* and answer directly some of the requests (i) if it is currently storing the required content. The replacement policy is an algorithm that tells the cache which content should be removed when a new content has to be stored and its storage capacity limit is reached. The most well-known of these policies are LFU (Least Frequently Used) and LRU (Least Recently Used). The choice of the right policy seems to be essential for a cache to be efficient. A considerable effort has already been spent to analyze, measure and model the efficiency of

J. Pahl, T. Reiners, and S. Voß (Eds.): INOC 2011, LNCS 6701, pp. 517–522, 2011.

a cache (the so-called *hit ratio*, i.e., the percentage of requests that can be answered by the cache) [1,5,6,8,13].

The effect of a cache is to reduce the load on upstream links and to reduce the delays to access to contents. Apart from its intrinsic efficiency, to be effective, a cache must also be deployed at the right place in a network (if it is too close to the server, it won't alleviate much links, and if it too close to the clients, then the number of caches to deploy might become huge). The location of caches can hence easily be modeled as an optimization problem, and cache and content location problems have also been quite intensively investigated [2,3,4,7,10,11]. Note that most of these optimization model extend a well-known location problem called p-median [14] and for which polynomial algorithms have been proposed in the case where the network is modeled by a tree [9,15].

In this paper, we propose an algorithmic approach to determine an optimal architecture where transparent caches are installed in a tree network. The root of the tree represents the central server which stores all contents, and each leaf represents a client (or an aggregated set of clients). We will consider probabilistic requests, weighted by the amount of clients performing the same demand. We will also include a capacity constraints on the links. Last and not least, each cache is supposed to contain the most popular contents taking into account the fact that some content might be already stored on other caches that are installed in the subtree rooted at the cache.

2 Problem Statement

We consider a telecommunication network that can be modeled by an arborescence (directed tree) $\mathscr{T} = (V,A)$ rooted at a special node $s \in V$ representing the central server. Without loss of generality, we will assume throughout the paper that the service is a Video-on-Demand (VoD) service and the data are hence videos. A subset $T \subset V$ of vertices, called terminals, represent the clients of a given data service. For each vertex $v \in V$, we denote $\delta^+(v)$ the set of sons of v in the arborescence (the vertices that come next to v in the paths from s to the clients). If $\delta^+(v) = \emptyset$, then v is a leaf. We denote by $L \subset V$ the set of leaves. We assume that $L = T$ (every leaf is a terminal and every terminal is a leaf). Let $S = V \setminus T$. We associate with each arc $a \in A$, a weight $w_a \geq 0$ which can either represent a distance or a unit cost (for transporting 1 Mb of data for instance). We assume that there is a finite set $\mathscr{D} = (d^1, \ldots, d^m)$ of data (=videos) available in the central server s. Each client $j \in T$ associates to each video $k \in K = \{1, \ldots, m\}$, a number $p_j^k \geq 0$, representing the popularity of the video k (depending on the number of requests of the video). Remember that a client can represent an aggregation of many clients.

In the cache location problem, we have to decide where to install caches in the network. We assume that each installed cache has a capacity of p (it can store up to p videos) and a cost of c_v if it is installed on vertex v. In some instances, we might assume that all installation costs are the same: $c_v = c$ for all $v \in V$. To model the behavior of the cache with respect to the dynamic arrival of requests, we will assume that the videos present in the cache are the p most popular ones, the popularity being measured at each potential location (a tree vertex) and taking into account the placement of other caches.

The problem is illustrated on a small 7 nodes instances in Figure 1.

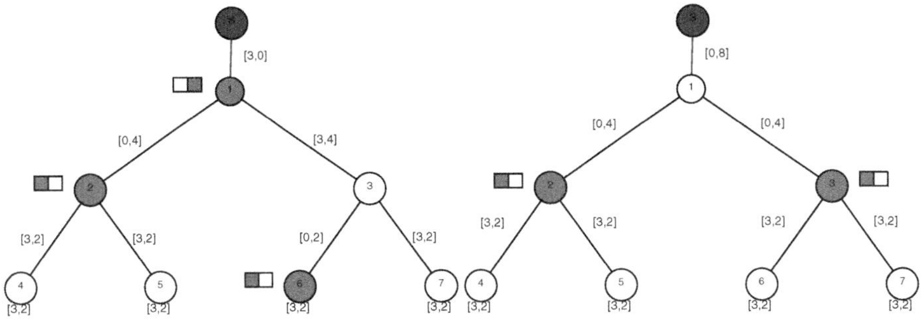

Fig. 1. A cache location problem instance with $m = 2$ and $p = 1$. The quantities below the leaves represent the popularities for both videos. The unit bandwidth cost is equal to 1 and each cache has a cost of 10. On the left, a feasible solution where caches are installed on nodes 1, 2 and 6. The most popular video is stored in the nodes 2 and 6 whereas the least popular one is stored in the root node. This solution has a cost of 61. An optimal solution, on the right, has a total cost of 56.

3 A Mixed Integer Programming Model

We use two sets of binary variables: the binary variable y_j is equal to 1 iff there is a cache located at vertex j and the binary variable x_j^k is equal to 1 if the video k is in a cache at node j. The flow variable f_v^k represent the popularity of video k at vertex v. This value results from popularities expressed by the sons of v. Our cache location problem can then be modeled by the following MIP:

$$\min \sum_{j \in V} c_j y_j + \sum_{k \in K} \sum_{i \in V \setminus \{s\}} w_{(\delta^-(i)i)} f_i^k, \tag{1}$$

$$\text{subject to:} \quad x_j^k \le y_j, \qquad j \in V, k \in K, \tag{2}$$

$$\sum_{k=1}^{m} x_j^k = p y_j, \qquad j \in V, \tag{3}$$

$$f_i^k = p_i^k (1 - x_i^k), \qquad i \in T, k \in K, \tag{4}$$

$$f_i^k = (1 - x_i^k) \sum_{j \in \delta^+(i)} f_j^k, \qquad i \in S, k \in K, \tag{5}$$

$$x_j^k, y_j \in \{0,1\}, f_j^k \ge 0, \qquad j \in V, k \in K. \tag{6}$$

The objective function integrates the installation cost and the access cost. We used $(\delta^-(i)i)$ to denote the arc from $\delta^-(i)$ to i. Constraints (2) and (3) are standard location constraints stating that a video can only be stored at a node where a cache is installed and a cache can contain at most p videos. Constraints (4) express the fact that the popularity for video k that is sent by a leaf node i to its parent is p_i^k if $x_i^k = 0$ and 0 otherwise. The next constraints (5) have a similar role for Steiner nodes thus allowing to propagate the popularities along the tree depending on the cache locations. Note that these constraints are non-linear, but they can easily be linearized using standard techniques.

Additional constraints should be introduced into the model to ensure that, at node $i \in V$, only the p most popular videos can be stored in a cache. To simplify the expressions, let $P_i^k = \sum_{j \in \delta^+(i)} f_j^k$ be the potential popularity for video k at node i. The new constraints should model the fact that, if $P_i^k > P_i^{k'}$ (video k is more popular than video k' at node i) and $x_i^{k'} = 1$ (video k' is stored in node i), then we must have $x_i^k = 1$. This can be achieved by the following constraints:

$$(x_i^k - x_i^{k'})(P_i^k - P_i^{k'}) \geq 0. \tag{7}$$

These constraints can also be linearized using standard techniques (For example Big M linearization).

4 A Dynamic Programming (DP) Approach

Several Dynamic Programming approaches have been proposed for location problems in trees (for the p-median problem [15] and for caches with a multicast protocol [11]). We propose an heuristic for our problem based on dynamic programming. Due to space limitation we only give a sketch of the algorithm. For each node i, the possible values of the popularities f_i^k depend on the popularities propagated by its sons. Then, if we know for each son of v all possible situations (depending on the caches installed in the subtree rooted at the son), we can build the set of all possible situations for i. To reduce the complexity of the procedure, many solutions are eliminated based on cost consideration and domination between solutions. Unfortunately, this can also eliminate the optimal solution. However, our numerical experiments show that the best solution obtained by dynamic programming is very close to the optimal solution.

5 Computational Experiments

In this section, we present some numerical results. The MIP formulation has been solved using MIP solver (Xpress-MP) and the DP algorithm has been implemented. Table 1 shows the computing times needed to solve various instances either using the MIP solver or the DP algorithm. In these tests $p = 2$ and $m = |K| = 5$.

Table 1. MIP and DP results

$\lvert V \rvert$	Xpress	DP	Gap (%)
99	14s	1s	0
106	10s	1s	0
132	73s	1s	0.0008
134	5s	1s	0
236	85s	1s	0
255	76s	1s	0
366	234s	4s	0.004
400	3684s	1s	0

All tree graphs have been generated with random parameters, following an uniform ditribution : Number of sons : [1,5], Edge capacity : [10,50], Edge cost :[1,20]. Moreover, cache cost and popularity have been chosen so that the optimal solution is non-trivial (cost cache : 195, popularity between 1 and 5).

The relative gap between the optimal solution found by the MIP solver and the approximate solution provided by DP is also shown on Table 1. We clearly see that this gap is really very small.

Notice the MIP formulation is solved in a brute way without any addition of valid inequalities. Even, the upper bound provided by DP was used to help the solver to find the optimal solution.

6 Conclusion

We have proposed a somewhat more accurate model for a the optimal location of transparent caches in a tree, taking into account the propagation of content popularities. We have given a MIP formulation for this problem and provided a dynamic programming algorithm. The preliminary computational experiments show that both approaches are promising and might solve problems with real-size instances. An enforcement of the MIP formulation using valid inequalities is the next research step. Capacity constraints can also be taken into account both in the MIP formulation and in the dynamic programming algorithm. Another generalization consists in considering the case where many servers are available with a rooted tree for each server. Then, a node might belong to different trees making the problem more complicated.

References

1. Asit, D., Towsley, D.: An approximate analysis of the lru and fifo buffer replacement schemes. SIGMETRICS Perform. Eval. Rev. 18(1), 143–152 (1990)
2. Avella, P., Boccia, M., Canonico, R., Emma, D., Sforza, A., Ventre, G.: Web cache location and network design in vpns (2003), `http://citeseer.ist.psu.edu/639771.html`
3. Cronin, E., Jamin, S., Danny, C., Yuval, R.: Constrained mirror placement on the internet (2002), `http://citeseer.ist.psu.edu/cronin02constrained.html`
4. Dantzig, P., Hall, R., Schwartz, M.: A case for caching file objects inside internetworks. In: SIGCOMM 1993, pp. 239–243 (1993)
5. Flajolet, P., Gardy, D., Thimonier, L.: Birthday paradox, coupon collectors, caching algorithms and self-organizing search. Discrete Appl. Math. 39(3), 207–229 (1992)
6. Gelenbe, E.: A unified approach to the evaluation of a class of replacement algorithms. IEEE Trans. Comput. 22(6), 611–618 (1973)
7. Hakimi, S., Schmeichel, E.: Locating replicas of a database on a network. Networks 30(1), 31–36 (1997)
8. Jelenkovic, P.R.: Asymptotic approximation of the move-to-front search cost distribution and least-recently-used caching fault probabilities. Annals of Applied Probability 9(2), 430–464 (1999)
9. Kariv, O., Hakimi, L.: An algorithmic approach to network location problems. ii: The p-median. SIAM J. Appl. Math. 37(3), 539–560 (1979)

10. Krishnan, P., Raz, D., Shavitt, Y.: The cache location problem. IEEE/ACM Transactions on Networking 8(5), 568–582 (2000), `citeseer.ist.psu.edu/635591.html`
11. Luss, H.: Optimal content distribution in video-on-demand tree networks. IEEE Transactions on systems, man, and cybernetics - part A: systems and humans 40(1) (2010)
12. Pathan, A.M.K., Buyya, R.: A taxonomy and survey of cdns. Tech. rep., The University of Melbourne (2007)
13. Podlipnig, S., Böszömenyi, L.: A survey of web cache replacement strategies. ACM Computing Surveys (CSUR) 35(4), 374–398 (2003)
14. Reese, J.: Solution methods for the p-median problem: An annotated bibliography. Networks 19, 125–142 (2006)
15. Tamir, A.: An $\mathcal{O}(pn^2)$ algorithm for the p-median and other related problems on tree graphs. Operations Research Letters 19, 59–64 (1996)

The Multi Terminal q-FlowLoc Problem: A Heuristic

Stephanie Heller and Horst W. Hamacher

Department of Mathematics, University Kaiserslautern, D-67663 Kaiserslautern, Germany
{heller,hamacher}@mathematik.uni-kl.de

Abstract. In this paper the multi terminal q-FlowLoc problem (q-MT-FlowLoc) is introduced. FlowLoc problems combine two well-known modeling tools: (dynamic) network flows and locational analysis. Since the q-MT-FlowLoc problem is NP-hard we give a mixed integer programming formulation and propose a heuristic which obtains a feasible solution by calculating a maximum flow in a special graph H. If this flow is also a minimum cost flow, various versions of the heuristic can be obtained by the use of different cost functions. The quality of this solutions is compared.

1 Introduction and Notations

FlowLoc problems combine two well studied modeling tools: network flow and locational analysis. Network flow models are often used to determine quickest flows or flow minimizing a given cost function (see [1] for an overview). Location theory on the other side is often used for finding "good" locations for facilities (see e.g. [5,6,10]). A field in which both problems occur is evacuation planning. A network flow represents people to be sent from a source to a sink. Depending on the objective function the overall evacuation time has to be minimized or the flow per time unit has to be maximized (see [2,3,4,7,8,9]). On the other hand facilities (like first aid wards, fire engines, fish&chip shops, etc.) have to be placed. Although the placement of the facilities and the according reduction of the capacity of some edges have influence on the optimum flow, the two methods have only been considered in an integrated fashion by the research group of the authors. FlowLoc problems combine network flows and locational analysis to obtain results (e.g. lower bounds for evacuation times) taking more factors into account and hence yielding more realistic results.

In this paper we first introduce notations and definitions. Then we give an IP formulation. In Section 2 a graph H is introduced and it is shown that flows with value q in graph H represent the feasible solutions of the q-MT-FlowLoc problem. By adding different cost functions and calculating a minimum cost flow, different solutions can be obtained. These are compared in Section 3.

Let an undirected network $G = (V, E)$ with capacity function $u : E \to \mathbb{N}$, a set $\mathbb{P}$ of facilities and a size function $r : \mathbb{P} \to \mathbb{N}$ be given. Furthermore let $nol : E \to \mathbb{N}$ be a function assigning to each edge the maximum number of facilities that can be located there. Let $\mathbb{L} = \{e \in E : nol(e) > 0\} \subseteq E$ be the subset of edges on which facilities can be placed. The q-MT-FlowLoc problem asks for an optimal allocation of all facilities $p \in \mathbb{P}$ to edges $e \in \mathbb{L}$ (see also [7]). This means an allocation maximizing $\sum_{v<w} f_{vw}$, where f_{vw} is the flow value between vertex v and w, has to be found. Here facilities can only be

J. Pahl, T. Reiners, and S. Voß (Eds.): INOC 2011, LNCS 6701, pp. 523–528, 2011.

placed on edges with capacity at least the size of the facility and the capacity u_e of edge e is reduced by the size of the largest facility $\tilde{p}$ placed on it to $u_e - r_{\tilde{p}}$. The number of facilities that has to be placed is denoted with $q = |\mathbb{P}|$ and the number of edges on which facilities can be placed with $L = |\mathbb{L}|$. The special case $q = 1$ is called 1-MT-FlowLoc problem or MT-FlowLoc single facility problem and is polynomial solvable (see [11]). For $q > 1$ the problem is NP-hard ([7]).

In IP 1 an integer programming formulation for the q-MT-FlowLoc problem is given. The objective function sums up the flow values of all vertex pairs. There are alternative objective functions (maximum difference, weighted sum) for the q-MT-FlowLoc problem but we will restrict ourselves to the sum objective function in this paper. This objective function can be used if it is not known in advance which vertex is source and which one sink.

IP 1 *q-MT-FlowLoc problem with sum objective function*

Variables

f_{vw}: *flow value of the flow with source v and sink w*
x_{ij}^{vw}: *flow on edge (i,j) of the flow from v to w*
y_{ijp}: *indicator variable: equal to one, if facility p is placed on edge (i,j), zero else*

Constants

nol_{ij}: *maximal amount of facilities that can be placed on edge (i,j)*
r_p: *size of facility p*
u_{ij}: *capacity of edge (i,j)*

$$\max \sum_{v<w \in V \times V} f_{vw} \tag{1}$$

$$\text{s.t.} \sum_{(i,w) \in E} x_{iw}^{vw} = f_{vw} \quad \forall (v,w) \in V \times V : v < w \tag{2}$$

$$\sum_{i:(i,j) \in E} x_{ij}^{vw} - \sum_{i:(j,i) \in E} x_{ji}^{vw} = 0 \quad \forall (v,w) \in V \times V : v < w, \forall i \in V \setminus \{v,w\} \tag{3}$$

$$\sum_{p \in \mathbb{P}} y_{ep} \le nol(i,j) \quad \forall (i,j) \in \mathbb{L} \tag{4}$$

$$\sum_{(i,j) \in \mathbb{L}: i<j} y_{ijp} = 1 \quad \forall p \in \mathbb{P} \tag{5}$$

$$x_{ij}^{vw} + x_{ji}^{vw} = 0 \quad \forall v < w \in V \times V, \forall (i,j) \in E \tag{6}$$

$$x_{ij}^{vw} + r_p \cdot y_{ijp} \le u(i,j) \quad \forall (i,j) \in \mathbb{L}, (v,w) \in V \times V : v < w, p \in \mathbb{P} \tag{7}$$

$$x_{ij}^{vw} \le u(i,j) \quad \forall (i,j) \in E \setminus \mathbb{L}, (v,w) \in V \times V : v < w \tag{8}$$

$$y_{ijp} = y_{jip} \quad \forall (i,j) \in \mathbb{L}, p \in \mathbb{P} \tag{9}$$

$$y_{ijp} \in \mathbb{B} \quad \forall (i,j) \in \mathbb{L}, p \in \mathbb{P} \tag{10}$$

Equation (2) stores the maximum flow value between vertex v and w in variable f_{vw}. Equation (3) ensures that the flow conservation constraints hold for flow between all vertex pairs. Inequality (4) guarantees that the number of facilities placed on edge (i,j) is less than or equal to $nol(i,j)$. Equations (5) and (9) ensure that every facility p is, indeed, located somewhere and that it influences the flow in both directions on the edge in which it is placed, resepectively. Equation (6) is the standard way to model the skew-symmetry in the undirected graph, and Equations (7) and (8) are capacity constraints. They guarantee that the flow on each edge and for each vertex pair does not exceed the capacity of the edge and - if a facility is placed on the edge - that the flow does not exceed the capacity reduced by the size of the facility.

2 Heuristic

In this section a heuristic is introduced that obtains feasible solutions by solving a maximum flow problem in a special network. This network represents all possible allocation of the facilities and all maximum flows correspond to feasible solutions for the q-MT-FlowLoc problem if the flow value is equal to q and vice versa. The main advantage of this heuristic is, that it always finds a feasible solution if one exists and that by adding a cost function to the edges in the network and calculating a minimum cost flow, it is possible to adapt the heuristic to different needs like objective functions or special graph classes.

Definition 1. *Given the network $G = (V,E)$ and the facilities $\mathbb{P}$ with size $r : \mathbb{P} \to \mathbb{N}$ of a (multi terminal) q-FlowLoc problem, the* q-FlowLoc feasible solution network *H with vertex set V_H and edge set E_H is defined as follows:*

$$V_H = \{s\} \cup \{t\} \cup \{p_i : 1 \le i \le L\} \cup \{e_j : 1 \le j \le L\}$$
$$E_H = \underbrace{\{(s,p_i)\ \forall 1 \le i \le q\}}_{E_H^0} \cup \underbrace{\{(e_j,t)\ \forall 1 \le j \le L\}}_{E_H^1} \cup \underbrace{\{(p_i,e_j)\ \forall i,j \text{ with } r(p_i) \le u(e_j)\}}_{E_H^2}$$

$$u(e) = \begin{cases} 1 & e \in E_H^0 \cup E_H^2 \\ nol(e_j) & e \in E_H^1 \text{ and } e = (p_i, e_j) \end{cases}$$

Using network H all feasible solutions for the q-MT-FlowLoc problem can be determined:

Theorem 1. *There exists a feasible solution for the (multi terminal) q-FlowLoc problem in G if and only if there exists a maximum s-t-flow in H with flow value q. Furthermore there is a one-to-one correspondence between the feasible solutions of the FlowLoc problem in G and the flows with flow value q in H.*

Proof. Let $l : \mathbb{P} \to \mathbb{L}$ be a feasible allocation of the facilities to the edges in G. Then define flow $x : E_H \to \mathbb{N}$ as follows:

$$x(e) = \begin{cases} 1 & e \in E_H^0 \text{ or } E_H^2 \text{ and } l(p_i) = e_j \\ \left|\{p \in \mathbb{P} : l(p) = e_j\}\right| & e = (e_j, t) \in E_H^1 \\ 0 & \text{else} \end{cases}$$

For x the capacity constraints are fulfilled since the allocation is feasible and hence the number of facilities placed on edge e_j is less or equal to $nol(e_j)$. Furthermore the flow conservation constraints hold for the vertices $v_i \in V_H$ since every facility has to be placed and for the vertices $e_j \in V_H$ since every facility has to be placed on exactly one edge.

On the other hand let x be a flow with value q in H, then define $l : \mathbb{P} \to \mathbb{L}$ as follows: $l(p_i) = e_j$, if $x(p_i, e_j) = 1$. This edge exists and is unique since the flow has flow value q and the flow conservation constraint is fulfilled for vertex v_i. On each edge $e_j \in \mathbb{L}$ at most $nol(e_j)$ facilities are placed because the capacity on edge $(e_j, t) \in E_H$ is equal to $nol(e_j)$. Facilities are only placed on edges having large enough capacity because of the definition of the edge set E_H. □

An immediate consequence of Theorem 1 is Heuristic 1, a generic heuristic to find a feasible solution for the q-MT-FlowLoc problem.

Heuristic 1. for the MTFLMFP with the sum objective function

Require: undirected graph $G = (V, E)$, capacities $u : E \to \mathbb{N}$, set of possible locations $\mathbb{L} \subseteq E$, set of facilities $\mathbb{P}$ with size $r(p)$, maximal number $nol(e)$ of facilities that can be placed on edge $e \in \mathbb{L}$

Ensure: allocation $l : \mathbb{P} \to \mathbb{L}$

1. construct the q-FlowLoc feasible solution network H
2. calculate a maximum s-t-flow x in H
3. construct allocation $l : \mathbb{P} \to \mathbb{L}$ from x (see proof of Theorem 1)
4. **return** l

The quality of the solution computed in Heuristic 1 can be influenced by computing special maximum s-t-flows in step (1). For instance a minimum cost s-t-flows sending q flow units can be calculated. For the cost function there are several choices, some of which are listed in Table 1. Any combination of the cost functions for edges (p_i, e_j) and (e_j, t) can be chosen. The idea of the cost functions is, that the larger the capacity of an edge, the larger the amount of facilities that can be placed on it and the smaller the facility itself, the smaller is the influence on the maximum flow values. It is not necessary to assign cost to edges (s, p_i) because every flow with flow value q has to use these edges, so in all cost combinations $c(s, p_i) = 0$.

Table 1. Cost functions for the edges in E_H

	a	b	c	d	e
$c(p_i, e_j)$	$-u(e_j)$	$-u(e_j) + r(p_i)$	$-u(e_j)nol(e_j)$	$-nol(e_j)(u(e_j) - r(p_i))$	0

	i	ii	iii	iv
$c(e_j, t)$	1	$-nol(e_j)$	$-u(e_j)$	0

3 Comparison of the Cost Functions

To indicate the influence of the cost selection, we list in Table 2 the performance of the heuristic for randomly generated test graphs (using random graph generator of BGL [12]) with $n = 30$ vertices and an edge density of 40%. A maximum of $nol_{\max} = 2$ facilities can be placed on a single edge. We tested all combinations of the cost functions given in Table 1. The number q of facilities and the ratio $L\%$ of edges in $\mathbb{L}$ to edges in E is given in the top row of Table 2. The entries in the table give the quotient of the heuristic solution and the optimal solution value computed by IP 1. All cost combinations for the tested heuristics yield solutions close to the optimum value. No heuristic achieved outstanding results for all tested parameter combinations. The "right" choice of the cost function depends on the observed problem and parameter setting. The heuristic is much faster in finding a feasible solution ($< 1sec$) then CPLEX (version 12.1.0, $> 5min$) in calculating an optimum solution. This follows from the fact that for obtaining the optimum solution $n(n-1)$ maximum flow problems have to be solved, while the heuristic only has to solve one maximum flow problem in a smaller graph. The advantage of graph H is that it is easy to construct and all feasible solutions are represented. Furthermore graph H is independent of the objective function of the FlowLoc problem and hence suitable for many problems. The cost function can be chosen corresponding to the objective function of the FlowLoc problem and the parameters of graph G. The special structure and the small size of H are further advantages.

Table 2. Comparison of Heuristics 1-3 and the different cost function combinations

$q, L\%$	8,40	16,40	5,100	10,100
(a,i)	0.8698	0.8943	0.9017	0.9104
(a,ii)	0.8698	0.8958	0.9041	0.9124
$(a,iii$	0.8737	0.8943	0.9005	0.9025
(a,iv)	0.8698	0.8943	0.9017	0.9104
(b,i)	0.8662	0.8785	0.9024	0.9061
(b,ii)	0.8662	0.8808	0.9047	0.9171
(b,iii)	0.8753	0.8785	0.9010	0.8995
(b,iv)	0.8662	0.8785	0.9024	0.9061
(c,i)	0.8712	0.8985	0.9017	0.9180
(c,ii)	0.8712	0.8985	0.9014	0.9169
(c,iii)	0.8777	0.8985	0.9005	0.9143
(c,iv)	0.8712	0.8985	0.9017	0.9180
(d,i)	0.8712	0.8985	0.9017	0.9180
(d,ii)	0.8712	0.8985	0.9041	0.9169
(d,iii)	0.8777	0.8985	0.9005	0.9143
(d,iv)	0.8712	0.8985	0.9017	0.9180
(e,i)	0.8730	0.8875	0.8989	0.9117
(e,ii)	0.8745	0.8967	0.9004	0.9122
(e,iii)	0.8775	0.8951	0.9051	0.9052
(e,iv)	0.8730	0.8875	0.8989	0.9117

4 Conclusion

The q-MT-FlowLoc problem is NP-hard and thus optimum solutions cannot be computed in polynomial time (unless P=NP). The q-FlowLoc feasible solution network incorporates all feasible solutions of the q-FlowLoc problem into the s-t-flows with flow value equal to q. By the use of different cost functions it is possible to obtain solutions for the q-MT-FlowLoc problem with objective value near the optimum. The time needed to calculate a maximum flow algorithm or a minimum cost flow in the small graph and the corresponding allocation of the facilities is negligible compared to solving the IP formulation to obtain a optimum solution. With the help of Heuristic 1 it is possible to find feasible solutions with reasonable objective values in short time also for large instances. The goal is now to quantify the performance of the heuristic by using the structure of the IP.

Acknowledgement

This paper is supported in part by the Federal Ministry for Education and Research (Bundesministerium für Bildung und Forschung, BMBF), Project REPKA, under FKZ 13N9961 (TU Kaiserslautern).

References

1. Ahuja, R., Magnanti, T., Orlin, J.: Network Flows. Massachusetts Institute of Technology, OR Center (1988)
2. Baumann, N.: Evacuation by earliest arrival flows. Ph.D. thesis, Universität Dortmund (2007)
3. Chalmet, L.G., Francis, R.L., Saunders, P.B.: Network models for building evacuation. Fire Technology 18(1), 90–113 (1982)
4. Choi, W., Hamacher, H.W., Tufekci, S.: Modeling of building evacuation problems by network flows with side constraints. European J. Oper. Res. 35(1988), 98–110 (1988)
5. Francis, R.L., McGinnis, L.F., White, J.A.: Locational analysis. European J. Oper. Res. 12(3), 220–252 (1983)
6. Hamacher, H.W.: Mathematische Lösungsverfahren für planare Standortprobleme. Vieweg, Braunschweig (1995)
7. Hamacher, H.W., Heller, S., Rupp, B.: Flow Location (FlowLoc) Problems and Evacuation Planning. Ann. Oper. Res (submitted)
8. Hamacher, H.W., Heller, S., Ruzika, S.: A Sandwich Approach for Evacuation Time Bounds. In: PED 2010 Conference Proceedings, pp. 503–513 (to appear, 2010)
9. Hamacher, H.W., Tjandra, S.A.: Mathematical modelling of evacuation problem: a state of the art. In: Schreckenberg, M., Sharma, S. (eds.) Pedestrian and Evacuation Dynamics, pp. 227–266. Springer, Heidelberg (2002)
10. Nickel, S., Puerto, J.: Location theory: a unified approach. Springer, Heidelberg (2005)
11. Rupp, B.: FlowLoc: Discrete Facility Locations in Flow Networks. Diploma thesis, TU Kaiserslautern (2010)
12. Siek, J., Lee, L., Lumsdaine, A.: The boost graph library. Addison-Wesley, Boston (2002)

Optimal Bandwidth Allocation in Mesh-Based Peer-to-Peer Streaming Networks

María Elisa Bertinat, Darío Padula, Franco Robledo Amoza, Pablo Rodríguez-Bocca, and Pablo Romero

Facultad de Ingeniería, Universidad de la República, Montevideo, Uruguay
dpadula@fing.edu.uy

Abstract. The design of efficient protocols for mesh-based Peer-to-Peer (P2P) networks has many challenges, one of them is the bandwidth allocation. On one hand, users (called *peers*) demand high Quality of Experience and network traffic when they watch their streaming contents. On the other, Internet Service Providers (ISPs) support their business with the capacity of their international links. A recent strategy considered in order to meet both peers and ISPs requirements is the Proactive Provider Participation, shortly named P4P [10]. This approach allocates the maximum total traffic in the network, reducing at the same time the percentage use of the most congested links. This paper addresses the bi-level P4P problem. We introduce a polytime solution which achieves any given accuracy when only one content is delivered in the network In addition, we design a greedy randomized technique when multiple contents are shared. Finally, we apply our algorithm to a real peer-to-peer live video-streaming platform, when a single content is delivered. The results highly outperform current strategies.

1 Introduction

An important amount of today's Internet traffic is due to live video streaming [2]. For this reason, several peer-to-peer streaming networks were developed in the last years. The most successful ones are PPlive, TVUnetwork, SopCast, with proprietary and unpublished mesh-based protocols [8]. Mesh-based P2P networks are virtual unstructured networks developed at the application layer, over the Internet infrastructure. Bittorrent is the best known mesh-based P2P protocol, developed for file sharing purposes [3]. The users, called *peers*, offer their resources(bandwidth in a streaming application) to others, basically because they share common interests. They can connect and disconnect freely. This makes P2P networks an attractive tool for them, but increases P2P's design challenges, because the resource availability depends on them.

In P2P, the cooperation is the key element in order to assure a certain quality of experience to end-users [8]. There are three main steps in all mesh-based P2P protocols for cooperation. First, when a peer enters the net it should discover other peers sharing the same content, which is called *swarm selection strategy*. Once a new peer knows other peers in his swarm, he must select the best ones to cooperate, what is called *peer selection strategy*. Once a new peer handshakes other peers, it should decide which pieces of the streaming content should be asked first, called the *piece selection strategy* [1]. This paper is focused on the swarm selection strategy and in the peer selection strategy. The

J. Pahl, T. Reiners, and S. Voß (Eds.): INOC 2011, LNCS 6701, pp. 529–534, 2011.

main issue is to locate the largest amount of traffic in the network without bottlenecks, and keeping the quality of experience between peers. In Section 2 the mathematical P4P model, based on [9,10], is explained. Related work on P4P can be found in [11,4,5].

Following the related work, in order to represent the complexity and scale of a real scenario with millions of peers, the peers are grouped in nodes. Each node is a geographical subset of Internet (for example: an autonomous system or an ISP point-of-presence), and they are interconnected by real links. Inside each node could be several peers sharing contents. Section 3 contains a polytime resolution for one content and a greedy randomized algorithm [7] for the general P4P problem. In Section 4 we show the performance of the single content polytime algorithm in GoalBit, which is the first open source real platform that widely offers live video streaming to final users in Internet [2]. Finally, Section 5 contains the main conclusions of this work.

2 Mathematical Model

Our model is inspired by [9,10], where the authors relax the model into a linear programming one. Consider a network with nodes set $V = \{v_1,\dots,v_n\}$ and two one-way links between each pair of nodes, whose respective capacities are represented by a non-negative matrix $C = (c_{i,j})_{1\leq i,j\leq n}$. The upload and download bandwidths for each node $v_i \in V$ are u_i^k and $d_i^k, i = 1,\dots,n$ respectively, where $k \in \{1,\dots,K\}$ represents different contents (each node v_i has K possible contents to download). Each link (i,j) uses a certain percentage of its capacity due to other applications, which is denoted by $b_{i,j}$, called the background traffic. Be $\mathscr{P} = \mathscr{P}_1^{k_1},\dots,\mathscr{P}_m^{k_m}$ a set of oriented paths in the network, where $\mathscr{P}_h^{k_h} = (x_h,\dots,y_h)$ (x_h is the uploader and y_h is the downloader). Be $t_1,\dots,t_m$ their respective traffic magnitudes. In words: x_h uploads a traffic magnitude t_h of content type k_h to y_h by the oriented path $\mathscr{P}_h^{k_h}$.

The objective function is to reduce the maximum link utilization ρ in the network. Constraint (1) imposes that the total traffic generated in the network must be maximized. Constraints (2) and (3) assure that each node does not upload (respectively download) more traffic than its capacity (for each content). Constraint (4) states that edge capacities must not be exceeded. Finally, constraint (5) classifies nodes as uploader or either downloader for each content.

$$\min_{\mathscr{P}} \max_{(i,j):i\neq j} \rho(\mathscr{P}) = b_{i,j} + \frac{\sum_{h:(i,j)\in\mathscr{P}_h^{k_h}} t_h}{c_{i,j}}, \qquad s.t.$$

P4P Model

$$\max_{\mathscr{P}} \sum_{h=1}^{m} t_h \tag{1}$$

$$\sum_{h:x_h=i,k_h=k} t_h \leq u_i^k, \qquad \forall i \in V, k \in \{1,\dots,K\} \tag{2}$$

$$\sum_{h:y_h=j,k_h=k} t_h \leq d_j^k, \qquad \forall j \in V, k \in \{1,\dots,K\} \tag{3}$$

$$b_{i,j}c_{i,j} + \sum_{h:(i,j)\in\mathscr{P}_h}^{k_h} t_h \leq c_{i,j}, \qquad \forall i \neq j \in V \tag{4}$$

$$u_i^k.d_i^k = 0 \qquad \forall i \in V, k \in K \quad (5)$$

In a real world scenario, the objective function has an economic interpretation: reduce the bottleneck of the most expensive edges in Internet at the same time fulfilling the peers demands (Constraint (1)) according with the available resources in the network (Constraints (2), (3) and (4)). In practice, once we have a set of oriented paths $\mathscr{P}$ and their respective magnitudes, it is possible to converge probabilistically to that traffic distribution in a real network. The swarm (list of peers) for a peer located at node v_i that asks for content k, is populated with the following probability of peers from node v_j:

$$w_{ji}^k = \frac{\sum_{h:(j,i)\in\mathscr{P}_h^k} t_h}{\sum_{x\in V}\sum_{h:(x,i)\in\mathscr{P}_h^k} t_h}. \quad (6)$$

Note that the numerator of w_{ji}^k represents, for content k, the traffic coming from node v_j to v_i, while the denominator adds all incoming traffics for the same content to node v_i. Equation (6) defines how to swarm must be created(i.e., the swarm selection strategy), where the peers in the swarm must be chosen randomly, according with the empirical probabilities w_{ji}^k. Moreover, the peer statistically takes in consideration these weights also in his peer selection strategy in order to have a faster converge. See [6] for details.

3 A Polytime Resolution and Its Generalization

Despite the high complexity of the general P4P formulation, we show that there exists a fully polynomial time approximation scheme (FPTAS) when one content is distributed in the network:

Theorem 1. *There is a FPTAS for the P4P Problem when $K = 1$.*

Proof. Connect two ideal nodes s and t (one transmitter and the other receiver) to every node v_i with corresponding capacities u_i and d_i. Find the mincut-maxflow ϕ_{max} via the classical Ford-Fulkerson [6] (FF) algorithm. Finally, reduce all capacities c_{ij} by a certain factor ρ. The minimum factor ρ_{min} that preserves ϕ_{max} can be found iteratively with a bipartition scheme in the closed interval $[\max_{1\leq i,j\leq n} b_{i,j}, 1]$ and calling FF. □

To the best of our knowledge, there is not an exact resolution in polytime for multiple contents [6]. A possible heuristic approach is detailed next.

Algorithm *RandomList* is very simple, and proposes a Greedy Randomized generalization for multiple contents. It receives the bandwidth matrices $U = (u_i^k)$ and $D = (d_i^k)$ (that store the bandwidth of every node for each content), the capacity and background matrices C and B respectively and a probability vector $(p_k)_{1\leq k\leq K}$, that measures the priority to the content type k. In each iteration, a content is chosen randomly according with the priority vector p. Immediately, FF is called in order to find the best bandwidth allocation for that content. The flows obtained so are added and the bandwidth and capacities updated (the entry $p_k = 0$). This process is repeated until there is no more capacity or after all contents were delivered.

Algorithm 1. $\mathscr{P} = RandomList(U,D,C,B,p)$

1. $\mathscr{P} = \emptyset$
2. **while** $(length(p) > 0$ AND $C \geq 0)$ **do**
3. $\quad k = ChooseContent(p)$
4. $\quad \mathscr{P} = \mathscr{P} \cup FF(U^k, D^k, C, B)$
5. $\quad Update(U,D,C,B,p)$
6. **end while**
7. **return** $\mathscr{P}$

4 Empirical Results

In this work we implemented Algorithm FF when one content is delivered to converge empirically to the optimum P4P bandwidth allocation in a real P2P video streaming platform, called GoalBit [2]. GoalBit maintains the BitTorrent's philosophy, trying to extend its success to video streaming. The P4P-based strategy we propose acts exactly in the moment of the peer list conformation, applying FF algorithm to skew routing to converge to the theoretical P4P solution. Emulations were carried out with information provided by The Uruguayan National Telephony Operator ANTEL from their GoalBit deployment live service. We contrast the swarm and peer selection strategy using our P4P algorithm(i.e. FF algorithm) versus the Classical GoalBit strategy (based on BitTorrent). For all emulations we evaluate the quality of experience of final users (buffering time), the total amount of exchanged traffic and the one which crosses international links. In particular, we show the results of two emulations for the cases of 60 and 100 simultaneous peers connected in average. Tables 1 and 2 show, for both strategies; the total traffic, incoming and outgoing international links[1] traffic. Also, shows the percentage growth of incoming and outgoing traffic $P_{G_{in}}$ and $P_{G_{out}}$ when the P4P model is applied, in relation with a Classical strategy. Specifically:

$$P_{G_{in}} = 100 \times (\frac{In_{P4P}}{In_{Classic}} - 1), P_{G_{out}} = 100 \times (\frac{Out_{P4P}}{Out_{Classic}} - 1),$$

where In_{P4P}, $In_{Classic}$ and Out_{P4P} and $Out_{Classic}$ represent the total incoming and outgoing traffic for P4P and Classic respectively. It is desirable to obtain negative percentage growth, interpreted as a reduction in the international links, and consequently, an improvement in relation with the Classical strategy.

It can be seen from Table 1 that the incoming reduction is 47.57%, while the outgoing reaches 74.96%, keeping the total traffic achieved by the classical strategy. Also, Table 2 shows important reductions for the 100 simultaneous peers case, and an increase in total traffic is perceived. Fig. 1 shows that the buffer time distributions are quite similar for both techniques. In both cases, 85% of peers perceived a buffering time lower than 55 seconds. In contrast, for the 100 simultaneous peers case, P4P present much lower buffering times when compared with the classical strategy (see Fig. 2). 68% of the peers wait at most 38 seconds to start playing when P4P is applied, but only 27% can start

[1] As we can only measure the traffic through the outgoing and incoming links to and from Uruguay, this will be our reference node.

Table 1. Link utilization for 60 peers

Model	Incoming	Outgoing	Total
Classic	31656	31366	183069
P4P	7927	16446	183067
% grow traffic	-47.57	-74.96	0.0

Table 2. Link utilization for 100 peers

Model	Incoming	Outgoing	Total
Classic	5681	10253	55078
P4P	3657	4451	58893
% grow traffic	-56.59	-35.63	6.93

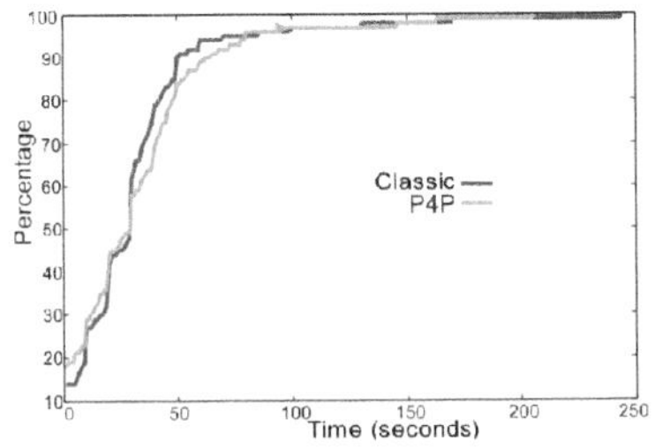

Fig. 1. Buffering time for 60 peers

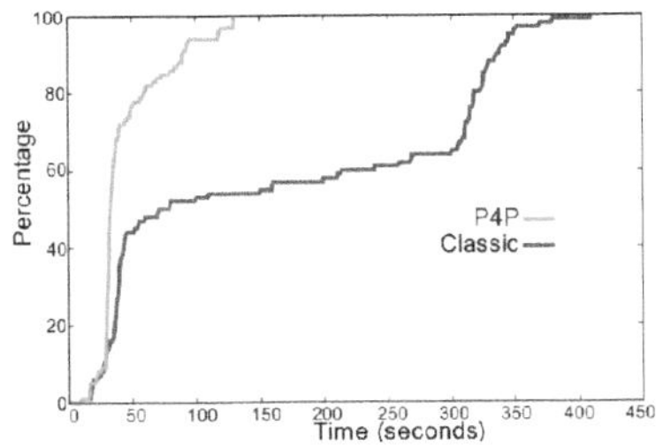

Fig. 2. Buffering time for 100 peers

playing the video during the same time. Many emulations were carried out for different inputs showing similar bandwidth savings, close to 30% on average [6].

5 Conclusions

In this work, the Proactive Participation Provider (P4P) performance was analyzed, and contrasted with the Classical GoalBit strategy. In a theoretical aspect, the P4P mathematical model can be solved with the desired precision when only one content is distributed in the network. However, when multiple contents are distributed the problem has not been solved exactly so far. Emulations of a real system indicate a 30% link utilization reduction in average with our P4P application, and at the same time the quality of experience seems to improve. Qualitatively, this highlights the competitiveness of the P4P optimum bandwidth allocation.

Acknowledgement

This paper was supported by the national telephony operator ANTEL (see `www.antel.com.uy`)

References

1. Bertinat, M.E., Vera, D.D., Padula, D., Robledo, F., Rodríguez-Bocca, P., Romero, P., Rubino, G.: A COP for Cooperation in a P2P Streaming Protocol. In: Proceedings of the IEEE International Conference on Ultra Modern Telecommunications (ICUMT 2009), October 12-14, pp. 1–7. IEEE Computer Society, Washington, DC (2009)

2. Bertinat, M.E., Vera, D.D., Padula, D., Robledo, F., Rodríguez-Bocca, P., Romero, P., Rubino, G.: Goalbit: The first free and open source peer-to-peer streaming network. In: Proceedings of the 5th international IFIP/ACM Latin American conference on Networking, September 2009, pp. 49–59. ACM Press, New York (2009)
3. Cohen, B.: Incentives build robustness in bittorrent 1, 1–5, (May 2003), `http://www.bramcohen.com`
4. Guo, Z., Min, L., Yang, H., Yang, S.: An Enhanced P4P-Based Pastry Routing Algorithm for P2P Network. In: 2010 IEEE International Conference on Granular Computing (GrC), August 2010, pp. 687–691 (2010)
5. Guo, Z., Yang, H., Yang, S.: P4P Pastry: A novel P4P-based Pastry routing algorithm in peer to peer network. In: 2010 The 2nd IEEE International Conference on Information Management and Engineering (ICIME), April 2010, pp. 209–213 (2010)
6. Padula, D.: Compromiso entre Pares e ISPs en el contexto P4P: Optimización en dos niveles. Master's thesis, Facultad de Ingeniería, Montevideo, Uruguay (July 2010), `http://premat.fing.edu.uy/IngenieriaMatematica`
7. Resende, M.G.C., Ribeiro, C.C.: Greedy Randomized Adaptive Search Procedures. In: Glover, F., Kochenberger, G. (eds.) Handbook of Methaheuristics. Kluwer Academic Publishers, Dordrecht (2003)
8. Rodríguez-Bocca, P.: Quality-centric design of Peer-to-Peer systems for live-video broadcasting. Ph.D. thesis, INRIA/IRISA, Université de Rennes I, France (April 2008)
9. Xie, H., Krishnamurthy, A., Silberschatz, A., Yang, Y.R.: P4P: Proactive Provider Participation for P2P. Tech. rep., Yale University, Department of Computer Science (2007)
10. Xie, H., Yang, R., Krishnamurthy, A., Liu, Y., Silberschatz, A.: P4P: Provider Portal for Applications. In: Conference on Data Communication, SIGCOMM (August 2008)
11. Yang, G., Sun, Y., Wu, H., Li, J., Liu, N., Dutkiewicz, E.: A Trust Model in P4P-integrated P2P Networks based on domain management. In: IEEE Symposium on Computers and Communications (ISCC), June 2010, pp. 824–829 (2010)

Hub Location Problems with Choice of Different Hub Capacities and Vehicle Types

Julia Sender and Uwe Clausen

Institute of Transport Logistics,
TU Dortmund, Leonhard-Euler-Str. 2, 44227 Dortmund, Germany
{sender,clausen}@itl.tu-dortmund.de

Abstract. Hub-and-spoke networks have wide (logistical) applications, e. g. , in air and cargo transportation, post delivery, and telecommunication systems. In these networks transport volume is specified as flow between several origins and destinations. Hubs are specific facilities that serve as sorting, handling, transfer, and distribution points. In hub-and-spoke networks, flows with different origins and destinations are consolidated in hubs. Consolidation of flows in hubs may decrease transportation costs but increase additional costs due to establishing and operating hub facilities. Hub location problems deal with the location of hubs and allocation of origin-destination nodes to hub nodes in order to route the demand of each origin-destination pair through the network. In this paper, we extend the classical capacitated hub location problem. For each potential hub node we consider a set of different capacity levels which can be chosen. Besides, we consider the choice of different vehicle types of different capacities and costs to model more realistic costs. We present a single and a multiple allocation version. We implement the resulting integer programs in GAMS and solve them with CPLEX.

1 Introduction

Hub location research is a very important area of location theory. This is due to the wide applications of hub-and-spoke networks in transportation and telecommunication systems [1,3]. In hub-and-spoke networks, origin-destination demand is served via hubs rather than directly. Hubs act simultaneously as collection, consolidation, transfer, and distribution points. Hub location problems consider the location of hubs and the allocation of origin-destination nodes to hub nodes in order to route the demand of each origin-destination pair through the network. These decisions determine the resulting network structure.

The hub location problem was first formulated mathematically by [12] as a quadratic integer program. Campbell [2] formulated this problem as a mixed integer LP. Later, [8,9] formulated a hub location problem as a multi-commodity flow problem. Recent reviews on hub location problems are given by [1,3]. Both give a review of classical hub location models, solution approaches and recent trends. They also outline different variants of basic models in terms of their objectives, network components, and topology, as well as several constraints (e.g., on nodes and arcs).

Hub location problems can be categorized into five basic variants [3,10]: p-Hub Median Problem, Hub Arc Location Problems, p-Hub Center Problems, Hub Covering Problems and Hub Location Problems. In particular, the *p-Hub Median Problem* has received

J. Pahl, T. Reiners, and S. Voß (Eds.): INOC 2011, LNCS 6701, pp. 535–546, 2011.

most attention in hub location research. It is characterized by a given number p of hubs to locate. The objective is to minimize the total flow costs between all pairs of nodes. Furthermore, direct connections between origins and destinations are usually not allowed. Similarly, in *p-Hub Center Problems* the number of hubs to locate is predisposed. The objective is to minimize the maximum distance (or costs) between each pair of nodes. Several variations of this model were considered [3]. In *Hub Covering Problems* demand is covered if both origin and destination nodes are within a particular distance around an allocated hub node. The objective of *Hub Arc Location Problems* is to locate a given number q of hub arcs instead of locating hubs [4,5]. The location of hub nodes is implicitly given by the location of hub arcs. The hub network may not be a complete graph. The objective is to minimize the total transportation costs. In *Hub Location Problems* the optimal number of hubs is determined as a part of the problem. In general, the constraint of using a given number of hubs is left out, when capacities in hubs are introduced. The objective is to minimize the total costs, whereas the total costs consist of the costs of hub facilities and transportation costs. Hence, the objective function reflects a trade-off between lower transportation costs and higher hub cost.

Aforementioned basic variants and in particular p-hub problems have received most attention in literature. Even these problems are usually $\mathcal{NP}$-hard [3]. The basic models are extended and modified in several ways. The models differ in their objective as well as in their constraints. Hub location problems can be classified in single and multiple allocation problems. In *single allocation* variants each non-hub node is allocated to a single hub node. In particular, this implies that a single hub node receives the whole transport flow of an assigned origin node and delivers the whole demand of an assigned destination node. In *multiple allocation* variants an origin-destination node may be allocated to several other hub nodes. In general, multiple allocation results in greater routing flexibility and hence is expected to have a lower objective function value. Furthermore, models may differ in the consideration of capacities on hub nodes or on arcs. Different types of capacity are considered in literature [3], e.g. there might be a restriction on the total flow through hub nodes or only on the collected flow from non-hub nodes. Most hub location problems take only one hub type into account. Two different hub types are considered by [11] in the context of air transportation, by [12] in the context of package delivery, and by [14] in the context of railway logistics. Further modifications and extensions are possible and result in a huge variety of hub location models. Surveys on different variations and classification schemes are given by [1,3,15]. The single allocation hub location problem with hub capacities is known as the capacitated single allocation hub location problem (CSAHLP). The multiple allocation version is known as the capacitated multiple allocation hub location problem (CMAHLP).

Usually in research a single discount factor between 0 and 1 for all inter-hub flows is considered. This factor is predetermined and independent of the actual flow volume. So far, only some research is dedicated to a flow-dependent transport cost structure [13,14]. In all aforementioned work, the capacities of potential hub nodes are predisposed and hence exogenous decisions. Nevertheless, a strategic network design for hub-and-spoke networks contains the strategic decision of dimensioning hub facilities, too. So far, only [6] has dealt with several capacity levels in hubs in a single allocation context with a predisposed discount factor for collection, transfer, and distribution arcs.

In this paper, we present a single and a multiple assignment hub location model with the choice of capacity levels. The number of hubs is not specified explicitly but is determined implicitly by the optimization. We limit the total flow through hub nodes, that is, the collected and transferred flow. Besides, we consider the choice of different vehicle types with different capacities and costs to model more realistic costs. Thus, the determination of the number of needed vehicles of each type to meet the demand becomes part of the problem. Hence, the impact of economies of scale on costs will vary across node pairs. To the best knowledge of the authors, these problems have not been considered in the literature.

The remainder of the paper is organized as follows. In Section 2, we develop a single and a multiple allocation version for hub location problems with the choice of different vehicle types and multiple capacity levels in hubs. We implement the resulting integer programs in GAMS and solve them with CPLEX for test data sets with 10 to 30 nodes. The results of our computational tests are presented in Section 3. The paper ends with conclusions on limitations and further research needs.

2 Formulations

In this paper, we extend the capacitated single assignment hub location problem with multiple capacity levels (CSAHLPM) introduced by [6]. Instead of predisposed discount factors we introduce the choice of different vehicle types to model more realistic costs. Besides, we restrict the total flow through hubs. We consider a single and a multiple allocation version. Before formulating both problems in Sections 2.2 and 2.3, we introduce some notation. The section ends with an extension of the formulation CSAHLPM of [6] to multiple allocation.

2.1 Notations

The following notation is introduced:

$N = \{1,\dots,n\}$	set of nodes
$\Gamma_k = \{1,\dots,s_k\}$	set of different capacity levels of a hub at node k $(k \in N)$
W_{ij}	transport volume of each origin-destination pair (i,j) $(i,j \in N)$
c_k^q	cost of a hub at location k with capacity level q $(k \in N, q \in \Gamma_k)$
γ_k^q	capacity of a hub k with capacity level q $(k \in N, q \in \Gamma_k)$
$U = \{1,\dots,\bar{u}\}$	set of different vehicle types on arcs
c_{kl}^u	cost of a vehicle of type u on arc (k,l) $(k,l \in N,\ u \in U)$
b_{kl}^u	capacity of a vehicle of type u on arc (k,l) $(k,l \in N, u \in U)$

Let $O_i = \sum_{j \in N} W_{ij}$ be the total flow originating at origin i and $D_j = \sum_{i \in N} W_{ij}$ the total demand of destination j $(i,j \in N)$. Costs c_k^q include fixed costs for establishing and operating a hub with capacity level q at node k $(k \in N, q \in \Gamma_k)$. Besides, we assume $\gamma_k^{q_1} < \gamma_k^{q_2}$ and $c_k^{q_1} < c_k^{q_2}$ for $q_1 < q_2$ and $q_1, q_2 \in \Gamma_k$. This assumption is helpful to do some preprocessing tests, cf. Table 2. Hence, it is not necessary to request $b_{kl}^{u_1} < b_{kl}^{u_2}$ with $c_{kl}^{u_1} < c_{kl}^{u_2}$ for $u^1 < u^2$ and $u^1, u^2 \in U$ in order to do further preprocessing tests. We assume $\sum_{k \in N} \gamma_k^{s_k} \geq \sum_{k \in N} O_k$ and $W_{ij} = 0$ for $i = j$ $(i,j \in N)$.

Each origin-destination path consists of three different components: a collection movement from an origin to the first hub, transfer movements from the first hub to the last hub, and a distribution movement from the last hub to the destination. A path with only one hub-stop is also possible. In this case, the transfer step can be omitted. We define three different sets of integer decision variables corresponding to the three components of an origin-destination path.

X_{ik} = flow from origin i to hub k $(i,k \in N)$
Y^i_{kl} = flow from hub k to hub l that originates at origin i $(i,k,l \in N)$
Z^i_{lj} = flow from hub l to destination j that originates at origin i $(i,j,l \in N)$

Each non-negative variable induces a single arc of the network. If the origin or destination is also a hub node, this arc might be from the particular node to itself. Transfer arcs connect hub nodes and do not exist if the origin-destination path is routed via exactly one hub. Besides, we have integer variables v^u_{pq} to consider the choice of different vehicle types u of different capacity on arc (p,q).

v^u_{pq} = number of vehicles of type u on arc (p,q) $(p,q \in N,\ u \in U)$

We consider binary decision variables for locating hubs and allocating capacity levels to the hubs:

$$H_k = \begin{cases} 1 \text{ if node } k \text{ is a hub } (k \in N) \\ 0 \text{ otherwise} \end{cases}$$

$$H^q_k = \begin{cases} 1 \text{ if node } k \text{ is a hub with capacity level } q\ (k \in N, q \in \Gamma_k) \\ 0 \text{ otherwise} \end{cases}$$

2.2 Multiple Allocation Formulation

In this section, we extend the classical capacitated multiple allocation hub location problem. Instead of discount factors on arcs we introduce the choice of different vehicle types on each arc. We also introduce the choice of different capacity levels in hubs. The formulation is given as follows:

$$(\mathrm{F_{MA}}) \qquad \min \sum_{k\in N}\sum_{q\in\Gamma_k} c^q_k H^q_k + \sum_{k\in N}\sum_{l\in N}\sum_{u\in U} c^u_{kl} v^u_{kl} \tag{1}$$

$$\text{s.t.} \sum_{k\in N} X_{ik} = O_i \qquad \forall\, i \in N \tag{2}$$

$$\sum_{l\in N} Z^i_{lj} = W_{ij} \qquad \forall\, i,j \in N \tag{3}$$

$$X_{ik} + \sum_{l\in N} Y^i_{lk} = \sum_{l\in N} Y^i_{kl} + \sum_{j\in N} Z^i_{kj} \qquad \forall\, i,k \in N \tag{4}$$

$$X_{ik} \le O_i H_k \qquad \forall\, i,k \in N \tag{5}$$

$$Y^i_{kl} \le O_i H_k \qquad \forall\, i,k,l \in N \tag{6}$$

$$Y^i_{kl} \le O_i H_l \qquad \forall\, i,k,l \in N \tag{7}$$

$$Z^i_{lj} \leq W_{ij} H_l \qquad \forall\, i,j,l \in N \qquad (8)$$

$$\sum_{i \in N} \left(X_{ik} + \sum_{l \in N} Y^i_{lk} \right) \leq \sum_{q \in \Gamma_k} \gamma^q_k H^q_k \qquad \forall\, k \in N \qquad (9)$$

$$\sum_{q \in \Gamma_k} H^q_k = H_k \qquad \forall\, k \in N \qquad (10)$$

$$X_{kl} + \sum_{i \in N} (Y^i_{kl} + Z^i_{kl}) \leq \sum_{u \in U} b^u_{kl} v^u_{kl} \qquad \forall\, k,l \in N, k \neq l \qquad (11)$$

$$X_{ik} \in \mathbb{Z}^+ \qquad \forall\, i,k \in N \qquad (12)$$

$$Y^i_{kl} \in \mathbb{Z}^+ \qquad \forall\, i,k,l \in N \qquad (13)$$

$$Z^i_{lj} \in \mathbb{Z}^+ \qquad \forall\, i,j,l \in N \qquad (14)$$

$$v^u_{kl} \in \mathbb{Z}^+ \qquad \forall\, k,l \in N,\ u \in U \qquad (15)$$

$$H_k \in \{0,1\} \qquad \forall\, k \in N \qquad (16)$$

$$H^q_k \in \{0,1\} \qquad \forall\, k \in N, q \in \Gamma_k \qquad (17)$$

The objective function (1) minimizes the total costs consisting of establishment and operation costs of hub facilities as well as transportation costs. Constraints (2), (3), and (4) are flow related constraints. Constraints (2) assure that all flow from each origin leaves the origin. Constraints (3) ensure that all origin-destination demand arrives at the proper destination and constraints (4) are the flow conservation constraints in hubs. If the origin or destination is also a hub node the corresponding arc might be from the particular node to itself. The flow conservation constraints still have to be valid for this node. Constraints (5), (6), (7), and (8) ensure that a hub is established for each collection, transfer or distribution movement. These constraints may be necessary to assure a feasible solution to the problem [7]. Constraints (9) are capacity constraints for all incoming flows at hubs, which includes the incoming flows from non-hub nodes and the flows transferred from other hubs. Constraints (10) ensure that for each hub node established a single capacity level is chosen. Constraints (11) assure that the flow on an arc cannot exceed the capacity of vehicles assigned to the corresponding arc. Therefore, the determination of the number of needed vehicles of each type to meet the demand becomes part of the problem. Hence, the impact of economies of scale on cost will vary across all node pairs. Finally, constraints (12)–(17) are domain constraints.

In Table 1 several preprocessing tests for formulation (F_{MA}) are proposed to reduce the size of the problem. Results of computational experiments are given in Subsection 3.2. In the next subsection we present a single allocation version of the problem above.

2.3 Single Allocation Formulation

The single allocation version is similar to the problem above. In addition, each non-hub node is restricted to be allocated to a single hub node. In the single allocation case, however, we can restrict the variables X_{ik} to be binary. Therefore, we can eliminate the Z^i_{lj} variables and replace the binary variables H_k by X_{kk}, whereas X_{kk} is one if node k becomes a hub and zero otherwise. Accordingly, we have the following new decision variables $X_{ik} = 1$ if node i is allocated to a hub node at k and 0 otherwise, for all$(i,k \in N)$.

Table 1. Preprocessing Tests for (F_{MA})

	Condition	Fixed Variables
Test 1	$k = l$ for $k, l \in N$	$v^u_{kl} = 0$ for all $u \in U$ $Y^i_{kl} = 0$ for all $i \in N$
Test 2	$i = l$ for $i, l \in N$	$Y^i_{kl} = 0$ for all $k \in N$
Test 3	$W_{ij} = 0$ for $i, j \in N$	$Z^i_{lj} = 0$ for all $l \in N$

The single allocation formulation is given as follows:

$$(\mathrm{F_{SA}}) \qquad \min \sum_{k\in N}\sum_{q\in\Gamma_k} c^q_k H^q_k + \sum_{k\in N}\sum_{l\in N}\sum_{u\in U} c^u_{kl} v^u_{kl} \tag{18}$$

$$\text{s.t.} \sum_{k\in N} X_{ik} = 1 \qquad \forall\, i \in N \tag{19}$$

$$O_i X_{ik} + \sum_{l\in N} Y^i_{lk} = \sum_{l\in N} Y^i_{kl} + \sum_{j\in N} W_{ij} X_{jk} \qquad \forall\, i,k \in N \tag{20}$$

$$X_{ik} \le X_{kk} \qquad \forall\, i,k \in N \tag{21}$$

$$Y^i_{kl} \le O_i X_{kk} \qquad \forall\, i,k,l \in N \tag{22}$$

$$Y^i_{kl} \le O_i X_{ll} \qquad \forall\, i,k,l \in N \tag{23}$$

$$\sum_{i\in N}\left(O_i X_{ik} + \sum_{l\in N} Y^i_{lk}\right) \le \sum_{q\in\Gamma_k} \gamma^q_k H^q_k \qquad \forall\, k \in N \tag{24}$$

$$\sum_{q\in\Gamma_k} H^q_k = X_{kk} \qquad \forall\, k \in N \tag{25}$$

$$O_k X_{kl} + \sum_{i\in N} Y^i_{kl} + D_l X_{lk} \le \sum_{u\in U} b^u_{kl} v^u_{kl} \qquad \forall\; k,l \in N, k \neq l \tag{26}$$

$$X_{ik} \in \{0,1\} \qquad \forall\, i,k \in N \tag{27}$$

$$Y^i_{kl} \in \mathbb{Z}^+ \qquad \forall\, i,k,l \in N \tag{28}$$

$$v^u_{kl} \in \mathbb{Z}^+ \qquad \forall\, k,l \in N,\ u \in U \tag{29}$$

$$H^q_k \in \{0,1\} \qquad \forall\, k \in N, q \in \Gamma_k \tag{30}$$

The objective (18) minimizes the cost for transportation and hubs. Constraints (19) assure that each non-hub node is allocated to a single hub node. Constraints (20) are the flow conservation equations at hubs. Constraints (21), (22), and (23) ensure that hub nodes are established for each collection, transfer, and distribution movement. Constraints (24) are capacity constraints for hub nodes. The following constraints (25) ensure that a single capacity level is chosen for each established hub node. Constraints (26) assure that the flow on each arc (k,l) cannot exceed the capacity of vehicles assigned to it. Finally, constraints (27)–(30) are domain constraints.

In Table 2 we present some reprocessing tests for the formulation above. Similar to the result of [6] we show that constraints (21) are not needed in the case of integer decision variables:

Table 2. Preprocessing Tests for (F_{SA})

	Condition	Fixed Variables
Test 1	$O_k > \gamma_k^q$ for $k \in N$ and $q \in \Gamma_k$	$H_k^q = 0$
Test 2	$O_i + O_k > \gamma_k^{s_k}$ for $i,k \in N, i \neq k$	$X_{ik} = 0$
Test 3	$O_k > \gamma_k^{s_k}$ for $k \in N$	$H_k^{s_k} = 0$
		$X_{ik} = 0$ for all $i \in N$
Test 4	$k = l$ for $k,l \in N$	$v_{kl}^u = 0$ for all $u \in U$
		$Y_{kl}^i = 0$ for all $i \in N$
Test 5	$i = l$ for $i,l \in N$	$Y_{kl}^i = 0$ for all $k \in N$
Test 6	$X_{ik} = 0$ for $i,k \in N$	$Y_{kl}^i = 0$ for all $l \in N$

Theorem 1. *Constraints (21) in formulation (F_{SA}) are redundant.*

Proof. Let $X_{kk} = 0,\ k \in N$. Due to constraints (25) we have $\sum_{q\in\Gamma_k} H_k^q = 0$ and hence $H_k^q = 0$ for all $q \in \Gamma_k$. Therefore, constraints (24) imply $\sum_{i\in N}\left(O_i X_{ik} + \sum_{l\in N} Y_{lk}^i\right) = 0$ and in particular $X_{ik} = 0$ for all $i \in N$. Let $X_{ik} = 1,\ i,k \in N$. Then $\sum_{i\in N} O_i X_{ik} > 0$ and also, $\sum_{i\in N}\left(O_i X_{ik} + \sum_{l\in N} Y_{lk}^i\right) > 0$. Hence, constraints (24) imply $\sum_{q\in\Gamma_k} \gamma_k^q H_k^q > 0$. Therefore, it is$\sum_{q\in\Gamma_k} H_k^q > 0$. Because of constraints (25) it holds $X_{kk} > 0$ and thus $X_{kk} = 1$. □

Due to this result, we can remove constraints (21) from (F_{SA}). Hence, this reduces the number of constraints, but the linear relaxation of the problem might be weaker.

2.4 Further Modifications

If we remove the constraints regarding vehicles, introduce costs concerning discount factors for collection, transfer, and distribution, limit only the incoming flow of non-hub nodes at hubs, and do not restrict the X, Y, and Z variables to be integer, we get a similar formulation to the one of [6] in the case of multiple allocation, the (CMAHLPM).

The following notation is introduced:

d_{ij}	distance from node i to node j $(i, j \in N)$
χ	unit costs for collection (origin-hub)
α	unit costs for transfer (hub-hub)
δ	unit costs for distribution (hub-destination)

In general, α is used as a discount factor to provide reduced costs per unit on arcs between hubs to reflect the economies of scale. We assume $\alpha < \chi$ and $\alpha < \gamma$. Besides, we suppose symmetric distances, i. e., $d_{ij} = d_{ji}(i, j \in N)$. The triangle inequality holds for the distances: $d_{ij} \leq d_{ik} + d_{kj}$ for all $i, j \in N$ and all $k \in N$. Correia et al. consider in [6] only non-processed flow incoming at the hubs, i. e. incoming flow from non-hub nodes and the flow originated at the hub itself. Hence, constraints (9) are replaced by constraints (32). Besides, the flow variables are not restricted to be integer.

The mixed integer formulation for a CMAHLPM version is the following:

$$\min \sum_{i\in N}\sum_{k\in N}\left(\chi d_{ik} X_{ik} + \sum_{l\in N} \alpha d_{kl} Y_{kl}^i + \sum_{j\in N} \delta d_{kj} Z_{kj}^i\right) + \sum_{q\in\Gamma_k} c_k^q H_k^q \tag{31}$$

$$\text{s. t. } (2)-(8),(10),(16),(17)$$

$$\sum_{i\in N} X_{ik} \le \sum_{q\in\Gamma_k} \gamma_k^q H_k^q \qquad \forall\, k\in N \qquad (32)$$

$$X_{ik} \ge 0 \qquad \forall\, i,k\in N \qquad (33)$$

$$Y_{kl}^i \ge 0 \qquad \forall\, i,k,l\in N \qquad (34)$$

$$Z_{kj}^i \ge 0 \qquad \forall\, i,j,k\in N \qquad (35)$$

3 Computational Experiments

In this section we present computational experiments for formulations (F_{MA}) and (F_{SA}). First, we describe the construction of the test data. Afterwards, we present the computational results. We tested both formulations with the same test data sets.

3.1 Test Data

The test data sets are generated randomly as follows: The integer origin-destination demand W_{ij} $(i,j\in N)$ is randomly chosen between 0 and 7. Let V be the total transport volume, $V=\sum_{i\in N}\sum_{j\in N} W_{ij}$. The capacity in hubs is constructed successively. For each potential hub node $k\in N$ the lowest capacity level γ_k^1 is randomly chosen between 40 and 100 % of the total transport volume: $\gamma_k^1 \in [0.4V, V]$. Each capacity level is reduced by 20 % of the capacity of the level directly above: $\gamma_k^q = 0.8\gamma_k^{q+1} \Leftrightarrow \gamma_k^{q+1} = 1.25\gamma_k^q$, for all $q,q+1\in\Gamma_k$. The costs of the highest capacity level at a hub is randomly chosen between 10 and 15 Euro multiplied with the corresponding hub capacity: $c_k^{S_k} \in \left[10\gamma_k^{S_k}, 15\gamma_k^{S_k}\right]$. The cost (per unit) of a lower capacity level is 10 % higher than the cost of the upper level: $c_k^q = 1.1\gamma_k^q \frac{c_k^{q+1}}{\gamma_k^{q+1}}$. In all test scenarios, we consider only two or three different vehicle types on arcs. In the case of two vehicle types the capacities are given by 20 for the first (smaller) vehicle and by 30 for the second one. In the case of three vehicle types the capacities of the vehicles are 15, 23, and 30. Hence, we have $b^u = b_{kl}^u$ for all $k,l\in N$ and $u\in U$, whereas b^u is the capacity of a vehicle of type u on each arc. In each case, the costs of the first vehicle are randomly chosen between 300 and 20,000 Euro. The costs from a node k to l are approximately the costs of l to k with a tolerance of 5 %: $c_{lk}^1 \in \left[0.95c_{kl}^1, 1.05c_{kl}^1\right]$. The costs of the other vehicles are constructed successively: $c_{kl}^{u+1} = 0.9b^{u+1}\frac{c_{kl}^u}{b^u}, k,l\in N, u,u+1\in U$. Hence, we have $c_{kl}^2 = 1.35c_{kl}^1$ in the case of two different vehicle types, and $c_{kl}^2 = 1.38c_{kl}^1$ and $c_{kl}^3 = 1.62c_{kl}^1$ in the case of three different vehicle types.

We generated several test scenarios in order to test and compare the influence of the different parameters. The test data sets differ in the number of hub types, the number of capacity levels and the number of vehicle types, cf. Table 3. These parameters influence the number of decision variables and constraints as well as the running time and gap, cf. Subsection 3.2.

Table 3. Test data sets

Scenario	Number nodes	Capacity levels	Vehicle types
01	10	2	2
02	10	2	3
03	10	3	2
04	10	3	3
05	20	2	2
06	20	2	3
07	20	3	2
08	20	3	3
09	30	2	2
10	30	2	3
11	30	3	2
12	30	3	3

3.2 Computational Results

We implemented the resulting integer programs in GAMS 23.3.3 and solved them with CPLEX 12.1.0. All tests were carried out on a PC under Windows 7 (64bit) with an Intel Core i5 650 CPU, 3,2 GHz and 8 GB RAM. We tested all constructed data sets for both formulations, whereas we set the time limit to 2 hours. The computational results of testing formulation (F_{MA}) are given in Table 4 and of testing formulation (F_{SA}) without constraints (21) in Table 5. In both tables we list the gap after a run-time of 2 hours as well as the gap of the first and last feasible solution found. Besides, we specify the number of constraints and decision variables for both formulations and all tested data sets, cf. Tables 4 and 5. The gap is calculated by $\frac{BF-BP}{BF}$, where BF is the objective function value of the current best integer solution and BP is the best possible integer solution (i.e. the current best lower bound).

In particular, the number of nodes has a great influence on the number of variables and constraints, cf. Tables 4 and 5. In both test series feasible solutions are found for each scenario, but finding optimal solutions is hardly possible for any test scenario. Only for two small data sets (scenario 03 and 04 in the single allocation case) optimal solutions are found. In all other tests no (provable) optimal solution is found.

In both test series we observe that the run time until a feasible solution is found increases with an increasing number of nodes. In the first four scenarios feasible solutions are found immediately, whereas the gap of the first feasible solutions found are relatively high. Nevertheless, the final gap in each of these scenarios is relatively small. An increase of the number of nodes by 10 results in larger run-times until feasible solutions are found. In the single allocation case, feasible solutions are found within 19 and 22 seconds for scenarios 05 - 08. In comparison to this, feasible solutions for the multiple allocation formulation are found relatively late. The run-time until a feasible solution is found is between 73 and 111 seconds. This disparity in the run-time is more significantly in the last four scenarios. In the multiple allocation case, feasible solution are found after run-times between 1,464 and 1,781 seconds, while in the single allocation case the run-time until feasible solutions are found is between 391 and 663 seconds.

Table 4. Computational results of solving(F_{MA}) optimally (run-time 2h)

Scenario	Decision variables	Constraints	Time(sec)/gap(%)	Time(sec)/gap(%) (*first feasible solution*)	Time(sec)/gap(%) (*last feasible solution*)
01	4,261	3,421	7,200 / 7.45	0 / 66.90	284 / 10.38
02	4,471	3,421	7,200 / 5.46	1/ 44.67	6,514 / 5.51
03	4,291	3,421	7,200 / 3.59	0 / 53.57	44 / 10.15
04	4,481	3,421	7,200 / 1.73	0 / 41.60	2,163 / 2.44
05	32,621	25,641	7,200 / 7.58	73 / 23.99	5,509 / 7.67
06	33,381	25,641	7,200 / 4.89	111 / 19.31	6,486 / 4.92
07	32,801	25,641	7,200 / 2.65	87 / 26.06	5,793 / 2.69
08	33,641	25,641	7,200 / 3.48	85 / 18.70	2,849 / 3.75
09	109,021	84,661	7,200 / 6.88	1,781 / 29.14	6,701 / 7.22
10	111,121	84,661	7,200 / 20.19	1,623 / 38.64	5,744 / 20.22
11	109,081	84,661	7,200 / 18.81	1,464 / 33.76	1,792 / 18.81
12	110,791	84,661	7,200 / 18.46	1,558 / 32.40	7,062 / 18.46

In scenario 05 of the single allocation case the gap of the first feasible solution found is relatively high with a gap of 79.60 % after a run-time of 19 seconds. However, the second feasible solution is found after 21 seconds with a gap of 8.70 %. Similar results can be observed for scenarios 06 and 07. In these cases, the first feasible solutions are found after 22 seconds with a gap of 25.25 % and after 21 seconds with a gap of 23.61 %, respectively. However, next feasible solutions are found after a run-time of 25 seconds with a gap of 10.96 % and after 26 seconds with a gap of 9.27 %, respectively.

In both test series and for all test data sets, only small improvements of the objective value are observed until the time-limit is reached. For example, in scenario 07 in Table 5 the last feasible solutions is found after 290 seconds with a gap of 0.92 %. Nevertheless, the gap decreases only by 0.53 % until the time-limit of 7,200 seconds is reached. In scenario 11 in Table 4 no improvement of the gap is observed after the last feasible solution is found after 1,792 seconds. Similar results, i. e. only minimal decrease of the gap, are observed in all test scenarios. Hence, verification of optimality is highly time-consuming. Remarkable is, that in the multiple allocation case an increase of the number of hub nodes by 10 results in a great increase of the gap, cf. scenario 09 - 12 in Table 4. In the single allocation case, similar results are observed. In these cases, however, the gap is relatively small in comparison to the multiple allocation case.

The comparison of the results of both formulations shows that the number of decision variables and constraints is notably larger in the multiple allocation case. The number of constraints is about 50 % and the number of decision variables is 70–90 % higher. This is due to the greater flexibility of multiple allocation formulations. Hence, the greater flexibility results in higher (final) gaps. In each scenario the gap after a run-time of 2 hours is (significantly) larger in the multiple allocation case, in particular in the last four scenarios. Nevertheless, a comparison of the optimal objective values shows that in almost all scenarios, except scenario 10, the objective value after a run-time of 2 hours is lower in the multiple allocation case while the final gap is significantly larger.

Table 5. Computational results of solving (F_{SA}) without constraints (21) optimally (run-time 2h)

Scenario	Decision variables	Constraints	Time(sec)/gap(%)	Time(sec)/gap(%) (*first feasible solution*)	Time(sec)/gap(%) (*last feasible solution*)
01	2,441	2,221	7,200 / 2.77	0 / 71.80	6,807 / 2.82
02	2,591	2,221	7,200 / 4.06	0 / 66.78	6,521 / 4.10
03	2,431	2,221	2,158 / 0.00	0 /65.39	949 / 1.94
04	2,611	2,221	4,838/ 0.00	0 / 67.10	4,754 / 0.07
05	17,621	16,841	7,200 / 0.86	19 / 79,60	3,370 / 1.93
06	18,381	16,841	7,200 / 1.53	22 / 25.25	1,874 /1.70
07	17,661	16,841	7,200 / 0.39	21 / 23.61	290 / 0.92
08	18,421	16,841	7,200 / 1.25	22 / 7.30	6,500 / 1.26
09	57,631	55,861	7,200 /1.74	427 / 9.59	6,897/1.80
10	59,371	55,861	7,200 / 7.13	419 / 13.05	7,146/ 7.14
11	57,691	55,861	7,200 / 7.04	391 / 12.89	6,597 /7.20
12	59,431	55,861	7,200 / 7.80	663/15.47	5,793 / 8.51

These results let assume, however, that greater (more realistic) test data sets would lead to higher computing times and gaps as well as an increasing complexity. Hence, refinements of the modeling approach and development of specific algorithms to solve larger problems efficiently are necessary in order to solve realistic test data sets (near-) optimal. In particular, further research should be done on alternative formulations of the problems, which should be compared (e.g., in terms of the bound provided by the linear relaxation). More research should also be done on redundant constraints as well as constraints which enhance the model.

4 Summary and Outlook

In this paper, we extend the classical hub location problem. We consider a single and a multiple allocation version of a capacitated hub location problem with several capacity levels in hubs and the choice of different vehicle types on arcs. We formulate the problem as a multi-commodity flow problem with additional constraints. Additionally, we present some preprocessing tests for both versions. We implemented the resulting integer programs in GAMS 23.3.3 and solved them with CPLEX 12.1.0 for test data sets with 10 to 30 nodes. We compare the computational results. For hardly any test scenario an optimal solution is found within a run-time of 2 hours. In general, only small improvements in the objective value are observed after first feasible solutions are found. The multiple allocation formulation increases the flexibility of the problem but also increases the number of constraints and decision variables significantly. Therefore, the tests of the multiple allocation formulation result in higher run-time until feasible solutions found and hence in higher gaps. A comparison of the optimal objective values shows that in almost all scenarios the objective values are lower in the multiple allocation than in the single allocation case although the final gaps are larger in the first test serial. These results lead to the conclusion that larger test data sets would result in higher computing times and gaps as well as in an increasing complexity. In order to find (near-) optimal solutions for realistic data sets refinements of the modeling approach and development of specific algorithms to solve larger problems efficiently are necessary. Hence, further research should be done on alternative formulations of the problems which should be compared (e.g., in terms of the bound provided by the linear relaxation). More research should also be done on redundant constraints as well as constraints which enhance the model.

Acknowledgement

The German Research Foundation (DFG) supported this research ("Diskrete Optimierungsmodelle und Algorithmen zur strategischen Planung von Transportnetzen im Einzelwagenverkehr"). This support is gratefully acknowledged.

References

1. Alumur, S., Kara, B.Y.: Network hub location problems: the state of the art. Eur. J. Oper. Res. 190, 1–21 (2008)
2. Campbell, J.F.: Integer programming formulations of discrete hub location problems. Eur. J. Oper. Res. 72, 387–405 (1994)
3. Campbell, J.F., Ernst, A.T., Krishnamoorthy, M.: Hub location problems. In: Drezner, Z., et al. (eds.) Facility Location: Applications and Theory, pp. 373–407. Springer, Berlin (2002)
4. Campbell, J.F., Ernst, A.T., Krishnamoorthy, M.: Hub arc location problems: part I – introduction and results. Management Science 51, 1540–1555 (2005)
5. Campbell, J.F., Ernst, A.T., Krishnamoorthy, M.: Hub arc location problems: part II – formulations and optimal algorithms. Management Science 51, 1556–1571 (2005)
6. Correia, I., Nickel, S., Saldanha-da-Gama, F.: Single-assignment hub location problems with multiple capacity levels. Transportation Research B 44, 1047–1066 (2010)
7. Correia, I., Nickel, S., Saldanha-da-Gama, F.: The capacitated single-allocation hub location problem revisited: A note on a classical formulation. European Journal of Operational Research 207, 92–96 (2010)
8. Ernst, A.T., Krishnamoorthy, M.: Efficient algorithms for the uncapacitated single allocation *p*-hub median problem. Location Science 4, 139–154 (1996)
9. Ernst, A.T., Krishnamoorthy, M.: Exact and heuristic algorithms for the uncapacitated multiple allocation *p*-hub median problem. European Journal of Operational Research 104, 100–112 (1998)
10. Gelareh, S.: Hub Location Models in Public Transport Planning. PhD-Thesis. Technische University Kaiserslautern (2008)
11. Hall, R.W.: Configuration of an overnight package air network. Transpn. Res.-A 23, 139–149 (1989)
12. O'Kelly, M.E.: A quadratic integer program for the location of interacting hub facilities. Eur. J. Oper. Res. 32, 393–404 (1987)
13. O'Kelly, M.E., Bryan, D.L.: Hub location with flow economies of scale. Transpn. Res.-B 32, 605–616 (1998)
14. Sender, J., Clausen, U.: Strategic network design for wagonload traffic in German railway logistics. In: Sumalee, A., Lam, W.H.K., Ho, H.W., Siu, B. (eds.) Proceedings of the 15th International Conference of Hong Kong Society for Transportation Studies (HKSTS), Hong Kong, pp. 255–262 (2010)
15. Wagner, B.: Hub&Spoke-Netzwerke in der Logistik. Modellbasierte Lösungsansätze für ihr Design. DUV, Wiesbaden (2006)

A Stochastic Optimization Model for Positioning Disaster Response Facilities for Large Scale Emergencies

Anurag Verma and Gary M. Gaukler

Texas A&M University, College Station, TX, USA
{anuragverma,gaukler}@tamu.edu

Abstract. In this paper, we present a two-stage stochastic program for locating disaster response facilities. Our modeling approach is unique in the literature in that it explicitly correlates the functioning of a facility with a particular disaster scenario. In particular, in our model the functioning of a facility is directly affected by its distance from the disaster epicenter. This represents an important modeling aspect of emergency facility location that to date has been ignored in the existing literature. To demonstrate the potential contributions of our model, we present a computational case study of (earthquake) disaster response facility location for the state of California. Our computational results show the distinct changes in the optimal location of facilities. Instead of placing facilities directly on top of some of the highest-risk areas (the traditional k-median solution), our model tends to place facilities still close to population centers, but farther away from high-risk areas.

1 Introduction

Recent years have brought forth a number of devastating large-scale emergency situations, such as Hurricanes Rita and Katrina that affected the United States as well as portions of the Carribbean in 2005, or the earthquakes in Sichuan, China in 2008, and in Haiti in 2010. In the United States, this has prompted policy makers at the federal and at the state level to coordinate the establishment of disaster response facilities. At these facilities, supplies (food rations, medical items, rescue equipment) are stored for use in the event of a large-scale emergency. An example of this pre-positioning strategy is the creation of the Strategic National Stockpile (SNS) [6]. A crucial aspect of such pre-positioning strategies is the judicious choice of the pre-positioning sites, which is a facility location problem.

In the existing literature, such emergency facility location has been subject to the application of traditional techniques using the k-center and k-median models. There exists literature on facility location with unreliable facilities [2,3,12]; however, these models are formulated for a conventional facility location setting, and not the scenario based setup that emerges in disaster response modeling. Emergency location models presented in [8,9,10,7,1,14] either do not model damage to facilities, or account for it by modeling the available capacity of a facility as a constant fraction of the total capacity, or account for the available capacity in a deterministic manner. We provide a model

J. Pahl, T. Reiners, and S. Voß (Eds.): INOC 2011, LNCS 6701, pp. 547–552, 2011.

that (1) acknowledges the direct effect a disaster can have on the facilities themselves; (2) includes the inherent stochastic nature of the occurrence and effects of large scale emergencies; and (3) considers the occurrence of a disaster to possibly encompass more than one node in the network.

A key consideration that makes emergency facility location significantly different from a regular k-median approach is the dependence of the availability of a facility on a particular disaster scenario. Under normal circumstances, there is no demand for emergency supplies. The occurrence of a disaster generates a localized demand, while simultaneously reducing the supplying capabilities of facilities in that area. We model this by correlating the functioning of a facility with each disaster scenario.

Furthermore, we distinguish between potential disaster locations (called epicenters), the points of interest like cities (the demand points) and the potential facility locations. As a result, we are able to frame a model in which a disaster at any location can affect nearby demand points and the disaster response facilities that were built. An implication of this framework is that now more than one demand point can be affected by a disaster scenario, unlike a regular k-median formulation [8].

We address the stochastic nature of the problem by formulating it as a 2-stage stochastic programming model, with the setup of the facilities being decided in the first stage. A number of transportation subproblems are solved in the second stage, depending on the demand for relief and the reduction in capabilities of the facilities due to structural damage caused by the disasters.

Some information that our model requires as input from a decision maker are the potential disaster locations with the probability of a disaster happening there, and an estimate of the effect a disaster will have on the nearby demand points and facilities. We provide a case study using our model for locating disaster response facilities to pre-position supplies for earthquake relief in the state of California.

2 The Stochastic Distance Dependent Model

This section formally presents the mathematical formulation of our problem as a 2-stage stochastic program with binary first stage. The first stage decision is to choose the locations of a given number (k) of facilities. The second stage decisions are the amounts to be routed from the opened facilities to the demand points for each disaster scenario, once the demands and reduced capability of the facilities are known.

For simplicity, and without loss of generality, the demands D_i at demand point $i \in I$ and capacities c_f of a facility $f \in F$ are represented in the same units. These units could represent, for example, the number of people whose needs have to be fulfilled, if there is one emergency supply packet stored per person to be served. The probability of a disaster occurring at and epicenter $e \in E$ is denoted by w_e. The quantity $\bar{d}_{if}$ denotes the travel distance from facility f to demand i, whereas d_{ex} denotes the Euclidean distance between e and $x \in I \cup F$, and is used to estimate the damage at x due to a disaster at e. The random variable $\tilde{p}_{ex}$, which is a function of d_{ex}, denotes the damage at $x \in I \cup F$ as a fraction of the total demand/capacity. It should be noted that given d_{ex}, we assume that we can obtain the distribution of $\tilde{p}_{ex}$.

In our model, the binary variable x_f denotes the decision to open a facility at location $f \in F$. The variable y_{eif} denotes the amount of service provided by facility f to demand point i when a disaster occurs at e. We now present our formulation,

$$\min \quad E_{\tilde{p}}[f(x,\tilde{p})] \tag{1}$$

$$s.t \quad \sum_{f \in F} x_f \leq k, \quad x \in \{0,1\}^{|F|} \tag{2}$$

Where for a particular realization p of damage $\tilde{p}$, we have

$$f(x,p) = \sum_{e \in E} \sum_{f \in F} \sum_{i \in I} w_e \bar{d}_{if} y_{eif} \tag{3}$$

$$s.t. \quad \sum_{i \in I} y_{eif} \leq (1-p_{ef}) c_f x_f \qquad \forall e \in E, \forall f \in F \tag{4}$$

$$\sum_{f \in F} y_{eif} \geq p_{ei} D_i \qquad \forall e \in E, \forall i \in I \tag{5}$$

$$y_{eif} \geq 0 \qquad \forall e \in E, \forall i \in I, \forall f \in F \tag{6}$$

Note that an expectation over all the possible disaster scenarios is already implicit in the objective function (3) of the subproblem. The only random variables we take an expectation over in the objective function (1) of the master problem are the damages to facilities and cities in each of the disaster scenarios.

3 Computational Results

For a computational study of our model, we solve the problem of locating facilities to pre-position earthquake response supplies such as potable water, tents, food packets for the state of California, USA. Earthquakes were chosen for their large geographical impact that can damage facilities as well as multiple cities. We chose California due to the availability of past earthquake data as well as the presence of significant research in forecasting the occurrence of earthquakes in California [11].

We use the 20 largest cities (by population) in California as the demand as well as potential facility locations, with demands proportional to their populations. In addition, 35 grid points with integer longitude and latitude values within Southern and Central California serve as additional potential facility locations. The earthquake scenarios are constructed by using historic data of the 23 earthquakes of magnitude larger than 6 on the Richter scale that occurred in and around California since 1973 [13]. An expectation over all these 23 scenarios (assumed equiprobable) is taken in the second stage of the stochastic programming formulation. The data used is available at `http://people.tamu.edu/~anuragverma/SEFLdata`.

The damage due to these earthquakes is modeled using the intensity-distance function provided in [4]. For the purpose of our computational study, we assume the damage to be proportional to the intensity, and be given as a function of the distance from the epicenter as $p_{ex}(d_{ex}) = 0.69e^{(0.364-0.130\ln(d_{ex})-0.0019d_{ex})}$ where d_{ex} is measured in kilometers [4]. To incorporate stochasticity of earthquake damages, $\tilde{p}_{ex}$ is modeled as a

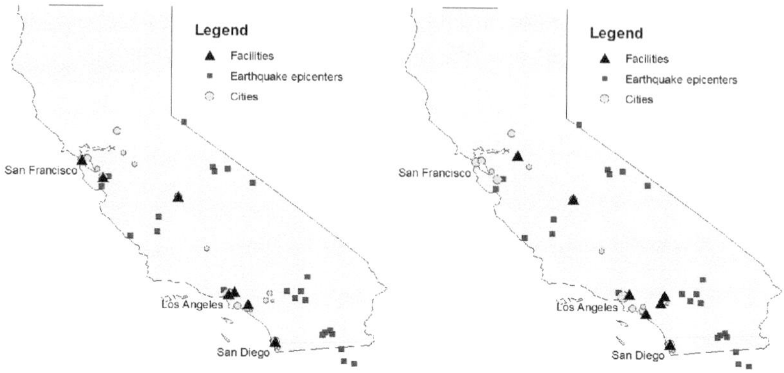

(a) k-median locations **(b)** Locations found by our model

Fig. 1. Locations of 7 facilities with capacity 800,000 as found by the k-median model and our model

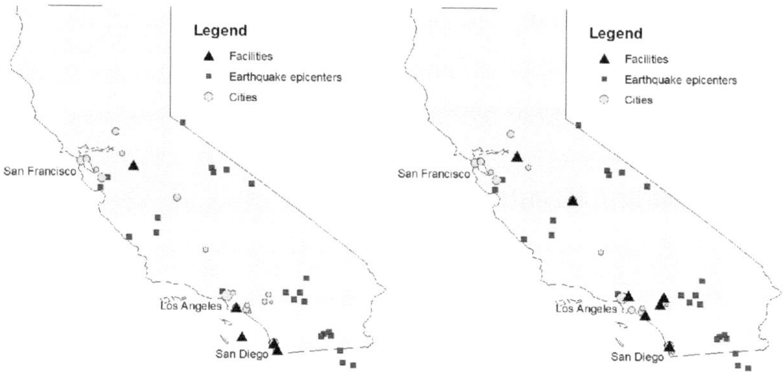

(a) 5 facilities **(b)** 7 facilities

Fig. 2. Locations of 5 & 7 facilities with capacity 800,000 as found by our model

random variable with values p_{ex}, $p_{ex}/2$ and $3p_{ex}/2$ with probabilities 0.6, 0.2 and 0.2 respectively.

To gauge the effects of capacity and number of facilities on their optimal location, we ran our model for placing 5,7, and 9 facilities with capacities 700,000, 800,000, 900,000, and 1,000,000 each. To compare the location of facilities suggested by our model to those in the existing literature, we implemented the regular k-median model presented in [8], where the damage to facilities is independent of distance from the disaster epicenter and the same in all scenarios. Due to computational time constraints, the findings we report for our stochastic model in this section are based on the incumbent solution found after running the L-shape algorithm [5] for 12,000 seconds on a workstation with a 2.66 GHz CPU with 4GB memory.

A snapshot of the results obtained is presented in Figures 1 and 2. From the figures, it can be observe that Los Angeles (LA) and the San Francisco (SF) Bay Area are potential high risk areas where a large city is located near an earthquake epicenter. From figure 1, which compares the placement of 7 facilities with capacities 800,000 using different models, we can observe that compared to the k-median model, our model places facilities at a safer distance from the epicenters. More specifically, a facility placed at an earthquake prone city near SF by the k-median model is moved farther away to a relatively safer city by our model. Furthermore, one of the two facilities in the SF Bay Area is moved near LA, which is earthquake prone and has a large population. Again, in our model facilities are moved from the immediate vicinity of LA to relatively safer but still close locations.

Figure 2 presents a sensitivity analysis of the solutions to our model. Figures 2a and 2b present the locations of 5 and 7 facilities with capacity 800,000 each. We observe that when a larger number of facilities is available to be placed, the model places facilities at more convenient, but riskier locations. This is because in such a situation, the impact of facility damage is less drastic due to facility redundancy.

4 Conclusion

This paper presents a 2-stage stochastic program for locating disaster response facilities. Our modeling approach is unique in the literature in that it explicitly correlates the functioning of a facility with a particular disaster scenario. This represents an important modeling aspect of emergency facility location that to date has been ignored in the existing literature.

Our research provides the quantitative tools to aid policy makers in the decision process for disaster response facility location. Our computational results show the distinct changes in the optimal facility locations under our new set of assumptions about damage to facilities. Furthermore, results of a sensitivity analysis on the number of facilities suggests a direct connection between available capacity and the proximity of optimal locations to high-risk areas. This suggests that our modeling approach will be especially useful in light of realistic, tight budgetary constraints.

Acknowledgement

We thank Preeti Shankar for preparing the graphics for visual representation of the results using Geographic Information System (GIS) software.

References

1. Balcik, B., Beamon, B.: Facility location in humanitarian relief. International Journal of Logistics: Research & Applications 11(2), 101–121 (2008)
2. Berman, O., Drezner, Z., Wesolowsky, G.: Locating service facilities whose reliability is distance dependent. Computers & Operations Research 30(11), 1683–1695 (2003)
3. Berman, O., Krass, D., Menezes, M.: Facility reliability issues in network p-median problems: Strategic centralization and co-location effects. Oper. Res. 55(2) (2007)

4. Howell Jr, B.F., Schultz, T.: Attenuation of modified mercalli intensity with distance from the epicenter. Bulletin of the Seismological Society of America 65, 651–665 (1975)
5. Birge, J., Louveaux, F.: Introduction to Stochastic Programming, 1st edn. Springer, Heidelberg (1997)
6. CDC: `http://www.bt.cdc.gov/stockpile/`
7. Duran, S., Gutierrez, M., Keskinocak, P.: Pre-positioning of emergency items worldwide for care international (2009)
8. Jia, H., Ordonez, F., Dessouky, M.: A modeling framework for facility location of medical services for large-scale emergencies. IIE Transactions 39(1), 41–55 (2007)
9. Paul, J., Batta, R.: Models for hospital location and capacity allocation for an area prone to natural disasters. International Journal of Operational Research 3(5), 473–496 (2008)
10. Rawls, C., Turnquist, M.: Pre-positioning of emergency supplies for disaster response. In: IEEE International Symposium on Technology and Society, ISTAS 2006, June 2006, pp. 1–9 (2006)
11. Rundle, J.B., Turcotte, D.L., Shcherbakov, R., Klein, W., Sammis, C.: Statistical physics approach to understanding the multiscale dynamics of earthquake fault systems. Rev. Geophys. 41 (February 2003)
12. Snyder, L.V., Daskin, M.S.: Reliability models for facility location: The expected failure cost case. Transportation Science 39(3) (2005)
13. USGS: `http://earthquake.usgs.gov/earthquakes/eqarchives/epic/`
14. Verma, A., Gaukler, G.: Locating disaster response facilities for large scale emergencies. Working paper (2010)

Efficient Robust Linear Optimization for Large Repositioning Problems

Haris Gavranović and Mirsad Buljubašić

Faculty of Natural Sciences, University of Sarajevo
haris.gavranovic@google.com, mirsad_bulj@yahoo.com

Abstract. The economic uncertainty and demand volatility put the maritime carriers under operational pressure and the obligation of better planning and better benchmarking of solutions. One of the ways to tackle uncertainty is to operate with robust plans. This paper investigates a robust optimization framework and uses well-known time-expanded networks and appropriate linear programs to study and solve three variants of a robust optimization empty container repositioning problem. We propose a new, dynamic generation constraints approach and effectively solve all these variants for a real world look-like set of instances. We use a list of global maritime ports to build artificial instances while exploring computational limits of the classic and dynamic approach. We are able to solve instances with hundreds of millions of additional, linear constraints. Our experimentation shows that this approach is efficient as well as indispensable for such big instances.

1 Introduction

Companies throughout the shipping world carefully forecast and plan the use of empty containers. Despite these efforts, everybody expects small changes on these forecasts due to uncertainty and volatility of demand. Even a single change can have considerable negative impact on daily operations of the company. A series of small changes on the demand can completely destroy the previous plan and the shipping company will fail to satisfy new demands. As a result, the company could loose new contracts and clients. One of the efficient ways to address these uncertainties is to operate under robust plans, slightly more expensive than nominal ones but able to recover from unexpected events.

In this paper we show how to effectively solve three different robust optimization problems. We construct large real world instances and practically test our approach. Similar instances of such size have not been solved before. When the number of constraints becomes untractable for any existing computer architecture, we show how to dynamically create and use additional constraints and find optimal solutions.

The rest of the paper is organized as follows. This introduction finishes with the presentation of related literature and experimental studies. Section 2 presents nominal and three robust optimization models and the theorems that characterize the additional constraints. Section 3 presents efficient algorithms to create additional constraints and report experimental results when the number of constraints is over several millions.

Related Studies: Empty repositioning problems often appear in the operational studies of railway or maritime traffic or the combination of the two and they received a lot

J. Pahl, T. Reiners, and S. Voß (Eds.): INOC 2011, LNCS 6701, pp. 553–558, 2011.

of attention both by practitioners and by theoreticians. The first studies appeared in the sixties and seventies and the topic never went out of focus (for an overview consult [6]). Most of these studies presume complete and accurate information about the future. The most popular models, used even today within transportation companies, result in dynamic programming solutions and/or linear or easy to solve integer programs.

Uncertainty and inaccuracy of data always existed in reality but models start to incorporate it for the first time in [7] and than again with works of Powell (e.g. [5]). A possible remedy for uncertainty, the approach we study here, is robustness of solutions, or how [3] call it, solutions "immunized against uncertainty". Beside these general studies about uncertainty and robustness there are also more practical, though theoretically less ambitious works, where only the right hand side is uncertain [1,4]. Erera et al. [4] consider computational properties of the approach. We continue here along this path studying and solving models for similar, real world instances resulting in huge number of additional linear constraints.

2 Integer Programming Models

2.1 The Nominal Model

In this section we introduce the nominal model together with all necessary notations. When possible we use existing notations from [4]. The nominal empty repositioning problem can be modeled using a time-expanded network $G = (N, A)$ where N is a set of all depot-timestep combinations. A node representing depot d at time t is noted by v_t^d. The vector $\mathbf{b}$ contains time-space net supply forecasts and represents the right-hand side values of the model. The set A contains two types of arcs: inventory that goes within the same depot and repositioning between two different depots. For each arc $a \in A$, $c(a)$ is a unit cost of flow on arc a. In many applications, as well as here, the unit holding costs at depots are negligible and therefore equal to zero.

The nominal repositioning problem may now be written as:

$$\mathbf{NP}\ \min_{\mathbf{x}}\{\mathbf{cx} : \mathbf{Ax} = \mathbf{b}, \mathbf{x} \in Z_+^{|A|}\} \tag{1}$$

where the decision vector $\mathbf{x}$ corresponds to the empty container flow on each arc and $\mathbf{A}$ is the node-arc incidence matrix implied by G and thus defining the typical network flow-balance constraints $\mathbf{Ax} = \mathbf{b}$.

Variation of the Nominal model: In [4] the authors add to G a sink node T with net supply $b(T) = \sum_{v_t^d \in V} b(v_t^d)$ and an arc connecting v_p^d to T for all $d \in D$. It is easy to construct unfeasible instances, e.g., when the net supply in the sink node is negative. We propose therefore a variant of the nominal problem, $\mathbf{NP}'$ introducing a special source node S and we add an arc connecting S to $v_0^d \forall d \in D$. In fact, the transportation company can decide to borrow, buy or rent certain quantity of containers at the initial phase of the operation (at 0 timestep) and this quantity is modeled by the flow on the arcs going from node S. The cost on the newly added arcs is considerably bigger than any other cost in the problem. Finally, we add the only backward arc in the model from T to S with cost zero. The linear program is still a typical network flow problem.

2.2 Robustness

The nominal estimate of the net supply $b(v)$ at each node v in V is just an estimate and therefore uncertain, and we experiment with robust framework developed in [2,4]. The uncertainty set comprises all sets with at most k nodes whose estimates $b(v)$ oscillate by at most $\hat{b}(v)$ units.

Robust repositioning plan is typical feasible flow plan which is also recoverable under a limited set of recovery actions and for some predefined uncertain outcome. We study here three different sets of recovery actions in TRP1, TRP2 and TRP3 as they are presented in [4].

The nominal model serves as a base for TRP models. The principle of solving other models is simply adding a set of constraints to ensure the robustness of the nominal solution.

TRP1 model: TRP1 also known under the name of Inventory Robust Repositioning Problem is the easiest among the three. Recovery actions set comprises only repositioning actions along the inventory arcs. In other words the company will keep sufficiently big reserve on every depot in the problem.

The set of additional constraints is described with the following theorem.

Theorem 1. *[4] A feasible solution x for the nominal problem 1 is also feasible for TRP1 if and only if*

$$x(a) \geq \vartheta(V_t^d, k), \qquad \forall a = (v_t^d, v_{t+1}^d) \in I \tag{2}$$

where $\vartheta(V_t^d, k)$ is node set vulnerability for set of vertices V_t^d and k. There is one new constraint for every depot-timestep. It turns out that $TRP1$ is polynomially solvable using standard minimum cost network flow algorithm.

TRP2 model: In this model we suppose that the company tackles the uncertainty at a given depot not only using its own inventory but also using containers at other depots. The following theorem describes the set of additional constraints.

Theorem 2. *[4] A feasible solution x for the nominal problem 1 is also feasible for TRP2 if and only if for every set of nodes $U \in \mathcal{U}_{W_2}$*

$$\sum_{a \in \Delta^{out}(U) \cap I} x(a) \geq \vartheta(U, k) \tag{3}$$

W_2 represents the subgraph that includes all arcs where the flow could change in the case of perturbation in the prevision. W_2 for $TRP2$ is a set of all arcs in the model, except the inventory arcs emanating from timestep zero nodes. Following definitions 3 and 4 in [4], the family $\mathcal{U}_{W_2}$ contains all inbound-closed sets U such that $\Delta^{out}(U) \cap I$ is competing.

TRP3 model: In this scenario there are two types of depots: those that serve as providers and those serving as recipients. For the sake of this study we use the formulation of additional constraints as defined for TRP2 at the appropriate network. There exists also another, more efficient way to define, and construct, these constraints [4].

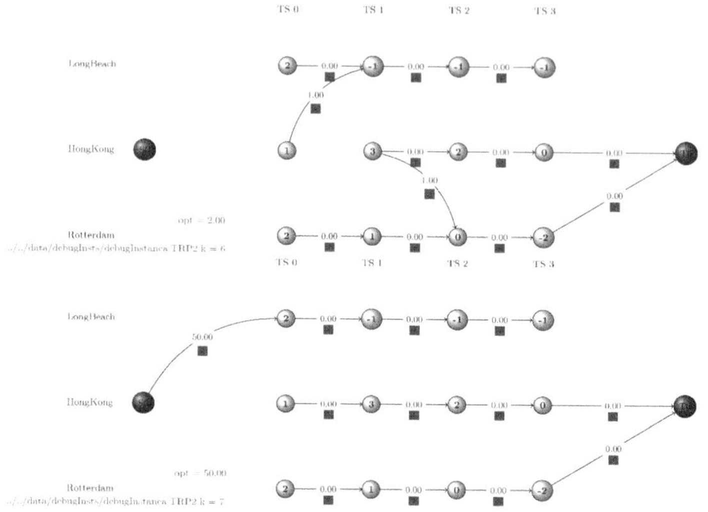

Fig. 1. The figure presents two optimal solutions for TRP2 problem. This toy problem has three ports : Long Beach, Hong Kong and Rotterdam and 4 timesteps. Only non-zero flows are presented. On the left is the solution with optimal value 2 for $k = 6$ while on the right is the one with optimal value 50 obtained for $k = 7$.

3 A Constraint Generation Linear Programming Approach

The proposed models result in linear programs that may have large, exponential number of constraints. This could be easily checked experimentally, as reported in figures 2 and 3. This is also already discussed in [4]. Huge number of constraints renders linear programs difficult, if not impossible to solve efficiently.

On the other hand, the ports in both real and experimental instances studied so far are naturally divided into several maritime regions resulting in small and tractable number of constraints. On our experimental machine (CPU:Intel Core i7 920 2.67GHz, RAM 6GB, Ubuntu) IBM-Cplex 11.2 routinely solves in several minutes instances having 60 ports divided in 8 regions with 50 timesteps.

The total number of constraints and therefore the time complexity of the solution is primarily influenced by the regional division of the ports but also partially by topology of the networks. The number of additional constraints is dominated by the number of constraints originated from the biggest region in the instance. This is why, in order to assess the computational limits of TRP2 and TRP3 models, it is enough to conduct tests with instances having a given number of ports all in one region. The number of constraint grows extremely fast and it becomes prohibitive for real world instances with more than 10 ports in a single region.

Note that all of the models have small, constant number of variables and fast growing number of additional constraints. We propose a simple yet computationally powerful and efficient way to solve such huge instances using constraint generation algorithm as follows

Algorithm 1. Constraint generation approach

Create Nominal model then choose and add a small subset of constraints and solve the resulting linear program
while there are unsatisfied constraints **do**
  Choose and add additional small subset of violated constraints
  optimize LP with the new set of constraints
  check if obtained solution satisfies all constraints
end while
{At the end of the while loop the algorithm checks if all additional linear constraints missing in the actual model are satisfied for the obtained solution. The constraints are checked one by one in the order of their creation and subset of violated constraints are kept and add to the model in the beginning of the next iteration.}

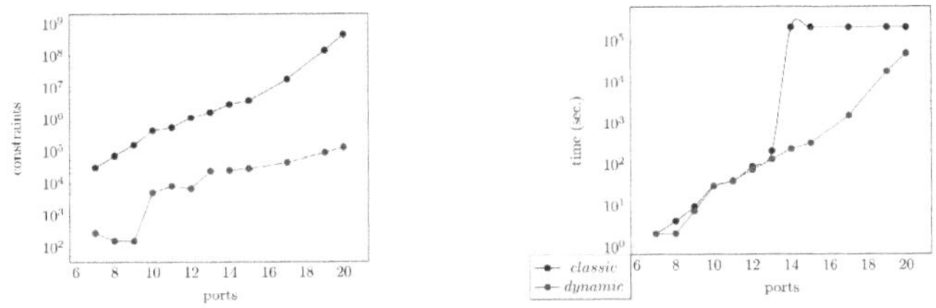

ports	iters
7	3
8	2
9	2
10	48
11	39
12	32
13	24
14	25
15	14
17	22
19	17
20	25

Fig. 2. The figures present number of constraints and solution time TRP2 model for instances with single region with 50 timesteps. Note that the graphics are semilog and it is not possible to solve, using classic approach, instances with more than 13 ports. The table presents the number of iterations in the while loop in the Algorithm 1.

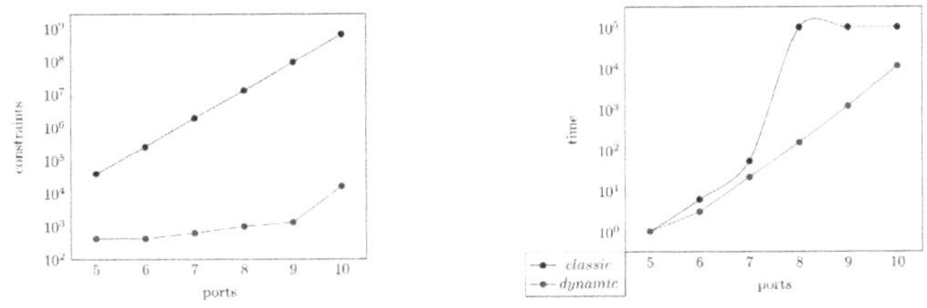

ports	iters
5	1
6	1
7	2
8	3
9	6
10	35

Fig. 3. These are results for TRP3 model. The number of additional constraints is really huge (in hundreds of millions). The instances have only 10 timesteps. It is not possible to solve, using classic approach, instances with more than 7 ports.

4 Conclusion

Despite the fact that the number of constraints for robust repositioning problems grows really fast it is still possible to solve real problems with a large number of ports using a dynamic generation constraints approach. We solved here the instances with hundreds of millions of additional, linear constraints. These instances are not solvable using a classical approach. Generation and checking of additional constraints is embarrassingly parallelizable procedure. It would be perspective to establish the computational limits for TRP problems using parallel approach.

References

1. Atamturk, A., Zhang, M.: Two-stage robust network flow and design under demand uncertainty. Operations Research 55, 662–673 (2007)
2. Ben-Tal, A., Goryashko, A., Guslitzer, E., Nemirovski, A.: Adjustable robust solutions for uncertain linear programs. Mathematical Programming A 99, 351–376 (2004)
3. Ben-Tal, A., Nemirovski, A.: Robust solutions of linear programming problems contaminated with uncertain data. Mathematical Programming A 88, 411–424 (2000)
4. Erera, A., Morales, J., Savelsbegh, M.: Robust optimization for empty repositioning problems. Operations Research 57, 468–483 (2009)
5. Powell, W.: A stochastic model of the dynamic vehicle allocation problem. Transportation Science 20, 117–129 (1986)
6. Powell, W.: Dynamic models of transportation operations. In: Graves, S., de Kok, T.A.G. (eds.) Handbooks in Operations Research and Management Science: Supply Chain Management, pp. 677–756. Elsevier, Amsterdam (2003)
7. Soyster, A.: Convex programming with set-inclusive constraints and applications to inexact linear programming. Operations Research 21, 1154–1157 (1973)

Robust Supply Vessel Planning

Elin E. Halvorsen-Weare[1] and Kjetil Fagerholt[2]

[1] Department of Industrial Economics and Technology Management,
Norwegian University of Science and Technology, Trondheim, Norway,
and Norwegian Marine Technology Research Institute (MARINTEK), Trondheim, Norway
elin.halvorsen-weare@iot.ntnu.no

[2] Department of Industrial Economics and Technology Management,
Norwegian University of Science and Technology, Trondheim, Norway
kjetil.fagerholt@iot.ntnu.no

Abstract. In the supply vessel planning problem, a set of offshore installations receives supplies from an onshore supply depot on a regular basis. This service is performed by a fleet of offshore supply vessels. The supply vessel planning problem then consists of determining the optimal fleet size and mix of supply vessels and the corresponding weekly voyages and schedules. This is a real planning problem faced by among others the energy company Statoil. In a previous study this problem was examined and a deterministic voyage based solution approach presented. In this study we address the problem of creating robust schedules to the supply vessel planning problem. Several approaches are tested and compared in a computational study, and the results show that there is an improvement potential if some robustness considerations are made when finding a solution to the supply vessel planning problem.

1 Introduction

Operators of offshore oil and gas installations need to have a reliable supply service from land. Interruptions of such service may in the worst case scenario result in temporarily shut-downs, which again will result in millions of USDs in lost income. The supply vessel planning problem consists of indentifying the optimal fleet of supply vessels to charter in to perform this service from one common onshore supply depot. At the same time the vessels' weekly routes and schedules need to be determined. A route is here defined as a combination of one or more voyages that a vessel sails during a week. Each voyage starts and ends at the supply depot and visits a number of offshore installations in between.

We study a real supply vessel planning problem faced by the energy company Statoil. In this problem, there are a number of practical constraints that must be considered. Each voyage has a minimum and maximum duration and a minimum and maximum number of offshore installations to visit. The offshore installations have an estimated weekly demand for goods (given in square meters of deck capacity onboard a supply vessel) and require a given number of visits per week. Some of the offshore installations have opening hours for when to perform unloading and loading operations of a supply vessel (typically closed for such operations at night). Each offshore installation has a service time which is the time it takes to unload and load a supply vessel. The departures

J. Pahl, T. Reiners, and S. Voß (Eds.): INOC 2011, LNCS 6701, pp. 559–573, 2011.

from the onshore supply depot to a given offshore installation should be evenly spread throughout the week.

The onshore supply depot is open for service eight hours a day (0800-1600), and has limited capacity so that only a limited number of supply vessels may be prepared for a new voyage on a given day. The turnaround time for a supply vessel is eight hours, which means that a vessel needs to be at the supply depot before 0800 to start on a new voyage the same day.

A supply vessel has a given deck capacity, service speed and time charter rate. The total demand for all offshore installations visited on a voyage cannot exceed the capacity of the vessel sailing the voyage.

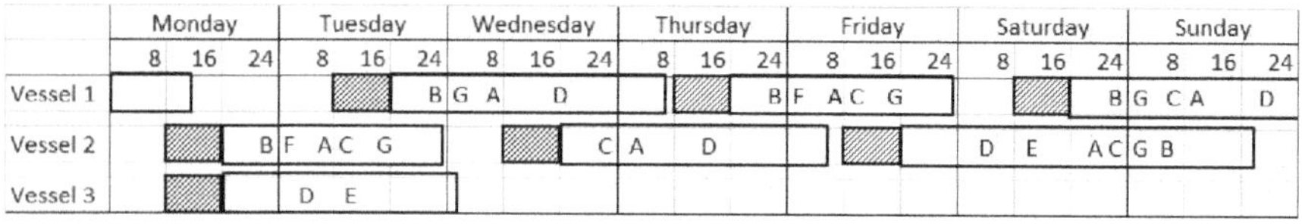

Fig. 1. A solution to the supply vessel planning problem

A solution to the supply vessel planning problem with three supply vessels and seven offshore installations (A, B,..., G) is illustrated in Figure 1. For a more thorough description of the supply vessel planning problem, we can refer to [15].

The supply vessel planning problem has similarities with other problems studied in the literature. It can be classified as a periodic fleet size and mix problem with time windows and multiple use of vehicles. Versions of the supply vessel planning problem are studied in [9] and [10]. The routing problem arising in the service of offshore installations is studied in [14]. Some references to the periodic vehicle routing problem are [5], [8], [16] and [24]. A recent survey on fleet composition and routing can be found in [17], and some studies of the vehicle routing problem with multiple use of vehicles are presented in [3], [4] and [23].

The energy company Statoil, that operates offshore installations in the North Sea and the Norwegian Sea, experiences that the supply service is highly affected by weather conditions. Especially during winter season there may be delays in the service due to rough weather conditions. In a previous study [15], a deterministic solution approach for the supply vessel planning problem was developed. A version of this model has been implemented and used by the planners in Statoil. This model has provided valuable decision support to the planning process, and savings of 3 million USDs were reported. However, the planners discovered that the planned schedules too often are difficult to execute in real life, resulting in replanning and extra costs involved when demand from installations need to be met in order to avoid production shut-downs. This revealed a need for creating schedules that are more robust and resilient to the prevailing weather conditions.

The purpose of this paper is to create robust solutions to the supply vessel planning problem. By robustness we mean the capability for a voyage or schedule to allow for unforeseen events during execution. Robust solutions will reduce the actual costs of the supply vessel service by avoiding expensive and unplanned means of bringing critical demand to the offshore installations. We aim to achieve robust schedules by combining simulation with the model from [15], and also testing other simpler approaches.

There are not many contributions in the literature that consider the uncertain elements in maritime transportation problems. Some references are [6] that studies a world-wide crude oil supply chain, [7] that tries to avoid vessels staying idle in weekend closed ports, [11] that combines simulation and optimization for strategic planning in shipping, and [18] that considers the risk of fluctuations in the spot market. A related topic is stochastic vehicle routing problems (SVPRs), were customers, demands and/or travel times are stochastic. A review of such problems can be found in [13], and the SVRP with stochastic travel and service times was studied in [20]. Some references on the combination of simulation and optimization are [2] and [12]. In [1] a solution approach combining optimization and simulation to an inventory routing problem is presented, and [19] and [21] present optimization and simulation approaches to variations of the SVRP.

The rest of this paper is organized as follows: A mathematical model for the supply vessel planning problem is presented in Section 2. Then Section 3 discusses the weather impact, and Section 4 describes different approaches to create robust schedules to the supply vessel planning problem. In Section 5 a computational study is provided. Finally the paper is concluded in Section 6.

2 Mathematical Model

In this section we present the mathematical formulation for the voyage based solution method from [15]. Let $\mathscr{V}$ be the set containing the supply vessels available for time charter, and let $\mathscr{N}$ be the set of offshore installations. Then set $\mathscr{R}_v$ is the set of pre-generated voyages that vessel $v \in \mathscr{V}$ may sail. Let $\mathscr{T}$ be the set of days in the planning horizon (a week), and $\mathscr{L}$ be a set containing all possible voyage durations in days (two or three) and let $\mathscr{H}$ be a set with all possible visit frequency values. Then set $\mathscr{R}_{vl}$ contains all candidate voyages of duration $l \in \mathscr{L}$ that vessel v may sail, and set $\mathscr{N}_k$ contains all offshore installations that require $k \in \mathscr{H}$ visits per week. C_v^{TC} represents the weekly time charter cost for vessel v, C_{vr}^S the sailing cost for vessel v sailing voyage $r \in \mathscr{R}_v$ while D_{vr} is the duration of voyage r sailed by vessel v in days (rounded up to nearest integer). S_i is the required weekly visit frequency to offshore installation $i \in \mathscr{N}$, F_v the number of days vessel v can be used during a week, B_t the number of supply vessels that may be serviced at the onshore supply depot on day $t \in \mathscr{T}$ and the binary parameter A_{vir} is 1 if vessel v visits offshore installation i on voyage r, and 0 otherwise. Further, $G_k \in [0, |\mathscr{T}|]$ is a number representing the length of a sub-horizon for the offshore installations with visit frequency k, and $\underline{P}_k$ and $\overline{P}_k$ are lower and upper bounds on the number of visits during the sub-horizon of length G_k an offshore installation with visit frequency k should receive. The variables are the binary variables δ_v that equals 1 if supply vessel v is chosen for time charter, 0 otherwise, and x_{vrt} that equals 1 if vessel v sails voyage r on day t, 0 otherwise.

The mathematical formulation then becomes:

$$min \sum_{v \in \mathscr{V}} C_v^{TC} \delta_v + \sum_{v \in \mathscr{V}} \sum_{r \in \mathscr{R}_v} \sum_{t \in \mathscr{T}} C_{vr}^S x_{vrt}, \tag{1}$$

subject to

$$\sum_{v \in \mathscr{V}} \sum_{r \in \mathscr{R}_v} \sum_{t \in \mathscr{T}} A_{vir} x_{vrt} \geq S_i, \quad i \in \mathscr{N}, \tag{2}$$

$$\sum_{r \in \mathscr{R}_v} \sum_{t \in \mathscr{T}} D_{vr} x_{vrt} - F_v \delta_v \leq 0, \quad v \in \mathscr{V}, \tag{3}$$

$$\sum_{v \in \mathscr{V}} \sum_{r \in \mathscr{R}_v} x_{vrt} \leq B_t, \quad t \in \mathscr{T}, \tag{4}$$

$$\sum_{r \in \mathscr{R}_{vl}} x_{vrt} + \sum_{r \in \mathscr{R}_v} \sum_{\nu=1}^{l-1} x_{vr,((t+\nu) \bmod |\mathscr{T}|)} \leq 1, \quad v \in \mathscr{V}, t \in \mathscr{T}, l \in \mathscr{L}, \tag{5}$$

$$\underline{P}_k \leq \sum_{v \in \mathscr{V}} \sum_{r \in \mathscr{R}_v} \sum_{\nu=0}^{G_k} A_{vir} x_{vr,((t+\nu) \bmod |\mathscr{T}|)} \leq \overline{P}_k, \quad k \in \mathscr{H}, i \in \mathscr{N}_k, t \in \mathscr{T}, \tag{6}$$

$$\delta_v \in \{0,1\}, \quad v \in \mathscr{V}, \tag{7}$$

$$x_{vrt} \in \{0,1\}, \quad v \in \mathscr{V}, r \in \mathscr{R}_v, t \in \mathscr{T}. \tag{8}$$

The objective function (1) minimizes the sum of the time charter costs and the sailing costs. Then constraints (2) ensure that all offshore installations get the required number of visits during the week, and constraints (3) ensure that a vessel is not in service more days than it is available during a week. These constraints also ensure that δ_v equals one if a vessel is in service. Constraints (4) limit the number of vessels to be serviced at the onshore supply depot on a given weekday, and constraints (5) ensure that a vessel does not start on a new voyage before it has returned to the onshore supply depot after the last voyage. Constraints (6) ensure that the visits to the offshore installations are evenly spread throughout the week. For example, if $k = 2$ then $G_k = 2$, $\underline{P}_k = 0$ and $\overline{P}_k = 1$, ensuring that there are no more than one visit to the offshore installation during a three day period. Constraints (7) and (8) set the binary requirements for the δ_v and x_{vrt} variables, respectively.

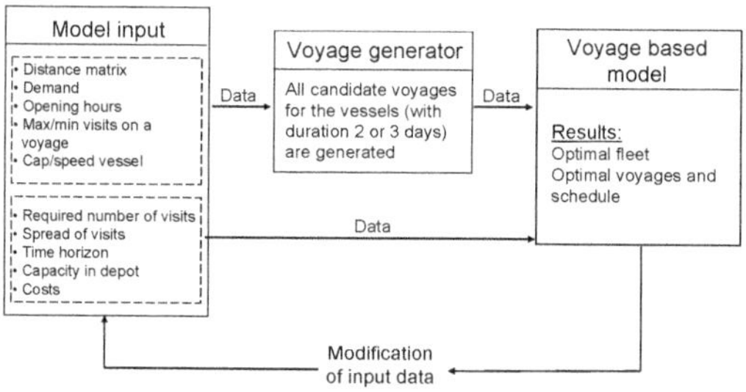

Fig. 2. Supply vessel planning model, schematic overview

A version of the mathematical model presented here has been implemented and is currently used by Statoil in their supply vessel planning process. Figure 2 gives a schematic overview over the solution method that contains a voyage generator and the voyage based model presented here. The voyage generator pregenerates all candidate voyages for a vessel taking into consideration constraints like opening hours at installations and loading capacities for the vessels. More information about the voyage generation procedure can be found in [15].

3 The Weather Impact

The prevailing weather conditions will affect the supply vessels' sailing speed and the unloading and loading operations at the offshore installations. This again may have severe consequences for the offshore supply service, especially during the winter season in the North Sea. The critical factor is the significant wave height (by definition the average wave height (trough to crest) of the one-third largest waves). If, for example, the significant wave height is more than 4.5 meters, it will not, due to safety regulations, be possible to perform offshore loading/unloading operations and a supply vessel will have to wait until the weather conditions improve before starting such operations. This is called waiting on weather (WOW). Rough weather conditions with significant wave heights of less than 4.5 meters can also make offshore loading/unloading operations challenging and can increase the time to perform such services. Rough conditions will also require a reduction in sailing speed. Table 1 defines four weather states based on the significant wave heights, and shows the reduction in sailing speed (in knots) and percentage increase in service time for loading/unloading operations at offshore installations. These are the conditions Statoil acts in accordance with in their supply vessel service.

Table 1. Weather states

Weather state	Wave height [m]	Sailing speed	Service time
1	< 2.5	0 kn	0%
2	$< 2.5,3.5$]	0 kn	20%
3	$< 3.5,4.5$]	- 2 kn	30%
4	≥ 4.5	- 3 kn	WOW

The weather conditions represent the major uncertain elements in the supply vessel planning problem and must be taken into consideration when we aim to create robust solutions to the supply vessel planning problem. We see from Figure 1 that a typical schedule may be very vulnerable for delays. Only a small decrease in sailing speed, or increase in service time, may result in delays that will affect later voyages. In the example in Figure 1, this is especially critical for the voyages sailed by vessel 1 on Tuesdays and vessel 2 on Wednesdays.

4 Robustness Approaches

The voyage based solution method from [15] does not take into account any robustness considerations, but finds the optimal schedule based on the supply vessels' service speeds and the given service times at the offshore installations. Here we present some robustness approaches that can be added to the model from Section 2.

4.1 Adding Slack to the Voyages and Schedule

We define *slack* as a supply vessel's idle time after finishing a voyage before it has to start preparing for the next voyage. Consequently, given that a supply vessel need to start preparing for a voyage at 0800, if it returns to the onshore supply depot at 0430 after last voyage, that voyage will have 3.5 hours slack. When voyages are joint in a weekly schedule, a voyage may have more than 24 hours slack if a supply vessel has one or more days idle before starting on a new voyage.

In [15] a robustness approach that required a given number of hours slack for each voyage was introduced. In this approach, all generated voyages that have less slack than required are either discarded (if the duration is maximum duration) or get transferred into a longer voyage with added slack of 24 hours. This is a simple way of assuring that every voyage allows for some delays due to reduced sailing speed or increased service time at the offshore installations.

Another simple approach considers the idle days for the supply vessels. For example, vessel 3 in Figure 1 only sails one voyage of duration two days during the week and has five idle days. The other two supply vessels do not have any idle days, and also sail some voyages with little slack that will not allow for much delay. For this schedule, it would probably be more beneficial if, for example, the voyage sailed by vessel 1 starting on Thursday was sailed by vessel 3. The solution method from [15] does not consider such consequences, and given that vessels 1 and 3 have the same sailing costs and both can sail that voyage, the two different schedules would be valued equally.

Based on the model formulation from [15] described in Section 2, the following may be added to create a weekly schedule that values idle days as a robustness consideration:

- Add a robustness profit in the objective function for each day a supply vessel is idle
- Add a robustness profit in the objective function for each supply vessel that has at least one idle day during the week
- Add a robustness profit in the objective function for each supply vessel that sails no more than two voyages during the week

All of the simple robustness approaches above can easily be implemented in the voyage based solution method from [15]. The purpose of adding robustness profit for idle days for supply vessels are to value idle days to some extent over longer voyages with more slack to each voyage. The robustness profit should not be too high so that it favors using more supply vessels than necessary. The first approach values each idle day equally independent on what supply vessel that has it, and may lead to one supply vessel having many idle days while others have none. The two other approaches give one fixed robustness profit if the statement occurs for a vessel.

4.2 An Optimization and Simulation Framework for Robust Schedules

Based on the model from [15] we have developed a solution method that combines optimization and simulation to provide robust and resilient schedules to the supply vessel planning problem.

Figure 3 illustrates the overall solution method. It consists of three steps:

Step 1. Generate all candidate voyages the vessels may sail
Step 2. Simulate each candidate voyage and assign a robustness measure

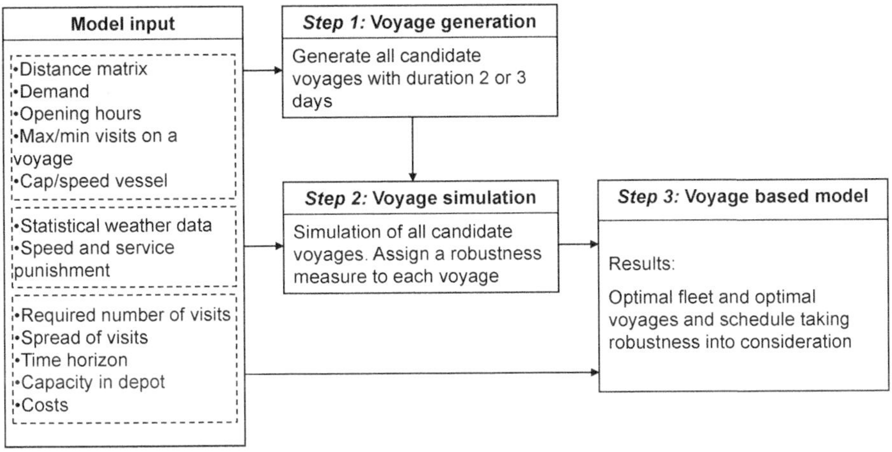

Fig. 3. Optimization and simulation for robust schedules, schematic overview

Step 3. Solve the voyage based model with robustness measures assigned to each voyage

Steps 1 and 3 are equivalent to phases 1 and 2 described in [15]. We now add some more input data: Statistical data about the uncertain elements of the problem, in this context weather data. This data is used in Step 2 to calculate a robustness measure for each candidate voyage. The robustness measure we use is not delivered volume. This is then used to create a robust weekly schedule for the supply vessel planning problem by giving it a cost in the objective function in the voyage based model.

Figure 4 shows a flow chart of the simulation procedure. For each simulation, a set of consecutive weather states are drawn from their respective probability distributions. There are four weather states as shown in Table 1. In the simulation procedure, we use

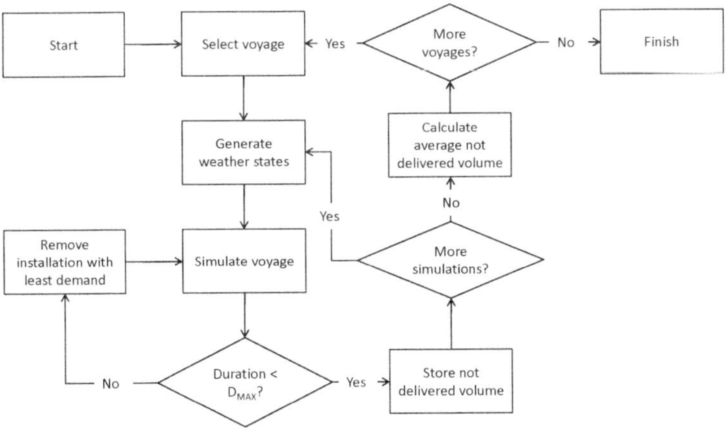

Fig. 4. Flow chart of the simulation procedure

a time interval of six hours, which means that each weather state lasts for six hours before the weather conditions may change. Each weather state has a given start state probability, as shown in Table 2. The next weather state will be dependent only on the current weather state, a random process recognized as a Markov chain, see e.g. [22]. Table 3 shows the transition probability matrix, which is the probability of moving from one weather state to another. For example, the probability of moving from weather state two in one time interval to weather state three in the next is 20.7%. All the probabilities are calculated from historical weather data provided by Statoil for the winter season in the North Sea.

Table 2. Start state probabilities

State	1	2	3	4
Probability	22.7%	27.1%	28.2%	22.0%

When the weather states are drawn, a voyage is simulated according to the necessary reduction in sailing speed and increase in service times the prevailing weather state demands. If the voyage cannot be completed within the maximum duration of that voyage (i.e. for a voyage with duration two days, the supply vessel needs to be back at the supply depot before these two days has passed), the offshore installation with the least demand is removed from the voyage. This procedure continues until the voyage can be completed. Then the total demand in square meters not delivered, calculated as the sum of the demand from the removed offshore installations, is stored and a new simulation started. The average demand not delivered over all simulations for each voyage is the output from the simulation procedure.

Table 3. Transition probability matrix

State	1	2	3	4
1	82.5%	16.9%	0.6%	0.0%
2	14.0%	60.6%	20.7%	4.7%
3	0.5%	23.9%	57.7%	17.9%
4	0.0%	0.6%	27.9%	71.5%

The objective function (1) from Section 2 is then replaced with:

$$min \sum_{v\in\mathscr{V}} C_v^{TC}\delta_v + \sum_{v\in\mathscr{V}}\sum_{r\in\mathscr{R}_v}\sum_{t\in\mathscr{T}} C_{vr}^{S}x_{vrt} + \sum_{v\in\mathscr{V}}\sum_{r\in\mathscr{R}_v}\sum_{t\in\mathscr{T}} C^{P}E_{vr}x_{vrt} \tag{9}$$

Here, E_{vr} is the average demand not delivered for voyage r sailed by vessel v, and C^P is the penalty cost for each square meter not delivered. This penalty cost is estimated based on the real cost of not delivered volume: This volume will, depending on the urgency, have to be delivered by a different supply vessel at a later time that could be one of the vessels in the fleet, a different vessel that needs to be chartered in on short term at a higher costs, or by helicopter at a much higher costs. To reflect these elements, C^P is calculated as the cost of a three day voyage sailed by a vessel chartered in on a somewhat higher time charter rate divided by the capacity of a vessel.

Based on the simulation procedure described here, a schedule simulation model was developed to test the schedules we get when using different solution approaches. In the schedule simulation model, a sequence of weather states for the whole time period of a schedule is drawn. Then every voyage sailed in the schedule is simulated. Extra slack in form of idle days for supply vessels is added to the voyage sailed before such an idle day, giving that voyage 24 hours (or more) of extra slack. The overall average square meters not delivered is then calculated and multiplied by the penalty cost.

5 Computational Study

The robustness approaches described in Section 4 have been tested on a number of problem instances based on the real supply vessel planning problem faced by Statoil. All test results were obtained on a 2.16 GHz Intel Core 2 Duo PC with 2 GB RAM. The voyage based model was implemented in Xpress-IVE 1.19.00 with Xpress-Mosel 2.4.0 and solved by Xpress-Optimizer 19.00.00. We set a time limit for solving the voyage based model to 3600 seconds. The voyage generator and the voyage and schedule simulating models were both written in C++ using Visual Studio 2005. A description of the problem instances is given in Section 5.1, followed by the computational results in Section 5.2.

5.1 Description of Problem Instances

Based on real data provided by Statoil for a supply vessel planning problem, 18 problem instances have been created. The instances are named based on how many offshore installations there are, and how many of these that have time windows for visits from supply vessels, i.e. how many that are closed for loading operations during nights between 1900 and 0700. As an example, problem instance 7-1 has seven offshore installations and one with such time windows.

The problem instances are created by taking the original supply vessel planning problem and removing or adding offshore installations. The offshore installations added to the problem are floating rigs that at some point in time were serviced from the onshore supply depot. An overview over the total number of offshore installation visits and the total weekly demand is shown in Table 4. The number of visits for each offshore installation vary from one to six, and the weekly demands varies from 250 to 960 m^2. These demands are divided by the number of weekly visits to get the demand for each visit. These again vary from 50 to 334 m^2. The service times at the offshore installations vary from 2.25 to 7 hours.

There are five supply vessels available that may be chartered. Their loading capacities vary from 900 to 1090 m^2, and weekly time charter rates are all somewhat above USD 100 000 (not necessarily reflecting today's market), highest for the vessels with the higher loading capacities. The service speeds for all the supply vessels are 12 knots.

The onshore supply depot has capacity to service three supply vessels every weekday, except for Sundays when it is closed. The duration of each voyage is two or three days,

Table 4. Number of visits and weekly demand

	# Offshore installations	# visits	Weekly demand [m^2]
Small	5	23	3248
instances	7	30	4402
Real sized	10	43	5995
instances	13	55	7429

and Mondays to Thursdays only two day voyages can start, three day voyages allowed on Fridays and Saturdays.

5.2 Numerical Results

The problem instances described in Section 5.1 have been tested using five different solution approaches:

Basic. Solution method from [15], i.e. solving the model from Section 2

4h. Solution method from [15] with four hours required slack at the end of each sailed voyage

Max2. Solution method from [15] with an extra robustness profit if a vessel sails no more than two voyages

SimSol. Solution method as described in Section 4.2 with robustness cost for each square meter not delivered

Combined. Solution method as described in Section 4.2 with robustness cost for each square meter not delivered (reduced to 1/3 of the cost used in SimSol) and robustness profit if a vessel sails no more than two voyages (same profit as for Max2)

We also tested the solution method from [15] with an extra robustness profit for each idle day for a supply vessel and with robustness profit for each supply vessel having at least one idle day, but this gave overall poorer results than the Basic solution approach. These results are therefore omitted from this presentation.

The robustness cost used in SimSol and Combined is calculated as described in Section 4.2. The robustness profit used in Max2 and Combined is set to an average time charter rate for half a day (weekly time charter rate divided by 14).

For the solution method from Section 4.2 (used in SimSol and Combined), each voyage was simulated 100 times for problem instances with 5, 7 and 10 offshore installations, and 20 times for the instances with 13 offshore installations due to time restrictions. The total CPU time for the simulations were 5-10 seconds, 2-5 minutes, 1-2.5 hours, and 33-50 hours for the instances with 5, 7, 10 and 13 offshore installations, respectively. To calculate the extra cost for not delivered volume for a schedule, we ran 10 000 simulations of each schedule. The total CPU time for 10 000 simulations varied from 10 seconds for the smaller problem instances to 90 seconds for the larger ones.

Table 5 shows the results from the small problem instances, with five and seven offshore installations. Listed in the table are the planned costs for the schedules (the cost of the schedule if there are no delays, robustness cost for not delivered volume

Table 5. Small problem instances

Problem instance		5-1	5-3	5-5	7-1	7-3	7-5	Total
Basic	PlanCost	460027	470500	815594	464452	471230	622331	3304134
	ExtraCost	73434	107290	80339	149977	157915	172871	741826
	Total	533461	577790	895933	614429	629145	795202	4045960
4h	PlanCost	101.40	100.00	100.00	101.34	100.00	100.00	100.38
	ExtraCost	105.05	99.36	99.17	67.17	87.66	94.54	89.78
	Total	101.90	99.88	99.93	93.00	96.90	98.81	98.44
Max2	PlanCost	100.00	100.00	100.00	100.00	100.00	100.00	100.0
	ExtraCost	103.95	101.19	109.40	66.92	87.23	80.43	87.61
	Total	100.54	100.22	100.84	91.93	96.79	95.75	97.73
SimSol	PlanCost	102.57	101.05	100.17	102.55	101.07	103.38	101.70
	ExtraCost	98.37	92.45	101.79	70.87	91.82	72.06	84.80
	Total	101.99	99.45	100.31	94.82	98.75	96.58	98.60
Combined	PlanCost	100.00	100.03	100.06	100.00	100.10	100.13	100.06
	ExtraCost	110.60	90.88	91.52	68.12	100.03	73.77	86.26
	Total	101.46	98.33	99.29	92.22	100.08	94.40	97.53

and/or robustness profit for idle days are not included), the extra cost for square meters not delivered (average cost over 10 000 simulations) and the total cost for the schedule: the sum of the planned cost and the extra cost. For the Basic solution approach, costs are reported in USD. The costs of the other solution approaches are given in percent relative to the cost of the Basic solution approach. The total costs over all problem instances are reported in the far right column.

We observe that for the smallest problem instances with only five offshore installations, the Basic solution approach gives good solutions compared with any of the suggested robustness approaches. This observation is related to that these schedules are not very constrained and have a lot of slack.

Although the Basic solution approach gives quite good results for the smaller problem instances, the overall best cost schedules was achieved using the Combined solution approach, 2.47 % less overall costs than the Basic approach. The Combined schedules also beat the Basic schedules for all problem instances except for 5-1.

Table 6 shows the results for the real sized problem instances, which are more interesting from a practical point of view. Where the solution cost is in bold, the Xpress-MP optimizer did not manage to prove optimal solution within the CPU limitation of 3600 seconds (proven gaps from optimal solution are, however, less than 1 % in all these cases). We observe that the Basic schedule is quite good for problem instance 10-1, but quite poor for the other instances. Overall for these problem instances, the SimSol schedules have the best results, 3.42 % better on average than the Basic schedules. The SimSol schedules beat the Basic schedules for all problem instances except for 10-1, while the Combined schedules beat the Basic schedules for all problem instances.

Table 6. Real sized problem instances

Problem instance		10-1	10-3	10-5	10-7	13-1	13-3	13-5	13-7	Total
Basic	PlanCost	626959	627097	628785	819845	**640273**	**641197**	**645143**	651859	5281158
	ExtraCost	140354	210929	276458	223263	277830	303550	338389	385235	2156008
	Total	767313	838026	905243	1043108	918103	944747	983532	1037094	7437166
4h	PlanCost	100.08	100.24	100.98	100.00	**102.14**	**102.09**	**101.57**	123.07	103.71
	ExtraCost	119.22	73.93	83.77	95.61	89.02	90.47	98.19	63.10	86.53
	Total	103.58	93.62	95.73	99.06	98.17	98.36	100.41	100.79	98.73
Max2	PlanCost	100.00	100.00	100.00	100.00	**100.08**	**99.92**	99.88	**100.00**	99.99
	ExtraCost	129.41	84.99	84.94	86.79	85.52	93.77	85.80	92.50	90.83
	Total	105.38	96.22	95.40	97.17	95.68	97.95	95.03	97.22	97.33
SimSol	PlanCost	103.25	103.48	103.71	101.23	103.00	103.54	102.74	101.97	102.81
	ExtraCost	101.46	79.25	70.28	76.97	81.39	82.46	82.36	83.78	81.33
	Total	102.93	97.38	93.50	96.04	96.46	96.77	95.73	95.21	96.58
Combined	PlanCost	100.27	100.36	101.19	100.41	**100.29**	101.32	100.98	100.87	100.70
	ExtraCost	96.49	89.05	74.08	77.77	91.50	90.19	82.44	96.02	87.13
	Total	99.58	97.51	92.91	95.56	97.63	97.74	94.60	99.07	96.77

We observe from the results in Tables 5 and 6 that the Basic schedules seem to be better when there are few offshore installations and few time windows. Table 7 shows the results from the problem instances with no time windows. For these instances, the

Table 7. Without time windows

Problem instance		5-0	7-0	10-0	13-0	Total
Basic	PlanCost	460000	464290	626781	**640273**	2191344
	ExtraCost	55193	118438	168062	217468	559161
	Total	515193	582728	794843	857741	2750505
4h	PlanCost	101.40	101.38	100.11	**102.18**	101.25
	ExtraCost	130.90	90.84	80.28	105.01	97.13
	Total	104.56	99.24	95.92	102.89	100.42
Max2	PlanCost	100.01	100.00	100.00	**100.00**	100.00
	ExtraCost	131.38	81.91	93.06	111.68	101.73
	Total	103.37	96.32	98.53	102.96	100.35
SimSol	PlanCost	102.57	102.60	103.25	**102.86**	102.86
	ExtraCost	127.71	87.54	84.01	107.53	98.22
	Total	105.27	99.54	99.18	104.04	101.91
Combined	PlanCost	100.01	100.04	100.26	100.21	100.15
	ExtraCost	135.16	89.51	87.07	113.48	102.60
	Total	103.77	97.90	97.47	103.58	100.65

Basic schedules were overall the best, with very good schedules for problem instances 5-0 and 13-0 compared with the other solution approaches. We also observe that the SimSol schedules have the overall worst result.

In general we see that addressing robustness when creating a schedule to the supply vessel planning problem give added value by having schedules that are more likely to be successfully executed. The planned cost of the schedule will be slightly higher than if we do not address robustness, but the simulated extra costs, representing a better estimate for the real costs, for not delivering goods to the offshore installation as planned will be less, resulting in a reduced total cost. However, the way the schedules are created by the voyage based solution method, there is a possibility that a schedule that does not address robustness will be better than one that does because we address robustness to individual voyages, and not the overall schedule.

6 Concluding Remarks

The supply vessel planning problem consists of determining optimal fleet size and mix of offshore supply vessels and their corresponding weekly routes and schedules to service a given set of offshore installations. The problem was addressed and a voyage based solution model was developed in [15]. A version of this model was implemented by the supply vessel planners at Statoil and savings of 3 million USDs was reported from this project. However, during the work with the model, it was realized that the weather conditions have a big impact on the execution of a schedule. A need for adding some robustness considerations into the model was thus discovered.

We have addressed the problem of creating robust schedules to the supply vessel planning problem. Several solution approaches have been implemented and tested:

- Requiring a number of hours of slack for each voyage sailed
- Benefiting schedules where vessels do not sail more than two voyages during a week
- Punishing voyages based on the average not delivered volume of goods calculated from simulating the voyages given certain probabilistic weather conditions

The solution approaches were then tested on several problem instances based on real data provided by Statoil.

From the computational results we see that adding some robustness criteria to the optimizing procedure of creating a schedule to the supply vessel planning problem gives a lower predicted cost than if no such criterion is considered. The effect is higher with increased problem size and more time windows. Variations of our simulation approach for robust schedules provide the overall best results, with a predicted cost saving of more than 3% for the real sized problem instances, but also simple approaches give good results and overall better schedules than not considering robustness.

Acknowledgement

This work was supported by the projects MARRISK and DESIMAL partly funded by the Research Council of Norway. The authors would like to thank the planners at Statoil Marine Operations for letting us get insight into their interesting planning problem. The authors are grateful to two former master students, Eirik Rosenberg and Pål Aastad, for developing an early version of the voyage simulating program.

References

1. Aghezzaf, E.-H.: Robust distribution planning for supplier-managed inventory agreements when demand rates and travel times are stationary. Journal of the Operational Research Society 59, 1055–1065 (2008)
2. April, J., Better, M., Glover, F., Kelly, J., Laguna, M.: Enhancing business process management with simulation optimization. In: Perrone, L.F., Wieland, F.P., Liu, J., Lawson, B.G., Nicol, D.M., Fujimoto, R.M. (eds.) Proceedings of the 2006 Winter Simulation Conference, pp. 642–649 (2006)
3. Brandão, J., Mercer, A.: A tabu search algorithm for the multi-trip vehicle routing and scheduling problem. European Journal of Operational Research 100, 180–191 (1997)
4. Brandão, J., Mercer, A.: The multi-trip vehicle routing problem. Journal of the Operational Research Society 49(8), 799–805 (1998)
5. Beltrami, E.J., Bodin, L.D.: Networks and vehicle routing for municipal waste collection. Networks 4(1), 65–94 (1974)
6. Cheng, L., Duran, M.A.: Logistics for world-wide crude oil transportation using discrete event simulation and optimal control. Computers and Chemical Engineering 28, 897–911 (2004)
7. Christiansen, M., Fagerholt, K.: Robust ship scheduling with multiple time windows. Naval Research Logistics 49(6), 611–625 (2002)
8. Cordeau, J.-F., Gendreau, M., Laporte, G.: A tabu search heuristic for periodic and multi-depot vehicle routing problems. Networks 30(2), 105–119 (1997)
9. Fagerholt, K.: Optimal fleet design in a ship routing problem. International Transactions in Operational Research 6(5), 453–464 (1999)
10. Fagerholt, K., Lindstad, H.: Optimal policies for maintaining a supply service in the Norwegian Sea. Omega 28, 269–275 (2000)
11. Fagerholt, K., Christiansen, M., Hvattum, L.M., Johnsen, T.A.V., Vabø, T.J.: A decision support methodology for strategic planning in maritime transportation. Omega 38, 465–474 (2010)
12. Fu, M.C.: Optimization for simulation: Theory vs. practice. INFORMS. Journal on Computing 14(3), 192–215 (2002)
13. Gendreau, M., Laporte, G., Séguin, R.: Stochastic vehicle routing. European Journal of Operational Research 88, 3–12 (1996)
14. Gribkovskaia, I., Laporte, G., Shlopak, A.: A tabu search heuristic for a routing problem arising in servicing of offshore oil and gas platforms. Journal of the Operational Research Society 59, 1449–1459 (2008)
15. Halvorsen-Weare, E.E., Fagerholt, K., Nonås, L.M., Asbjørnslett, B.E.: Fleet size and mix and periodic routing of offshore supply vessels. Working paper submitted. Norwegian University of Science and Technology (2010)
16. Hemmelmayr, V.C., Doerner, K.F., Hartl, R.F.: A variable neighborhood search heuristic for periodic routing problems. European Journal of Operational Research 195, 791–802 (2009)
17. Hoff, A., Andersson, H., Christiansen, M., Hasle, G., Løkketangen, A.: Industrial aspects and literature survey: Fleet composition and routing. Computers & Operations Research 37, 2041–2061 (2010)
18. Hwang, H.-S., Visoldilokpun, S., Rosenberger, J.M.: A branch-and-price-and-cut method for ship scheduling with limited risk. Transportation Science 42(3), 336–351 (2008)

19. Juan, A., Faulin, J., Grasman, S., Riera, D., Marull, J., Mendez, C.: Using safety stocks and simulation to solve the vehicle routing problem with stochastic demands. Transportation Research Part C (2010), doi:10.1016/j.trc.2010.09.007
20. Li, X., Tian, P., Leung, S.C.H.: Vehicle routing problems with time windows and stochastic travel and service times: Models and algorithm. International Journal of Production Economics 125, 137–145 (2010)
21. Sörensen, K., Sevaux, M.: A practical approach for robust and flexible vehicle routing using metaheuristics and Monte Carlo sampling. Journal of Mathematical Modelling and Algorithms 8(4), 387–407 (2009)
22. Ross, S.M.: Introduction to Probability Models, 9th edn. Academic Press, San Diego (2007)
23. Taillard, E.D., Laporte, G., Gendreau, M.: Vehicle routeing with multiple use of vehicles. Journal of the Operational Reserach Society 47(8), 1065–1070 (1996)
24. Tan, C.C.R., Beasley, J.E.: A heuristic algorithm for the period vehicle routing problem. Omega 12(5), 497–504 (1984)

A Liner Shipping Network Design – Routing and Scheduling Impacted by Environmental Influences

Volker Windeck and Hartmut Stadtler

Institute for Logistics and Transportation, University of Hamburg,
Von-Melle-Park 5, 20146 Hamburg, Germany
volker.windeck@uni-hamburg.de,
hartmut.stadtler@uni-hamburg.de

Abstract. Not only attempts to reduce costs but also emissions are driving shipping companies to operate their fleet in slow steaming mode. We show a strategic liner shipping network design decision support system which takes into account three environmental influences: waves, currents and wind. Our model will answers the question, whether environmental influences and the use of additional propulsion systems can influence the cost structure or the on-time delivery of a liner shipping network and its schedule. We will present a variable neighborhood solution approach, that can solve this this network design problem in reasonable time.

1 Introduction

Increasing fuel costs and growing ecological concerns are forcing ship owners, operators as well as forwarding companies, to operate their fleet in a so called slow steaming mode or even to consider the use of alternative propulsion techniques. The tremendous rise of bunker fuel costs (over 28% in the last 12 months; [2]) now represents an increasing portion of freight charges. To reduce bunker fuel costs per trip, operators can decrease their fleets average speed resulting in reduced fuel consumption. Because ship operators guarantee port visits at a fixed frequency, the increase in travel time due to a slower speed can lead to a need for additional ships. The majority of liner shipping companies ensure port visits at a weekly frequency.

The optimal selection of speed combined with the number of vessels needed to perform a permanent route roundtrip at minimum costs, is addressed. In addition growing ecological concerns have an influence on these decisions. Two different sides put pressure on ship operators to reduce emissions. On the one side governments, caused by the MARPOL Annex VI (International Convention for the Prevention of Marine Pollution from Ships) controlled by the IMO (International Maritime Organisation) force ships to use marine diesel fuel that emits less NOX (nitrogen oxides) and SOX (sulphur oxides). This marine diesel is sold due to its higher quality at a higher price than regular bunker fuel. On the other side companies (e.g. Tchibo, a German coffee shop chain also offering other goods such as clothing, household items and electronics) intend to reduce emissions for improving the carbon footprint, when shipping their goods [12]. Due to the fact that these items are mainly produced in East Asia and therefore transported by ship to Europe, a large amount of the emissions will be generated by sea transport.

J. Pahl, T. Reiners, and S. Voß (Eds.): INOC 2011, LNCS 6701, pp. 574–576, 2011.

Besides using fuel that emits less NOX and SOX, reducing the speed of ships will reduce emissions most. This is due to the fact that fuel consumption increases with speed almost to the power of three.

Other approaches to reduce fuel consumption or sulphur emission rely on using alternative propulsion techniques. Recent ideas include the use of solar power, wave propelled ships or wind dependent propulsion systems. Wind propulsion systems are the most practical and advanced techniques. The system we are considering in this paper is a kite system like the one manufactured by SkySails [11]. Depending on the wind direction and speed, related to the ships direction and speed, these kites can lower the engine power output when travelling at a given speed and thus reduce fuel consumption. Other techniques use the effect of Flettner rotors (e.g. E-Ship, [5]).

2 Literature Review and Problem Statement

Little research has been done on ship routing and scheduling and especially few contributions on the liner shipping network design can be found. A recent classification approach and literature review on liner shipping problems is given by Kjeldsen 2009 [8].

Several research contributions on fleet design and ship routing make use of a predefined sets of routes from which the most promising routes are selected via a set partitioning problem, as done in Fagerholt [6,7] and Cho and Perakis [4]. Fagerholt [6] formulates the liner shipping task as a multi-trip vehicle routing problem, where cargos are picked up in production harbours and dropped off at a central depot. The size of a heterogeneous fleet of vessels is then determined.

A proposal on designing a strategic container liner shipping network and simultaneously solving an empty container redistribution planning problem is given by Shintani et al. [10]. Spare space on ships is used to transport empty containers.

A mathematical model allowing for transhipment and multiple visits of one harbour in each round trip is presented by Agarwal and Ergun [1]. One round trip is assumed and speed variation is not accounted for. Three algorithms for this simultaneous ship-scheduling and cargo-routing problem are given: a greedy heuristic, Column Generation and Benders decomposition. These are compared with regard to computational efficiency and solution quality.

The idea to model a liner shipping network problem with a vessel being able to stop at a port on its inbound and outbound journey was first introduced by Rana and Vickson [9]. In our approach speed is a variable for each port to port relation. For the first time we will account for monthly mean weather dependent travel times and resulting travel costs as well as variable speed settings on arcs between two consecutive harbours for a strategic liner network design problem.

3 Solution Approach

Our mathematical model for a strategic liner network design problem is capable of creating round trips for each type of ship specifying the ports to be visited by this type of ship and its corresponding schedule. This is done for multiple types of ships simultaneously leading to multiple parallel round trips. Our objective is to maximize the profit gained by revenues when transporting cargo from its pick-up port to its destination port and subtracting all variable speed and type of ship dependent travel costs as well as fixed charter costs per ship. Not all cargo demand has to be transported. Only cargo that contributes to the profit will be serviced.

Our mixed integer programming model will addresses the following questions:

- Which harbours on a ships' route should be visited on a liner round trip?
- Which cargo can be transported from the cargos' pick-up port to its destination port along the same round trip?
- How many ships of a specific type are needed to guarantee an e.g. weekly visit to all ports along a round trip?
- Which average speed should be selected on a trip between two consecutive harbours?

For solving this model we created a hybrid algorithm, consisting of a Variable Neighborhood Search (VNS) and a relaxed MIP-model. This approach is comparable to a Matheuristic as described in Caserta and Voß [3].

References

1. Agarwal, R., Ergun, O.: Ship scheduling and network design for cargo routing in liner shipping. Transportation Science 42(2), 175–196 (2008)
2. BWI: Bunkerworld index (2011), `http://www.bunkerworld.com/prices/index/bwi`
3. Caserta, M., Voß, S.: Metaheuristics: Intelligent problem solving. In: Maniezzo, V., Stützle, T., Voß, S. (eds.) Matheuristics, Annals of Information Systems, vol. 10, pp. 1–38. Springer, US (2010)
4. Cho, S.C., Perakis, A.N.: Optimal liner fleet routeing strategies. Maritime Policy & Management 23(3), 249–259 (1996)
5. Enercon: Segelrotor-Schiff e-ship 1 in der Erprobung. Windblatt Enercon Magazin, p. 6 (February 2010)
6. Fagerholt, K.: Optimal fleet design in a ship routing problem. International Transactions in Operational Research 6(5), 453–464 (1999)
7. Fagerholt, K.: Designing optimal routes in a liner shipping problem. Maritime Policy & Management 31(4), 259–268 (2004)
8. Kjeldsen, K.H.: Liner shipping network design, routing and scheduling. Ph.D. thesis, CORAL - ASB (2009)
9. Rana, K., Vickson, R.G.: Routing container ships using lagrangean relaxation and decomposition. Transportation Science 25(3), 201–214 (1991)
10. Shintani, K., Imai, A., Nishimura, E., Papadimitriou, S.: The container shipping network design problem with empty container repositioning. Transportation Research Part E: Logistics and Transportation Review 43(1), 39–59 (2007)
11. SkySails: Skysails (2011), `http://www.skysails.info/`
12. Tchibo: Sustainable development report (2009), `http://www.tchibo-nachhaltigkeit.de/controller.aspx?n=120\&l=2`

A VND-ILS Heuristic to Solve the RWA Problem

Alexandre Xavier Martins[1], Christophe Duhamel[2], Mauricio Cardoso de Souza[3], Rodney Rezende Saldanha[4], and Philippe Mahey[2]

[1] Programa de Pós-Graduação em Engenharia Elétrica , Universidade Federal de Minas Gerais, Av. Antônio Carlos, 6627, Belo Horizonte, MG, Brasil
xmartins@decea.ufop.br

[2] Laboratoire LIMOS, CNRS-UMR6158, Université Blaise Pascal, Campus des Cézeaux, BP 10125, 63173 Aubiere CEDEX, France
{duhamel,mahey}@isima.fr

[3] Departamento de Engenharia de Produção, Universidade Federal de Minas Gerais, Av. Antônio Carlos, 6627, Belo Horizonte, MG, Brasil
mauricio.souza@pq.cnpq.br

[4] Departamento de Engenharia Elétrica, Universidade Federal de Minas Gerais, Av. Antônio Carlos, 6627, Belo Horizonte, MG, Brasil
rodney@cpdee.ufmg.br

Abstract. The Routing and Wavelength Assignment problem occurs in optical networks where communication requests fulfilled without inducing wavelength conflicts. We propose a VND-ILS algorithm which is a hybridization of both the ILS metaheuristic and the VND local search. Three neighborhood structures are defined for the VND as well as a perturbation step in the ILS. The efficiency of our approach is illustrated on classical realistic RWA instances. The computational results show our method outperforms some of the best existing methods in the literature.

1 Introduction

Consider a directed graph $G = (V, E)$, where V denotes the set of nodes and E denotes the set of communication links. Let R be the set of requests. Each request $i \in R$ is defined by an origin and destination pair $(s_i, d_i) \in V \times V$. The Routing and Wavelength Assignment (RWA) consists in determining the paths in which each attended request will be routed and its associated wavelength. A single wavelength must be assigned along the entire route if no converters are available, this is called continuity constraint. Besides, lightpaths that share a common physical link cannot be assigned the same wavelength. This is called the wavelength clash constraint.

Generally the RWA problem can be classified into two types: static, in which the requests between the node pairs are given in advance; dynamic, in which the on-line establishment of lightpaths is involved for connection requests that occur dynamically. This paper addresses the static min-RWA offline variant, where the objective is to minimize the number of wavelengths used satisfying all the traffic requests in R.

An instance of the static RWA is illustrated in Fig. 1a, along with the set of traffic requests. A feasible associated solution is shown in Fig. 1b. The arc $(1,7)$ is used to

J. Pahl, T. Reiners, and S. Voß (Eds.): INOC 2011, LNCS 6701, pp. 577–582, 2011.

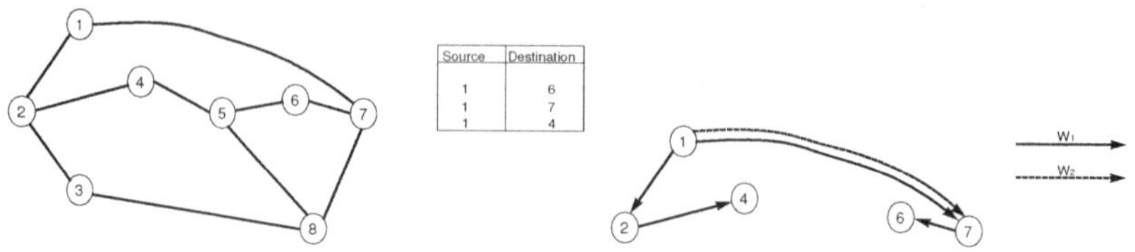

(a) Instance of static RWA **(b)** Feasible Solution

Fig. 1. RWA Problem Example

attend the requests $\{1,6\}$ and $\{1,7\}$. Therefore we only need to use two wavelengths on it.

According to [2] the min-RWA is NP-hard. Thus, it is very difficult to get an optimal solution for problems of practical sizes. Approximation algorithms and heuristics are thus usually applied to compute a good solution in a reasonable amount of time for real size instances.

In this paper, we present a combination of both the VND and the ILS heuristics for min-RWA. The VND heuristic performs a local search in the neighborhood of the current solution while the ILS heuristic is used to disturb an existent solution, allowing a new search to be done, instead of generating a new solution.

The rest of this paper is organized as follows. Section 2 is dedicated to the heuristics to solve the min-RWA problem. The implemented VND and ILS are depicted, respectively, in Sections 3 and 4. The experimental results are shown in Section 5. Final remarks are done in Section 6.

2 Related Work

Two strategies are commonly used to solve min-RWA. They both decompose the RWA into two subproblems: the routing problem and the wavelength assignment problem. The first strategy solves the wavelength problem after having solved the routing problem ([1], [6]). The second strategy solves both subproblems at the same time ([9], [4]).

Four constructive heuristics are proposed in [9] based on methods used to solve the Bin Packing problem: FF-RWA is based on the first fit heuristic; BF-RWA is based on the best fit heuristic; FFD-RWA is based on the first fit decreasing heuristic; and BFD-RWA is based on the best fit decreasing heuristic. Computational results show that FFD-RWA and BFD-RWA outperform the Greedy-EDP-RWA [4]. An efficient implementation of these heuristics is presented in [7] along with a new set of test instances. The best results were obtained by BFD-RWA.

One of the drawbacks of the BFD-RWA algorithm is that there are many requests with the same min-length value. To handle this problem [8] proposed a genetic algorithm with random keys to order those requests. The requests are ordered according to their min-length and to their keys values. The key value to each request is a real number within $[0,1]$ that is randomly generated in the initial population. The fitness of the chromosome is given by the cost of the solution found by a decoding heuristic that receives as input the random-keys vector and outputs a feasible solution with its corresponding cost. The decoding consists of two steps. First, the requests are sorted in non-increasing order of the sum of their min-length and key values. Then, the relative order between requests with the same min-length value is defined by their keys. The objective function of the resulting order is computed using the BFD-RWA algorithm [9].

3 Variable Neighborhood Descent

The VND technique [5] consists in systematically exploring different neighborhoods of a solution within a local search in order to reduce the risk of becoming trapped in local optima. During the exploration, only improving movements are accepted. Whenever a new improving solution is found, the method restarts with the first neighborhood structure at the new current solution.

Here, given a solution, the general idea is to choose a graph (ie. a wavelength) with fewer requests to be eliminated. Instead of trying to remove the graph at once, the idea is to transfer, one by one, the requests included in this graph.

Let λ_t be the chosen graph. We must now choose the request to be transferred. The requests are sorted in non-increasing order of the lengths of their shortest path. Then the first element, r_t, is selected to be transferred.

The first neighborhood consists of a simple reallocation. For each graph, different from λ_t, the method tries to shift the request $r_t \in \lambda_t$ into another graph. The search is stopped as soon as a feasible shift is found. Upon success, the search continues by selecting another request belonging to λ_t. If λ_t is empty, it can be removed. Otherwise, the method proceeds to the next neighborhood.

The second neighborhood is called when a simple reallocation is not possible. Therefore, a graph λ_i is selected and we try to reallocate each of its requests to the other graphs, different from λ_t. If at least one change is done it means we can try to reallocate the request r_t to the graph λ_i. For each iteration of this neighborhood a single graph λ_i is selected. The choice is done sequentially. Upon success, the search continues by selecting another request belonging to λ_t and returns to the first neighborhood. Otherwise, the method proceeds to the next neighborhood.

Finally, if no reallocation is possible, the method performs the search in the third neighborhood. The goal here is to exchange the request r_t with another request. The swap is accepted only for requests whose shortest path length is lower than the length of the shortest path of r_t. Upon success, the search continues by selecting another request belonging to λ_t and returns to the first neighborhood. Otherwise, the method stops and the ILS steps goes on.

4 Iterated Local Search

The ILS [3] method aims to disturb a current solution without undermining the gains already achieved. In this sense we have developed a method that changes the current solution without adding any other wavelength and without allocating any requests in λ_t.

The perturbation is done as follows: for each graph, excluding λ_t, we pull out a chosen randomly request. The arcs of each request are reactivated. Then the method constructs an assignment problem (graphs $\times$ requests) by checking all possible allocations between requests and graphs. The costs of edges in the assignment problem are as follow: $c_{ij} = -2$ if the request i belongs to the graph j, $c_{ij} = -\infty$ if there is no path for the request i in the graph j. Otherwise there is a path and the cost is $c_{ij} = -(1.0 - SP(i)/SP(i,j)) + max(SP(i,i) - SP(i,j), 0.0)$, where $SP(i)$ is the length of shortest path of request i and $SP(i,j)$ is the length of shortest path of request i in the graph j.

The assignment problem can be solved in polynomial time. After the problem has been solved, we check whether there has been any change in the current solution. If so, the method returns to the VND, otherwise another ILS iteration is done.

5 Results

In order to assess the performance of the proposed VND-ILS, one set of computational experiments is designed. All of the computational tests have been carried out on a computer with a Core 2 Duo processor, 1.97 GHz and 4 GB of RAM memory, running the MS Windows XP operating system. All of the algorithms have been implemented in C++.

Three sets of test instances were used in the computational experiments. The first set is a collection of the most studied realistic instances in the literature. The other two tests (sets Y and Z) correspond to the largest and most difficult instances proposed in [7].

The experiment addresses the performances of the GA-RWA algorithm proposed in [8] with a time limit of 10 min and the VND-ILS we propose with a time limit of 5 min in order to take into account the processor difference. The VND-ILS algorithm starts with a solution provided by the BFD algorithm [9]. For each instance, the tables report the value of the best found solution and the average gap obtained with GA-RWA over five runs and VND-ILS over thirty runs (each run with a different seed for the random number generator) for statistical consistency. The last column presents the lower bound obtained by solving the linear relaxation of a multicommodity flow formulation equivalent to min-RWA without the wavelength continuity constraints.

The average gap over the realistic instances, the set Z and the set Y are shown in Table 1, 2 and 3, respectively. For GA-RWA these gaps are 4.1%, 6.0% and 13.8% while for VND-ILS these gaps are 0.0%, 3.4% and 9.9%. The column σ shows the standard deviation for the VND-ILS method. We can see that for realistic instances our method is able to find all the optimal solutions. Moreover, the VND-ILS has been able to improve all the upper bounds that could still be improved (on the sets Z and Y, but also on the first set).

Table 1. GA-RWA vs. VND-ILS, Realistic

Method	GA-RWA		VND-ILS			
Instance	λ_{min}	$gap(\%)$	λ_{min}	$gap(\%)$	σ	LB
ATT	24	20.0	20	0.0	0.0	20
ATT2	113	0.0	113	0.0	0.0	113
Finland	46	0.4	46	0.0	0.0	46
NSF.3	22	0.9	22	0.0	0.0	22
NSF.12	39	2.6	38	0.0	0.0	38
NSF2.12	35	0.6	35	0.0	0.0	35
Average		4.1		0.0		

Table 2. GA-RWA vs. VND-ILS, Set Z

Method	GA-RWA		VND-ILS			
Instance	λ_{min}	$gap(\%)$	λ_{min}	$gap(\%)$	σ	LB
Z.10x10.20	31	15.6	29	7.4	0.0	27
Z.6x17.40	87	4.0	85	1.2	0.0	84
Z.10x10.60	87	13.2	85	10.4	0.0	77
Z.4x25.60	195	2.0	193	0.6	0.3	192
Z.10x10.80	115	12.4	112	9.7	0.3	103
Z.8x13.80	134	3.9	131	1.6	0.2	129
Z.6x17.80	176	3.0	172	0.6	0.0	171
Z.5x20.80	209	2.0	206	0.5	0.0	205
Z.4x25.80	260	1.3	258	0.6	0.5	257
Z.5x20.100	257	2.8	253	1.6	0.3	250
Average		6.0		3.4		

Table 3. GA-RWA vs. VND-ILS, Set Y

Method	GA-RWA		VND-ILS			
Instance	λ_{min}	$gap(\%)$	λ_{min}	$gap(\%)$	σ	LB
y.4.20.4	20	6.3	19	0.0	0.0	19
y.3.40.5	59	12.8	56	7.6	0.3	53
y.3.60.5	86	12.5	83	7.9	0.3	77
y.4.60.5	58	18.4	55	12.8	0.4	49
y.5.60.1	36	9.7	35	7.6	0.5	33
y.3.80.1	122	15.5	116	9.5	0.2	106
y.3.80.5	113	8.8	110	4.2	0.3	104
y.4.80.1	73	55.3	70	49.2	0.3	47
y.4.80.5	75	16.0	72	11.3	0.5	65
y.5.80.1	47	11.2	46	8.1	0.5	43
y.5.80.2	60	1.7	59	0.1	0.3	59
y.4.100.1	90	18.4	87	14.5	0.0	76
y.5.100.1	58	5.5	57	4.2	0.5	55
y.5.100.2	74	1.6	73	1.1	0.4	73
Average		13.8		9.9		

6 Conclusions

The objective of this work was to propose an efficient VND-ILS with respect to the computational time and to the solution quality when compared with well known heuristics available in the literature for min-RWA problem.

We take advantage of an existing method to generate initial solutions (the BFD-RWA). This method was combined with a VND algorithm with three different neighborhood structures. Moreover, an ILS scheme was developed. This strategy proved being quite successful. It not only finding the optimal solutions for all realistic instances, but it also improved the upper bounds for all instances in which it was still possible.

Acknowledgement

The authors are grateful to Thiago Ferreira de Noronha, the coauthor of [7,8], for providing datasets used. The first author would like to thank the Brazilian Agency CAPES (Coordenação de Aperfeiçoamento de Pessoal de Nível Superior, process 5117-09-7).

References

1. Banerjee, D., Mukherjee, B.: A practical approach for routing and wavelength assignment in large wavelength-routed optical networks. IEEE Journal on Selected Areas in Communications 5, 903–908 (1996)
2. Erlebach, T., Jansen, K.: The complexity of path coloring and call scheduling. Theoretical Computer Science 255, 33–50 (2001)
3. Lourenço, H.R., Martin, O., Stuetzle, T.: Iterated Local Search. In: Glover, F., Kochenberger, G. (eds.) Kluwer Academic Publishers, Dordrecht (2002)
4. Manohar, P., Manjunath, D., Shevgaonkar, R.K.: Routing and wavelength assignment in optical networks from edge disjoint path algorithms. IEEE Communications Letters 6(5), 211–213 (2002)
5. Mladenovic, N., Hansen, P.: A Variable Neighborhood Search. Computers and Operations Research 24, 1097–1100 (1997)
6. Noronha, T.F., Ribeiro, C.C.: Routing and wavelength assignment by partition colouring. European Journal of Operational Research 171(3), 797–810 (2006)
7. Noronha, T.F., Resende, M.G.C., Ribeiro, C.C.: Efficient implementations of heuristics for routing and wavelength assignment. In: McGeoch, C.C. (ed.) WEA 2008. LNCS, vol. 5038, pp. 169–180. Springer, Heidelberg (2008)
8. Noronha, T.F., Resende, M.G.C., Ribeiro, C.C.: A biased random-key genetic algorithm for routing and wavelength assignment. J. Glob. Optim. (2010), doi: 10.1007/s10898-010-9608-7
9. Skorin-Kapov, N.: Routing and wavelength assignment in optical networks using bin packing based algorithms. European Journal of Operational Research 177(2), 1167–1179 (2007)

Recoverable Robust Knapsacks: Γ-Scenarios*

Christina Büsing[1], Arie M.C.A. Koster[2], and Manuel Kutschka[2]

[1] Technische Universität Berlin, Institut für Mathematik,
Straße des 17. Juni 136, D-10623 Berlin, Germany
cbuesing@math.tu-berlin.de

[2] RWTH Aachen University, Lehrstuhl II für Mathematik,
Wüllnerstr. 5b, D-52062 Aachen, Germany
{koster,kutschka}@math2.rwth-aachen.de

Abstract. In this paper, we investigate the recoverable robust knapsack problem, where the uncertainty of the item weights follows the approach of Bertsimas and Sim [3,4]. In contrast to the robust approach, a limited recovery action is allowed, i.e., up to k items may be removed when the actual weights are known. This problem is motivated by the assignment of traffic nodes to antennas in wireless network planning. Starting from an exponential min-max optimization model, we derive an integer linear programming formulation of quadratic size. In a preliminary computational study, we evaluate the *gain of recovery* using realistic planning data.

1 Introduction

An important problem in the design of wireless networks is the assignment of traffic nodes, e.g., aggregations of users, to antennas. Each antenna has a limited bandwidth capacity to be partitioned among the users in the area covered by the antenna. Users, in general, do not generate a constant traffic rate. Depending on their needs in data traffic or web browsing the requested bitrate fluctuates (e.g., 64 kbps, 384 kbps, 2 Mbps). In the network capacity planning phase usually an average traffic volume is considered. However, during operation, individual users with their actual bitrates need to be admitted to the antenna.

The actual bitrates are difficult to predict in advance, but using historical data average and peak values can be derived. It is also observed that not all peaks occur simultaneously. Therefore, we may assume that the bitrates of only a limited number Γ of users deviate from their average at the same time. Such behavior is captured by the Γ-scenario set introduced by Bertsimas and Sim [3,4].

To ensure a good quality of service for all users at any point in time, a robust assignment is appropriate. Such a robust solution is static over time, neglecting the possibility to reassign (a limited number of) users to other antennas, according to the current traffic volume. Focusing on a single antenna, the robust approach reduces to a classical knapsack problem with uncertainty in the weights. Including the possibility to change the

* This work was supported by the Federal Ministry of Education and Research (BMBF grant 03MS616A, project *ROBUKOM - Robust Communication Networks*, www.robukom.de).

J. Pahl, T. Reiners, and S. Voß (Eds.): INOC 2011, LNCS 6701, pp. 583–588, 2011.

assignment during runtime yields a *recoverable robust knapsack problem* (rrKP): Compared to the planning phase, up to k users can be refused a connection at this antenna (and should be reassigned to another one).

In this paper, we study the rrKP with Γ-scenarios. In Section 2, we describe the problem in detail and give an overview on previous work. In the next section, we derive an integer linear programming formulation. In Section 4, we present the results of preliminary computational experiments on the gain of recovery using data from wireless network planning. We close with concluding remarks in Section 5.

2 Recoverable Robust Knapsack Problem

Despite its simple structure, the knapsack problem (KP) is weakly **NP**-hard [10] but solvable in pseudo-polynomial time [2]. Alternatively, branch-and-cut algorithms can be used to solve the KP. For a detailed introduction see [11, 13].

Yu [15] defined a robust version of the knapsack problem by introducing uncertainty in the profit values via a discrete set of scenarios. For sets with an unbounded number of scenarios, the decision version of the problem is strongly **NP**-complete and can not be approximated, mentioned by Aissi et al. [1]. On the other hand, if the set contains a constant number of scenarios, the problem is only weakly **NP**-complete, solvable in pseudo-polynomial time [15] and there exists an FPTAS [1]. A recoverable robust knapsack problem with discrete scenarios is investigated by [6]. Here, up to k items can be removed and ℓ items added to a first stage solution.

Recently, Klopfenstein and Nace [12] considered robust knapsacks with uncertainty in the weights based on Γ-scenarios [3, 4]. We will in the following extend this model by a recovery action of deleting up to k items.

Definition 1 (Recoverable Robust Knapsack (rrKP)). *Let N be a set of n items with profits p_i, nominal (or default) weight $\underline{w}_i$, and maximum deviation $\hat{w}_i$, $i \in N$. For a given $\Gamma \in \mathbb{N}$, the set $\mathcal{S}_\Gamma$ consists of all scenarios S which define a weight function $w^S : N \to \mathbb{N}$ s.t. $w_i^S \in [\underline{w}_i, \underline{w}_i + \hat{w}_i]$ for all $i \in N$ and $|\{i \in N : w_i^S > \underline{w}_i\}| \leq \Gamma$. For $k \in \mathbb{N}$ and a subset $X \subseteq N$ the* recovery set *$\mathcal{X}_X^k$ consists of all subsets of X with at least $|X| - k$ elements, i.e., $\mathcal{X}_X^k = \{X' \subseteq X : |X \setminus X'| \leq k\}$. Given a knapsack capacity $c \in \mathbb{N}$, the rrKP is to find a set $X \subseteq N$ with maximum profit $p(X) := \sum_{j \in X} p_j$ s.t. for every scenario $S \in \mathcal{S}_\Gamma$ there exists a set $X' \in \mathcal{X}_X^k$ with $w^S(X') \leq c$.*

Note that k models the quality of service: for $k = 0$ (the robust case), every user granted connection is connected, whereas for $k = n$ no service guarantee is given. We now focus on a compact formulation of an rrKP instance with Γ-scenarios.

3 A Compact ILP Formulation

In this section we present an ILP-formulation for the rrKP. To this end, we define binary variables $x_i \in \{0, 1\}$, $i \in N$, denoting the items in the knapsack. Any 0-1 point x satisfying the (exponential many) inequalities

$$\sum_{i \in N} \underline{w}_i x_i + \max_{\substack{\bar{X} \subseteq N \\ |\bar{X}| \leq \Gamma}} \Big(\sum_{i \in \bar{X}} \hat{w}_i x_i - \max_{\substack{Y \subseteq N \\ |Y| \leq k}} \big(\sum_{i \in Y} \underline{w}_i x_i + \sum_{i \in \bar{X} \cap Y} \hat{w}_i \big) \Big) \leq c \tag{1}$$

represents a feasible solution. In the following, we characterize the same polytope by a linear number of constraints. First, we consider a subproblem of finding a scenario $S \in \mathscr{S}_\Gamma$ that imposes the maximum weight on a chosen subset $X \subseteq N$. For given parameters $\Gamma \in \mathbb{N}$ and $k \in \mathbb{N}$ we define the *weight* of a subset $\overline{X} \subseteq X$ as

$$\text{weight}(\overline{X},X) = \sum_{i\in\overline{X}} \hat{w}_i - \max_{\substack{Y\subseteq X\\|Y|\leq k}} \Big(\sum_{i\in Y} \underline{w}_i + \sum_{i\in Y\cap\overline{X}} \hat{w}_i\Big).$$

A *maximum weight set* X_Γ^k is a subset of X with $|X_\Gamma^k| \leq \Gamma$ and with maximum weight. The *maximum weight set problem* (MWSP) is to find for a given set X, and parameters Γ and k, a maximum weight set X_Γ^k.

As the following example indicates, there is no inclusion relation between optimal solutions of an MWSP for different Γ values, i.e., in general $X_\Gamma^k \subsetneq X_{\Gamma+1}^k$.

Example 1. Consider the set $X = \{1,\ldots,4\}$ with nominal weights $\underline{w} = \{3,3,10,10\}$ and deviations $\hat{w} = \{2,2,5,5\}$ and $k = 1$. For $\Gamma = 1$, the sets $X' = \{1\}$ and $X'' = \{2\}$ are the maximum weight sets for this instance with weight$(X',X) = -8$. But, $\bar{X} = \{3,4\}$ is the maximum weight set with weight$(\bar{X},X) = -5$ for $\Gamma = 2$, whereas the sets $\{1,3\}$, $\{1,4\}$, $\{2,3\}$ and $\{2,4\}$ have a weight of -8.

Yet, the MWSP can be solved in polynomial time by exploiting linear programming duality. Computing a maximum weight set is formulated by the following ILP

$$\max_{y\in\{0,1\}^{|X|}} \Big\{ \sum_{i\in X} \hat{w}_i y_i - \max_{z\in\{0,1\}^{|X|}} \big\{ \sum_{i\in X} (\underline{w}_i + \hat{w}_i y_i) z_i \;:\; \sum_{i\in X} z_i \leq k \big\} \;:\; \sum_{i\in X} y_i \leq \Gamma \Big\} \tag{2}$$

The variables y_i represent the choice, whether an item i is in the maximum weight set X_Γ^k, and z_i, whether the item i is removed due to its high weight.

Given a vector y, (2) can be solved by its linear relaxation, since the matrix is totally unimodular [14, Sec. 3.2]. By duality, we obtain a compact ILP reformulation:

$$\begin{aligned}
\max \quad & \sum_{i\in X} \hat{w}_i y_i - k\cdot u - \sum_{i\in X} v_i & & & \text{(3a)}\\
s.t. \quad & \sum_{i\in X} y_i & \leq \Gamma & & \text{(3b)}\\
& \hat{w}_i \cdot y_i - u - v_i & \leq -\underline{w}_i & \quad \forall i \in X & \text{(3c)}\\
& u, v_i & \geq 0 & \quad \forall i \in X & \text{(3d)}\\
& y_i & \in \{0,1\} & \quad \forall i \in X & \text{(3e)}
\end{aligned}$$

where the dual variable u corresponds to $\sum_{i\in X} z_i \leq k$ and v_i to $z_i \leq 1$ for $i \in X$. Next, we parametrize (3a)–(3e) by the possible u-values and denote with $z(u)$ the value of (3a)–(3e).

Lemma 1. *For a fixed parameter u', let $w_i(u') = \min\{\hat{w}_i, -\underline{w}_i + u'\}$ for all $i \in \{1,\ldots,n'\}$, $X^-(u') = \{i \in X \mid w_i(u') < 0\}$, and $X(u') \subseteq X \backslash X^-(u')$ maximizing $\sum_{i\in X(u')} w_i(u')$ with $|X(u')| \leq \Gamma$. Then*

$$z(u') = \sum_{i\in X(u')} w_i(u') + \sum_{i\in X^-(u')} w_i(u') - k\cdot u'$$

holds. Furthermore, there always exists an optimal solution (u^*, y^*, v^*) *of* (3a)–(3e) *with* $u^* \in U := \{0\} \cup \{\underline{w}_i : i \in X\} \cup \{\underline{w}_i + \hat{w}_i : i \in X\}$.

See [5] for the omitted proof. By Lemma 1, inequality (1) is equivalent to

$$\sum_{i \in N} \underline{w}_i x_i + \max_{u \in U} \Big(\sum_{i \in X^-(u)} w_i(u) \cdot x_i - k \cdot u + \max_{\substack{X' \subseteq N \\ |X'| \le \Gamma}} \sum_{i \in X} w_i(u) \cdot x_i \Big) \le c. \tag{4}$$

This inequality can be transformed into the following set of constraints

$$\begin{aligned} \sum_{i \in N} \underline{w}_i x_i + \sum_{i \in X^-(u)} w_i(u) \cdot x_i + \max \sum_{i \in N} w_i(u) \cdot x_i \cdot y_i^u &\le c + ku && \forall u \in U \\ \sum_{i \in N} y_i^u &\le \Gamma && \forall u \in U \\ y_i^u &\in \{0,1\} && \forall i \in N, \forall u \in U. \end{aligned}$$

By dualizing the last part, which is totally unimodular, we get the following ILP

$$\max \sum_{i \in N} p_i x_i \tag{6a}$$

$$s.t. \sum_{\substack{i \in N: \\ \underline{w}_i < u}} \underline{w}_i x_i + \sum_{\substack{i \in N: \\ \underline{w}_i \ge u}} u x_i + \Gamma \xi^u + \sum_{i \in N} \theta_i^u \le c + ku \quad \forall u \in U \tag{6b}$$

$$\min\{\hat{w}_i, -\underline{w}_i + u\} x_i - \xi^u - \theta_i^u \le 0 \quad \forall i \in N, \forall u \in U \tag{6c}$$

$$x_i \in \{0,1\}, \quad \xi^u, \theta_i^u \ge 0 \quad \forall i \in N, \forall u \in U \tag{6d}$$

with new dual variables ξ^u and θ_i^u. The model contains $\mathcal{O}(n^2)$ variables and $\mathcal{O}(n^2)$ constraints depending on the number of different values of $\underline{w}_i, \hat{w}_i, i \in N$.

4 Computational Experiments

In this section, we present some preliminary results of computational experiments on the gain of recovery for the rrKP with Γ-scenarios. As test instances, we consider a wireless network planning problem based on [8]. Given the planning instances, rrKP instances were generated for all 51 antennas with 15 to 221 traffic nodes (geometric mean: 87). Uncertain demands are modeled as in [7].

We implemented formulation (6a)–(6d) of the rrKP in C++ using IBM ILOG CPLEX 12.2 [9] as MIP solver. All computations were carried out on a Linux machine with 2.93 GHz Intel Xeon W3540 CPU, 12 GB RAM, and a time limit of one hour per instance. All instances could be solved to optimality.

We investigate the *gain of recovery*, i. e., the (percentual) increase in the objective value by allowing recovery. As values for k and Γ we consider (rounded-up) relative values of 0 %, 5 %, ..., 25 % of the number of traffic nodes.

Comparing all test instances, Figure 1 shows the geometric mean resp. maximum gain of recovery achieved in these instances (normalized to $k = 0$). Further, the added value for each value k is shown.

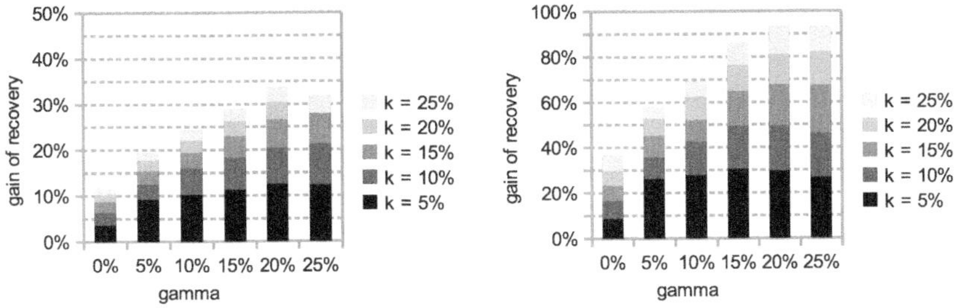

(a) geometric mean (b) observed maximum

Fig. 1. Gain of Recovery. For each instance, Γ,and k, the gain of recovery is determined by the objective value normalized to the corresponding case with $k = 0\,\%$

Fixing k, we observe that in geometric mean a higher gain of recovery is obtained by increasing Γ (e. g., $k = 20\,\%, \Gamma = 5\,\%$ yields 18 %, while $k = 20\,\%, \Gamma = 20\,\%$ yields 30 %). By evaluating the maximum observed gain of recovery, we estimate the potential added value by recovery. It ranges from 25 % ($k = 5\,\%$) to 71 % ($k = 25\,\%$) in geometric mean with an absolute maximum of 93 % ($\Gamma = k = 25\,\%$).

In summary, the results of our preliminary study show that the recoverable robust approach gives a promising added value to the robust approach for small k already.

5 Concluding Remarks

In this paper, we considered the recoverable robust knapsack problem (rrKP) with Γ-scenarios which is a subproblem in wireless network planning under traffic uncertainties. In detail, we introduced a compact ILP-formulation for this problem which is linear in the input size. Using realistic application-based data, we presented the results of a preliminary computational study evaluating the gain of recovery.

In the future, the polyhedral structure of the rrKP with Γ-scenarios should be studied to improve the overall solving process in a branch-and-cut approach.

Acknowledgement

The authors would like to thank Alexander Engels for providing us with realistic network data to evaluate our optimization model.

References

1. Aissi, H., Bazgan, C., Vanderpooten, D.: Min-max and min-max regret versions of some combinatorial optimization problems: a survey (2007), `http://hal.archives-ouvertes.fr/docs/00/15/86/52/PDF/AN7LAMSADE_1-32.pdf`
2. Bellman, R.: Notes on the theory of dynamic programming IV - maximization over discrete sets. Naval Research Logistics Quarterly 3, 67–70 (1956)

3. Bertsimas, D., Sim, M.: Robust discrete optimization and network flows. Mathematical Programming Ser. B 98, 49–71 (2003)
4. Bertsimas, D., Sim, M.: The Price of Robustness. Operations Research 52(1), 35–53 (2004)
5. Büsing, C.: Recoverable Robustness in Combinatorial Optimization. Ph.D. thesis, Technische Universität Berlin (to appear, 2011)
6. Büsing, C., Koster, A.M.C.A., Kutschka, M.: Recoverable robust knapsacks: the discrete scenario case. Optimization Letters, Online First (2011), `http://dx.doi.org/10.1007/s11590-011-0307-1`
7. Claßen, G., Koster, A.M.C.A., Schmeink, A.: Planning wireless networks with demand uncertainty using robust optimization. Optimization Online Eprint server (2011), `http://www.optimization-online.org/DB_HTML/2011/03/2954.html`
8. Engels, A., Reyer, M., Mathar, R.: Profit-oriented combination of multiple objectives for planning and configuration of 4G multi-hop relay networks. In: 7th International Symposium on Wireless Communication Systems (IEEE ISWCS), York, UK, pp. 330–334 (2010)
9. ILOG: CPLEX version 12.2 (2010), `http://www.ibm.com`
10. Karp, R.: Reducibility among combinatorial problems. Complexity of Computer Computations, pp. 85–103 (1972)
11. Kellerer, H., Pferschy, U., Pisinger, D.: Knapsack Problems. Springer, Heidelberg (2004)
12. Klopfenstein, O., Nace, D.: Valid inequalities for a robust knapsack polyhedron - Application to the robust bandwidth packing problem. Networks (to appear, 2010)
13. Martello, S., Toth, P.: Knapsack Problems: Algorithms and Computer Implementations (1990)
14. Wolsey, L.A.: Integer Programming. Wiley, Chichester (1998)
15. Yu, G.: On the max-min 0-1 knapsack problem with robust optimization applications. Operations Research 44, 407–415 (1996)

A Tabu Search Heuristic Based on k-Diamonds for the Weighted Feedback Vertex Set Problem

Francesco Carrabs[1], Raffaele Cerulli[1], Monica Gentili[2], and Gennaro Parlato[3]

[1] University of Salerno, Department of Mathematics
fcarrabs@unisa.it, raffaele@unisa.it
[2] University of Salerno, Department of Computer Science
mgentili@unisa.it
[3] Liafa, CNRS and University Paris Diderot
gennaro@liafa.jussieu.fr

Abstract. Given an undirected and vertex weighted graph $G = (V,E,w)$, the Weighted Feedback Vertex Problem (WFVP) consists of finding a subset $F \subseteq V$ of vertices of minimum weight such that each cycle in G contains at least one vertex in F. The WFVP on general graphs is known to be NP-hard and to be polynomially solvable on some special classes of graphs (e.g., interval graphs, co-comparability graphs, diamond graphs). In this paper we introduce an extension of diamond graphs, namely the *k-diamond* graphs, and give a dynamic programming algorithm to solve WFVP in linear time on this class of graphs. Other than solving an open question, this algorithm allows an efficient exploration of a neighborhood structure that can be defined by using such a class of graphs. We used this neighborhood structure inside our Iterated Tabu Search heuristic. Our extensive experimental results show the effectiveness of this heuristic in improving the solution provided by a 2-approximate algorithm for the WFVP on general graphs.

1 Introduction

Given an undirected graph $G = (V,E)$, a *Feedback Vertex Set* (fvs) of G is a subset $F \subseteq V$ of vertices such that each cycle in G contains at least one vertex in F, i.e. the residual graph induced by the set of vertices $V \setminus F$ is acyclic. The *Feedback Vertex Problem* (FVP) consists of finding an fvs of minimum cardinality. When a weight $w(v)$ is associated with each vertex v of G then we have a *vertex weighted graph*. The Weighted Feedback Vertex Problem (WFVP) on a weighted graph G consists of finding an fvs of minimum weight, where the weight of the set is the sum of the weights of its elements. Both FVP and WFVP are NP-complete problems and have application in several areas of computer science such as circuit testing, deadlock resolution, placement of converters in optical networks, combinatorial cut design. This problem becomes polynomial when addressed on diamond graphs [5], co-comparability graphs [6], convex bipartite graphs [6], permutation graphs [14], interval graphs [15]. The best known approximation algorithm for WFVP has approximation ratio 2. The MGA algorithm introduced in [3] was the first one having such an approximation ratio. Other approximation algorithms for the WFVS are proposed in [1,18] for general graphs and in [2,8,13] for special graph classes. There are also exact algorithms finding a minimum FVS in a graph on n vertices in time $O(1.9053^n)$ [17] and in time $O(1.7548^n)$ [9].

J. Pahl, T. Reiners, and S. Voß (Eds.): INOC 2011, LNCS 6701, pp. 589–602, 2011.

In this paper we focus on the weighted feedback vertex set problem (WFVP). In particular, we introduce an extension of diamond graphs, namely the *k-diamond* graphs, and give a linear time algorithm to solve WFVP on it based on a dynamic programming approach. Moreover, we show how this new class of graphs can be used to define a neighborhood structure (namely, the *k-diamond Neighborhood*) of a given feasible solution and, successively, we show how to solve the problem on general graphs by means of a tabu search technique using the k-diamond neighborhood. Such a class of neighborhood was already introduced in [4], where, however, the computational complexity of finding an optimum WFVP on a k-diamond graph was left open and a heuristic approach was used to solve the problem. We solve such an open problem (by giving a linear time algorithm) and also show the effectiveness of the chosen neighborhood in improving a given initial feasible solution when explored by means of our exploration strategy. In order to do this, experimental results are given to show how our Iterative Tabu Search can improve the initial feasible solution, returned by the 2-approximate MGA algorithm [3], when the k-diamond neighborhood is defined and efficiently explored.

The sequel of the paper is organized as follows. Section 2 introduces the basic notation. Section 3 describes the class of k-diamond graphs and contains the main properties to solve WFVP in linear time on this class. The role of k-diamonds to define a neighborhood structure is described in Section 4, together with the proposed Iterated Tabu Search heuristic. Computational results are reported in Section 5. Finally, concluding remarks are discussed in Section 6.

2 Definitions and Notation

Let $G = (V, E, w)$ be an undirected and vertex weighted graph, where V is the set of n vertices, E is the set of m edges, and, $w(v)$ is a positive weight associated with each vertex $v \in V$. Given a subset $X \subseteq V$ of vertices, let us define its weight $W(X)$ as the sum of the weights of its elements, i.e. $W(X) = \sum_{v \in X} w(v)$ and $\bar{X} = V \setminus X$ its complementary set. If $X = \emptyset$ then $W(X) = 0$. We denote by $G[X]$ the subgraph of G induced by the set of vertices $X \subseteq V$. Formally, $G[X] = (X, E_{[X]}, w)$ where $E_{[X]} = \{(x,y) \in E : x, y \in X\}$. A *tree* T_r rooted in r is an acyclic and connected graph. We define a *forest* $\mathscr{F}$ as a graph where any connected component is a tree. A subset of vertices X is a feedback vertex set of G if and only if $G[\bar{X}]$ is a forest. From now on we denote by $F(G)$ and $F^*(G)$, any feedback vertex set and the minimum weight feedback vertex set of G, respectively. When no confusion may arise we simply denote these sets by F and F^* respectively. Moreover, we define $F_{\bar{v}}$ a feedback vertex set of G not containing vertex v. A vertex $v \in F$ is *redundant* if and only if $F \setminus \{v\}$ is a feedback vertex set of G. Any vertex $v \in V$ is said to be *appended* if it is not included in any cycle of G. Obviously, a set of vertices is an fvs of G if and only if it is an fvs of the graph G' obtained from G after deleting all the appended vertices. We say a graph is *reduced* if it does not contain any appended vertex. The reduction operation of a graph can be performed in linear time. W.l.o.g., from now on we suppose graph G to be a reduced graph. For any additional definition and notation we refer to [7].

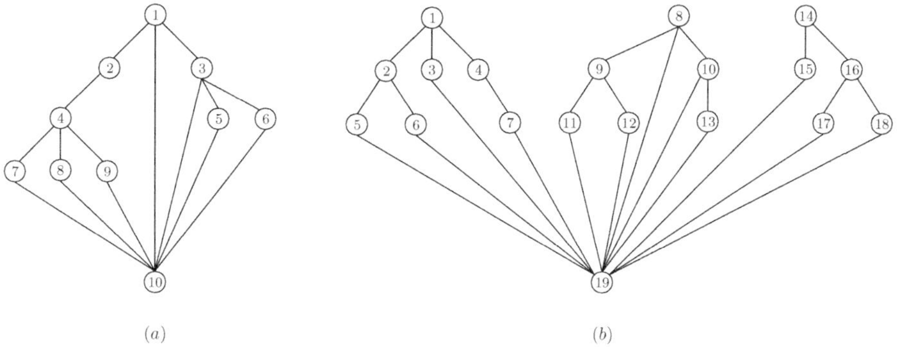

Fig. 1. *(a) A diamond with upper apex $r = 1$ and lower apex $z = 10$. (b) A 3-diamond with upper apices $R = \{1, 8, 14\}$ and lower apex $z = 19$. Note that, as stated by property 1, it is composed by the three diamonds D_1, D_8 and D_{14}.*

3 The Class of k-Diamond Graphs

In this section we first recall the definition of the class of diamond graphs introduced in [5], and successively we formally describe the extended class of k-diamond graphs. Then, we prove the basic properties that are useful to optimally solve WFVP on this new class in linear time.

A weighted diamond $D_{r,z} = (V_r, E_r, w)$ is an undirected and vertex weighted graph where (i) each vertex $v \in V_r$ is included in at least one simple path between r and z and (ii) $D_{r,z}[\bar{z}]$ is a tree. The two vertices r and z are called the *upper* and *lower* apex of a diamond $D_{r,z}$, respectively, and, the subgraph $D_{r,z}[\bar{z}]$ is referred to as the tree T_r rooted in r associated with $D_{r,z}$. In Figure 1(a) the diamond $D_{1,10}$ with upper apex $r = 1$ and lower apex $z = 10$ is shown. Note that by deleting vertex z we obtain the tree $T_1 = D_{1,10}[\bar{10}]$.

As shown in [5], WFVP can be solved in linear time on a diamond graph by a dynamic programming algorithm. Let us refer to such an algorithm as DP. In the sequel we show how to use DP to solve WFVP in linear time on a k-diamond. A k-diamond is a generalization of a diamond where multiple upper apices are allowed, formally:

Definition 1. *A weighted k-diamond $D_{R,z} = (V_R, E_R, w)$, where $k \geq 1$, $R = \{r_1, r_2, \ldots, r_k\} \subseteq V_R$ and $z \in V_R$, is an undirected and vertex weighted graph where (i) each vertex $v \in V_R$ is included in a simple path between exactly one of the k apices $r_i \in R$ and z and (ii) $D_{R,z}[\bar{z}]$ is a forest with k connected components.*

Following the definition introduced for diamond graphs, we refer to the set of vertices R and to vertex z of $D_{R,z}$ as the set of *upper apices* and the *lower apex* of $D_{R,z}$, respectively. The subgraph $D_{R,z}[\bar{z}]$ is referred to as the forest $\mathscr{F}_R$, associated with $D_{R,z}$, whose connected components are the k trees T_{r_i} rooted in $r_i \in R$. Figure 1(b) shows a 3-diamond. The set of upper apices is composed of the three vertices $R = \{1, 8, 14\}$, while the lower apex is vertex $z = 19$. The graph obtained from $D_{R,z}$, after deleting the lower apex, is a forest with the three connected components T_1, T_8 and T_{14} . To keep notation simple, in the sequel of the paper and when no confusion may arise, we denote a k-diamond $D_{R,z}$, with $R = \{r_1, \ldots, r_k\}$, just by D_R and a diamond graph $D_{r,z}$ by D_r. Note that for $k = 1$ a k-diamond is a diamond. Moreover, it is easy to see that the following property holds:

Property 1 (Decomposition). Any k-diamond D_R is composed by k distinct diamonds D_{r_i}, with $r_i \in R$, having all the same lower apex z.

For instance, the 3-diamond depicted in Figure 1(b) is composed by three diamonds D_1, D_8 and D_{14}. We will see in the following how to use the decomposition property to solve WFVP on D_R. By definition of k-diamond, the following properties obviously hold.

Property 2. The singleton $\{z\}$ is an *fvs* of D_R.

Property 3. Every cycle of D_R contains vertex z and vertices belonging to the same connected component of $\mathscr{F}_R$.

Observe that, by property 2, a minimum weight feedback vertex set $F^*(D_R)$ of D_R either contains vertex z or not. Therefore, to find $F^*(D_R)$ we can proceed as follows: (i) compute the minimum weight feedback vertex set $F^*_{\bar{z}}(D_R)$ that does not contain z; (ii) if $W(F^*_{\bar{z}}(D_R)) < w(z)$ then set $F^*(D_R) = F^*_{\bar{z}}(D_R)$ otherwise set $F^*(D_R) = \{z\}$. The computation of $F^*_{\bar{z}}(D_R)$ can be carried out by finding the fvs $F^*_{\bar{z}}(D_{R_i})$ of minimum weight on each of the k diamonds D_{r_i} that compose D_R as proven by the following lemma:

Lemma 1. *Given the k-diamond D_R, let $F^*_{\bar{z}}(D_{r_i})$, $\forall r_i \in R$, be a minimum weight feedback vertex set of diamond D_{r_i} not containing vertex z. Then:* $F^*_{\bar{z}}(D_R) = \bigcup_{r_i \in R} F^*_{\bar{z}}(D_{r_i})$.

Proof. Let $X = \bigcup_{r_i \in R} F^*_{\bar{z}}(D_{r_i})$. We need to prove that X is a minimum fvs of D_R not containing z, i.e. $X = F^*_{\bar{z}}(D_R)$. From property 3 and by definition of $F^*_{\bar{z}}(D_{r_i})$, it is evident that X is an fvs of D_R therefore we have only to prove that X is minimum. Let us suppose, by contradiction, there exists another fvs, say Y, such that $z \notin Y$ and $W(Y) < W(X)$. Let $Y_i = Y \cap D_{r_i}$. By property 3, each set Y_i is an fvs of D_{r_i} that does not contain vertex z. Therefore, since $Y = \bigcup_{r_i \in R} Y_i$ and $W(Y) < W(X)$, there must exist at least a set, say Y_h such that $W(Y_h) < W(F^*_{\bar{z}}(D_{r_h}))$: a contradiction. □

Corollary 1. *Given the k-diamond D_R, a minimum weight feedback vertex set $F^*(D_R)$ is either the set $F^*_{\bar{z}}(D_R)$ or the singleton $\{z\}$.*

From Corollary 1, the problem of finding an optimum WFVS on a k-diamond is reduced to compute $F^*_{\bar{z}}(D_{r_i})$ on each of the k diamonds that compose D_R. These fvs's can be computed using the *DP* algorithm given in [5]. Fig. 2 reports the pseudo-code of our algorithm DP_{multi} that solves WFVP on k-diamonds. Theorem 1 to follow proves that this algorithm runs in linear time.

Theorem 1. *Given a k-diamonds $D_R = (V_R, E_R, w)$, the DP_{multi} algorithm computes $F^*(D_R)$ in $O(|V_R|)$ time.*

Proof. The computation of $F^*_{\bar{z}}(D_{r_i})$ carried out in step 1 of DP_{multi} algorithm takes $O(|V_{r_i}|)$ time (see [5]). Since this computation is repeated for each root $r_i \in R$, then the

Procedure: DP_{multi}

Step 1. **for all** D_{r_i} **compute** $F^*_{\bar{z}}(D_{r_i})$;
Step 2. **Set** $F^*_{\bar{z}}(D_R) \leftarrow \bigcup_{r_i \in R} F^*_{\bar{z}}(D_{r_i})$;
Step 3. **if** $W(F^*_{\bar{z}}(D_R)) < w(z)$ **then** $F^*(D_R) \leftarrow F^*_{\bar{z}}(D_R)$ **otherwise** $F^*(D_R) \leftarrow \{z\}$;
Step 4. **Return** $F^*(D_R)$;

Fig. 2. Pseudo code of algorithm DP_{multi}

total cost of step 1 is equal to $O(|V_R|)$ time. The joining operation carried out at step 2 requires $O(k)$ time, while step 3 and step 4 require constant time. Consequently, DP_{multi} runs in $O(|V_R|)$ time. □

The next sections contain a description of a general neighborhood structure based on the class of k-diamonds and introduce an operator that, using DP_{multi}, efficiently explores such a neighborhood. This operator will be later embedded into our Iterated Tabu Search.

4 The Neighborhood Structures and the Iterative Tabu Search

The basic paradigm of tabu search is to use information about the search history to guide local search approaches to overcome local optimality. Based on some sort of memory certain moves may be forbidden, we say they are set tabu (and appropriate move attributes are put into a list, the so-called tabu list). The search may imply acceptance deteriorating moves when no improving moves exist or all improving moves of the current neighborhood are set tabu. We implemented an extension of the standard Tabu Search [10,11,12] (TS), namely the *Iterated Tabu Search* [16] (ITS), whose central idea is based on the concept of *intensification* and *diversification*. The intensification phase is focused in finding a better (locally optimal) solution in "surroundings", i.e. neighborhood, of the current solution. The ITS method uses the classical TS to achieve such an improvement. The Diversification phase is used whenever the tabu memory indicates that one is trapped in a certain basin of attraction and then allows to escape from the current local optimum and to move towards new regions in the solution space.

In the following subsections the main components of the algorithm are described: (i) the neighborhood structures (namely, the k-diamond neighborhood and the 2-exchange neighborhood), (ii) the corresponding exploration strategies (namely, the $Single-Insert$ and the $Double-Insert$ operators, respectively); (iii) the tabu list and (iv) the diversification phase. The pseudo-code of our Iterative Tabu Search (ITS) is given in Fig. 5.

4.1 The k-Diamond Neighborhood

Given a graph G, let F be any not redundant fvs of G and $\mathscr{F} = G[\bar{F}]$ the forest induced by vertices not in F. By inserting a vertex $z \in F$ in $\mathscr{F}$ a k-diamond D_R is obtained. Let I_z

be an fvs of D_R not containing z, then the set $F' = I_z \cup \{F \setminus \{z\}\}$ is a new fvs of G. Note that, F' could contain redundant vertices. Let O_z be the set of the redundant vertices of F'. Note that, by construction of I_z, we have $O_z \subseteq F \setminus \{z\}$. Add z to O_z and consider the vertex set $F_{new} = I_z \cup \{F \setminus O_z\}$. F_{new} is a not redundant fvs of G and: if $W(I_z) < W(O_z)$, its weight is lower than the weight of F. Given a vertex $z \in F$, we define the couple (I_z, O_z) an *exchange set* of z, formally:

Definition 2. *Given a vertex $z \in F$, the couple (I_z, O_z), where $I_z \subseteq V \setminus F$, $O_z \subseteq F$ and $z \in O_z$, is an exchange set of z if the set $I_z \cup \{F \setminus O_z\}$ is a not redundant* fvs *of G.*

Let us denote by $\mathscr{E}(F,z)$ the collection of all the exchange sets associated with $z \in F$, i.e. $\mathscr{E}(F,z) = \{(I_z, O_z) : I_z \cup \{F \setminus O_z\}$ *is a not reduntant fvs of* $G\}$. The k-diamond neighborhood is defined as follows:

Definition 3. *Given a graph G and an fvs F, the k-diamond neighborhood $\mathscr{N}(F)$ is the set of all not redundant fvs of G that can be obtained from F through the exchange sets associated with each vertex $z \in F$:*

$$\mathscr{N}(F) = \left\{ I_z \cup \{F \setminus O_z\} : (I_z, O_z) \in \mathscr{E}(F,z), \forall z \in F \right\}$$

Note that, given a vertex $z \in F$, finding the minimum cost set I_z associated with it corresponds to find the minimum weight feedback vertex set on the k-diamond associated with z. Hence, by applying the DP_{multi} algorithm we can perform an implicit exhaustive exploration on the neighborhood to find a local optimum in polynomial time. This exploration is carried out by our first operator $Single-Insert$ that is described next.

4.1.1 The $Single-Insert$ Operator

Given a not redundant fvs F of G and the incumbent solution F^*, the $Single-Insert$ operator builds, for each $z \in F$, the k-diamond D_R by introducing z in $\mathscr{F} = G[\bar{F}]$. Successively, it computes an exchange set (I_z, O_z) where $I_z = F_{\bar{z}}^*(D_R)$, i.e I_z is the minimum feedback vertex set of D_R not containing z. The operator selects the best exchange set (I_z^*, O_z^*) such that $W(I_z^*) - W(O_z^*) = min_{(I_z,O_z):z \in F}\{W(I_z) - W(O_z)\}$. More in detail (see Fig. 3), the operator builds the k-diamond D_R (step 1), finds the fvs $F_{\bar{z}}^*(D_R)$ by applying algorithm DP_{multi} and sets $I_z \leftarrow F_{\bar{z}}^*(D_R)$ (step 2). The operator (step 3) finds redundant vertices (if any) of the new fvs $F_{new} = F \setminus \{z\} \cup \{I_z\}$ to be inserted in O_z (initially $O_z = \{z\}$). To this end, $Single-Insert$ builds the forest $\mathscr{F}' = G[\bar{F}_{new}]$ and reintroduces, one by one, each vertex $z' \in F \setminus \{z\}$ to check whether z' is redundant or not. If z' is redundant then it is moved from F_{new} to O_z. The final fvs $F_{new} = I_z \cup \{F \setminus O_z\}$ is then obtained after all the vertices in $F \setminus \{z\}$ are checked for redundancy. Note that the pair (I_z, O_z) is the move corresponding to the transition from solution F to its neighbor F_{new}.

The weight of the new set F_{new} is then compared with the weight of the incumbent solution F^* found so far. If $W(F_{new}) < W(F^*)$, then the operator sets the best move (I_z^*, O_z^*) equal to (I_z, O_z) even if this move is tabu (this represents the application of an aspiration criterion [12]). Otherwise, if (I_z, O_z) is not tabu and the corresponding solution is better than the solution associated with (I_z^*, O_z^*), the algorithm sets (I_z^*, O_z^*) equal to (I_z, O_z). Finally, if both previous cases do not hold, (I_z, O_z) is neglected.

Procedure: *Single − Insert(G,F, F*)*

Set $W(I_z^*) \leftarrow \infty$, $W(O_z^*) \leftarrow 0$
for all $z \in F$ **do**

Step 1. Insert z in $G[\bar{F}]$ and reduce the obtained graph to produce the k-diamond D_R;
Step 2. Set $I_z \leftarrow F_{\bar{z}}^*(D_R)$;
Step 3. Find the set of redundant nodes O_z, add z to O_z, and set $F_{new} \leftarrow I_z \cup \{F \setminus O_z\}$;
Step 4. **if** $W(F_{new}) < W(F^*)$ **do** *// aspiration criterion //*
$I_z^* \leftarrow I_z, O_z^* \leftarrow O_z$;
else if $W(I_z) - W(O_z) < W(I_z^*) - W(O_z^*)$ **and** (I_z, O_z) *is not tabu* **do**
$I_z^* \leftarrow I_z, O_z^* \leftarrow O_z$;

end for
return $(I_z^* \cup \{F \setminus O_z^*\})$;

Fig. 3. *Pseudo-code of operator Single − Insert*

4.2 The 2-Exchange Neighborhood and the *Double − Insert* Operator

Additional neighborhoods similar to the k-diamond neighborhood above described can be considered if more than one vertex of F is selected to be introduced in $\mathscr{F} = G[\bar{F}]$. Indeed, a drawback of the *Single − Insert* operator concerns the diversification of the explored solutions. In fact, when there are not redundant vertices, only one vertex (the lower apex z) is moved from F to $\mathscr{F} = G[\bar{F}]$. Hence, in the worst case, several applications of the operator *Single − Insert* are necessary to remove more than one vertex from F. In order to overcome this issue, we consider a new neighborhood, namely the 2-exchange neighborhood, to diversify the explored solutions, that is, we considered the case when two vertices $\{z_i, z_j\}$ are selected to be inserted in $\mathscr{F}$. This neighborhood is explored by the operator *Double − Insert* (see Fig. 4) that differs from *Single − Insert* since it inserts two lower apices $\{z_i, z_j\}$ into $\mathscr{F}$, and finds the fvs I_{z_i,z_j} by applying algorithm MGA. MGA is a greedy algorithm that selects at each iteration the vertex v such that the ratio $w(v)/d(v)$ is minimum, where $d(v)$ is the degree of the vertex. When a vertex is selected, it is removed from G and G is then reduced to obtain the subgraph G'. The degree of each vertex v in G' is updated and for each edge (u,v) that was removed during the reduction process, the weight of its endpoints is decreased by the quantity $w(v)/d(v)$. The selection of a new vertex is then carried out on G' until it is not empty. For more details on MGA the reader can refer to [3].

The *Single − Insert* operator and the *Double − Insert* operator will be used during the intensification phase of Iterative Tabu Search metaheuristic.

4.3 The Tabu List

At iteration t, after a relocation of vertices is carried out according to a resulting exchange set (I_z, O_z), the inverse move (O_z, I_z) cannot be carried out for the next Δ iterations, where Δ is the tabu list size. To implement a fast way for storing each move (O_z, I_z) we use a bit mask and an hash-table. Since both I_z and O_z are vertex sets and each vertex has a distinct ID, we allocate two bit-mask *bl* and *br* whose size is $|V|$. We set the bits in *bl* and *br* corresponding to the vertices in O_z and I_z, respectively, equal to

Procedure: $Double-Insert(G,F,F^*)$

Set $W(I_C^*) \leftarrow \infty$, $W(O_C^*) \leftarrow 0$
for all pair $C = (z_i, z_j)$ with $z_i, z_j \in F$ **do**

Step 1. Insert z_i and z_j in $G[\bar{F}]$ and reduce it to obtain G';
Step 2. Apply MGA to find an fvs I_C of G';
Step 3. Find the set of redundant nodes O_C, add z_i and z_j to O_C and set $F_{new} \leftarrow I_C \cup \{F \setminus O_C\}$;
Step 4. **if** $W(F_{new}) < W(F^*)$ **do** *// aspiration criterion //*
$I_C^* \leftarrow I_C$, $O_C^* \leftarrow O_C$;
else if $W(I_C) - W(O_C) < W(I_C^*) - W(O_C^*)$ **and** (I_C, O_C) is not tabu **do**
$I_C^* \leftarrow I_C$, $O_C^* \leftarrow O_C$;

end for
return $I_C^* \cup \{F \setminus O_C^*\}$;

Fig. 4. *Pseudo-code of operator* $Double-Insert$

1. The two strings are then concatenated to generate the string of bits *bl-br*, that is the *key* associated with the move. This key, that is unique for each move, is given to the hash function to save the move. To verify if a move is tabu it is sufficient to generate its key and check whether it is inside the hash table. The key generation, the insertion into the hash table and the checking operations require $O(|I_z| + |O_z|)$ time. The keys are saved inside a FIFO queue whose size is Δ, hence when the queue is full and a new key has to be inserted, the key on the head is removed from the queue and from the hash table. This operation requires constant time. We used a reactive tabu list, that uses a list whose size is dynamically updated during the computation according to the evolution of the search. The value of Δ ranges between a lower bound β^- and an upper bound β^+ that are fixed at the beginning of the computation and never change. Given an initial fvs F, we set $\beta^- = 5$, $\beta^+ = \max\{3\beta^-, \frac{|F|}{3}, \frac{|\bar{F}|}{3}\}$ and $\Delta = \beta^- + \frac{(\beta^+ - \beta^-)}{2}$. After each iteration t, if the new solution F' found during the intensification phase (steps 4-17 in Fig. 5) is better than F^*, then Δ is increased by one. Otherwise, if F' is worse than F^* but better than the solution found at the previous iteration then Δ is not changed. Finally, if F' is worse than the previous one then Δ is decreased by one.

4.4 The Diversification Phase

The diversification phase is implemented using a modified version of the $Double-Insert$ operator (namely the $Multi-Insert$ operator). Given a solution F, $Multi-Insert$ differs from $Double-Insert$ since a subset of vertices $P \subset F$ with $|P| > 2$ is inserted into the forest $\mathscr{F} = G[\bar{F}]$ to obtain a new graph G'. There are three main aspects to take into account in the diversification phase: (i) when to apply the diversification and on which solution, (ii) the cardinality of the set P, and, (iii) which vertices to introduce in P. We apply the diversification either to the best solution F^* found so far (step 23 in Fig. 5) or to the solution F' (step 25 in Fig. 5) computed during the intensification phase. We keep a counter q that ranges from 1 to θ (that is the maximum number of diversification operations performed by the algorithm) and, as soon as this bound is reached, the ITS stops. The cardinality of P is computed according to the following

Procedure: $ITS(G,\theta,\sigma)$

```
F ← F* ← MGA(G);
for q = 1 to θ do
    // Intensification Phase
    F' ← V;
    for h = 1 to σ do
        F1 ←Single-Insert(G,F,F*);
        F2 ←Double-Insert(G,F,F*);
        if W(F1) < W(F2) then
            F ← F1;
        else
            F ← F2;
        end if
        Save the inverse move into the tabu list.
        if W(F) < W(F') then
            F' ← F; h ← 1;
        end if
    end for
    if W(F') < W(F*) then
        F* ← F';
    end if
    // Diversification Phase
    if q is even then
        F ← Diversification(F*);
    else
        F ← Diversification(F');
    end if
end for
return F*;
```

Fig. 5. *Pseudo-code of Iterated Tabu Search*

formula: $\max\left\{5, \frac{|F|\times(20+5q)}{100}\right\}$. Finally, to remove vertices from F we consider the last iteration $it^+(v)$ when v has been inserted in F: the vertices of F are sorted in increasing order according to $it^+(v)$ and the first $|P|$ vertices of F are selected.

5 Computational Results

The ITS algorithm was coded in C and run on a 2.33 GHz Intel Core2 Q8200 processor. Since there are no available benchmark instances for the WFVP, we generated instances for the following class of graphs: random graphs, squared and not squared grids, taurus and hypercube. Each instance is characterized by the number of vertices, the number of edges, a seed and a range of values for the weight of the vertices. The weight ranges are: 10-25, 10-50 and 10-75. For each combination of parameters we generated five instances with the same characteristics except for the seed. The results reported in the tables are average values over these five instances. Small instances have 25, 50 and 75 vertices. Large instances have 100, 200, 300, 400 and 500 vertices.

Tables 1-2-3 report the results of the MGA algorithm and of our ITS algorithm. The first and second columns in each table report the id and characteristics of each instance, respectively. For random graphs (Table 1): number of vertices (n), number of edges (m), the lower (low) and upper (up) bounds for the weight of each vertex. For the hypercube graphs (Table 2a and 3a) the number of edges is omitted because it depends on the number of vertices. For the remaining graph classes: x is the number of rows and y is the number of columns. The third column in each table reports the solution

Table 1. *(a) Test results on random graphs: (a) small instances and (b) large instances*

(a)

RANDOM GRAPHS: Small Instances								
ID	Instance				MGA	ITS		GAP
	n	m	low	up	Value	Value	Time	
1	25	33	10	25	70.8	63.8	0.00	-9.89%
2	25	33	10	50	105.4	99.8	0.00	-5.31%
3	25	33	10	75	133.6	125.2	0.00	-6.29%
4	25	69	10	25	166.8	157.6	0.00	-5.52%
5	25	69	10	50	294.8	272.2	0.00	-7.67%
6	25	69	10	75	455	409.4	0.00	-10.02%
7	25	204	10	25	286.4	273.4	0.02	-4.54%
8	25	204	10	50	527	507	0.01	-3.80%
9	25	204	10	75	829.8	785.8	0.01	-5.30%
10	50	85	10	25	191.4	175.4	0.03	-8.36%
11	50	85	10	50	298.2	280.8	0.03	-5.84%
12	50	85	10	75	377.2	348	0.02	-7.74%
13	50	232	10	25	409	389.4	0.07	-4.79%
14	50	232	10	50	746.8	708.6	0.06	-5.12%
15	50	232	10	75	1018.4	951.6	0.04	-6.56%
16	50	784	10	25	612.6	602.2	0.11	-1.70%
17	50	784	10	50	1204.2	1172.2	0.15	-2.66%
18	50	784	10	75	1685.2	1649.4	0.14	-2.12%
19	75	157	10	25	347.2	321	0.13	-7.55%
20	75	157	10	50	571.2	526.2	0.14	-7.88%
21	75	157	10	75	815	757.2	0.11	-7.09%
22	75	490	10	25	654.2	638.6	0.16	-2.38%
23	75	490	10	50	1286.6	1230.6	0.27	-4.35%
24	75	490	10	75	1870.8	1793.6	0.13	-4.13%
25	75	1739	10	25	903.2	891	0.40	-1.35%
26	75	1739	10	50	1681	1664.8	0.35	-0.96%
27	75	1739	10	75	2479.8	2452.8	0.33	-1.09%
AVG								**-5.18%**

(b)

RANDOM GRAPHS: Large Instances								
ID	Instance				MGA	ITS		GAP
	n	m	low	up	Value	Value	Time	
1	100	247	10	25	536.4	501.4	0.33	-6.52%
2	100	247	10	50	910.4	845.8	0.37	-7.10%
3	100	247	10	75	1279.2	1223.8	0.28	-4.33%
4	100	841	10	25	846	828.2	0.27	-2.10%
5	100	841	10	50	1793.2	1729.6	0.60	-3.55%
6	100	841	10	75	2512.2	2425.6	0.35	-3.45%
7	100	3069	10	25	1151.2	1134	0.59	-1.49%
8	100	3069	10	50	2218	2179	0.69	-1.76%
9	100	3069	10	75	3284	3228.8	0.77	-1.68%
10	200	796	10	25	1547.8	1488.4	3.48	-3.84%
11	200	796	10	50	2544.2	2442.6	2.50	-3.99%
12	200	796	10	75	3277.4	3157	2.78	-3.67%
13	200	3184	10	25	2035.6	2003.6	2.78	-1.57%
14	200	3184	10	50	3775.2	3683.6	2.67	-2.43%
15	200	3184	10	75	5259	5158.6	2.76	-1.91%
16	200	12139	10	25	2467.4	2450	11.31	-0.71%
17	200	12139	10	50	4182.2	4149.4	8.91	-0.78%
18	200	12139	10	75	5568.8	5531.4	6.98	-0.67%
19	300	1644	10	25	2136.6	2072.6	10.19	-3.00%
20	300	1644	10	50	4384.6	4239.4	9.12	-3.31%
21	300	1644	10	75	6411.2	6154.4	11.09	-4.01%
22	300	7026	10	25	3267.6	3231	19.59	-1.12%
23	300	7026	10	50	6368.4	6261.4	21.12	-1.68%
24	300	7026	10	75	8825.2	8660.6	17.21	-1.87%
25	300	27209	10	25	3749.2	3729.2	44.74	-0.53%
26	300	27209	10	50	5774.2	5738	29.26	-0.63%
27	300	27209	10	75	10514	10469.6	50.88	-0.42%
28	400	2793	10	25	3097	3015.2	29.99	-2.64%
29	400	2793	10	50	6726.8	6528	35.82	-2.96%
30	400	2793	10	75	9006.8	8730	35.36	-3.07%
31	400	12369	10	25	4514.4	4451.8	55.14	-1.39%
32	400	12369	10	50	6896	6837.4	35.88	-0.85%
33	400	12369	10	75	10788.8	10661.8	48.12	-1.18%
34	400	48279	10	25	5090	5060.8	123.27	-0.57%
35	400	48279	10	50	7142.6	7109.2	85.15	-0.47%
36	400	48279	10	75	15202.4	15114.6	127.31	-0.58%
37	500	4241	10	25	4197.4	4102.8	68.35	-2.25%
38	500	4241	10	50	7447.6	7285	70.14	-2.18%
39	500	4241	10	75	11619.6	11285.6	63.93	-2.87%
40	500	19211	10	25	5817.2	5745.8	99.12	-1.23%
41	500	19211	10	50	7819.2	7725	89.63	-1.20%
42	500	19211	10	75	14335.4	14167.8	80.09	-1.17%
43	500	75349	10	25	6388.6	6366.4	181.71	-0.35%
44	500	75349	10	50	8709	8671.2	155.18	-0.43%
45	500	75349	10	75	16994.6	16939.2	201.96	-0.33%
AVG								**-2.09%**

Table 2. *(a) Test results on small instances:(a) hypercube graphs, (b) taurus graphs, (c) squared grid graphs and (d) not squared grid graphs*

(a)

HYPERCUBE GRAPHS: Small Instances							
ID	Instance			MGA	ITS		GAP
	n	low	up	Value	Value	Time	
1	16	10	25	77.4	72.2	0.00	-6.72%
2	16	10	50	99.8	93.8	0.00	-6.01%
3	16	10	75	99.8	97.4	0.00	-2.40%
4	32	10	25	177.2	170	0.01	-4.06%
5	32	10	50	249.4	241	0.00	-3.37%
6	32	10	75	286.2	277.6	0.00	-3.00%
7	64	10	25	377.6	354.6	0.13	-6.09%
8	64	10	50	486.2	476	0.05	-2.10%
9	64	10	75	514	503.8	0.05	-1.98%
AVG							**-3.97%**

(b)

TAURUS GRAPHS: Small Instances								
ID	Instance				MGA	ITS		GAP
	x	y	low	up	Value	Value	Time	
1	5	5	10	25	113.2	101.4	0.00	-10.42%
2	5	5	10	50	135.2	124.4	0.00	-7.99%
3	5	5	10	75	167.4	157.8	0.00	-5.73%
4	7	7	10	25	206	197.4	0.03	-4.17%
5	7	7	10	50	243.4	234.2	0.02	-3.78%
6	7	7	10	75	282.6	269.6	0.02	-4.60%
7	9	9	10	25	324.8	310.4	0.20	-4.43%
8	9	9	10	50	388.4	370	0.17	-4.74%
9	9	9	10	75	448.4	432.2	0.16	-3.61%
AVG								**-5.50%**

(c)

SQUARED GRID GRAPHS: Small Instances								
ID	Instance				MGA	ITS		GAP
	x	y	low	up	Value	Value	Time	
1	5	5	10	25	122.4	114	0.00	-6.86%
2	5	5	10	50	208.4	199.8	0.00	-4.13%
3	5	5	10	75	335.2	312.6	0.00	-6.74%
4	7	7	10	25	270.8	252.4	0.03	-6.79%
5	7	7	10	50	464.6	439.8	0.03	-5.34%
6	7	7	10	75	749.4	718.4	0.03	-4.14%
7	9	9	10	25	466	444.2	0.22	-4.68%
8	9	9	10	50	805.8	754.6	0.29	-6.35%
9	9	9	10	75	1209.6	1138	0.13	-5.92%
AVG								**-5.66%**

(d)

NOT SQUARED GRID GRAPHS: Small Instances								
ID	Instance				MGA	ITS		GAP
	x	y	low	up	Value	Value	Time	
1	8	3	10	25	104.8	96.8	0.00	-7.63%
2	8	3	10	50	174.8	157.4	0.00	-9.95%
3	8	3	10	75	246.6	220	0.00	-10.79%
4	9	6	10	25	326.4	295.8	0.07	-9.37%
5	9	6	10	50	512	489.4	0.04	-4.41%
6	9	6	10	75	801	755	0.04	-5.74%
7	12	6	10	25	431.6	399.8	0.15	-7.37%
8	12	6	10	50	717.2	673.4	0.12	-6.11%
9	12	6	10	75	1092.8	1017.4	0.10	-6.90%
AVG								**-7.59%**

value returned by MGA. We do not report the computational time of MGA since it is always negligible. Fourth and fifth columns in the tables report the solution value and the computational time (in seconds) of our ITS algorithm. Finally, last column reports the percentage gap between the solution values returned by the two algorithms. This gap is positive if MGA finds a better solution than ITS and negative otherwise. The last line of the tables reports the average value of this gap computed on all the instances of the table. On small instances of random graphs (Table 1) we can see from the gap column that ITS always finds a better solution than MGA and the CPU time is less than half of a second. On the 27 instances of Table 1a, this gap is greater than 5% for 15 instances and in one case (instance 6) it is greater than 10%. On average, the improvement obtained by ITS is around 5%. It is interesting to observe that as the density of graph increases the gap decreases. This reveals that the selection criterion applied by MGA (the ratio between weight and degree of a vertex) is less effective on sparse graphs. This trend is evident on large instances (Table 1b) where (see for example instances with 500 vertices) the gap for sparse instances is more that 2% and it is less that 0.4% on more dense instances. The computational time of ITS is less than 1 minute for the first 33 instances and is less that 4 minutes for the remaining large instances.

Consider Table 2 (for small instances) and Table 3 (for large instances) to compare the algorithms on the other types of graph. Let us analyze the small instances for the hypercube graphs. From the gap column, we can see that in three cases (instances 1,

Table 3. *(a) Test results on large instances: (a) hypercube, (b) taurus, (c) squared grid and (d) not squared grid graphs*

(a)

HYPERCUBE GRAPHS: Large Instances							
ID	Instance			MGA	ITS		GAP
	n	low	up	Value	Value	Time	
1	128	10	25	784.8	740	1.09	-5.71%
2	128	10	50	1125.4	1071	0.40	-4.83%
3	128	10	75	1196.4	1163.6	0.34	-2.74%
4	256	10	25	1641.2	1542.6	9.41	-6.01%
5	256	10	50	2429.4	2311.4	6.45	-4.86%
6	256	10	75	2673.4	2590.8	3.94	-3.09%
7	512	10	25	3416.4	3240.8	73.51	-5.14%
8	512	10	50	5147.4	4921.8	67.58	-4.38%
9	512	10	75	5789.2	5588.6	51.74	-3.47%
AVG							**-4.47%**

(b)

TAURUS GRAPHS: Large Instances								
ID	Instance				MGA	ITS		GAP
	x	y	low	up	Value	Value	Time	
1	10	10	10	25	413	388.8	0.38	-5.86%
2	10	10	10	50	476.4	458.6	0.37	-3.74%
3	10	10	10	75	523	504.8	0.25	-3.48%
4	14	14	10	25	793.8	750.8	5.96	-5.42%
5	14	14	10	50	908.2	875.6	3.68	-3.59%
6	14	14	10	75	1062.4	1017.2	3.59	-4.25%
7	17	17	10	25	1167.4	1110.2	21.98	-4.90%
8	17	17	10	50	1364.8	1307.6	20.93	-4.19%
9	17	17	10	75	1551.4	1502.4	23.18	-3.16%
10	20	20	10	25	1621.2	1548.6	88.75	-4.48%
11	20	20	10	50	1867.2	1803.4	81.03	-3.42%
12	20	20	10	75	2109.6	2042.6	55.19	-3.18%
13	23	23	10	25	2136.4	2043.4	278.08	-4.35%
14	23	23	10	50	2520	2412.2	177.53	-4.28%
15	23	23	10	75	2818.8	2705.4	184.99	-4.02%
AVG								**-4.15%**

(c)

SQUARED GRID GRAPHS: Large Instances								
ID	Instance				MGA	ITS		GAP
	x	y	low	up	Value	Value	Time	
1	10	10	10	25	613	570.6	0.54	-6.92%
2	10	10	10	50	1002	948.8	0.41	-5.31%
3	10	10	10	75	1657.4	1566	0.51	-5.51%
4	14	14	10	25	1273.6	1209.4	8.07	-5.04%
5	14	14	10	50	2103	2008.6	8.06	-4.49%
6	14	14	10	75	3618.6	3401.2	7.23	-6.01%
7	17	17	10	25	1917	1834.2	42.63	-4.32%
8	17	17	10	50	3231	3070.6	29.71	-4.96%
9	17	17	10	75	5380.8	5089.8	29.68	-5.41%
10	20	20	10	25	2781	2619.8	85.42	-5.80%
11	20	20	10	50	4516.8	4321.2	103.84	-4.33%
12	20	20	10	75	7650.4	7272.6	127.81	-4.94%
13	23	23	10	25	3626.8	3462.8	371.23	-4.52%
14	23	23	10	50	6171.4	5865.4	291.52	-4.96%
15	23	23	10	75	10195.6	9723.4	240.50	-4.63%
AVG								**-5.14%**

(d)

NOT SQUARED GRID GRAPHS: Large Instances								
ID	Instance				MGA	ITS		GAP
	x	y	low	up	Value	Value	Time	
1	13	7	10	25	552	513	0.36	-7.07%
2	13	7	10	50	870	803.4	0.31	-7.66%
3	13	7	10	75	1471	1390.8	0.34	-5.45%
4	18	11	10	25	1284.6	1208	6.78	-5.96%
5	18	11	10	50	2149.2	2049.8	8.77	-4.62%
6	18	11	10	75	3643.6	3431	5.79	-5.83%
7	23	13	10	25	2049.4	1930.6	42.54	-5.80%
8	23	13	10	50	3366.2	3194.8	43.01	-5.09%
9	23	13	10	75	5653	5286.6	34.27	-6.48%
10	26	15	10	25	2690.6	2532.8	104.81	-5.86%
11	26	15	10	50	4387.4	4164.8	82.30	-5.07%
12	26	15	10	75	7427.6	7063.4	85.79	-4.90%
13	29	17	10	25	3443.2	3270	236.94	-5.03%
14	29	17	10	50	5716.6	5430.4	251.17	-5.01%
15	29	17	10	75	9451.8	8993.2	196.66	-4.85%
AVG								**-5.65%**

2 and 7) the gap is greater than 5% while on average it is around 4%. This difference becomes more significant on the other three classes of graphs: for taurus graph the average gap is around 5.5%, for the squared grid graphs the average gap is 5.66% and on the not squared grid graphs it is 7.59% (and, except for instance 5, it is always greater than 5%). The CPU time of ITS on these four classes of graphs is negligible being always less than half of a second. Note that, since in these graphs several vertices have the same degree, the selection criterion applied by MGA is essentially led by the weight of the vertices and this probably causes its poor results.

On large instances, there is a sensible reduction of the gap between ITS and MGA for taurus, squared grid and not squared grid graphs, while this gap increases on hypercube graphs. In detail, on the hypercube, the gap is greater than 3% for three instances (instances 1, 4 and 7) with an average value of 4.47%. The computational time of ITS on this class of graphs is, in the worst case, slightly more than 1 minute. On taurus graphs the average gap is equal to 4.15% and on two instances (1 and 4) it is greater than 5%. ITS computational time increases to 5 minutes in the worst case. For half of the squared grid instances, the gap is greater than 5% while the average gap is equal to 5.14%. These graphs ended to be more expensive for ITS in terms of computational time. Finally, as already observed for small instances, the not squared grid graphs are the hardest instances for MGA. Indeed, only in 3 cases (instances 5, 12 and 15) the gap is less than 5% while the average gap is equal to 5.65%.

6 Conclusions

We addressed a well known NP-complete problem in the literature (the Weighted Feedback Vertex Set Problem) with application in several areas of computer science such as circuit testing, deadlock resolution, placement of converters in optical networks, combinatorial cut design. In this paper we presented a polynomial time exact algorithm (the DP_{multi} algorithm) to solve the problem on a special class of graphs, namely the k-diamond graphs. In addition, we proposed an Iterative Tabu Search algorithm considering two different neighborhood structures one of which is based on the k-diamond graphs where the DP_{multi} algorithm was hugely used for a better exploration. We carried out an extensive experimentation to show the effectiveness of our approach when compared with the well known 2-approximation algorithm MGA. Our approach shows a very good trade-off between solution quality and computational time: our ITS solves the problem in less than 1 second for instances up to 100 vertices with an improvement of the quality of the solution when compared to those returned by MGA. This makes ITS suitable to be embedded on an exact approach, that is object of our future research.

References

1. Bafna, V., Berman, P., Fujito, T.: A 2-Approximation Algorithm for the Undirected Feedback Vertex Set Problem. SIAM J. Discrete Math. 12(3), 289–297 (1999)
2. Bar-Yehuda, R., Geiger, D., Naor, J., Roth, R.M.: Approximation Algorithms for the Feedback Vertex Set Problem with Applications to Constraint Satisfaction and Bayesian Inference. SIAM J. Comput. 27(4), 942–959 (1998)
3. Becker, A., Geiger, D.: Approximation Algorithms for the Loop Cutset Problem. In: Proceedings of the 10th Conference on Uncertainty in Artificial Intelligence, pp. 60–68 (1994)
4. Brunetta, L., Maffioli, F., Trubian, M.: Solving The Feedback Vertex Set Problem On Undirected Graphs. Discrete Applied Mathematics 101, 37–51 (2000)
5. Carrabs, F., Cerulli, R., Gentili, M., Parlato, G.: A Linear Time Algorithm for the Minimum Weighted Feedback Vertex Set on Diamonds. Information Processing Letters 94, 29–35 (2005)
6. Chang, M.S., Liang, Y.D.: Minimum feedback vertex sets in cocomparability graphs and convex bipartite graphs. Acta Informatica 34, 337–346 (1997)
7. Cormen, T.H., Leiserson, C.E., Rivest, R.L.: Introduction to Algorithms. MIT Press, Cambridge (2001)
8. Even, G., Naor, J., Schieber, B., Zosin, L.: Approximating Minimum Subset Feedback Sets in Undirected Graphs with Applications. SIAM J. Discrete Math. 13(2), 255–267 (2000)

9. Fomin, F.V., Gaspers, S., Pyatkin, A.V.: Finding a minimum feedback vertex set in time $\mathcal{O}(1.7548^n)$. In: Bodlaender, H.L., Langston, M.A. (eds.) IWPEC 2006. LNCS, vol. 4169, pp. 184–191. Springer, Heidelberg (2006)
10. Glover, F.: Tabu Search. ORSA Journal on Computing 1, 190–206 (1989)
11. Glover, F., Laguna, M.: Tabu search. Kluwer, Dordrecht (1997)
12. Glover, F.: Tabu Search: A Tutorial. Interfaces 20, 74–94 (1990)
13. Kleinberg, J., Kumar, A.: Wavelength Conversion in Optical Networks. Journal of Algorithms, 566–575 (1999)
14. Liang, Y.D.: On the feedback vertex set problem in permutation graphs. Information Processing Letters 52, 123–129 (1994)
15. Lu, C.L., Tang, C.Y.: A Linear-Time Algorithm for the Weighted Feedback Vertex Problem on Interval Graphs. Information Processing Letters 61, 107–111 (1997)
16. Misevicius, A., Lenkevicius, A., Rubliauskas, D.: Iterated tabu search: an improvement to standard tabu search. Information Technology and Control 35(3), 187–197 (2006)
17. Razgon, I.: Exact computation of maximum induced forest. In: Arge, L., Freivalds, R. (eds.) SWAT 2006. LNCS, vol. 4059, pp. 160–171. Springer, Heidelberg (2006)
18. Vazirani, V.V.: Approximation Algorithms. Springer, Heidelberg (2001)

Cuts, c-Cuts, and c-Complexes over the n-Cube

M. Reza Emamy-Khansary

Dept. of Mathematics, University of Puerto Rico, Rio Piedras, Puerto Rico
sanjuancube@yahoo.com

Abstract. A cut-complex over the geometric n-cube is an induced subgraph of the cube whose vertices are strictly separated from the rest of the cube by a hyperplane of R^n. A cut is the set of all the edges connecting the cut-complex to its node-complement subgraph. Here, we will extend the concepts of hyperplane cuts and cut-complexes to a larger class of cubical graphs called c-cuts, and c-complexes respectively. Then, we prove connectivity of the c-complexes that is essential for their characterization. Finally, we outline new open problems regarding the c-cuts and c-complexes over the n-cube.

1 Introduction

The cut number $S(n)$ of the n-cube is the minimum number of hyperplanes in R^n that slice (i.e. cutting the edges but missing vertices) all the edges of the n-cube. The cut number problem for the hypercube of dimensions $n \geq 4$ was posed by P. O'Neil about forty years ago [10], however two different proofs of $S(4) = 4$ by Emamy [2,3] and a set of 5 hyperplanes that slice all the edges of the 6-cube (M. Paterson) [11], are the only early results for the lower dimensional cubes. Since O'Neil's paper in 1971 and after about 30 years, Sohler-Ziegler [13] obtained a computational solution to the 5-cube problem that shows $S(5) = 5$. This settles down the problem for $n \leq 6$, but the problem remains open for $n \geq 7$. Of course, finding a short and computer-free proof for the 5-cube will remain to be a challenging problem. More of Paterson type of covering cuts can be found in Ziegler [14]. The cut number problem has also been raised by Grünbaum [8], Klee [9] and as a comprehensive exposition in Saks [11]. In the computational proof of Sohler-Ziegler [13] and in the other computational results of Ziegler [14], the authors have used advanced parallel computing to overcome the technical complexities of the computational process. The parallel algorithms have recently been improved by T. Schumacher, E. Lübbers, P. Kaufmann, M. Platzner [12]. For more on related problems and parallel computational results see Emamy-Ziegler [5,6]. On the other hand, N. Calkin et.al. [1] have applied probabilistic methods on the cuts and slicing hyperplanes in order to find $S(7)$. A special class of induced subgraphs of the n-cube called cut-complexes play an important role in any solution to this problem. In fact Sohler-Ziegler [13] have applied the characterization of all the cuts and cut-complexes of the 5-cube to prove that $S(5) = 5$. In this note, we will extend the concepts of hyperplane cuts and cut-complexes, to a larger class of cubical graphs called c-cuts, and c-complexes respectively. And we prove a basic connectivity lemma for the latter class that is essential to characterize c-complexes. Finally, we outline new open problems regarding these larger class of cuts over the n-cube.

J. Pahl, T. Reiners, and S. Voß (Eds.): INOC 2011, LNCS 6701, pp. 603–606, 2011.

2 Terminologies

The n-dimensional cube Q_n will refer to the geometric cube embedded in R^n whose vertex set is $B^n = \{0,1\}^n$ and the adjacent vertices are connected by the straight line segments in the usual manner. The cut number problem in a covering format can be stated as follows. Any slicing hyperplane divides the set of 2^n vertices of the n-cube into two disjoint sets S, S' of vertices that stay in the opposite sides of the hyperplane. In fact, we have a 2-vertex coloring of the cube in which vertices in S are red and those of S' are blue. In this vertex coloring, an edge is called colorful if it has two blue and red end vertices. A cut is the set of all the colorful edges, that is, they have one end vertex in S and the other one in S', and then the subgraph induced by any of the sets S (or S') is defined to be a cut-complex. The cut number problem can be redefined as the minimum number of distinct cuts that cover all the edges of the n-cube. In the latter format, the class of cuts are determined by the class of slicing hyperplanes. So evidently, for any specific class of 2-vertex colorings, a new type of cut number problem can be defined naturally according to the cuts associated with the 2-colorings. In the following, some specific set of planar vertices (cubical rectangles) will be defined that play an important role in the definition of this new class of cuts and complexes.

A cubical rectangle is any set $\{x,y,w,z\}$ of the hypercube vertices that form a plane rectangle. In the latter definition, any square will be treated as a regular rectangle, and thus the set of vertices of any 2- dimensional face of the hypercube is considered to be a cubical rectangle. These cubical rectangles combined with a 2-vertex coloring of the n-cube will lead us to a new class of cuts that is larger and in fact contains the class of hyperplane cuts in the following manner. For a 2-vertex coloring, a cubical rectangle is called alternating if its vertices are colored alternately, i.e., two opposite vertices of the rectangle are red and the other two are blue. A 2-vertex coloring is called feasible if it has no alternating cubical rectangle over the n-cube. Let a feasible 2-coloring be given, then the set of all the colorful edges is called a c-cut. Similarly a c-complex is defined to be the subgraph of the n-cube induced by the vertices of the same color (either red or blue). Finally, $cS(n)$ is defined to be the minimum number of c-cuts that can cover all the edges of the cube. Every cut-complex is a c-complex but not conversely. A computational characterization of cut-complexes can be found in [2,13], where a basic search algorithm has been implemented to find all the nonisomorphic cut-complexes of lower dimensional cubes. These algorithms apply the fact that cut-complexes are connected subgraphs of the n-cube. Thus, it is natural and useful to show that c-complexes are also connected and so they can be characterized by some search algorithms.

Remark

Evidently, if C is a c-complex over Q_n then its restriction to any face Q_m of the $n-$cube with $2 \leq m \leq n$, and its graph node-complement C' both are also c-complexes. On the other hand, to characterize c-complexes (or cut-complexes) it is more convenient that nonisomorphic complexes to be considered, however this is not exactly the same as graph isomorphism. Two complexes are isomorphic if there is a hypercube symmetry (the hypercube isometries that are generated by rotations and reflections) that takes one complex to the other one. The characterization methods that have been applied in [2,13] include search algorithms that are based on the following Lemma 1. It is an interesting observation that the proof of this lemma applies the separability of cut-complexes by slicing hyperplanes and does not work for c-complexes. So, any attempt to characterize c-complexes should begin with such a proof for the c-complexes. In this context, Lemma 2 is as essential for those characterization algorithm to work. For the proof and application of Lemma 1, see [2,13].

Lemma 1. *For any cut-complex C_k of k vertices, there exists a vertex $w \in vert(Q_n) - vert(C_k)$ such that the subgraph induced by $C_k \cup \{w\}$ is also a cut-complex.*

Let x, y be two given vertices of the n−cube. In the following lemma $cub(x,y)$ denotes the smallest face (subcube) containing x, y, and $d_H(x,y)$ stands for the hamming distance between x, y. It is clear that the dimension of the face $cub(x,y)$ is exactly $d_H(x,y)$.

Lemma 2. *Any c-complex is a connected subgraph of Q_n.*

Proof. In a given feasible coloring of Q_n , let $\mathscr{B}$ denote the set of blue and $\mathscr{R}$ be the set of red vertices. To show that the subgraph induced by $\mathscr{R}$ is connected, we suppose it is not connected and obtain a contradiction. Then there exist $x, y \in \mathscr{R}$ such that any path (in Q_n) connecting these two vertices has at least one interior vertex in $\mathscr{B}$. Let us choose $x, y \in \mathscr{R}$ such that $k = dim(cub(x,y))$ is minimum, that is $d_H(x,y)$ is minimum among all $x, y \in \mathscr{R}$ with such property. Let $x_1, x_2, ..., x_k$ and $y_1, y_2, ..., y_k$ be the k neighbors of x, y in $cub(x,y)$ respectively. If all of these neighbors are in $\mathscr{B}$, then we have many alternating rectangles, for instance the one that contains the vertices x, y and two opposite vertices of $cub(x,y)$ from the list of neighbors. Otherwise, suppose $x_1 \in \mathscr{R}$. Then, consider $x_1, y \in \mathscr{R}$ with $d_H(x_1,y) = k-1$, and evidently the property that each path connecting x_1, and y has at least one vertex in $\mathscr{B}$. This contradicts the minimality of $d_H(x,y)$ and the proof is complete.

Finally we conclude the note with a few research problems:

Problem 1. *Give a complete characterization of c-complexes by a computer-free method or by an efficient algorithm over the 5-cube.*

Problem 2. *In Lemma 1, prove a similar result for those of c-complexes.*

Problem 3. *By a geometric or combinatorial approach, prove that the class of cut-complexes and c-complexes coincide for the 5-cube.*

Problem 4. *Find a computer-free proof that shows $cS(n) = n$ for $4 \leq n \leq 5$.*

We proceed with the terminology that is needed to state the rest of the problems. For any given set of vertices $S \subset vert\, Q_n$, $cub\, S$ denotes the smallest face of Q_n containing S. In particular for any given induced subgraph C of Q_n, $cub\, C$ stands for the $cub(vert\, C)$. An edge uv of C is said to be a leaf of C if the edge has a vertex of degree one in C, the other vertex will be called the root of the leaf. Other than a trivial case when the complex is only one single edge, leaves have always unique roots. A vertex $w \in C$ is called an internal vertex for C, if all of its neighbors in $cubC$ are in C. The proofs of the following theorems can be found in [4].

Theorem 3. *Let uv be a leaf of a cut-complex C in Q_n with $deg(v) = 1$ and $n \geq 2$, then the root u of the leaf is an internal vertex of $C' = C - v$ in $cubC'$.*

Theorem 4. *Let a cut-complex C be given with $n \geq 2$. Then, $dim(cubC) = max\{deg_C(v) : v \in C\}$.*

Problem 5. *Do Theorems 3 and 4 hold for the c-complexes?*

Problem 6. *Solve Problems 1 and 3 for dimensions* $6 \leq n \leq 8$.

Problem 7. *Find the values of $cS(n)$ for dimensions* $7 \leq n \leq 8$.

References

1. Calkin, N., James, K., Bowman, L.J., Myers, R., Riedl, E., Thomas, V.: Cuts of the Hypercube. Dept. of Mathematical Sciences, Clemson University (July 3, 2008) (manuscript)
2. Emamy-K, M.R.: On the cuts and cut number of the 4-cube. J. Combin. Theory, Ser. A 41(2), 221–227 (1986)
3. Emamy-K, M.R.: On the covering cuts of the d-cube, $d \leq 5$. Disc. Math. 68, 191–196 (1988)
4. Emamy-K, M.R.: Geometry of cut-complexes and threshold logic. J. Geom. 65, 91–100 (1999)
5. Emamy-K., M.R., Ziegler, M.: New Bounds for Hypercube Slicing Numbers. In: Discrete Mathematics and Theoretical Computer Science Proceedings AA (DM-CCG), pp. 155–164 (2001)
6. Emamy-K., M.R., Ziegler, M.: On the Coverings of the d-Cube for d $\leq$ 6. Discrete Applied Mathematics 156(17), 3156–3165 (2008)
7. Grünbaum, B.: Convex Polytopes. In: Kaibel, V., Klee, V., Ziegler, G.M. (eds.) Springer, Heidelberg (2003)
8. Grünbaum, B.: Polytopal graph. MAA Studies in Math. 12, 201–224 (1975)
9. Klee, V.: Shapes of the future. Some unresolved problems in high-dimensional intuitive geometry. In: Proceedings of the 11th Canadian Conference on Computational Geometry, vol. 17 (1999)
10. O'Neil, P.E.: Hyperplane cuts of an n-cube. Disc. Math. 1, 193–195 (1971)
11. Saks, M.E.: Slicing the hypercube. In: Surveys in Combinatorics, pp. 211–255. Cambridge University Press, Cambridge (1993)
12. Schumacher, T., Lübbers, E., Kaufmann, P., Platzner, M.: Accelerating the Cube Cut Problem with an FPGA-Augmented Compute Cluster. In: Advances in Parallel Computing, vol. 15, IOS Press, Amsterdam (2008); ISBN 978-1-58603-796-3
13. Sohler, C., Ziegler, M.: Computing Cut Numbers. In: 12th Annual Canadian Conference on Computational Geometry, CCCG 2000, pp. 73–79 (2000)
14. Ziegler, G.: `http://www.uni-paderborn.de/cs/cubecuts`
15. Ziegler, G.: Lectures on Polytopes. Springer, Heidelberg (1994)

Exact and Metaheuristic Approaches to Extend Lifetime and Maintain Connectivity in Wireless Sensors Networks

Andrea Raiconi[1] and Monica Gentili[2]

[1] University of Salerno,Via Ponte Don Melillo, 84084 Fisciano (SA) Italy
araiconi@unisa.it
[2] University of Salerno,Via Ponte Don Melillo, 84084 Fisciano (SA) Italy
mgentili@unisa.it

Abstract. Wireless sensor networks involve a large area of real-world contexts, such as national security, military and environmental control applications, traffic monitoring, among others. These applications generally consider the use of a large number of low-cost sensing devices to monitor the activities occurring in a certain set of target locations. One of the most important issue that is considered in this context is maximizing network lifetime, that is the amount of time in which this monitoring activity can be performed by opportunely switching the sensors from active to sleep mode. Indeed, the lifetime of the network can be maximized by individuating subset of sensors (i.e., *covers*) and switching among them. Two important aspects need to be taken into account among others: (i) coverage: each determined cover has to cover the entire set of targets, and (ii) connectivity: each cover should provide satisfactory network connectivity so that sensors can communicate for data gathering or data fusion (*connected* covers). In this paper we consider the problem of determining the maximum network lifetime to monitor all the targets by means of connected covers. We analyze the problem and propose an exact approach based on column generation and two heuristic approaches, namely a greedy algorithm and a GRASP algorithm, to solve it. We analyze the performance of the heuristic approaches by comparing the obtained solutions with those provided by the exact approach when available. Our preliminary experimental results show the proposed solution algorithms to be promising in terms of tradeoff between quality of solutions and computational effort.

1 Introduction

Wireless sensor networks have met a growing interest in the last years due to their applicability to a large class of contexts, such as traffic control, military and environmental applications. These networks are generally characterized by a large number of small sensing devices (*sensors*), often randomly disposed all over the region of interest in order to perform a monitoring activity on a set of target points. One of the key issues in this scenario involves the maximization of the amount of time during which this activity can be carried out taking into account the limited power of the battery of each sensor. This problem is usually known as *Maximum Network Lifetime Problem (MLP)*. The lifetime of the network can be maximized by individuating subsets of sensors (i.e., *covers*) and switching among them. Two important aspects need to be taken

J. Pahl, T. Reiners, and S. Voß (Eds.): INOC 2011, LNCS 6701, pp. 607–619, 2011.

into account among others: (i) coverage: each determined cover has to cover the entire set of targets, and (ii) connectivity: each cover should provide satisfactory network connectivity so that sensors can communicate for data gathering or data fusion (*connected* covers). The MLP is extensively studied in the literature by mainly considering the coverage requirement: exact and heuristic methods to build the set of covers and to assign them appropriate activation times have been presented in several works ([2],[3],[10]), and also some variants of the problem solved in a similar way have been proposed ([4], [7],[11]).
In this paper we consider the problem of determining the maximum network lifetime to monitor all the targets by means of connected covers. To take into account communication between sensors, we consider an additional node (*root*) representing a central processing station and we assume there exists a *communication link* among each couple of sensors (or sensor and root) if they are close enough to communicate directly between each other. A solution composed by connected covers is such that the sensors in each cover are connected to the root through a path that uses communication links.

The connectivity requirement has also been addressed in the literature ([1], [6], [9], [12], [13], [14]). In [12] and [13] the authors propose sufficient conditions in the coverage level of each sensor to imply connectivity. The connectivity issue is also considered in [9] for an unreliable wireless sensor grid network, and a necessary and sufficient condition for coverage and connectivity is presented. In [6], an approximated algorithm is presented to maximize the lifetime of the network where it is assumed that each sensor only needs to know the distances between adjacent nodes in its transmission range and their sensing radii. Existing papers that are nearer to our study are [1] and [14]. The authors in [14] describe an energy consumption model of the sensors that takes into account their different role (either a relay role, a source role or both) in the cover. They define the problem as the Maximum Tree Cover problem and present a greedy heuristic and an approximation algorithm to solve it. Two solution approaches are presented in [1] to maximize network lifetime and maintain connectivity: an exact approach based on column generation and a heuristic algorithm aiming at a distributed implementation. To our knowledge, there is a lack of contributions regarding metaheuristic approaches related to the problem.

In this paper we present a GRASP metaheuristic, as well as a greedy heuristic (that is used as a subroutine inside the GRASP) and an exact method based on column generation that is used to evaluate the performance of our heuristics. Our column generation differs from the one proposed in [1] in the subproblem definition as it will be better clarified in Section 3. We evaluate the performance of our algorithms by comparing them on some benchmark instances. Our preliminary experimental results show the proposed solution algorithms to be promising in terms of trade-off between solution quality and computational effort.

The paper is organized as follows. The next section gives the needed notation, introduces the problem and contains the mathematical formulation. The Column Generation approach is presented in Section 3. Our greedy algorithm and GRASP algorithm are object of Section 4. Experimental results are shown in Section 5. Further research is discussed in Section 6.

2 Notation and Problem Definition

Let $N = (T,S)$ be a wireless sensor network where $T=\{t_1,..,t_n\}$ is the set of target points and $S=\{s_0,s_1,..,s_m\}$ is the set composed by the sensor nodes and the special root node s_0. For each $s_i \in S$, let $T_{s_i} \subseteq T$ be the subset of targets covered by s_i; we assume $T_{s_0} \equiv \emptyset$ since the root node does not have coverage purposes. Given a subset $C \subseteq S$, we define the set of targets covered by C as $T_C = \bigcup_{s_i \in C} T_{s_i}$. By extension, each target in T_C is said to be covered by C. Moreover, for each $s_i \in S$, let $neigh(s_i) \subseteq S$ be the set of elements of S that are close enough to s_i to allow direct communication. We consider an energy consumption model of the sensors where it is assumed that the energy requirement to send data to each communication link is constant. Note that if $s_i \in neigh(s_j)$, then $s_j \in neigh(s_i)$. Given a subset $C \subseteq S$, we define $neigh(C) = \bigcup_{s_i \in C} neigh(s_i)$. Now, consider the induced undirected graph $G = (S,E)$ such that the communication link $(s_i,s_j) \in E$ if and only if $s_j \in neigh(s_i)$ $\forall s_i,s_j \in S$ (see Figure 1A for an example). We define a set $C \subseteq S$ to be a connected cover of N if: (i) $s_0 \in C$, (ii) $T_C \equiv T$, and (iii) C is such that there exists a path of communication links connecting each of its sensors to s_0 (see Figure 1B-1C).

The **Connected Maximum Network Lifetime Problem (CMLP)** is defined as follows: *Find a collection of pairs* (C_j,w_j), $j = 1,2,\ldots,p$, *where* C_j *is a connected cover and* w_j *is its corresponding activation time, such that the sum of the activation times* $\sum_{j=1}^{p} w_j$ *is maximized, and, each sensor is used for a total time that does not exceed its battery:* $\sum_{j\in\{1,..p\}|s_i\in C_j} w_j \leq 1$ *for each* $s_i \in S \setminus \{s_0\}$.

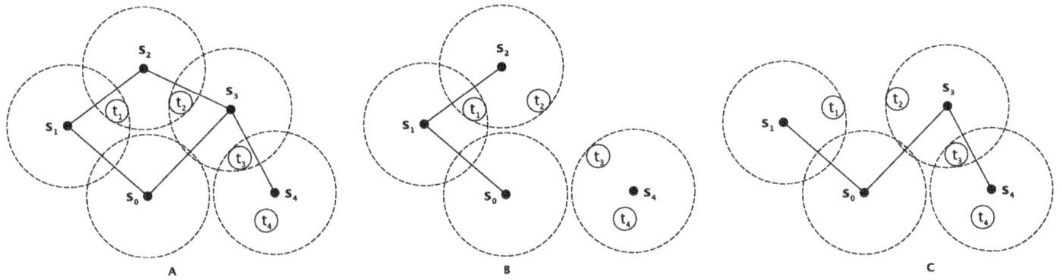

Fig. 1. A: An example network N and the corresponding induced graph G. **B:** Unfeasible (disconnected) cover. **C:** Feasible connected cover.

The problem can be formulated as the classical MLP [2] with the only difference that each cover considered in the formulation is connected. Let $C_1,\ldots,C_M$ be the collection of *all* the feasible connected covers in the network, let parameters a_{ij} be equal to 1 if sensor s_i belongs to cover C_j and equal to 0 otherwise, and, $w_1\ldots,w_M$ be the decision variables denoting the activation times for each cover. The mathematical formulation of the problem is as follows:

$$\max \sum_{j=1}^{M} w_j \tag{1}$$

subject to

$$\sum_{j=1}^{M} a_{ij} w_j \le 1 \qquad \forall s_i \in S \setminus \{s_0\} \tag{2}$$

$$0 \le w_j \le 1 \qquad \forall j = 1, ..., M \tag{3}$$

The objective function (1) maximizes the network lifetime. The set of constraints (2) state that the total energy consumption for each sensor cannot exceed its battery life that we normalized to be equal to 1. Finally, constrains (3) are non-negativity constraints and they define an upper bound for the activation time of each cover.

3 Column Generation Approach

Delayed Column Generation, or simply Column Generation is an efficient technique to solve linear programming formulations when the set of variables is too large to consider all of them explicitly. Since most of them will be nonbasic and, therefore, they assume a value of zero in the optimal solution, the method aims to generate only variables which have potential to improve the objective function, while the others are implicitly discarded. In particular, the general iteration of the Column Generation considers a primal problem restricted only to a subset of variables (the so-called *Restricted Primal*) and optimally solves it. In order to determine whether the returned solution is optimal for the entire problem, all the reduced costs of the nonbasic variables should be computed and it should be verified whether they satisfy the optimality conditions. If this is the case the algorithm stops, otherwise a new variable (column) is generated, it is added to the restricted primal and the algorithm iterates. To check for the optimality condition an additional problem is solved (the so-called *Separation Problem*) whose solution either returns a new column to be added to the restricted primal or verifies the optimality of the current solution.

Let us consider the mathematical formulation given in the previous section restricted only to a subset p of feasible covers, and, let B and NB be the index sets of the optimal basic and non basic variables corresponding to the optimal solution of the restricted primal, respectively. Moreover, let π_i, $i = 1, 2, \dots, m$, be the set of dual optimal multipliers associated with each primal constraint (that is, with each sensor s_i). The current primal solution is optimal if the reduced costs associated with the non basic variables are all non negative, that is, $\sum_{i:s_i \in C_j} \pi_i - c_j \ge 0$ for each $j \in NB$, where c_j is the coefficient of variable w_j in the objective function (1) of the primal problem. Note that, all the coefficients of the variables in the objective function (1) are equal to 1, therefore the optimality conditions reduce to $\sum_{i:s_i \in C_j} \pi_i - 1 \ge 0$ for each $j \in NB$. Instead of computing all the reduced costs, we could compute the minimum among all of them and check whether it is greater than or equal to zero. Such a minimum reduced cost can be computed solving the separation problem (described next), that returns the non basic connected cover with minimum reduced cost. The proposed formulation for the separation problem ensures the connectivity of each generated cover by selecting sensors and communication links that define a tree rooted in s_0. We consider three types of variables: the binary variable act_i associated with sensor s_i that is equal to 1 if s_i is activated in the new cover, and it is equal to 0 otherwise; the binary variable x_{ij} associated with the communication link (s_i, s_j) that is equal to 1 if sensor s_i and sensor s_j are both active in the cover and the communication link (s_i, s_j) is used in the cover and it is equal to 0 otherwise; non-negative variable f_{ij} associated with each communication link (s_i, s_j) that denotes the flow passing through (s_i, s_j). Parameters of the model are the dual prices π_i associated with each sensor, and the binary parameters g_{ki} equal to 1 if target point t_k is covered by s_i and is equal to 0 otherwise.

The formulation is a single-commodity formulation to find a tree connecting all the activated sensors and covering all the targets such that the sum of the cost of the sensors in the tree is minimized. Such a tree can be found by sending from the root node s_0 a unit of flow to each active sensor.

$$\min \sum_{s_i \in S \setminus \{s_0\}} \pi_i act_i \tag{4}$$

subject to

$$\sum_{s_i \in S \setminus \{s_0\}} g_{ki} act_i \geq 1 \qquad \forall t_k \in T \tag{5}$$

$$\sum_{(s_0, s_i) \in E} f_{0i} = \sum_{s_i \in S} act_i \tag{6}$$

$$\sum_{(s_i, s_j) \in E} f_{ij} - \sum_{(s_j, s_i) \in E} f_{ji} = act_j \qquad \forall s_j \in S \setminus \{s_0\} \tag{7}$$

$$\sum_{(s_i, s_j) \in E} x_{ij} = act_j \qquad \forall s_j \in S \setminus \{s_0\} \tag{8}$$

$$x_{ij} \leq f_{ij} \leq |S| x_{ij} \qquad \forall (s_i, s_j) \in E \tag{9}$$

$$act_i \in \{0, 1\} \qquad \forall s_i \in S \setminus \{s_0\} \tag{10}$$

$$x_{ij} \in \{0, 1\} \qquad \forall (s_i, s_j) \in E \tag{11}$$

The objective function (4) minimizes the sum of the cost of the active sensors (that is, the sensors that are selected to be in the cover). Constraints (5) ensure all the targets are covered by the selected sensors. If a sensor is selected to be in the cover then exactly one (entering) communication link has to be selected, this is ensured by constraints (8). Constraint (6) guarantee that the total amount of flow that is sent from the root node s_0 is equal to the total number of active sensors in the cover. To ensure connectivity among the active sensors, flow conservation constraints are imposed by constraints (7). Constraints (9) guarantee there is a positive flow only on communication links that are selected. Finally, constraints (10) and (11) are binary constraints on the variables.

After solving the above separation problem, if its optimal objective function value is ≥ 1, we can safely deduce that the solution found by the master problem was optimal; otherwise, the returned new column (defined by the optimal solution value of variables act_i) is introduced into the master problem that is solved again and the process is iterated.

We solved the separation problem by solver CPLEX. It is easy to check that there always exists an optimum solution composed of minimal covers, therefore we added the following set of constraints ensuring only minimal covers are generated. Let

$\{C_1, C_2, \ldots, C_u\}$ be the set of connected covers generated by the algorithm so far, then the set of constraints added to the separation problem is the following:

$$\sum_{s_i \in S \setminus \{s_0\}} a_{ij} act_i \leq \sum_{s_i \in S \setminus \{s_0\}} a_{ij} - 1 \qquad \forall j = 1, \ldots, u \qquad (12)$$

These inequalities ensure that the new connected cover returned by the separation problem differs from the already generated covers in at least one sensor, and, therefore it is not contained in none of them. One last issue concerns the initialization of the column generation with the first initial set of connected covers. In the preliminary results presented in Section 5 we used two different approaches. For small instances, we determined an initial set of covers by running the subproblem several times associating random weights with each sensor, while, for large instances we used a heuristic initializations provided by the GRASP procedure described in Section 4.

The above column generation approach differs from that presented in [1] in the definition of the subproblem that results from a different energy consumption model associated with each sensor. The authors in [1] consider all the sensors to be connected to each other (i.e., the underlying graph is complete) but with different costs depending on the role of the sensors and on the distance between them. Such an energy consumption model could be introduced in our definition of the problem and an analysis of such a case is object of further study as outlined in the concluding section.

4 Greedy and GRASP Algorithms

Here we present a greedy heuristic (namely, the CMLP-Greedy) and a GRASP metaheuristic (namely, CMPL-GRASP) designed to solve the CMLP problem.

The greedy algorithm is based on the greedy heuristic presented in [4] for the classical Maximum Network Lifetime Problem and in [7] for the α-MLP variant. The algorithm iteratively produces new covers. Each cover is initialized with the root node and new sensors, chosen among the ones with positive residual energy, are added. Each new sensor to be included is also chosen such that it is directly connected to at least one element of the cover, to guarantee connectivity. The algorithm terminates when the sensors with positive residual energy cannot guarantee either coverage or connectivity.

Algorithm 1 contains a description of our CMLP-Greedy. Line 1 contains the input parameters. Granularity factor $gf \in (0,1]$ represents a maximum amount of activation time assigned to each generated cover during the algorithm execution. The S_R set initialized in line 2 contains the list of sensors with a residual energy > 0. Parameters R_{s_i} initialized in lines 3-5 represent the amount of residual energy for each sensor s_i and the SOL set initialized in line 6 will contain the final solution. Line 7 checks whether the remaining sensors can cover the whole set of targets T. Lines 8-9 create a new cover C_l and initialize it with the root node s_0. The T_U set initialized in Line 10 contains the targets that still need to be covered in C_l. The loop in lines 11-26 ensures that new sensors are added to C_l until either all targets are covered or a new cover cannot be achieved. In particular, lines 12-13 check whether there exist sensors with residual energy > 0 that are not in C_l yet but are connected to at least one of its sensors or the root; let S_{sel} be the set of these selectable sensors. If this set is empty, it is impossible to complete C_l and the current solution SOL is returned. Otherwise, one of the sensors in S_{sel} will be selected and included in C_l. As shown in lines 16-21, if S_{sel} covers targets that are uncovered in C_l so far, a target t_c is selected as *critical* and a sensor s_c covering t_c with the *greatest contribution* is selected (we will explain such concepts in more detail in the following), otherwise if S_{sel} does not cover new targets in C_l but a sensor in S_{sel} is needed for connection then s_c is directly selected. In lines 22-25, C_l and T_U are updated according to the selection of s_c. Line 27 sets an activation time for C_l equal to gf if each sensor of C_l has a residual energy $\geq gf$, otherwise the activation time of C_l is equal to the minimum of their residual energies. Lines 28-33 update the residual energies and check if the set R_S must be updated. Cover C_l is added to SOL in line 34. The final set of covers is returned in line 36. In our experimentations, we refined the network lifetime associated with the solution returned by CMLP-Greedy by solving the mathematical model given in Section 2 restricted to the set of generated covers.

Algorithm 1. CMLP-Greedy algorithm

1\. input: wireless network $N = (T, S)$, granularity factor $gf \in (0, 1]$
2\. $S_R \leftarrow S \setminus \{s_0\}$
3\. **for** each $s_i \in S_R$ **do**
4\. $R_{s_i} \leftarrow 1$
5\. **end for**
6\. $SOL \leftarrow \emptyset$
7\. **while** $\bigcup_{s_i \in S_R} T_{s_i} \equiv T$ **do**
8\.   Create a new empty cover C_l
9\. $C_l \leftarrow \{s_0\}$
10\. $T_U \leftarrow T$
11\.   **while** $T_U \not\equiv \emptyset$ **do**
12\. $S_{sel} \leftarrow (S_R \cap neigh(C_l)) \setminus C_l$
13\.   **if** $S_{sel} \equiv \emptyset$ **then**
14\.   **return** SOL
15\.   **end if**
16\.   **if** $T_U \cap T_{S_{sel}} \not\equiv \emptyset$ **then**
17\.   Find a **critical target** $t_c \in T_U \cap T_{S_{sel}}$
18\.   Select $s_c \in S_{sel}$ s.t. $t_c \in T_{s_c}$ and s_c has the **maximum contribution**
19\.   **else**
20\.   Select $s_c \in S_{sel}$ s.t. s_c has the **maximum contribution**
21\.   **end if**
22\. $C_l \leftarrow C_l \cup \{s_c\}$
23\.   **for** each $t_j \in T_U$ s.t. $t_j \in T_{s_c}$ **do**
24\. $T_U \leftarrow T_U \setminus \{t_j\}$
25\.   **end for**
26\.   **end while**
27\. $w_l =$ **max feasible activation time** $\leq gf$ for C_l
28\.   **for** each $s_i \in C_l \setminus \{s_0\}$ **do**
29\. $R_{s_i} \leftarrow R_{s_i} - w_l$
30\.   **if** $R_{s_i} = 0$ **then**
31\. $S_R \leftarrow S_R \setminus \{s_i\}$
32\.   **end if**
33\.   **end for**
34\. $SOL \leftarrow SOL \cup \{C_l\}$
35\. **end while**
36\. **return** SOL

Critical Target: A critical target is the target that, among the ones which are covered by sensors in S_{sel}, is covered by sensors with the least positive residual energy; that is

$$t_c = \arg \min_{t_j \in T_U \cap T_{S_{sel}}} \{ \sum_{s_i \in S_R | t_j \in T_{s_i}} R_{s_i} \} \quad (13)$$

Sensor Contribution: If S_{sel} covers targets that are in T_U, and therefore we chose a critical target t_c, we consider the contribution of each selectable sensor s_i covering t_c as the number of uncovered targets in T_{s_i}, that is,

$$Contr(s_i) = |T_U \cap T_{s_i}| \quad \forall s_i \in S_{sel} | t_c \in T_{s_i} \quad (14)$$

Otherwise, since none of these sensors covers new targets (as in line 20 of CMLP-Greedy), the contribution of each selectable sensor s_i is given by the sum of the number of targets in T_U that can be covered by its neighbors with residual energy, i.e.

$$Contr(s_i) = \sum_{s_{i'} \in S_R \cap neigh(s_i)} |T_U \cap T_{s_{i'}}| \quad \forall s_i \in S_{sel} \quad (15)$$

If even these contributions are all equal to 0, a sensor belonging to S_{sel} is chosen randomly.

We embedded CMLP-Greedy into a GRASP scheme (CMLP-GRASP). GRASP is a metaheuristic consisting in a multi-start iterated local search. At each iteration, a new starting solution is generated according to a randomized heuristic; this solution is then refined by the local search phase. The best solution among those computed during the execution is then returned (for more detail the reader can refer to [5]). In order to implement these two steps, we developed two variants of CMLP-Greedy. CMLP-Greedy' takes as additional input parameter a positive integer rcl. Each time that a critical target or a sensor with the greatest contribution has to be selected, instead of performing the best greedy choice, we create a Restricted Candidate List (RCL) of the best rcl choices (or less, in case that the number of available choices is smaller than rcl). One element of RCL is then selected at random. In the local search phase, we build solution neighborhoods using a second variant, CMLP-Greedy". This variant imposes the inclusion of a cover (passed as input parameter) as part of the solution, while the others are generated using CMLP-Greedy.

A high level outline of CMLP-GRASP is given in Algorithm 2. Line 1 describes the input parameters. Besides the input network and the value rcl, we consider a set of granularity factor values $\{gf_1, \ldots, gf_k\}$ and a maximum number of iterations it_{max} from the last improvement in the objective function value (that is in the network lifetime). In lines 2-4 we initialize the best found solution SOL, its value lt and the current iteration number it. In order to evaluate the objective function lt of each solution generated throughout the algorithm execution, we solve the mathematical formulation given in Section 2 restricted to the set of covers composing the solution. The GRASP loop is contained in lines 5-22. In lines 7-8 a randomized solution is generated using a predefined granularity factor and evaluated. The local search is performed as described in lines 9-16; a new neighbor is generated and evaluated for each cover in the current solution and for each granularity factor value gf_i. The best solution found and its value are stored using a steepest descent approach, as shown in lines 12-14. If the incumbent optimum was improved during a GRASP iteration, it is updated and the it value is reset, as can be seen in lines 17-20. Finally, the best solution found is returned in line 23.

Algorithm 2. CMLP-GRASP algorithm

```
input: wireless network N = (T,S), granularity factors {gf_1,...,gf_k}, gf_i ∈ (0,1] ∀i = 1,...,k,
    max iterations parameter it_max, RCL size rcl > 0
it ← 0
SOL ← ∅
lt ← 0
while it < it_max do
    it ← it + 1
    SOL' ← CMLP-Greedy'(N, gf_1, rcl)
    lt' ← Evaluation(SOL')
    for each C'_l ∈ SOL' and each i ∈ 1,...,k do
        SOL'' ← CMLP-Greedy"(N, gf_i, C'_l)
        lt'' ← Evaluation(SOL'')
        if lt'' > lt' then
            SOL' ← SOL''
            lt' ← lt''
        end if
    end for
    if lt' > lt then
        SOL ← SOL'
        lt ← lt'
        it ← 0
    end if
end while
return SOL
```

5 Computational Results

In this section we compare our CG exact procedure and our two heuristic approaches on a preliminary testbed. As in [4], we generated instances with a small population of targets and a high number of sensors. Specifically, we consider the sets $n = 15$ and $m = 25, 50, 75, 100, 150, 200$. For each value of m, we generated 5 different test instances. In each instance, targets, sensors and the root node are randomly disposed on a grid and communication links as well as target coverages are determined accordingly.

Table 1 contains average instance characteristics, in terms of connectivity among sensors and number of targets covered by each sensor. In particular, the columns labeled with Sensor Connectivity report the degree of the sensor nodes and of the root node in the induced graph. The connectivity is expressed in percentage, that is, an element of S with m communication links would have 100% connectivity. For each scenario (that is, for each value of m) we report average values of the maximum, minimum and average

Table 1. Instance characteristics (average values)

	Sensor Connectivity (%)			Sensor Coverage		
m	max	min	avg	max	min	avg
25	40.0	7.2	21.6	7.4	1.2	2.8
50	36.4	6.0	21.6	7.4	1.0	2.8
75	37.6	6.1	22.6	7.4	1.0	3.0
100	36.0	7.0	22.0	7.8	1.0	3.0
150	34.9	7.2	22.6	7.8	1.0	3.0
200	35.2	7.4	22.9	7.8	1.0	3.0

Table 2. Computational results

	CMLP-Greedy		CMLP-GRASP		CG	
m	sol	time	sol	time	sol	time
	1	0.01	1	1.58	1	0.07
	0	0.01	0	0.13	0	0
25	1	0.01	1	1.5	1	0.03
	2	0.01	2	7.93	2	0.25
	1	0.01	1	1.5	1.66	0.64
	2	0.03	2	4.87	2	1.28
	2	0.03	2.5	6	2.5	2.49
50	3	0.03	3	5.39	3	1.57
	4	0.04	4	9.99	4	3.83
	3	0.03	4	12.44	4.66	452.35
					CMLP-GRASP + CG	
	5	0.08	6	39.4	7	2200.16
	4	0.09	4	19.78	4	23.1
75	3	0.04	3	6.63	3	7.56
	6	0.13	7	42.92	7	50.89
	6	0.07	6	31.21	7	143.61
	8	0.18	9	53.67	9.68[**10**]	TL
	6	0.37	7	53.53	7	64.28
100	6	0.1	7	28.26	7	40.64
	9	0.27	9	74.71	9	88.2
	8	0.32	8	139.83	8	156.76
	12	0.4	15	111.9	15*(20)*	TL
	11	0.54	13	108.84	13.43[**14**]	TL
150	10	0.27	12	49.1	14.99[**16**]	TL
	12	1.01	13	202.98	13	297.05
	13	0.46	14	211.81	14	466.74
	18	0.94	20	259.49	20.69*(32)*	TL
	17	1.37	18	228.29	18.32 *(20)*	TL
200	13	0.44	14.95	252.2	18.52[**19**]	TL
	16	2.1	18	363.79	18	1150.66
	18	1.13	19	352.99	19	894.09

Optimal values found after the time limit are reported in square brackets and bold. Upper bounds are reported in round brackets and italic.

degree of each sensor node. For example, instances with $m = 25$ are such that the element of S with the maximum degree has on average $25 \times 0.4 = 10$ links, the one with the minimum degree has on average 1.8 links, and finally the average degree is equal to 5.4. Sensor Coverage columns give information about the number of targets covered by each sensor; that is, they give information about the quantity $|T_{s_i}|$, $s_i \in S$. In particular, for each scenario column *max* gives the average value (over the five instances) of the size of the set T_{s_i} whose cardinality is maximum, column *min* gives the average value of the size of the set T_{s_i} whose cardinality is minimum, while the last column gives the average value of the size of all the sets T_{s_i}.

After a tuning phase, we determined the values to be assigned to the parameters used by the algorithms. For the CMLP-Greedy heuristic, we set gf equal to 0.2. For CMLP-GRASP, we used the set of granularity factors $\{0.2, 0.3, 0.5, 0.7\}$, and chose $rcl = 5$ and $it_{max} = 20$. We also considered a one hour time limit for each execution of the Column Generation algorithm. As already mentioned in Section 3, for larger instances (those where $m \geq 75$), we initialized the Column Generation with heuristic solutions provided by the GRASP procedure: since good starting solutions help to speed up the convergence of the method. Computational times reported for the Column Generation on these instances will be therefore the sum of the times of both the procedures. For instances with $m \leq 50$, instead, we obtained better computational times by initializing our CG with covers generated by repeated executions of the subproblem where a random weight is associated with each sensor. All tests were performed on a workstation with Intel Core 2 Duo processor at 2.4Ghz and 3GB of RAM. All the procedures have been coded in C++, using IBM ILOG CPLEX 11.2 and the Concert Technology library to solve the mathematical formulations. Detailed results and computational times can be found in Table 2, while average values are shown in Table 3.

The Column Generation algorithm is able to find certified optimal solutions for every instance with up to 75 sensors within the considered time limit; it reaches the time limit once for $m = 100$, three times for $m = 150$ and three times for $m = 200$. We run again the procedure for up to 6 hours on these instances, obtaining 4 new optimal solutions (1 for $m = 100$, 2 for $m = 150$ and 1 for $m = 200$). We reported these new optima in square brackets and bold in Table 2. For the three remaining instances, we reported naive upper bounds in round brackets and italic in Table 2. Upper bounds are computed considering the coverage level of the least covered target of the network. For $m \geq 100$, the average values reported in brackets in Table 3 are evaluated considering optimal solutions where available and upper bounds otherwise. On average, CG solutions found within the 1-hour time limit are 0.73% worse than this value for $m = 100$, 8.57% worse for $m = 150$ and 12.45% worse for $m = 200$.

As could be easily expected, CMLP-Greedy is the fastest algorithm, computing the solutions in less than 0.1 seconds on average for instances with up to 75 sensors, and

Table 3. Computational results (average values)

	CMLP-Greedy		**CMLP-GRASP**		**CG**	
m	sol	time	sol	time	sol	time
25	1	0.01	1	2.53	1.13	0.2
50	2.8	0.03	3.1	7.74	3.23	92.3
					CMLP-GRASP + CG	
75	4.8	0.08	5.2	27.99	5.6	485.06
100	7.4	0.25	8	70	8.14 **[8.2]**[1]	TL
150	11.6	0.54	13.4	136.93	14.08 *(15.4)*[2]	TL
200	16.4	1.19	17.99	291.35	18.91 *(21.6)*[3]	TL

[1] *1 optimum found after TL used in the average.*
[2] *2 optima found after TL and 1 upper bound used in the average.*
[3] *1 optimum found after TL and 2 upper bounds used in the average.*

in 1.19 seconds on average for $m = 200$. It finds the optimum in 11 instances out of 30. On datasets where certified optimal solutions could be computed for each instance ($25 \leq m \leq 100$), the returned solutions are on average smaller than the optimum of a percentage between 9.76% ($m = 100$) and 14.29% ($m = 75$). While much slower than CMLP-Greedy, the CMLP-GRASP algorithm retains reasonable computational times, running on average in less than 5 minutes in the worst case ($m = 200$). It also provides better solutions, finding the optimum on 19 instances. On the datasets where we have certified optimal solutions for every instance, the returned solutions are on average smaller than the optimum of a percentage between 2.02% ($m = 50$) and 11.5% ($m = 25$).

6 Conclusions

We addressed the problem of maximizing network lifetime of a wireless sensor network when all the targets need to be covered and connectivity among sensors needs to be ensured. The problem is studied in the literature but an efficient metaheuristic algorithm still is missing. In this paper we present a GRASP metaheuristic, as well as a greedy heuristic (that is used as a subroutine inside the GRASP) and an exact method, based on column generation, that is used to evaluate their performances.

Our preliminary computational results show the proposed solution algorithms to be promising in terms of tradeoff between solution quality and computational effort. Our very next step in this research topic is a wider experimentation of the approaches aimed at evaluating their performances on different datasets (namely scattering and design datasets [8]). We also intend, in order to speed up our exact approach, to solve the subproblem by means of a heuristic approach for generating covers with positive reduced cost, and use the MIP only when the heuristic fails. Moreover, we will try to enhance the MIP formulation by improving constraints (9) by means of better upper bounds on the flows.

Additional steps will focus on applying the same approaches to solve the problem when the energy consumption of the sensors for transmission and connectivity activities is more detailed. Moreover, we will also investigate how our approaches can be modified to solve the connected variant of α-MLP [7] when the connected covers are not required to cover all the targets but only a portion α of them.

References

1. Alfieri, A., Bianco, A., Brandimarte, P., Chiasserini, C.F.: Maximizing system lifetime in wireless sensor networks. European Journal of Operational Research 181, 390–402 (2007)
2. Berman, P., Calinescu, G., Shah, C., Zelikovsky, A.: Power Efficient Monitoring Management in Sensor Networks. In: Proceedings of the Wireless Communications and Networking Conference, pp. 2329–2334 (2004)
3. Cardei, M., Du, D.-Z.: Improving Wireless Sensor Network Lifetime through Power-Aware Organization. ACM Wireless Networks 11, 333–340 (2005)
4. Cardei, M., Thai, M.T., Li, Y., Wu, W.: Energy-Efficient Target Coverage in Wireless Sensor Networks. In: Proceedings of the 24th conference of the IEEE Communications Society (INFOCOM), vol. 3, pp. 1976–1984 (2005)

5. Festa, P., Resende, M.G.C.: GRASP: Basic components and enhancements. Telecommunication Systems 46, 253–271 (2011)
6. Kasbekar, G.S., Bejerano, Y., Sarkar, S.: Lifetime and coverage guarantees through distributed coordinate-free sensor activation. In: Proceedings of the 15th annual international conference on Mobile computing and networking, pp. 169–180 (2009)
7. Gentili, M., Raiconi, A.: α-Coverage to Extend Network Lifetime on Wireless Sensor Networks. Technical Report 2-2010, Department of Mathematics and Computer Science, University of Salerno (2010)
8. Lopes, L., Gentili, M., Efrat, A., Ramasubramanian, S.: Scheduling Redundant Sensors Optimally for Maximum Lifetime. Technical report 11-2010, Department of Mathematics and Computer Science, University of Salerno (2010)
9. Shakkottai, S., Srikant, R., Shroff, N.: Unreliable sensor grids: coverage, connectivity and diameter. In: IEEE INFOCOM, pp. 1073–1083 (2003)
10. Slijepcevic, S., Potkonjak, M.: Power Efficient Organization of Wireless Sensor Networks. In: IEEE International Conference on Communications, vol. 2, pp. 472–476 (2001)
11. Thai, M.T., Wang, F., Du, D.H., Jia, X.: Coverage Problems in Wireless Sensor Networks: Designs and Analysis. International Journal of Sensor Networks 3, 191–200 (2008)
12. Wang, X., Xing, G., Zhang, Y., Lu, C., Pless, R., Gill, C.: Integrated coverage and connectivity configuration in wireless sensor networks. In: Proceedings of the 1st international conference on Embedded networked sensor systems, pp. 28–39 (2003)
13. Zhang, H., Hou, J.: Maintaining Sensing Coverage and Connectivity in Large Sensor Networks. Ad Hoc & Sensor Wireless Networks 1
14. Zhao, Q., Gurusamy, M.: Lifetime maximization for connected target coverage in wireless sensor networks. IEEE/ACM Transactions on Networking 16, 1378–1391 (2008)

Computing Upper Bounds for a LBPP with and without Probabilistic Constraints

Hugo Rodríguez[1], Pablo Adasme[1,2], Abdel Lisser[2], and Ismael Soto[1]

[1] Departamento de Ingenieria Eléctrica,
Universidad de Santiago de Chile, Av. Ecuador 3769 Santiago, Chile
{hugo.rodriguez,pablo.adasme,ismael.soto}@usach.cl
[2] Laboratoire de Recherche en Informatique(LRI),
Université Paris Sud, Paris XI 91405 Orsay Cedex, France
{abdel.lisser,pablo.adasme}@lri.fr

Abstract. In this paper, we compute upper bounds for a generic linear bilevel programming problem (LBPP) using the iterative min-max (IMM) algorithm proposed in [4,5]. Neither in [4] nor in [5] the authors give optimal solutions for this problem or gaps to measure IMM efficiency. To this purpose, we implement the construction method proposed by Jacobsen [3] to generate valid test instances with their respective optimal solutions. Afterward, we add to these valid test instances knapsack probabilistic constraints in the upper-level sub-problem as in [4,5]. The latter allows us to compute upper bounds for stochastic bilevel instances as well. Our numerical results show average relative gaps of 43.97% for the valid test instances and 28.93% while adding probabilistic constraints.

1 Introduction

In mathematical programming, the bilevel programming problem (BPP) is a hierarchical optimization problem. It consists in optimizing an objective function subject to a constrained set in which another optimization problem is embedded. The first level optimization problem (upper-level problem) is known as the leader's problem while the lower-level is known as the follower's problem. Applications concerning BPP include transportation, networks design, management and planning among others. It has been shown that BPPs are strongly NP-hard even for the simplest case in which all the involved functions are affine [1]. In both papers [4,5], neither optimal solutions nor gaps have been reported. In this paper, we implement the construction method proposed by Jacobsen [3] to obtain valid test instances with their respective optimal solutions for our LBPP. Then, we compute upper bounds by solving these instances with the iterative min-max (IMM) algorithm proposed in [4]. Additionally, we add knapsack probabilistic constraints in the upper-level subproblem same as in [4,5]. The probabilistic constraints are added in such a way each of them does not cut off the optimal point previously found by Jacobsen method. As a consequence, we obtain upper bounds for stochastic bilevel instances as well. The paper is organized as follows. In Section 2, we state the LBPP and briefly explain Jacobsen construction method. We also introduce the probabilistic constraints. For sake of clarity in Section 3, we briefly explain how IMM algorithm computes the upper bounds. In Section 4, we provide optimal solutions

J. Pahl, T. Reiners, and S. Voß (Eds.): INOC 2011, LNCS 6701, pp. 620–625, 2011.

and upper bounds for the valid test instances with and without probabilistic constraints. Finally, in Section 5 we give the main conclusions of the paper.

2 Problem Formulation and the Jacobsen Construction Method

In this section, we present a generic LBPP, the knapsack probabilistic constraints and briefly explain Jacobsen construction method [3]. We consider the generic LBPP [3,4]:

$$\begin{aligned}
\text{LBP1:} \quad & \max_x \; c_1^T x + d_1^T y && (1)\\
& \text{s.t. } A^1 x + B^1 y \leq b^1 && (2)\\
& 0 \leq x \leq \mathbb{1}_{n_1} && (3)\\
& y \in \arg\max_y \{c_2^T x + d_2^T y\} && (4)\\
& \text{s.t. } A^2 x + B^2 y \leq b^2 && (5)\\
& 0 \leq y \leq \mathbb{1}_{n_2} && (6)
\end{aligned}$$

where $x \in \mathbb{R}^{n_1}$ and $y \in \mathbb{R}^{n_2}$ are the decision variables. Vectors $\mathbb{1}_{n_1}$ and $\mathbb{1}_{n_2}$ are vectors of size n_1 and n_2 with entries equal to one. Matrices A^1, B^1, A^2, B^2 and vectors $c_1, c_2, d_1, d_2, b_1 \in \mathbb{R}^{m_1}, b_2 \in \mathbb{R}^{m_2}$ are input real matrices/vectors defined accordingly. In LBP1, (1)-(3) correspond to the leader's problem while (4)-(6) represent the follower's problem. In mathematical terms, a solution of a BPP is achieved when both; the leader and follower find an optimum equilibrium point such that for each valued vector x found by the leader, vector y corresponds to the optimal solution of the follower problem. Knapsack probabilistic constraints can also be added to the upper-level problem of LBP1. Let $w = w(\omega) \in \mathbb{R}_+^{n_1}$ and $S = S(\omega) \in \mathbb{R}_+$ be two random variables distributed according to a discrete probability distribution Ω. The authors in [4,5] define the following knapsack probabilistic constraints

$$\mathbb{P}\left\{w^T(\omega)x \leq S(\omega)\right\} \geq (1-\alpha) \tag{7}$$

where α represents the risk we take while not satisfying some of the constraints. Since Ω is discrete, one may suppose that $w = w(\omega)$ and $S = S(\omega)$ are concentrated in a finite set of scenarios such as $w(\omega) = \{w_1, .., w_K\}$ and $S(\omega) = \{s_1, ..s_K\}$, respectively with probability vector $P^T = (p_1, .., p_K)$ for all k where $\sum_{k=1}^K p_k = 1$ and $p_k \geq 0$. According to [4], constraints in (7) can be transformed into the following pair of deterministic constraints

$$w_k^T x \leq s_k + M_k z_k, k = 1:K, \quad \text{and} \quad P^T Z \leq \alpha \tag{8}$$

where vector $Z^T = (z_1, .., z_K,)$ is composed of binary variables. If $z_k = 0$ the constraint is included, otherwise it is not activated. M_k for each $k = 1:K$ is defined as $M_k = \sum_{i=1}^{n_1} w_k^i - s_k$, where w_k^i denotes the ith component of vector w_k. In this paper, we generate and randomly add the pair of constraints in (8) in the upper-level problem of LBP1. Putting it altogether, it yields the following deterministic mixed integer LBPP

$$\text{LBP2:} \quad \max_x \; (1) \quad \text{s.t.} \quad (2)-(3),(8), \quad y \in \arg\max_y \{c_2^t x + d_2^t y\} \quad \text{s.t.} \quad (5)-(6)$$

Although LBP2 contains binary variables, it can be converted into an equivalent continuous LBPP (see ref. [33] in [4]).

Jacobsen construction method: an important step within this paper is the generation of the input data with known optimal solution for LBP1. This step is crucial since it allows the evaluation of IMM algorithm [4]. We implement Jacobsen construction method as described in [3]. The method uses linear programming (LP) together with some neighbor local vertex search. It mainly consists of selecting a random vertex situated approximately in the middle of P as the global solution of a LBPP, where $P = \{Ax + By \leq b, x, y \geq 0\}$ with $A = (A^1, A^2)^T$, $B = (B^1, B^2)^T$, and $b = (b^1, b^2)^T$. Subsequently, the method systematically designates a predetermined number of constraints to be assigned to the lower level problem. As local vertex search strategy, we randomly interchange basic with non-basic variables until we find a neighbor basic feasible solution.

3 The Iterative Min-Max Algorithm

In order to apply IMM algorithm, model LBP1 (resp. LBP2) should be transformed into the so called global linear complementary problem (GLCP). In case of LBP2, we have to reformulate it first as a continuous LBPP due to its binary variables [4]. The GLCP is a single level formulation. The idea is to replace the follower's problem with its initial constraints and complementary slackness conditions. The decision variables of GLCP are: the leader, the follower and the follower's dual variables. To show how IMM algorithm works, we derive the GLCP for LBP2 as

$$\begin{aligned}
\text{LBP2}_G: \quad & \max_{x,y,z,\lambda,\mu_1,\mu_2,\mu_3} \quad (1) \\
\text{s.t.} \quad & (2)-(3), (5)-(6), (8) \\
& (B^2)^T\lambda + \mu_3^T \mathbb{I}_{n_2} \geq d_2 && (9) \\
& \mathbb{I}_K \mu_1 + \mathbb{I}_K \mu_2 \geq \mathbb{1}_K && (10) \\
& \lambda^T(b^2 - A^2x - B^2y) = 0 && (11) \\
& \mu_1^T z = 0; \quad \mu_2^T(\mathbb{1}_K - z) = 0; \quad \mu_3^T(\mathbb{1}_{n_2} - z) = 0 && (12) \\
& y^T((B^2)^T\lambda + \mathbb{I}_{n_2}\mu_3 - d_2) = 0 && (13) \\
& 0 \leq z \leq \mathbb{1}_K; \quad \lambda, \mu_1, \mu_2, \mu_3 \geq 0 && (14)
\end{aligned}$$

where $\mathbb{I}_{n_2}, \mathbb{I}_K$ represent identity matrices of size n_2 and K, respectively. $\lambda, \mu_1, \mu_2, \mu_3$ and (9)-(10) are dual variables and constraints of the follower. This model is a quadratic problem since the complementary constraints (11)-(13) are quadratic, and thus it is hard to solve directly. The first step of IMM consists in relaxing these quadratic constraints into the following Lagrangian function

$$\begin{aligned}
\mathscr{L}(x,y,z,\lambda,\mu_1,\mu_2,\mu_3) = {} & c_1^T x + d_1^T y + \lambda^T(b^2 - A^2x - B^2y) + \\
& + \mu_1^T z + \mu_2^T(\mathbb{1}_K - z) + \mu_3^T(\mathbb{1}_{n_2} - z) + \\
& + y^T((B^2)^T\lambda + \mathbb{I}_{n_2}\mu_3 - d_2)
\end{aligned} \tag{15}$$

This allows writing a min-max relaxation for LBP2$_G$ as follows

$$\text{LGN:} \quad \min_{\{\lambda,\mu_1,\mu_2,\mu_3\}} \max_{\{x,y,z\}} \; (15) \quad \text{s.t.}(2)-(3),(5)-(6),(8),(9)-(10),(14)$$

The second step of IMM consists of decomposing LGN into two linear programming subproblems: LGNs and LGNd as

$$\text{LGNs:} \max_{\{x,y,z,\varphi\}} \quad \varphi \quad \text{s.t.} \quad (2)-(3),(5)-(6),(8), \quad 0 \leq z \leq \mathbb{1}_K$$
$$\varphi \leq \mathscr{L}(x,y,z,\lambda^q,\mu_1^q,\mu_2^q,\mu_3^q) \quad \forall q = 0,1,..,N-1 \tag{16}$$

and

$$\text{LGNd:} \quad \min_{\{\lambda,\mu_1,\mu_2,\mu_3,\beta\}} \quad \beta \quad \text{s.t.} \quad (9)-(10), \quad \lambda,\mu_1,\mu_2,\mu_3 \geq 0$$
$$\beta \geq \mathscr{L}(x^q,y^q,z^q,\lambda,\mu_1,\mu_2,\mu_3) \quad \forall q = 1,..,N \tag{17}$$

where φ and β are defined as free real variables. Finally, the third step of the algorithm consists in solving iteratively both LGNs and LGNd. At iteration q, the auxiliary constraint (16) (resp. (17)) is added to LGNs (resp. LGNd) in order to enforce the convergence of their optimal solution values towards the optimal solution value of LGN. The iteration process stops when either $\beta - \varphi < \delta$ or $(\beta - \varphi)/\beta < \varepsilon$ for small $\delta > 0$ and $\varepsilon > 0$. The convergence of IMM is proven in [4].

4 Numerical Results

In this section we present numerical results which measure the performance of IMM algorithm when applied to both; LBP1 and LBP2 respectively. The input data is generated according to Jacobsen algorithm [3], except for the objective functions of the leader and follower problems. These data and the data for the probabilistic constraints is generated exactly as in [4]. This generation procedure ensures that the inducible region of the generated LBPP is bounded, but it does not guarantee non-emptiness. Jacobsen and IMM algorithms are implemented using Matlab 7.8. The linear programs are solved using Mosek [2]. The simulations are run in a 2100 MHz computer with 2 Gb Ram under windows XP. Table 1 shows the upper bounds for LBP1. Similarly, table 2 shows numerical results for LBP2. The averages are computed over 100 sample runs for each instance in both tables. In table 1, columns 2-4 give the size of the instances. Columns 5, 6 and 7 give the average optimal values, average for the upper bounds and the standard deviation for these upper bounds, respectively. In column 8-10, we provide the minimal gap, maximal gap and the average relative gaps. The relative gaps are computed, for each run and for each instance as $\left(\frac{UB-Opt}{UB}\right) * 100$. Finally, columns 11 and 12 give the averages regarding the number of LP problems IMM algorithm solves and the cpu time in seconds it takes to converge. We mainly observe from table 1, that the gaps do not increase with the size of the instances and that they are not near optimal. On the other hand, the cpu times show that the algorithm is fast. In table 2, we provide the same information as in table 1, except for the new added column 2. In this column, we give

Table 1. Upper bounds for LBP1 using IMM

#	Instance Size			Avg	UBs		Gaps			Avg	Avg Cpu
	m_1+m_2	n_1	n_2	Opt	Avg	Std	Min	Max	Relative(%)	# LP	Time
1	5	5	5	24.96	43.24	9.87	0.494	95.81	42.85	2.42	0.004
2	5	10	10	42.71	84.51	14.85	7.385	97.14	49.01	2.90	0.008
3	10	15	15	66.44	123.21	19.88	8.627	95.36	46.07	3.86	0.012
4	10	20	20	88.01	170.54	21.22	13.405	91.44	48.28	4.61	0.015
5	15	25	25	99.82	120.72	22.73	6.592	91.61	42.50	5.39	0.021
6	15	30	30	136.18	254.57	28.08	10.561	91.47	46.70	5.71	0.024
7	20	35	35	162.46	300.70	25.45	3.540	84.50	46.19	7.41	0.038
8	20	40	40	180.33	334.57	27.74	15.642	83.56	45.89	7.28	0.048
9	30	45	45	239.52	385.72	31.35	8.153	74.57	37.71	10.41	0.101
10	30	50	50	240.31	421.04	30.67	7.722	79.34	42.86	9.49	0.101
11	40	60	60	315.74	506.87	36.97	9.442	73.01	37.57	14.01	0.220
12	50	100	100	500.18	866.59	52.19	16.523	83.00	42.00	16.81	0.549

Table 2. Upper bounds for LBP2 using IMM

#	K	Instance Size			Avg	UBs		Gaps			Avg	Avg Cpu
		m_1+m_2	n_1	n_2	Opt	Avg	Std	Min	Max	Relative(%)	# LP	Time
1	5	5	10	10	54.26	85.27	13.71	1.536	87.02	35.73	5.61	0.013
2	10	5	10	10	59.79	86.47	14.15	1.880	62.35	30.36	13.61	0.034
3	20	5	10	10	57.26	83.25	16.12	0	91.53	30.24	22.00	0.069
4	50	5	10	10	59.01	85.87	16.11	0.250	80.12	30.53	68.58	0.543
5	5	10	15	15	88.49	127.25	19.71	2.515	90.36	30.14	8.05	0.024
6	10	10	15	15	94.99	127.01	19.84	0.740	83.25	24.56	13.55	0.040
7	20	10	15	15	91.66	127.63	16.93	0.288	89.96	28.21	25.08	0.103
8	50	10	15	15	89.91	127.02	20.07	0.247	79.20	28.70	57.87	0.530
9	5	15	20	20	118.95	169.87	20.05	1.722	84.19	29.66	8.19	0.029
10	10	15	20	20	130.45	170.86	24.88	0.641	78.54	23.29	26.33	0.164
11	20	15	20	20	128.49	172.09	20.01	1.131	83.42	25.13	23.85	0.127
12	50	15	20	20	119.19	173.49	21.18	0.615	86.20	31.04	55.82	0.635

the number of probabilistic constraints we add to each valid test instance previously generated with Jacobsen algorithm. In table 2, we observe slightly better upper bounds when compared to table 1. On the opposite, the number of LPs and the cpu times are larger. This can be explained by the number of constraints which in LBP2 are larger than in LBP1. Finally, the gaps for LBP2 are not near optimal either.

5 Conclusions

In this paper, we computed upper bounds for a LBPP using IMM algorithm proposed in [5,4]. We generated valid test instances with Jacobsen construction method [3] together with their respective optimal solutions. Then, we added to these valid test instances linear probabilistic constraints in the upper-level sub-problem. Our numerical results show average relative gaps of 28.93% and 43.97% with and without using probabilistic constraints. As future research, valid cuts such as knapsack valid inequalities or cover ones could be added to reduce these gaps.

Acknowledgement

The authors are grateful for the financial support given by Conicyt through the projects: Inserción de Capital Humano en la Academía, cod: 79100020 and Dicyt-Usach, cod: 0611175.

References

1. Hansen, P., Jaumard, B., Savard, G.: New Branch and Bound Rules for Linear Bilevel Programming. SIAM Journal on Scientific and Statistical Computing 13, 1194–1217 (1992)
2. MOSEK is an optimization software designed to solve large-scale mathematical optimization problems, http://www.mosek.com/
3. Moshirvaziri, K., Amouzegar, M., Jacobsen, S.: Test Problems Construction for Linear Bilevel Programming Problems. Journal of Global Optimization 8, 235–243 (1996)
4. Kosuch, S., Lebodic, P., Leung, J., Lisser, A.: On Stochastic Bilevel Programming Problem. Networks (accepted paper) (to appear, 2011)
5. Kosuch, S., Lebodic, P., Leung, J., Lisser, A.: On a Stochastic Bilevel Programming Problem with Knapsack Constraints. In: International Network Optimization Conference INOC 2009, Pisa, Italy (2009)

Mixed Integer Programming Model for Pricing in Telecommunication

Mustapha Bouhtou[1], Jean-Robin Medori[1], and Michel Minoux[2]

[1] Mustapha Bouhtou, Orange Labs, 38-40 rue du General Leclerc, 92794 Issy-Les-Moulineaux Cedex 9, France
{mustapha.bouhtou, jeanrobin.medori}@orange-ftgroup.com
[2] UPMC-LIP6, 4 place Jussieu 75252 Paris Cedex 5, France
michel.minoux@poleia.lip6.fr

Abstract. Yield management has been successfully applied in the context of airline companies. However, so far, application to telecommunication industry have been scarce. Using Yield management principles, this paper investigates the new problem of maximizing revenue of telecommunications operator by setting prices on voice services. This pricing is based on available resource i.e. network load. We first propose a Mixed Integer Program to model this problem. Then we study the particular case where demand and load are linear functions of price. Using numerical results, we also study the impact of a Big M constant on computational effort required to solve the problem.

1 Introduction

For the last three decades, yield management (YM) has attracted interest from theoretical as well as operational point of view. Generally, YM is defined as the application of information systems and pricing strategies to allocate the right capacity to the right customer at a right price and the right time in order to maximize revenue.

Airlines were among the first to implement YM [5] and some other industries have chosen later to implement such a pricing system. A state of the art can be found in [4].

Despite this interest, so far, only a few attempts have been conducted to apply YM to the telecommunication industry. Humair [2] propose to define basis and model a framework for telecommunications. Several papers propose YM techniques to control the Internet congestion and mobile network congestion. Manaffar [3] proposes to integrate pricing with call admission control in mobile networks. Bouhtou et al. [1] studie pricing for telecommunications and propose bi-level optimization formulations to model competition between operators. Viterbo et al. [6] investigate revenue maximization in the context of a real-time pricing model for cellular networks.

In this article, we deal with the new problem of pricing a voice telecommunication service provided by an operator according to available resource. In a way similar to airlines, we apply YM principles to sell voice communication on a perishable and limited resource. The main objective is revenue optimization (versus congestion control). First we formulate the problem as a Mixed Integer Program (MIP), then we study the particular case where demand and load functions are linearly dependent on price which leads to a Mixed Integer Quadratic Program (MIQP). Some preliminary numerical results are presented in Section 4.

J. Pahl, T. Reiners, and S. Voß (Eds.): INOC 2011, LNCS 6701, pp. 626–630, 2011.

2 Pricing Problem in Voice Telecommunication Networks

We consider an operator providing voice telecommunication services using a volume pricing strategy. We apply YM principles, as in airlines, and propose to set dynamic prices throughout the day, based on available resource capacity, in order to maximize daily revenue. We split a day in several time slots, with possibly different durations.

We adopt a pricing strategy based on network load. Indeed, we see in fig.1 an example of load throughout a day at two different fixed prices. This chart shows that load is price dependent, and so the importance of choosing a pricing strategy based on network load.

Fig. 1. Load vs time

Table 1. Price Grid

Load levels	Load intervals	Prices
0	$[0;Th_0]$	r_0
1	$[Th_0;Th_1]$	r_1
$\vdots$	$\vdots$	$\vdots$
I	$[Th_{I-1};\infty[$	r_I

We consider several load thresholds Th_i, and the last one correspond to the network capacity. Congestion is allowed, but not in two consecutive time slots.

In order to set a price for each time slot based on load, we assign a price to each load level. So, the daily pricing problem is reduced to finding an optimal price grid such as table 1. Price r_i is applied at time slot t if load of the previous time slot $t-1$ was in level i. Thus, load used to price time slot t depend on price set at $t-1$. This rule makes our problem dynamic.

3 MIP Formulation

3.1 Notation

The following notation is used throughout the paper :

1. Input Data :
   - T : Number of time slots indexed by t
   - I : Number of load thresholds indexed by i
   - Th_i : Load thresholds
   - p_{-1} : Input representing price of the last time slot of the day before
   - $\rho_t(p_{t-1})$: Function representing load at t
   - $D_t(p_t)$: Function representing demand (number of seconds consumed) at t
   - $R_t(p_t)$: Function representing revenue obtained in period t
2. Variables :
   - $p_t \in [P_{min};P_{max}]$: Variable representing price assigned to time slot t
   - $r_i \in [P_{min};P_{max}]$: Variable representing price assigned to load level i, i.e. between Th_{i-1} and Th_i

- β_i^t : Boolean variable equals to 1 if load at t is in load level i
- α_i^t : Variable equals to load at t if it is in load level i, 0 otherwise.

3.2 General Problem Formulation

Using previous notation, we propose the general voice pricing problem $Pb(M)$ based on resource capacity, and modeled as a MIP.

The objective function is taken as sum of revenues $R_t(p_t)$ on the successive periods. In constraint (2), we use α_i^t to model load at t in load level i. Constraint (3) allows to set the load level in which the load ρ_t is. Constraint (4) ensures each period to be assigned only to one load level. Constraint (5) assigns a time slot to a load level. It allows the price r_i to be equals to price p_t if and only if load at t is in the level i (i.e. $\beta_i^t = 1$). M represents a Big M constant. Constraint (6) prevents congestion phenomenons to occur in two consecutive time slots.

$$
Pb(M)\begin{cases}
\max \sum_{t=0}^{T-1} R_t(p_t) & & (1)\\
\textbf{subject to} & & \\
\sum_{i=0}^{I} \alpha_i^t = \rho_t(p_{t-1}) & \forall t & (2)\\
\beta_i^t Th_{i-1} \le \alpha_i^t \le \beta_i^t Th_i & \forall i,t & (3)\\
\sum_{i=0}^{I} \beta_i^t = 1 & \forall t & (4)\\
(\beta_i^t - 1)M \le p_t - r_i \le (1-\beta_i^t)M & \forall i,t & (5)\\
\beta_I^{t-1} + \beta_I^t \le 1 & \forall t \ne 0 & (6)\\
\beta_i^t \in \{0;1\} & \forall i, \forall t & \\
\alpha_i^t \ge 0 & \forall i, \forall t & \\
r_i \in [P_{min};P_{max}] & \forall i & \\
p_t \in [P_{min};P_{max}] & \forall t &
\end{cases}
$$

3.3 Linear Demand and Load Model

We focus on the special case where load and demand are linear and decreasing functions of price, and so revenue is a quadratic function of price.

$$\rho_t(p_{t-1}) = d_t - c_t \cdot p_{t-1} \tag{7}$$

$$R_t(p_t) = D_t(p_t) \cdot p_t = b_t \cdot p_t - a_t \cdot p_t^2 \tag{8}$$

where a_t, b_t, c_t and d_t are positive input provided by statistical data. In this case, $Pb(M)$ becomes a MIQP with $T(I+1)$ binary variables, $T(I+1)+T+I+1$ continuous variables, and $4T(I+1)+3T-1$ linear constraints.

4 Experimentation and Numerical Results

We have built several instances with different numbers of time slots, using data measured every 30 minutes on a real network. Several tests have been performed to solve those instances using Mixed Integer quadratic programming solver from Cplex 12.1 on a workstation with 4 processors Intel(R) Xeon(R) 1.6-GHz, and 4096-KB memory size.

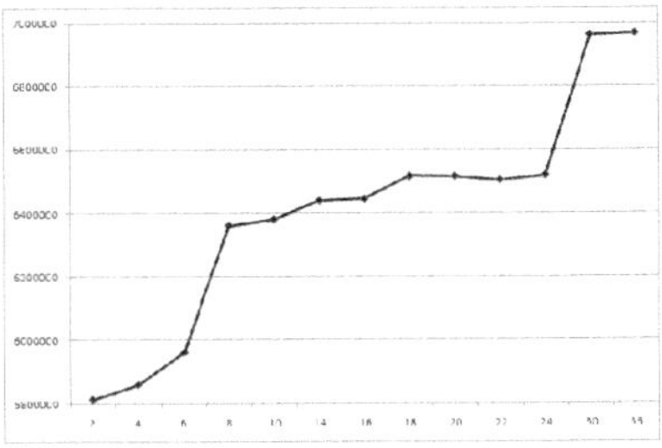

Fig. 2. Revenue vs Number of periods

In this section, we discuss the interest to consider a high number of time slots. As can be seen in fig. 2, revenue tends to increase as the number of time slots increases. The increase in optimal solution value is about 17%. This confirms the importance to consider the problem for a sufficiently large number of time slots. However, the problem is hard to solve. Results show that for $T = 8$ to $T = 48$, integrity gaps become very high, between 34% and 400%. Moreover, with $M = 10000$, we only obtain in 4 hours, approximations with respectively for the instances with $T = 42$ and $T = 48$, a 1.41% and a 2.95% gaps. Thus, we have investigate ways to improve efficiency by choosing a more accurate value of the Big M constant.

Let (p_t^*, r_i^*) be an optimal solution. we note :

$$P_{max}^* = \max_t \{p_t^*\} = \max_i \{r_i^*\} \tag{9}$$

$$P_{min}^* = \min_t \{p_t^*\} = \min_i \{r_i^*\} \tag{10}$$

When $\beta_i^t = 0$, equation (5) leads to $|p_t - r_i| \leq M$. As $P_{max}^* - P_{min}^*$ represents the highest difference between two prices, M just have to be greater than $P_{max}^* - P_{min}^*$. So, the idea in the following is to derive an upper bound to $P_{max}^* - P_{min}^*$, i.e. upper and lower bounds respectively to P_{max}^* and P_{min}^*. As a price is always positive, we consider 0 as the lower bound of P_{min}^*.

Let us define $\widehat{P_{t,i}}$, $\overline{P_t}$ and $\widetilde{P_t}$ such as

$$\rho_t(\widehat{P_{t,i}}) = Th_i \tag{11}$$

$$\rho_t(\overline{P_t}) = 0 \tag{12}$$

$$\widetilde{P_t} = \arg\max_p \{R_t(p)\} = \frac{b_t}{2a_t} \tag{13}$$

Proposition 1.

$$P^*_{max} \leq \min\left\{\max_t\{\overline{P_t}\}\ ;\ \max_t\{\widehat{P_{t,0}}\ ;\ \widetilde{P_t}\}\right\} \tag{14}$$

Proof. Because ρ_t is positive and decreasing, P^*_{max} is obviously lower than each $\overline{P_t}$. Let us consider an optimal solution such as $P^*_{max} = r_i \geq \max_t\{\widehat{P_{t,0}}\ ;\ \widetilde{P_t}\}$. By contradiction, it is easy to prove that this solution is not optimal, because the solution with the same prices but $\max_t\{\widehat{P_{t,0}}\ ;\ \widetilde{P_t}\}$ instead of r_i, is better, and still feasible (any time slot is assigned to another load level).

We performed tests on the same instances but with $M = P^*_{max}$. Results show that the choice of the M value does not have any impact on integrity gap. Table 2 represents computing time and number of generated nodes according to M value for some instances. Computing time and number of generated nodes are globally smaller using M smaller which validate the interest of our proposition. With an accurate M the instance with $T = 42$ has been solved in less than 3 hours, and we obtain for the instance with $T = 48$ a better approximate solution at 1.82% (vs 2.95) in about 3 hours (vs 4 hours).

Table 2. Big M influence on computing time and number of nodes

Number of time slots	Integrity gap	Computing time in seconds		Number of generated nodes	
		$M = 10000$	$M = 430$	$M = 10000$	$M = 430$
12	339.65%	0.64	0.33	509	408
20	177.15%	11.52	4.51	9000	7500
24	175.84%	26.95	9.98	19400	18600
30	160.34%	353.91	157.11	200100	199000
42	396.5%	14400[1.41%]	9986.65.96	32572500	11000400
48	183.01%	14031.41[2.95%]	7639.57[1.82%]	8006100[2.95%]	7810300[1.82%]

5 Conclusion and Future Research Work

This paper deals with a new problem of yield management in the telecommunication industry. A MIP formulation has been proposed to model this problem of pricing voice services based on available resource. We focused our study on the particular case, where demand and load are functions linearly dependent on price resulting in a MIQP. A study on the impact of a big M constant has also been carried on and led to improve the computational efficiency. In future work, we intend to extend this model, considering for example more general models for load and demand depending on several variables.

References

1. Bouhtou, M., Erbs, G., Minoux, M.: Joint optimization of pricing and resource allocation in competitive telecommunications networks. Networks 50, 37–49 (2007)
2. Humair, S.: Yield Management for Telecommunication Networks: Defining a New Landscape. PhD thesis, Massachusetts Institue Of Technology (2001)
3. Manaffar, M., Bakhshi, H., Pilevari, M.: A new dynamic pricing scheme with call admission control to reduce network congestion. In: Advanced Information Networking and Applications - Workshops, pp. 347–352 (2008)
4. Bitran, R., Caldentey, R.: An Overview of Pricing Models for Revenue Management. MIT Sloan Working Paper No. 4433-03 (2002)
5. Smith, B., Leimkuhler, J., Darrow, R.: Yield Management at American Airlines. Interfaces (1992)
6. Viterbo, E., Chiasserini, C.F.: Dynamic Pricing in Wireless Networks (2001)

UL RSSI as a Design Consideration for Distributed Antenna Systems, Using a Radial Basis Function Model for UL RSSI

Sarel Roets[1], Praven Reddy[1], and Poovendren Govender[2]

[1] Mobile Telephone Network, Johannesburg, South Africa
{roets_sa,goven_po}@mtn.co.za
[2] Ericsson, Ericsson South Africa, Woodmead, South Africa
praven.reddy@ericsson.com

Abstract. This paper aims to develop a non-linear model for the Uplink Received Signal Strength Indicator (UL_RSSI) during high capacity stadium events. Radial Basis Functions (RBF's) are used to aid in developing the model for the UL_RSSI as the amount of users in the stadium changes. Air Speech Equivalent (ASE) is used to model users on the uplink. Furthermore a model is derived to assist planning engineers to determine a suitable amount of sectors needed for Distributed Antenna Systems (DAS) that will not cause degradation in the noise floor. Results for the UL_RSSI model prove to follow trending of actual noise floor characteristics during high capacity stadium events with reasonable accuracy. The model was used along with a design example and results if implemented should provide Enhanced Uplink/High Speed Uplink Packet Access (EUL/HSUPA) users with an acceptable Quality of Service.

1 Introduction

With the advent of both social network applications and the smart phone revolution, it has become increasingly important for operators to handle the increased capacity and provide higher data rates in the uplink for users through WCDMA's EUL/HSUPA technologies. In the WCDMA standard, the User Equipment (UE) can have a minumum of two radio links, one in the uplink the other in the downlink. The downlink is nearly orthogonal with users well separated, however the power in the downlink is shared by the users. The uplink is non-orthogonal, interference limited as every UE has a power budget and interferes with each other. Improving the uplink has now become more critical as it dictates at what power level a UE should transmit to be measured/detected by the NodeB.

During high capacity stadium events it has been seen that the noise on the uplink increases non-linearly and subsequently the UE has to transmit at higher power levels. Higher noise levels on the uplink now however tend to decrease the quality of the uplink and EUL services.

From data acquired during high capacity events it came to light that the model used for UL_RSSI does not indicate the non-linear output from actual data. The motivation was now to use Neural Networks (NN) to obtain a non-linear model for UL_RSSI during high capacity stadium events. Our work improves on the model in [1]. An improved

J. Pahl, T. Reiners, and S. Voß (Eds.): INOC 2011, LNCS 6701, pp. 631–642, 2011.

model representing actual behavior of the noise floor can assist planning engineers in designing stadium solutions to provide the best possible service to the users. An accurate model can predict what changes in the noise floor for various conditions.

2 WCDMA Network Theory

2.1 WCDMA Network Architecture Overview

According to 3GPP the radio access network for UMTS Radio Access Network (WCDMA RAN), provides a connection between the Core Network (CN) and the UE while also interfacing towards the external Network Management Systems (NMS) [2]. The Network Elements (NE's), RNC, RXI, and Radio Base Station (RBS) manage the data links between WCDMA RAN and the UE. An architecture diagram for the WCDMA network is shown in figure 1. Radio Access Bearers (RAB) are setup between the CN and the UE, as shown in figure 1, for different traffic classes, conversational (voice), streaming class (video), interactive (internet browsing) and background (email). The fore-mentioned services utilise a signalling connection between the NE's. RRC (radio resource control) signalling protocol messages occur between the UE and RNC. Non Access Spectrum (NAS) signalling protocol messages occur between the UE and CN. The RAB contains an Iu bearer from the RNC to the CN, a user plane Transport Channel (TRCH) on the Iub which forms a radio link between the UE and RNC [2].

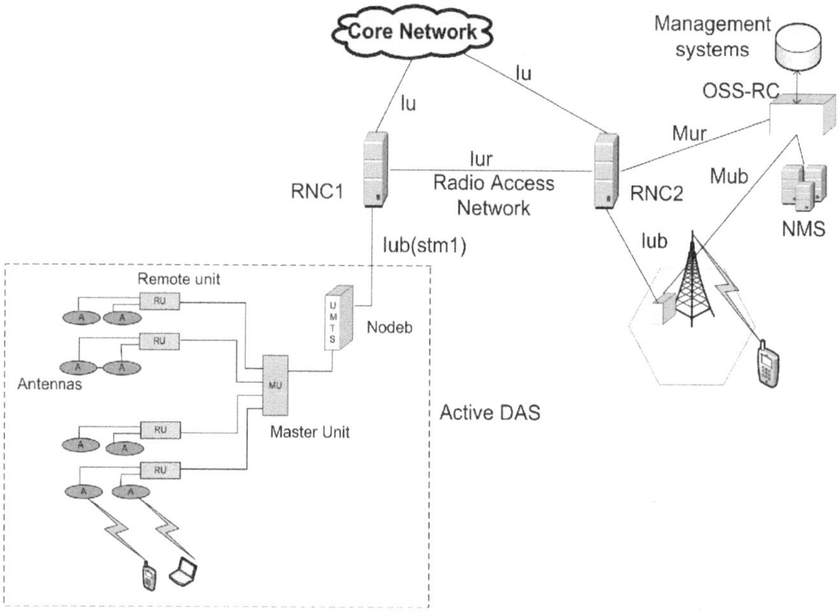

Fig. 1. WCDMA network architecture layout. Architecture is drawn from the Subscriber side up to the Core Network.

In the case of an active DAS in figure 1 , minimal RF gain is used from the Node-B. RF signals are converted to optical frequency via the Master Unit (MU) for distribution over fiber and converted back to RF when terminating at wideband Remote Units (RU's). RU's are installed close to the antennas to ensure that lost energy is minimized. The active DAS is ideal for multisector/multioperator applications, offering the most uniform coverage and optimal trunking efficiency with the best distribution of RF for HSPA services. The active DAS has a lower noise figure in the uplink when compared to the high system noise figure on the high loss passive DAS system. Passive DAS designs require the Bidirectional Amplifiers (BDA's) as close as possible to the antenna to circumvent the passive attenuation which impacts the noise figure, limiting uplink performance. In the case of a hybrid DAS, the uplink in the passive portion is the limiting factor and careful consideration should be taken in the choice of value for the UL Attenuator, as the noise power from the active portion may degrade the passive portion [3].

2.2 Discussion on HSUPA/EUL

After the Release99 version of WCDMA access networks it was realized that higher data throughputs was required on the downlink. This new access release for WCDMA entailed the enhanced downlink capabilities and saw users obtaining speeds of up to 14.4 Mbps. Release5 entailed enhanced capabilities for downlink and was named, High Speed Data Packet Access (HSDPA). User tendencies after Release5 indicated users requiring faster speeds on the uplink leading to Release6 and 7 with enhanced uplink (EUL) capabilities. Enhancements on Release5 uplink to obtain HSUPA are an increase in peak data rate and cell throughput as well as a shorter round trip time. Various aspects (for example 2ms TTI and Hybrid Automatic Repeat Request (HARQ) with soft combining, dynamic power allocation and fast link adaptation) were introduced to achieve the above mentioned performance. Higher modulation schemes are used for HSDPA users with good channel quality to receive resources (code/power) and not be bandwidth limited. EUL uses fast power control in the uplink for link adaptation and does not require a higher modulation like 16QAM, rather QPSK as it is interference limited.

A higher modulation scheme introduces a higher Peak to Average ratio (PAR), which means a higher Electromagnetic Interference (EMI) being generated by the UE. Soft handover reduces intercell interference and provides macro diversity. Improved system response time is realized as the scheduling and HARQ functionality are present on the node b instead of the RNC as in R99. In HSUPA/EUL various channels where introduced as extras like the E-DCH, E-DPDCH, E-DPCCH, E-HICH, E-AGCH, and E-RGCH. These channels control throughput, signaling and admission control. For the admission control the UE TX Power and Rise over Thermal (RoT) is considered. RoT is a measure of the capacity on the Uplink channel as perceived by the Node-B. Each UE that transmits on the uplink is seen by the UE as interference decreasing the bandwidth on the uplink.

2.3 Theory on Uplink Noise

The radio link on the uplink is used to communicate from the UE towards the RAN. To communicate on the uplink a UE is required to transmit at a power larger than the thermal noise floor, discussed in section 4.1. When a single UE is present in the coverage of a cell, the UE is required to transmit at a power just above the thermal noise floor. If a second UE is however introduced it contributes to the noise floor in the environment and the first UE needs to increase its uplink transmission power.

Modelling and measuring the amount of users on the uplink is done by using ASE. An ASE is used to reserve air-interface resources in both the uplink and downlink [4]. In the case of a speech user, the Orthogonal Variable Spreading Factor, (OVSF), of 128 is used according to [5]. Utilising the OVSF code for a speech user can now indicate the bits per symbol or data rate used for a spesific service. In [6] the OVSF is expressed as in (1) where the chip rate is known to be 3.84 $Mcps$.

$$OVSF = \frac{ChipRate}{DataRate} \tag{1}$$

Obtaining the data rate from (1) results in (2).

$$DataRate = \frac{ChipRate}{OVSF} = \frac{3.84 \times 10^6}{128} = 30kbps \tag{2}$$

A date rate of 30 $kbps$ is found for one ASE/Speech user on the uplink according to [5]. Furthermore accroding to [4] one speech user is equivalent to 1.61 ASE. ASEs are used to provide a reservation for air-interface resources in both uplink and downlink. This monitor is based on estimation of the air interface usage per radio link type (Radio Connection Type) in a cell, separately for the uplink and downlink. The ASE of a radio link is a relative value, defined as the air-interface load relative to a speech radio link (12.2 $kbps$, 50 % activity). The general method of estimating the ASE value for a specific service is as follows:

$$ASE = \frac{maxrateradiolink}{maxrateradiolinkspeech} \times \frac{activityfactorradiolink}{activityfactorradiolinkspeech} \tag{3}$$

If the UE is in Soft Handover the UL ASE is distributed evenly between the cells in the active set, since UL interference in reduced in Soft Handover. The DL ASE values are naturally unchanged and counted in each cell.

3 Introduction to RBF Neural Networks

A short introduction behind the use and workings of RBF are discussed in this section. RBF's can be used to approximate linear or non-linear functions. A RBF has the feature of monotonicaly decreasing/increasing as the distance from the centre point changes, according to the radial function. The most widely used radial function is a Gaussian RBF [7].

$$h(x) = e^{\frac{-(x-c)^2}{r^2}} \tag{4}$$

In (4), c is the centre of the radial function and r the radius from the centre for a Gaussian RBF. The input to the radial function is, x, and the output is $h(x)$. When performing some function or pattern estimation the trained values for c and r are used to estimate

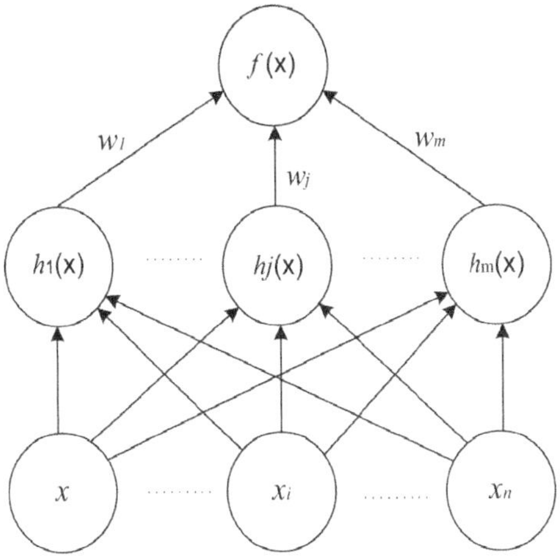

Fig. 2. Neural Network layout for a single layered Radial Basis Function

$h(x)$ for any given value for x. A RBF can be used for either a single layer or multi-layer neural network. For the single layered neural network the network architecture can be viewed as in figure 2. According to figure 2, a function approximation for a RBF Neural Network can thus mathematically be expressed as in equation (5).

$$f(x) = \sum_{j=1}^{m} \omega_j h_j(x) \tag{5}$$

The term $h(x)$ in (5) refers to (4) which is the Gaussian RBF. Weights, ω, for RBF functions are representative of the, r and c terms in (4).

Training the weights can be done by using a training set and least squares estimation. A training set consists of known input and output values for some function, thus x and $h(x)$ is known in (4) but for various ranges changing x into a vector $\mathbf{x}$ [7]. Rewriting the input x into a vector results in (6).

$$f(\mathbf{x}) = \sum_{\mathbf{j=1}}^{\mathbf{m}} \omega_{\mathbf{j}} \mathbf{h}_{\mathbf{j}}(\mathbf{x}) \tag{6}$$

Training the RBF now uses least squares estimation to minimise the sum-squared error with respect to the weights of the model [7]. In (7) the sum-squared error is represented by S.

$$S = \sum_{i=1}^{p} (\hat{y}_i - f(\mathbf{x}))^2 \tag{7}$$

When using ridge-regression as in [7], a cost function is introduced into (7) and the new equation (8) is obtained where C is represents the sum squared error with the additional cost function.

$$C = \sum_{i=1}^{p} (\hat{y}_i - f(\mathbf{x}_i))^2 + \sum_{j=1}^{m} \lambda_j \omega_j^2 \tag{8}$$

In (8), λ indicates the regularisation parameter, which is used to minimise the cost function. According to [7] minimising the cost function results in m linear equation to be solved. To determine, m, the *Optimal Weight Matrix* is used which determines the optimum amount of weights to be used. To find the optimum number of weights, m, one starts of by differentiating the cost function in (8) as follows:

$$\frac{\partial C}{\partial \omega_j} = s\sum_{i=1}^{p}(f(\mathbf{x}_i) - \hat{y}_i)\frac{\partial f}{\partial \omega_j}(\mathbf{x}_i) + 2\lambda_j \omega_j \tag{9}$$

The model, f also needs to be differentiated as seen in (10).

$$\frac{\partial f(\mathbf{x}_i)}{\partial \omega_j} = h_j(\mathbf{x}_i) \tag{10}$$

By substituting (10) into (11) and equating to zero the equation in (11) is found.

$$\sum_{i=1}^{p} f(\mathbf{x}_i)h_j(x_i) + \lambda_j \hat{\omega}_j = \sum_{i=1}^{p} \hat{y}_i h_j(\mathbf{x}_i) \tag{11}$$

Equation (11) has $1 \leq j \leq m$ constraints, where for each constraint there also exists an unknown. One can conclude that there is, m, unique solutions. By rewriting (11) into algebraic notations one can solve (11).

$$\mathbf{h}_j^T \mathbf{f} + \lambda_j \hat{\omega}_j = \mathbf{h}_j^T \hat{y} \tag{12}$$

For each value of, j, in (12) from 1 to m a relation can be established between the vector quantities of $\mathbf{h}_j^T \mathbf{f}$ and $\mathbf{h}_j^T \hat{y}$ as in (13).

$$\begin{bmatrix} \mathbf{h}_1^T \mathbf{f} \\ \mathbf{h}_2^T \mathbf{f} \\ \vdots \\ \mathbf{h}_m^T \mathbf{f} \end{bmatrix} + \begin{bmatrix} \lambda_1 \hat{\omega}_1 \\ \lambda_2 \hat{\omega}_2 \\ \vdots \\ \lambda_m \hat{\omega}_m \end{bmatrix} = \begin{bmatrix} \mathbf{h}_1^T \hat{y} \\ \mathbf{h}_2^T \hat{y} \\ \vdots \\ \mathbf{h}_m^T \hat{y} \end{bmatrix} \tag{13}$$

Equation (13) can then be rewritten with matrix notation to obtain the following expression:

$$\mathbf{H}^T \mathbf{f} + \Lambda \hat{\omega} = \mathbf{H}^T \hat{\mathbf{y}} \tag{14}$$

In (14), Λ and $\mathbf{H}$ can be represented as follows:

$$\Lambda = \begin{bmatrix} \lambda_1 & 0 & \dots & 0 \\ 0 & \lambda_2 & \dots & 0 \\ \vdots & \vdots & \ddots & \vdots \\ 0 & 0 & \dots & \lambda_m \end{bmatrix} \quad \text{and} \tag{15}$$

$$\mathbf{H} = \begin{bmatrix} h_1(\mathbf{x}_1) & h_2(\mathbf{x}_1) & \dots & h_m(\mathbf{x}_1) \\ h_1(\mathbf{x}_2) & h_2(\mathbf{x}_2) & \dots & h_m(\mathbf{x}_2) \\ \vdots & \vdots & \ddots & \vdots \\ h_1(\mathbf{x}_p) & h_2(\mathbf{x}_p) & \dots & h_m(\mathbf{x}_p) \end{bmatrix} \tag{16}$$

Considering, **f** to be the composition of two terms, the design matrix (**h**) and the weight vector (Λ).

$$f_i = f(x_i) = \sum_{j=1}^{m} \hat{\omega}_j h_j(\mathbf{x}_i) = \mathbf{H}\hat{\mathbf{w}} \tag{17}$$

By now substituting (14) into (12) one obtains the solution for $\hat{\mathbf{w}}$ in (18).

$$\begin{aligned} \mathbf{H}^T\hat{\mathbf{y}} &= \mathbf{H}^T\mathbf{f} + \Lambda\hat{\mathbf{w}} \\ &= \mathbf{H}^T\mathbf{H}\hat{\mathbf{w}} + \Lambda\hat{\mathbf{w}} \\ \hat{\mathbf{w}} &= (\mathbf{H}^T\mathbf{H} + \Lambda)^{-1}\mathbf{H}^T\hat{\mathbf{y}} \end{aligned} \tag{18}$$

Both the number of weights/neurons and the values for each of them are determined through the process discussed above to obtain $\hat{\mathbf{w}}$ in (18). The process to obtain $\hat{\mathbf{w}}$ is called training and hereafter the values for $\hat{\mathbf{w}}$ is used as a constant vector in (6).

4 Modelling Uplink Noise

4.1 Uplink Noise Floor

In a mobile environment with zero users there exists a noise floor. The noise floor represents the minumum power at which a UE should transmit to be heard by the network. Using the ambient temperature as well as the Boltzmann constant a value for the noise floor on the uplink can be determined as in (19) as discussed in [5].

$$\begin{aligned} N_{floor} &= T \times B \\ &= 290 \times 1.3806503 \times 10^{-23} \\ &= -105dBm \end{aligned} \tag{19}$$

In (19), T represents the temperature in Kelvin and B the Boltzman Constant in units of $m^2kg^{-2}K^1$. For purposes in this paper it is assumed that the ambient temperature remains constant at $290 \pm 3K$. The actual stadium installation had the cabinets contained in a temperature controlled room with similiar specifications.

4.2 Current RoT Model

Currently models for RoT indicate a parabolic relationship between the amount of users versus the increase in uplink noise. A model is found in [1], indicating the relationship to be parabolic. When considering a single UE transmitting to the NodeB at a constant power, the UE can be situated anywhere in the coverage of the NodeB. However if another UE is now also introduced into the coverage of the NodeB and at a distance closer to the NodeB than UE 1, noise is introduced into the system. Noise caused by more than one UE in the coverage area causes cell-breathing leading to UE's having to transmit at higher power to be received by the NodeB. According to [1], the effect of cell breathing can be modelled by (20):

$$\frac{E_b}{I_0 + N_0} = \frac{\frac{C}{R_b}}{N_0 + \frac{\alpha(1+i)(N-1)C}{W}} = \frac{G_p}{\frac{N_b W}{C} + \alpha(1+i)(N-1)} \tag{20}$$

In (20), C indicates the required received power from the UE, N_0 represents the power spectral density of the noise, W is the CDMA chip rate, N is representative of the amount of active users, α indicates the transmitter's duty cycle, G_p is the processing gain and i represents the ratio of the interference. If C is now solved from (20) an expression for the required received power or UL_RSSI is found in (21).

$$C = \frac{N_0 W\left(\frac{E_b}{I_0+N_0}\right)}{G_p - \left(\frac{E_b}{I_0+N_0}\right)\alpha(1+i)(N-1)} \tag{21}$$

For various sets of users the UL_RSSI versus users can be plotted according to (21) producing the results in figure 6. The relationship between the noise rise and uplink noise is thus a quadratic relationship. To obtain the results in figure 6, the following assumptions were made with regards to the values used for terms in (21). Assuming the OVSF is equal to the processor gain a value of 256 for speech users are used in (21) for G_p. A value of 7 dB for E_b/I_0+N_0 is used and 50 % for α. For i a value of 55 % was used and for the receiver noise, N_0, a value of 5 dB.

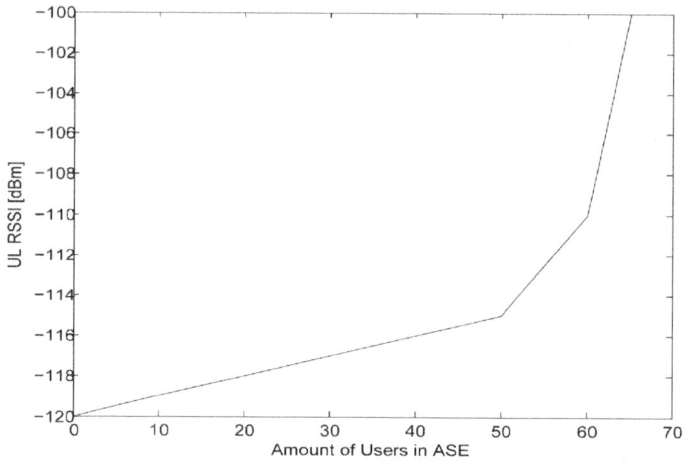

Fig. 3. Graphical results for 0 to 65 users using the expression for UL_RSSI in (21)

4.3 New Proposed Model for RoT

During an event at a stadium with high subscriber capacity it was noticed that UL_RSSI on the DAS in the stadium was limiting the performance of EUL. Data for the amount of users, ASE, was used along with the data for RoT providing insight into the equivalent noise rise attributed per voice user on the 3G network. A graphical representation of the average results for UL_RSSI versus ASE or users was found as in figure 4.

Figure 4 results in a 6-th order polynomial function approximation for the UL_RSSI as a function of users. However it was decided not to use this results for benchmarking further models as it is based on actual data for a full capacity stadium event. The idea

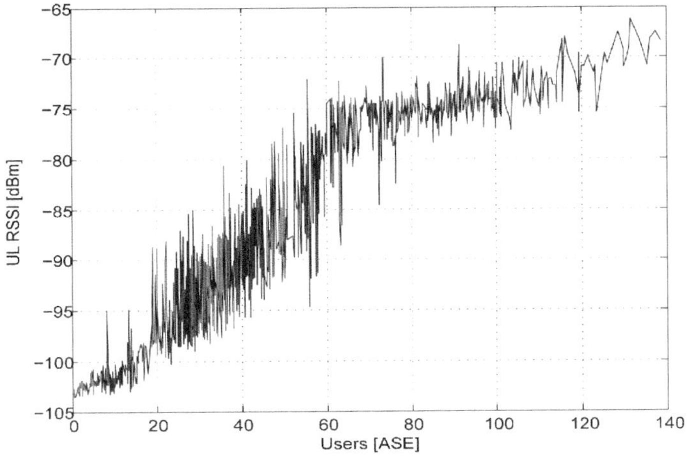

Fig. 4. UL_RSSI for a high capacity stadium during an event as a function of the amount of equivalent voise users

is to rather use a RBF Neural Network to create a model of the non-linear relationship between the UL_RSSI and the amount of users. Using the average UL_RSSI and average amount of users for the data from figure 4 results in obtaining the (noise per user factor), $k = -1.866\ dBm/user$.

Using actual data along with the non-linear function abilities associated with neural networks may now provide an improved model to estimate the effect of the amount of users on the UL_RSSI.

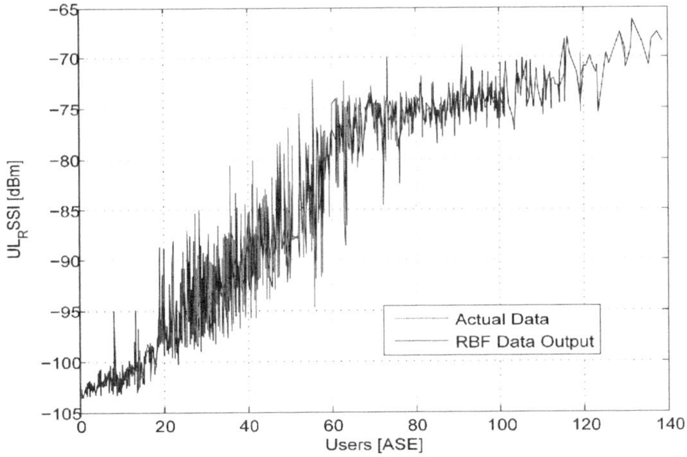

Fig. 5. Actual data plotted over estimated data for a stadium at full capacity during an event. An average error between the actual data and estimated data is -0.0000048 dBm.

RBF Approximation of the UL_RSSI as a function of ASE. Using both the theory in sections II/III and the data for figure 4, a RBF was trained with the amount of users as the input data. The resulting UL_RSSI was used as the target data. The training set used is for a full high capacity stadium during an event. Testing the RBF was done by using ASE data from the same stadium but from a different event (event data 2). The actual data was then compared to the UL_RSSI estimation from event data 2. An average error of $-0.000048\ dBm$ is observed between the actual data and the RBF estimated UL_RSSI. The model was tested for various other cases with similiar results and was therefore considered to be valid as a model. The trained neural network consists of 1175 neurons and has an error of $\pm 3\ \%$ after using least squared estimation for training the algorithm.

5 Design Model for DAS System

The main aim of this paper is to design a model to assist Radio Planning Engineers in effectively designing DAS's for stadiums. As mentioned in several sections the amount of users does affect the UL_RSSI which in turn influences the performance of EUL/HSUPA. The aim now is to use the data which can be provided by the new model for UL_RSSI as a function of users to be able to design the layout for a DAS for stadiums.

In this section we derive an expression for the number of subscribers that should be covered with an acceptable degradation in uplink noise. Stadiums are designed to house up to 100 000 people and usually a country has several mobile operators. Therefore this derivation will start by considering the amount of market share a network operator has as a percentage expression. We are then able to deduce the amount of possible users in a stadium based on market share.

$$S \times M = N \tag{22}$$

In (22), S is the stadium capacity, M indicates the network operator's market share and N is the amount of possible users in the stadium. Either figure 4 or the RBF model can now be used to determine how many users one wants to be covered for an acceptable UL_RSSI level on a sector, this amount of users is represented by N_d. Expanding on (22) to include the amount of desired users for acceptable UL_RSSI results in (23).

$$\frac{S \times M}{N_d} = \frac{N}{N_d} = N_o \tag{23}$$

where N_o indicates the amount of sectors required for the design. One determines the amount of required NodeB's, R_s to cover the stadium as in (24), by dividing the amount of sectors by the count of sectors per NodeB, designated by the symbol B in (24). It should be noted that one NodeB sector may contain several DAS antennas.

$$\frac{\frac{N}{N_d}}{B} = \frac{N_o}{B} = R_s \tag{24}$$

To test and evaluate if the number of sectors will provide coverage with suitable levels of UL_RSSI one can multiply the results of (22) with k. The results provide the noise rise per user and this should then indicate if suitable design levels were chosen.

Also one can expand to include the amount of 3G users in the network. The market share, M in (22) indicates the total market share of the network operator. If the percentage of 3G users, G, in the network is included a more accurate design can be realized. The result of including the percentage of 3G users means that (22) can be substituted by (25).

$$S \times M \times G = N \tag{25}$$

The model derived in this section should now allow a radio planner to design adequate coverage without degrading EUL services.

6 Design Results

This paper already provided various results in terms of models and actual data from major events in stadiums. A design example is presented to highlight the use of our model in this paper.

Using Figure 4 it was determined that a maximum number of users to be covered is 45 per cell, thus $N_d = 45$. The RBF model was then used to determine the effect of a maximum of 45 users on the UL_RSSI, resulting in the relationship in figure 6. At this stage the planner should recognise that the noise rise due to 45 simultaneous users has an UL_ RSSI of -80 dBm, which is in accordance to figure 4. If the degradation of UL_RSSI is acceptable to the planner he/she may continue with the design. Further inputs to the planner might be that the stadium capacity is 50,000, with a network operator market share of 45 % and 14 % of the network's traffic is on 3G. Using (25):

$$50000 \times 0.45 \times 0.14 = 3150 \tag{26}$$

According to (26) a possible total of up to 3150 voice users can be expected in the stadium using the resources of the network operator. Using (23) the amount of required cells can be determined using $N_d = 45$.

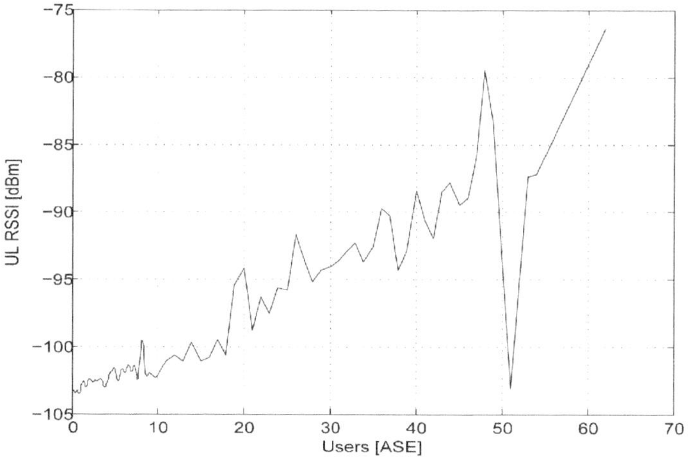

Fig. 6. Estimated degradation of the UL_RSSI with a maximum of up to 45 simultaneous users

$$\frac{3150}{45} = 70 \tag{27}$$

The result in (27) indicates that 70 sectors is required to cover users within the acceptable noise rise limits. Using (24) the amount of necessary NodeB's with three sectors can be determined as in (28).

$$\frac{70}{3} = 23 \tag{28}$$

The design requires that 23 NodeB's be installed to provide coverage with acceptable power levels on the uplink.

7 Conclusion

Data from actual events in a high capacity stadium was used to model the relationship between UL_RSSI and the amount of speech users on 3G. The model proved accurate and estimations for various stadium designs can now be made. A RBF was used to create the model for the UL RSSI as it provides non-linear system estimation with acceptable accuracy and ease of implementation. A design model was derived as the main objective of this paper and uses data from the UL_RSSI model to determine the amount of sectors required to fulfill design requirements for a stadium DAS. The results of a design scenario provided the planner with a guideline for the number of users per sector as well as the amount of NodeBs that are required to meet the design requirements. The model enables planners to now use the UL_RSSI as a design consideration when designing a DAS for stadium events. Meeting the ever demanding requirements of users now also focusses on providing superior quality on the uplink to fulfill various needs of the user. Both the model for UL_RSSI and the design model proved efficient and adequate. The authors carry experience from the 2010 Soccer World Cup and maintain if such design considerations are followed, improved EUL/HSUPA performance can be provided on a stadium DAS.

References

1. Nawrocki, M., Dohler, M.: Understanding UMTS Radio Network Modelling Planning and automated Optimisation. John Wiley & Sons, Chichester (2006)
2. 3GPP TS 25.331: Radio Resource Control Protocol for the UE-Utran Radio Interface
3. Tolstrup, M.: Indoor Radio Planning A practical guide for GSM, UMTS and HSPA. John Wiley & Sons, Chichester (2008)
4. Ericsson Course Material. Wcdma Ran Workshop
5. Laiho, J., Wacker, A., Novostad, T.: Radio Planning and Optimisation for UMTS. John Wiley & Sons, Chichester (2005)
6. Castle, W.: UMTS Air Interface. Course Material (2005)
7. Orr, M.: Introduction to Radial Basis Function Neural Networks (1996), `http://www.anc.ed.ac.uk/mjo/papers/intor.html` (cited, May 2010)

Handling Rest Requirements and Preassigned Activities in Airline Crew Pairing Optimization

Michael Römer and Taïeb Mellouli

Institute of Business Information Systems and Operations Research,
Martin-Luther-University, Halle-Wittenberg
{michael.roemer,mellouli}@wiwi.uni-halle.de

Abstract. For the complex task of scheduling airline crews, this paper discusses the integration of rostering requirements into the crew pairing optimization process. Our approach is based on a network flow model which uses a state expanded network to represent pairing chains for crew members at different domiciles. We enhance this model by proposing a refined representation of rest requirements along with preassignments such as pairings from the previous planning period, office and simulator activities as well as vacation and part-time leaves. In particular, we introduce the concept of availability blocks to mitigate the loss of information following from the aggregated anonymous flow of crew members in the network model. Experimental results with real world data sets show that the refined model remains tractable in practical settings.

1 Introduction

Due to its difficulty and due to the amount of costs which can be saved by finding good solutions, airline crew scheduling is a planning task which has received a lot of attention in the research literature. In crew scheduling, the crews are assigned schedules (*rosters*) for a planning period of usually two weeks or a complete month. Mainly because of the complexity of the planning task, the planning process is typically divided into at least two steps which are performed in a sequential order.

In the *crew pairing* step, an optimal set of pairings covering all flights is generated. A *pairing* is a legal sequence of typically up to five duties which starts and ends at a *crew domicile* (*home base*). A *duty* is a sequence of flights which forms a working day. After a duty, a crew member has to be awarded a *daily rest*. The daily rest can occur either at the domicile of the crew member or at a remote airport within a pairing in which case expenses for hotel overnights are incurred. At the beginning and at the end of a pairing as well as between the duties of a pairing, a *proceeding* (*deadhead*) can be used to position the crew.

In the subsequent *crew assignment* (*crew rostering*) step, these anonymous pairings are assigned to crew members. When the rosters are constructed, *preassignments* such as simulator activities or vacation blocks have to be respected as well as the need for *weekly rest* requirements. A common rule for weekly rests is that a crew member has to be assigned a weekly rest block within every period of seven or eight days.

The way in which rosters are constructed varies from airline to airline, but it typically falls in one of the following three categories: In the *bidline approach*, the airline generates a set of legal anonymous schedules (bidlines) for which crew members can bid,

J. Pahl, T. Reiners, and S. Voß (Eds.): INOC 2011, LNCS 6701, pp. 643–656, 2011.

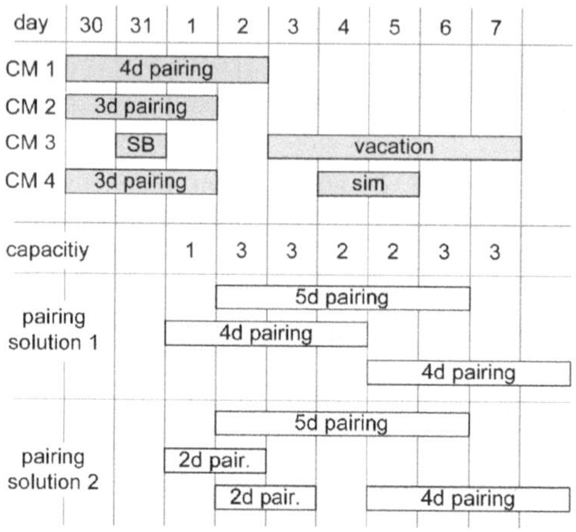

Fig. 1. A roster with preassignments and two pairing solutions

typically in seniority order. In the *preferential bidding* approach, crew members bid for certain properties of the schedule and the rosters are then assigned in a way that crew satisfaction is maximized in seniority order. In the *personalized rostering* approach, the schedules are constructed based on fair share with respect to preferences and to an equal distribution of the workload among all crew members.

In this paper, we focus on crew scheduling in the personalized rostering process mainly found at European airlines. In particular, we deal with the interdependencies between the generation of crew pairings and the following rostering step. To provide an idea of the problems addressed in this work, we give a simple example in Figure 1.

Figure 1 depicts the rosters for a small crew domicile with four crew members (CM 1... CM 4) before crew scheduling is carried out. The rosters contain several activities at the last days of the previous planning period as well as some preassignments such as standby duties (*SB*), vacations and simulator trainings (*sim*) within the current planning period. The line below the rosters depicts the aggregated available capacities for every day in the planning period. In addition, there are two sets of pairings which could result from the pairing optimization step (*2d pairing* denotes a pairing with a length of two days). Both solutions are feasible with respect to the aggregated available capacities. However, it is easy to see that the pairings of the first set do not fit into the gaps available in the rosters above. In contrast, the second set of pairings exhibits no conflicts with respect to the preassignments. Yet, a feasible roster also has to fulfill the weekly rest requirements. When considering the weekly rest rule mentioned above, there also exists no feasible rostering solution for the second set of pairings.

If such a situation arises in practice, the pairing solution has to be modified in order to achieve a feasible rostering solution. Such modifications can be achieved by swapping pairings between crew domiciles and/or by breaking and maybe recombining pairings. These local modification steps typically cause higher costs, e.g. since additional expenses for proceedings are incurred.

In this paper, we propose an approach to integrate characteristics of the crew rostering step into crew pairing optimization and we investigate the sensitivity of this integration on the overall crew scheduling process. In the following section, we provide a description of the relevant planning problems and give a short review of related work. Our network-based modeling and solution approach is described in Section 3. In Section 4, we present the results of our computational experiments based on real world data sets of a medium-sized German airline. In the last section, we give a short summary and an outlook on further research opportunities.

2 Related Work

In this section, we describe the planning problems arising in short term airline crew scheduling and review existing work on the topic. To be more specific, we shortly review the crew pairing problem, the crew rostering problem and approaches for integrated crew scheduling.

2.1 Crew Pairing Optimization

In the *crew pairing problem* (CPP), the flights which fall into the planning period are sequenced to form an optimal set of legal pairings. The main goal in the CPP is to generate a cost-minimal set of pairings covering all flights. Depending on the airlines' crew payment structure, the cost function of a pairing can be a complex nonlinear function of the flights and duties it contains. In most European airlines, crews are payed a fixed salary plus an overtime payment if a threshold of monthly or yearly working time is exceeded. In this case, the most important pairing-related variable costs are those caused by proceedings and hotel stays, per diem expenses for working time and rests outside the domicile and overtime payments. In airlines operating from more than one crew domicile, a second goal is to distribute the flight hours evenly among the domiciles based on the available capacity. An uneven distribution over several periods would cause bottlenecks in some domiciles with crew members reaching their yearly maximum flight and working times. An important difficulty in the CPP stems from the multitude of rules such as company and governmental rules governing the legality of duties and pairings. While some rules depend on a simple accumulation of measures such as flight time, there are also complex path-dependent rules which cannot be expressed linearly in terms of the flights and their connections and thus are difficult to model.

The CPP has been studied extensively, for surveys on the topic see, e.g., [2,7]. It is mostly stated as a generalized set partitioning problem in which pairings form the columns and there is one row for every flight to be covered. Most state of the art approaches solve the set partitioning model by branch and price (column generation) approaches in which the variables of the set partitioning problem are generated dynamically within the solution procedure by solving a subproblem, using dual information from the solution of the set partitioning problem. This solution approach was first presented in [9] and later refined by several authors.

In most cases, the column generation subproblem is formulated as a variant of a shortest path problem on a time space network. While every legal pairing forms a path in such a time-space subproblem network, there are many paths in the network which violate the complex rule set governing pairing legality. Due to this important circumstance, solving a standard shortest path problem is not a feasible approach to generate columns (pairings). Instead, most authors propose to solve a resource constrained shortest path problem in which resource variables and resource extension functions are used to model the complex legality rules (see e.g. [4]). Another approach considered, e.g., in [1] is to look for multiple shortest paths on the network and check each path for legality until a legal pairing is found.

2.2 Crew Rostering

The *crew rostering problem* (CRP) consists of assigning pairings (and in some cases additional activities such as standby duties) to individual crew members in a way that all roster legality rules are fulfilled. In the personalized rostering approach considered in this paper, the goal is to achieve fairness and equality among the crew members. Among the many regulations governing the legality of a crew roster, the weekly and monthly rest requirements can be considered two of the most important rules. A typical example for a weekly rest requirement we encountered at the airline in our practical case study is the following rule: every crew member has to have at least two rest days within every seven day period and one rest block of two sequential days within every eight day period. In each calendar month, a crew member of the same airline needs to be granted at least eleven rest days.

Another important aspect in rostering are the preassigned activities such as simulator, office, vacation and part-time leaves: they define the gaps in which the work activities and rest blocks can be inserted. Some types of preassigned activities are counted as working time which is the case for flight activities overlapping from the previous planning period, simulator trainings or office activities. Other activities such as vacation or part-time leave represent rest time. A detailed description of the CRP as well as an overview of solution approaches is provided by [8]. Like the CPP, the CRP can be modeled as a set partitioning problem and solved by column generation (see, e.g. [6]). Alternative approaches are presented in [3], where a multicommodity flow problem is solved and in [11], where the authors employ a heuristic approach based on the scatter search metaheuristic.

2.3 Integration in Crew Scheduling

As described in the introduction, the sequential proceeding in crew scheduling can lead to problems since important rostering constraints are not accounted for in the pairing optimization step. Particularly at airlines with small crew domiciles, preassigned activities and the rest requirements can lead to the situation that the generated pairings are not directly assignable to form legal rosters. As a consequence, the optimal pairing solution has to be modified. Often, these mainly local modifications negatively affect the quality of the pairing solution.

To overcome these problems caused by the sequential planning process, some authors propose approaches for integrated procedures in which rosters are constructed directly from the flights. An approach based on a branch and price algorithm is presented in [5] where the authors report significant advantages in solving the integrated problem in contrast to using a sequential procedure. The biggest instance they solve contains 521 flights within a solution time of 27 hours. In [16], the authors propose a mixed integer programming model using pre-generated duties for the integrated crew scheduling problem which is solved with a standard solver after a clique-based reformulation. However, the instances solved in that work are relatively small (the maximum number of flights considered was 210). Proceedings required in a multiple crew domicile case are not considered. Recently, [15] have provided an approach in which they integrate crew pairing and anonymous bidline generation. By applying column generation in combination with dynamic constraint aggregation, they are able to solve instances with up to 7527 flights within 44 hours. The authors report significant savings with respect to the number of bidlines generated in comparison to a sequential approach.

Instead of solving the integrated crew scheduling problem, [12] proposes a partially integrated approach which consists of solving the *crew pairing chain problem* (CPCP) to generate pairings. The author defines a *pairing chain* as a sequence of pairings spaced by weekly periods which can be performed by the same crew member. The result of the CPCP is an optimal set of pairing chains which also accounts for the availability of crew members at the beginning of the planning period as well as the preassigned activities and weekly rest requirements. The model proposed in [12] is an aggregated state expanded multicommodity network model in which a commodity corresponds to a crew member from a certain crew domicile. The states correspond to the number of sequential working days in a pairing and flights are represented by pre-generated duty arcs. A solution of the aggregated network flow model can be interpreted as a set of possible pairing chains. A concrete solution can be obtained by decomposing the flow into paths which represent feasible pairing chains.

While the model presented in [12] captures the weekly rest requirements quite precisely due to the fact that every working block of at most five days has to be followed by a weekly rest arc, it is not capable of modeling all complex path dependent pairing legality rules which incorporate more than one duty. In a subsequent work which is presented in [13] and later refined in [14] the author introduces a different network-based model for the CPCP in which flights are represented in state-expanded subnetworks in which all relevant rules are enforced by construction of the network. The weekly rest requirements are approximately modeled by incorporating rest arcs as well as constraints which enforce that for every seven day period, the number of rest periods is greater or equal to the number of available crew members at every crew domicile. Since this model forms the basis for the approach chosen in this paper, we describe it in more detail in the following section.

3 Modeling and Solution Approach

The approach we propose in this section is based on the second network flow model for the CPCP of [13,14]. In the first two subsections, we describe the underlying network structure and give the formulation of the basic model. In the following subsections, we describe our new modeling approach to handle rest requirements and preassigned activities and discuss our approach to solve the resulting mixed integer programming model. Note that while the problem which we model is taken from our concrete case study, structurally similar problems can be found at many airlines, especially at airlines from Europe.

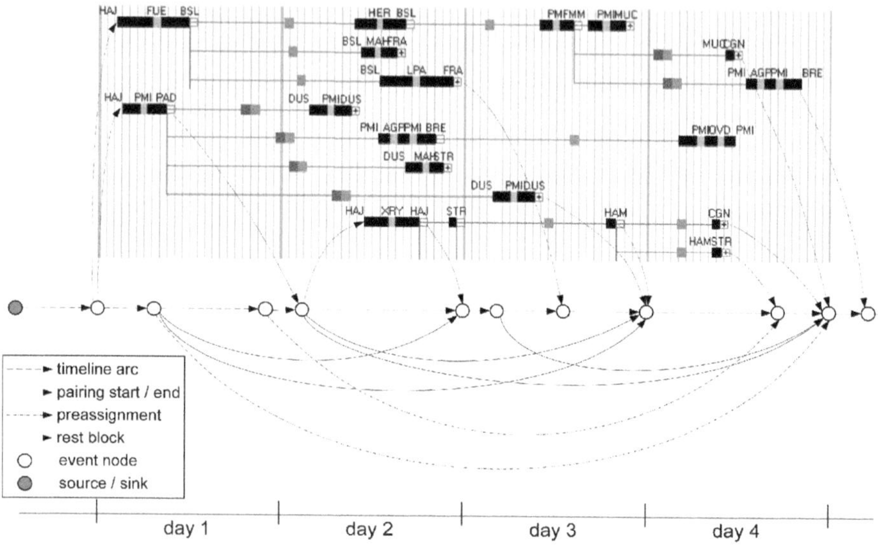

Fig. 2. Sketch of the network structure for a crew domicile

3.1 Network Construction

For every crew domicile k from the set of all domiciles K, a model network $G^k = (N^k, A^k)$ is constructed. A sketch for such a network is provided in Figure 2. The connections between activities which start and end at the domicile are modeled by a *timeline* (*connection line*) which starts at a source node and ends at a sink node. The supply / demand of a node is denoted as b_i^k. For the source / sink node of domicile k, the supply corresponds to the number of crew members cm^k / $-cm^k$. All other nodes in the network have 0 demand. Besides the source and the sink, the timeline contains event nodes which represent points in time at which activities start and end. These timeline nodes are connected by waiting arcs.

The flight activities are modeled in state-expanded pairing subnetworks which are connected to the timeline via starting and ending event nodes. In these subnetworks (loosely depicted in the upper part of Figure 2), every flight can occur several times in different states. As a simple example, the state of a flight which forms the first flight in a duty is different from the state of the same flight if it is the third flight within a duty. Another simple example is that the state of a flight depends on the number of days already worked within a pairing. In general, the states depend on the rules that govern the legality of a pairing. As a consequence, the state-expanded subnetworks have the important property that every path in the subnetwork corresponds to a legal pairing respecting all contractual and governmental legality rules. The paths can be one to five days long and can contain proceeding activities as well as overnight stays at hotels. The times of the end events from the pairing networks correspond to the ready-times of the crew members after the pairings, that is, the daily rest time is included. The arcs in the pairing subnetwork alternate between arcs representing flight activities and arcs representing connections between flights.

Preassigned activities are modeled as arcs with a fixed flow starting and ending at the domicile timeline. For every block of assigned working days as well as for each block of assigned days which count as rest time, such a preassignment arc is introduced.

In addition to the arcs which represent assigned rests, the network contains arcs without upper bound for the flow which represent possible rest blocks: for every possible two-day and three-day rest block in the planning period, a weekly rest arc is connected to the timeline. A flow on these arcs (which blocks capacities which cannot be used for flying) is enforced by the rest-related constraints presented in Section 3.3.

3.2 Basic Model

The flow variables on the arcs $(i,j) \in A^k$ are denoted with X_{ij}^k. Arcs representing activities (or rests) are integer variables whereas arcs modeling connections of activities (including the timeline arcs) are continuous variables. For every node i in the network, the flow balance constraints have to hold:

$$\sum_{j:(i,j)\in A^k} X_{ij}^k - \sum_{j:(j,i)\in A^k} X_{ji}^k = b_i \qquad \forall k \in K, \forall i \in N^k \tag{1}$$

The set A_f^k contains all arcs in the pairing subnetwork of domicile k which represent flight f from the set F of all flights. Each flight has to be covered by exactly one flow unit:

$$\sum_{k\in K} \sum_{(i,j)\in A_f^k} X_{ij}^k = 1 \qquad \forall f \in F \tag{2}$$

In addition, there are constraints which govern the distribution of flight hours among the crew domiciles. With $a_{fh(ij)}^k$, we denote the flight hours arising at arc (i,j) and with $b_{fhGoal)}^k$ the flight hour goal of domicile k. By introducing variables U_{fh}^k and O_{fh}^k for the under- and overcovering of the flight hour goals, the following soft constraints can be formulated:

$$\sum_{(i,j)\in A^k} a_{fh(ij)}^k X_{ij}^k = b_{fhGoal}^k - U_{fh}^k + O_{fh}^k \qquad \forall k \in K \tag{3}$$

Analogously to the flight hour base constraints, additional base constraints are introduced to approximately model the flight overtime worked in the planning period measured by the variable O_{wh}^k. Overtime is caused if crew members exceed their paid monthly flight time including compensation times for non flying activities.

The objective function consists of a combination of cost minimization and goal programming to achieve an even distribution of flight hours among the crew domiciles. It contains costs for proceedings and accommodation as well as per diem expenses which are represented in the cost coefficients c_{ij}^k of the flow variables, and overtime costs and penalties for the deviation from the flight hour goals:

$$\min \sum_{k\in K} \Big(\sum_{(i,j)\in A^k} c_{ij}^k X_{ij}^k + c_{fh-}^k U_{fh}^k + c_{fh+}^k O_{fh}^k + c_{wt+}^k O_{wt}^k \Big) \tag{4}$$

3.3 Weekly and Monthly Rests

The weekly rest requirements are fulfilled if every crew member has at least two rest days in every seven-day period and at least one rest block of at least two days in each eight-day period. In the first set of constraints, we aggregate this requirement for each seven day period. Let P_{rest} be the set of seven day periods within the planning period T for which weekly rest requirements have to hold. $A^k_{restIn(p)}$ denotes the set of all arcs representing a rest block with at least two days overlapping with period p and cm^k is the number of crew members available at domicile k. The aggregated set of weekly rest constraints can then be stated as follows:

$$\sum_{(i,j)\in A^k_{restIn(p)}} X^k_{ij} + B^k_p \geq cm^k \qquad \forall k \in K \forall p \in P_{rest} \tag{5}$$

In these constraints, the variable B^k_p is introduced. It represents the number of rest requirements in the seven-day period p which are fulfilled by starting and ending rest blocks at the borders of period p. We further define $A^k_{restStart(t)}$ and $A^k_{restEnd(t)}$ as the sets of all arcs representing a rest block starting and ending at day t. Using this notation, we constrain B^k_p to be the minimum of the number of rest blocks ending and starting at the borders of period p:

$$B^k_p \leq \sum_{(i,j)\in A^k_{restEnd(t)}} X^k_{ij} \qquad \forall k \in K \forall p \in P_{rest}, t = Startday(p) \tag{6}$$

$$B^k_p \leq \sum_{(i,j)\in A^k_{restStart(t)}} X^k_{ij} \qquad \forall k \in K \forall p \in P_{rest}, t = Endday(p) \tag{7}$$

In addition, we formulate constraints which enforce weekly rests before and after long working blocks of five days. To this end, we introduce the following sets and constraints. The set $A^k_{workEnd(t,d)}$ contains the set of all arcs which end at day t, with d days of work in sequence. This set includes arcs representing preassigned work as well as arcs from the pairing subnetworks ending at the domicile timeline. Note that for every arc representing the end of a pairing it is known how many days the pairing contains. We then formulate the respective constraints:

$$\sum_{(i,j)\in A^k_{workEnd(t,5)}} X^k_{ij} \leq \sum_{(i,j)\in A^k_{restStart(t+1)}} X^k_{ij} \qquad \forall k \in K, \forall t \in T \tag{8}$$

$$\sum_{(i,j)\in A^k_{workEnd(t,5)}} X^k_{ij} \leq \sum_{(i,j)\in A^k_{restEndt(t-5)}} X^k_{ij} \qquad \forall k \in K, \forall t \in T \tag{9}$$

A similar set of constraints is introduced for working blocks with a length of four days.

The monthly rest requirements are modeled via the following constraints which guarantee that at least sum of all the monthly rest requirements of all crew members, denoted with $b^k_{restInMonth}$, is awarded in the planning period.

$$\sum_{(i,j)\in A^k_{rest}} a^k_{restDays(ij)} X^k_{ij} \geq b^k_{restInMonth} \qquad \forall k \in K \tag{10}$$

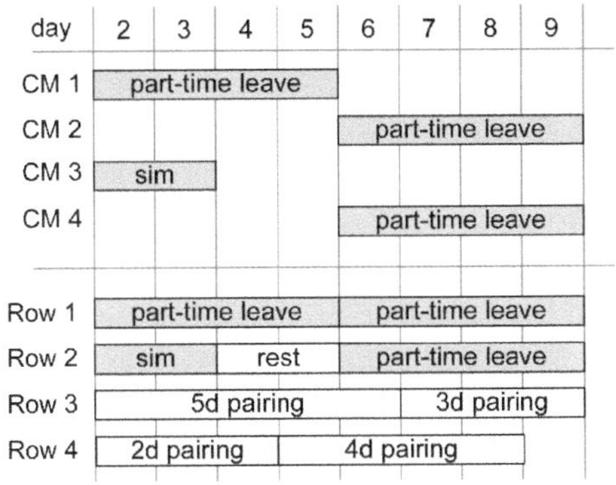

Fig. 3. Unassignable flow decomposition

3.4 Availability Blocks

The network model described so far incorporates necessary conditions formulated on the aggregated homebase level which have to hold for every feasible rostering solution. However, due to the loss of identity of the crew members within the aggregated flow model, there can still arise feasible pairing solutions which cannot be assigned with respect to the preassignments and the weekly rest requirements.

Such a situation is depicted for a small example in Figure 3: In the upper part, the preassignment situation before the rostering step is shown, consisting of three part-time leave blocks and one simulator training, each assigned to a different crew member. In the network flow model, however, all crew members from the same domicile are aggregated and form an anonymous flow in which the information that the four activities are assigned to different crew members is lost. A (decomposed) feasible solution to the flow model is depicted in the in the lower part of the figure. In the decomposition shown there, all preassignments are assigned to only two rows. In addition, the only two-day rest block is assigned to a row in which there is already a part-time block that counts as rest while the last two rows do not contain any rest. It can easily be seen that the activities cannot be rearranged in a way that they form feasible rostering solution.

In order to overcome problems of this kind, we introduce the idea of *availability blocks*. Availability blocks are periods in which a crew member is available for work without interruption. The main idea is that the set of pairings which are generated in the pairing optimization step has to fit into the availability blocks of the crew members. The number of crew members which are available for a certain availability block depends on the preassigned activities and on the weekly rest blocks planned within the pairing optimization model.

Due to the anonymity of the crew members in the aggregated flow, we cannot use the history of a flow unit to define the available working blocks. Instead, we use the observation that a crew member will not have two rest blocks shortly after another to characterize the availability in a period: If we assume that a crew member will be available for work for at least four days after a weekly rest, we can indirectly calculate the number of crew members available within a period by subtracting the number of rest blocks which touch this period from the number of crew members available at the homebase cm^k. Preassignments such as vacation or part-time off which count as rest time are also included in this calculation. As an example, the capacity available for period $[5;6]$ of Figure 3 would be 0.

Arcs which represent rest blocks that have at least one day which overlaps with period p are defined to be contained in the set $A_{restTouch(p)}$. If an individual crew member has two preassigned rest blocks touching period p (and thus violates the assumption stated above), only one rest block is counted.

The capacity provided by the availability blocks defined in the way described above can be used by *working blocks*. Note that every working block does not only need at least one availability block for its complete period, but also consumes the availability capacity from the blocks corresponding to all periods which are contained in its working period. For example, a five-day pairing starting at day t does not only need capacity from an availability block of five days starting at t, but also reduces the availability in all one-, two-, three- and four-day blocks it contains. Note that the solution presented in the lower part of Figure 3 violates this condition: since the available capacity period $[5;6]$ is 0, no pairing that contains this two-day block can be part of a feasible solution.

We define $A^k_{work(p)}$ as the set of all arcs which represent working blocks that contain the complete period p. This set includes the preassigned work activities as well as arcs from the pairing subnetwork. Here, it is important that for each possible path in the pairing subnetwork, only one arc is counted. This is achieved by using only the arcs contained in the sets $A^k_{workEnd(t,d)}$ defined above for which the period defined by the days $t-d+1$ and t completely contains the period p.

Using these definitions, we can formulate the following set of constraints which enforce the consistency between availability blocks and working blocks. They are imposed for every crew domicile k and for the set P_{avail} of all periods with a length between one and five days which are completely contained in the planning period T:

$$\sum_{(i,j)\in A^k_{work(p)}} X^k_{ij} + \sum_{(i,j)\in A^k_{restTouch(p)}} X^k_{ij} \leq cm^k \qquad \forall k \in K, \forall p \in P_{avail} \tag{11}$$

3.5 Solution Approach

The model described above forms a mixed integer linear program. Since pairings are modeled implicitly in the network structure instead of being explicitly enumerated like in the set partitioning modeling approach, there is no need to resort to a decomposition approach such as column generation. Instead, owing to the effective aggregated flow formulation, it can be solved directly using a standard mixed integer linear programming solver. In order to speed up the solution process and to solve large scale instances, we employ an iterative fixing heuristic which can be characterized as a variant of LP plunging (see, e.g. [10]): After having solved the LP relaxation of the problem, we fix variables in the pairing subnetwork which have an integer or near-integer value on the whole path. Note that the fixation of a single flight arc to one on a path affects several variables since all other variables which represent the same flight can be fixed to zero. We repeat this fixing step until certain thresholds with respect to the relative gap to the first LP solution is reached. Afterwards, the modified mixed integer program is solved by a standard MIP solver. In case that the relative gap of the integer solution to the first LP solution exceeds a certain threshold, we backtrack by undoing the fixation of the last step and solving the resulting MIP.

4 Computational Results

The basic model described above forms the key part of the *crew optimizer*, a decision support system for crew scheduling which has been productive use at a medium-sized German airline with 38 aircrafts and 11 crew domiciles for several years. One of the motivations for the refined model described in this work comes from the fact that the planners at the airline often have to modify the pairing solution in order to obtain a feasible and balanced rostering solution. Such problems primarily arise at small crew domiciles with a small number of crew members and at the beginning of the planning period. In order to mitigate these problems, in the current standard configuration of the crew optimizer only pairings with a duration of up to three days are generated for small domiciles and at the first few days of the planning period. This is due to the fact that it is more likely to find a feasible rostering solution for short pairings than for long ones.

The system also contains a component which can be used to manipulate a given set of pairings in order to prepare the crew assignment step. This manipulation can either be carried out manually or by an automatic heuristic. The most important task of the heuristic is to identify and resolve certain types of capacity conflicts which affect the assignability of the pairings. It resolves these conflicts by local modification steps such as moving pairings between domiciles or breaking and recombining pairings. Note, however, that it is not capable of providing a solution which is guaranteed to be assignable. In addition to the conflict resolution, the heuristic tries to improve the (possibly modified) pairing solution with respect to the measures of costs and goal variance. Goal variance is defined here as the sum of the quadratic deviations from the domicile flight hour goals.

We use this postprocessing heuristic and its performance measures to compare the quality of the pairing solution generated with the new refined model with the current productive version of the model. In the productive version, the fixed preassignment arcs are included as well as a simplified version of the weekly and monthly rest constraints.

The experiments were conducted with real world problem instances from our partner airline including all relevant characteristics such as legality rule sets for duties and pairings (e.g. EU Ops rules, and airline-specific contractual rules). Table 1 presents the main characteristics of the problem instances on which we performed our tests.

The experiments were conducted on a personal computer (Intel Core 2, 2.4 GHz, 3GB RAM, Win XP Prof. 32-bit). To solve the linear relaxations within the fixing

Table 1. Characteristics of the solved instances

Name	Function	# Aircrafts	# Flights	# Domiciles	# Crew Members	# Days
CPJul	CP	38	2459	11	278	12
FOJul	CP	38	2459	11	255	12
CPAug	CP	38	2376	11	278	12
FOAug	FO	38	2376	11	255	12

heuristic and for the MIP phase, we used CPLEX 12.2 with two threads and default parameter settings. Within the final MIP solution phase, we stopped at MIP-gap of 0.3%. The experimental results are shown in Table 2. The column *Gen.* shows the time needed to build the model in minutes. In particular, the construction of the pairing networks including the calculation of proceeding opportunities, legality checks of the paths representing pairings and the aggregation of the network by consolidating equivalent states consumes a considerable amount of time. The three figures presented in the following column are the solution time of the first LP, the solution time of the last MIP as well as the overall time spent for solving the mathematical model. The gap listed in the next column is the gap between the optimal solution of the first LP relaxation and the final MIP solution expressed in percent. The following four columns show the costs and goal deviation before and after the postprocessing heuristic was carried out.

Table 2. Comparison of models and results

Instance	Model	# Cols	# Rows	# Nonzeros	Gen.	Solution time	Gap	Costs Pre	Costs Post	GoalDev Pre	GoalDev Post
CPJul	standard	448286	192622	1231487	25	6.7 / 1.2 / 12.3	0.8	59056	59242	1571	1108
	refined	462628	199629	1360321	25	11.3 / 2.1 / 17.2	0.93	59720	59753	812	616
FOJul	standard	469597	199325	1282676	25	7.1 / 0.8 / 12.1	0.91	53047	53544	2467	111
	refined	462628	199629	1360321	25	11.3 / 2.1 / 17.2	0.93	52708	52522	1613	279
CPAug	standard	443927	187817	1160869	23	3.8 / 1.8 / 15.4	1.4	62136	62478	3079	1358
	refined	470273	202124	1381174	23	13.2 / 0.7 / 21.8	0.6	64179	64287	1439	904
FOAug	standard	446614	187908	1166918	23	4.5 / 1.1 / 9.5	0.82	53831	53871	1003	799
	refined	447488	190833	1261031	23	9.1 / 1.2 / 15.1	0.85	52348	52355	1018	328

The experiments show that the mathematical model based on our aggregated state-expanded network flow approach for crew scheduling can be solved by a state-of-the-art solver in acceptable times and exhibits a low LP-IP gap. The technique of iteratively solving the LP relaxation and fixing whole pairings which form paths of the pairing networks in the domicile layers preserves this low gap and helps to solve multi-domicile crew scheduling problems of practical size. This remains true for the refined model which handles rest requirements and preassignments in a more precise way - the solution times are only slightly higher in this case. With regard to the quality of the solutions measured in costs and goal deviation after the postprocessing step, both models behave similarly, that is, none of the models dominates the other. Surprisingly, for the instance FOAug, the refined model shows a least costly solution even before the postprocessing step. We think this is due to the refined and more exact handling of the rest requirements in this model.

In order to find an explanation for the similarity of the results obtained by both models, we performed some additional, more qualitative analyses of the pairing solutions. One observation we made was that in many cases, the solution of the standard model contained a considerable amount of relatively short pairings which reduces the probability that unassignabilities arise. In addition, by qualitatively analyzing the pairing solutions in bottleneck situations (e.g. at small domiciles), we found that the postprocessing heuristic did not recognize all assignability problems which mainly occurred in the solutions of the standard model. We will explore this issue further by employing an automatic assignment heuristic which is under development at the moment and which will permit a more detailed evaluation of the quality of a pairing solution with respect to assignability.

5 Conclusions and Outlook

As shown in our computational experiments, the network flow based approach to optimizing crew pairings is able to solve real world instances within an acceptable amount of time. One aspect contributing to the efficiency of the model is the aggregation of the crew members in the anonymous flow of the domicile network layers. However, this aggregation leads to a loss of information which can lead to problems in the subsequent crew assignment step. In this paper, we proposed the concept of availability blocks to model preassignments and rest requirements more exactly in order to mitigate these problems. In our experiments, the refinements of the model only lead to a moderate increase with regard to the size of the model and to solution times. Since our goal is to improve the overall crew scheduling process, the next step in our investigations will be to evaluate the effects of the refined crew pairing model presented on the whole process by employing an automatic assignment module.

References

1. Andersson, E., Housos, E., Kohl, N., Wedelin, D.: Crew pairing optimization. In: Yu, G. (ed.) Operations Research in the Airline Industry, pp. 228–258. Kluwer, Boston (1998)
2. Barnhart, C., Cohn, A., Johnson, E.L., Klabjan, D., Nemhauser, G.L., Vance, P.: Airline crew scheduling. In: Hall, R.W. (ed.) Handbook of Transportation Science, pp. 517–560 (2003)
3. Cappanera, P., Gallo, G.: A multicommodity flow approach to the crew rostering problem. Operations Research 52(4), 583–596 (2004)
4. Desaulniers, G., Desrosiers, J., Dumas, Y., Marc, S., Rioux, B., Solomon, M.M., Soumis, F.: Crew pairing at air france. European Journal of Operational Research 97(2), 245–259 (1997)
5. Freling, R., Lentink, R.M., Wagelmans, A.P.: A decision support system for crew planning in passenger transportation using a flexible Branch-and-Price algorithm. Annals of Operations Research 127(1), 203–222 (2004)
6. Gamache, M., Soumis, F., Marquis, G., Desrosiers, J.: A column generation approach for Large-Scale aircrew rostering problems. Operations Research 47(2), 247–263 (1999)
7. Gopalakrishnan, B., Johnson, E.: Airline crew scheduling: State-of-the-Art. Annals of Operations Research 140(1), 305–337 (2005)
8. Kohl, N., Karisch, S.E.: Airline crew rostering: Problem types, modeling, and optimization. Annals of Operations Research 127(1), 223–257 (2004)
9. Lavoie, S., Minoux, M., Odier, E.: A new approach for crew pairing problems by column generation with an application to air transportation. European Journal of Operational Research 35(1), 45–58 (1988)
10. Löbel, A.: Solving large-scale multiple-depot vehicle scheduling problems. Lecture notes in economics and mathematical systems, pp. 193–220 (1997)
11. Maenhout, B., Vanhoucke, M.: A hybrid scatter search heuristic for personalized crew rostering in the airline industry. European Journal of Operational Research 206(1), 155–167 (2010)

12. Mellouli, T.: A network flow approach to crew scheduling based on an analogy to a Train/Aircraft maintenance routing problem. In: Voss, S., Daduna, J. (eds.) Computer-Aided Scheduling of Public Transport. LNEMS, vol. 505, pp. 91–120. Springer, Berlin (2001)
13. Mellouli, T.: Scheduling and routing processes in public transport systems. Professorial dissertation, University of Paderborn (2003)
14. Mellouli, T.: A network flow optimization model for integrated capacity-based crew scheduling. In: 23rd EURO conference on Operational research, Bonn (2009)
15. Saddoune, M., Desaulniers, G., Elhallaoui, I., Soumis, F.: Integrated airline crew pairing and crew assignment by dynamic constraint aggregation. Working Paper G-2010-05, Montréal (2010)
16. Zeghal, F., Minoux, M.: Modeling and solving a crew assignment problem in air transportation. European Journal of Operational Research 175(1), 187–209 (2006)

On the Cover Scheduling Problem in Wireless Sensor Networks

André Rossi[1,2], Marc Sevaux[1,3,⋆], Alok Singh[2], and Martin Josef Geiger[3]

[1] Université de Bretagne-Sud, Lab-STICC, Lorient, France
[2] University of Hyderabad,
Department of Computer and Information Sciences, Hyderabad, India
[3] Helmut Schmidt University, Logistics Management Department, Hamburg, Germany
marc.sevaux@univ-ubs.fr

Abstract. Cover scheduling for wireless sensor networks has attracted a lot of attention during the last decade. Many heuristic approaches and exact methods have been proposed for producing covers, i.e. subsets of sensors to be used at the same time. However, the actual schedule of the generated covers has never been explicitly addressed to the best of our knowledge. Though, this problem is of particular relevance when coverage breach is allowed in the aforementioned problems, i.e., when a full coverage of targets at any time is not mandatory. In that case, the objective of the wireless sensor network cover scheduling problem (WSN-CSP) is to schedule the covers so as to minimize the longest period of time during which a target is not covered in the schedule. In this paper, this problem is proved $\mathcal{NP}$-Hard, a MILP formulation is provided along with a greedy heuristic and a genetic algorithm based approach. Computational results show the effectiveness of the last approach.

1 Introduction

The use of wireless sensor networks (WSNs) has been increasing at a rapid pace in remote or hostile environments for data gathering [1]. This includes battlefield surveillance, fire monitoring in forests, or undersea tsunami monitoring. In such environments, sensors are usually deployed in an *ad hoc* manner or at random when it is not possible to place them precisely. To compensate for this random deployment, a greater number of sensors are used than what is actually required. This also increases the fault tolerance as some targets are redundantly covered by multiple sensors.

In the aforementioned situation, sensors are gathered into a number of subsets (not necessarily disjoint) such that sensors in each subset cover the targets. Such subsets are referred to as *covers*. Covers are activated sequentially in a mutually exclusive manner, *i.e.*, at any instant of time only sensors belonging to the active cover are used, whereas all other sensors are not. Using covers significantly increases network lifetime for two main reasons. First, sensors consume much more energy in an active state than in an inactive state [12]. Second, a sensor battery has been shown to last longer if it oscillates frequently between active and inactive states [2, 3].

⋆ Corresponding author.

J. Pahl, T. Reiners, and S. Voß (Eds.): INOC 2011, LNCS 6701, pp. 657–668, 2011.

One among the most popular objectives is the maximization of lifetime under maximum coverage breach limitation [8, 9, 5]. This problems is very efficiently solved in a previous framework based on column generation [14] and in that work, the effect of bandwidth is also taken into account: the number of sensors in any cover has to be at most w, where w is a given integer. The problem of minimizing the coverage breach under bandwidth constraints is referred to as MCBB, and the problem of maximizing the network lifetime under bandwidth constraints is referred to as MNLB, as in [4, 15].

The output of the previous framework [14] is the design of the covers and the time these covers should be used to satisfy one of the two objectives. Once the covers are determined, they should be scheduled. In the literature, the term 'scheduling' mostly means cover generation as in [5], and the actual schedule of the covers is not addressed. This is fully justified when the targets have to be continuously covered: in that case, the covers can be scheduled in any order. However, when coverage breach is allowed, then some targets may not be covered at any time. Scheduling the covers is then to minimize the longest period of time during which a target is not covered.

1.1 Introductory Example

In the example shown in Figure 1, five covers have been generated. Gray boxes represent the coverage of the targets. In the current scheduling (in natural order), target T_4 is not covered for 8 consecutive time units. Of course, this can be improved by changing the schedule. Figure 2 gives, for the same example, another order for which the objective value is 5 (due to targets T_2 and T_3).

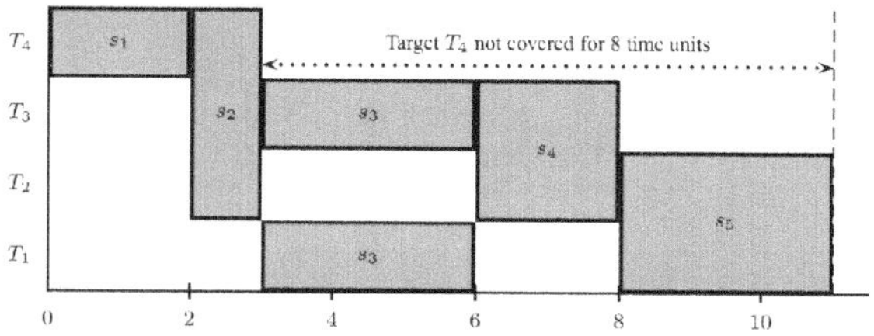

Fig. 1. A first schedule whose objective value is 8. Target T_4 is not covered for 8 consecutive time units because of successive scheduling of covers s_3, s_4 and s_5 (in any order).

1.2 Problem Definition and Notations

Given a set of covers, an usage duration and the set of targets covered for each cover, the wireless sensor network cover scheduling problem (WSN-CSP) is to find a schedule that minimizes the longest period of time during which a target is not covered.

More formally, WSN-CSP can be defined as follows. Let m be the number of targets, and q be the number of covers. For all $j \in \{1, \dots, q\}$ the duration of cover s_j is denoted by p_j: we refer to it as a processing time like in a scheduling problem [11]. Processing times are not necessarily integers and preemption is not allowed. Moreover, the starting time of cover j is denoted by $start_j$ for all $j \in \{1, \dots, q\}$. For all $k \in \{1, \dots, m\}$, C_k is the set of all the covers that cover target k. For all $j \in \{1, \dots, q\}$, D_j is the set of the

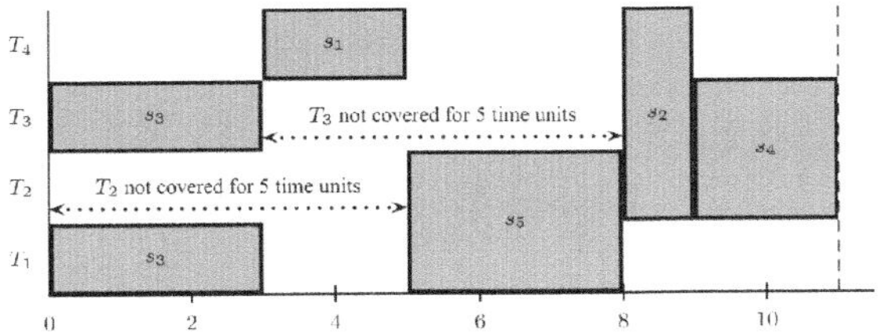

Fig. 2. An improved schedule of the same covers has an objective value of 5. Targets T_2 and T_3 have the same maximum non-coverage.

targets that are covered by cover s_j. The cover scheduling problem is to schedule the covers so as to minimize Δ_{max}, the maximum period of time during which a target is not covered.

In the previous example, $m = 4$, $q = 5$, $p = \{2, 1, 3, 2, 3\}$, $C_1 = \{s_3, s_5\}$, $C_2 = \{s_2, s_4, s_5\}$, $C_3 = \{s_2, s_3, s_4\}$ and $C_4 = \{s_1, s_2\}$. Moreover $D_1 = \{4\}$, $D_2 = \{2, 3, 4\}$, $D_3 = \{1, 3\}$, $D_4 = \{2, 3\}$ and $D_5 = \{1, 2\}$.

1.3 Organization of the Paper

The rest of the paper is organized as follows. Section 2 addresses WSN-CSP in a particular case, that is used for proving it to be $\mathcal{NP}$-Hard in the strong sense. A lower bound is derived for WSN-CSP in Section 3 and a MILP formulation is proposed in Section 4. A heuristic approach is presented in Section 5, whereas Section 6 describes a genetic algorithm based approach. Section 7 reports computational results and the last section concludes the paper and provides directions for future work.

2 Complexity Analysis

2.1 The Cover Scheduling Problem in a Particular Case

The following particular case of the cover scheduling problem is considered: $m = 2$ targets, the q sensors are partitioned in $O_1 + O_{12}$ where O_1 is the set of all covers that cover target t_1 only, and O_{12} is the set of all covers that cover targets t_1 and t_2. We assume that $O_1 \cup O_{12} = \{1, \ldots, q\}$, with $O_1 \cap O_{12} = \emptyset$. The cardinality of these sets is defined by $q_1 = |O_1|$ and $q_{12} = |O_{12}|$, so $q = q_1 + q_{12}$.

Moreover, p_j is equal to 1 for all the covers in O_1, and is a strictly positive integer for the covers in O_{12}. As a consequence, the network lifetime LT is also integer and is equal to

$$LT = q_1 + \sum_{j \in O_{12}} p_j$$

Target t_1 is covered by all the covers, but t_2 is covered by the covers in O_{12} only. The total amount of time during which t_2 is not covered is equal to q_1. This duration can be split in at most $q_{12} + 1$ time intervals. Since all the covers duration are integer, the minimum duration of the maximum time interval during which t_2 is not covered is $\left\lceil \frac{q_1}{q_{12}+1} \right\rceil$.

Lemma 1. *The covers in O_{12} are assumed to be numbered in $\{1, \ldots, q_{12}\}$. An optimal schedule to WSN-CSP in this particular case is returned by setting the starting time of cover s_j in O_{12} as follows:*

$$start_j = j \left\lceil \frac{q_1}{q_{12}+1} \right\rceil + \sum_{z<j} p_z \qquad \forall j \in \{1, \ldots, q_{12}\}$$

The covers in O_1 are scheduled in between, in any order.

Proof. $\left\lceil \frac{q_1}{q_{12}+1} \right\rceil$ is the minimum value for the maximum duration during which t_2 is not covered. It can be seen that the first cover in O_{12} stats at time $\left\lceil \frac{q_1}{q_{12}+1} \right\rceil$, and that the time elapsed between two consecutive covers in O_{12} is $\left\lceil \frac{q_1}{q_{12}+1} \right\rceil$ by construction. Now, we can check that the time elapsed between the completion time of the last cover in O_{12} and LT is also less than or equal to $\left\lceil \frac{q_1}{q_{12}+1} \right\rceil$. This duration can be written as

$$\begin{aligned} LT - q_{12} \left\lceil \tfrac{q_1}{q_{12}+1} \right\rceil + \sum_{z=1}^{q_{12}} p_z &= q_1 - q_{12} \left\lceil \tfrac{q_1}{q_{12}+1} \right\rceil \\ &= (q_{12}+1) \tfrac{q_1}{q_{12}+1} - q_{12} \left\lceil \tfrac{q_1}{q_{12}+1} \right\rceil \\ &= q_{12} \left(\tfrac{q_1}{q_{12}+1} - \left\lceil \tfrac{q_1}{q_{12}+1} \right\rceil \right) + \tfrac{q_1}{q_{12}+1} \end{aligned}$$

Since $\left(\frac{q_1}{q_{12}+1} - \left\lceil \frac{q_1}{q_{12}+1} \right\rceil \right)$ is negative or zero, we have

$$\begin{aligned} LT - q_{12} \left\lceil \tfrac{q_1}{q_{12}+1} \right\rceil + \sum_{z=1}^{q_{12}} p_z &\leq \tfrac{q_1}{q_{12}+1} \\ &\leq \left\lceil \tfrac{q_1}{q_{12}+1} \right\rceil \end{aligned}$$

□

Remark 1. It may happen that the computed starting time of the last cover in O_{12} is equal to LT. In that case this cover should start at any time $t \leq LT - p_{q_{12}}$ provided it does not overlap another cover.

Finally, as all the covers in O_1 have a duration of one unit of time, they can be scheduled in the $q_{12}+1$ time intervals, that all have an integer duration that is less than or equal to $\left\lceil \frac{q_1}{q_{12}+1} \right\rceil$.

2.2 WSN-CSP Is $\mathcal{NP}$-Hard in the Strong Sense

In order to show that WSN-CSP is $\mathcal{NP}$-Hard in the strong sense, we consider the following instance: $m = 2$ targets, the covers in O_{12} all have a one duration, the network lifetime is $LT = q_{12} + (q_{12}+1)B$, where B is a given integer with $B > 1$. The covers in O_1 have an integer duration $p_j \leq B$ for all $j \in O_1$, such that $\sum_{j \in O_1} p_j = (q_{12}+1)B$. This implies that the minimum value for the longest period of time during which t_2 is not covered is B.

Determining whether or not the covers in O_1 can be scheduled in those $q_{12}+1$ time intervals which duration is B is equivalent to address the decision problem version of the bin packing problem which is: "does there exist a partition of O_1 in $q_{12}+1$ sets such that the sum of the durations of the covers in each set does not exceed B?"

Since a particular instance of the cover scheduling problem is equivalent to the bin packing problem, the WSN-CSP is $\mathcal{NP}$-Hard in the strong sense.

3 A Lower Bound for WSN-CSP

First, it can be observed that Δ_{max} is lower bounded by the duration of any cover that does not cover all the targets:

$$\Delta_{max} \geq \max_{\substack{j \in \{1,\ldots,q\} \\ |D_j| < m}} p_j$$

This lower bound is likely to be efficient for instances in which the coverage breach rate is low. We remind that the breach rate is the ratio of non-covered targets at any time in a solution. A breach rate equal to zero means that all the targets are covered.

Furthermore, for a specific target, the total amount of time during which it is not covered is equal to the lifetime of the network minus the time this target is covered. This amount of time can be split into up to $|C_k|+1$ time intervals, by interspersing the covers that cover that target in an equally distributed separation. In Figure 1, target T_1 is not covered for 5 units of time. Hence, by positioning appropriately covers s_3 and s_5, the maximum period of time during which target T_1 is not covered is $5/3 \approx 1.67$ time units. If we repeat this computation for all targets and take the maximum, we can deduce that Δ_{max} is lower bounded by

$$\Delta_{max} \geq \max_{k \in \{1,\ldots,m\}} \left(\frac{LT - \sum_{j \in C_k} p_j}{|C_k|+1} \right)$$

This bound is likely to perform well for instances in which the breach rate is high.

So in conclusion, Δ_{max} is lower bounded by $\overline{\Delta}$

$$\overline{\Delta} = \max \left(\max_{\substack{j \in \{1,\ldots,q\} \\ |D_j| < m}} p_j, \max_{k \in \{1,\ldots,m\}} \left(\frac{LT - \sum_{j \in C_k} p_j}{|C_k|+1} \right) \right) \tag{1}$$

Naturally, this bound is not tight as the problem is $\mathcal{NP}$-Hard in the strong sense.

4 A MILP Formulation of WSN-CSP

For solving this problem, we have developed a non-linear programming formulation which is too difficult to use in practice since even the smallest instances cannot be solved [13].

The solution to WSN-CSP can be modeled as a permutation of $\{1,\ldots,q\}$. For all $(j,i) \in \{1,\ldots,q\} \times \{1,\ldots,q\}$, let $x_{j,i}$ be a binary variable that is set to one if and only if cover s_j is in position i in the solution, and which is set to zero otherwise. The numerical value for Δ_{max} is fixed by the total duration of a subset of consecutive covers such that none of them is covering at least one target.

WSN-CSP can be stated as follows:

$$\text{Minimize } \Delta_{max} \tag{2}$$

$$\sum_{i=1}^{q} x_{j,i} = 1 \qquad \forall j \in \{1,\ldots,q\} \tag{3}$$

$$\sum_{j=1}^{q} x_{j,i} = 1 \qquad \forall i \in \{1,\ldots,q\} \tag{4}$$

$$\Delta_{max} \geq \overline{\Delta} \tag{5}$$

$$\sum_{j\in S}\sum_{i=1}^{|S|} p_j x_{j,i+r} \leq \Delta_{max} \qquad \forall S \subseteq \{1,\ldots,q\}, |S| > 1, \; |\cup_{j\in S} D_j| < m, \forall r \in \{0,\ldots,q-|S|\} \tag{6}$$

$$x_{j,i} \in \{0,1\} \qquad \forall (j,i) \in \{1,\ldots,q\} \times \{1,\ldots,q\} \tag{7}$$

$$\Delta_{max} \geq 0 \tag{8}$$

The objective function is to minimize Δ_{max}. Constraints (3), (4) and (7) are those of the assignment problem. They enforce that x is a valid permutation of $\{1,\ldots,q\}$. Constraint (5) is enforcing the lower bound on Δ_{max} defined by Equation (1), and (8) is a non-negativity constraint on Δ_{max}.

Constraint (6) states that the duration of any subset of covers such that there exists at least one target that is not covered by any of them, and that are scheduled consecutively, is less than Δ_{max}. More precisely, any subset S of covers that do not cover at least one target and that are scheduled consecutively can occupy positions 1 to $|S|$, or 2 to $|S|+1$, etc, up to positions $q-|S|$ to q. Note that the actual order of the covers in S does not matter. In constraint (6), r is an offset that allows to track the starting position of the corresponding subset of consecutive covers in the solution: $i+r$ is in $\{1,\ldots,q-|S|\}$ for $i=1$. The case where $|S|=1$ is dealt with, with constraint (5). In other words, this constraint tracks all possible pairs, triplets, or larger cardinality subsets of covers that do not cover at least one target, at any position it can possibly occupy in the solution.

A practical way of using the above formulation is described in [13].

5 A Greedy Heuristic Approach to WSN-CSP

The proposed cover scheduling heuristic (CSH) for WSN-CSP takes its inspiration from the lower bound introduced in Section 3. Targets are sorted by decreasing minimum non coverage time, i.e. the targets that are the most likely to be responsible for the longest non coverage period of time are processed first. In this heuristic approach, the non-overlapping constraint on covers is first relaxed. So the covers are allocated unfeasible starting times for minimizing the longest period of non coverage time of targets. More precisely, for each non-processed target k, the covers that cover it but that are still unscheduled are scheduled so as to minimize the longest non coverage duration. When all the covers have been scheduled, they are sorted by increasing (unfeasible) starting times, and are finally scheduled according to that sequence.

The reminder of this section presents CSH in more detail. Let $Q = \{0, \ldots, q+1\}$ be the set of covers where two dummy covers s_0 and s_{q+1} have been added to the set of covers. They both have a zero duration, s_0 is scheduled at time 0 and s_{q+1} is scheduled at time LT. Moreover, $D_0 = D_{q+1} = \{1, \ldots, m\}$ as they cover all the targets. These dummy covers are used for indicating that all targets are considered to be covered at time zero, and also at time LT. Finally, these two dummy covers are considered to be already scheduled.

First, the targets are sorted by decreasing uncovered time. For all targets, indexed by $k \in \{1, \ldots, m\}$, the uncovered time UT_k is the total time during which it is not covered divided by the number of covers that cover it minus one.

$$UT_k = \frac{\sum\limits_{j \notin C_k} p_j}{|C_k| - 1} \quad \forall k \in \{1, \ldots, m\}$$

Note that $|C_k| \geq 2$ for all k because of covers s_0 and s_{q+1}.

Algorithm 1 describes CSH, it relies on the following variables. Let $sched_k$ be the number of covers in C_k that have already been scheduled (when processing target $k = 1$, $sched_1 = 2$ because of the two dummy covers). SC_k is the set of the scheduled covers in C_k. Initially for $k = 1$, $SC_1 = \{0, q+1\}$. UC_k is the set of the unscheduled covers in C_k. Initially for $k = 1$, $UC_1 = C_1 \backslash \{0, q+1\}$. C_k is then partitioned in two disjoint subsets $C_k = SC_k \cup UC_k$, but UC_k might possibly be empty with no inconvenience. By definition, $sched_k = |SC_k|$, for all $k \in \{1, \ldots, m\}$. The introduction of $sched_k$ covers in the schedule creates $sched_k + 1$ time intervals during which target k is not covered. Some of these time intervals may have a negative duration because of overlapping. Let w_z be the magnitude of these time intervals for all $z \in \{1, \ldots, sched_k - 1\}$, computed in Equation (9).

As SC_k is sorted by increasing starting times, $SC_k(sched_k - 1)$ is the last element of SC_k, so it is cover $q+1$ as this cover is the last one. Thus, $start_{SC_k(sched_k - 1)} = LT$. Moreover, Add_z is the set of all the covers that are to be added to time interval z, for all $z \in \{1, \ldots, sched_k - 1\}$ (initially, i.e. at the beginning of the process of target k, Add_z is empty for all z). Note that $start_{SC_k(z)}$ is the starting time of the cover in position z in the ordered set SC_k.

$$\begin{aligned} w_z = {} & start_{SC_k(z)} - \left(start_{SC_k(z-1)} + p_{SC_k(z-1)}\right) \\ & - \sum_{g \in Add_z} p_g \qquad \forall z \in \{1, \ldots, sched_k - 1\} \end{aligned} \tag{9}$$

In Algorithm 1, in line 1, $Add_z(h)$ refers to the cover index of the element in position h in Add_z, and UC_k are sorted by increasing duration of the corresponding covers in

Algorithm 1. Cover Scheduling Heuristic (CSH)

Input: the set of covers Q and their duration
Output: a sequence of the covers

// **Initialization**
1 Compute UT_k for all targets
2 Sort the targets in decreasing values of UT_k and re-index
// **Main loop over all targets**
3 **for** $k = 1 \rightarrow m$ **do**
4   Sort SC_k by increasing order of the starting times of these covers
5   Sort UC_k by increasing duration of these covers
6   **while** $UC_k \neq \emptyset$ **do**
7   uc is the last cover index in UC_k // i.e. with maximum duration
8   compute w_z as stated by equation (9)
9   compute $r = \underset{z\in\{1,\ldots,sched_k-1\}}{\arg\max} \left(\frac{w_z}{1+|Add_z|}\right)$
10   allocate $uc \in UC_k$ such that $Add_r \leftarrow Add_r \cup \{uc\}$
11         $UC_k = UC_k \setminus \{uc\}$
    // **Schedule the covers in C_k by setting starting times**
12   **forall** $z \in \{1,\ldots,sched_k - 1\}$ **do**
13   **forall** $g \in \{1,\ldots,|Add_z|\}$ **do**
14             $\sigma = Add_{z,g}$ // i.e. the cover in position g in Add_z
15             $start_\sigma = \left(start_{SC_k(z-1)} + p_{SC_k(z-1)}\right) + g\frac{w_z}{1+|Add_z|} + \sum_{h=1}^{g-1} p_{Add_z(h)}$
16         $SC_k \leftarrow SC_k \cup Add_z$
17         $Add_z \leftarrow \emptyset$
// **Repairing phase**
18 Since some of the covers may overlap, sort them in increasing order of starting times
19 Output the final sequence (will be used for building a semi-active schedule of covers)

order to use the longest covers for splitting the longest time interval during which target k is not covered (see lines $7-11$ in Algorithm 1).

Each time interval z is allocated $|Add_z|$ unscheduled covers, and these covers are scheduled in lines $12-17$ so as to minimize the maximum amount of time during which target k is not covered. Finally, lines $18-19$ allow to build a feasible solution by sorting the starting times of the covers by increasing order.

6 A Genetic Algorithm Based Approach to WSN-CSP

This section introduces a genetic algorithm based approach for WSN-CSP. The proposed approach is referred to as CSGA (Cover Scheduling Genetic Algorithm) in the sequel.

CSGA uses permutation encoding, i.e., a chromosome consists of a linear permutation of covers. A permutation of covers specifies the order in which the covers are scheduled, i.e., a value of j at the position i in the permutation indicates that cover j will be the ith cover in the schedule.

The fitness function is same as the objective function of WSN-CSP. In order to compute the fitness of a chromosome, we have to find the maximum duration for which a target is uncovered in the schedule represented by that chromosome. The complexity of this fitness evaluation is $\mathcal{O}(mq)$.

The probabilistic binary tournament selection is used for selecting the two parents for crossover, where the candidate with better fitness is selected with probability p_b. The reason for using probabilistic binary tournament selection is that probabilistic binary tournament is similar to rank selection as far as selection pressure is concerned. At the same time, it is much more computationally efficient [7]. However, with very small probability p_r, we have selected a parent randomly instead of selecting it through probabilistic binary tournament selection. This has been done with the motive of increasing the diversity of the population.

The crossover operator used is the cycle crossover operator CX as initially described in [10]. This crossover operator preserves the absolute positions of the covers in the schedule from one parent or the other and causes a proper mix of the schedules of the two parents. Uniform order based crossover [6] was also tried, but, CX gave better results. Three swap mutations are used inside a single mutation operator to mutate the chromosomes. We have applied the crossover operator always to generate an offspring, whereas mutation operator has been used only with probability p_m.

CSGA relies on the steady-state [6] population replacement model instead of the commonly used generational model. Unlike generational replacement, where the entire parent population is replaced with an equal number of newly created children every generation, in the steady-state population replacement method a single child is produced in every generation and it replaces a less fit member of the population. In comparison to the generational method, the steady-state population replacement method generally finds better solutions faster. This is because of permanently keeping the best solutions in the population and the immediate availability of a child for selection and reproduction. Another advantage of the steady-state population replacement method is the ease with which duplicate copies of the same individuals are prevented in the population. In the generational approach, duplicate copies of the highly fit individuals may exist in the population. Within few generations, these highly fit individuals can dominate the whole population. When this happens, the crossover becomes totally ineffective and the mutation becomes the only possible way to improve solution quality. Under this situation, improvement in solution quality, if any, is very slow. Such a situation is called the premature convergence. In the steady-state approach we can easily avoid this situation by simply checking each newly generated child against current population members and discarding it if it is identical to any member.

Initial population is generated randomly with the restriction that each member of the initial population should be unique. The offspring created during each generation is checked for uniqueness with respect to the existing population members, and, if it is unique, it is inserted into the population replacing the worst member, otherwise it is discarded.

With the intent of improving the solution quality even further, we have applied a local search on the best solution obtained through genetic algorithm. This local search follows an iterative process. During each iteration, it tries to swap a cover involved in the longest breach with a cover which covers the target involved in the longest breach. If such a swap can decrease the longest breach then it is accepted, otherwise original schedule is restored. This process is repeated till no swap move is possible that can decrease the longest breach. We have tried this local search inside the genetic algorithm but the resulting approach was found to be too slow. Actually, each swap move requires evaluating the fitness of the schedule from scratch which takes $\mathcal{O}(mq)$ time as mentioned above. This time complexity of a single swap move makes this local search inappropriate for use inside the genetic algorithm.

As far as parameter settings of the CSGA are concerned, the population size is fixed to 400 individuals, p_b is set to 0.8, p_r is set to 0.01, and p_m is taken to be equal to 0.05. The CSGA terminates when the best solution does not improve over 20,000 iterations. CSGA can also terminate when it fails to find a solution different from current population members in 20 consecutive trials. All these parameter settings are chosen empirically. Though these settings provide good results, they are in no way optimal for all instances.

7 Computational Results

The set of instances used in this work is taken from the output of the framework presented in [14]. In order to produce instances for WSN-CSP, MNLB has been solved beforehand for different size and parameter values. The number of targets m is taken in $\{50, 100, 150, 200\}$. For each value of m a corresponding value of the number of sensors is given in $\{30, 60, 90, 120\}$ and the sensing range will be fixed to 150. Note that the number of sensors and the sensing range value are not parameters of WSN-CSP, but have an impact on the number of covers and their duration. For each value of m, the bandwidth constraint is set to either 5, 10 or equal to the number of targets (i.e. no bandwidth constraint is enforced) and the breach rate is either set to 0.1 or 0.2. For each combination of the parameters, 30 randomly generated instances are considered, and the grand total is 720 instances.

All experiments were carried out on an Intel Core 2 Quad computer with 4 Gb of RAM running at 2.83 Ghz under Ubuntu 9.04. The code is developed in C and compiled with gcc version 4 without any special multi-core utilization features. The running times are not reported in details but the heuristic CSH running time is always less than 0.02s. The average and maximum CPU time for the CSGA are 5.83s and 40.61s when $\alpha = 0.1$ and 14.72s and 60.71s when $\alpha = 0.2$. All our approaches are executed once on each instance.

Table 1 presents the results obtained by CSH and by CSGA in terms of solution quality. The first column defines the instance parameters (namely, the number of targets, the bandwidth constraint and $\overline{q}$, the average number of covers). Each row corresponds to 30 instances for $\alpha = 0.1$ and 30 instances for $\alpha = 0.2$. The next four columns provide the results for the instances where the breach rate is $\alpha = 0.1$ and the last four columns correspond to $\alpha = 0.2$. For each group of four columns, the first figure is the deviation of CSH over LB in percentage, the second figure is the number of proved optimal solutions found by CSH over 30 instances (CSH=LB), the third figure is the deviation of CSGA over LB, and the fourth figure is the number of proved optimal solutions over 30 instances. A solution can be proved optimal whenever the objective value is equal to the lower bound. The last two lines of the Table summarize the results.

Table 1. Results of CSH and CSGA

	$\alpha = 0.1$				$\alpha = 0.2$			
Parameters	% CSH	# CSH	% CSGA	# CSGA	% CSH	# CSH	% CSGA	# CSGA
$m = 50, w = 5, \overline{q} = 41$	56.6	2	1.8	23	90.5	0	3.5	12
$m = 50, w = 10, \overline{q} = 47$	64.7	0	1.2	22	78.5	0	2.7	14
$m = 50, w = 50, \overline{q} = 47$	50.3	2	0.2	24	94.5	0	3.5	14
$m = 100, w = 5, \overline{q} = 90$	76.1	0	0.2	28	133.9	0	9.2	8
$m = 100, w = 10, \overline{q} = 96$	88.7	1	0.7	26	126.1	0	7.8	10
$m = 100, w = 100, \overline{q} = 96$	78.4	0	0	30	131.0	0	6.8	10
$m = 150, w = 5, \overline{q} = 137$	96.6	0	0.5	26	158.4	0	12.2	3
$m = 150, w = 10, \overline{q} = 140$	102.1	0	1.0	26	164.3	0	18.9	2
$m = 150, w = 150, \overline{q} = 142$	95.2	0	0.6	27	162.8	0	11.4	4
$m = 200, w = 5, \overline{q} = 182$	106.7	0	1.1	26	182.0	0	22.2	0
$m = 200, w = 10, \overline{q} = 188$	133.9	0	0	27	201.3	0	16.2	3
$m = 200, w = 200, \overline{q} = 189$	131.4	0	1.3	24	187.7	0	17.6	2
Total proved opt. sol.		**5**		**309**		**0**		**82**
Average Deviation to LB	**90.1%**		**0.7%**		**142.6%**		**11.1%**	

Table 1 suggests that the problem difficulty increases with α. Indeed, WSN-CSP is trivial when $\alpha = 0$ as the targets are continuously covered by all the covers, and the gap to the lower bound is maximum for $\alpha = 0.2$. However, solution quality is assessed through a lower bound that may not be tight, especially when α is large. Consequently, the solutions for $\alpha = 0.2$ may be closer to optimality than they appear in Table 1. This situation arises in the example introduced in Section 1.1. The optimal objective value is 5 whereas the lower bound is only 3 in that case. Nevertheless, this bound allows to prove the optimality of 391 solutions returned by CSGA out of a grand total of 720 instances ($\approx 54\%$).

8 Conclusions and Perspectives

To the best of our knowledge, this paper is the first attempt to address the Cover Scheduling Problem in Wireless Sensor Networks (WSN-CSP). This problem appears once the covers have been generated in an earlier stage of the a wireless sensor network problem (like MCBB or MNLB) in which the breach is non zero and the covers are non-disjoint. This problem is shown to be $\mathcal{NP}$-Hard in the strong sense, and a MILP formulation is proposed.

Our practical contribution consists of a greedy heuristic and a genetic algorithm with local search. Both are tested on a large set of instances and give convincing results. Our future work will be oriented in two directions. First, we plan to improve the genetic algorithm with dedicated operators for the WSN-CSP. Second, we also intend to propose more efficient lower bounds for this problem.

References

1. Akylidiz, I., Su, W., Sankarasubramaniam, Y., Cayirci, E.: A survey of sensor network. IEEE Communication Magazine 40(8), 102–116 (2002)
2. Benini, L., Castelli, G., Macii, A., Macii, E., Poncino, M., Scarsi, R.: A discrete-time battery model for high-level power estimation. In: Proceedings of the IEEE Design Automation and Test in Europe conference, Paris, France, pp. 35–39 (2000)
3. Benini, L., Bruni, D., Macii, A., Macii, E., Poncino, M.: Discharge current steering for battery lifetime optimization. IEEE Transaction on computers 52(8), 985–995 (2003)
4. Cheng, M.X., Ruan, L., Wu, W.: Achieving minimum coverage breach under bandwidth constraints in wireless sensor networks. In: INFOCOM 2005. 24th Annual Joint Conference of the IEEE Computer and Communications Societies, pp. 2638–2645 (2005)
5. Chow, K., Lui, K., Lam, E.: Wireless sensor networks scheduling for full angle coverage. Multidimensional Systems and Signal Processing 20, 101–119 (2009)
6. Davis, L.: Handbook of Genetic Algorithms. Van Nostrand Reinhold (1991)
7. Goldberg, D.E., Deb, K.: A comparative analysis of selection schemes used in genetic algorithms. In: Proceedings of the 1991 Conference on Foundations of Genetic Algorithms, pp. 69–93. Morgan Kaufmann, San Francisco (1991)
8. Huang, C., Tseng, Y.: The coverage problem in a wireless sensor network. Mobile Networks and Applications 10, 519–528 (2005)
9. Lin, J., Chen, Y.: Improving the coverage of randomized scheduling in wireless sensor networks. IEEE Transaction on Wireless Communications 7(12), 4807–4812 (2008)
10. Oliver, I.M., Smith, D.J., Holland, J.R.C.: A study of permutation crossover operators on the travelling salesman problem. In: Proocedings of the second international conference on genetic algorithms, pp. 224–230. Erlbaum, Mahwah (1987)
11. Pinedo, M.: Scheduling: theory, Algorithms and Systems. Prentice-Hall, Englewood Cliffs (2008)
12. Raghunathan, V., Schurgers, C., Park, S., Srivastava, M.: Energy aware wireless microsensor networks. IEEE Signal Processing Magazine 19(2), 1007–1023 (2002)
13. Rossi, A., Sevaux, M., Singh, A., Geiger, M.J.: The cover scheduling problem arising in wireless sensor networks. Research report RR-11-02-01, Université de Bretagne-Sud, Lorient, France (2011), `http://www-labsticc.univ-ubs.fr/or/`
14. Rossi, A., Singh, A., Sevaux, M.: Génération de colonnes et réseaux de capteurs sans fil. In: Proceedings of the ROADEF, 11ème congrès de la Société Française de Recherche Opérationnelle et d'Aide à la Décision, Toulouse, France, February 24-26 (2010)
15. Wang, C., Thai, M.T., Li, Y., Wang, F., Wu, W.: Optimization scheme for sensor coverage scheduling with bandwidth constraints. Optimization letters 3(1), 63–75 (2009)

Author Index